TYPHOON TURNING ATLAS

TYPHOON TURNING ATLAS

Weihong Qian
Xiaolong Shan
Haoyuan Liang

Peking University, China

World Scientific

NEW JERSEY · LONDON · SINGAPORE · BEIJING · SHANGHAI · HONG KONG · TAIPEI · CHENNAI

Published by

World Scientific Publishing Co. Pte. Ltd.
5 Toh Tuck Link, Singapore 596224
USA office: 27 Warren Street, Suite 401-402, Hackensack, NJ 07601
UK office: 57 Shelton Street, Covent Garden, London WC2H 9HE

Library of Congress Cataloging-in-Publication Data
Qian, Weihong
Typhoon turning atlas / Weihong Qian, Peking University, China, Xiaolong Shan, Peking University, China, Haoyuan Liang, Peking University, China.
pages cm
Includes bibliographical references and index.
ISBN 978-9814531818 (pbk. : alk. paper)
1. Typhoons--Tracks--Observations. 2. Typhoons--Forecasting. 3. Cyclones--Tropics--Forecasting. 4. Decomposition method. I. Shan, Xiaolong, author. II. Liang, Haoyuan, author. III. Title.
QC948.Q253 2014
551.55'20916460223--dc23

2013023792

British Library Cataloguing-in-Publication Data
A catalogue record for this book is available from the British Library.

Typeset by Stallion Press
Email: enquiries@stallionpress.com

Printed by FuIsland Offset Printing (S) Pte Ltd Singapore

Preface

Typhoons, or severe tropical cyclones, with most destructive forces are frequently formed in the South China Sea (SCS) to the western Pacific regions. Typhoons generally move westwards or northwestwards in summer and autumn, following the steering flows associated with the dominant Pacific subtropical high. But, there are often some unusual tracks, turning to the left and/or to the right for one or more times. It is difficult to forecast whether, when, and where the turning occurs based on the traditional synoptic charts and numerical weather forecast model products, because atmospheric variables include multiscale physical components and observational/model errors. *Typhoon Turning Atlas* provides an additional tool to deal with the problem. The global atmospheric variables of observations and model products can be decomposed physically into four components: (i) the zonal-mean symmetric climate, (ii) the asymmetric climate (monsoon troughs), (iii) the planetary-scale anomaly, and (iv) the synoptic-scale anomaly (anomalous vortices). This atlas, with thousands of plots, covers important information of 322 historical typhoons occurring in the last three decades in the period of 1979–2011 and demonstrates how the typhoons' motions and intensities were influenced by climatic monsoon troughs and anomalous vortices, respectively.

Classification of the positions from tropical disturbance to typhoon's peak intensity as well as diagnostic method of typhoon motions in this atlas will be surely useful as a good reference for researchers and forecasters in the practice of typhoon prediction. The method of physical decomposition used in this atlas demonstrates how to extract useful information from observations (or model products), which is unlike most mathematical decompositions based on orthogonal basis functions, such as spherical harmonic expansion and empirical orthogonal functions. The atlas consists of seven chapters as follows.

The principle and method of physical decomposition of atmospheric variables, as well as some distribution features of typhoons occurring in the western Pacific and the SCS are described in Chapter 1. In order to understand this atlas, three typhoons with unusual motions and intensity changes are also illustrated as examples in this chapter.

In Chapter 2, typhoon motions without turning are divided into five sections according to their geneses in the SCS, intensity peaked in the SCS, the Philippines, the east seas of China, and the western Pacific, respectively. Typhoons achieving their peak intensities in the SCS, the Philippines, the equatorial Pacific, and the tropical Pacific are illustrated in Chapter 3. Typhoons generated in the Philippines, the equatorial western Pacific, the equatorial central Pacific, and the tropical Pacific and also achieving their peak intensities in the Philippines are discussed in Chapter 4. Most typhoons with their peak intensities located in the Taiwan area but generated in the equatorial western Pacific, the equatorial central Pacific, the tropical western Pacific, and the tropical central Pacific are given in Chapter 5. Some typhoons with their intensities peaked in the east seas of China but generated in the equatorial western Pacific, the equatorial central Pacific, the tropical western Pacific, the east seas of China, and the Northwest Pacific are plotted in Chapter 6. Typhoons with their intensities peaked in the equatorial Pacific and in the Northwest Pacific, typhoons generated in the western/central Pacific and entered the east seas of China, as well as twin typhoons and complex tracks of typhoons are described in Chapter 7.

We would like to thank Professor Chou Jifan for his useful suggestions and Dr Zhang Sheng for his revision to the atlas.

Weihong Qian
28 December 2012

Contents

Preface v

Chapter 1: Physical Decomposition 1

1.1. Typhoons in the Northwest Pacific and the South China Sea . . . 1
1.2. Physical Decomposition Principles . . . 2
1.3. Statistical Tracks . . . 5
1.4. Unusual Motions of Three Typhoons . . . 8
1.5. Climatic Monsoon Troughs . . . 22
1.6. Two Close Tropical Cyclones . . . 29
References . . . 35

Chapter 2: Typhoons Without Obvious Turning Motions 37

2.1. Geneses in the South China Sea and the Philippines . . . 37
2.2. Intensity Peaked in the South China Sea . . . 49
2.3. Intensity Peaked in the Philippines . . . 76
2.4. Intensity Peaked in the East Seas of China . . . 109
2.5. Intensity Peaked in the Western Pacific . . . 131

Chapter 3: Intensity Peaked in the South China Sea 151

3.1. Geneses in the South China Sea . . . 151
3.2. Geneses in the Philippines . . . 166
3.3. Geneses in the Equatorial Pacific . . . 171
3.4. Geneses in the Tropical Pacific . . . 177

Chapter 4: Intensity Peaked in the Philippines 195

4.1. Geneses in the Philippines . . . 195
4.2. Geneses in the Equatorial Western Pacific . . . 198
4.3. Geneses in the Equatorial Central Pacific . . . 213
4.4. Geneses in the Tropical Pacific . . . 235

Chapter 5: Intensity Peaked in the Taiwan Area 253

5.1. Geneses in the South China Sea . . . 253
5.2. Geneses in the Philippines . . . 256
5.3. Geneses in the Equatorial Western Pacific . . . 262
5.4. Geneses in the Equatorial Central Pacific . . . 267
5.5. Geneses in the Tropical Western Pacific . . . 271
5.6. Geneses in the Tropical Central Pacific . . . 291

Chapter 6: Intensity Peaked in the East Seas of China 297

6.1. Geneses in the Equatorial Western Pacific . . . 297
6.2. Geneses in the Equatorial Central Pacific . . . 301
6.3. Geneses in the Tropical Western Pacific . . . 309

6.4. Geneses in the East Seas of China 326
6.5. Geneses in the Northwest Pacific 333

Chapter 7: Intensity Peaked in the Pacific 345

7.1. Intensity Peaked in the Equatorial Pacific 345
7.2. Intensity Peaked in the Northwest Pacific 359
7.3. Geneses in the Western Pacific and Entering the East Seas of China 382
7.4. Geneses in the Central Pacific and Entering the East Seas of China 386
7.5. Twin Typhoons 406
7.6. Complex Tracks 423

Chapter 1

Physical Decomposition

1.1. Typhoons in the Northwest Pacific and the South China Sea

A tropical cyclone is a rotating, organized system of clouds and thunderstorms that generates over tropical or subtropical oceanic basins and has a closed low-level circulation. Tropical cyclones rotate in a counterclockwise manner in the Northern Hemisphere (NH) and in a clockwise manner in the Southern Hemisphere (SH). They are classified as follows:

- **Tropical Disturbance (TB):** A TB with apparently organized convection (generally 100 to 300 miles in diameter) originates in the tropics or subtropics, with a non-frontal migratory character, and maintains its identity for 24 h or more.
- **Tropical Depression (TD):** A tropical cyclone with maximum sustained winds of 38 mph (33 knots or 17 m/s) or less.
- **Tropical Storm (TS):** A tropical cyclone with maximum sustained winds of 39 to 73 mph (34 to 63 knots or 18 to 32 m/s).
- **Typhoon (TY) or Hurricane (HC):** A tropical cyclone with maximum sustained winds of 74 mph (64 knots or 33 m/s) or higher. The terms "typhoon" and "hurricane" are used in the Northwest Pacific (NW Pacific or NWP) and the Atlantic Ocean, respectively. Similar storms in the Indian Ocean and South Pacific Ocean are called "tropical cyclones."
- **Super Typhoon (ST) or Major Hurricane (MH):** A tropical cyclone with maximum sustained winds of 148 mph (129 knots or 67 m/s) or higher.

Most tropical cyclones form between 5° and 30° of latitude away from the equator in the NH (SH) and move typically westwards. Sometimes, the winds at the lower, middle and upper troposphere change and steer the cyclone toward the north and northwest in the NH and toward the south and southwest in the SH. When tropical cyclones reach latitudes near 30°N in the NH, they often move northeastwards. Figure 1.1 shows 543 TYs generated in the NW Pacific and in the South China Sea (SCS).

Seasonally, most typhoons form from May to December in the NW Pacific and from May to October in the SCS (Fig. 1.2). The maximum numbers of TYs in the NW Pacific and in the SCS are 104 and 11 in September, respectively. In the NW Pacific, TYs develop in all seasons. In contrast, no TYs occur in January, February, and March in the SCS.

During the lifetime of a TY, there are locations of formation, peak intensity, and dissipation. To better describe the positions of generation and peak intensity, the following nine regions are shown in Fig. 1.3: (i) the east seas of China, (ii) the Taiwan area, (iii) the subtropical Northwest Pacific, (iv) the SCS, (v) the Philippines, (vi) the tropical western Pacific, (vii) the equatorial western Pacific, (viii) the tropical central Pacific, and (ix) the equatorial central Pacific. The east seas of China consist of the Bohai Sea, the Yellow Sea, the East China Sea, and the Taiwan area.

1.2. Physical Decomposition Principles

Previous studies have shown that tropical cyclone motions are mainly governed by the tropospheric average steering flow (Chan and Gray 1982), generally defined as the vertical weighted average of the flow in the troposphere. During summer and autumn, tropical cyclones in the NWP generally move westwards, northwestwards, or northwards, mainly driven by large-scale steering flow associated with the dominant NWP subtropical high and westerly trough–ridge distribution. After passing the ridge line of the NWP subtropical high, tropical cyclones accelerate northeastwards owing to the influence of the large-scale westerly steering flow. Thus, most tropical cyclones moving westwards in the NWP would eventually turn northwards or northeastwards. This kind of track is designated as the recurvature (George and Gray 1977).

There were, however, some unusual tracks turning left and/or right for one or several times, but not following the large-scale steering flows. It is, thus, difficult to forecast whether, when, and where the turning motion occurs based on the traditional synoptic chart and the numerical weather forecast model products, because atmospheric variables include multiscale physical components and observational/model errors.

Some unusual motions of TYs were caused by the binary interaction with other tropical cyclones in the vicinity. The concept of binary interaction was well described by a Japanese meteorologist,

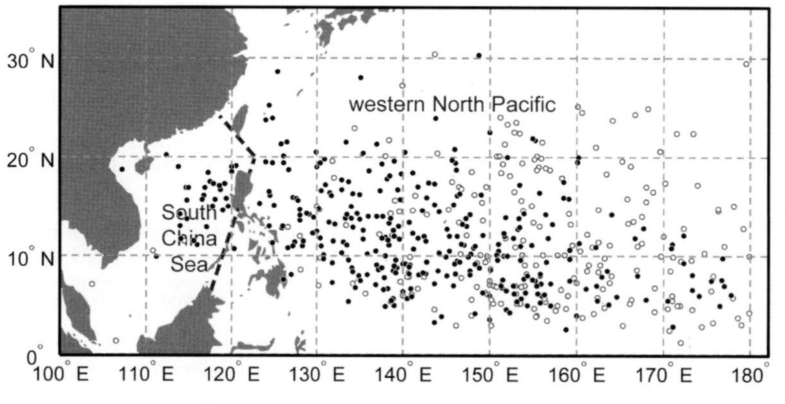

Figure 1.1. Locations of 543 TYs generated in the period of 1979–2011. Among them, tracks of 330 TYs (solid dots) finally reached the north of 15°N and the west of 130°E during their lifetime.

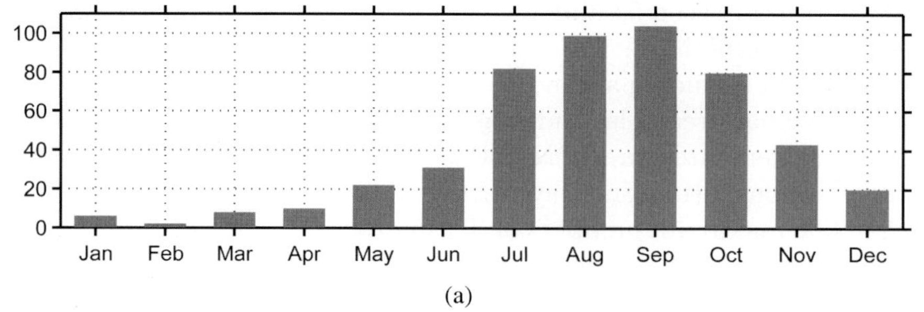

(a)

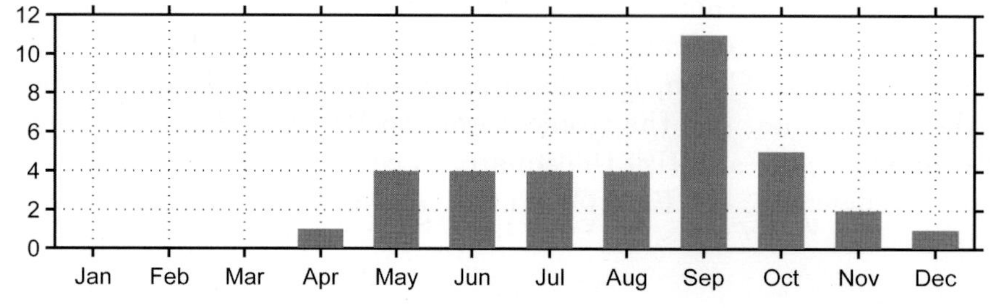

(b)

Figure 1.2. The monthly number of TYs generated in (a) the NW Pacific and (b) the SCS in the period of 1979–2011.

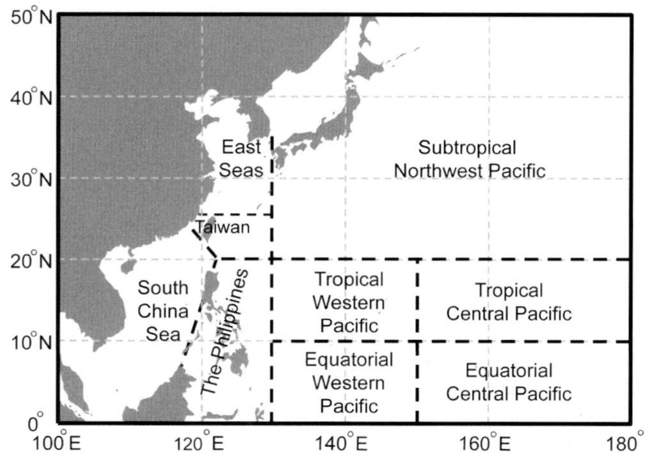

Figure 1.3. Nine regions near the China coasts and in the NWP described in this atlas.

named Fujiwhara (1921; 1931), in the early 20th century. This concept has now been referred to as the "Fujiwhara effect." A previous study indicated that the Fujiwhara effect tends to be masked by environmental circulations (Willoughby 2007). Studies indicated that two tropical cyclones can interact with each other through approaching, merging, and escaping (Lander and Holland 1993). Specifically, Brand (1970) showed that two tropical cyclones begin to circle around each other when their distance is less than 750 nautical miles (nmi), and attract each other when their distance is less than 400 nmi.

Most unusual tracks in this atlas are those of tropical cyclones that did not encounter other tropical cyclones in the vicinity but their turning motions cannot be explained by large-scale steering flows. In order to resolve this problem, we need to find another explanation.

Usually, total flows consist of several large-scale and small-scale circulation systems. These large-scale circulation systems are the steering flows that are associated with the inter-tropical convergence zone (ITCZ), subtropical highs, southwesterly monsoon flow, and westerly flow. Small-scale circulation systems include vortices in the upper and lower troposphere. Logically, large-scale steering flows provide a background of tropical cyclone motion, while the interaction can only be observed in small-scale circulation systems, such as two TY vortices and other vortices in the vicinity. Thus, extracting small-scale anomalous flow from total flow may be a possible method to examine this interaction and explain unusual tracks of TY motion (Qian *et al.* 2014).

Since TYs or tropical cyclones are embedded in the large-scale environmental flow, the decomposition of total or observed wind field is necessary. There are many mathematical decomposition methods based on orthogonal basis functions, such as spherical harmonic expansion globally and empirical orthogonal functions regionally. In a limited region, the harmonic-sine/cosine series expansion was used in decomposing and reconstructing wind field (Chen and Kuo 1992a; 1992b). The non-divergent and non-rotational wind components can be separated by using this method and used to describe the large-scale steering flow and the small-scale TY vortex, respectively. The spatial distributions of two components rely on the choice of domain. These two components are referred to as the large-scale spatial steering flow and small-scale spatial anomalous flow. This decomposition method is independent of historical observations.

On the other hand, we introduce another decomposition method that relies on climatology. Traditionally, the climate is defined, to a large extent, as the average conditions over certain time intervals, such as monthly, seasonal, and annual averages (Peixoto and Oort 1992). In contrast to the method mentioned above, we decompose total flow (or other variables such as geopotential height and temperature) into four components physically as follows.

$$\boldsymbol{V}(\lambda,\varphi,t)_{d,y} = [\overline{\boldsymbol{V}}(\varphi,t)]_d + \boldsymbol{V}^*(\lambda,\varphi,t)_d + [\boldsymbol{V}(\varphi,t)]'_{d,y} + \boldsymbol{V}(\lambda,\varphi,t)'^*_{d,y}. \tag{1}$$

The term $\boldsymbol{V}(\lambda,\varphi,t)_{d,y}$ on the lefthand side of Eq. (1) is the historical or current meteorological observations of wind vector $\boldsymbol{V}(\lambda,\varphi,t)_{d,y} = \boldsymbol{i}u(\lambda,\varphi,t)_{d,y} + \boldsymbol{j}v(\lambda,\varphi,t)_{d,y}$ at a position of longitude λ and latitude

φ, and at a diurnal time t (24 h a day), for example at 0000 or 0100 UTC Coordinated Universal Time (UTC) on a calendar date d in a year y.

The first component $[\bar{\boldsymbol{V}}(\varphi, t)]_d$ on the righthand side of Eq. (1) is the zonal-mean symmetric climate of wind averaged from total N years at a diurnal time t on a calendar date d. This component changes with latitude φ and diurnal time t in hours and calendar date d, depending on the inclination of incoming solar ray and the surface condition. This zonal-averaged symmetric climate of wind can be estimated by observational winds in the total N years

$$[\bar{\boldsymbol{V}}(\varphi, t)]_d = \sum_{y=1}^{N}\sum_{\lambda=1}^{P} \boldsymbol{V}_t(\lambda, \varphi, t)_{d,y}/(N \times P), \tag{2}$$

where y is the years from 1 to N and λ is longitude-grid numbers along a zonal circle from the first to the last P grids. $N = 30$ is the total years (1981–2010) of observational wind data used, which must be at a diurnal time t, i.e. 0000 UTC or 0600 UTC, 1200 UTC, and 1800 UTC in this atlas, used to calculate the climate.

The second component $\boldsymbol{V}^*(\lambda, \varphi, t)_d$ is the asymmetric climate of wind at a diurnal time t on a calendar date d, namely

$$\boldsymbol{V}^*(\lambda, \varphi, t)_d = \sum_{y=1}^{N} \boldsymbol{V}_t(\lambda, \varphi, t)_{d,y}/N - [\bar{\boldsymbol{V}}(\varphi, t)_d]. \tag{3}$$

It is caused by the sea–land topographical distribution and sea–land thermal contrast. The asymmetric component can be used to describe climatic monsoon circulation and widely used in the atlas.

Two components $[\bar{\boldsymbol{V}}(\varphi, t)]_d$ and $\boldsymbol{V}^*(\lambda, \varphi, t)_d$ consist of the standard climate of wind at a diurnal time t on a calendar date d,

$$\begin{aligned}\tilde{\boldsymbol{V}}(\lambda, \varphi, t)_d &= \sum_{y=1}^{N} \boldsymbol{V}(\lambda, \varphi, t)_{d,y}/N \\ &= [\bar{\boldsymbol{V}}(\varphi, t)]_d + \boldsymbol{V}^*(\lambda, \varphi, t)_d.\end{aligned} \tag{4}$$

It means that the standard climate of wind (or climatic wind) varies every hour t on a calendar date d.

The third component

$$[\boldsymbol{V}(\varphi, t)]'_{d,y} = \sum_{\lambda=1}^{P} \boldsymbol{V}_t(\lambda, \varphi, t)_{d,y}/P - [\bar{\boldsymbol{V}}(\varphi, t)_d] \tag{5}$$

is the zonal-averaged wind anomaly at a diurnal time t on a calendar date d in a year y. This component is directly linked with the westerly index cycle.

The fourth component $\boldsymbol{V}(\lambda, \varphi, t)'^{*}_{d,y}$ at a diurnal time t on the calendar date d in a year y can be extracted from observational total wind after filtering out the climatic wind and the zonal-averaged wind anomaly

$$\begin{aligned}\boldsymbol{V}(\lambda, \varphi, t)'^{*}_{d,y} &= \boldsymbol{V}(\lambda, \varphi, t)_{d,y} - \tilde{\boldsymbol{V}}(\lambda, \varphi, t)_d \\ &\quad - [\boldsymbol{V}(\varphi, t)]'_{d,y}.\end{aligned} \tag{6}$$

The term $\boldsymbol{V}(\lambda, \varphi, t)'^{*}_{d,y}$ refers to the anomalous flow describing anomalous systems, such as anomalous vortices (anomalous high or anomalous low). The environmental flow or large-scale steering flow consists of the climatic winds $\tilde{\boldsymbol{V}}(\lambda, \varphi, t)_d$ and the zonal-averaged wind anomaly $[\boldsymbol{V}(\varphi, t)]'_{d,y}$. Anomalous pressure systems and anomalous temperature centers are extracted by decomposing the observational geopotential height and temperature, respectively. This method obtaining the four components is referred to as the physical decomposition (Qian 2012a).

Physically, the asymmetric climatic wind reflects the monsoon circulation in the global monsoon region (Qian and Tang 2010; Qian 2011a). The zonal-averaged anomaly of variables is mainly forced by anomalous thermal contrasts between polar and equatorial zones (Qian and Liang 2012). The synoptic-scale anomalies of variables can indicate extreme weather events (Qian 2011b; 2012a; 2012c). Each component of the decomposition has its impact on the intensity and the track of TYs. The percentage variances of the four components of daily meridional wind at 850 hPa are 6%, 20%, 1%, and 73%, respectively (Qian 2012b). Other variables, such as total wind as well as temperature and geopotential height, also show that the components of zonal-averaged climate and zonal-averaged anomaly are smaller than that of the asymmetric climate and the synoptic-scale anomaly. In this atlas, we focus on examining how the climatic monsoon trough and anomalous circulation systems (anomalous vortices) influence the tracks and intensities of the TYs. In practice, total wind can be directly decomposed into climatic wind and anomalous wind (Qian *et al.* 2014).

Two data sets of atmospheric variables are used in this decomposition. The first is the NCEP/NCAR Reanalysis (National Centers for Environmental Prediction/National Center for Atmospheric Research) global grid daily atmospheric wind and geopotential height products in the period 1981–2010 obtained from the website (http://www.esrl.noaa.gov/psd/data/gridded/data.ncep.reanalysis.html), of which the horizontal resolution is 2.5 × 2.5 grid degrees. Another is the European Centre for Medium Range Weather Forecasts (ECMWF) Interim Reanalysis (ERA Interim) data (http://www.ecmwf. int/research/era/do/get/index) in the period 1981–2010, of which the horizontal resolution is 0.75×0.75 grid degrees. The data of the best TY track are obtained from the Joint Typhoon Warning Center (JTWC) (http://www.usno.navy.mil/ NOOC/nmfc-ph/RSS/jtwc/best_tracks/). This atlas plots 330 TYs in the SCS to the NW Pacific in the period of 1979–2011. The peak intensity of a TY is defined by the maximum wind speed (or the deepest surface pressure depression) reached in its lifetime. The initial TB refers to a tropical cyclone of which the wind speed reaches 15 knots and the dissipation stage to that of which the wind speed reduces to less than 20 knots.

1.3. Statistical Tracks

A TY usually originates from a TB supported by favorable conditions, including warm sea surface temperature, high tropospheric relative humidity, weak vertical shear of the horizontal winds, as well as low atmospheric static stability and favorable patterns of synoptic-scale atmospheric circulation (e.g. Chen 2012). After a TY genesis, it generally moves westwards or northwestwards along large-scale steering flows associated with a NW Pacific subtropical high. Figure 1.4 shows lines connecting the positions where each TY formed from initial disturbance to where its intensity peaked and to where it dissipated. In September, most disturbances formed in the tropical NW Pacific and moved northwestwards. Interestingly, in September, if their intensities peaked in the SCS or near the Philippines, TYs usually continued to move northwestwards and made landfall along the China and Vietnam coasts. If their intensities peaked in the east of 130°E, however, most TYs tended to turn the northeastward. In particular, if their intensities peaked in east seas of Taiwan and the Philippines, TYs could head toward Japan.

In October, statistical positions of peak intensities moved southwards to the central SCS and the east sea of the Philippines, though most of the TYs still originated from TBs in the equatorial NW Pacific and headed northwestwards. In this month, most TYs peaked in the SCS and would make landfall along the South China and Vietnam coasts. On the other hand, those peaked in the NW Pacific would move in the northeastward direction.

In November, statistical positions of peak intensities shifted to the south of 15°N. TBs originated near 10°N in the equatorial NW Pacific and headed northwestwards. In this month, a number of TYs peaked in the SCS and the Philippines and tended to make landfall in the vicinity of the SCS. Other TYs peaking in the NW Pacific mainly moved northeastwards.

Track and intensity are two key forecast elements during a TY's lifetime, especially when it approaches offshore regions. Variations in motion and intensity are influenced by many factors. Figure 1.4 shows only some statistical characteristics of TY tracks and positions of peak intensities.

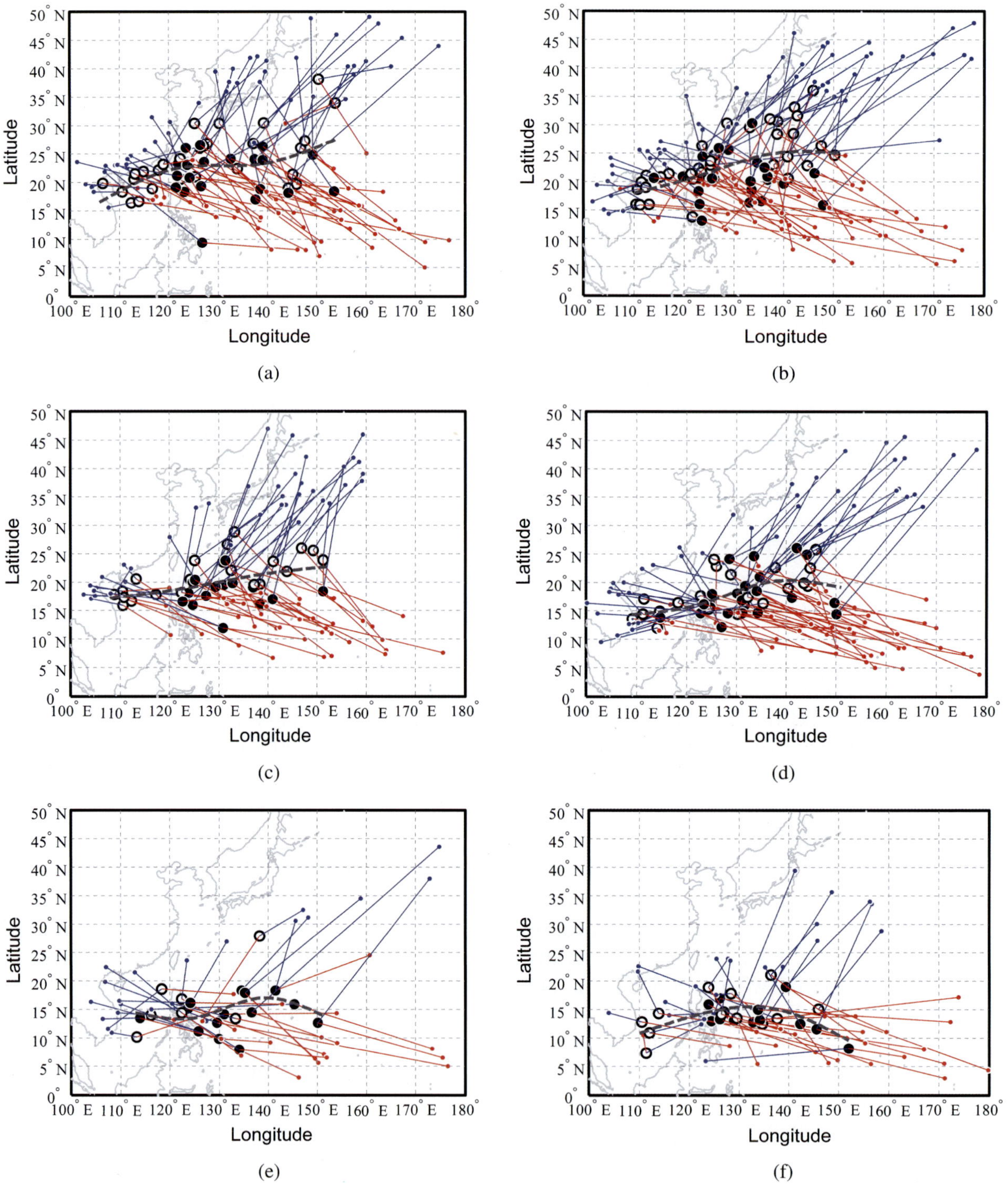

Figure 1.4. Initial positions of TB (small red dots), positions of peak intensity (STs are denoted by large black dots and generic TYs are denoted by open dots), and dissipating positions (small blue dots). Black dashed lines are the cubic splines fitting the positions of peak intensities in (a) 1–15 September, (b) 16–30 September, (c) 1–15 October, (d) 16–31 October, (e) 1–15 November, and (f) 16–30 November, respectively. The three positions for each individual TY are joined by red and blue lines.

Most TYs follow along the large-scale steering flows associated with the subtropical high, and they, consequently, move northwestwards in the NW Pacific and then northwards or northeastwards.

Using the method of physical decomposition of atmospheric variables, the impact of the monsoon trough on a TY's motion and intensity is statistically illustrated in Fig. 1.5. Statistical positions of peak intensities shifted, following the climatic monsoon troughs southwards seasonally from September to October and November. For individual TYs shown in Fig. 1.4, its unusual motion and irregular intensity

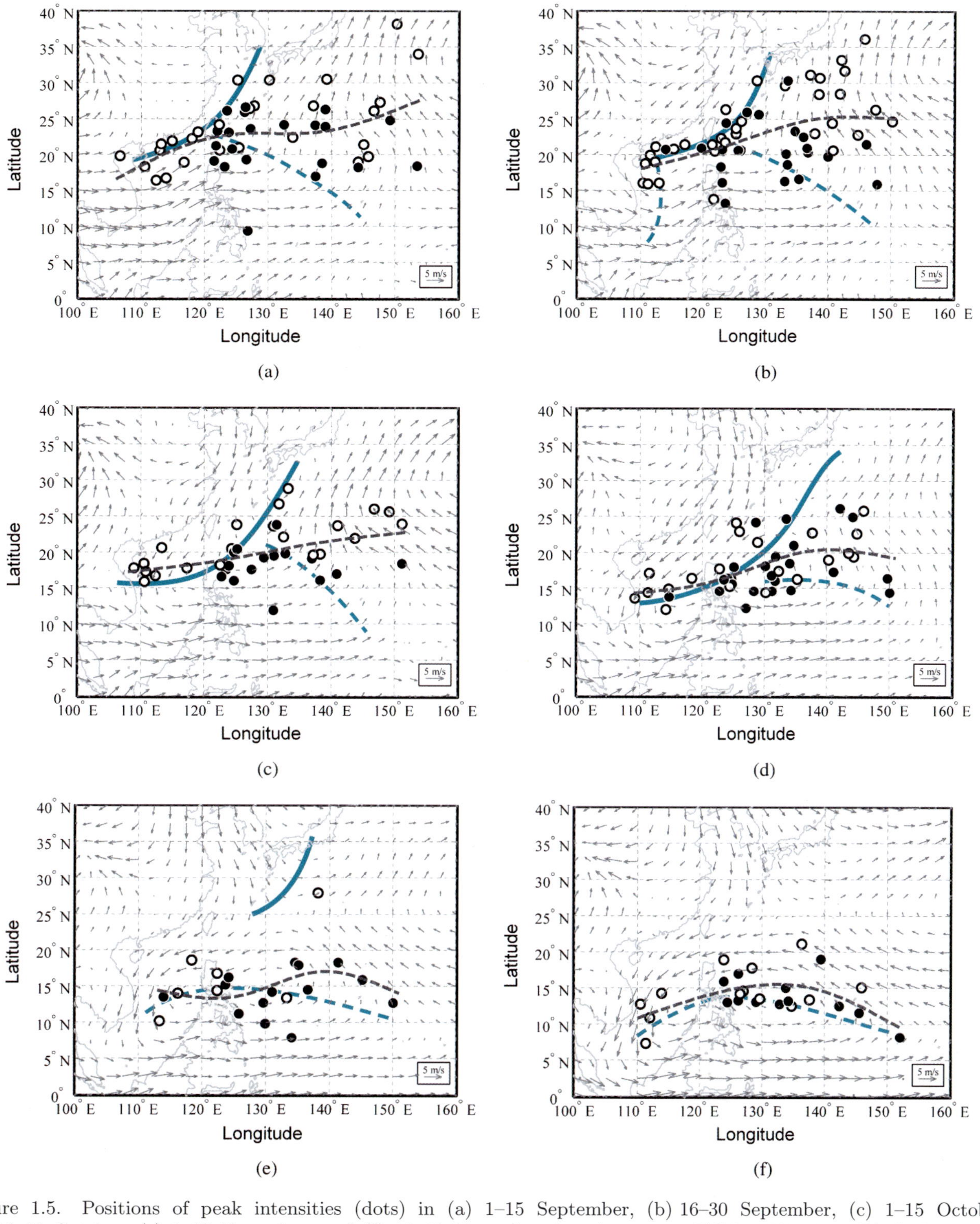

Figure 1.5. Positions of peak intensities (dots) in (a) 1–15 September, (b) 16–30 September, (c) 1–15 October, (d) 16–31 October, (e) 1–15 November, and (f) 16–30 November, based on every TY in 1980–2011. Black dots represent STs ($\boldsymbol{V}_{\max} \geq 115$ knots) and open dots represent generic TYs ($64 \leq \boldsymbol{V}_{\max} < 115$ knots). Black dashed lines are the cubic splines fitting the positions of peak intensities. Arrows indicate the pentad-mean asymmetric climatic winds (monsoon flow) averaged over the period 1981–2010 in pentads 50 (3–7 September), 53 (18–22 September), 56 (3–7 October), 59 (18–22 October), 62 (2–6 November), and 65 (17–21 November) at 850 hPa, respectively. Thick solid blue lines in each panel indicate the East Asian subtropical monsoon trough and blue dashed lines denote the peninsula monsoon trough in the SCS and the Pacific tropical monsoon trough.

Source: The data used is obtained from NCEP/NCAR Reanalysis.

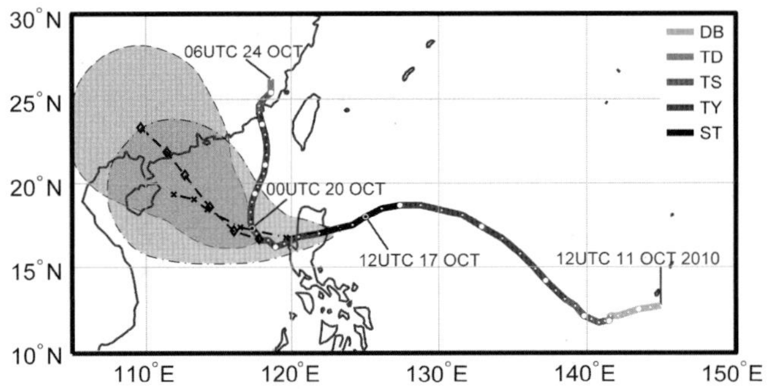

Figure 1.6. The track of TY Megi (2010_No.15) and two forecasts issued by China NMC at 0000 UTC 18 October 2010 and at 1200 UCT 19 October 2010. The white spots show the locations of Megi at every 6 h from 1200 UTC 11 October 2010 to 0600 UTC 24 October 2010. Two fine dashed lines with shadings indicate the track forecasts. Megi's intensity is indicated by disturbance (DB), tropical depression (TD), tropical storm (TS), typhoon (TY), and super typhoon (ST).

variation were influenced by climatic monsoon troughs and synoptic-scale anomalies (or anomalous vortices).

1.4. Unusual Motions of Three Typhoons

As shown in Figs. 1.4 and 1.5, many TYs turned right when they arrived at climatic monsoon troughs. All of them are shown in the above figures without detailed tracks. Three cases of unusual TY tracks are illustrated in this section.

The first case of TY is named Megi (2010). It was first identified by the JTWC on 12 October 2010 as a TB located in the southeast of Guam (Fig. 1.6). Before 16 October 2010, the TY gradually strengthened and moved northwestward along the easterly flows to the south of the NWP high. On 17 October 2010, it strengthened to a ST. It later peaked with its wind velocity 160 knots (296 km/h, 184 mph) and lowest surface pressure of 903 hPa. It made landfall on the northern Philippines on 18 October 2010. On 19 October, it reduced its speed and turned northwards abruptly. On 20 October, it temporarily strengthened with its maximum wind velocity reaching 115 knots (212 km/h, 132 mph) and surface pressure reducing to 937 hPa. On October 22, it weakened into a severe TS approaching the Taiwan Strait. At 0:55 p.m. Beijing Time on October 23, it made landfall over Zhangpu County in Fujian Province, Southeast China.

Following its track, it is observed that Megi experienced the following three processes: (i) moving west-southwestwards after formation, (ii) turning left on 14–17 October before its intensity peaked in the northeast sea of the Philippines, (iii) and turning right after it moved into the SCS on 19–22 October 2010. Two forecasts from China National Meteorological Center (NMC) are also plotted to compare with the track of Megi. At 0000 UTC 18 October, operational forecast from China NMC showed that Megi would pass the northern Philippines, then enter the SCS, and eventually approach the Hainan Island in three days. At 1200 UCT 19 October, NMC announced that Megi would move northwestwards and make landfall in the west coast of South China (Guangdong Province) on 23 October 2010. Meanwhile, ECMWF and JTWC made the same prediction (Xu 2011; Qian *et al.* 2012). However, none of them correctly predicted the rapid turning.

To examine what actually caused the left turning of Megi from 14–17 October, height anomalies are extracted along two longitudes at all isobaric levels from 1000 to 100 hPa using Eq. (6). Vertical-meridional section of geopotential height anomaly along 135.75°E longitude at 1200 UTC 15 October 2010 shows a negative center of height anomaly at the lower troposphere around Megi in the 15°N (Fig. 1.7a). Another vertical-meridional section of geopotential height anomaly along 130.50°E longitude at 1200 UTC 15 October 2010 is given in Fig. 1.7b. There was another anomalous vortex in the southwest of the Megi vortex in the lower troposphere. The distance between the two anomalous vortices was under 750 nmi.

To examine whether there were interactions between anomalous circulations that caused the right turning of Megi, Fig. 1.8 shows the vertical-meridional sections of geopotential height anomaly and temperature anomaly along 117.00°E longitude at 1200 UTC 20 October 2010. No other negative centers of height anomaly are found near Megi in the lower troposphere and at around 500 hPa (Fig. 1.8a).

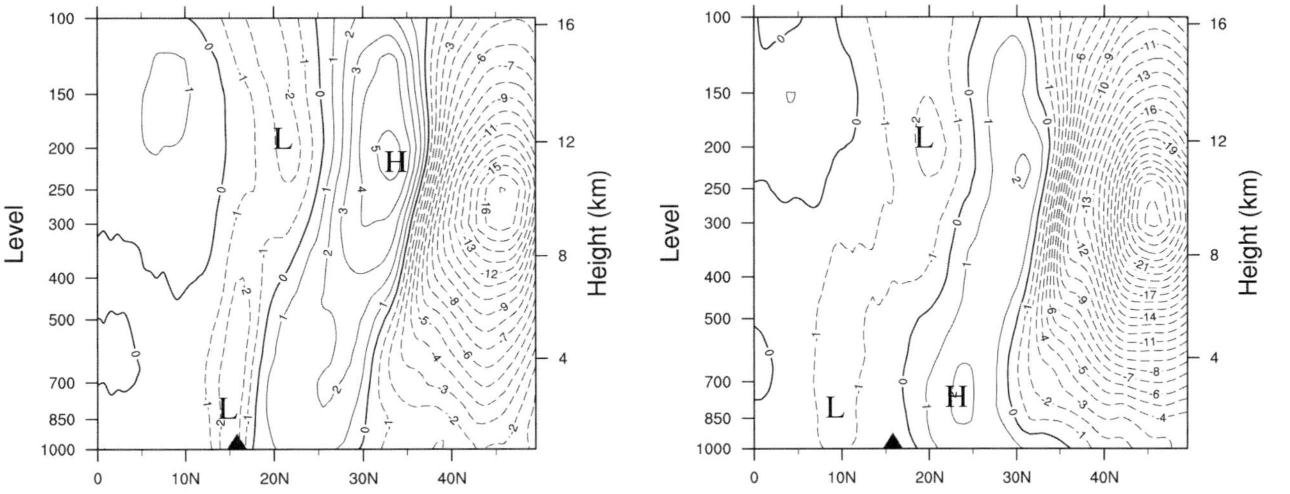

(a) Along 135.75°E

(b) Along 130.50°E

Figure 1.7. Vertical-meridional sections of geopotential height anomalies (10 gpm) along two longitudes of (a) 135.75°E and (b) 130.50°E at 1200 UTC 15 October 2010. The position of Megi is indicated by the symbol ▲. The low and high centers of height anomalies are denoted by letters L and H, respectively.

Source: The data used is obtained from the ERA Interim.

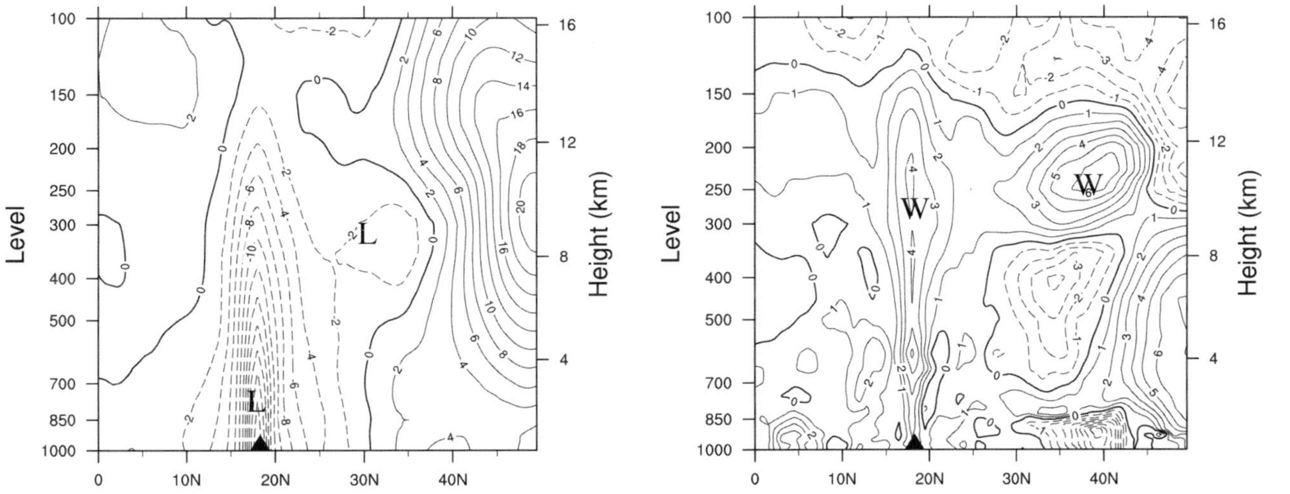

(a) Height anomaly

(b) Temperature anomaly

Figure 1.8. Vertical-meridional sections of (a) geopotential height anomaly (10 gpm) and (b) temperature anomaly (K) along 117.00°E longitude at 1200 UTC 20 October 2010. The position of Megi is indicated by the symbol ▲. The low centers of height anomaly and warm centers of temperature anomaly at troposphere are denoted by letters L and W, respectively.

Source: The data used is obtained from the ERA Interim.

But two negative centers of height anomaly exist in the upper troposphere at around 300 hPa. One was the Megi vortex in the northeastern SCS and another was the subtropical anomalous vortex over the central China continent near the 30°N. This subtropical anomalous vortex is baroclinic with an anomalous warm center at the upper troposphere and an anomalous cold center at the middle troposphere (Fig. 1.8b). The Megi warm vortex extended from 1000 to 150 hPa in the troposphere.

To find whether there were other vortices interacting with the Megi vortex apart from that shown in Fig. 1.8, the vertical-meridional sections of geopotential height anomaly and temporal temperature anomaly along 109.50°E longitude at 1200 UTC 20 October 2010 are shown in Fig. 1.9. There was a baroclinic anomalous vortex in the northwest of the Megi vortex. Two anomalous vortices were centered at around 300 hPa and their distance was less than 750 nmi.

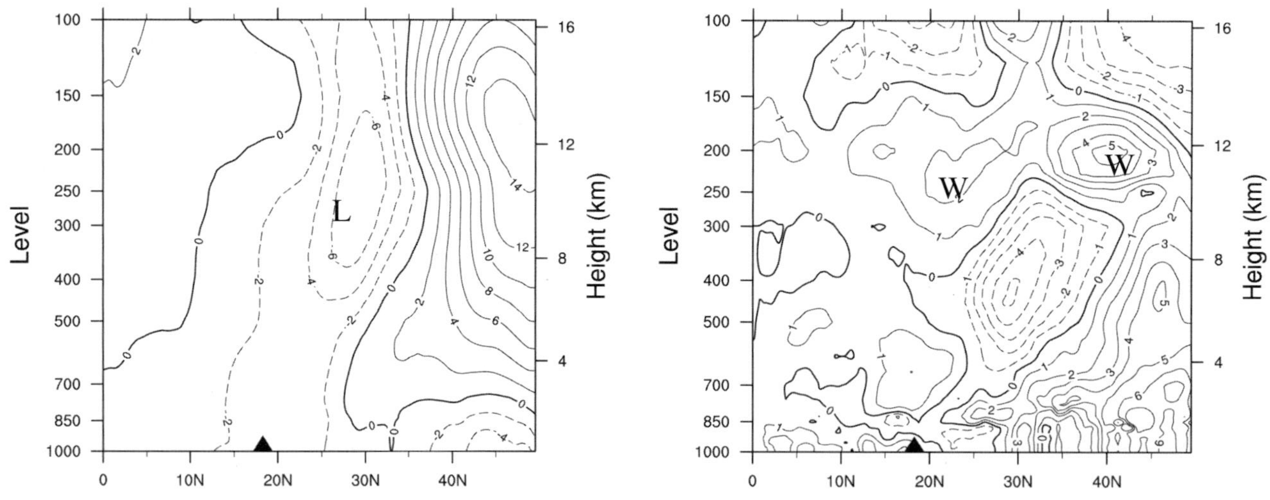

(a) Height anomaly

(b) Temperature anomaly

Figure 1.9. Vertical-meridional sections of (a) geopotential height anomaly (10 gpm) and (b) temperature anomaly (K) along 109.50°E longitude at 1200 UTC 20 October 2010. Symbols used are same as in Fig. 1.8.

Source: The data used is obtained from the ERA Interim.

According to the distributions of height and temperature anomalies in Figs. 1.7–1.9, traditional and anomalous synoptic charts should be drawn at 850 hPa and 300 hPa, respectively. Traditional and anomalous synoptic charts at 850 hPa are shown in Fig. 1.10. From 11 to 13 October 2010, a TB (Megi) formed and moved west-southwestwards basically along the easterly steering flow in the south of the subtropical high in the NW Pacific at the traditional synoptic chart. In the first three days, no other TBs (cyclones) interacted with Megi. An anomalous anti-cyclonic center in its north moved westwards in the first three days. Another anomalous anti-cyclonic center in its south moved eastwards and weakened. The motion of Megi vortex in the first three days might be influenced by the large-scale easterly steering flow and the two anomalous anti-cyclonic centers.

Megi was slowly moving northwestwards on 14 October but it moved southwestwards on 17 October 2010. Its intensity was weak before 16 October and the negative height anomaly was clear in the low troposphere over the TY's position as shown in Fig. 1.7. The traditional synoptic charts with total flow and height at 850 hPa are shown in Fig. 1.11. There was another vortex in the left of Megi on 14–16 October and their distance was less than 750 nmi. The two vortices circled each other in a counterclockwise manner so that the Megi vortex is expected to turn left according to the Fujiwhara effect. On 17 October, two vortices merged together and Megi's intensity reached its maximum.

After removing the climatic wind (or flow) and the zonal-mean wind (or flow) anomaly, the two vortices are clearer than that in the total flow. They circled each other in a counterclockwise manner (Fig. 1.12) and merged in the east sea of the Philippines on 17 October when Megi reached the peak intensity. During this circling process, only the Megi vortex moved and surrounded, while the anomalous vortex was staying still located in the southeast sea of the Philippines.

The flows are different while comparing Figs. 1.11 and 1.12. Easterly steering flow clearly existed in the north of Megi as shown in the traditional synoptic charts. But its left turn cannot be explained simply by this easterly steering flow. The interaction between two vortices can be clearly observed in the traditional and anomalous synoptic charts. In this case, the steering flow and the interaction between two vortices influenced the motion of Megi.

The track and the intensity of Megi were also influenced by the asymmetric climatic wind (or the climatic monsoon flow). The asymmetric climatic wind is averaged from 16 to 20 October at 850 hPa (Fig. 1.13). Megi tended to move toward the climatic monsoon trough before 16 October. It followed the climatic monsoon trough from 16 to 19 October 2010 and the intensity peaked on 1200 UTC 17 October 2010 along the trough before making landfall in the Philippines. The TY Megi is expected to move westwards continually in the northern SCS if there are no influence of other anomalous systems.

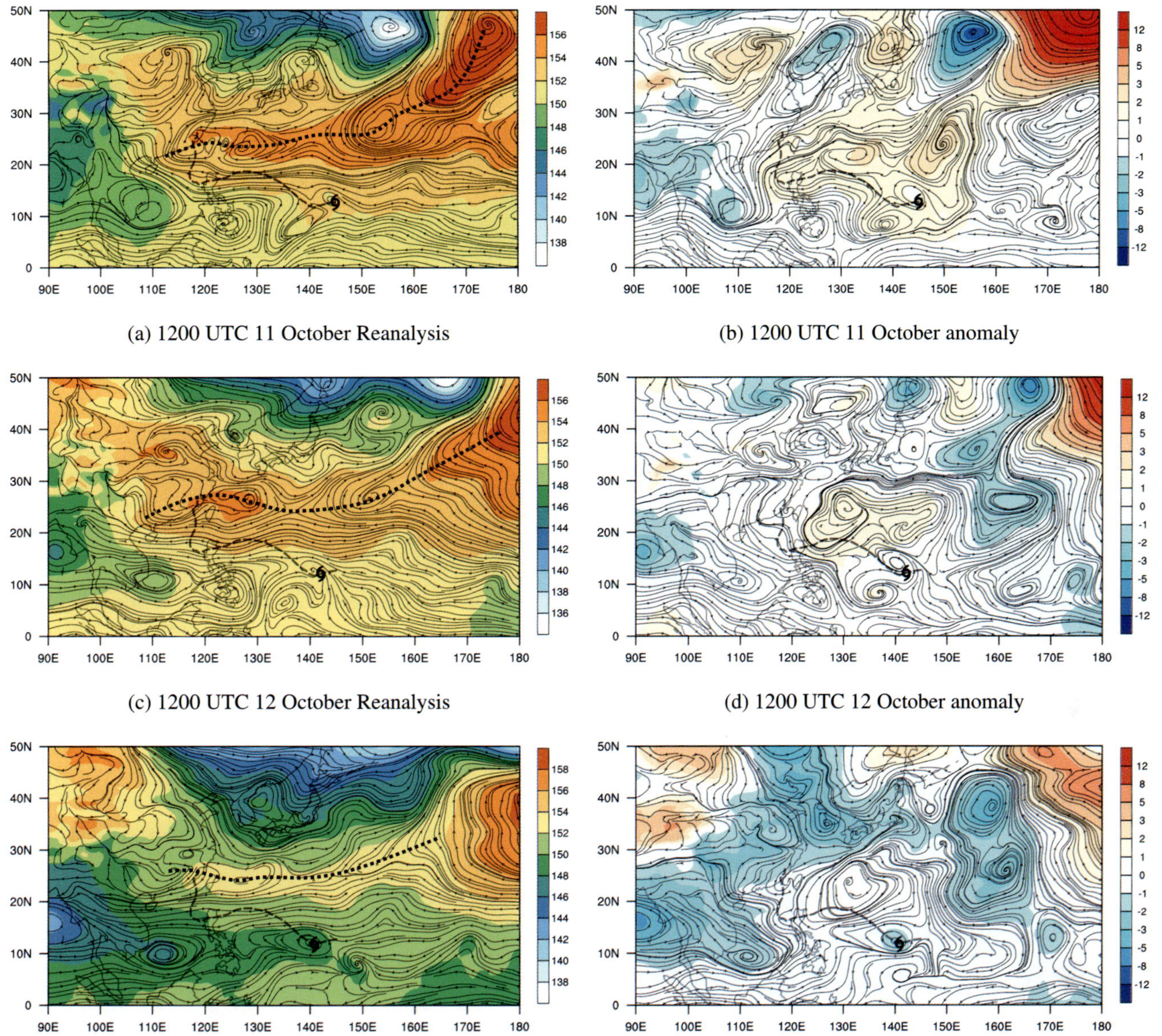

(a) 1200 UTC 11 October Reanalysis

(b) 1200 UTC 11 October anomaly

(c) 1200 UTC 12 October Reanalysis

(d) 1200 UTC 12 October anomaly

(e) 1200 UTC 13 October Reanalysis

(f) 1200 UTC 13 October anomaly

Figure 1.10. Traditional synoptic charts (shading: height; unit: 10 gpm) with total flow in left panels (a), (c), (e) and anomalous synoptic charts (shading: height anomaly; unit: 10 gpm) with flow anomaly in right panels (b), (d), (f) at 850 hPa and at 1200 UTC 11 October, 12 October, and 13 October 2010. The position of TY Megi is indicated by the symbol 𝟔 along its track with black-dashed line and the ridge line of subtropical high is denoted by a dotted line.

Source: The data used is obtained from the ERA Interim.

The unexpected right turn of Megi occurred after it entered the SCS. From 17 to 19 October, the subtropical high ridge was located in the north of Megi with two highs or anti-cyclonic circulation centers in its northeast and northwest, respectively. The easterly flow was in the tropical zone and the westerly flow prevailed in the China continent and Japan. At 1200 UTC 19 October, Megi was located in the easterly steering flow, thus it is expected to continue moving westwards (Fig. 1.14a). Since no tropical cyclone was in the vicinity of Megi, the binary interaction of Fujiwhara effect cannot be used to explain its right turn, based on the traditional synoptic charts.

The right turn of Megi occurred on 19 October 2010. Two troughs at 300 hPa can be traced in the large-scale westerly total flow over the middle-high latitudes from 19 to 22 October 2010 (Fig. 1.14). At 1200 UTC 19 October, the northern and the southern troughs were located in Mongolia and Northwest China, respectively. Then, the northern trough had moved eastwards to North China and the Japan

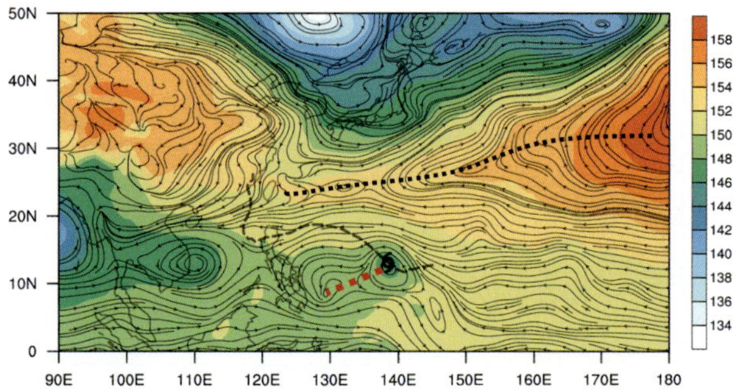

(a) 1200 UTC 14 October

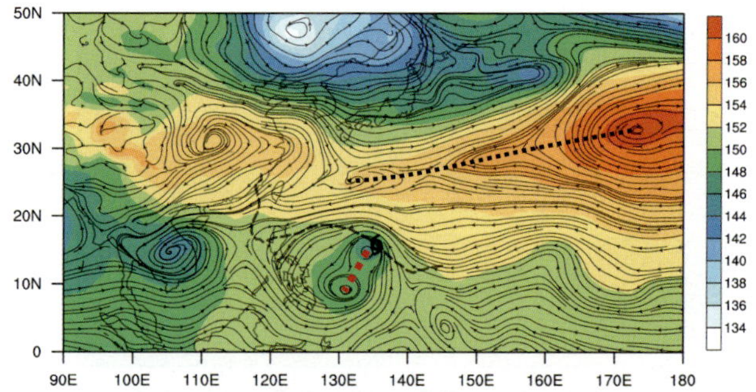

(b) 1200 UTC 15 October

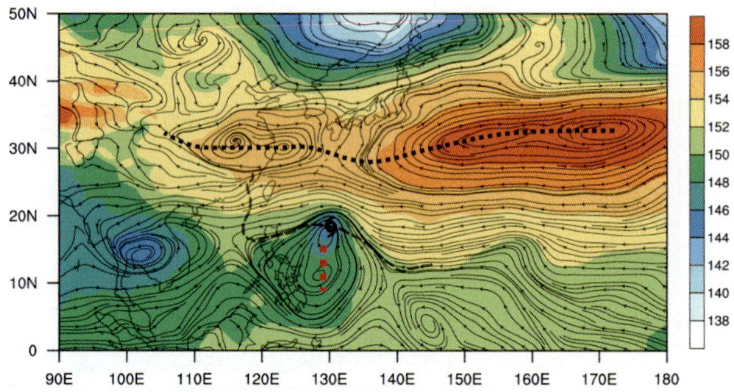

(c) 1200 UTC 16 October

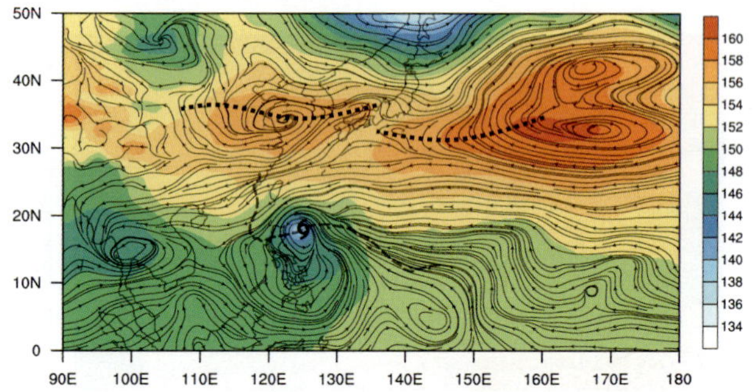

(d) 1200 UTC 17 October

Figure 1.11. Traditional synoptic charts (shading: height; unit:10 gpm) at (a) 1200 UTC 14 October, (b) 1200 UTC 15 October, (c) 1200 UTC 16 October, (d) 1200 UTC 17 October 2010 at 850 hPa. Symbols used are same as in Fig. 1.10. The position of TY Megi is indicated by the symbol 6 and the ridge line of subtropical high is denoted by a dotted line. The two vortices are joined by a red-dotted line.

Source: The data used is obtained from the ERA Interim.

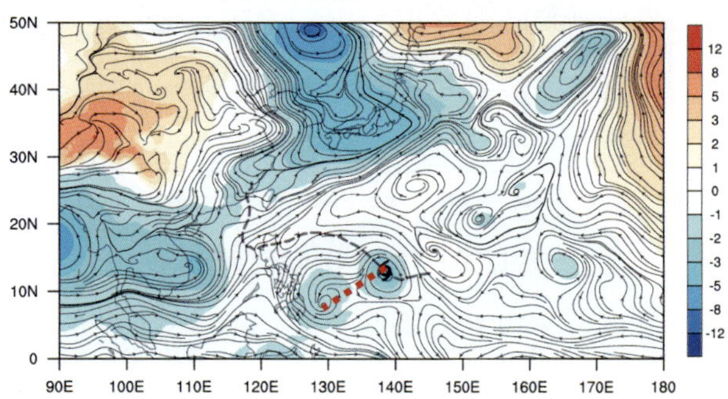

(a) 1200 UTC 14 October

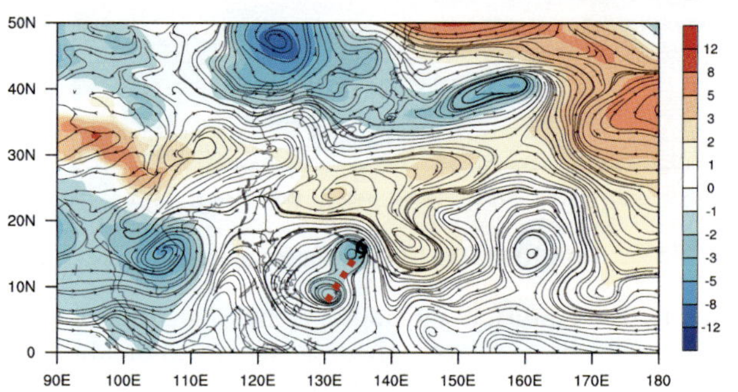

(b) 1200 UTC 15 October

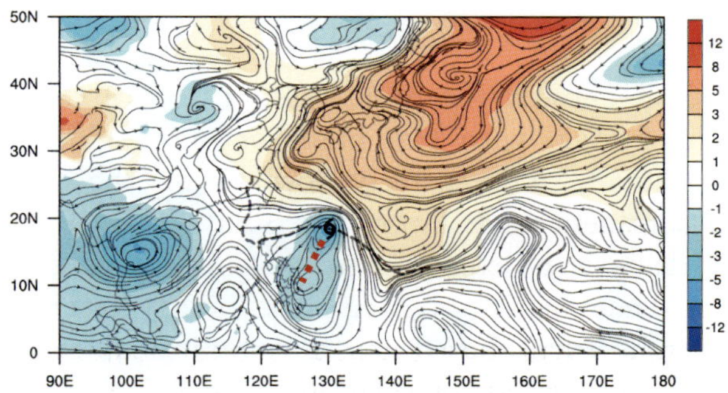

(c) 1200 UTC 16 October

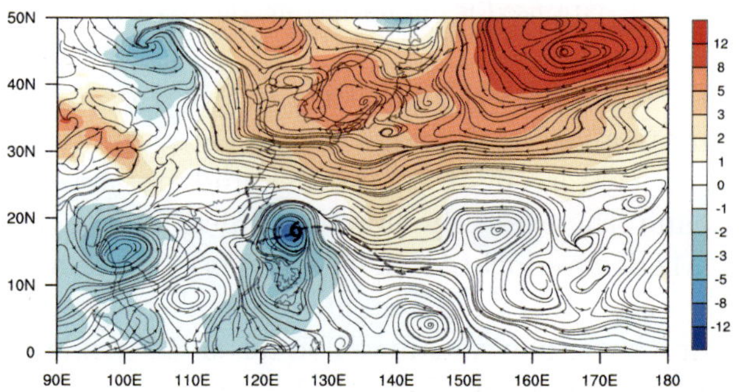

(d) 1200 UTC 17 October

Figure 1.12. Anomalous synoptic charts from (a) 1200 UTC 14 October, (b) 1200 UTC 15 October, (c) 1200 UTC 16 October, (d) 1200 UTC 17 October at 850 hPa. Shading is the height anomaly (unit: 10 gpm). The position of TY Megi is indicated by the symbol 6. The two vortices are joined by a red-dotted line.

Source: The data used is obtained from the ERA Interim.

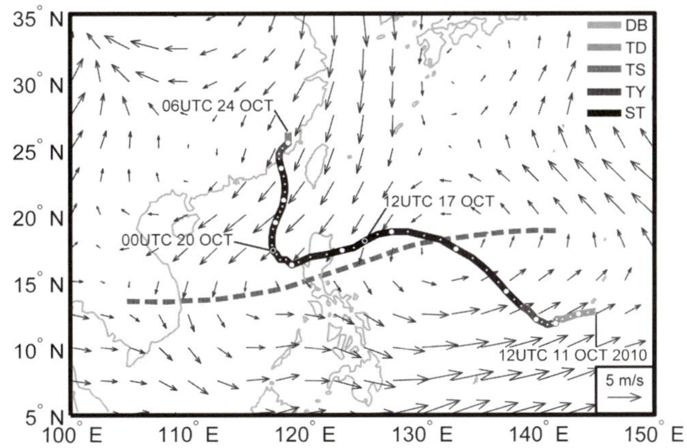

Figure 1.13. The track of TY Megi (2010_No.15) and the climatic monsoon flow (m/s) averaged from the period of 16–20 October at 850 hPa. The climatic monsoon trough is denoted by a heavy-dashed line.

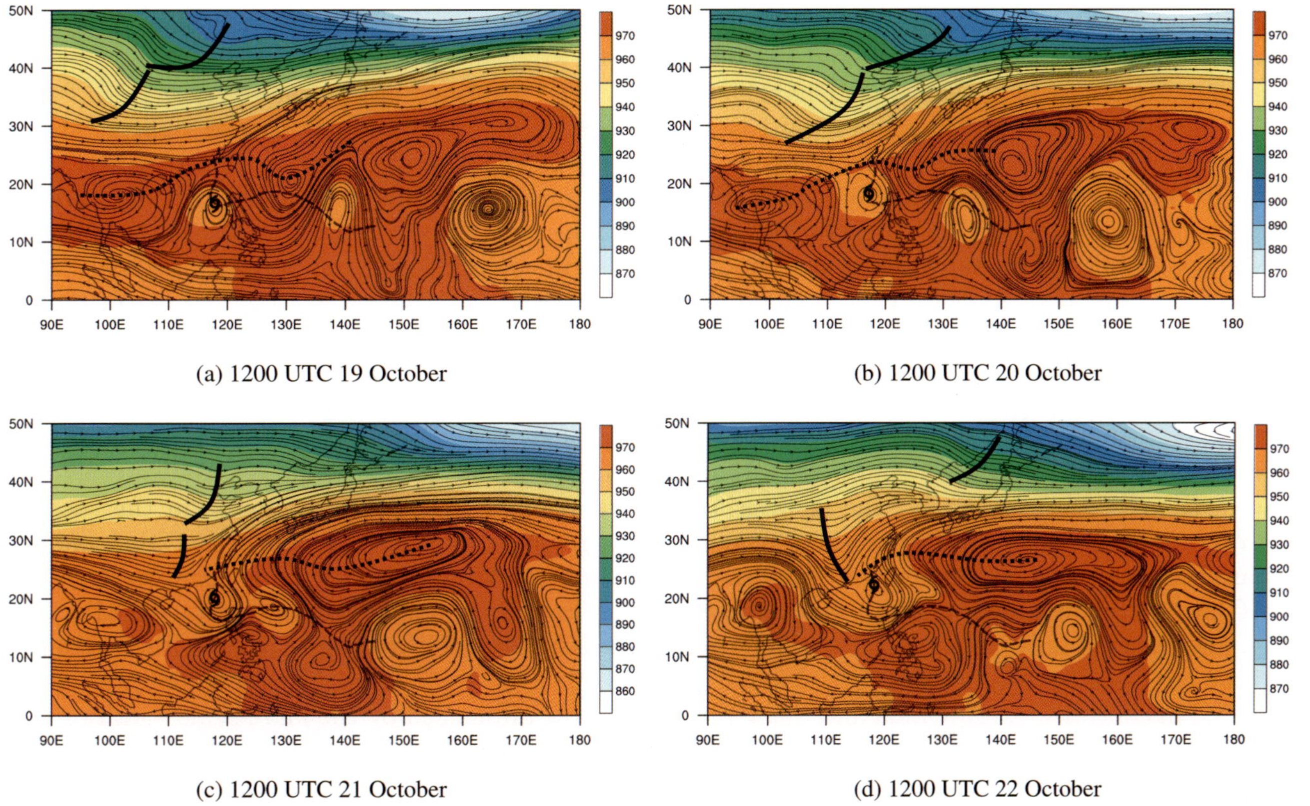

Figure 1.14. Traditional synoptic charts (shading: height; unit:10 gpm) from (a) 1200 UTC 19 October, (b) 1200 UTC 20 October, (c) 1200 UTC 21 October, (d) 1200 UTC 22 October at 300 hPa. The trough and the ridge line of the subtropical high are denoted by a heavy-curved line and a black-dotted line, respectively.

Source: The data used is obtained from the ERA Interim.

Sea, while the southern trough had moved eastward to central and South China from 20 to 22 October. In these four days, the ridge line of the subtropical high was still located in the north of the TY. The trough, the ridge, and the TY are three systems between which there must be interactions but it is difficult to judge whether the TY would continue moving westwards or turn northwards based on the total flow.

The anomalous flow is used to explain the turning right of Megi in the northern SCS as shown in Fig. 1.15. At 0000 UTC 19 October (Fig. 1.15a), two anomalous vortices marked by I and II existed in Mongolia and western China, respectively. Negative height anomalies partially indicated the positions of troughs. Anomalous vortex I moved eastwards in North China at 1200 UTC 19 October and disappeared in Northeast China after 0000 UTC 20 October 2010. Anomalous vortex II moved southeastwards and deepened gradually from 19 to 21 October 2010. Meanwhile, the Megi vortex turned northward rapidly on 19 October 2010. Although there were no other tropical cyclones that interacted with Megi, the anomalous vortex in Southwest China and the Megi vortex in the northern SCS were both evident. The distance between the two vortices was less than 750 nmi since 20 October 2010. They circled around each other in a counterclockwise manner and became closer. When the distance reduces to less than 400 nmi on 22 October 2010 (Figs. 1.15g and 1.15h), the two vortices eventually merged together at 1200 UTC 22 October 2010.

In comparison of Figs. 1.14 and 1.15, the influences of both middle-latitude trough and anomalous vortex on the motion of TY Megi are shown by different circulation patterns. In Fig. 1.14, there are large-scale (or steering flows and the ridge of the subtropical high) and small-scale circulation systems. In Fig. 1.15, however, only small-scale circulation systems, such as anomalous vortices, are observed without the mask of large-scale steering flows or the climatic ridge of the subtropical high. Therefore, the interaction between the Megi vortex and another anomalous vortex is apparent in Fig. 1.15. They experienced a rotating and merging process, which is very similar to the Fujiwhara effect. It is believed that this effect can only be determined after the environmental flow or climatic flow is filtered out from observations as described by Dong and Neumann (1983).

The second case of TY is named Chanchu (2006_No.02). It formed in the equatorial western Pacific on 7 May 2006 and moved northwestward to the central SCS (Fig. 1.16a). It turned right at around 1800 UTC 14 May before its intensity peaked at 0000 UTC 15 May 2006. From 11 to 14 May 2006, Chanchu moved northwestwards basically along the climatic monsoon trough (Fig. 1.16b) and its intensity peaked at 0000 UTC 15 May 2006 before turning right.

The vertical-meridional sections of geopotential height anomaly and temperature anomaly along 20.25°N latitude at 0000 UTC 17 May 2006 are shown in Fig. 1.17. The warm-core vortex of the TY Chanchu was obviously located in the upper troposphere. Therefore, the flow at 300 hPa is chosen to illustrate the interaction of different systems.

The relationship between total flow and the TY Chanchu during the turning right period from 14 to 17 May 2006 is shown in Fig. 1.18. At 1200 UTC 14 May 2006 (Fig. 1.18a), there was a large-scale westerly flow from the China continent to Japan in middle latitude and a large-scale southwesterly flow in the equatorial SCS and the equatorial western Pacific. The TY was located in the south of the ridge line of subtropical high. Two troughs were, respectively, located in North China and the Japan Sea within the large-scale westerly flow, and a southern trough was in the southwestern SCS to the Gulf of Thailand. The three troughs and the ridge line in its north still existed on 15 May 2006 (Fig. 1.18b). It is hard to explain the right turn of the TY Chanchu by these troughs, the ridge line, and the large-scale steering flows. On 16 and 17 May 2006 (Figs. 1.18c and 1.18d), a trough was clearly observed in central-western China and the TY Chanchu was influenced by the large-scale southwesterly steering flow when it entered the northernmost SCS and made landfall on the South China coast. The prediction of its right turn in the central SCS from 14 to 15 May 2006 is also difficult by analyzing these troughs, the ridge line, and the large-scale steering flows (Figs. 1.18a and 1.18b).

After removing the climatic flow and the zonal-averaged flow anomaly, anomalous circulation systems are clearly observed in Fig. 1.19. At 1200 UTC 14 May 2006 (Fig. 1.19a), there were two anomalous vortices (V1 and V2) in the east and west of the Chanchu vortex. However, their distances were greater than 750 nmi so they might not have any interaction. An anomalous anti-cyclone in the north of Chanchu is identified in Fig. 1.19a. This anomalous subtropical high separated the TY Chanchu from the anomalous vortex in northern China. At 1200 UTC 15 May 2006 (Fig. 1.19b), two anomalous vortices in the middle latitude, marked by I and II, appeared in western China. From 16 to 17 May 2006 (Figs. 1.19c and 1.19d), anomalous vortex I disappeared in northern China but anomalous vortex II moved southeastwards day by day. Anomalous vortex II and the Chanchu vortex both circled around each other in a counterclockwise manner for about three days. The interaction between the two anomalous vortices is similar to the Fujiwhara effect.

The third case of TY is named Linfa (2009_No.03). It originated in the southeast sea of the Philippines at 0600 UTC 13 June 2009 and moved northwestwards basically along the climatic monsoon trough (Fig. 1.20). It became stagnant from June 17 to 19 and then turned right rapidly. Its intensity

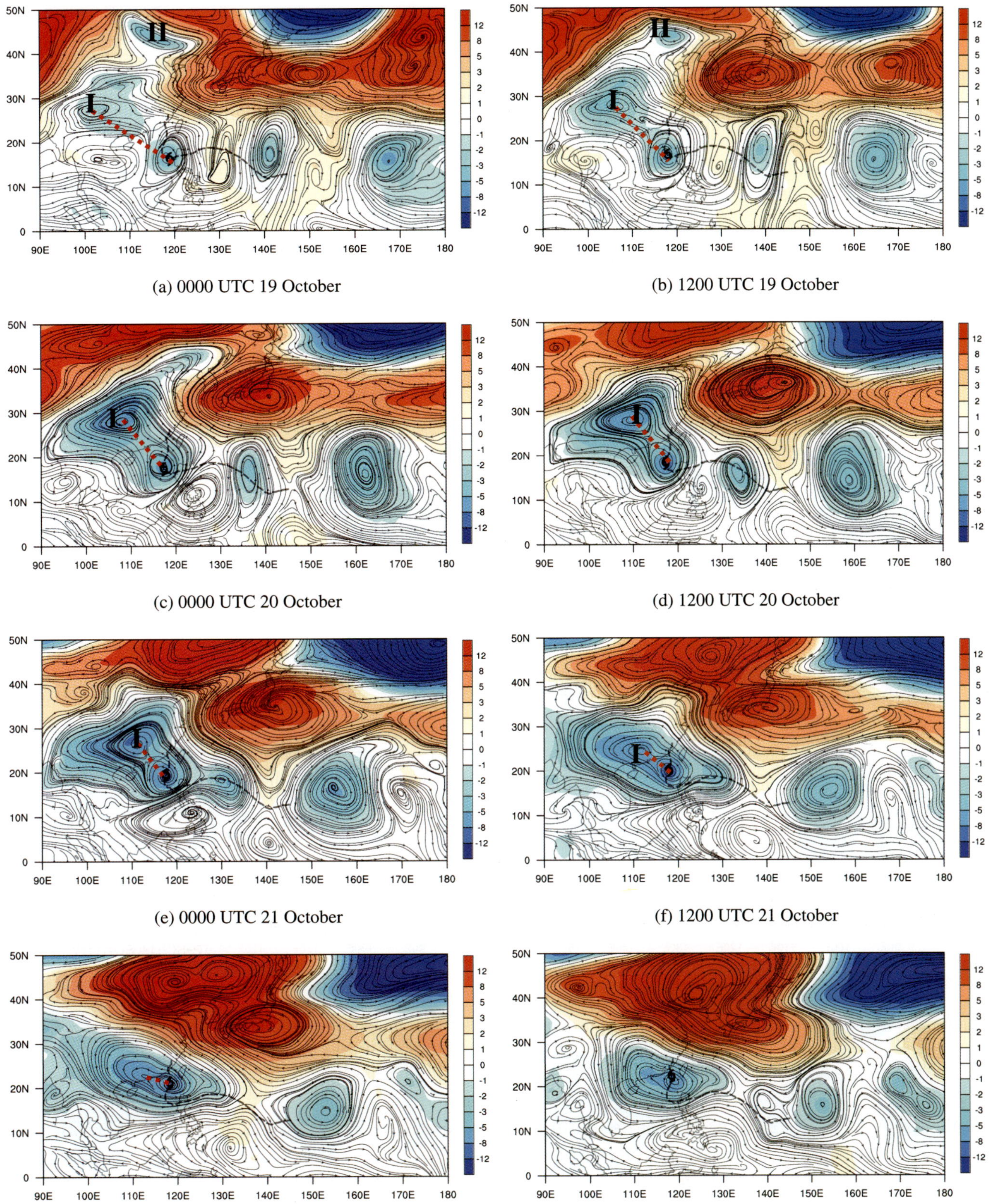

(a) 0000 UTC 19 October (b) 1200 UTC 19 October

(c) 0000 UTC 20 October (d) 1200 UTC 20 October

(e) 0000 UTC 21 October (f) 1200 UTC 21 October

(g) 0000 UTC 22 October (h) 1200 UTC 22 October

Figure 1.15. Anomalous synoptic charts with flow anomaly and height anomaly (shading; unit: 10 gpm) at (a) 300 hPa at 0000 UTC and (b) 1200 UTC 19 October, (c) 0000 UTC and (d) 1200 UTC 20 October, (e) 0000 UTC and (f) 1200 UTC 21 October, (g) 0000 UTC and (h) 1200 UTC 22 October 2010, respectively. Two anomalous vortices are indicated by I and II, respectively. The red-dashed lines connect two vortices.

Source: The data used is obtained from the ERA Interim.

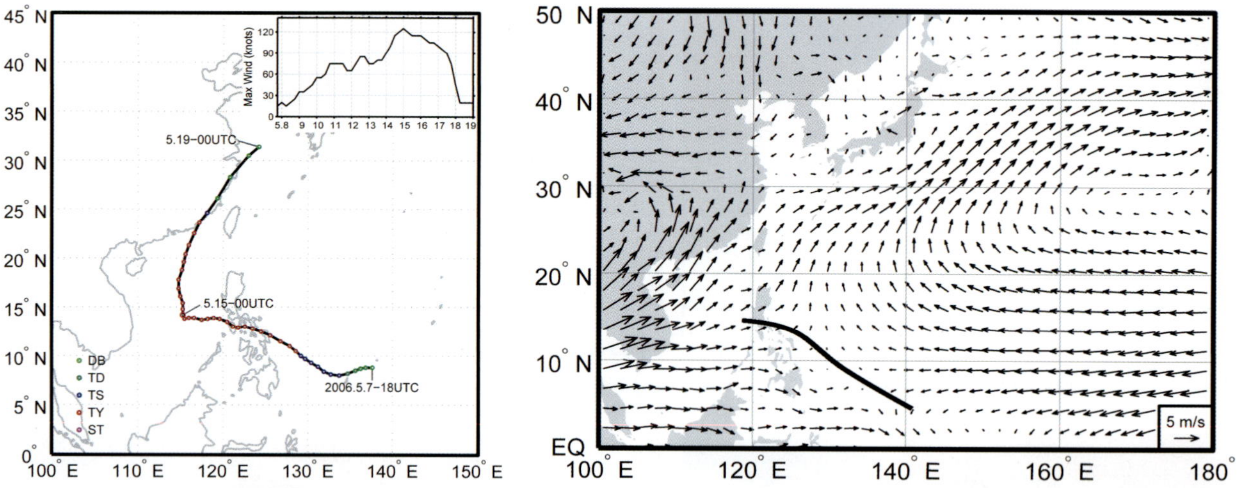

(a) 2006_02_TC Track

(b) Pentad 27 (11–15 May)

Figure 1.16. (a) The track and intensity (maximum wind: knot in the upper panel) of the TY Chanchu (2006_No.02) and (b) the climatic monsoon flow (m/s) averaged in the period of pentad 27 (11–15 May). In (a), the TY position is marked by open dots at 6-h interval, and linked by a heavy line. Its intensity is indicated by disturbance (DB), tropical depression (TD), tropical storm (TS), typhoon (TY), and super typhoon (ST) with various colors labeled in the bottom left corner. In (b), the climatic monsoon trough is denoted by a black line.

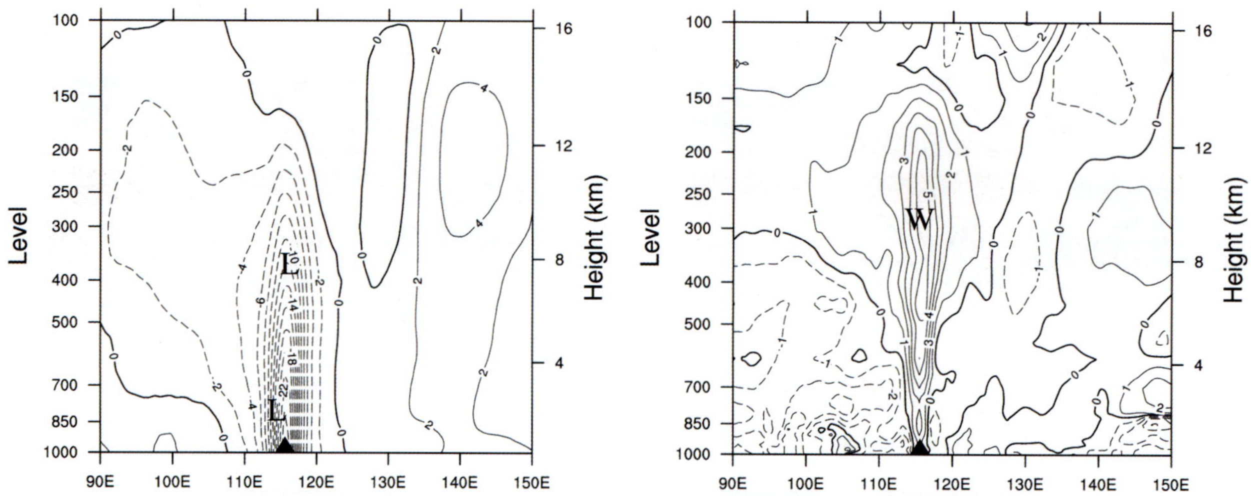

(a) Height anomaly

(b) Temperature anomaly

Figure 1.17. Vertical-meridional sections of (a) geopotential height anomaly (10 gpm) and (b) temperature anomaly (K) along 20.25°N latitude at 0000 UTC 17 May 2006. Symbols used are same as in Fig. 1.8.

peaked at 1200 UTC 20 June before its landfall on the southeast coast of China at 1200 UTC 21 June 2009.

Same as previous two cases, vertical-meridional sections of geopotential height anomaly and temperature anomaly along 18.00°N latitude at 0000 UTC 19 June 2009 for the TY Linfa is shown in Fig. 1.21. An anomalous warm-core vortex with two warm centers at around 850 hPa and 300 hPa respectively, is clearly observed. The anomalous vortex also extended from the lower troposphere to the upper troposphere. For this TY, the circulation systems at 850 hPa and 300 hPa are examined, respectively.

Total flows at 850 hPa and the motion of TY Linfa are shown in Fig. 1.22. On 16 and 17 June 2009 (Figs. 1.22a and 1.22b), the ridge line of the subtropical high was in the north of the TY so it moved westwards. From 18 to 19 June 2009 (Figs. 1.22c and 1.22d), the large-scale steering flow was

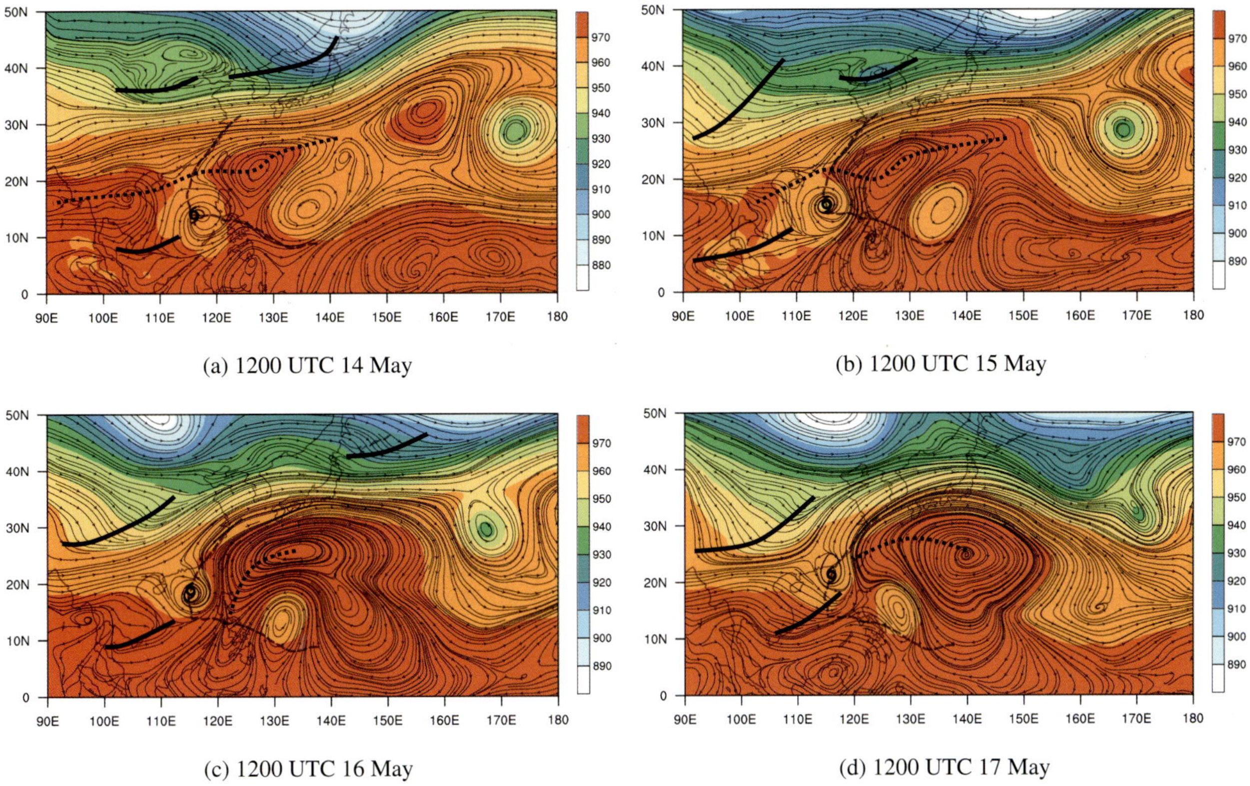

Figure 1.18. Total flows at 300 hPa at 1200 UTC on (a) 14, (b) 15, (c) 16, and (d) 17 May 2006. Symbols used are same as in Fig. 1.14.

Source: The data used is obtained from the ERA Interim.

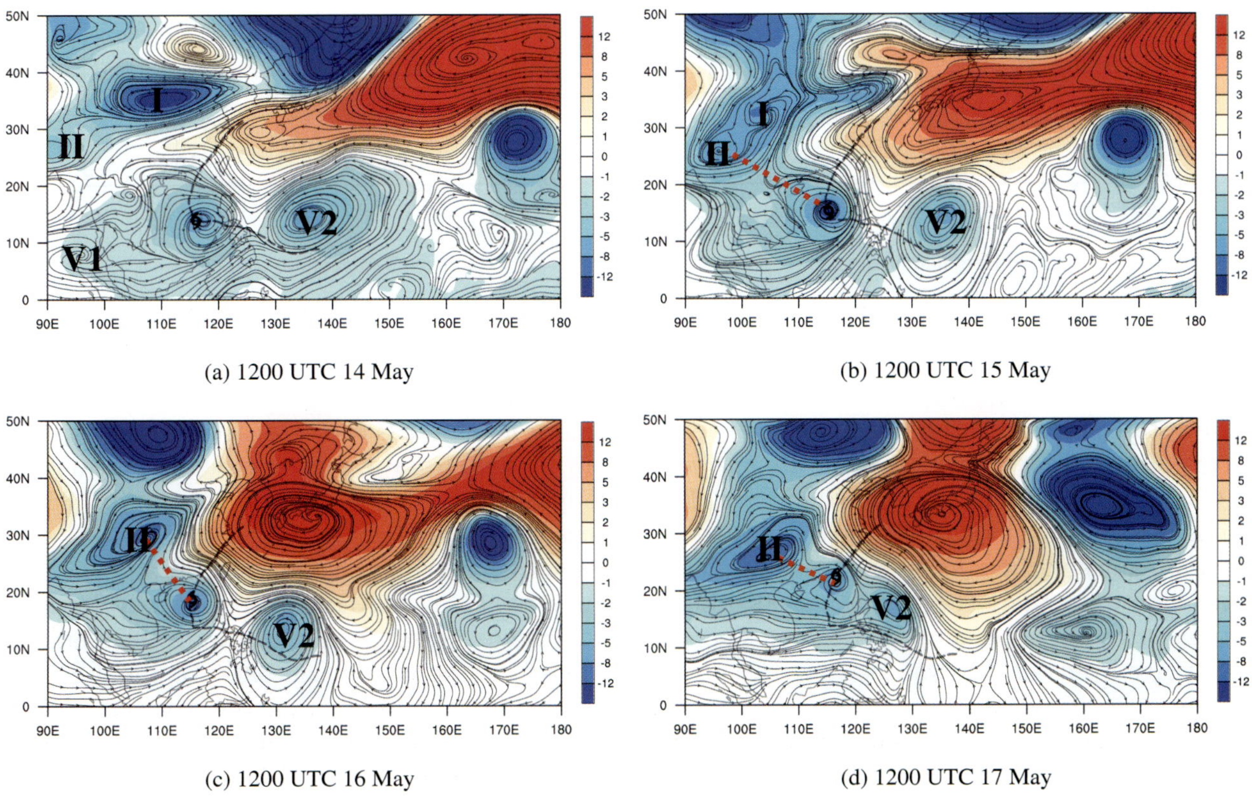

Figure 1.19. Anomalous synoptic charts with flow anomaly and height anomaly (shading; unit: 10 gpm) at 300 hPa at 1200 UTC on (a) 14, (b) 15, (c) 16, and (d) 17 May 2006. Two anomalous vortices in the low latitude are indicated by V1 and V2, respectively. Symbols used are same as in Fig. 1.15.

Source: The data used is obtained from the ERA Interim.

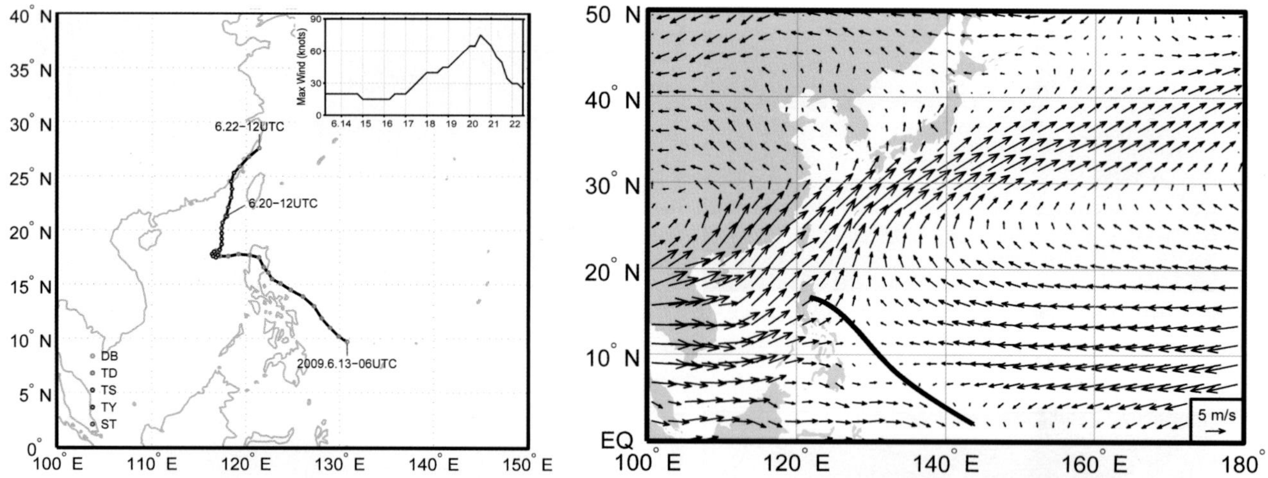

(a) 2009_03_TC Track

(b) Pentad 34 (15–19 June)

Figure 1.20. The track and intensity (maximum wind: knot in the upper panel) of the TY Linfa (2009_No.03) and (b) the climatic monsoon flow (m/s) averaged in the period of pentad 34 (11–15 May). Symbols used are same as in Fig. 1.16.

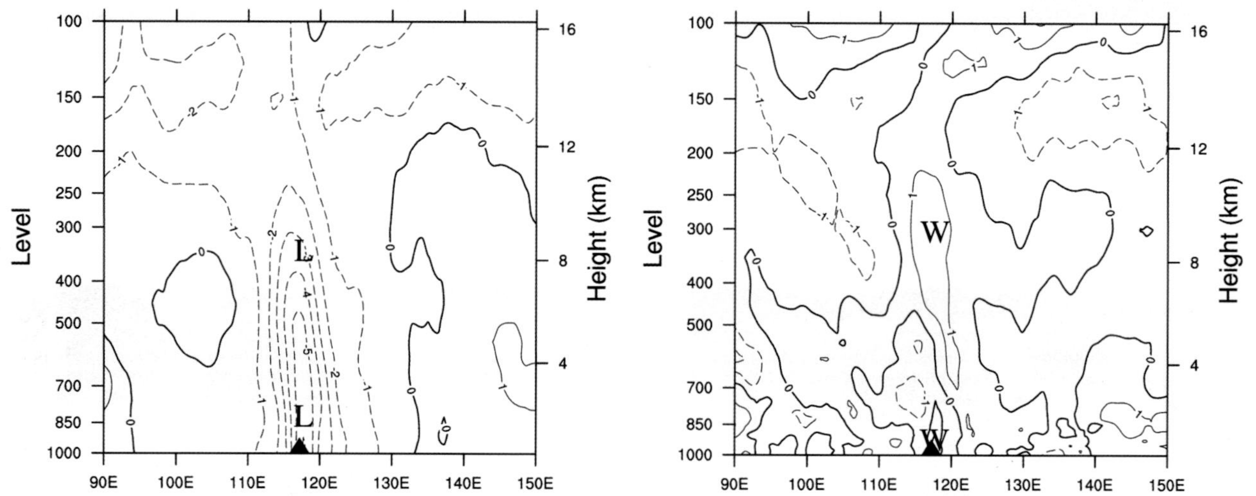

(a) Height anomaly

(b) Temperature anomaly

Figure 1.21. Vertical-meridional sections of (a) geopotential height anomaly (10 gpm) and (b) temperature anomaly (K) along 18.00°N latitude at 0000 UTC 19 June 2009. Symbols used are same as in Fig. 1.8.

southwesterly wind and the TY, thus, is expected to move northwards or northeastwards. However, it had experienced a stagnant state for about two days. In these four days, several troughs were active in the China continent and the East China Sea. These troughs and the large-scale steering flow at 850 hPa cannot explain the changes of the TY motion in this case.

Anomalous vortices are clearly observed at 850 hPa after removing the climatic flow and the zonal-averaged flow anomaly (Fig. 1.23). At 0000 UTC 16 June 2009 (Fig. 1.23a), there were two anomalous vortices in the Southeast Asian region. The larger anomalous vortex I was located in Thailand but its distance from the TY was greater than 750 nmi. Therefore, the interaction between the TY vortex and the eastern anomalous vortex II should be examined. In these four days, the TY vortex in the northeastern SCS and the anomalous vortex II in the southeast sea of the Taiwan Island circled around each other in a counterclockwise manner, i.e. the eastern anomalous vortex II moved northwards and the TY vortex moved westwards.

Total flows at 300 hPa are also used to examine the motion of TY Linfa (Fig. 1.24). At 0000 UTC 16 June 2009 (Fig. 1.24a), the TY appeared in the easterly total flow. Since the ridge line of the subtropical high was in its north, it moved westwards. The Asian continental subtropical high was strengthened

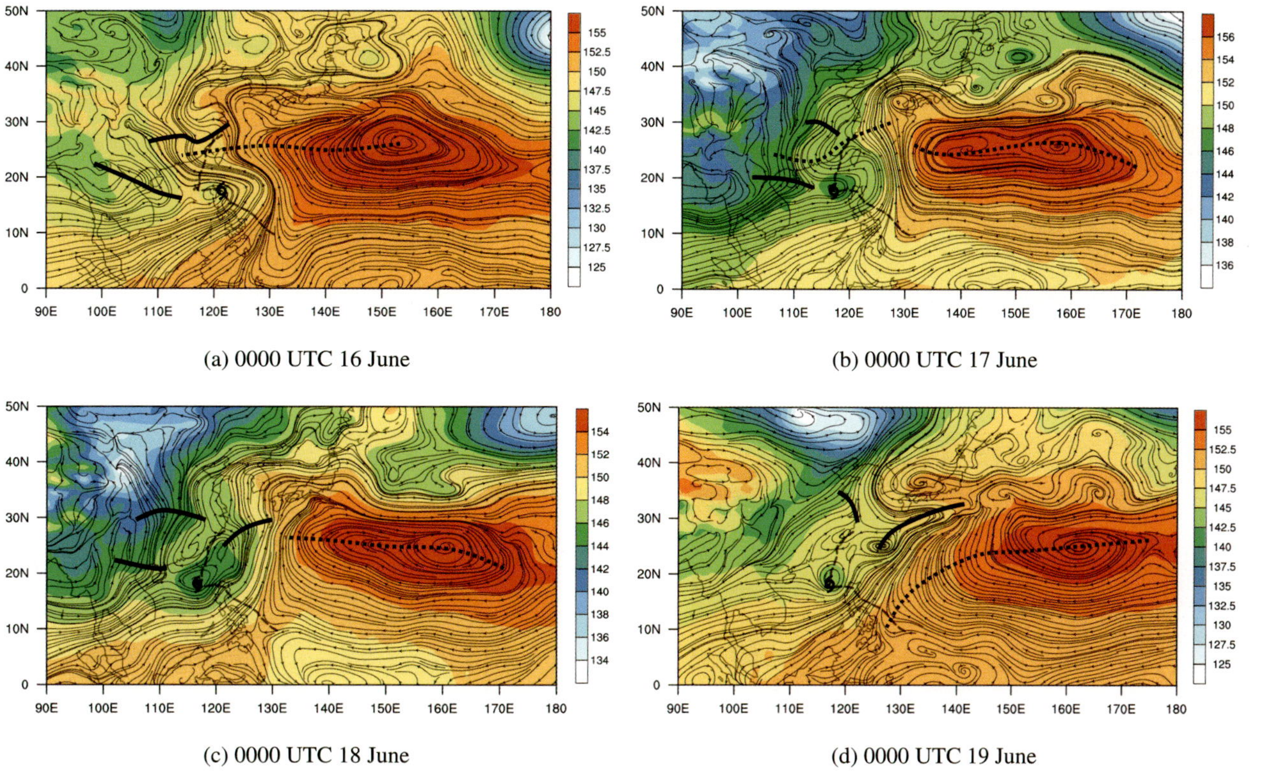

Figure 1.22. Total flows at 850 hPa at 0000 UTC on (a) 16, (b) 17, (c) 18, and (d) 19 June 2009. Symbols used are same as in Fig. 1.14.

Source: The data used is obtained from the ERA Interim.

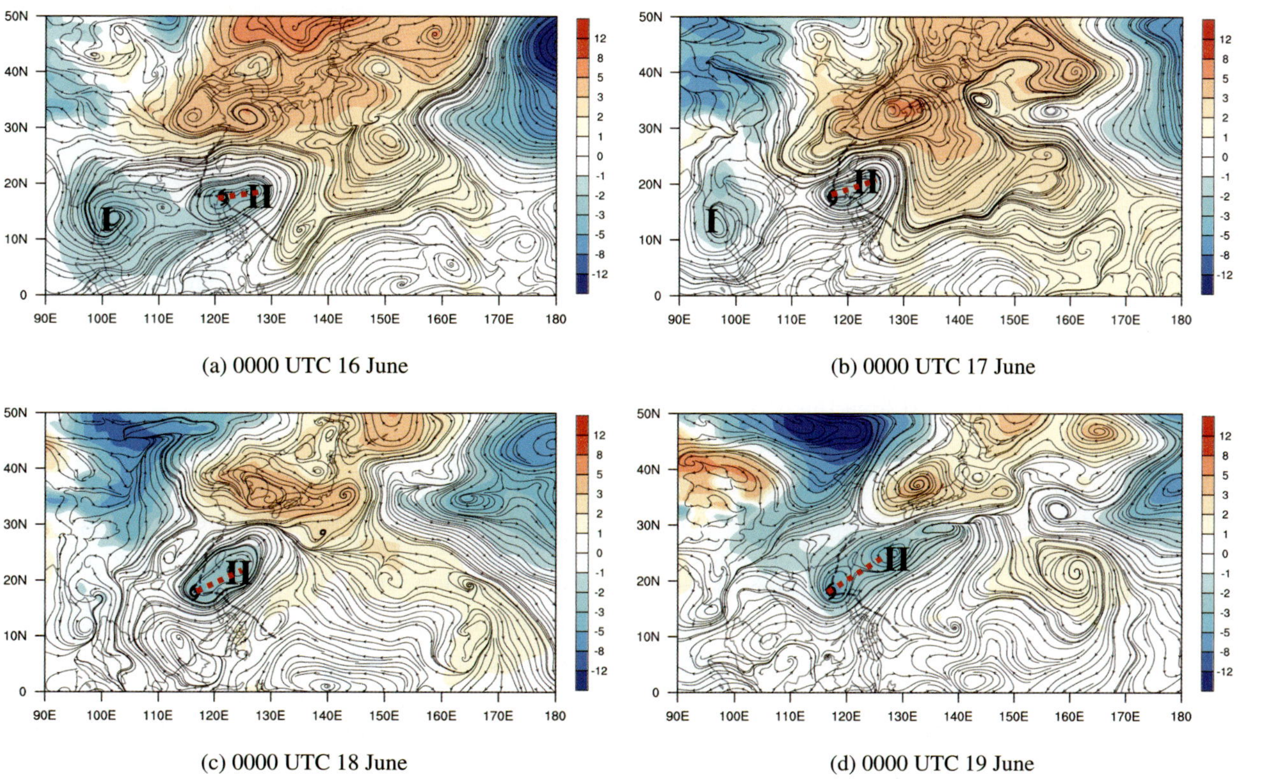

Figure 1.23. Anomalous synoptic charts with flow anomaly and height anomaly (shading; unit: 10 gpm) at 850 hPa at 0000 UTC on (a) 16, (b) 17, (c) 18 and (d) 19 June 2009. Symbols used are same as in Fig. 1.15.

Source: The data used is obtained from the ERA Interim.

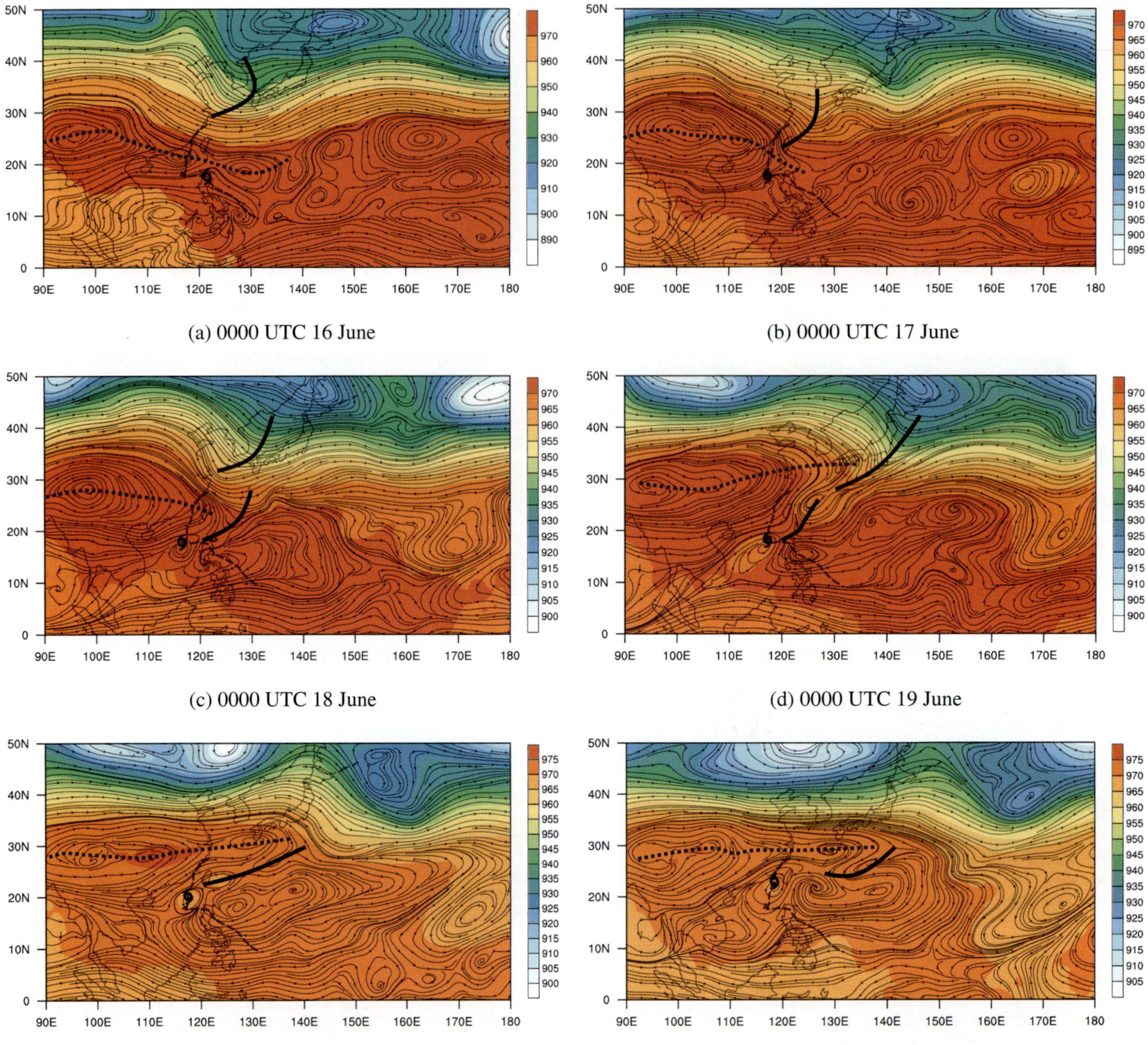

(a) 0000 UTC 16 June

(b) 0000 UTC 17 June

(c) 0000 UTC 18 June

(d) 0000 UTC 19 June

(e) 0000 UTC 20 June

(f) 0000 UTC 21 June

Figure 1.24. Total flows at 300 hPa at 0000 UTC on (a) 16, (b) 17, (c) 18, (d) 19, (e) 20, and (f) 21 June 2009. Symbols used are same as in Fig. 1.14.

Source: The data used is obtained from the ERA Interim.

on 17 June 2009 (Fig. 1.24b). Under the easterly steering flow in the south of the ridge line of the subtropical high, the TY is expected to move westwards continually. However, it experienced a stagnant period for about two days instead. Thus, this stagnant period cannot be explained by the large-scale easterly total flow or easterly steering flow. From 19 to 21 June 2009, the TY appeared in the large-scale easterly total flow or easterly steering flows and it is expected to move westwards, but it actually turned right in the northeastern SCS. Thus, the large-scale easterly total flow or easterly steering flows cannot explain the right turn of this TY motion in the northeastern SCS. Linfa's motion was associated with troughs, shears, and a small vortex. In the first four days, a westerly trough moved eastwards from the East China Sea to the east sea of Japan, while Linfa moved westwards and then with a stagnant state for about two days. A shear line was active in the east sea of the Taiwan Island from 18 to 21 June 2009. On 20 June, a small closed vortex was formed along the shear line and located near the Taiwan Island. After only one day, this vortex and the TY merged together. Thus, the shear line and the small

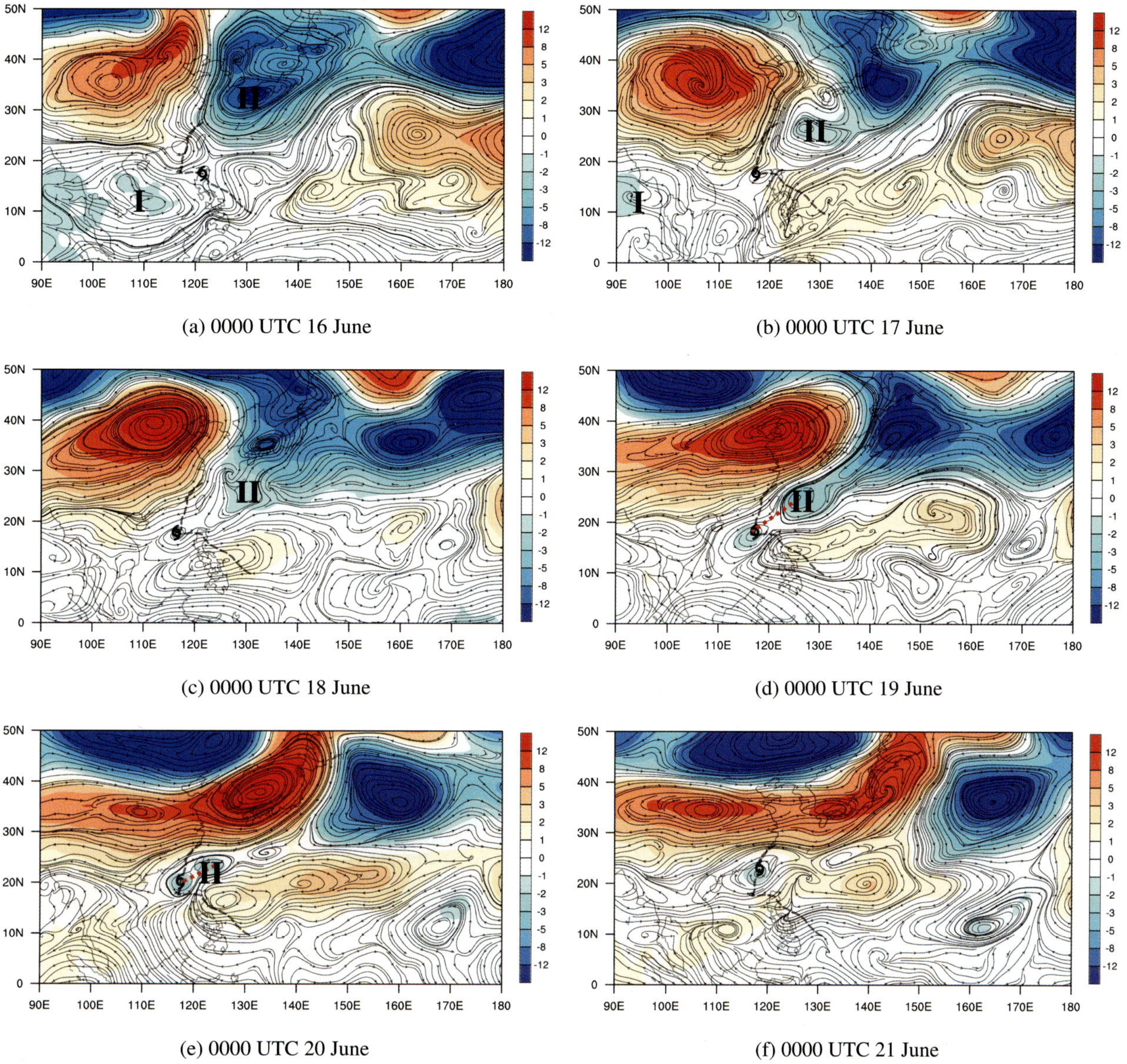

Figure 1.25. Anomalous synoptic charts with flow anomaly and height anomaly (shading; unit: 10 gpm) at 300 hPa at 0000 UTC on (a) 16, (b) 17, (c) 18, (d) 19, (e) 20, and (f) 21 June 2009. Symbols used are same as in Fig. 1.15.

Source: The data used is obtained from the ERA Interim.

vortex were linked with the unusual motion of TY while the westerly trough cannot explain this unusual motion.

The small-scale anomalous vortex is clearly observed when the climatic flow and the zonal-averaged flow anomaly are removed from the total flow (Fig. 1.25). At 0000 UTC 16 June 2009 (Fig. 1.25a), two small-scale anomalous high circulations were located in the east and northwest of the TY. Anomalous vortex I existed in the southwest of the TY. The TY is expected to move westwards in this situation. The distance between the anomalous vortex I and the Linfa vortex was less than 400 nmi, so they attracted each other. From 17 to 18 June 2009 (Figs. 1.25b and 1.25c), the reversed anomalous flows appeared in southeast and northwest of the TY. The anomalous vortex I had moved away from the Linfa vortex, so it is expected to have a stagnant period. On 19 June 2009 (Fig. 1.25d), another small-scale anomalous vortex II appeared in the northeast of the TY. This small-scale anomalous vortex II shortened its distance from the Linfa vortex to less than 400 nmi on 20 June 2009 (Fig. 1.25e). Finally, they attracted each other and merged on 21 June 2009 (Fig. 1.25f). It is clear that the attraction and mergence of the TY vortex and the small-scale anomalous vortex II led to the right turn.

It is noted that the motion of TY is determined by the climatic monsoon flow and the synoptic-scale flow anomaly. The climatic monsoon trough provides one of the favorable conditions for TY developing and moving. Usual motion of TY basically follows the climatic monsoon flow and particularly follows the climatic monsoon trough while its unusual motion is influenced by the synoptic-scale flow anomaly. The interaction between a TY vortex and an anomalous vortex in the vicinity can be clearly identified after the total flow is decomposed into four physical components. This identification is useful to us developing a model to forecast the unusual typhoon track.

1.5. Climatic Monsoon Troughs

The global atmospheric centers of action and global monsoon system have been calculated from daily geopotential heights or daily winds based on Eq. (3). Figure 1.26 shows the distributions of atmospheric centers of action and global monsoon troughs based on the 850 hPa geopotential height. Globally, there are 22 climatic troughs including 9 climatic monsoon troughs (Qian and Tang 2010). Among them, there are three planetary-scale monsoon troughs: (i) the ITCZ in the tropical Pacific, (ii) the tropical Indian Ocean, (iii) and the tropical Atlantic Ocean, respectively. Around the Tibet Plateau, there are five peninsula-scale troughs. In the region of the SCS and the NW Pacific, the peninsula-scale monsoon trough (No. 8) is located in the SCS, the subtropical monsoon trough (No. 9) reaches the northernmost part in East Asia in early August, and the Pacific tropical monsoon trough (No. 1) is active in the tropical NW Pacific.

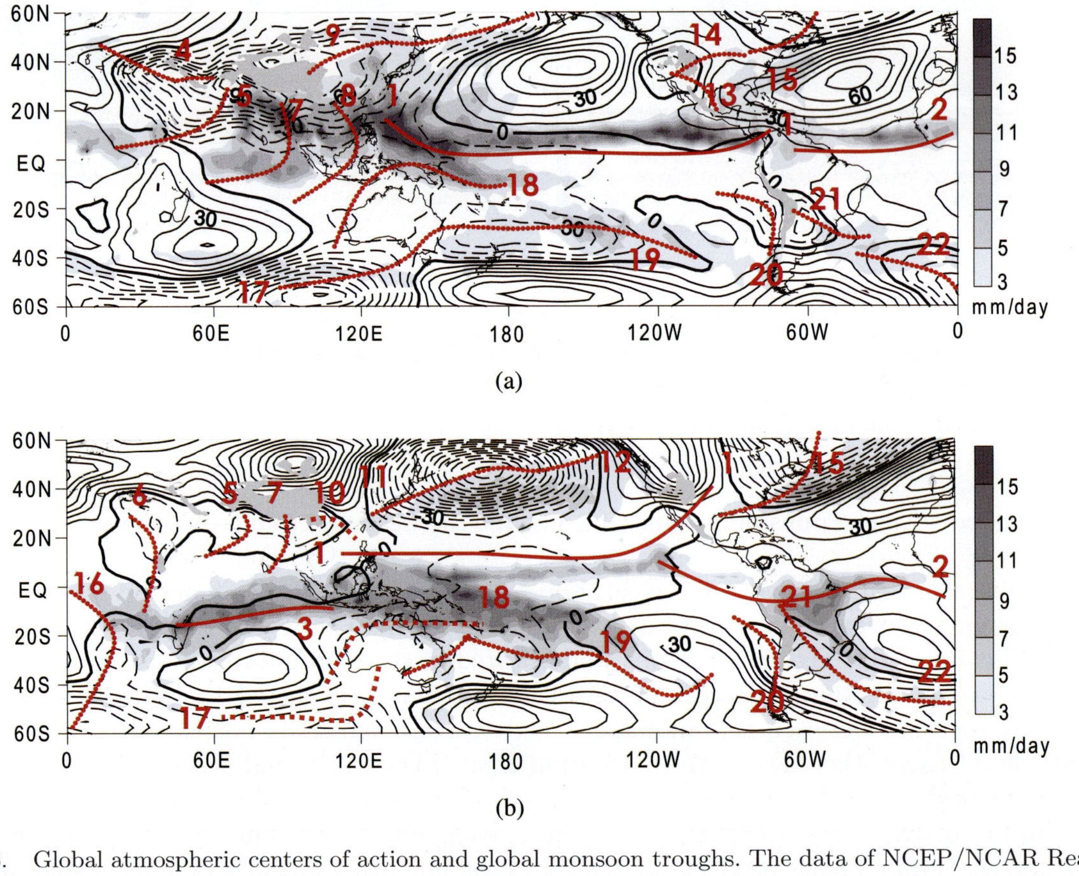

Figure 1.26. Global atmospheric centers of action and global monsoon troughs. The data of NCEP/NCAR Reanalysis is used to calculate the asymmetric climate of geopotential height (Qian and Tang 2010). The shading represents the pentad-mean precipitation (mm/day) and the contour lines denote the asymmetric climate of 850 hPa geopotential height (gpm) at pentad 44 (a) and pentad 8 (b). Heavy solid red lines indicate the ITCZ and dotted red lines represent peninsula-scale (or basin-scale) troughs. A total of 22 troughs are marked by numbers. Among them, nine monsoonal troughs occupy monsoonal precipitation. The numbers 1, 2, and 3 represent the ITCZ in Pacific, Atlantic, and Indian Ocean, respectively.

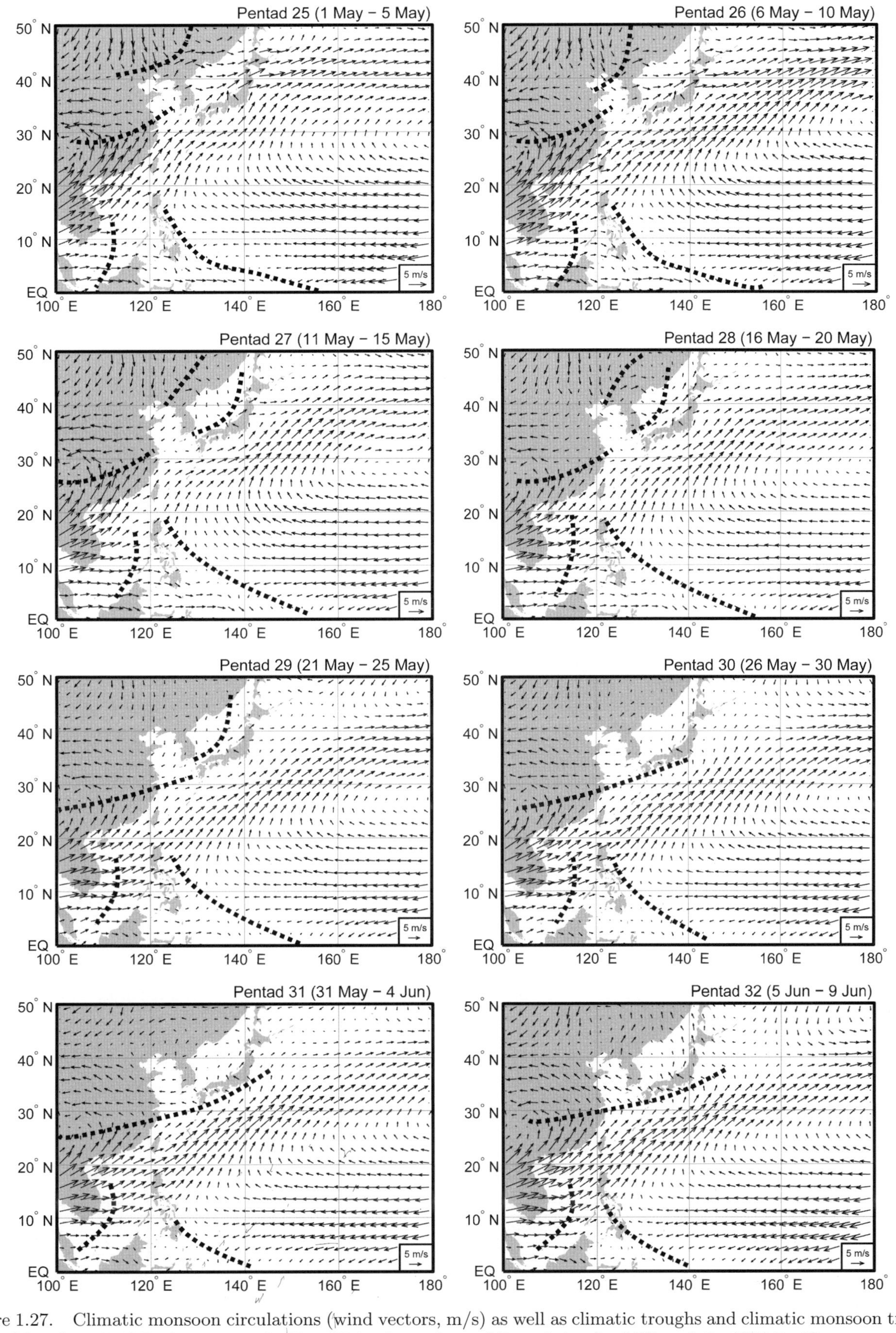

Figure 1.27. Climatic monsoon circulations (wind vectors, m/s) as well as climatic troughs and climatic monsoon troughs (dotted lines) at 850 hPa from pentads 25 to 73 in the regions of East Asia, the SCS, and the NW Pacific.

Source: The data of the ERA Interim with the 2.25° resolution is used.

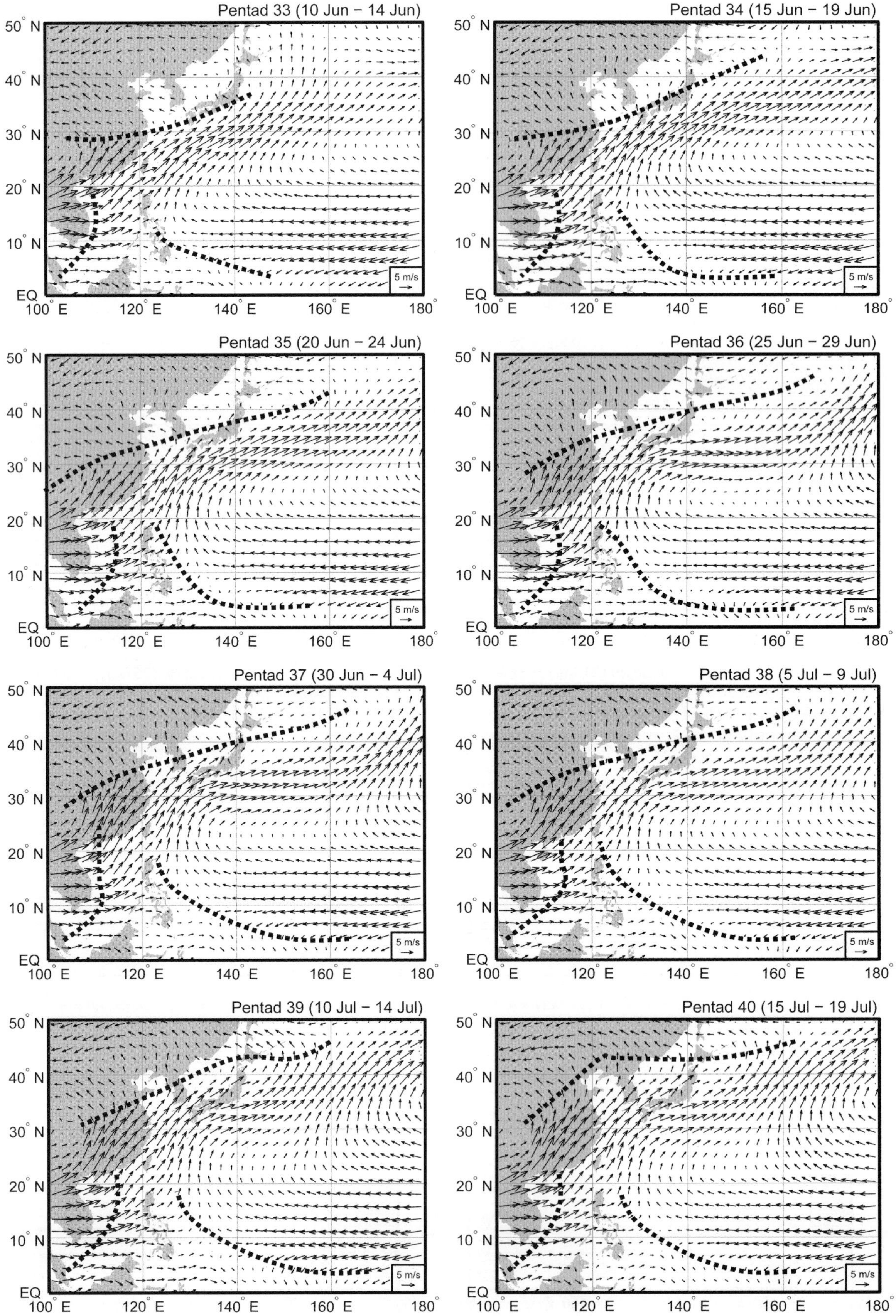

Figure 1.27. (*Continued*)

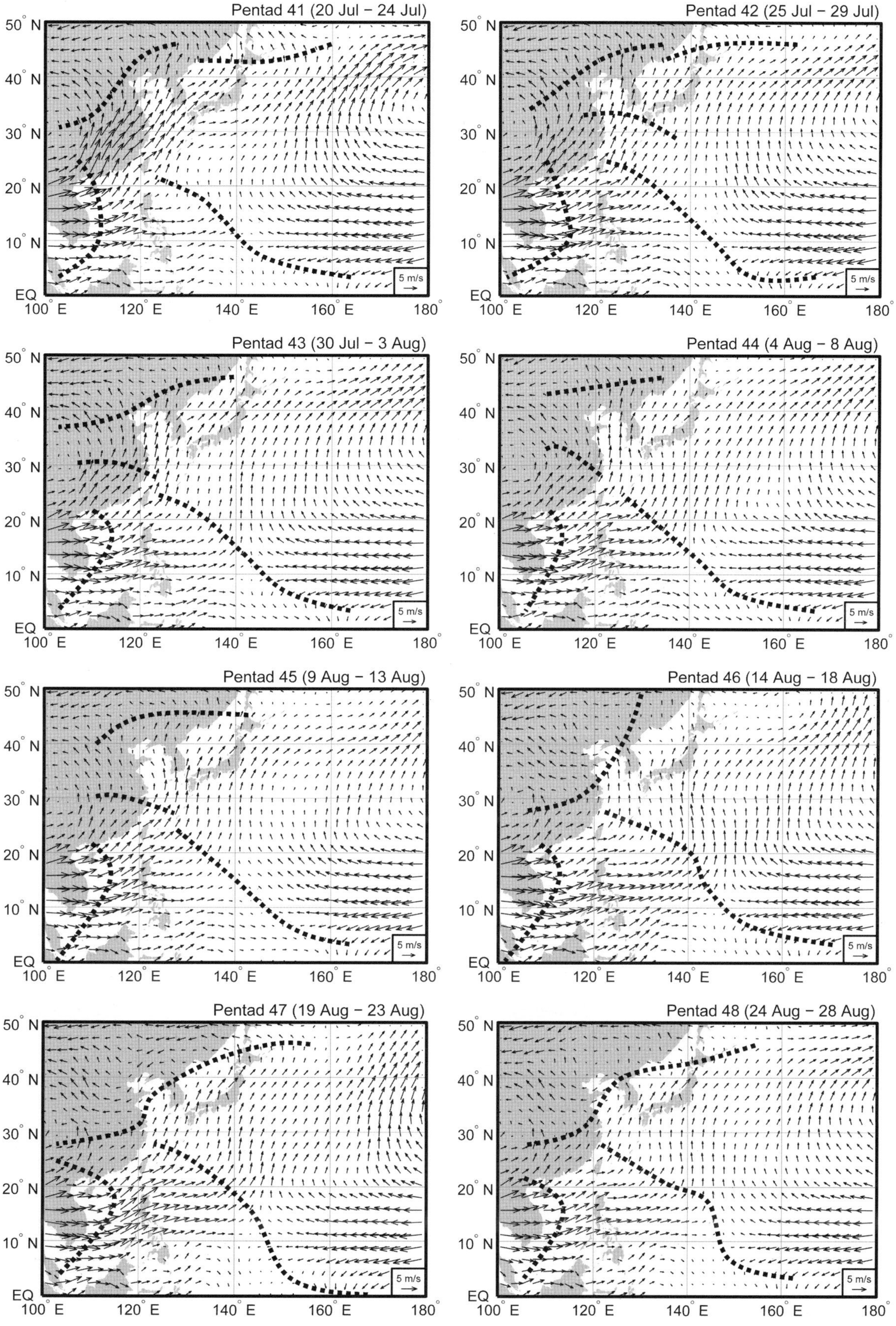

Figure 1.27. (*Continued*)

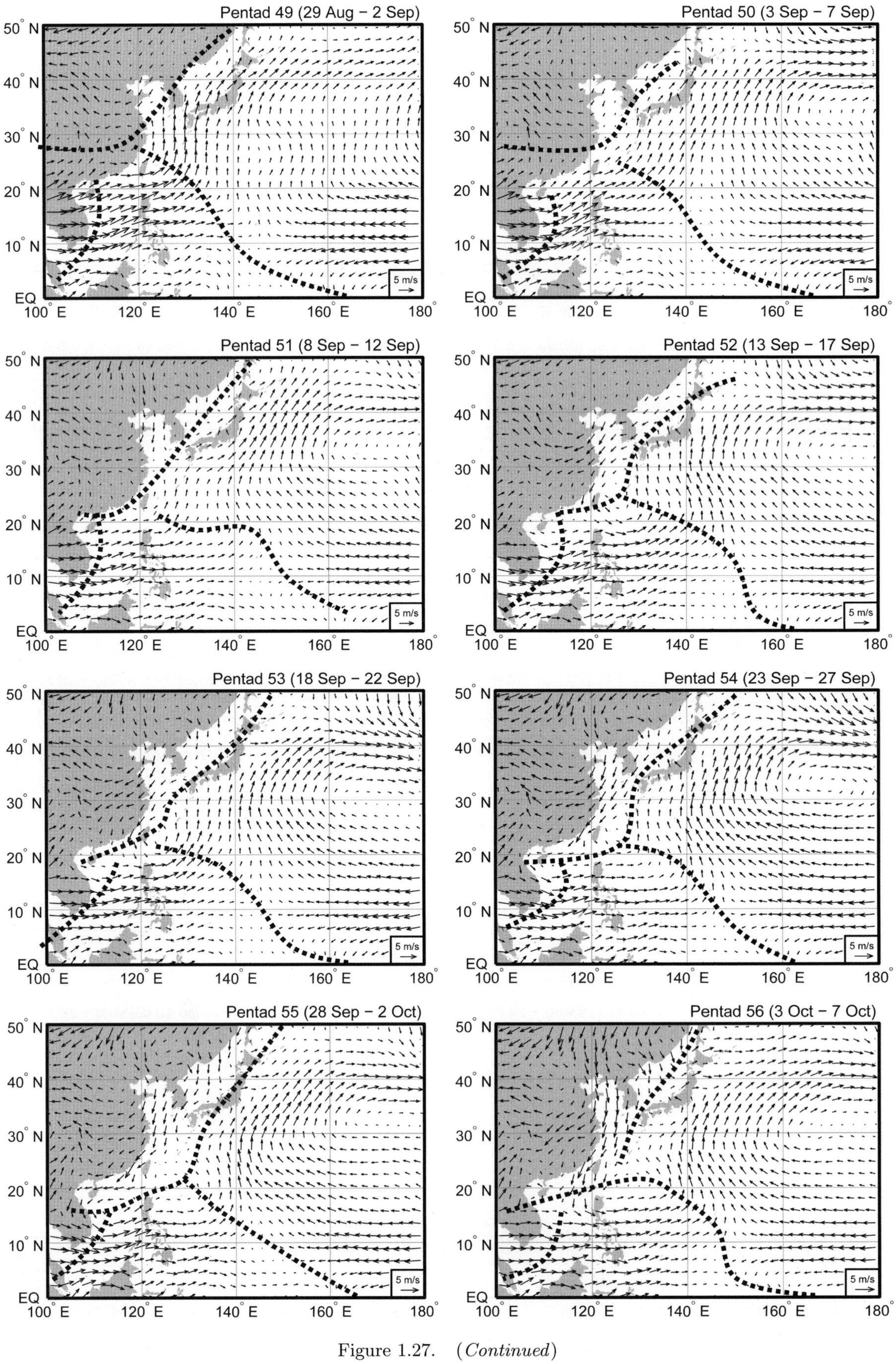

Figure 1.27. (*Continued*)

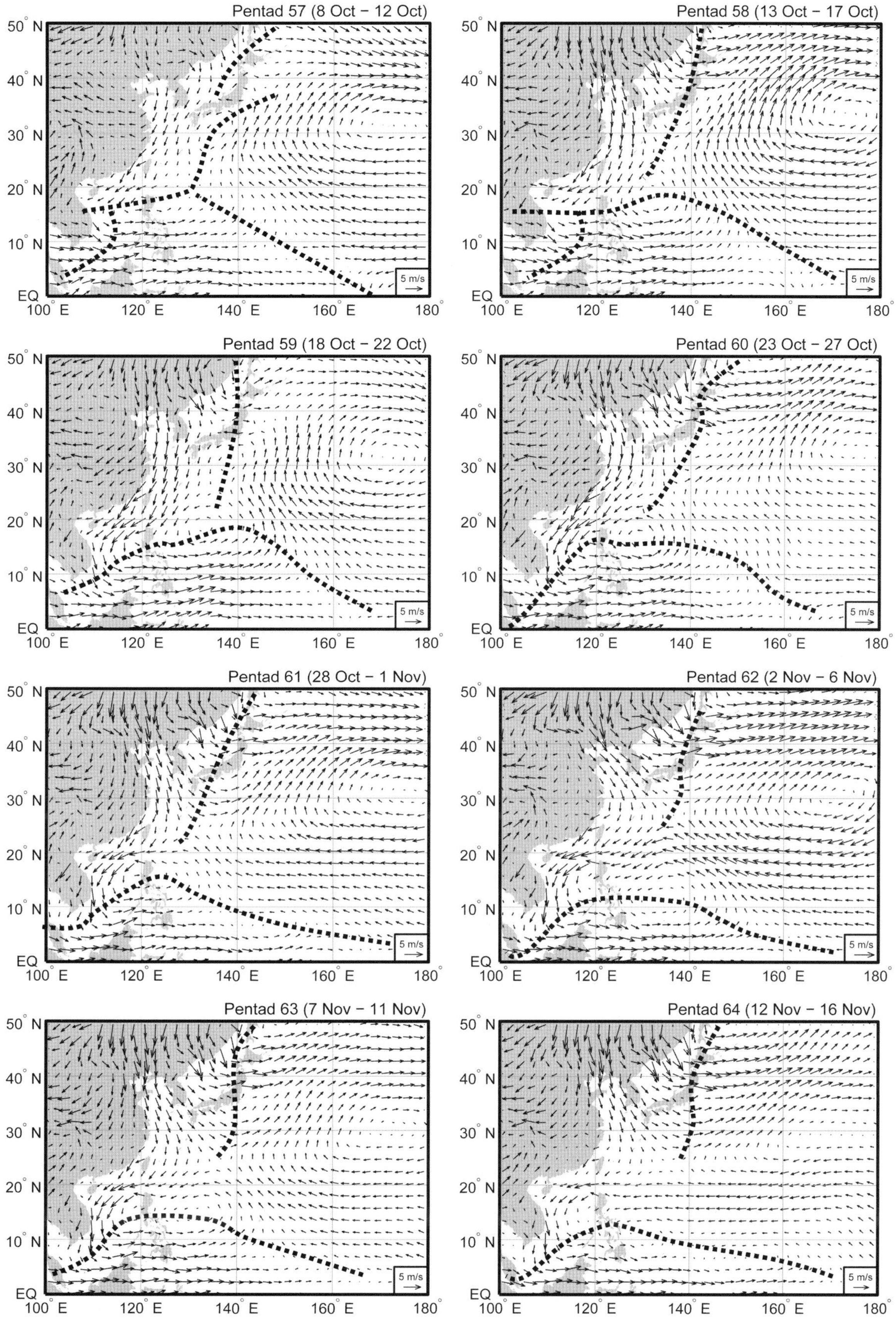

Figure 1.27. (*Continued*)

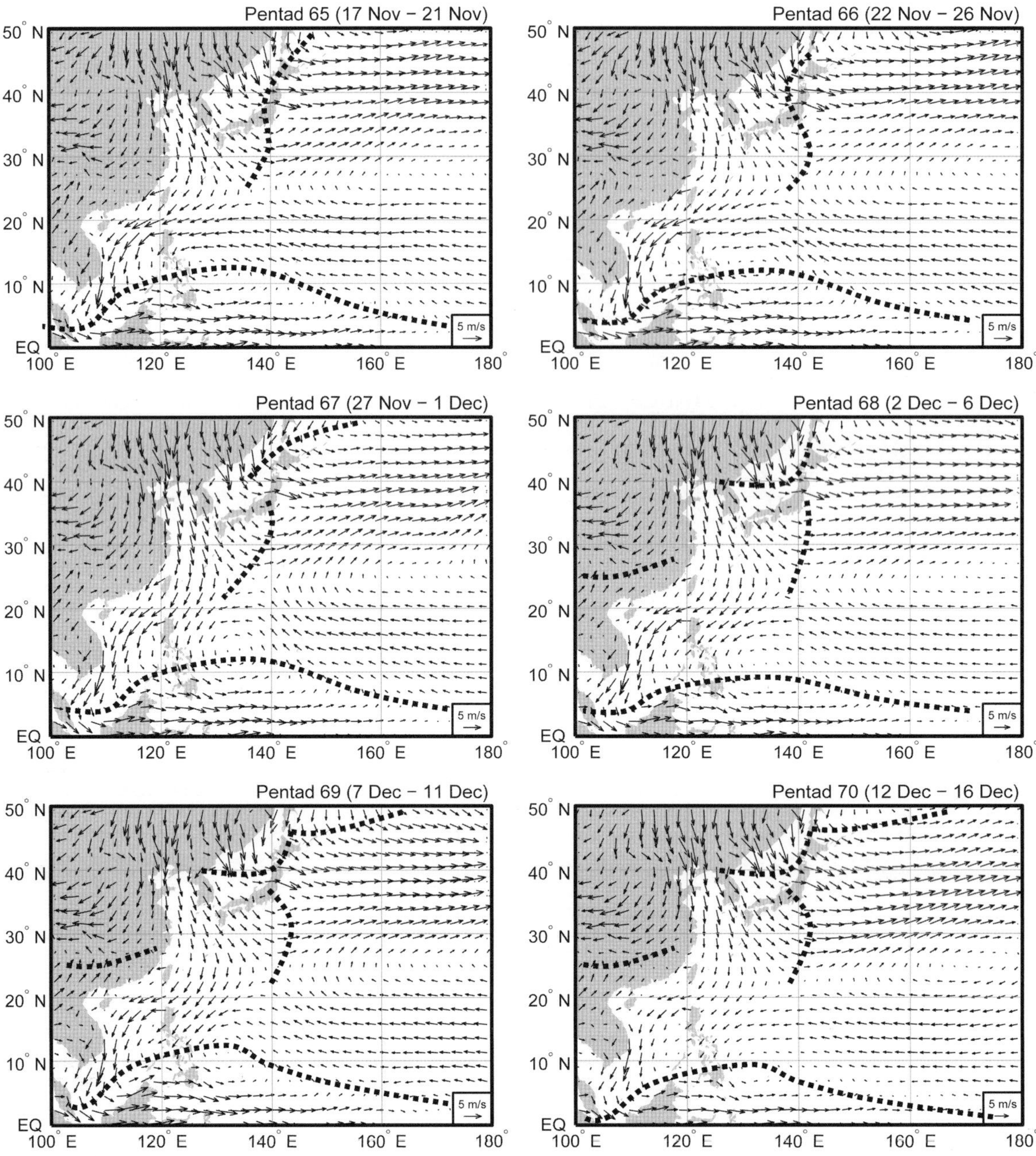

Figure 1.27. (*Continued*)

The East Asian subtropical monsoon trough, the SCS monsoon trough, and the Pacific tropical monsoon trough are active and shift their locations with season. The subtropical monsoon trough from the East Asian continent retreats southwards to the northernmost SCS in middle September. The data of original ERA Interim with the horizontal resolution 0.75° from 1981 to 2010 is used to reconstruct the 2.25° resolution. Each panel shown in Fig. 1.27 is calculated by Eq. (3) and averaged for a pentad (five days). Detail location changes of these troughs are shown in each pentad from May to December.

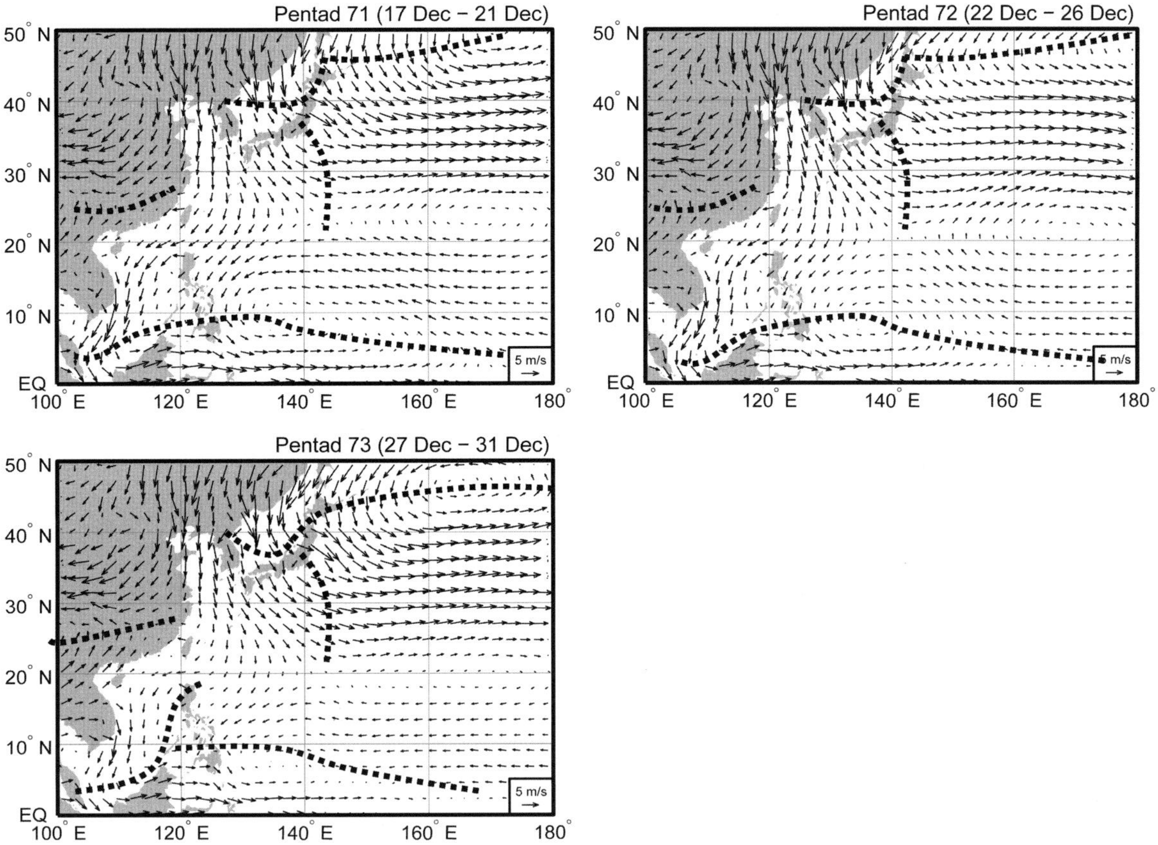

Figure 1.27. (*Continued*)

Table 1.1. The frequency (times) of two (or three) close tropical cyclones (or TYs) appearing in the NW Pacific in each year from 1980 to 2011.

Year	Times	Year	Times	Year	Times	Year	Times
1980	1	1988	1	1996	3	2004	0
1981	1	1989	3	1997	3	2005	1
1982	2	1990	3	1998	6	2006	2
1983	2	1991	3	1999	2	2007	4
1984	3	1992	2	2000	1	2008	2
1985	1	1993	5	2001	0	2009	1
1986	2	1994	5	2002	2	2010	3
1987	0	1995	3	2003	0	2011	1

1.6. Two Close Tropical Cyclones

Two tropical cyclones are defined to be "close" when their distance is less than 2000 km for at least two consequent days of observations. There were in total 62 times that two close tropical cyclones and six times that three close tropical cyclones appeared at the same time during the period from 1980 to 2011 (Table 1.1). Among them, there were only 15 times that the TBs (or cyclones) reached the TY intensity. All 68 times of tropical cyclones and their tracks are plotted in Fig. 1.28.

The positions of peak intensity of tropical cyclones shift southwards with season. Monthly positions of two close tropical cyclones with one reached its peak intensity during its lifetime in 1980–2011 are plotted in Fig. 1.29.

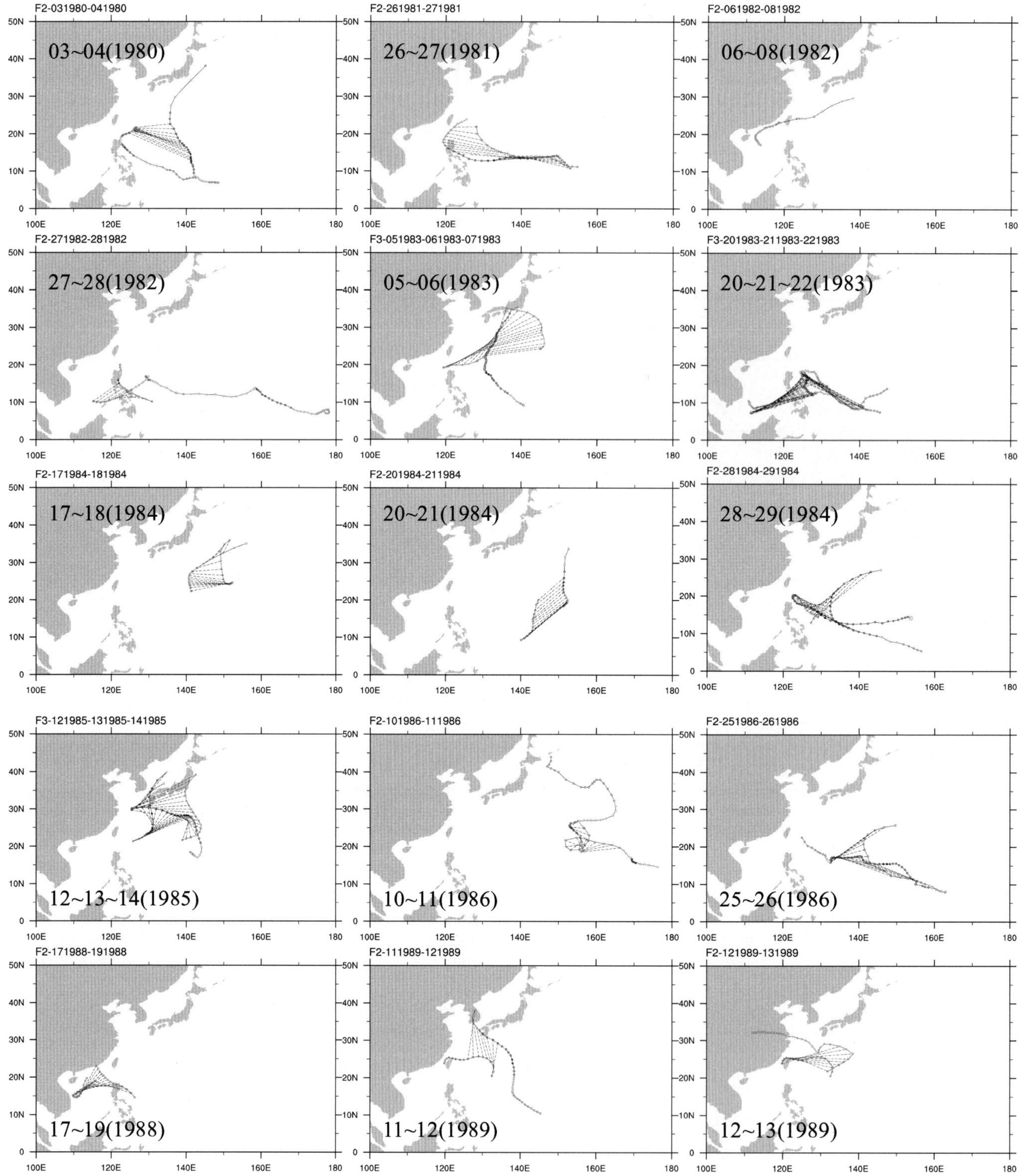

Figure 1.28. The tracks of two (or three) close tropical cyclones or tropical turbulences from 1980 to 2011. Each dashed line connects two tropical cyclones (turbulences) when their distance is less than 2000 km.

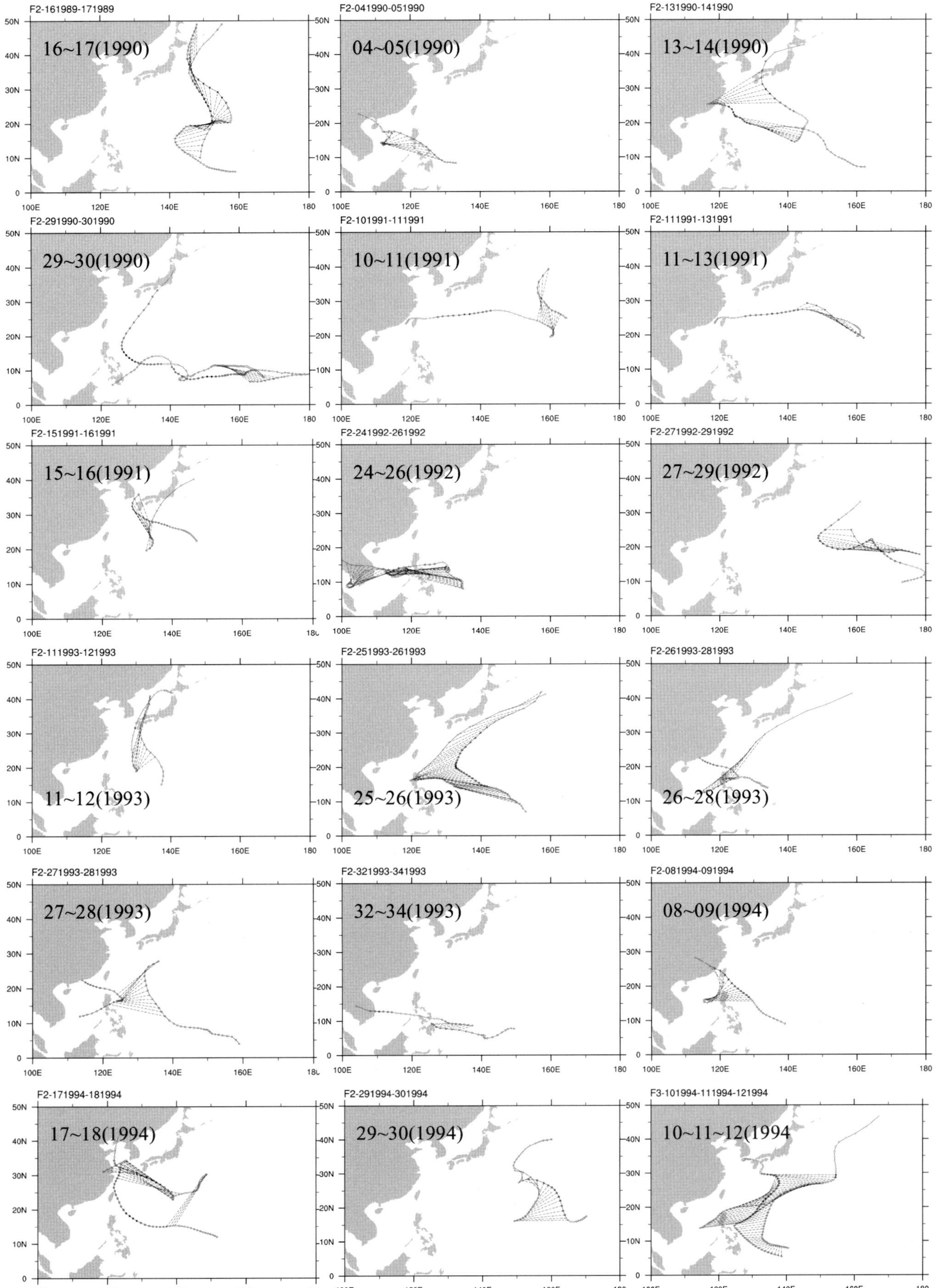

Figure 1.28. *(Continued)*

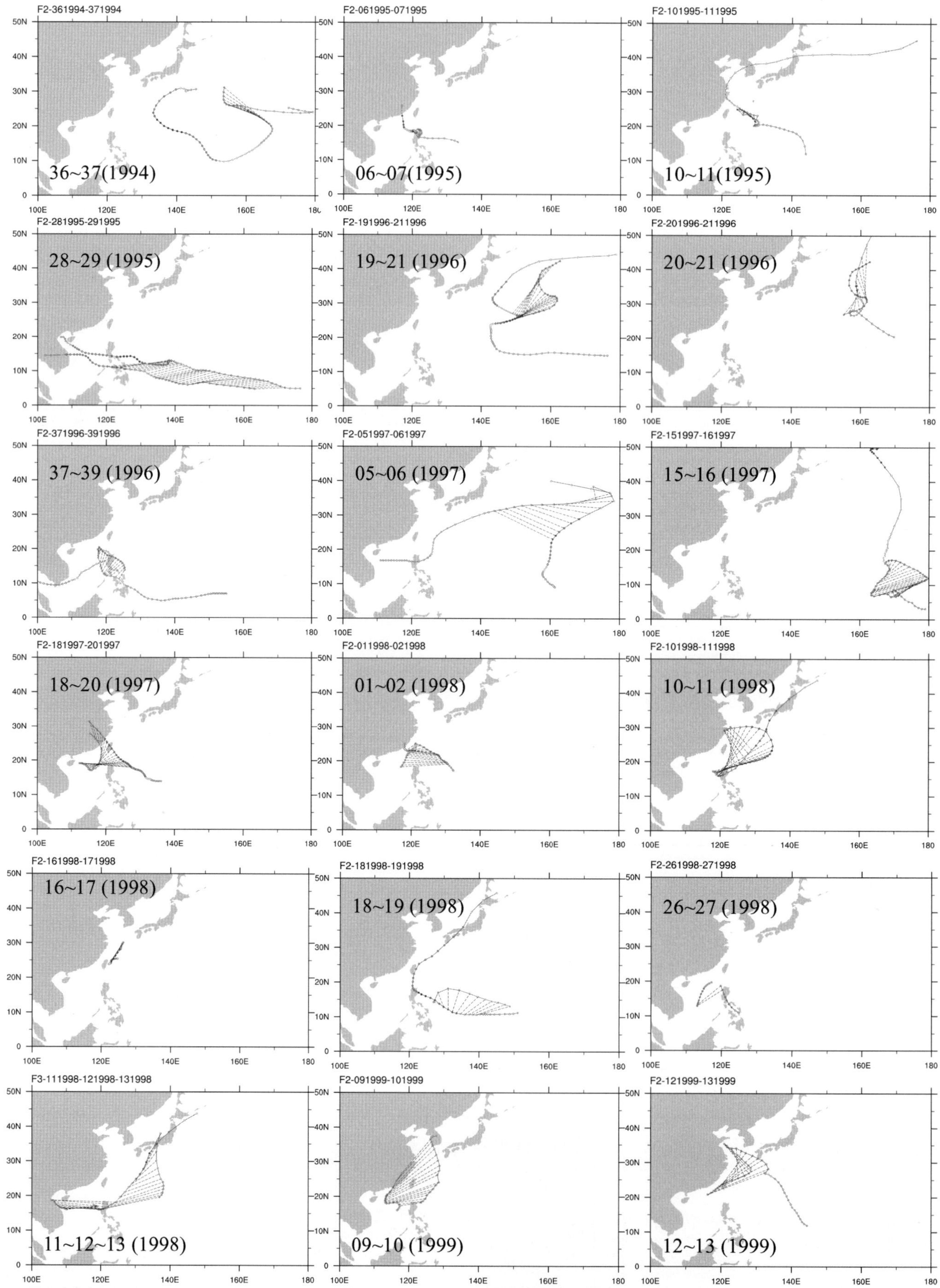

Figure 1.28. (*Continued*)

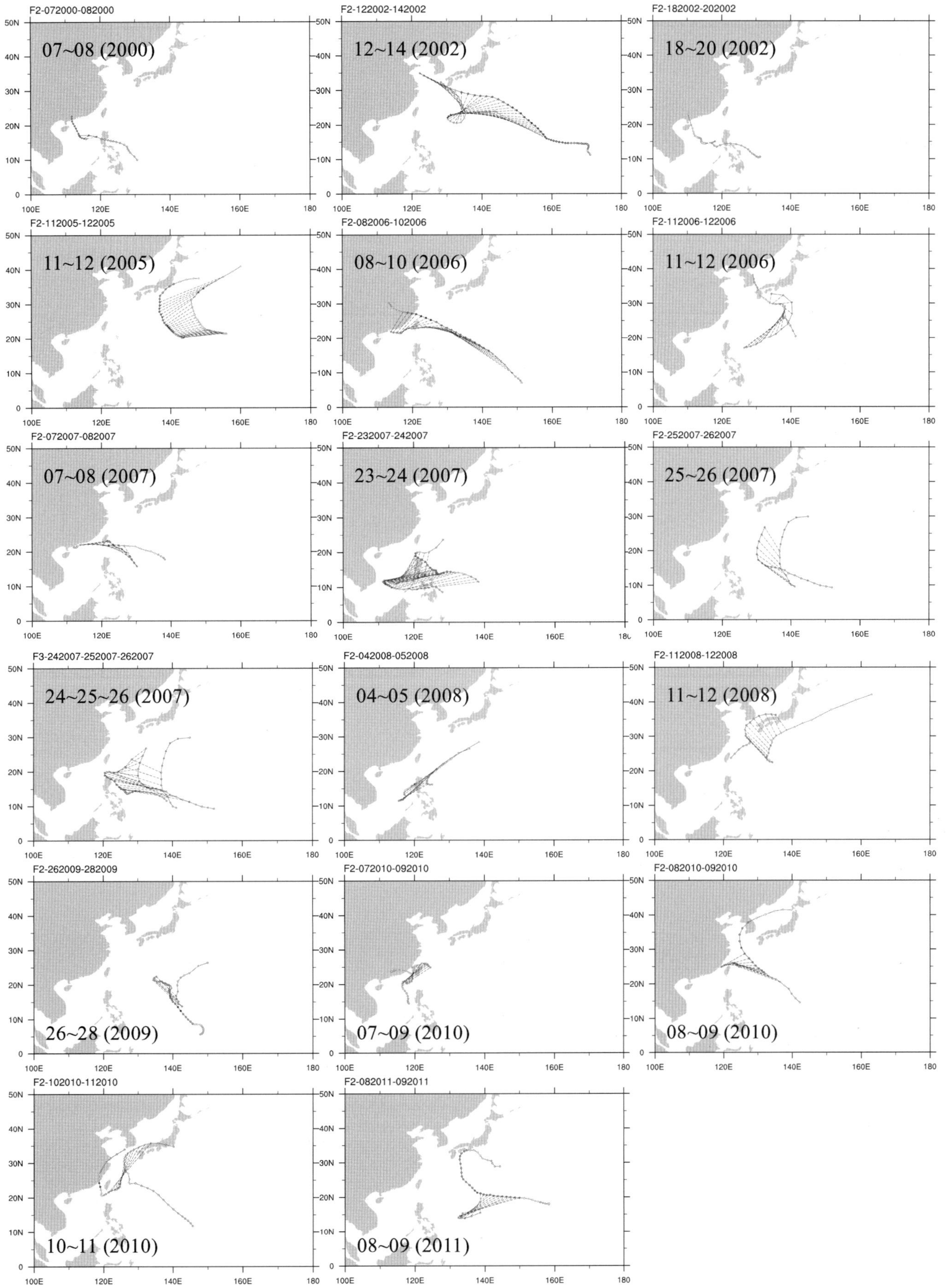

Figure 1.28. (*Continued*)

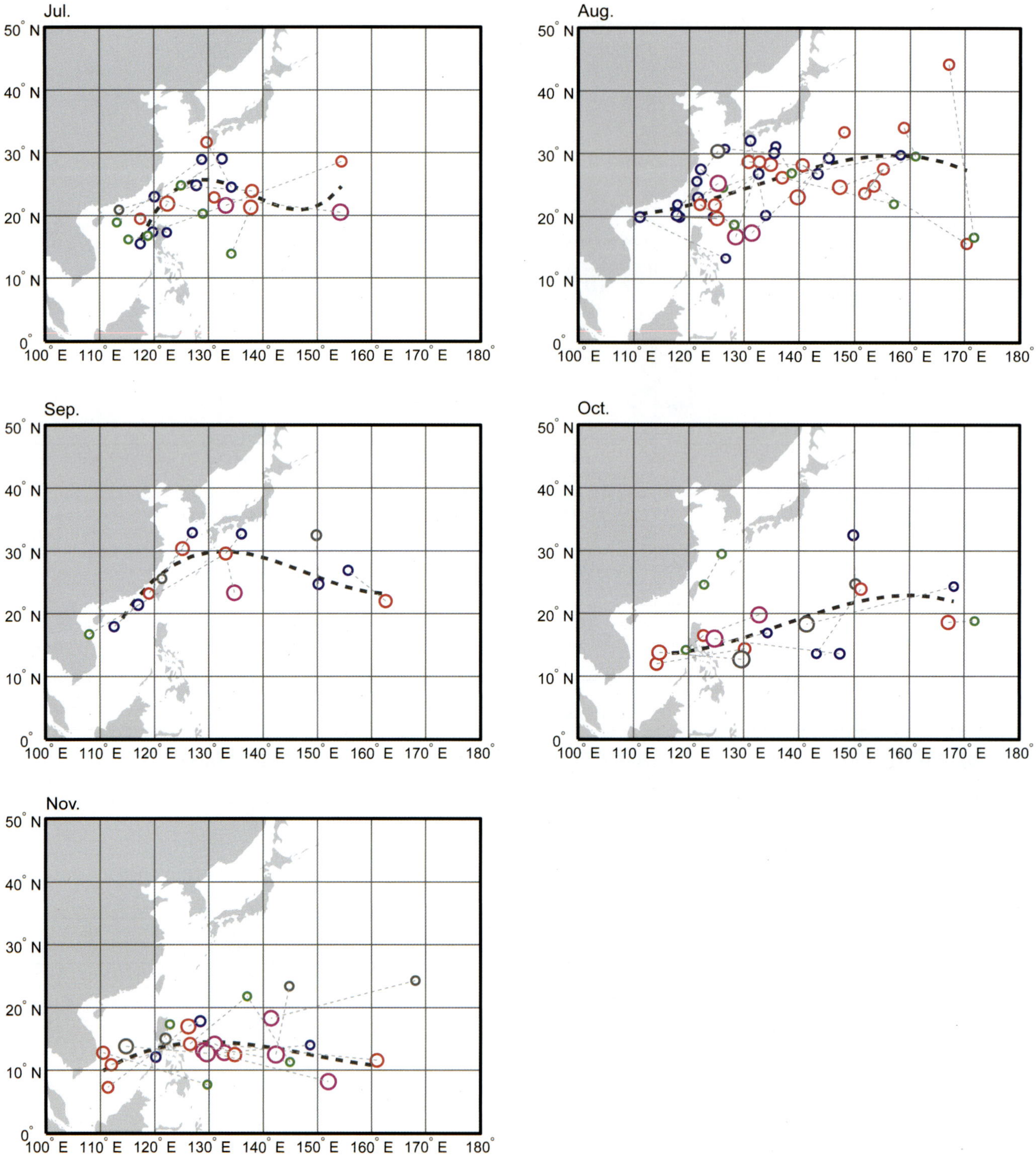

Figure 1.29. Monthly positions of two close tropical cyclones with one reached the peak intensity during its lifetime in 1980–2011. Open dots colored in pink, red, blue, and green indicate the STs, TYs, TSs, and TDs, respectively. Black dashed lines are the cubic spline fitting the positions of peak intensity of tropical cyclones.

References

Brand, S., 1970: Interaction of binary tropical cyclones of the western North Pacific Ocean, *J. Appl. Meteorol.*, **9**, 433–441.

Chan, J. C. L. and W. M. Gray, 1982: Tropical cyclone movement and surrounding flow relationship, *Mon. Wea. Rev.*, **110**, 1354–1374.

Chen, L. S., 2012: Research progress on the structure and intensity change for the landfall tropical cyclones, *J. Tropical Meteorol.*, **18**, 113–118.

Chen, Q. S. and Y. H. Kuo, 1992a: A harmonic-sine series expansion and its application to reconstruction problems in a limited area, *Mon. Wea. Rev.*, **120**, 91–112.

Chen, Q. S. and Y. H. Kuo, 1992b: A consistency condition for wind-field reconstruction in a limited area and a harmonic-cosine series expansion, *Mon. Wea. Rev.*, **120**, 2653–2670.

Dong, K. and C. J. Neumann, 1983: On the relative motion of binary tropical cyclones, *Mon. Wea. Rev.*, **111**, 945–953.

Fujiwhara, S., 1921: The mutual tendency towards symmetry of motion and its application as a principle in meteorology, *Q. J. R. Meteorol. Soc.*, **47**, 287–293.

Fujiwhara, S., 1931: Short note on the behavior of two vortices, *Proc. Phys. Math. Soc. Japan* (Ser. 3), **13**, 106–110.

George, J. E. and W. M. Gray, 1977: Tropical cyclone recurvature and nonrecurvature as related to surrounding wind-height fields, *J. Applied Meteorol.*, **16**, 34–42.

Lander, M. and G. J. Holland, 1993: On the interaction of tropical cyclone-scale vortices. I: Observations, *Q. J. R. Meteorol. Soc.*, **119**, 1347–1361.

Peixoto, J. P. and A. H. Oort, 1992: *Physics of Climate.* Springer-Verlag, American Institute of Physics Press, New York, 1–520.

Qian, C. H., Y. H. Duan, S. H. Ma, S. and Y. L. Xu, 2012: The current status and future development of China operational typhoon forecasting and its key technologies, *Adv. Meteor. Sci. Technol.*, **2**(5), 36–43. (in Chinese)

Qian, W. H., 2011a: *Atlas of Climate Change and China Extreme Climate Events.* Beijing, China Meteorological Press, 1–259. (in Chinese)

Qian, W. H., 2011b: Middle-range forecast system and method of the low-temperature freezing rain based on the physical decomposition of atmospheric variables. State Intellectual Property Office of P.R.C., No. ZL 2011 1 0061438.7.

Qian, W. H., 2012a: Making methods of anomalous synoptic chart and low-frequency synoptic chart and their applications in weather forecast. State Intellectual Property Office of P.R.C., CN 201210134358.4, 2012-05-02.

Qian, W. H., 2012b: Physical decomposition principle of regional-scale atmospheric transient anomaly, *Chinese J. Geophys.*, **55**, 1439–1448. (in Chinese)

Qian, W. H., 2012c: *Principles of Medium to Extended Range Weather Forecasts.* Science Press, Beijing, 1–410.

Qian, W. H. and H. Y. Liang, 2012: Propagations of planetary-scale zonal mean wind anomalies and polar oscillations, *Chinese Sci. Bull.*, **57**, 2606–2614.

Qian, W. H., X. L. Shan, H. Y. Liang, J. Huang, and C. H. Leung, 2014: A generalized beta-advection model to improve unusual typhoon track prediction by decomposing total flow into climatic and anomalous flows. *J. Geophys. Res. Atmos.*, **119**: doi:10.1002/2013JD020902.

Qian, W. H. and S. Q. Tang, 2010: Identifying global monsoon troughs and global atmospheric centers of action on a pentad scale, *Atmos. Oceanic Sci. Lett.* **3**, 1–6.

Willoughby, H. E., 2007: Forecasting hurricane intensity and impacts, *Science* **315**, 1232–1233.

Xu, Y. L., 2011: Forecast analysis on the abrupt northward recurvature of super typhoon Megi (2010), *Meteorol. Mon.*, **37**, 821–826.

Chapter 2

Typhoons Without Obvious Turning Motions

Chapter 2 displays typhoons without obvious turning motions during their lifetime but their tracks were modified by anomalous flows or anomalous vortices. These typhoons were generated in the South China Sea (SCS) and the Philippines (Sec. 2.1). Their intensities peaked in the SCS (Sec. 2.2), in the Philippines (Sec. 2.3), in the east seas of China (Sec. 2.4), and in the western Pacific (Sec. 2.5), respectively. In this chapter, anomalous flows or anomalous vortices are shown in anomalous synoptic charts but traditional synoptic charts with steering flows have not been shown.

2.1. Geneses in the South China Sea and the Philippines

In the period of 1979–2011, 10 typhoons were generated in the SCS and moved without obvious turning motions. Their tracks were only modified by anomalous vortices (Figs. 2.1–2.10).

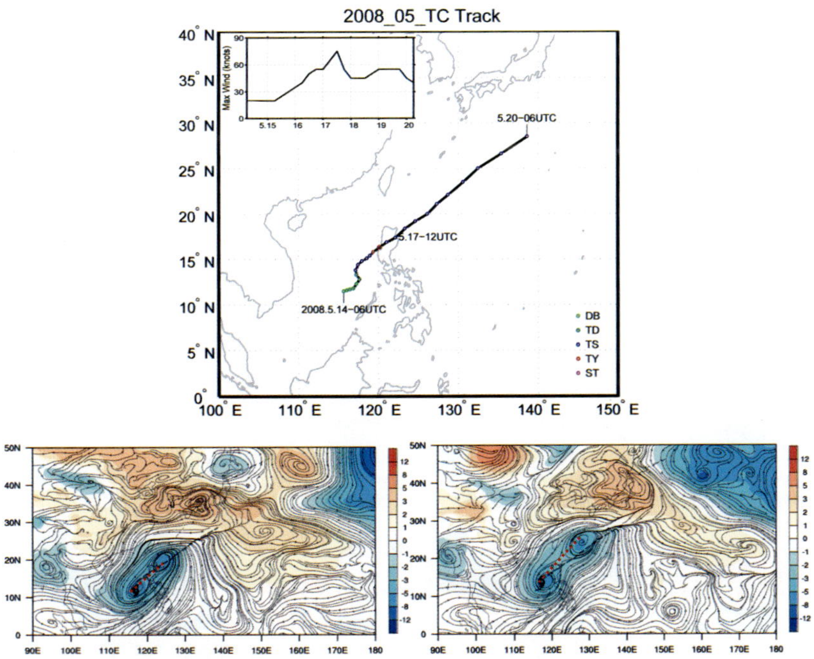

850 hPa flow/height (shading, 10 gpm) anomalies at 0000 UTC 15 and 16 May 2008.

Figure 2.1. The typhoon (2008_No.05) generated in the central SCS on 14 May 2008 and moved northeastward. Its intensity peaked near the climatic monsoon trough on 17 May 2008 before landfall on the northern Philippines. Two anomalous vortices existed at 850 hPa along the direction of typhoon motion with an escaping process during the typhoon lifetime period. In the typhoon track figures: (i) the typhoon position is marked by open dots at 6-h interval, and linked by heavy line; (ii) the intensity is shown by the curved line in the upper small panel and indicated also by disturbance (DB), tropical depression (TD), tropical storm (TS), typhoon (TY), and super typhoon (ST) with various colors in the bottom corner; (iii) the climatic (or monsoon) flow and the climatic monsoon trough may be referenced from Fig. 1.27. In the anomalous synoptic charts, shadings denote geopotential height anomaly (10 gpm) and the red-dotted line connects the two centers of anomalous vortices indicated by flow/height anomalies.

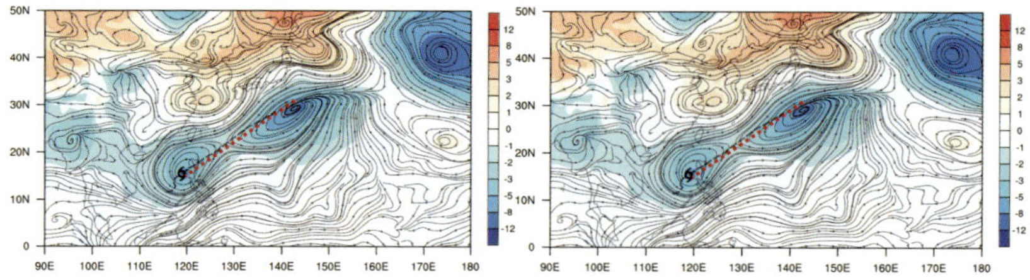

850 hPa flow/height anomalies at 0000 UTC 17 and 18 May 2008.

Figure 2.1. (*Continued*)

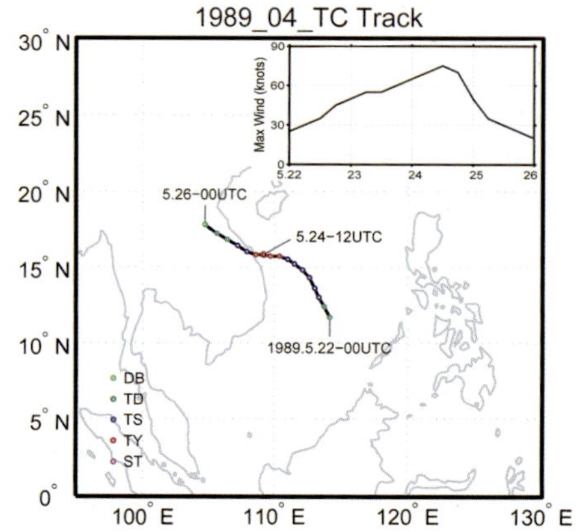

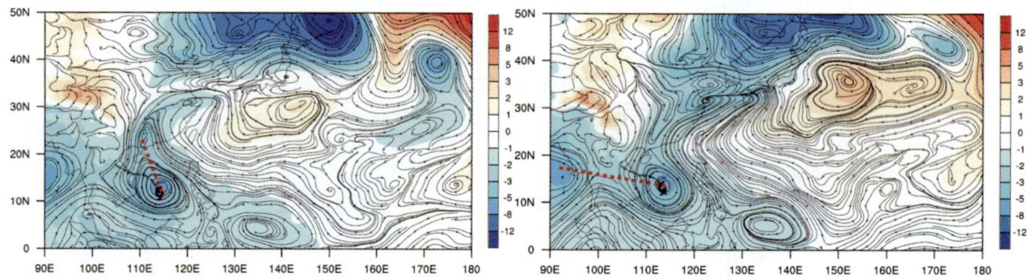

850 hPa flow/height anomalies at 0000 UTC and 1200 UTC 22 May 1989.

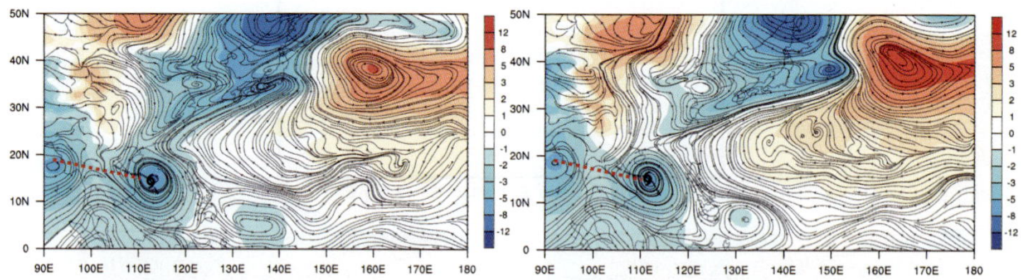

850 hPa flow/height anomalies at 0000 UTC and 1200 UTC 23 May 1989.

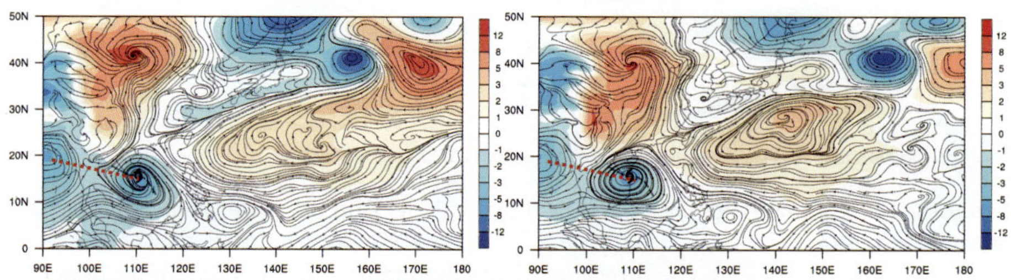

850 hPa flow/height anomalies at 0000 UTC and 1200 UTC 24 May 1989.

Figure 2.2. The typhoon (1989_No.04) generated in the central SCS on 22 May 1989 and moved northwestward. Its intensity peaked near the climatic monsoon trough in the western SCS on 24 May 1989 before landfall on the central east coast of Vietnam. Its track was modified by an anomalous vortex at 850 hPa.

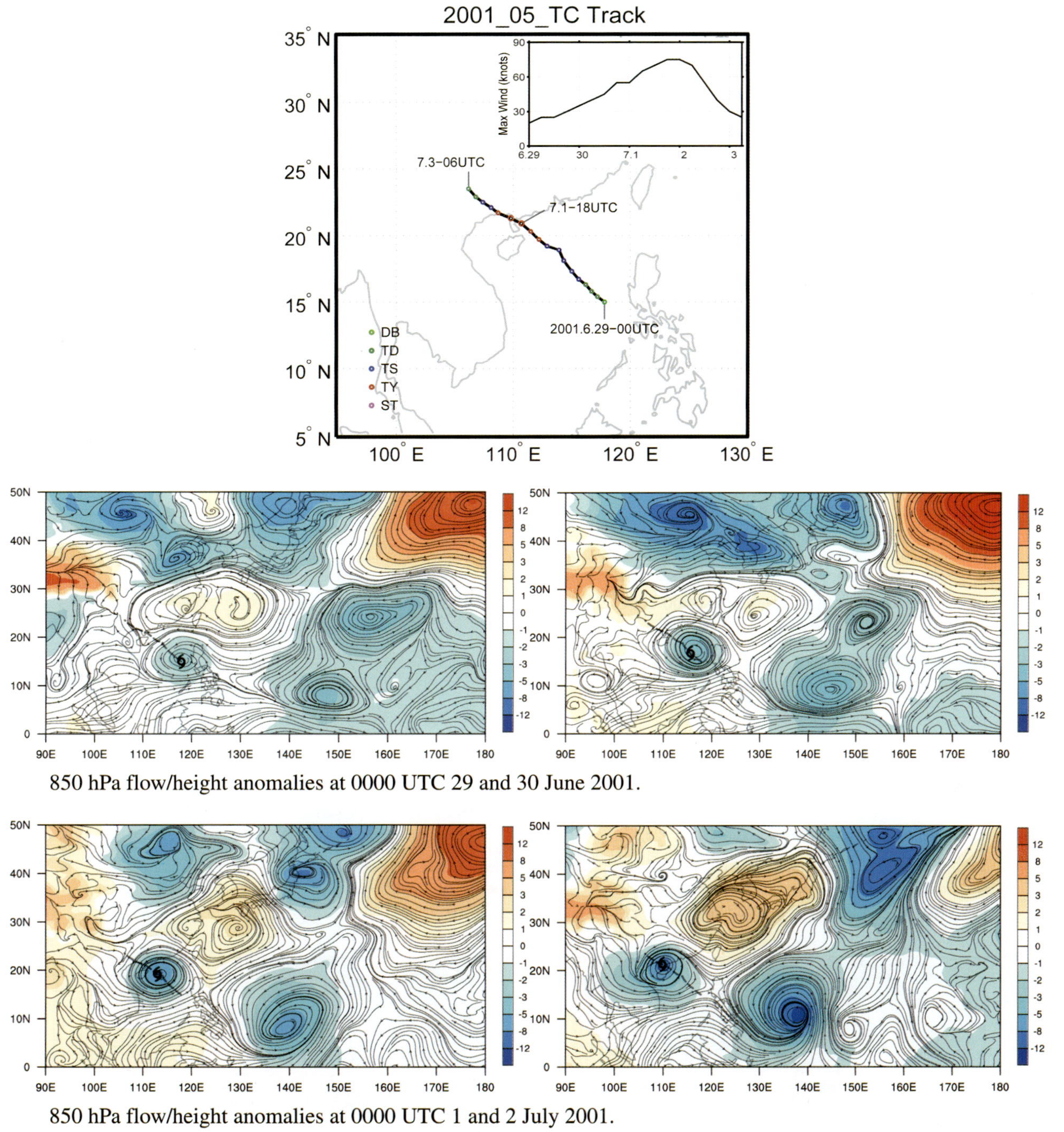

Figure 2.3. The typhoon (2001_No.05) generated in the central eastern SCS on 29 June 2001 and moved northwestward. Its intensity peaked near the climatic monsoon trough on 1 July 2001 before landfall on the South China coast. No other anomalous vortices interacted with the typhoon vortex at 850 hPa.

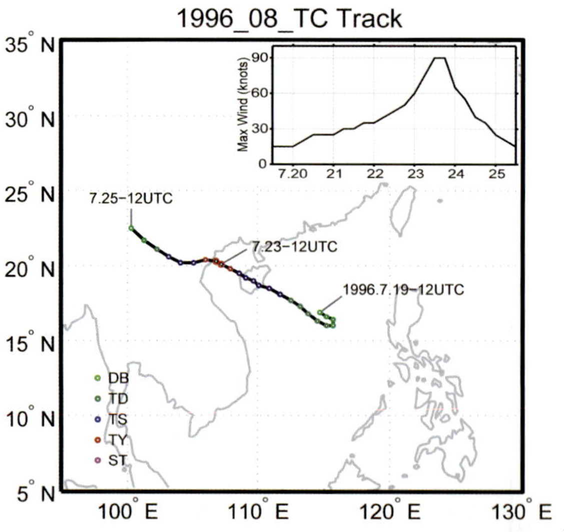

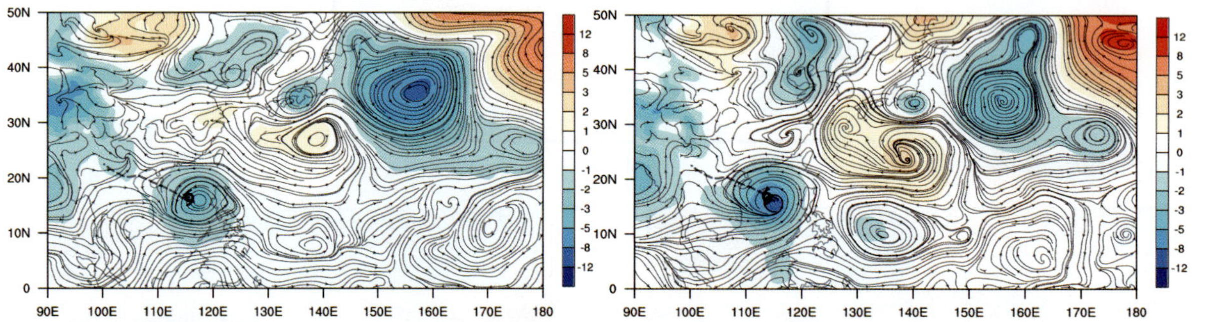

850 hPa flow/height anomalies at 0000 UTC 20 and 21 July 1996.

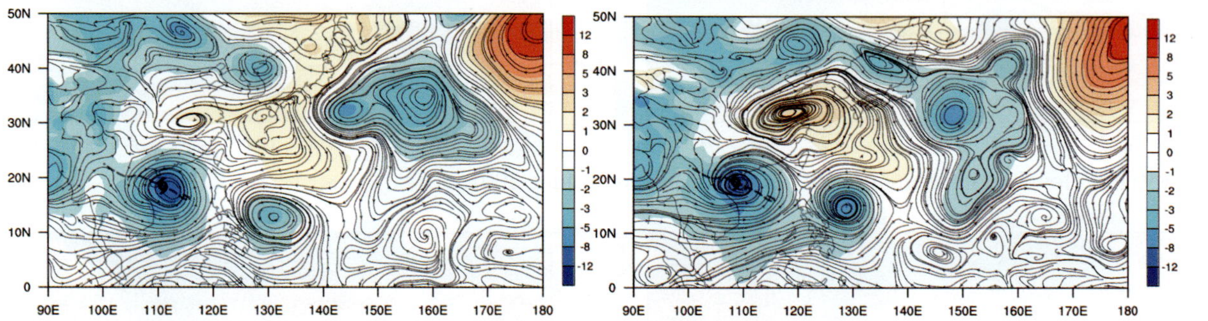

850 hPa flow/height anomalies at 0000 UTC 22 and 23 July 1996.

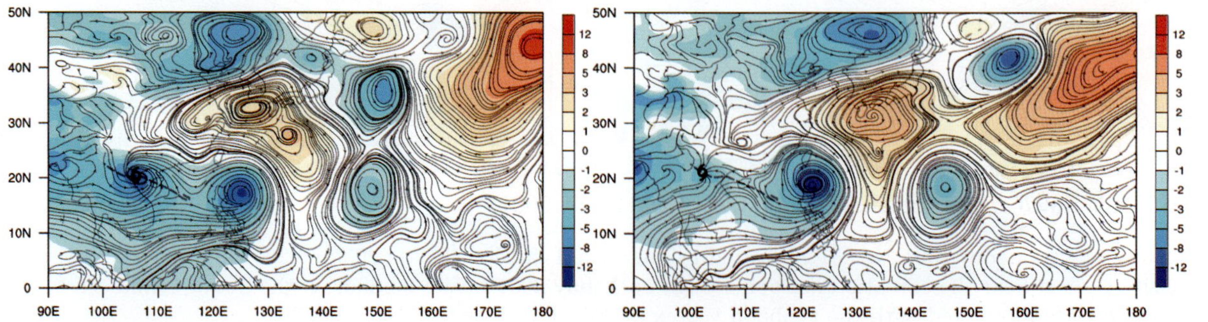

850 hPa flow/height anomalies at 0000 UTC 24 and 25 July 1996.

Figure 2.4. The typhoon (1996_No.08) generated in the central SCS on 19 July 1996 and moved northwestward. Its intensity peaked near the climatic monsoon trough in the north-westernmost SCS on 23 July 1996 before landfall on the northeast coast of Vietnam. No other anomalous vortices interacted with the typhoon vortex at 850 hPa.

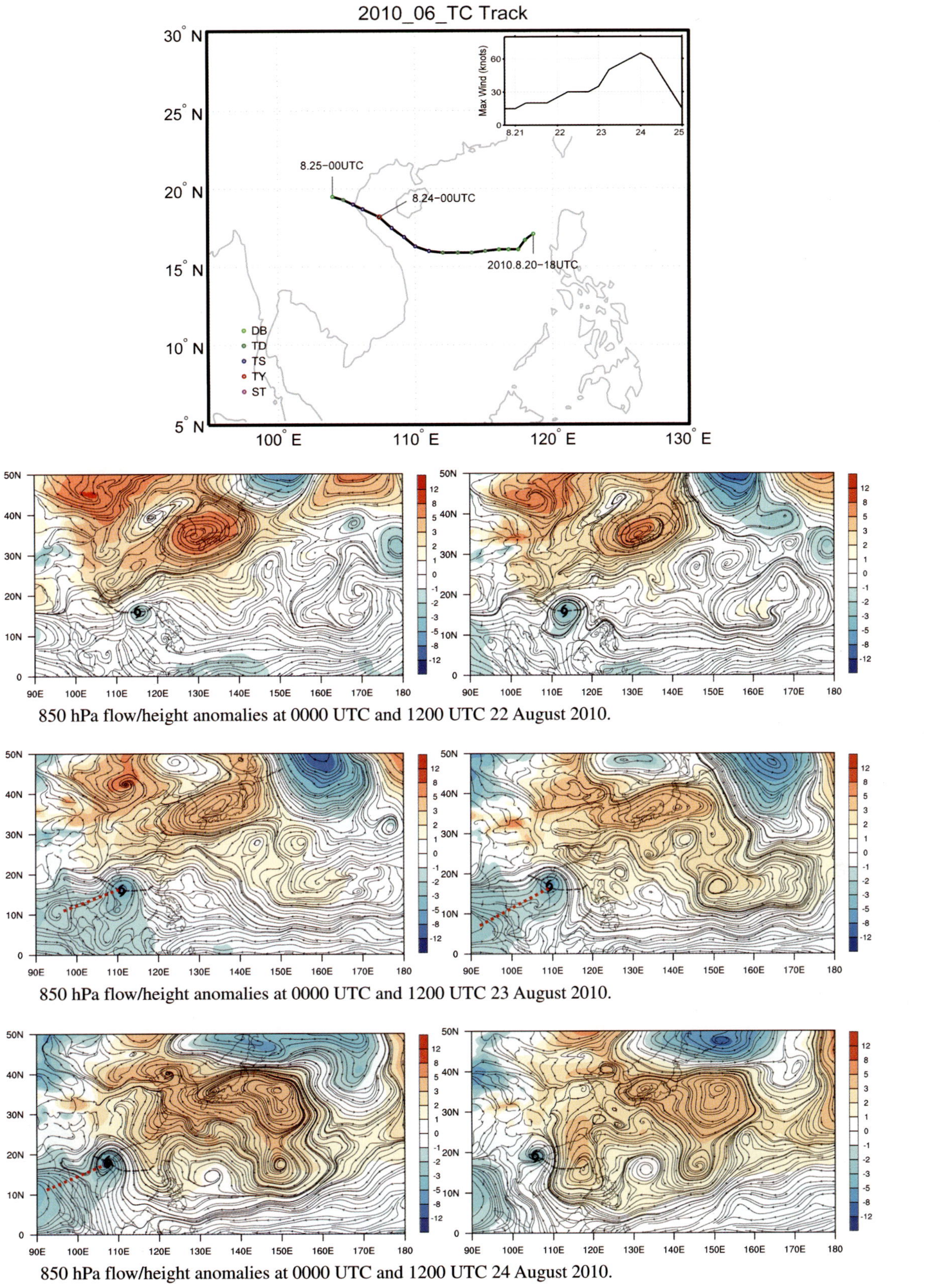

Figure 2.5. The typhoon (2010_No.06) generated in the eastern SCS near the northern Philippines on 20 August 2010. Its intensity peaked near the climatic monsoon trough in the northwestern SCS on 24 August 2010 before landfall on the northeastern coast of Vietnam. Its northwestward motion was modified by an anomalous vortex at 850 hPa so that there was a small change in its track.

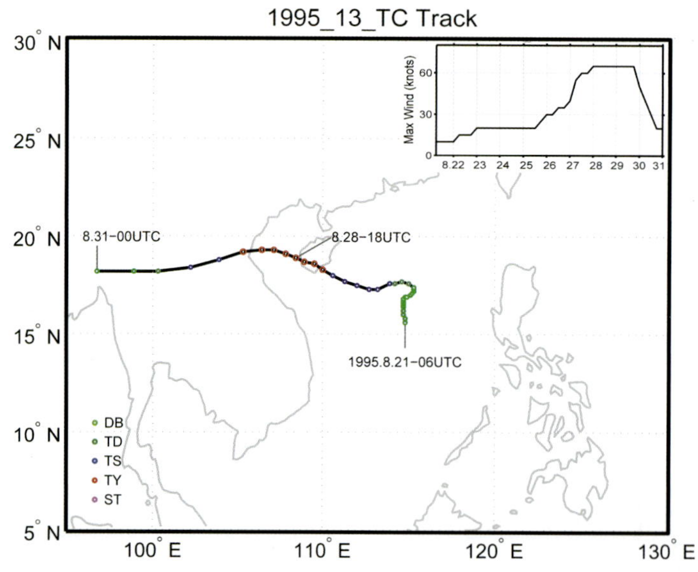

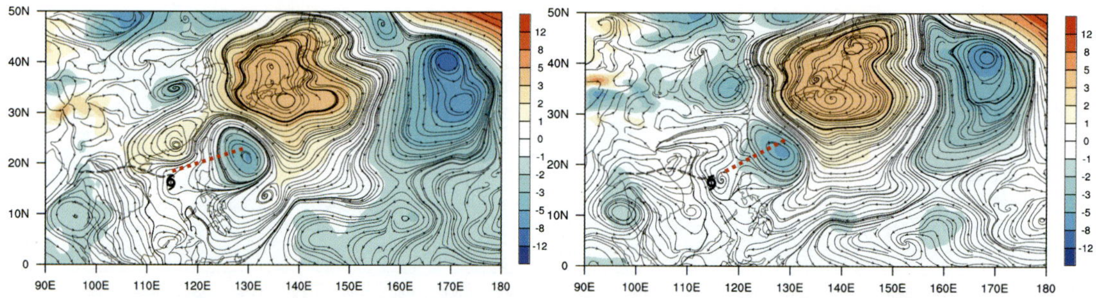

850 hPa flow/height anomalies at 0000 UTC and 1200 UTC 22 August 1995.

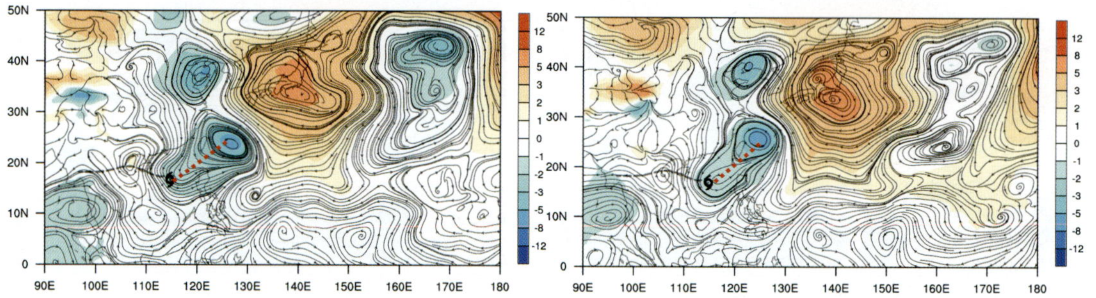

850 hPa flow/height anomalies at 0000 UTC and 1200 UTC 23 August 1995.

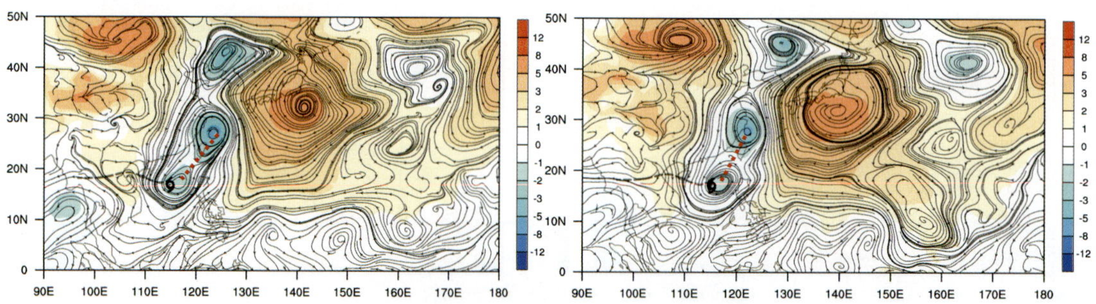

850 hPa flow/height anomalies at 0000 UTC and 1200 UTC 24 August 1995.

Figure 2.6. The typhoon (1995_No.13) generated in the central SCS on 21 August 1995. Its intensity peaked near the climatic monsoon trough near the Hainan Island on 28 August 1995 before landfall on the northeastern coast of Vietnam. Its motion was modified by an anomalous vortex at 850 hPa.

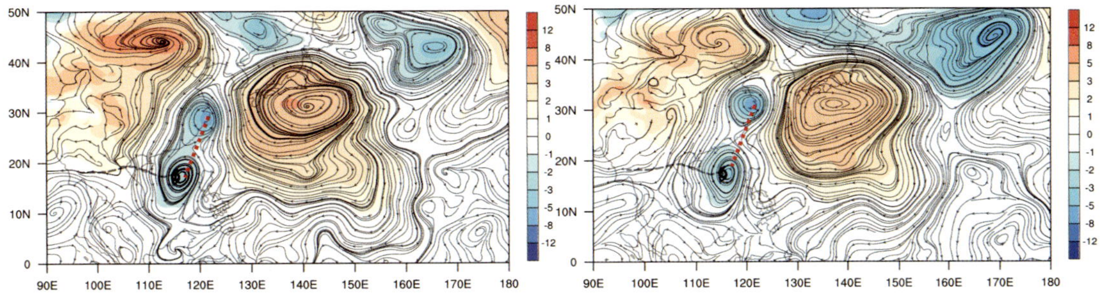

850 hPa flow/height anomalies at 0000 UTC and 1200 UTC 25 August 1995.

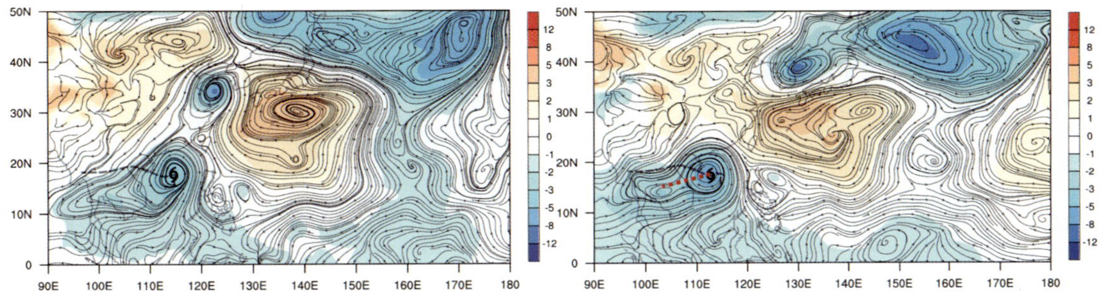

850 hPa flow/height anomalies at 0000 UTC 26 and 27 August 1995.

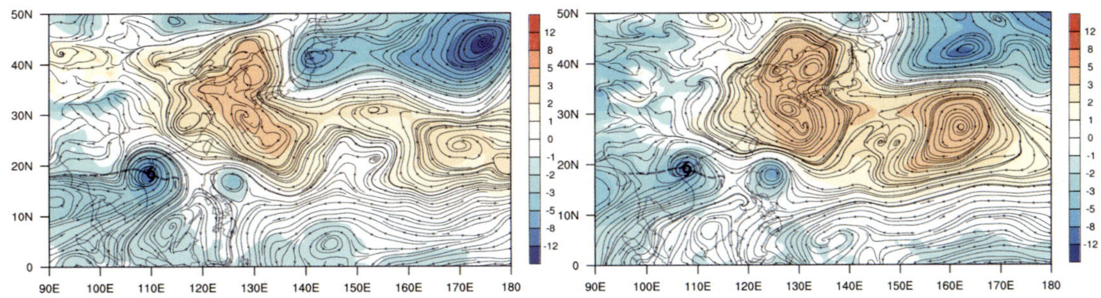

850 hPa flow/height anomalies at 0000 UTC 28 and 29 August 1995.

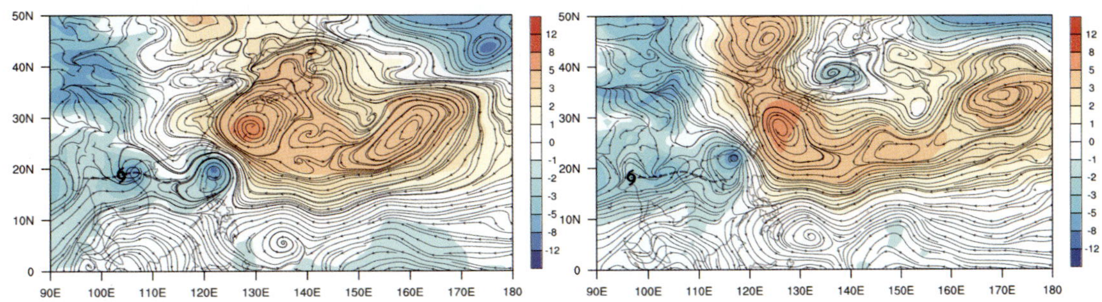

850 hPa flow/height anomalies at 0000 UTC 30 and 31 August 1995.

Figure 2.6. *(Continued)*

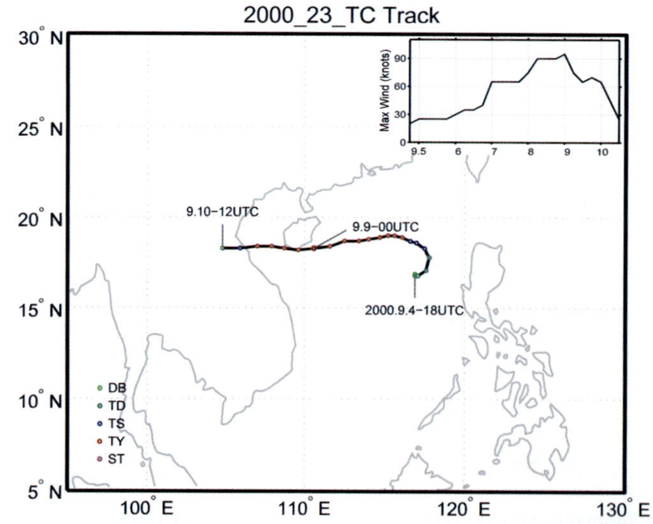

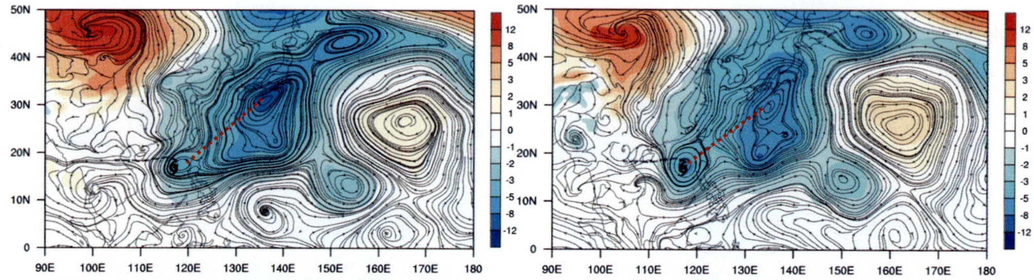

850 hPa flow/height anomalies at 0000 UTC and 1200 UTC 5 September 2000.

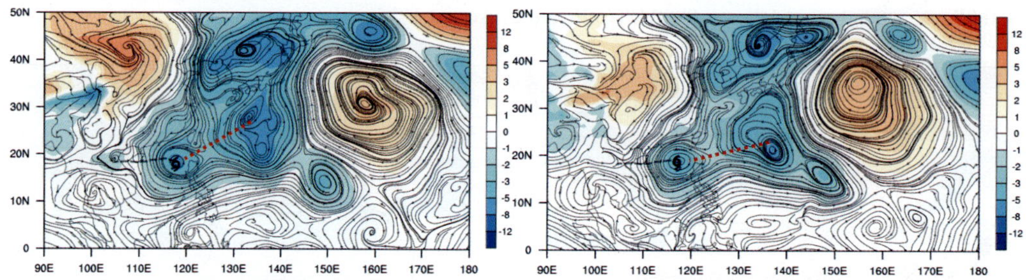

850 hPa flow/height anomalies at 0000 UTC and 1200 UTC 6 September 2000.

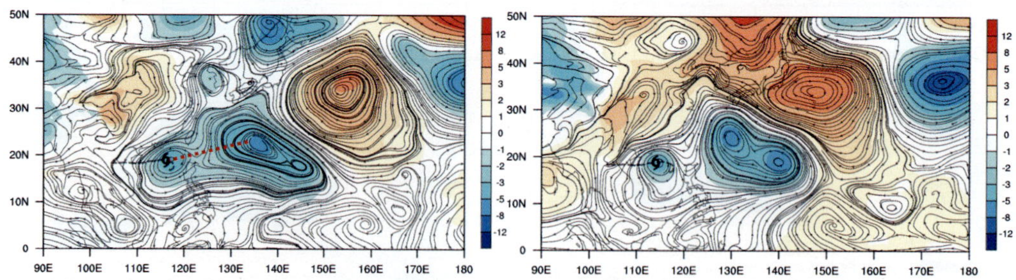

850 hPa flow/height anomalies at 0000 UTC 7 and 8 September 2000.

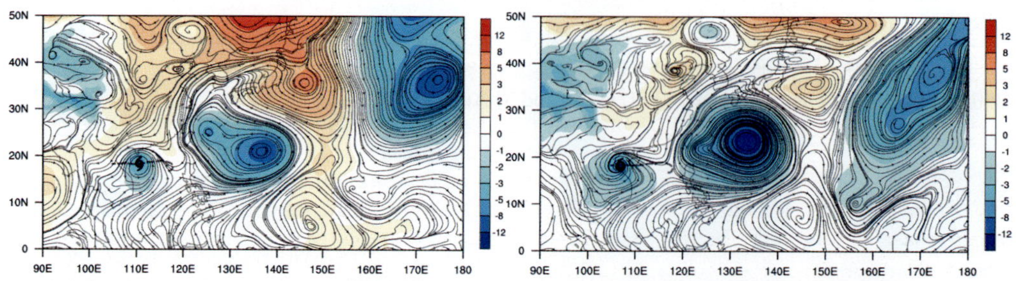

850 hPa flow/height anomalies at 0000 UTC 9 and 10 September 2000.

Figure 2.7. The typhoon (2000_No.23) generated in the eastern SCS on 4 September 2000 and had a semicircle track in the first three days due to the influence of an anomalous vortex at 850 hPa. Its intensity peaked near the climatic monsoon trough near the Hainan Island on 9 September 2000. Its westward motion was not influenced by any anomalous vortices at 850 hPa.

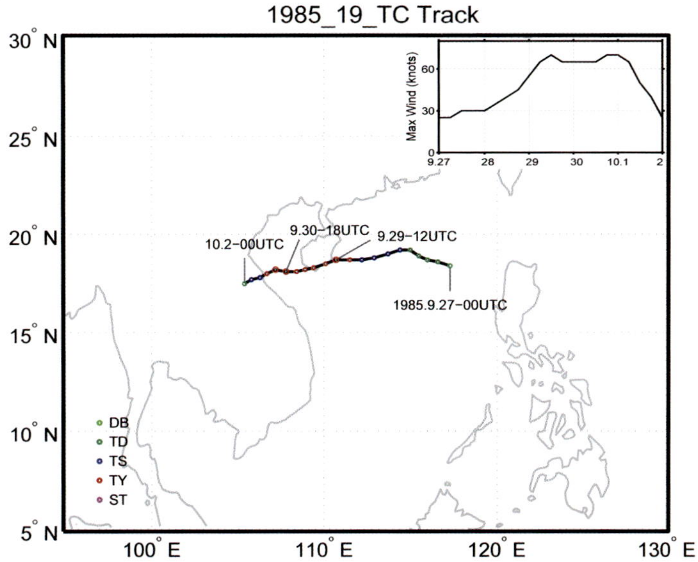

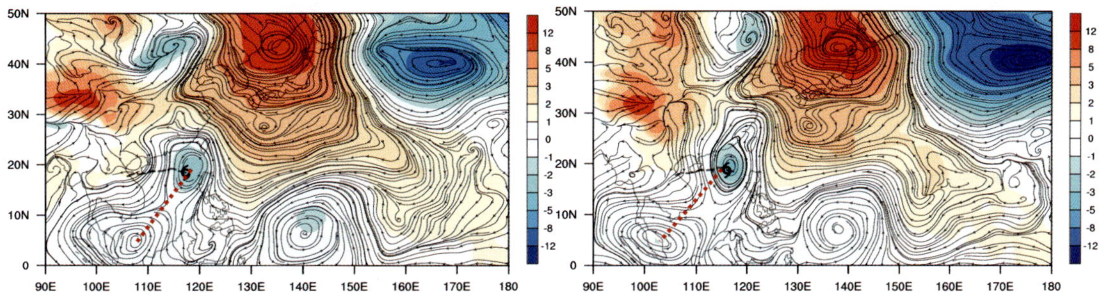

850 hPa flow/height anomalies at 0000 UTC and 1200 UTC 27 September 1985.

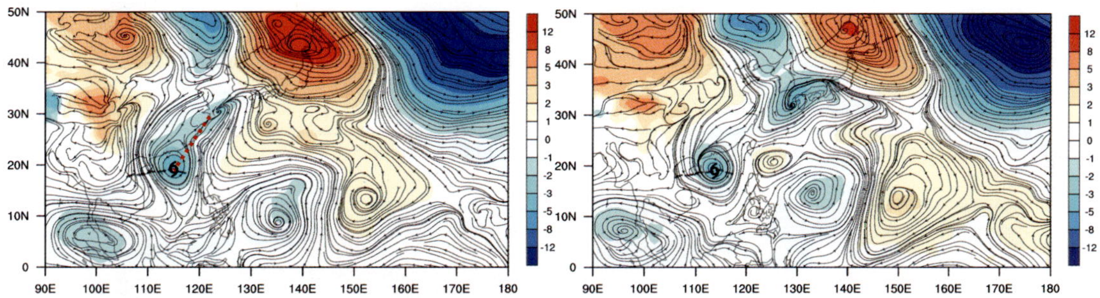

850 hPa flow/height anomalies at 0000 UTC and 1200 UTC 28 September 1985.

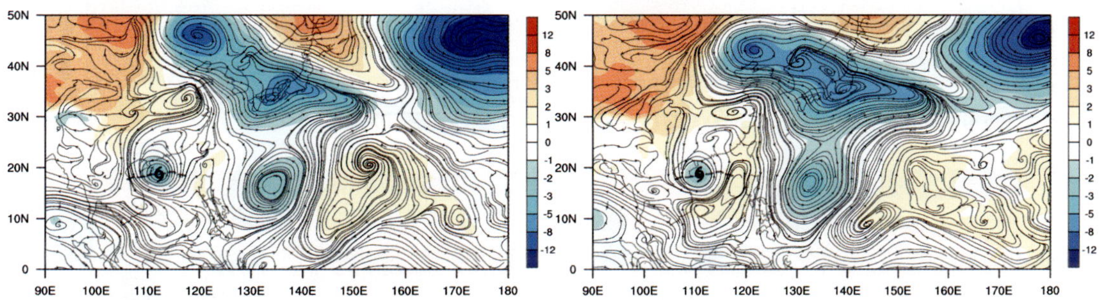

850 hPa flow/height anomalies at 0000 UTC and 1200 UTC 29 September 1985.

Figure 2.8. The typhoon (1985_No.19) generated in the northeastern SCS on 27 September 1985 and moved westward. Its intensity peaked near the climatic monsoon trough in the northwestern SCS on 29–30 September 1985 before landfall on the northeastern coast of Vietnam. In the first three days, its track had a small change due to the influence of an anomalous vortex at 850 hPa.

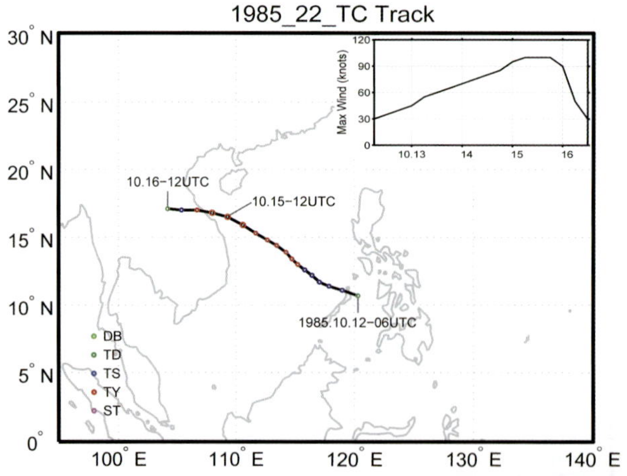

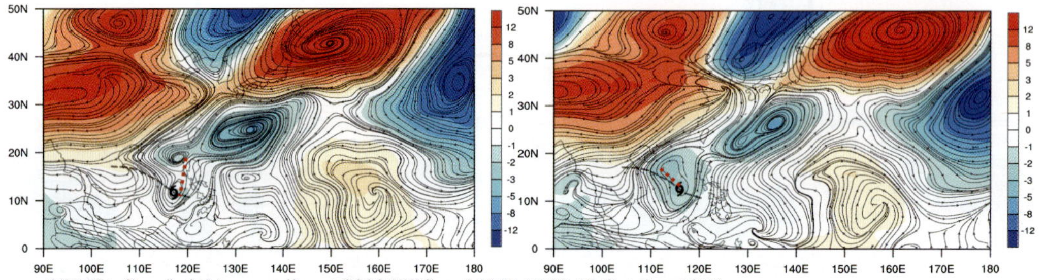

300 hPa flow/height anomalies at 0000 UTC and 1200 UTC 13 October 1985.

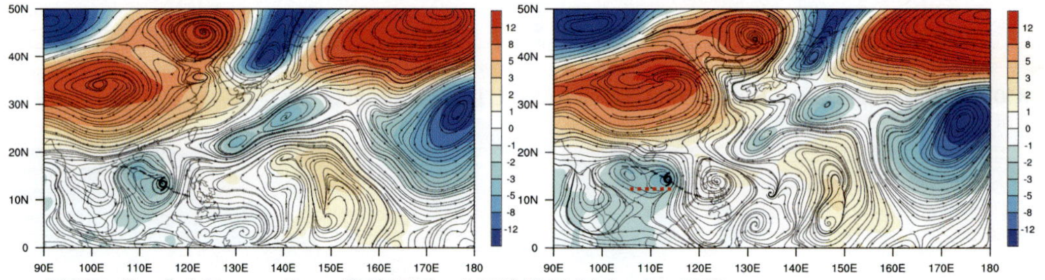

300 hPa flow/height anomalies at 0000 UTC and 1200 UTC 14 October 1985.

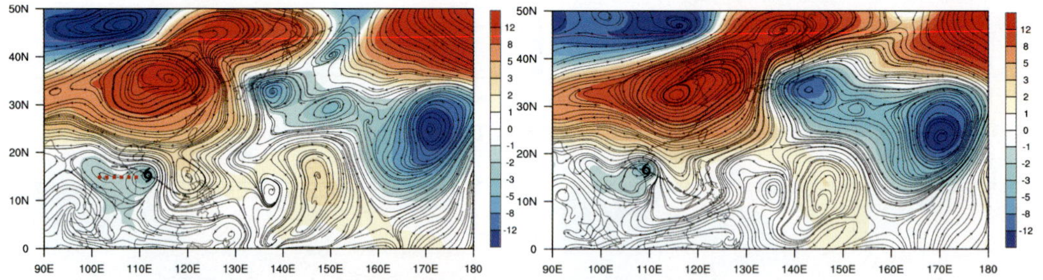

300 hPa flow/height anomalies at 0000 UTC and 1200 UTC 15 October 1985.

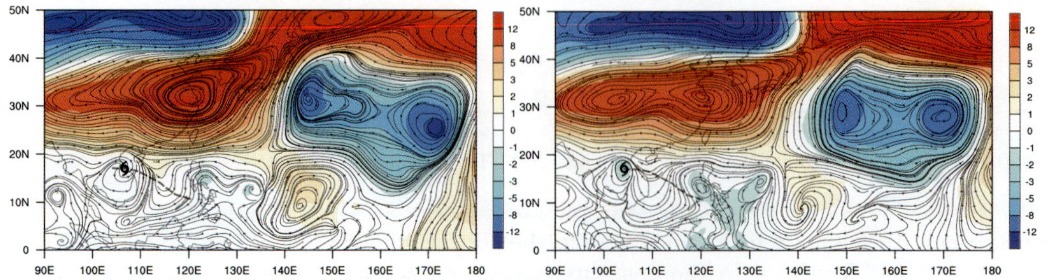

300 hPa flow/height anomalies at 0000 UTC and 1200 UTC 16 October 1985.

Figure 2.9. The typhoon (1985_No.22) generated in the central Philippines on 12 October 1985 and moved northwestward. Its intensity peaked near the climatic monsoon trough in the south sea of the Hainan Island on 15 October 1985 before landfall on the northeastern coast of Vietnam. Its track had a small change due to the influence of two small anomalous vortices at 300 hPa.

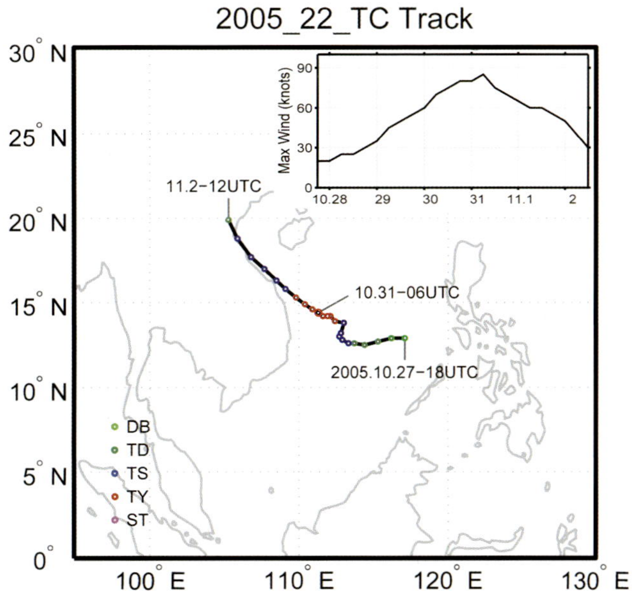

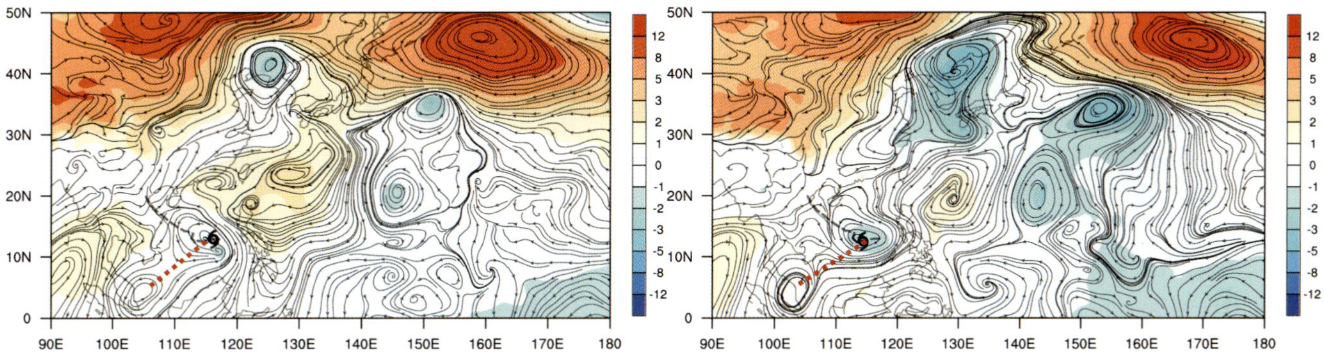

850 hPa flow/height anomalies at 0000 UTC and 1200 UTC 28 October 2005.

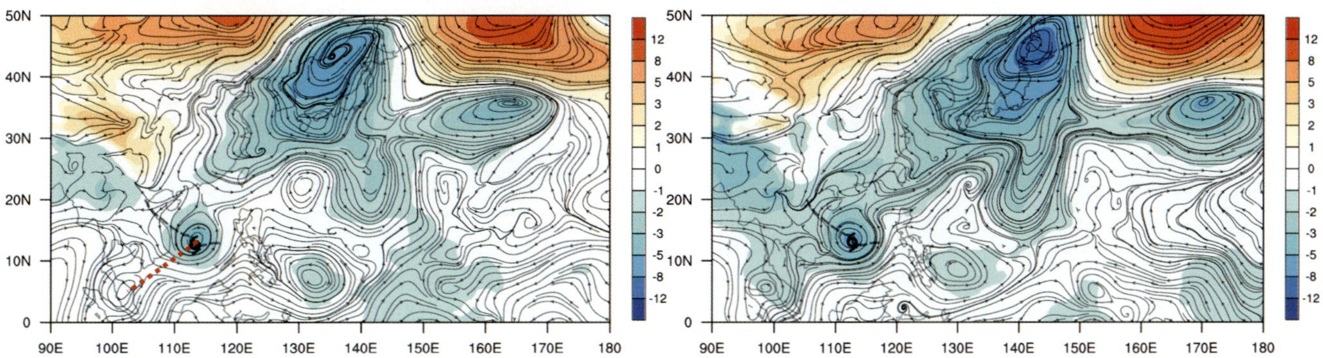

850 hPa flow/height anomalies at 0000 UTC and 1200 UTC 29 October 2005.

Figure 2.10. The typhoon (2005_No.22) generated in the central eastern SCS on 27 October 2005 and moved northwestward. Its intensity peaked near the climatic monsoon trough in the central western SCS on 31 October 2005 before landfall on the Vietnam coast. Its track had a small change due to the influence of an anomalous vortex at 850 hPa in the first two days.

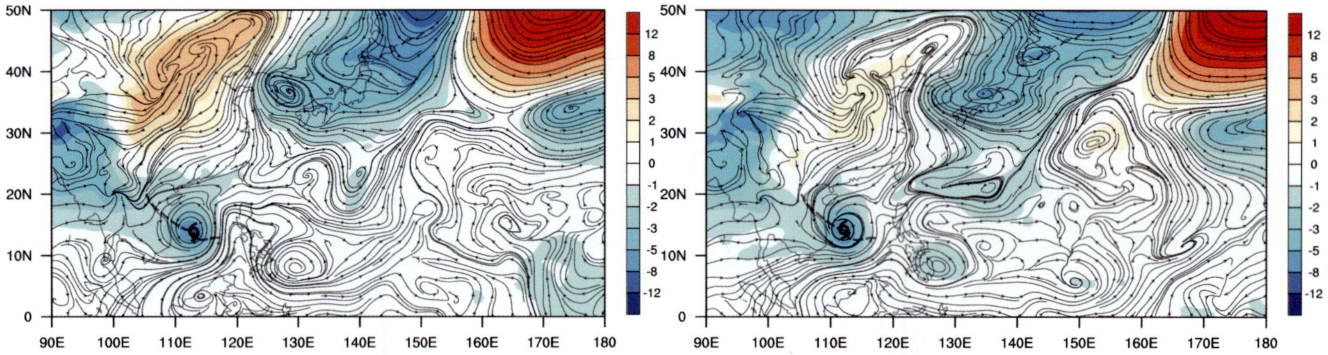

850 hPa flow/height anomalies at 0000 UTC and 1200 UTC 30 October 2005.

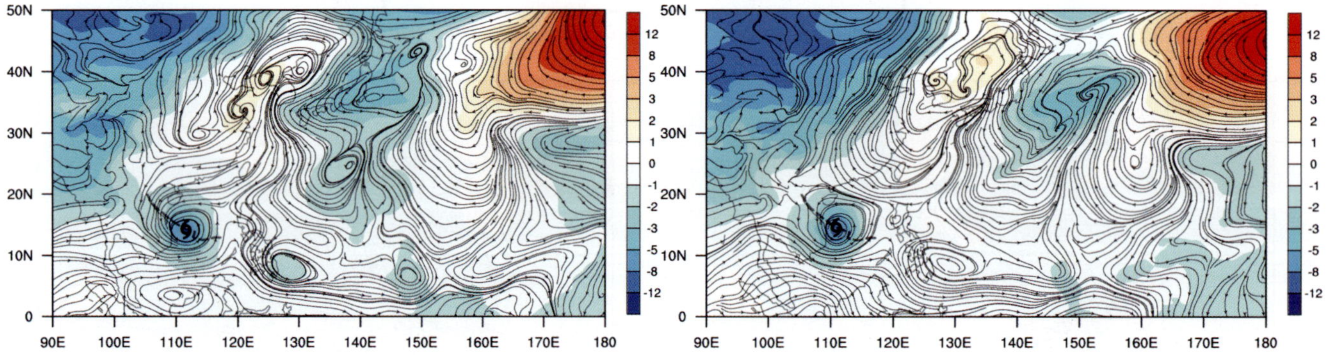

850 hPa flow/height anomalies at 0000 UTC and 1200 UTC 31 October 2005.

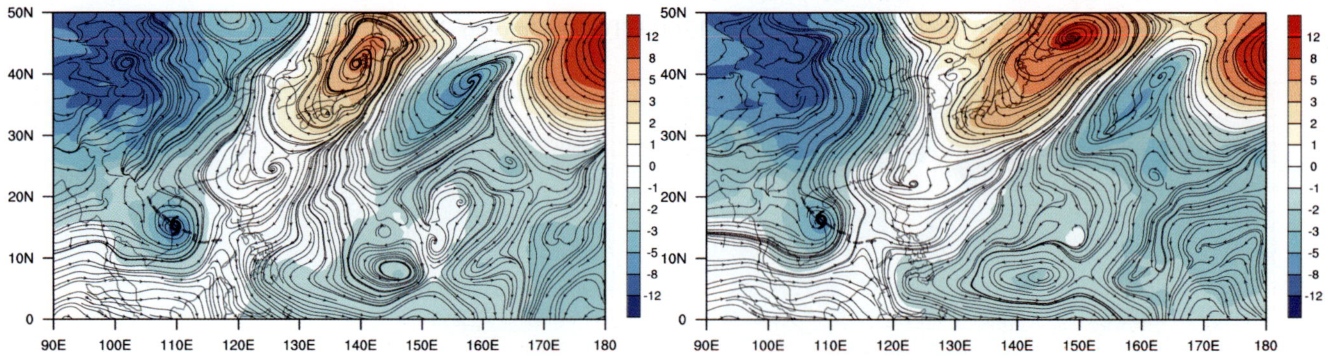

850 hPa flow/height anomalies at 0000 UTC and 1200 UTC 1 November 2005.

Figure 2.10. *(Continued)*

2.2. Intensity Peaked in the South China Sea

This section lists 26 typhoons generated in other regions without obvious turning motions during their lifetime periods and their intensities peaked in the SCS (Figs. 2.11–2.36).

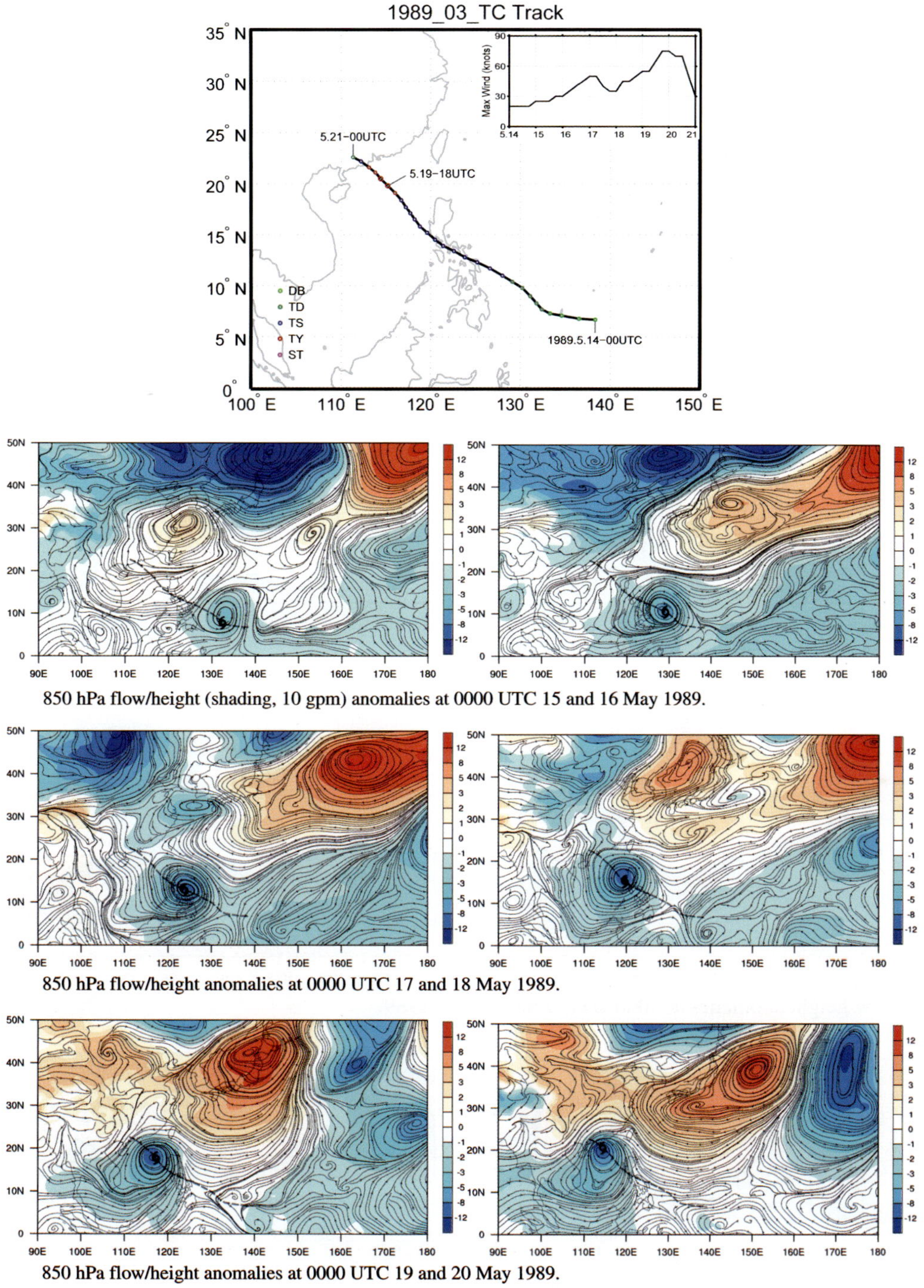

Figure 2.11. The typhoon (1989_No.03) generated in the equatorial western Pacific on 14 May 1989 and moved northwestward along the climatic monsoon trough without the influence of other anomalous vortices at 850 hPa. Its intensity peaked on 19 May 1989 before landfall on the South China coast.

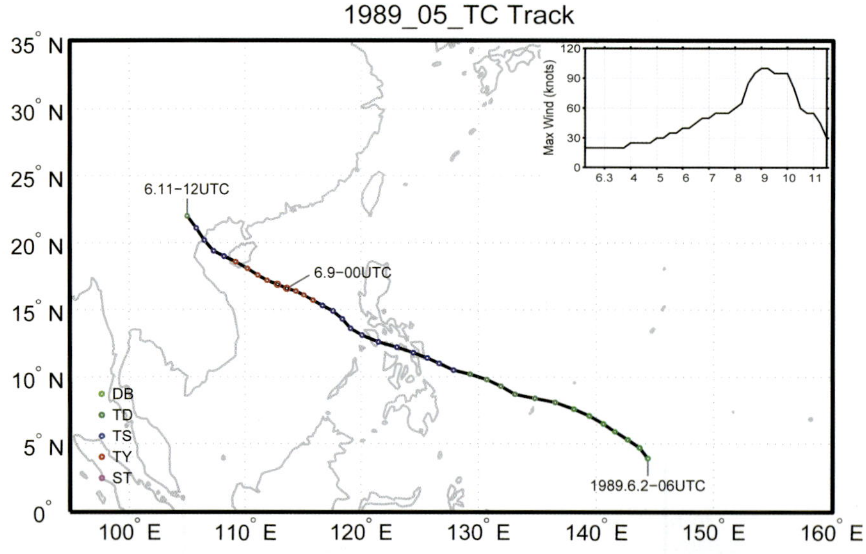

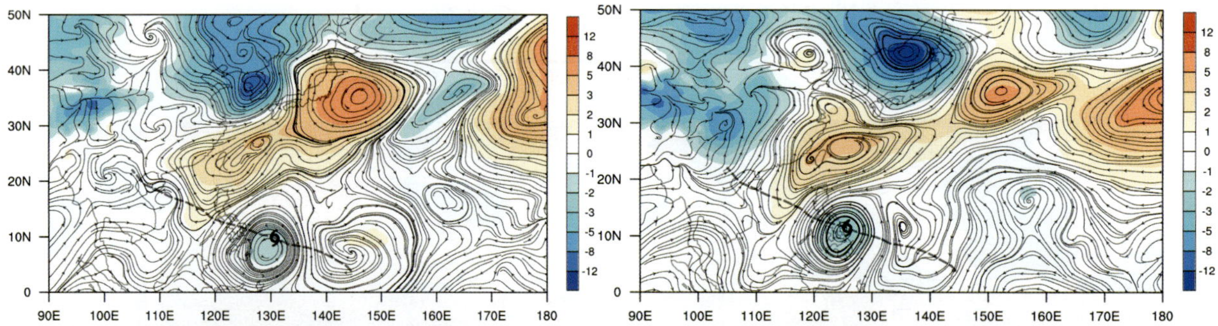

850 hPa flow/height anomalies at 0000 UTC 5 and 6 June 1989.

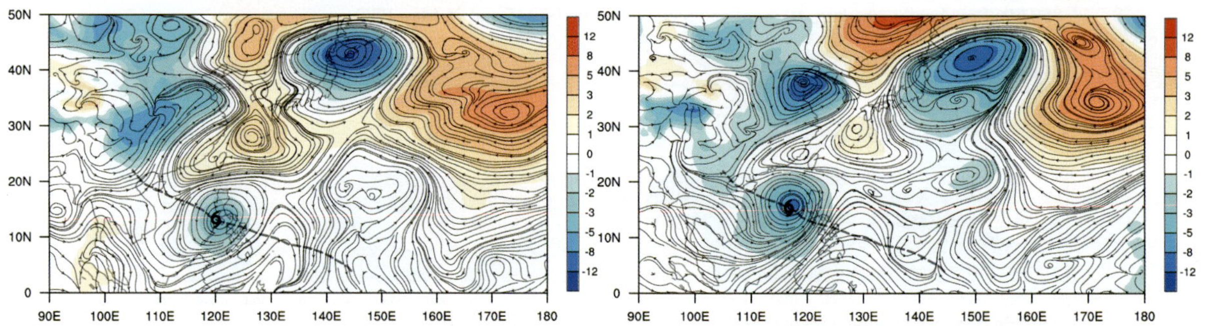

850 hPa flow/height anomalies at 0000 UTC 7 and 8 June 1989.

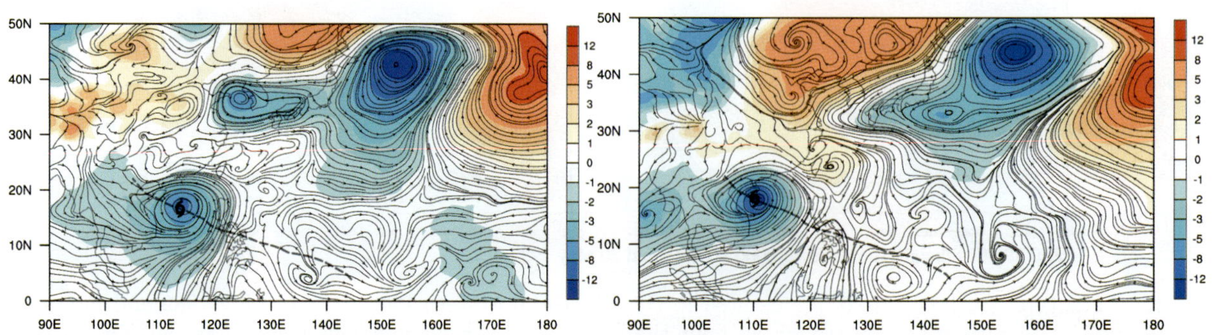

850 hPa flow/height anomalies at 0000 UTC 9 and 10 June 1989.

Figure 2.12. The typhoon (1989_No.05) generated in the equatorial western Pacific on 2 June 1989 and moved northwestward without the influence of other anomalous vortices at 850 hPa. Its intensity peaked near the climatic monsoon trough in the central northern SCS on 9 June 1989 before landfall on the northeast coast of Vietnam.

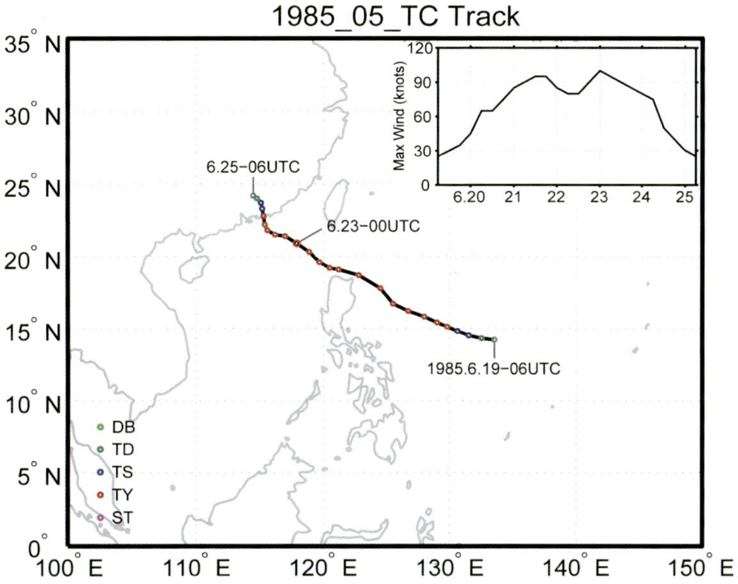

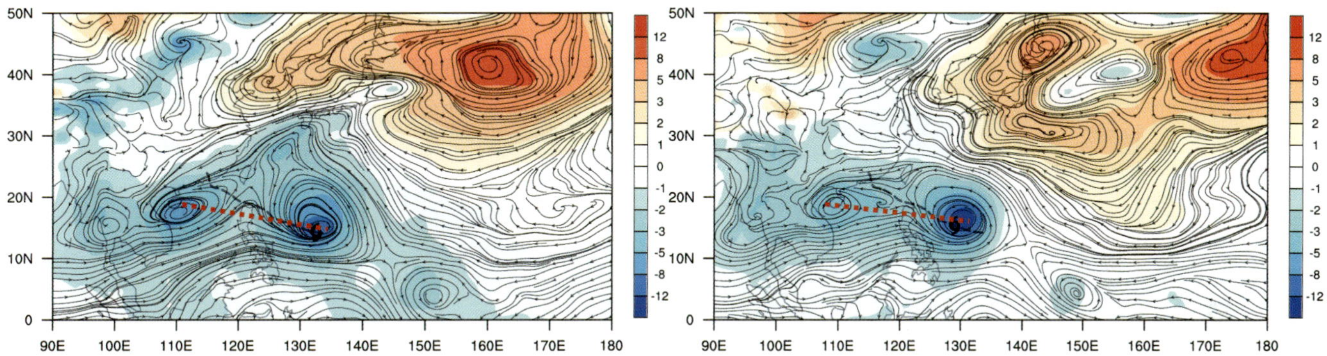

850 hPa flow/height anomalies at 1200 UTC 19 and 20 June 1985.

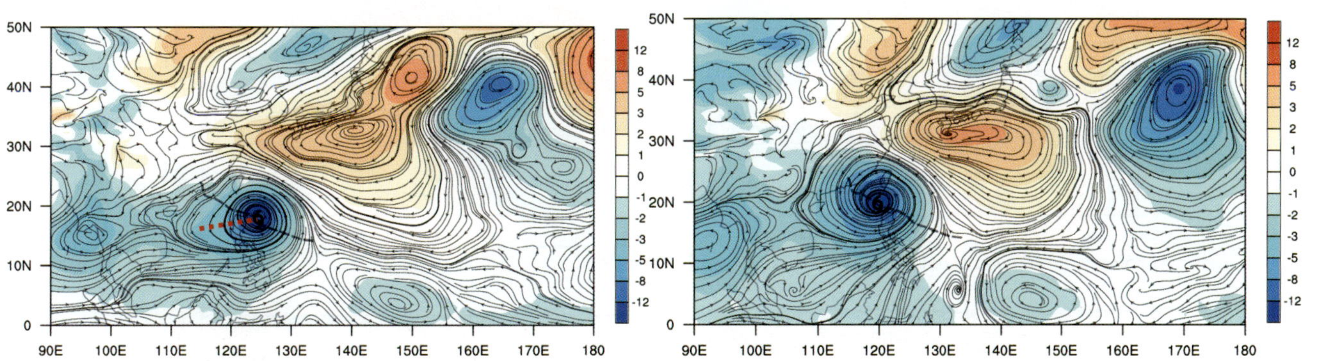

850 hPa flow/height anomalies at 1200 UTC 21 and 22 June 1985.

Figure 2.13. The typhoon (1985_No.05) generated in the tropical western Pacific on 19 June 1998. Its intensity peaked in the northern SCS on 23 June 1998 due to the combination of two anomalous vortices at 850 hPa before landfall on the South China coast. It moved northwestward along the climatic monsoon trough without the influence of other anomalous vortices.

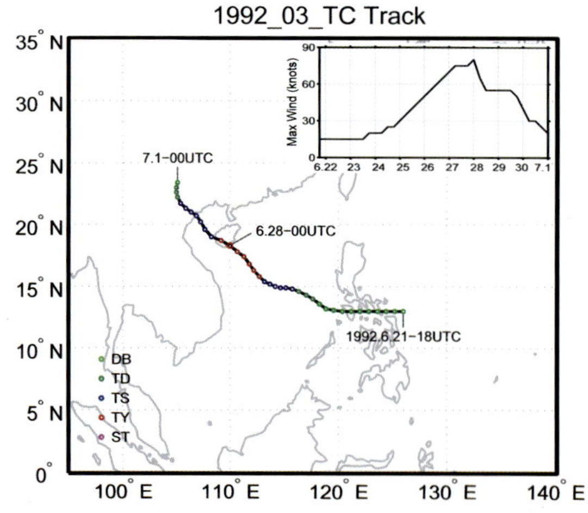

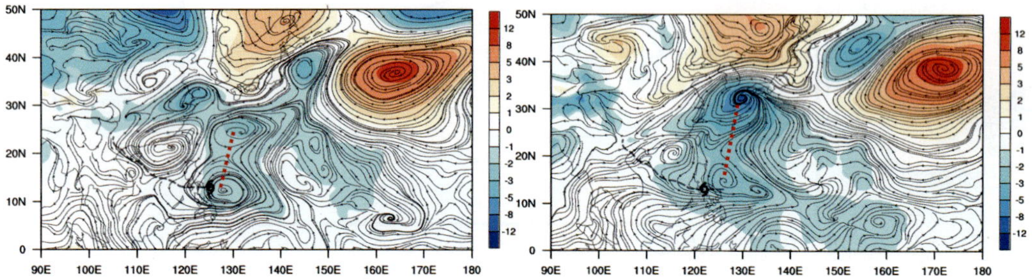

850 hPa flow/height anomalies at 0000 UTC 22 and 23 June 1992.

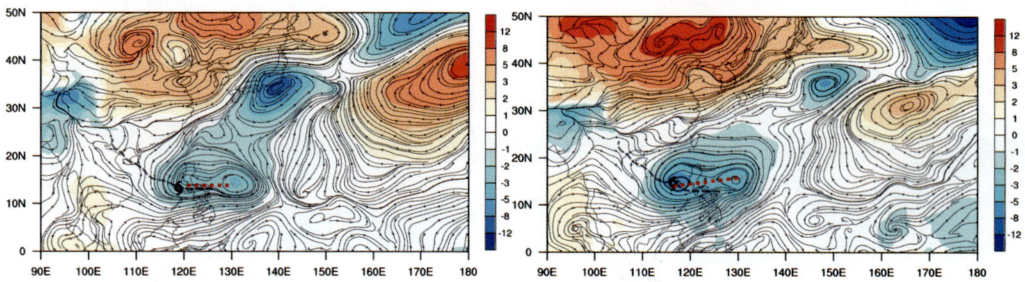

850 hPa flow/height anomalies at 0000 UTC 24 and 25 June 1992.

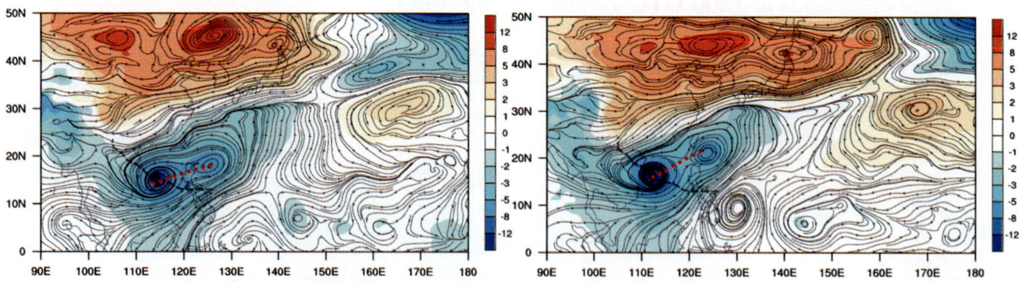

850 hPa flow/height anomalies at 0000 UTC 26 and 27 June 1992.

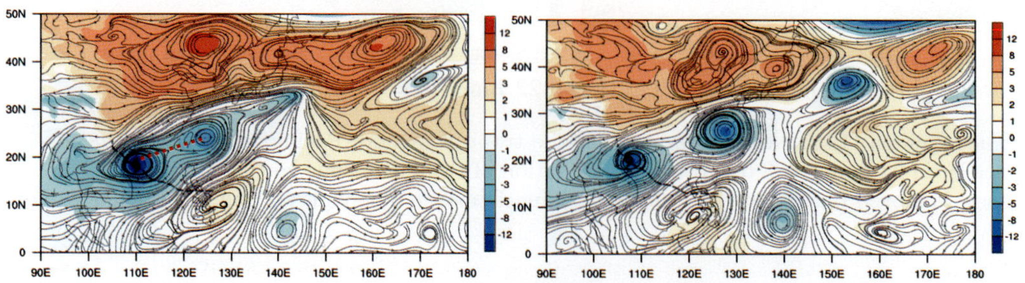

850 hPa flow/height anomalies at 0000 UTC 28 and 29 June 1992.

Figure 2.14. The typhoon (1992_No.03) generated in the east sea of the Philippines on 21 June 1992 and moved northwestward. Its intensity peaked near the climatic monsoon trough in the Hainan Island on 28 June 1992 before landfall on the island. Its motion was modified by the interaction of anomalous vortices at 850 hPa.

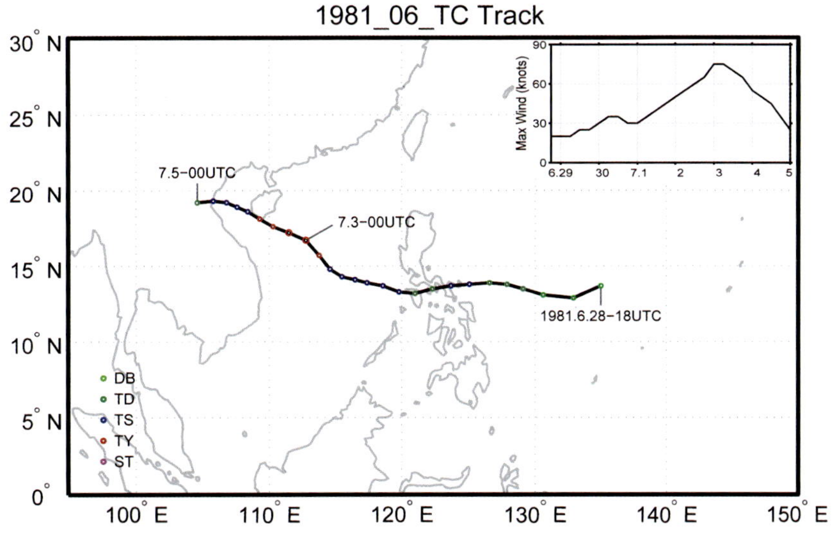

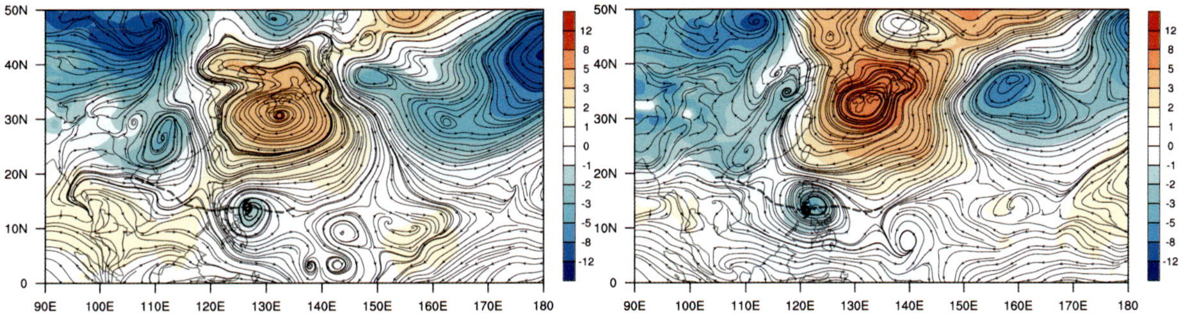

850 hPa flow/height anomalies at 0000 UTC 30 June and 1 July 1981.

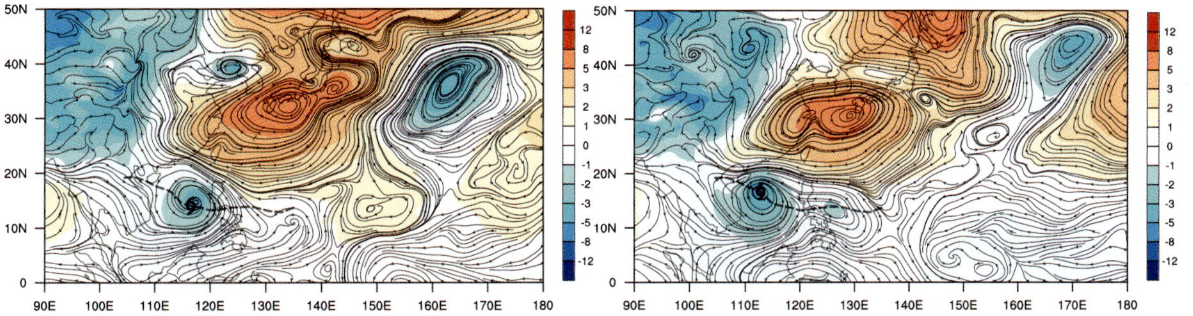

850 hPa flow/height anomalies at 0000 UTC 2 and 3 July 1981.

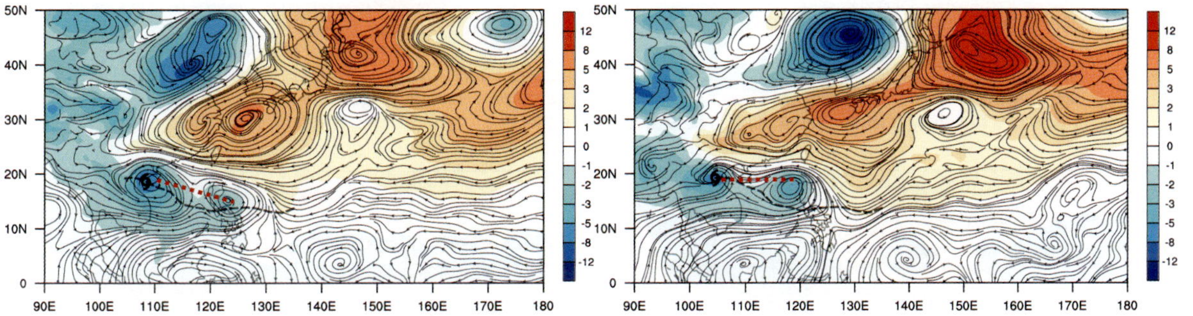

850 hPa flow/height anomalies at 0000 UTC 4 and 5 July 1981.

Figure 2.15. The typhoon (1981_No.06) generated in the tropical western Pacific on 28 June 1981 and moved westward. Its intensity peaked near the climatic monsoon trough in the southeast sea of the Hainan Island on 3 July 1981 before landfall on the island and its waving track was influenced by anomalous flows at 850 hPa.

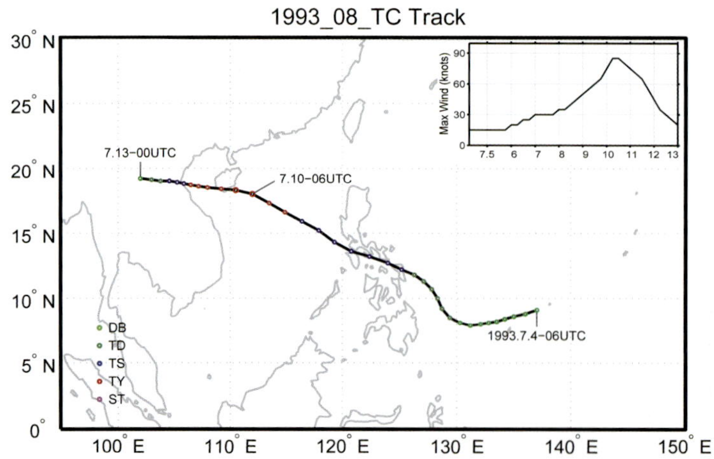

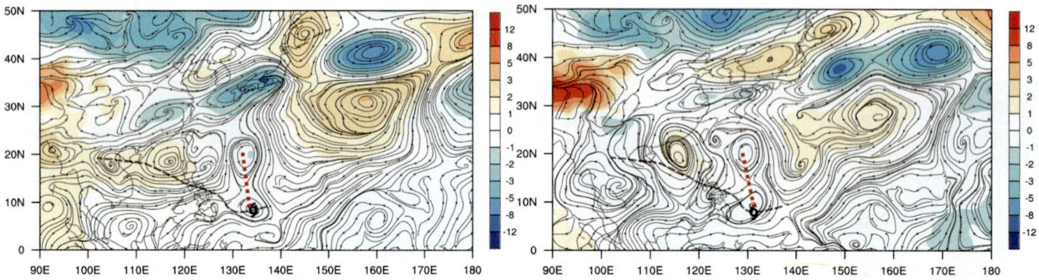

850 hPa flow/height anomalies at 0000 UTC 5 and 6 July 1993.

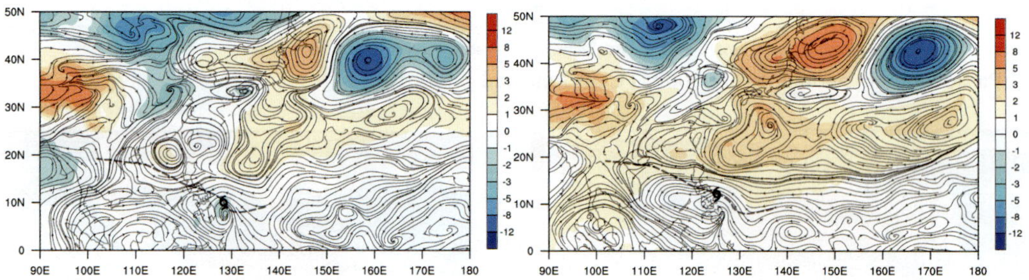

850 hPa flow/height anomalies at 0000 UTC 7 and 8 July 1993.

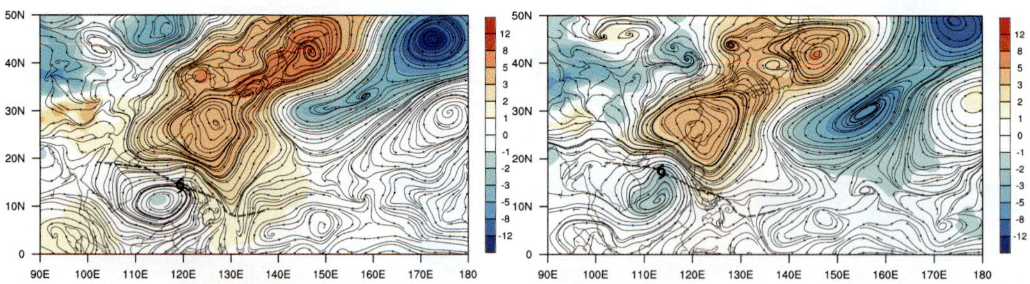

850 hPa flow/height anomalies at 0000 UTC 9 and 10 July 1993.

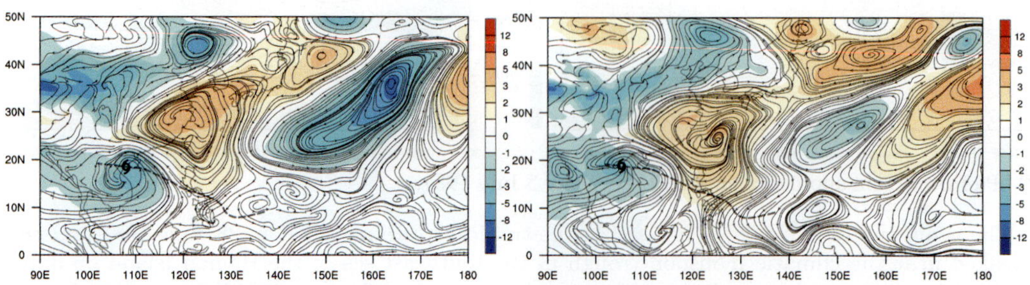

850 hPa flow/height anomalies at 0000 UTC 11 and 12 July 1993.

Figure 2.16. The typhoon (1993_No.08) generated in the equatorial western Pacific on 4 July 1993 and moved northwestward. Its intensity peaked near the climatic monsoon trough near the Hainan Island on 10 July 1993 before landfall on the island and the northeast coast of Vietnam. At early stage, its motion was modified by anomalous vortices at 850 hPa.

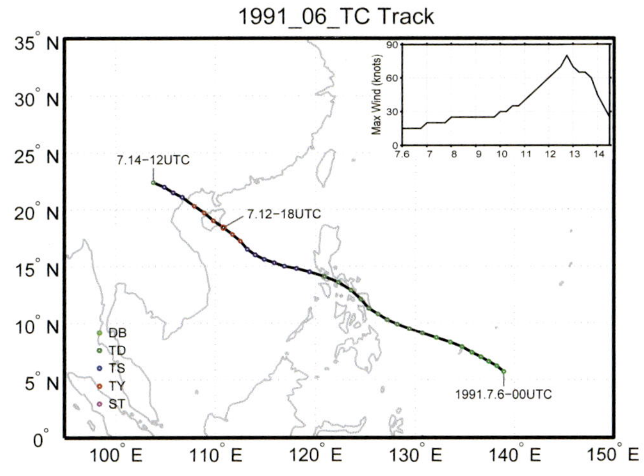

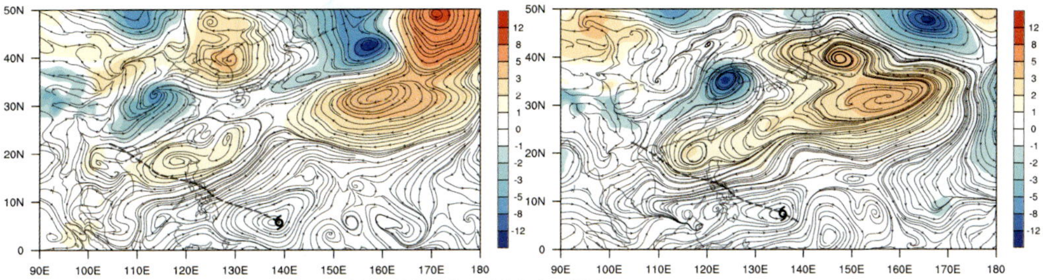

850 hPa flow/height anomalies at 0000 UTC 6 and 7 July 1991.

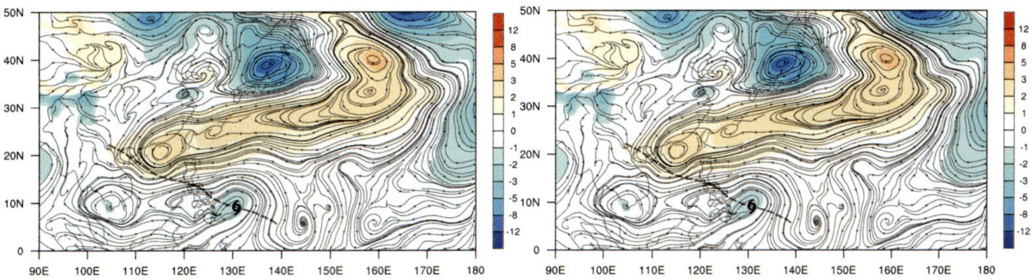

850 hPa flow/height anomalies at 0000 UTC 8 and 9 July 1991.

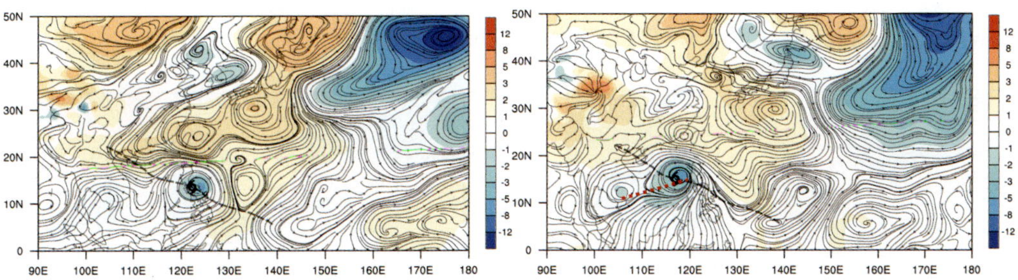

850 hPa flow/height anomalies at 0000 UTC 10 and 11 July 1991.

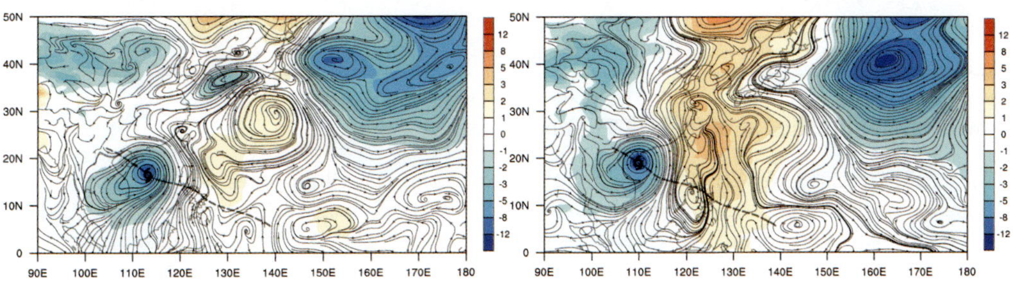

850 hPa flow/height anomalies at 0000 UTC 12 and 13 July 1991.

Figure 2.17. The typhoon (1991_No.06) generated in the equatorial western Pacific on 6 July 1991 and moved northwestward along the climatic monsoon trough but modified by another anomalous vortex at 850 hPa. Its intensity peaked near the climatic monsoon trough in the northwestern SCS on 12 July 1991 before landfall on the Hainan Island.

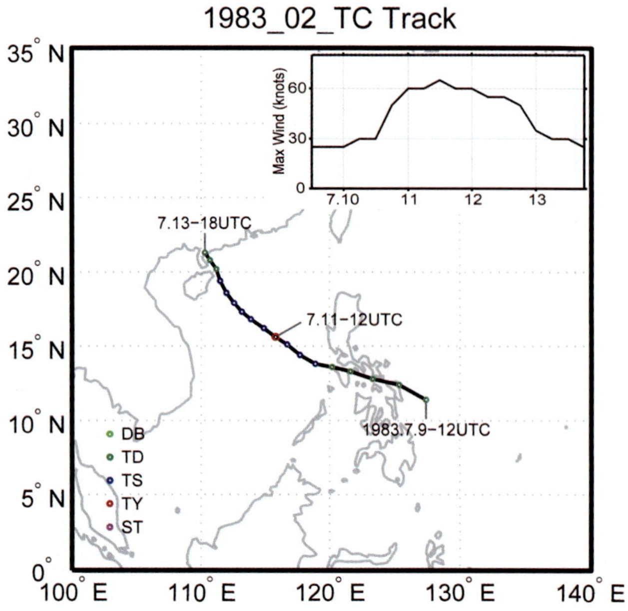

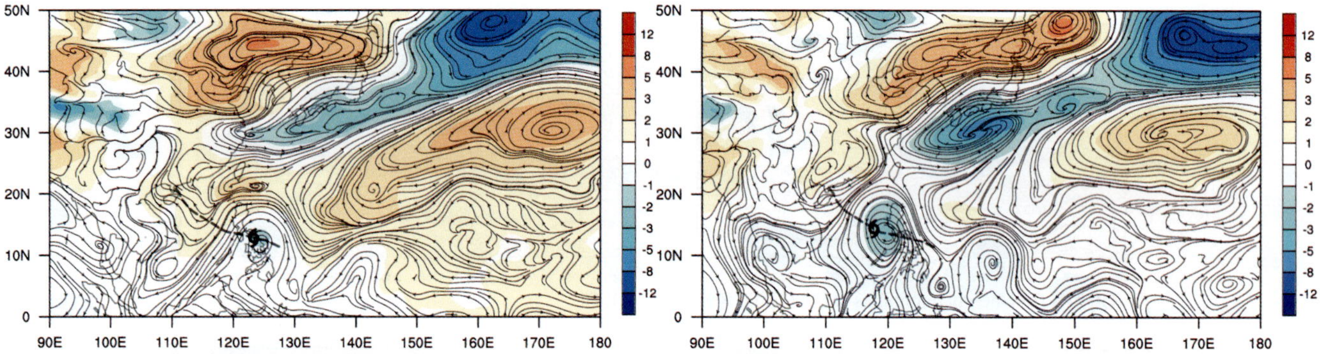

850 hPa flow/height anomalies at 0000 UTC 10 and 11 July 1983.

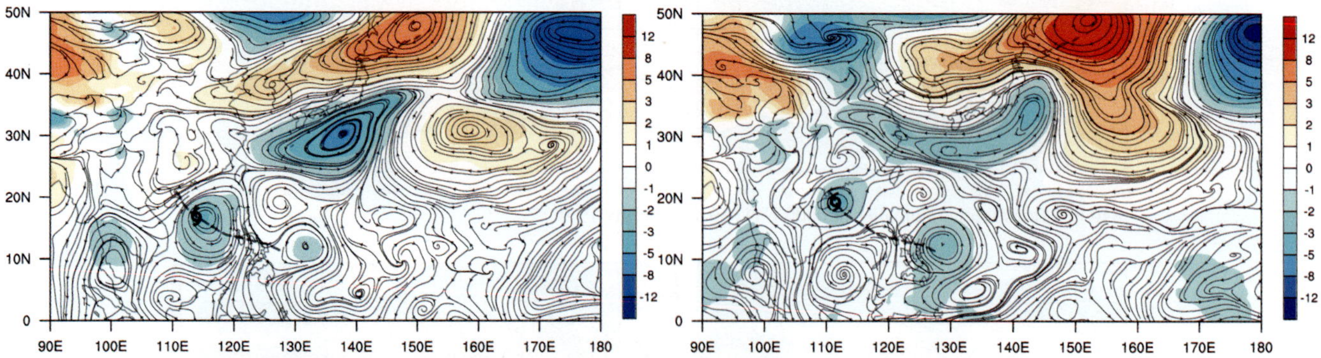

850 hPa flow/height anomalies at 0000 UTC 12 and 13 July 1983.

Figure 2.18. The typhoon (1983_No.02) generated in the east sea of the Philippines on 9 July 1983 and moved northwestward. Its intensity peaked near the climatic monsoon trough in the central SCS on 11 July 1983 and disappeared when it made landfall on the South China coast and its motion was not affected by other anomalous vortices at 850 hPa.

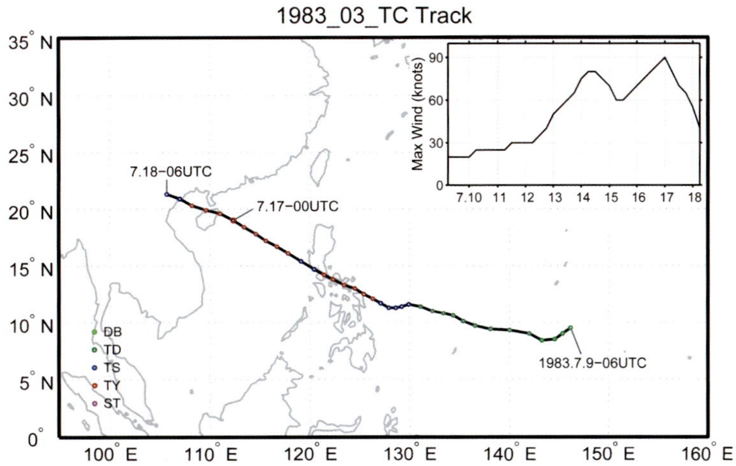

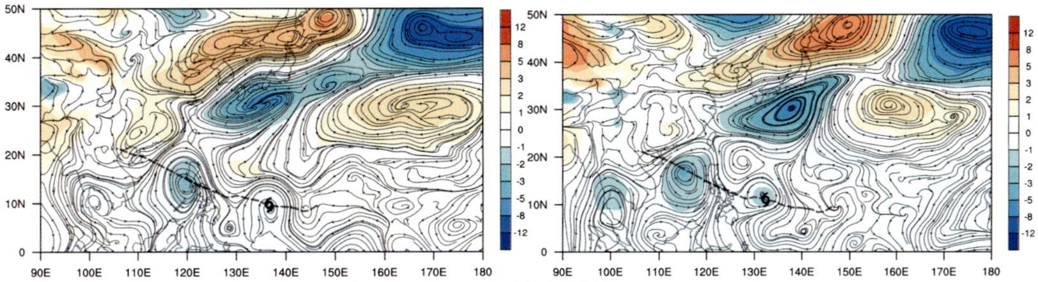

850 hPa flow/height anomalies at 0000 UTC 11 and 12 July 1983.

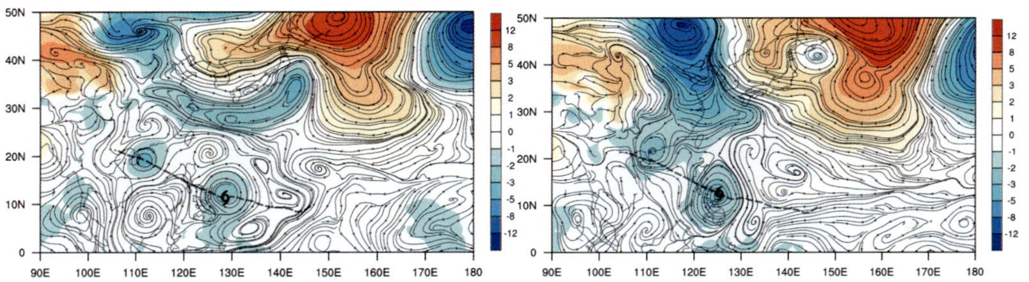

850 hPa flow/height anomalies at 0000 UTC 13 and 14 July 1983.

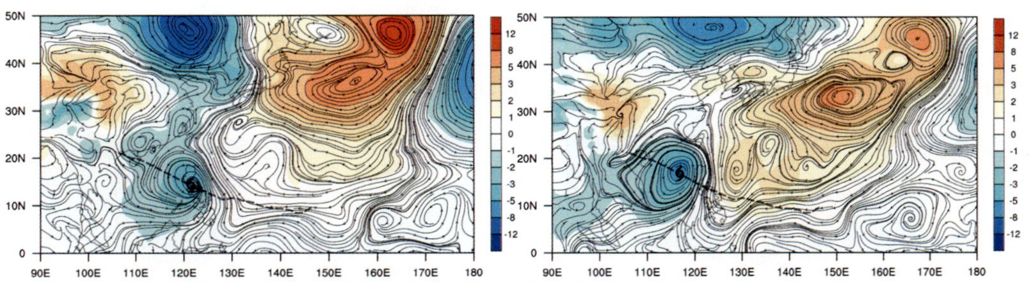

850 hPa flow/height anomalies at 0000 UTC 15 and 16 July 1983.

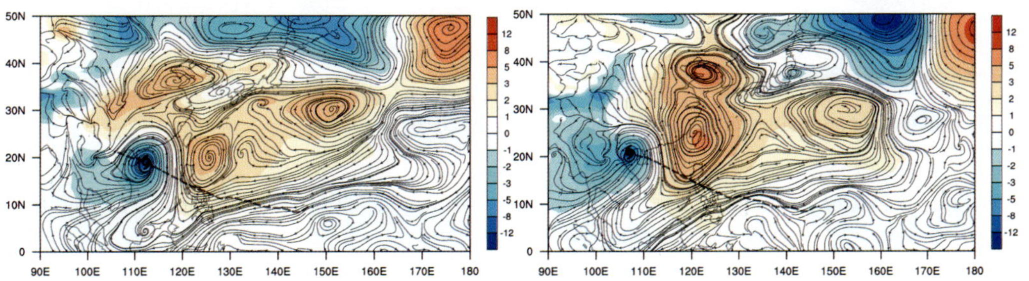

850 hPa flow/height anomalies at 0000 UTC 17 and 18 July 1983.

Figure 2.19. The typhoon (1983_No.03) generated in the equatorial western Pacific on 9 July 1983 and moved west-northwestward without the influence of other anomalous vortices at 850 hPa. Its intensity peaked near the climatic monsoon trough on 17 July 1983 before landfall on the Hainan Island.

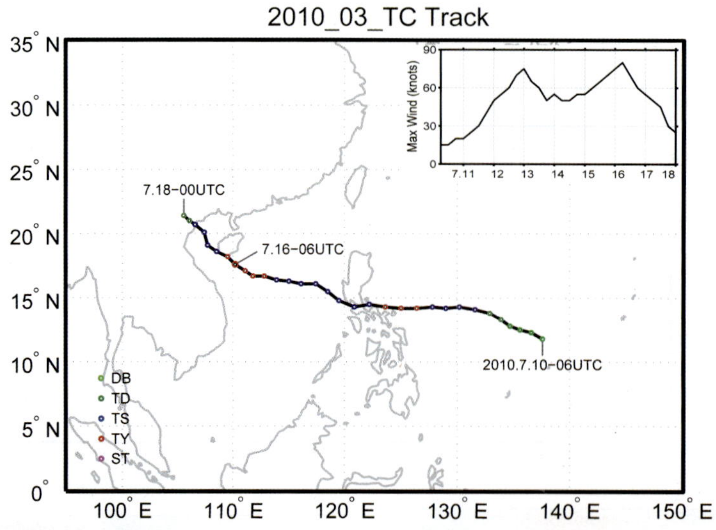

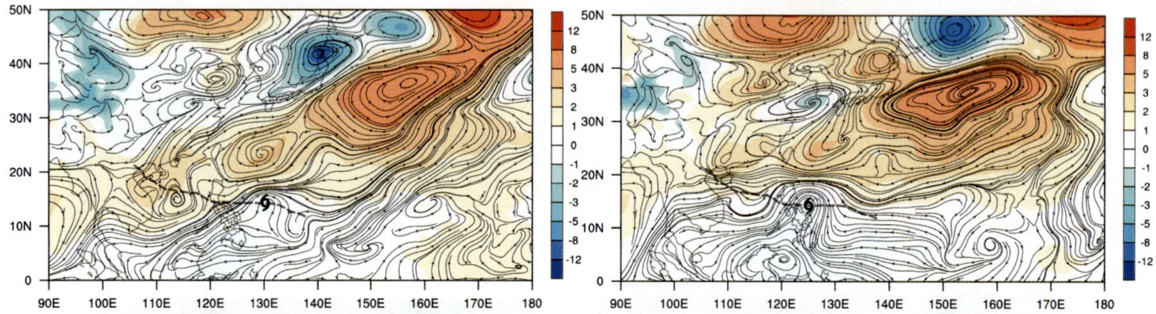

850 hPa flow/height anomalies at 0000 UTC 12 and 13 July 2010.

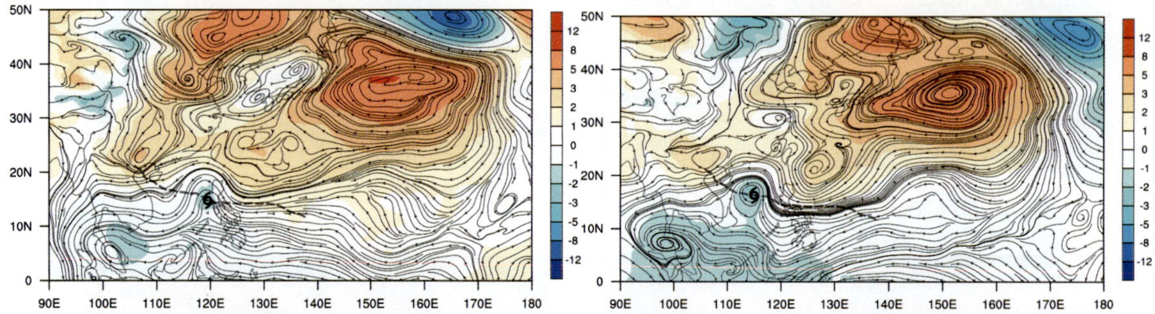

850 hPa flow/height anomalies at 0000 UTC 14 and 15 July 2010.

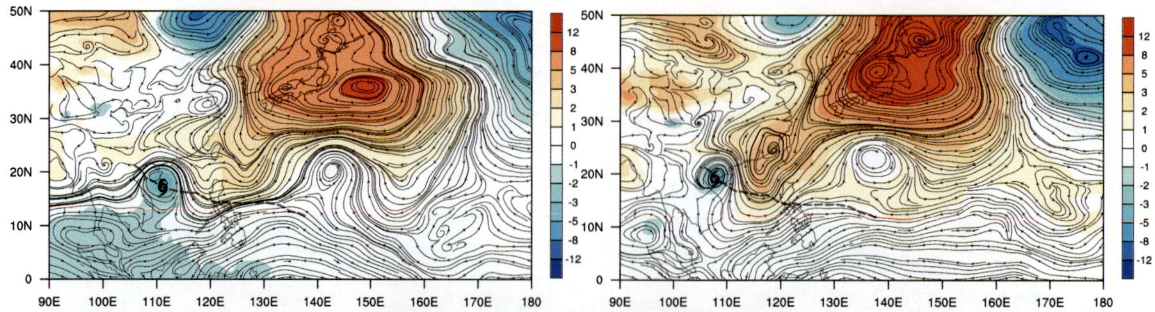

850 hPa flow/height anomalies at 0000 UTC 16 and 17 July 2010.

Figure 2.20. The typhoon (2010_No.03) generated in the tropical western Pacific on 10 July 2010 and moved west-northwestward without the direct influence of other anomalous vortices at 850 hPa. Its intensity peaked near the climatic monsoon trough on 16 July 2010 before landfall on the Hainan Island.

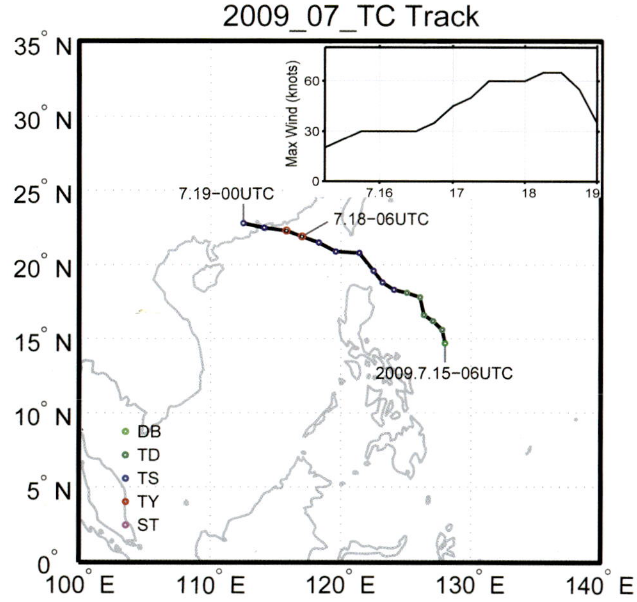

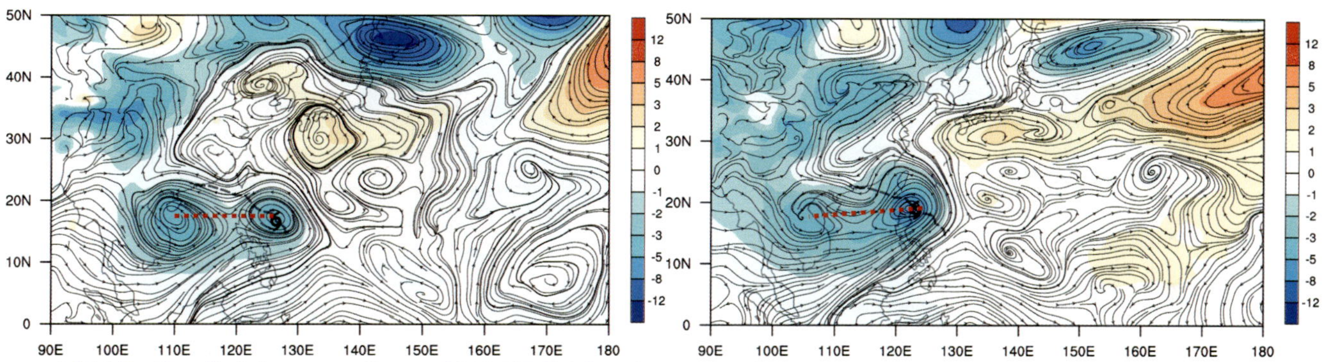

850 hPa flow/height anomalies at 0000 UTC 16 and 17 July 2009.

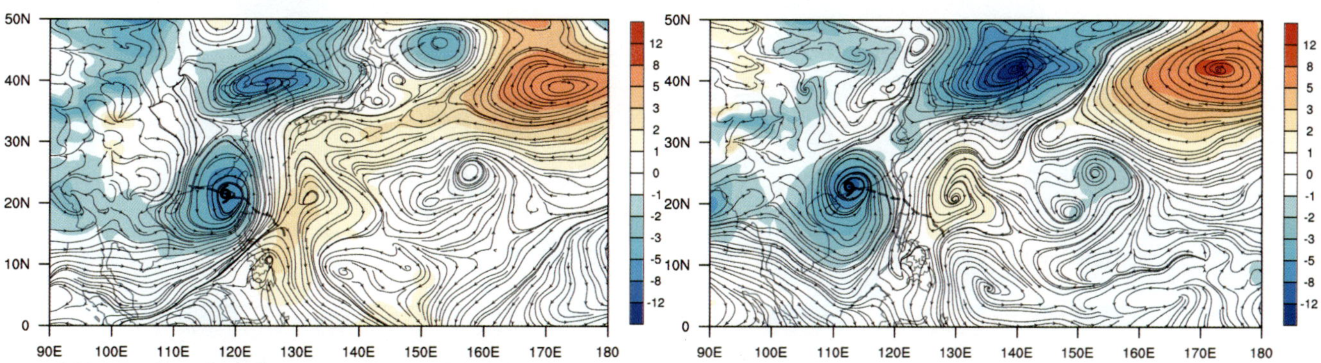

850 hPa flow/height anomalies at 0000 UTC 18 and 19 July 2009.

Figure 2.21. The typhoon (2009_No.07) generated in the east sea of the Philippines on 15 July 2009 and moved northwestward. Its intensity peaked on 18 July 2009 before landfall on the South China coast. Its track was modified by another anomalous vortex at 850 hPa.

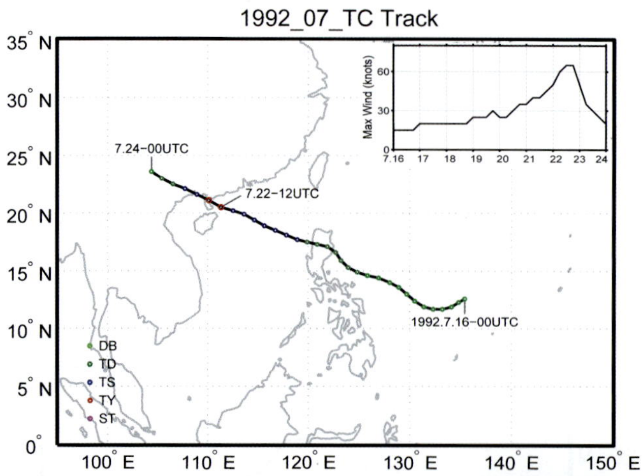

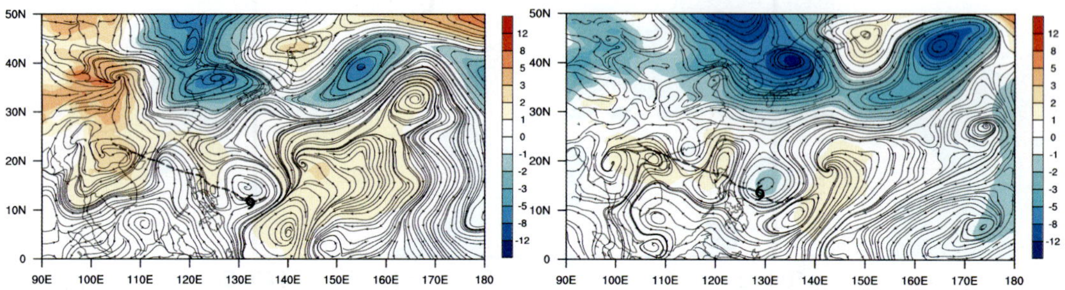

850 hPa flow/height anomalies at 0000 UTC 17 and 18 July 1992.

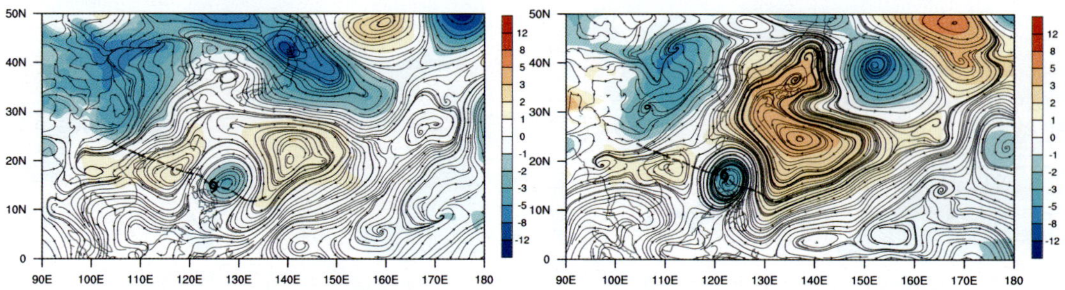

850 hPa flow/height anomalies at 0000 UTC 19 and 20 July 1992.

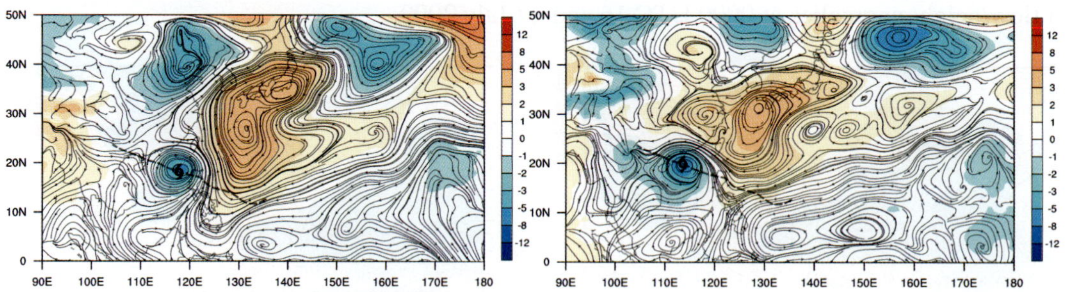

850 hPa flow/height anomalies at 0000 UTC 21 and 22 July 1992.

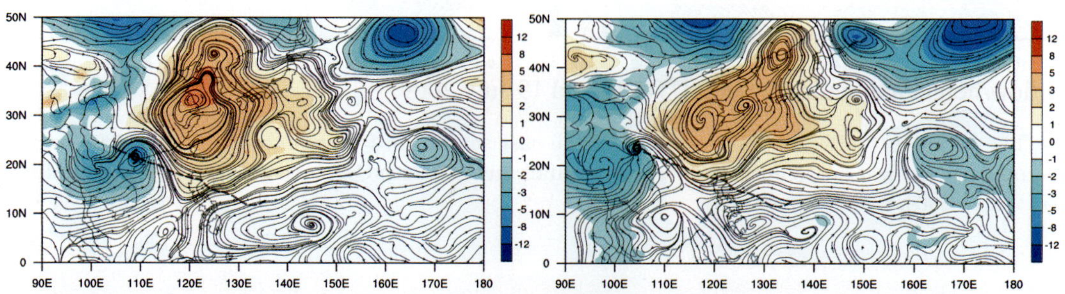

850 hPa flow/height anomalies at 0000 UTC 23 and 24 July 1992.

Figure 2.22. The typhoon (1992_No.07) generated in the tropical western Pacific on 16 July 1992 and moved northwestward. Its intensity peaked near the climatic monsoon trough on 22 July 1992 before landfall on the South China coast and its motion was not affected by any anomalous vortices at 850 hPa.

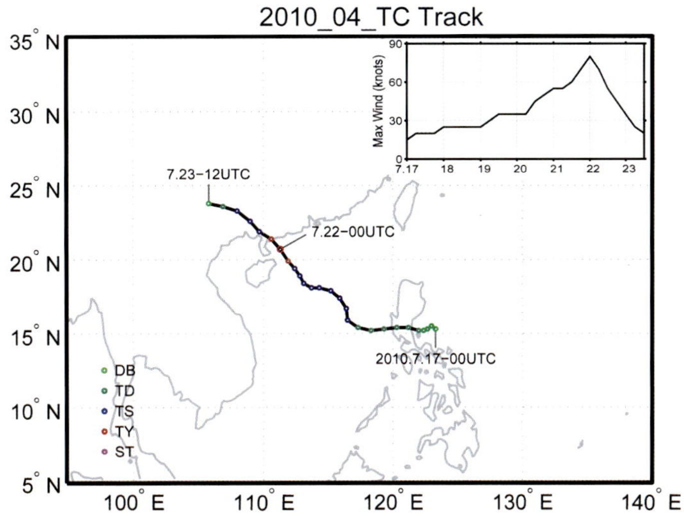

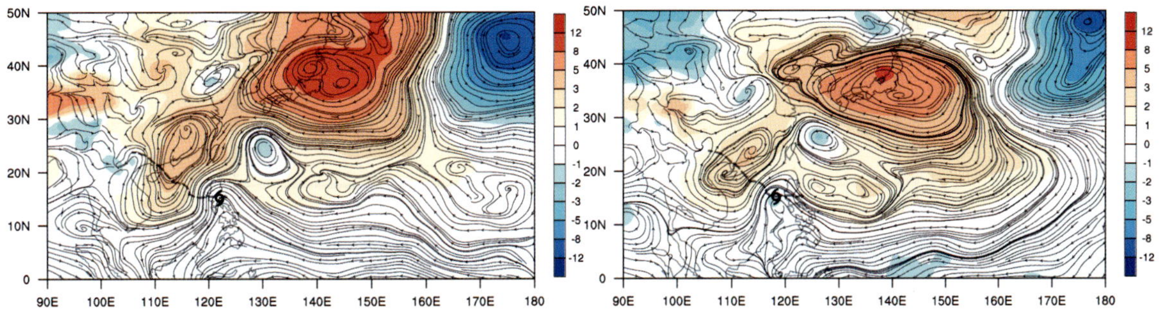

850 hPa flow/height anomalies at 0000 UTC 18 and 19 July 2010.

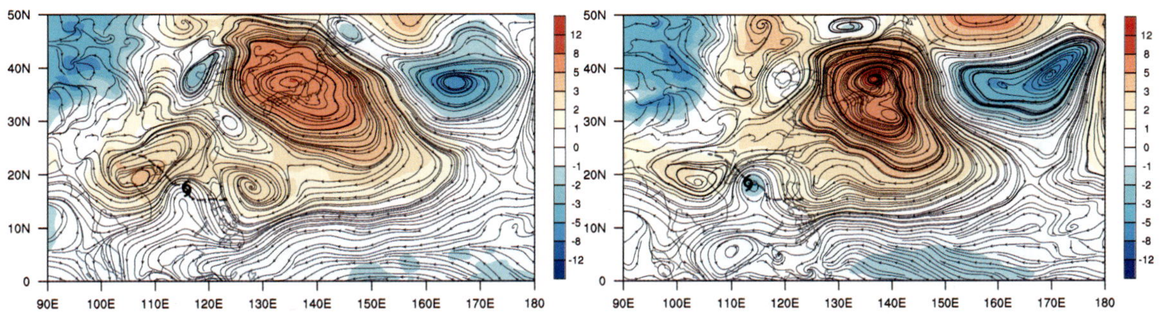

850 hPa flow/height anomalies at 0000 UTC 20 and 21 July 2010.

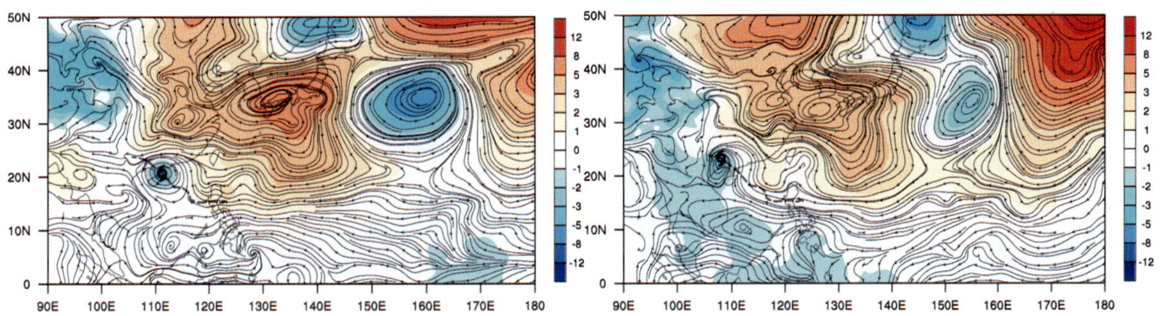

850 hPa flow/height anomalies at 0000 UTC 22 and 23 July 2010.

Figure 2.23. The typhoon (2010_No.04) generated in the east sea of the Philippines on 17 July 2010 and moved northwestward. Its intensity peaked near the climatic monsoon trough on 22 July 2010 before landfall on the South China coast and its waving path was influenced by anomalous flows at 850 hPa.

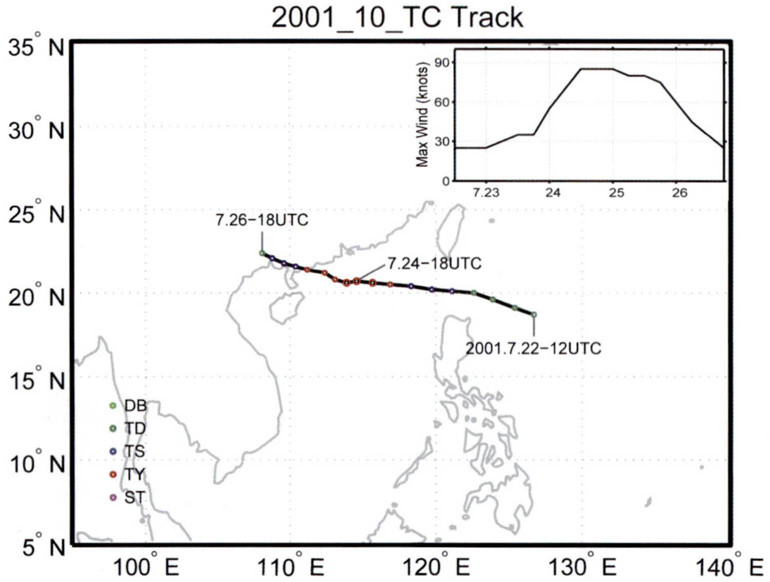

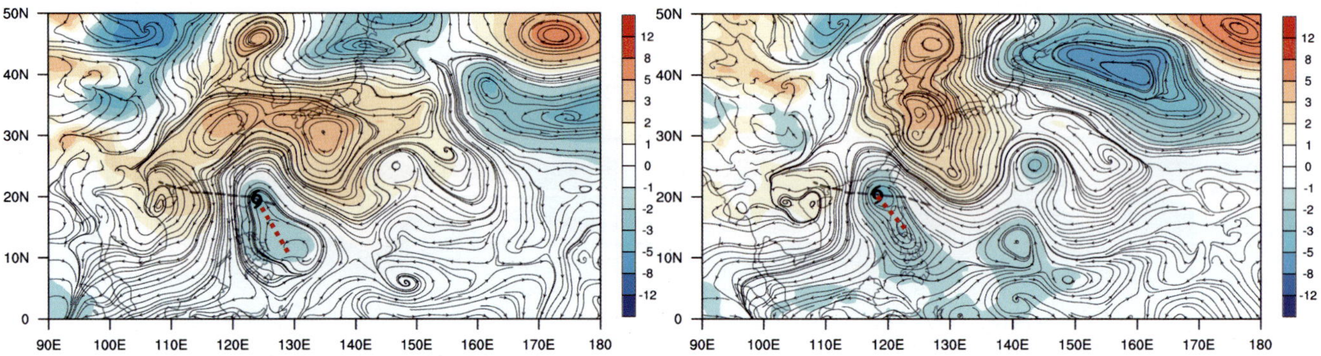

850 hPa flow/height anomalies at 0000 UTC 23 and 24 July 2001.

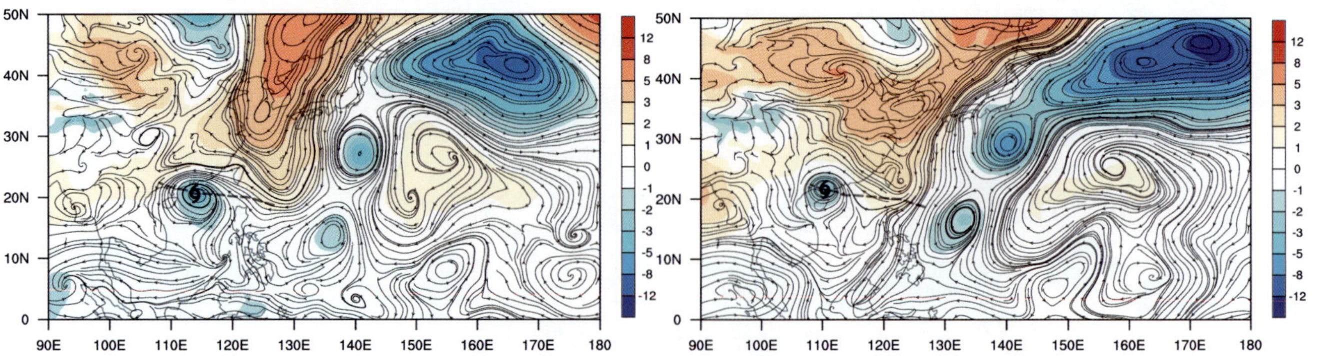

850 hPa flow/height anomalies at 0000 UTC 25 and 26 July 2001.

Figure 2.24. The typhoon (2001_No.10) generated in the northeast sea of the Philippines on 22 July 2001 and moved westward. Its intensity peaked near the climatic monsoon trough in the northernmost SCS on 24 July 2001 before landfall on the South China coast and its motion was modified by an anomalous vortex at 850 hPa.

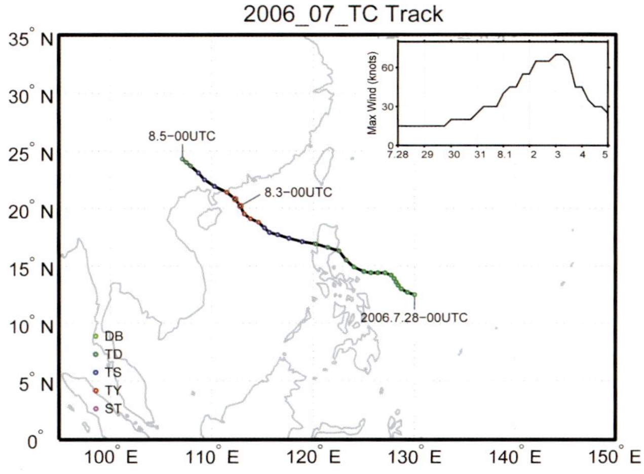

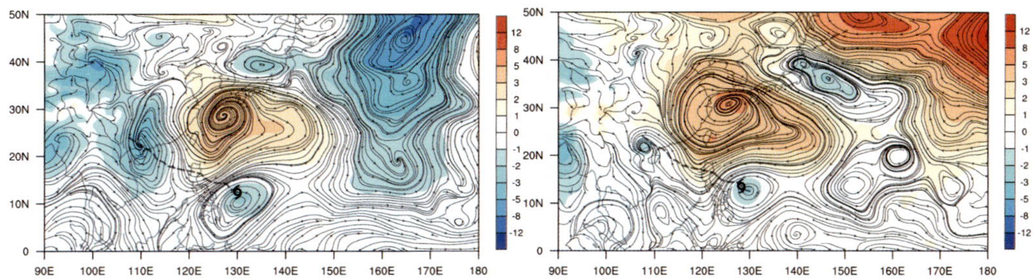

850 hPa flow/height anomalies at 0000 UTC 28 and 29 July 2006.

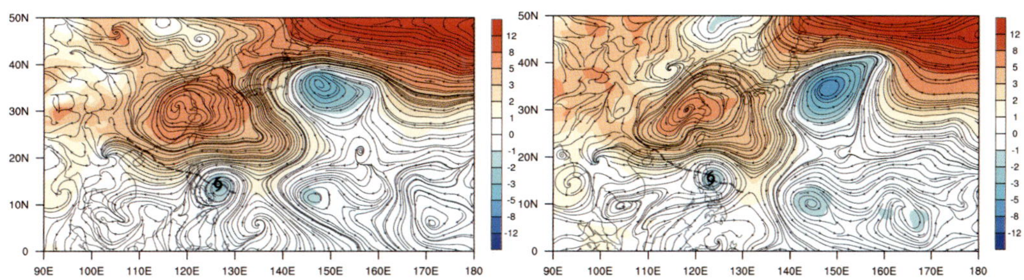

850 hPa flow/height anomalies at 0000 UTC 30 and 31 July 2006.

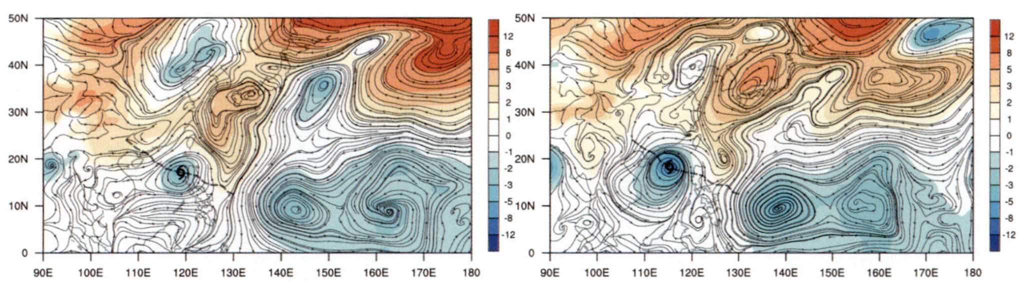

850 hPa flow/height anomalies at 0000 UTC 1 and 2 August 2006.

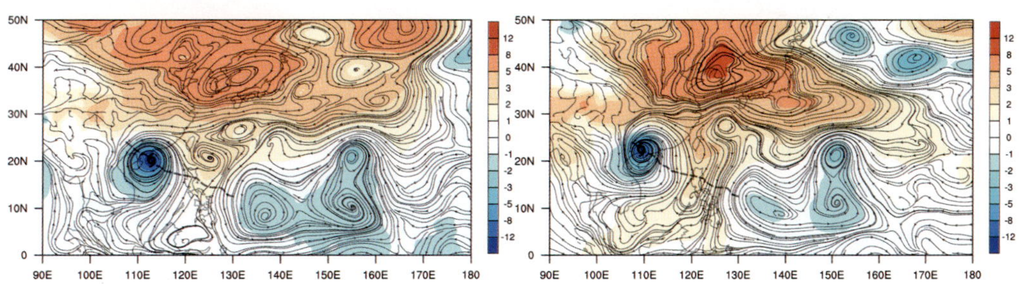

850 hPa flow/height anomalies at 0000 UTC 3 and 4 August 2006.

Figure 2.25. The typhoon (2006_No.07) generated in the east sea of the Philippines on 28 July 2006 and moved northwestward. Its intensity peaked near the climatic monsoon trough in the northernmost SCS on 3 August 2006 before landfall on the South China coast and its track was not affected by other anomalous vortices at 850 hPa.

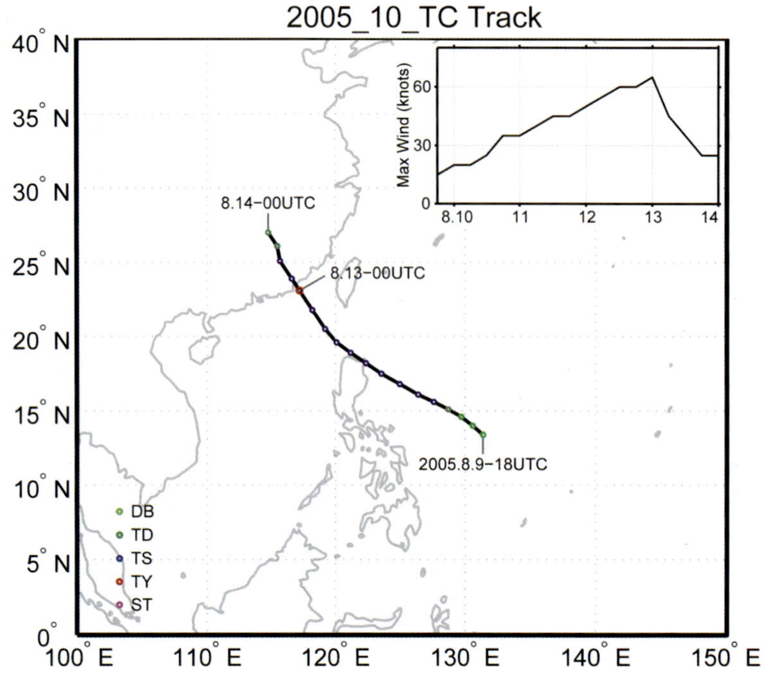

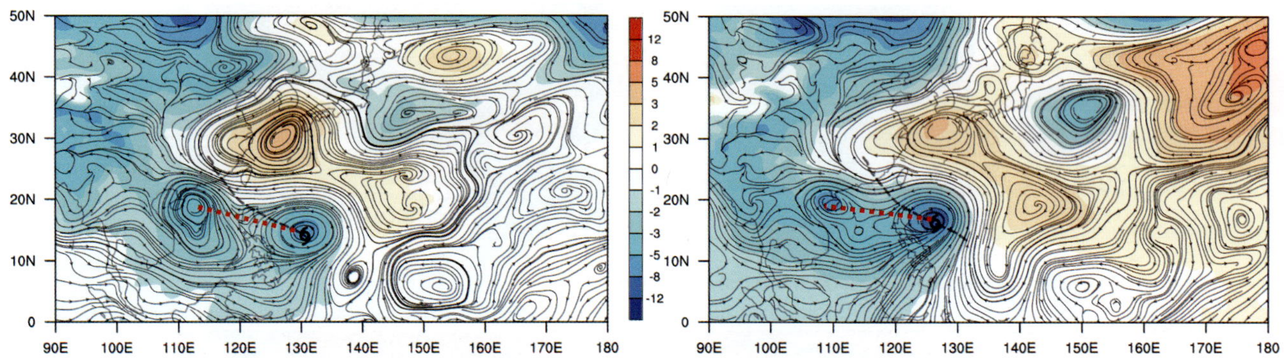

850 hPa flow/height anomalies at 0000 UTC 10 and 11 August 2005.

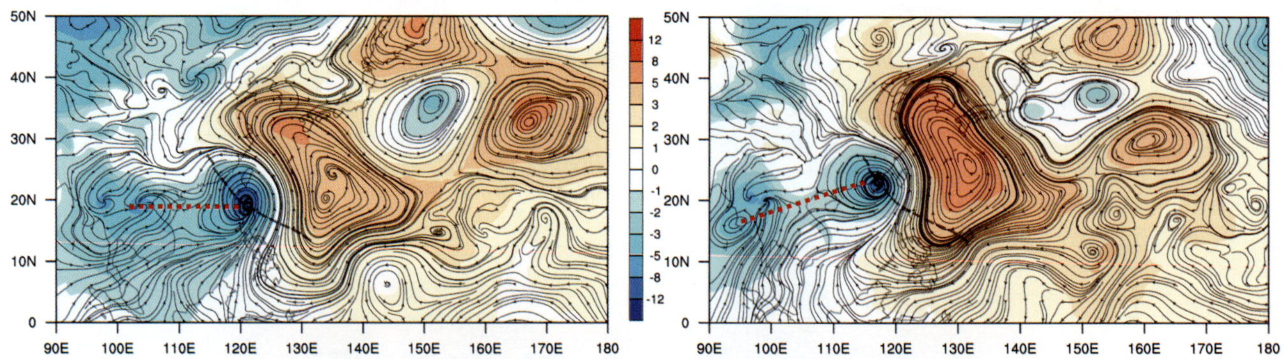

850 hPa flow/height anomalies at 0000 UTC 12 and 13 August 2005.

Figure 2.26. The typhoon (2005_No.10) generated in the tropical western Pacific on 9 August 2005 and moved northwestward. Its intensity peaked on 13 August 2005 before landfall on the South China coast. Its motion was modified by an anomalous vortex at 850 hPa.

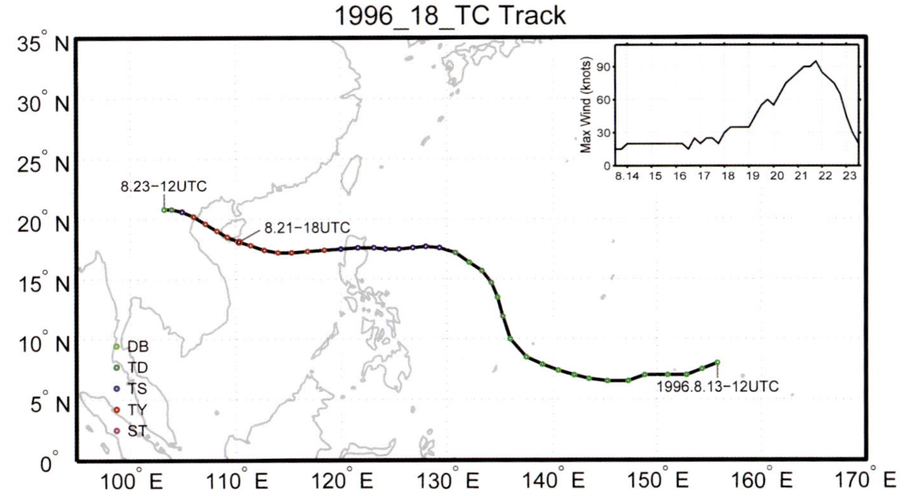

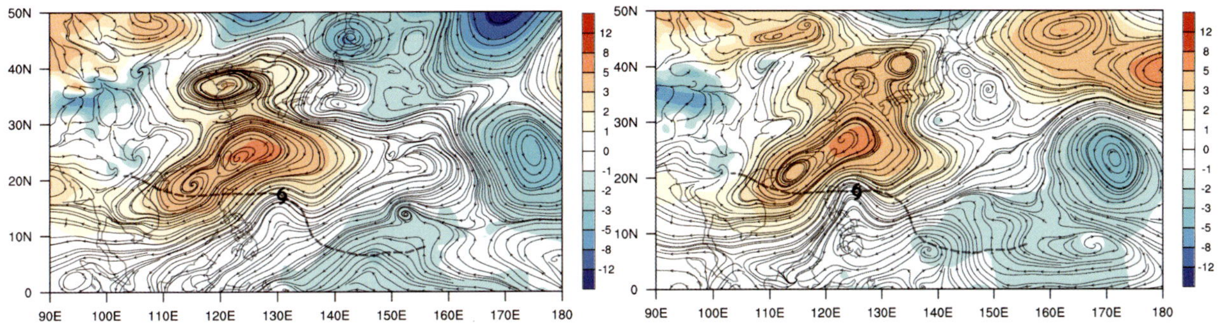

850 hPa flow/height anomalies at 0000 UTC 18 and 19 August 1996.

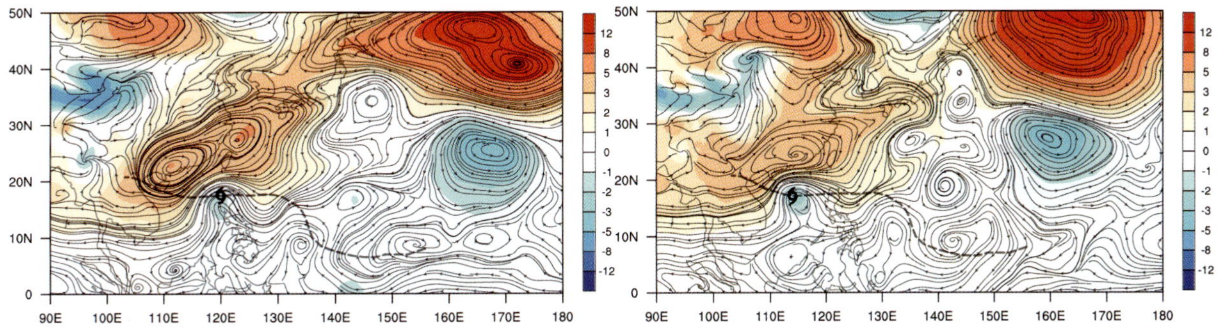

850 hPa flow/height anomalies at 0000 UTC 20 and 21 August 1996.

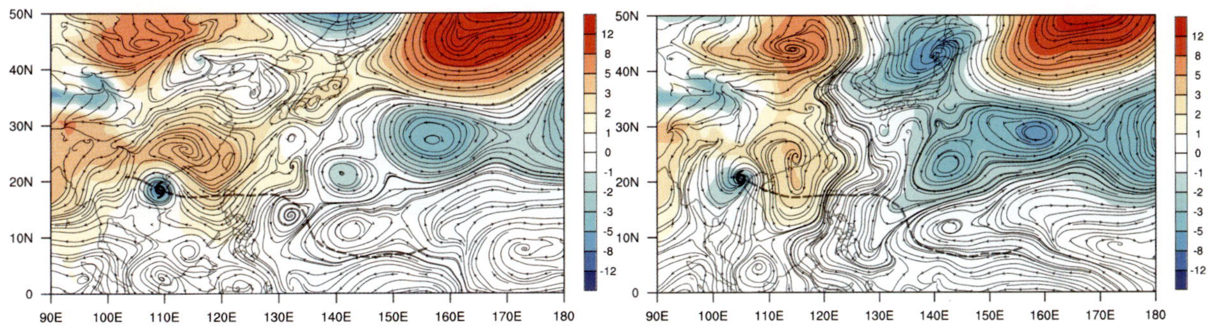

850 hPa flow/height anomalies at 0000 UTC 22 and 23 August 1996.

Figure 2.27. The typhoon (1996_No.18) generated in the equatorial western Pacific on 13 August 1996 and moved westward in the first five days. Its intensity peaked near the climatic monsoon trough on 21 August 1996 before landfall on the Hainan Island and its track was not affected by other anomalous vortices at 850 hPa in the Philippines and SCS.

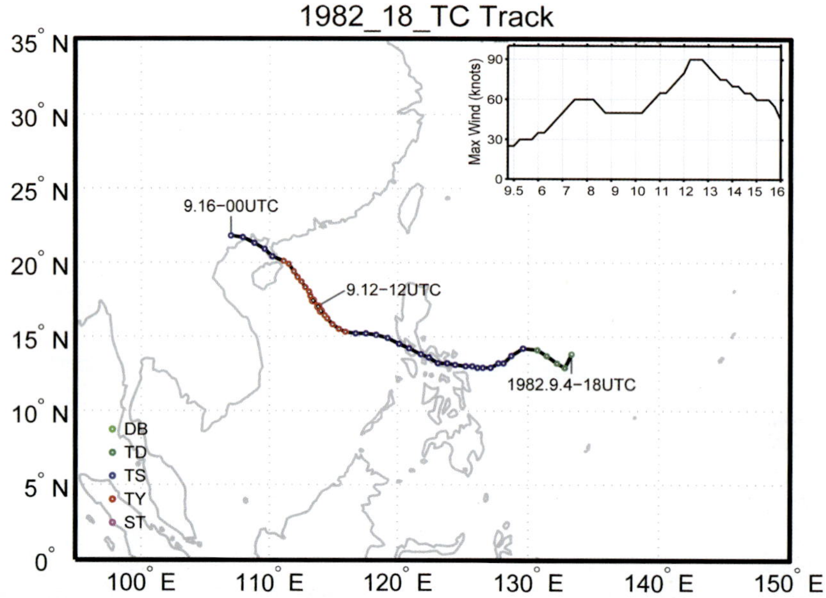

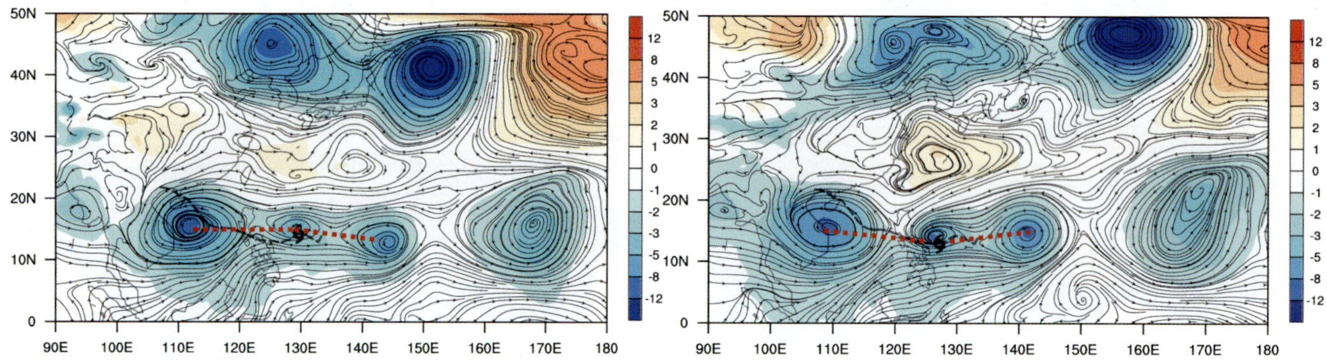

850 hPa flow/height anomalies at 0000 UTC 6 and 7 September 1982.

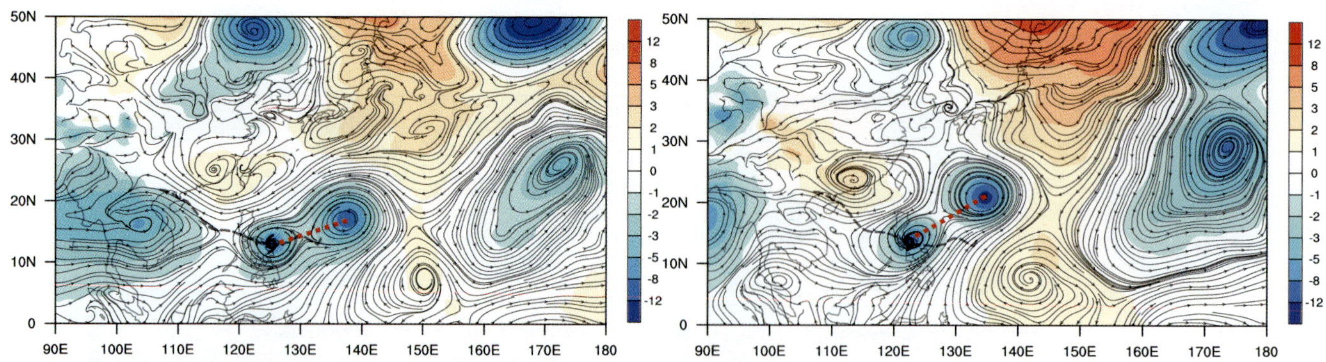

850 hPa flow/height anomalies at 0000 UTC 8 and 9 September 1982.

Figure 2.28. The typhoon (1982_No.18) generated in the tropical western Pacific on 4 September 1982 and moved west-northwestward. Its intensity peaked near the climatic monsoon trough in the northern SCS on 12 September 1982 before landfall on the South China coast and its motion was modified by two anomalous vortices at 850 hPa.

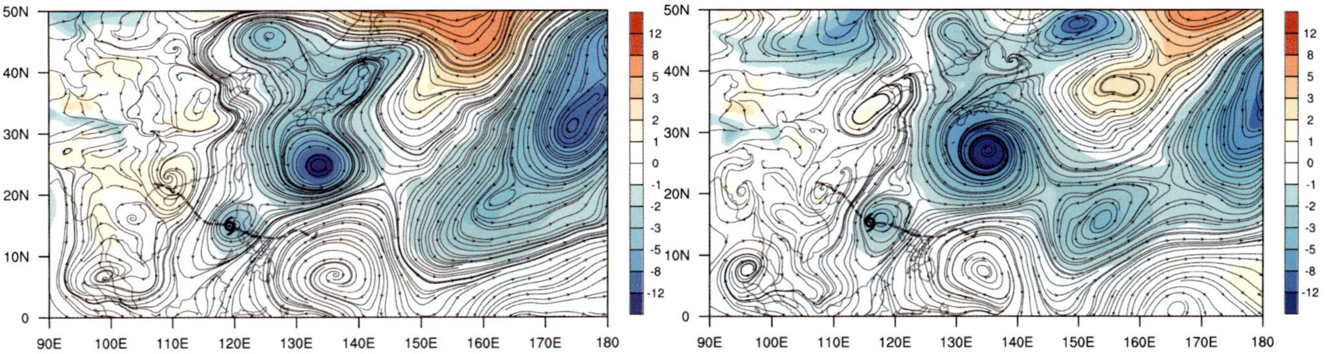

850 hPa flow/height anomalies at 0000 UTC 10 and 11 September 1982.

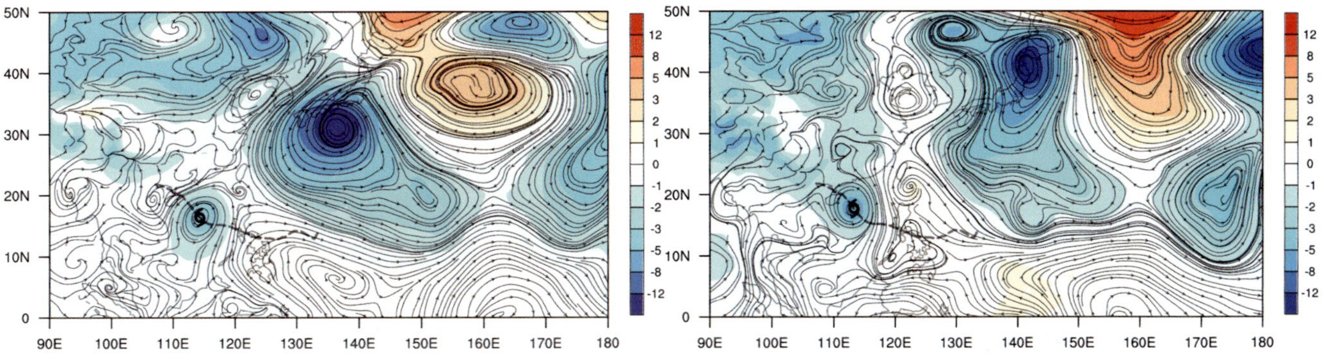

850 hPa flow/height anomalies at 0000 UTC 12 and 13 September 1982.

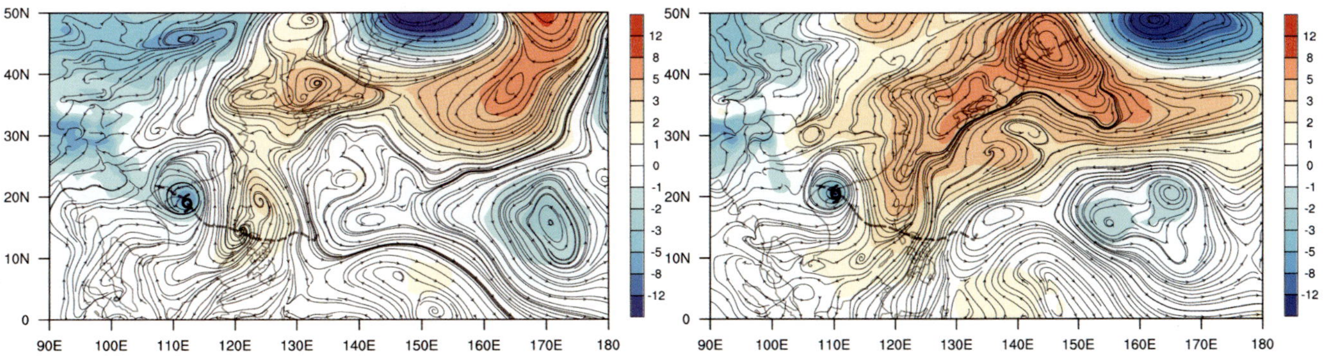

850 hPa flow/height anomalies at 0000 UTC 14 and 15 September 1982.

Figure 2.28. *(Continued)*

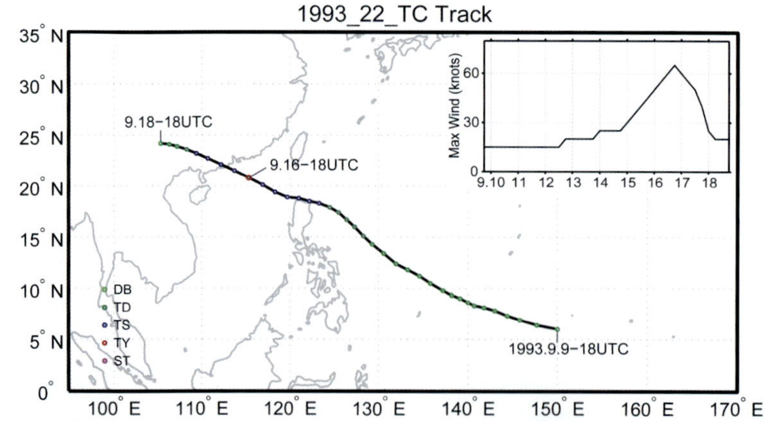

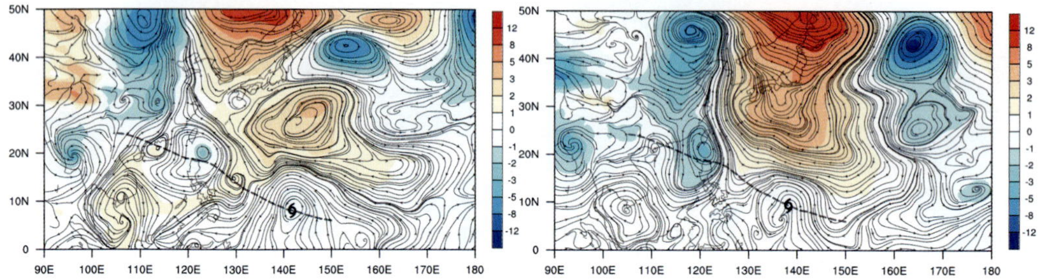

850 hPa flow/height anomalies at 0000 UTC 11 and 12 September 1993.

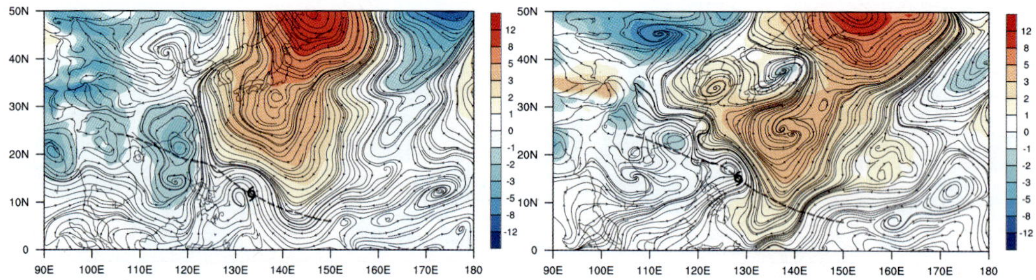

850 hPa flow/height anomalies at 0000 UTC 13 and 14 September 1993.

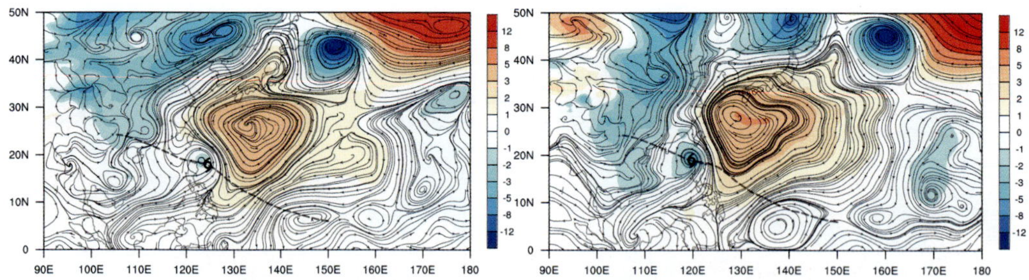

850 hPa flow/height anomalies at 0000 UTC 15 and 16 September 1993.

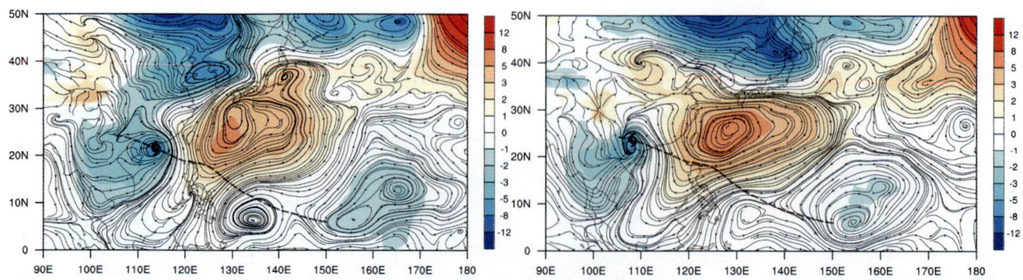

850 hPa flow/height anomalies at 0000 UTC 17 and 18 September 1993.

Figure 2.29. The typhoon (1993_No.22) generated in the equatorial western Pacific on 9 September 1993 and moved northwestward. Its intensity peaked near the climatic monsoon trough in the northernmost SCS on 16 September 1993 before landfall on the South China coast and its track was not affected by other anomalous vortices at 850 hPa.

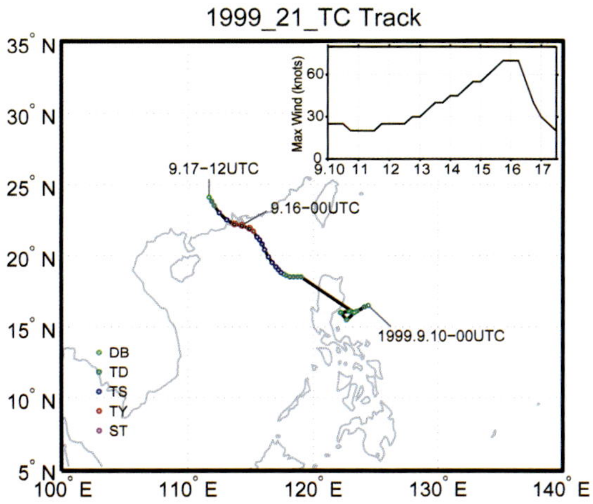

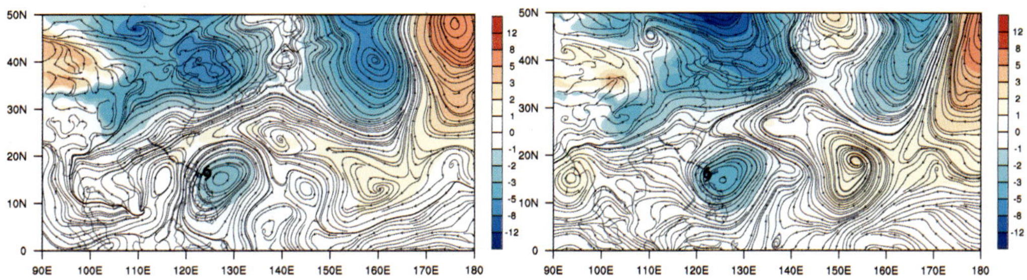

850 hPa flow/height anomalies at 0000 UTC 10 and 11 September 1999.

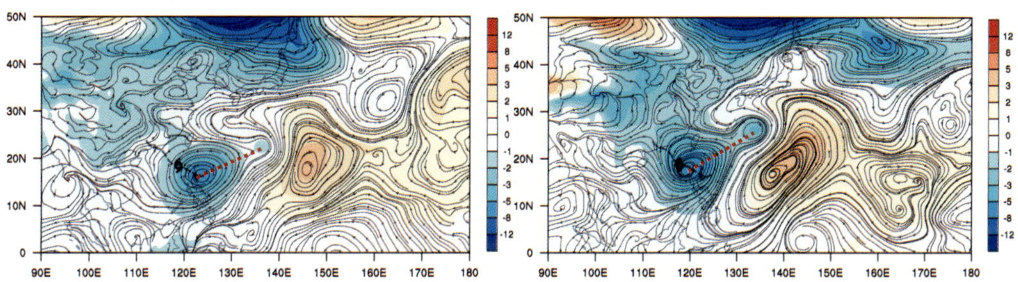

850 hPa flow/height anomalies at 0000 UTC 12 and 13 September 1999.

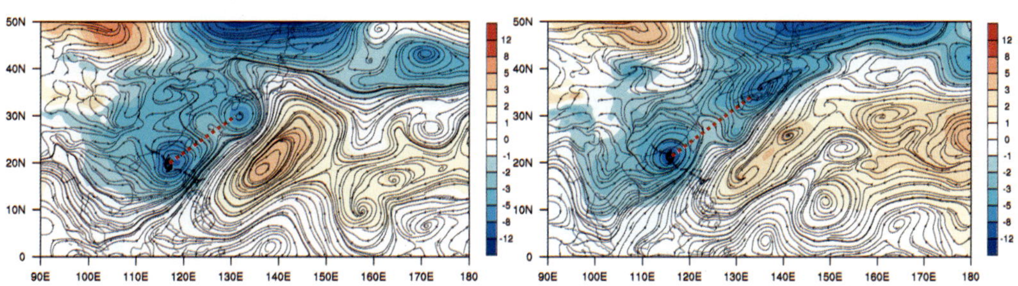

850 hPa flow/height anomalies at 0000 UTC 14 and 15 September 1999.

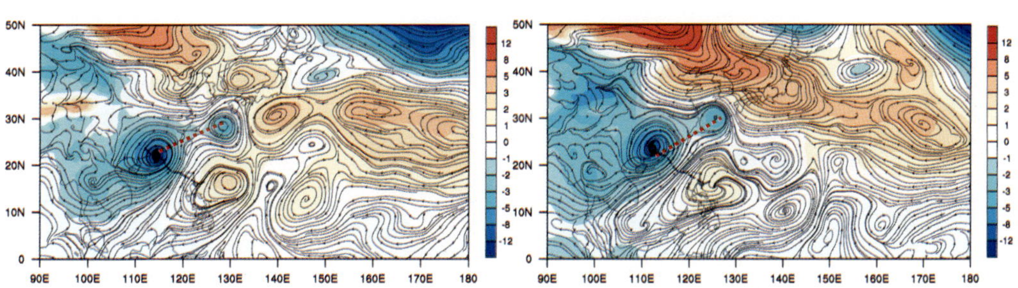

850 hPa flow/height anomalies at 0000 UTC 16 and 17 September 1999.

Figure 2.30. The typhoon (1999_No.21) generated in the east sea of the Philippines on 10 September 1999 and moved northwestward. Its intensity peaked near the climatic monsoon trough in the northernmost SCS on 16 September 1999 before landfall on the South China coast and its track was modified by an anomalous vortex at 850 hPa.

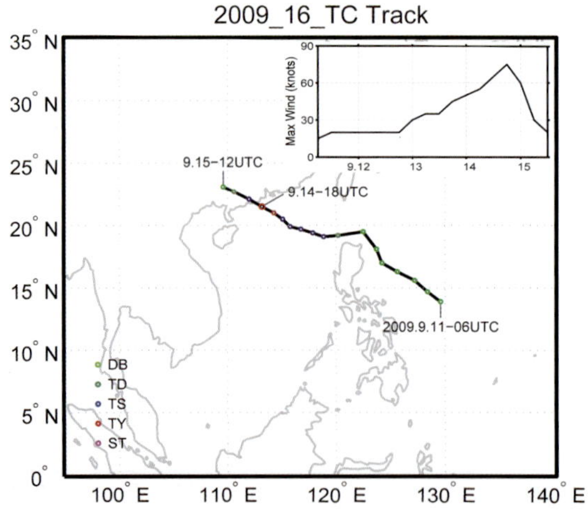

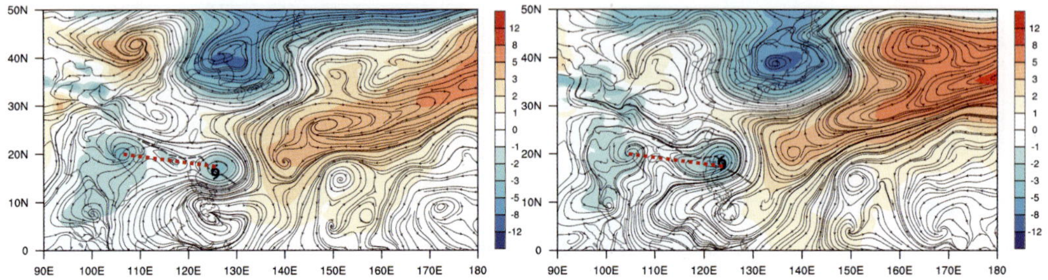

850 hPa flow/height anomalies at 0000 UTC and 1200 UTC 12 September 2009.

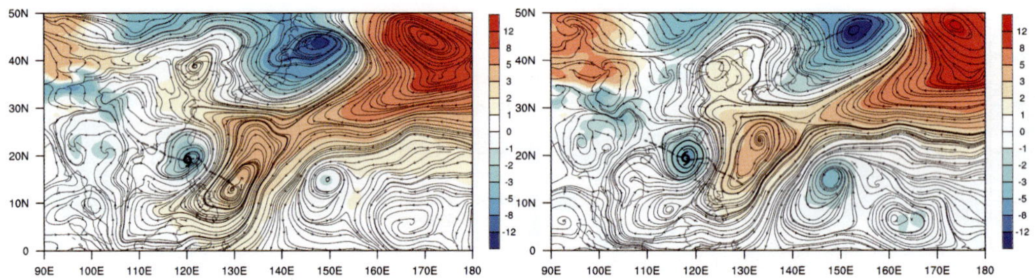

850 hPa flow/height anomalies at 0000 UTC and 1200 UTC 13 September 2009.

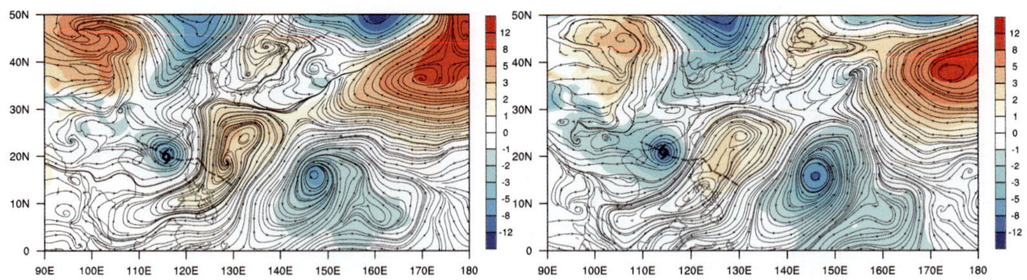

850 hPa flow/height anomalies at 0000 UTC and 1200 UTC 14 September 2009.

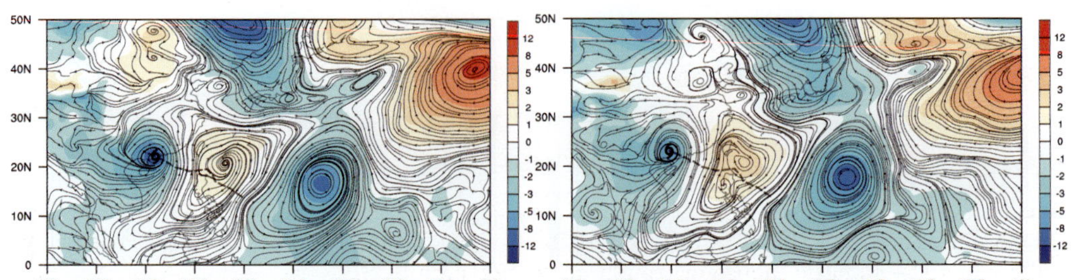

850 hPa flow/height anomalies at 0000 UTC and 1200 UTC 15 September 2009.

Figure 2.31. The typhoon (2009_No.16) generated in the east sea of the Philippines on 11 September 2009 and moved northwestward. Its intensity peaked near the climatic monsoon trough in the northernmost SCS on 14 September 2009 before landfall on the South China coast and its track was modified by an anomalous vortex at 850 hPa.

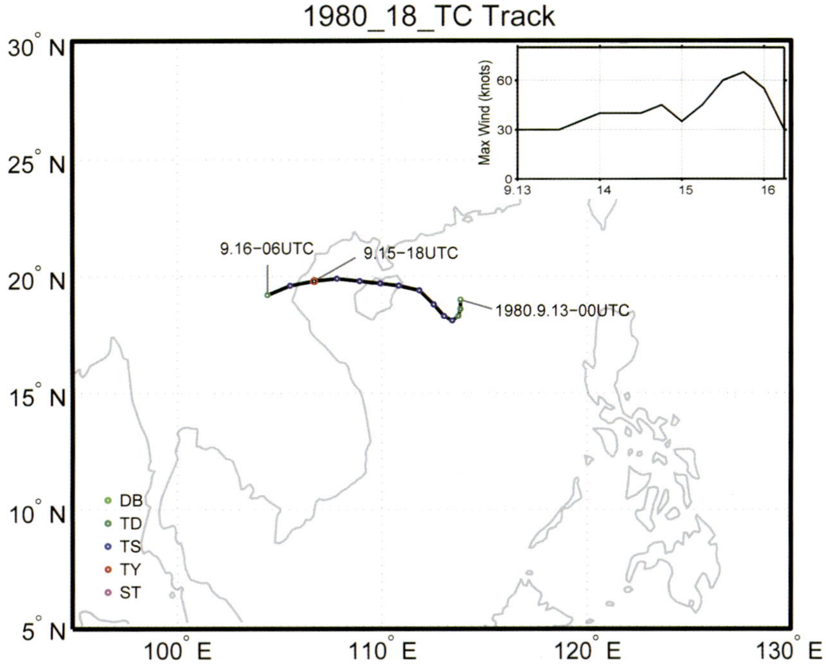

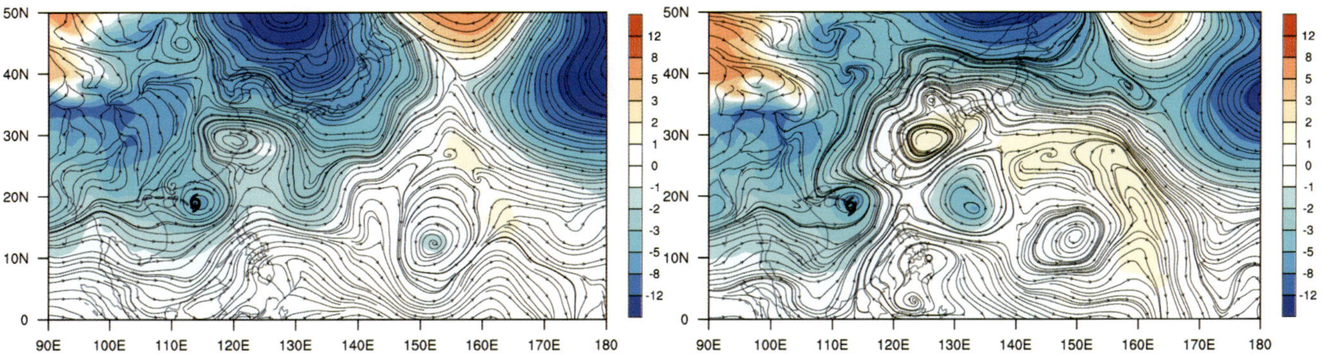

850 hPa flow/height anomalies at 0000 UTC 13 and 14 September 1980.

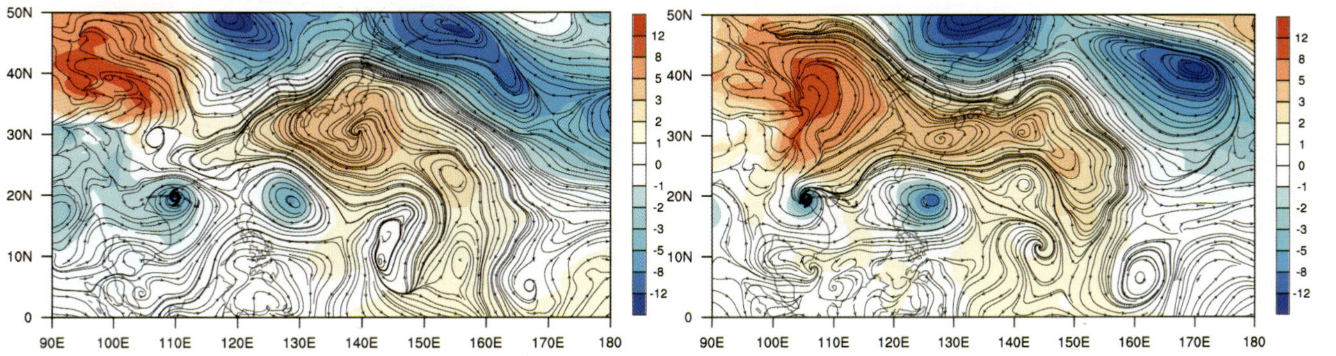

850 hPa flow/height anomalies at 0000 UTC 15 and 16 September 1980.

Figure 2.32. The typhoon (1980_No.18) generated in the east sea of the Hainan Island on 13 September 1980. Its intensity peaked in the north-westernmost SCS on 15 September 1980 before landfall on the northeast coast of Vietnam and its track was basically along the climatic monsoon trough in the northern SCS.

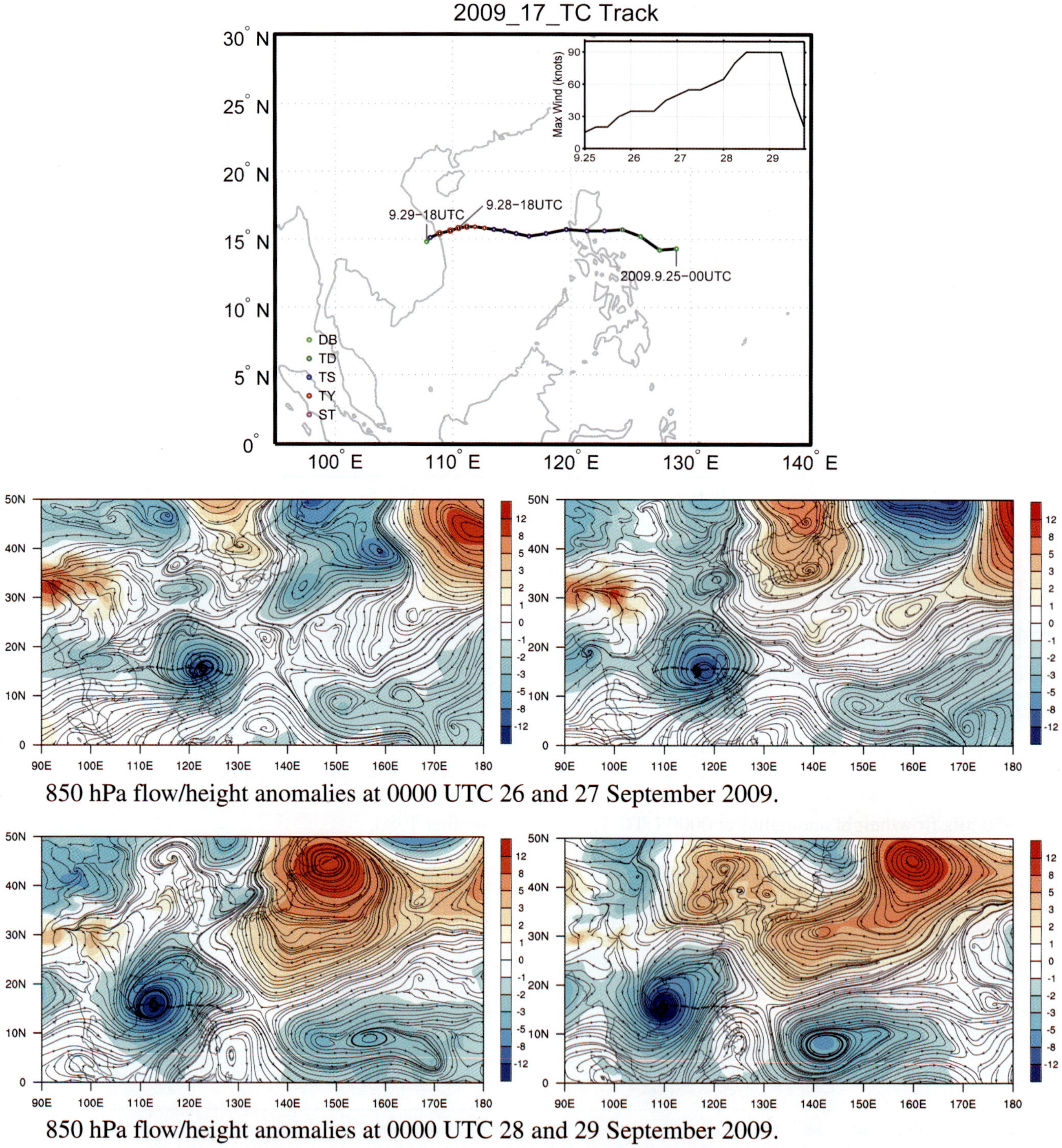

Figure 2.33. The typhoon (2009_No.17) generated in the east sea of the Philippines on 25 September 2009. Its intensity peaked on 28 September 2009 before landfall on the central east coast of Vietnam and its track was basically along the climatic monsoon trough and without the influence of anomalous vortices at 850 hPa.

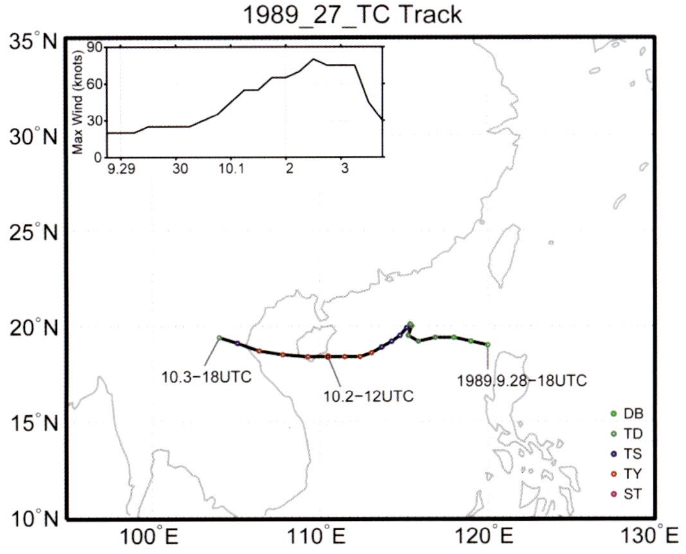

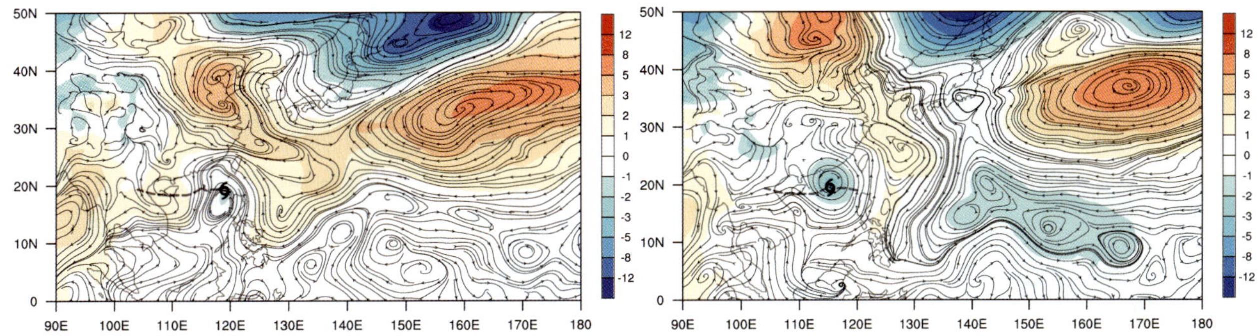

850 hPa flow/height anomalies at 0000 UTC 29 and 0000 UTC 30 September 1989

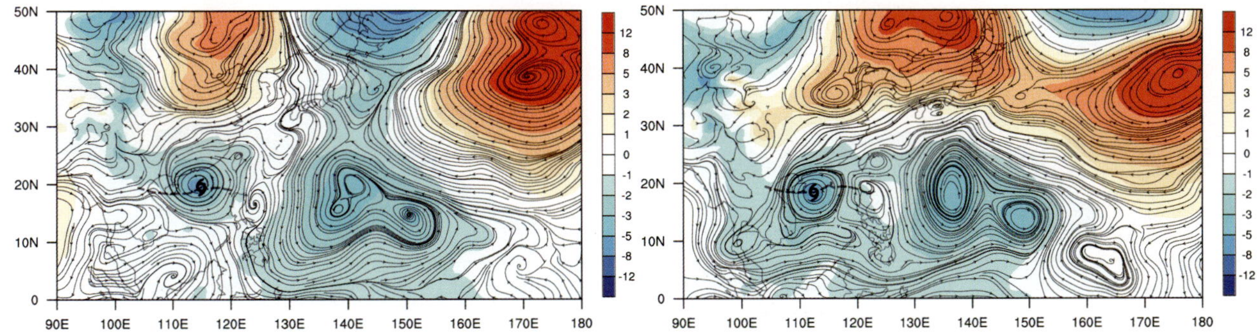

850 hPa flow/height anomalies at 0000 UTC 1 and 0000 UTC 2 October 1989

Figure 2.34. The typhoon (1989_No.27) generated in the South China Sea on 28 September 1989 and its motion was basically along the climatic monsoon trough. Its intensity peaked on 2 October 1989 before landfall on the Hainan Island.

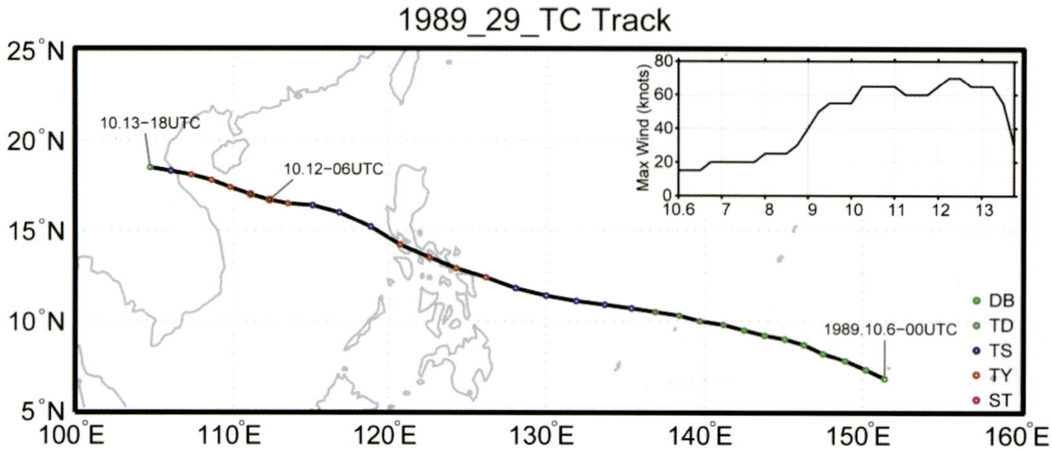

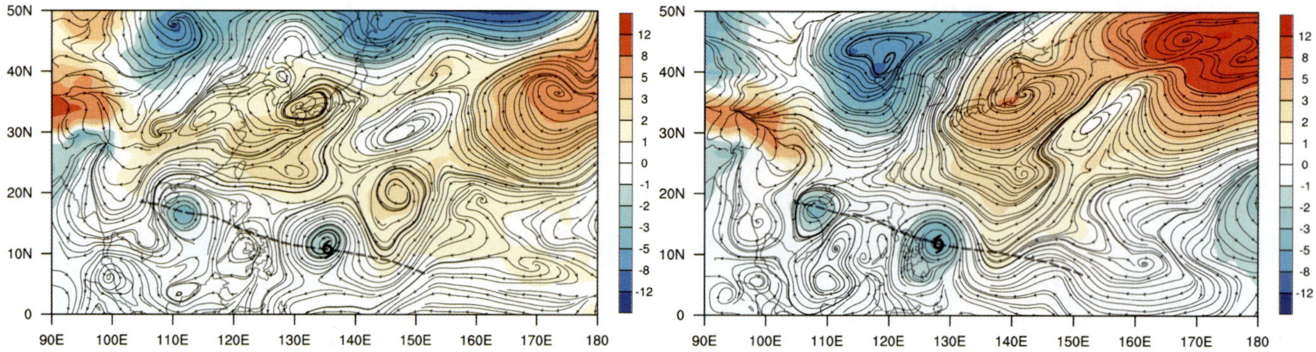

850 hPa flow/height anomalies at 0000 UTC 9 and 10 October 1989

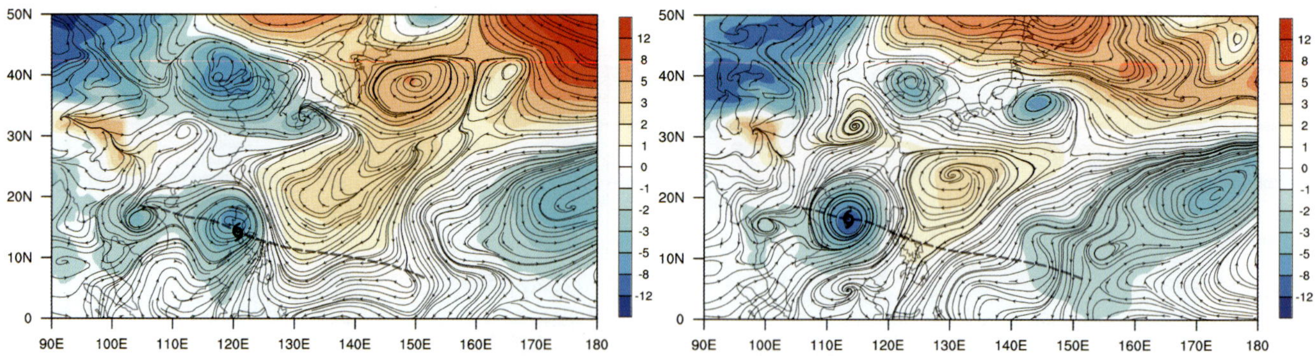

850 hPa flow/height anomalies at 0000 UTC 11 and 12 October 1989

Figure 2.35. The typhoon (1989_No.29) generated in the tropical western Pacific on 6 October 1989 and moved northwest ward. Its intensity peaked on 12 October 1989 when it was approaching the climatic monsoon trough in the central South China Sea.

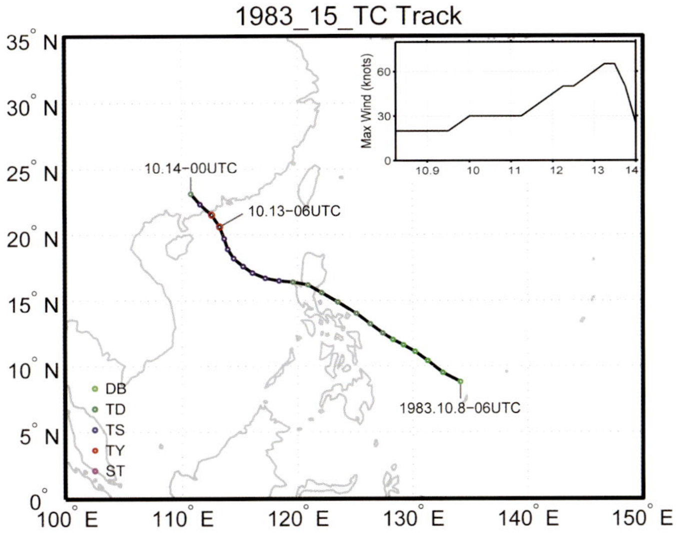

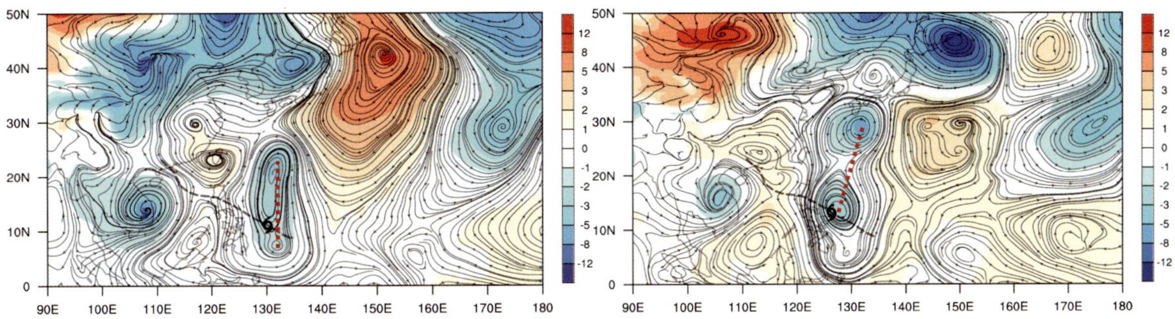

850 hPa flow/height anomalies at 0000 UTC 9 and 10 October 1983.

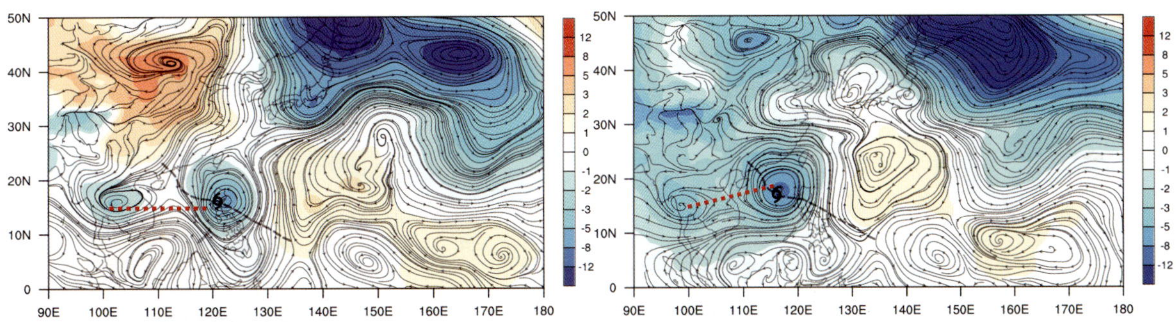

850 hPa flow/height anomalies at 0000 UTC 11 and 12 October 1983.

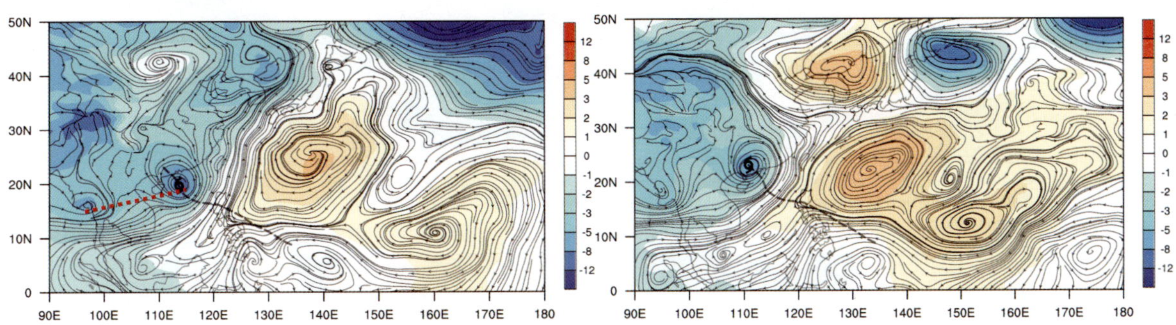

850 hPa flow/height anomalies at 0000 UTC 13 and 14 October 1983.

Figure 2.36. The typhoon (1983_No.15) generated in the equatorial western Pacific on 8 October 1983 and moved northwestward. Its intensity peaked on 13 October 1983 before landfall on the South China coast and its motion in the western Pacific was basically along the climatic monsoon trough and modified by two anomalous vortices at 850 hPa.

2.3. Intensity Peaked in the Philippines

Many typhoons generated from original tropical disturbances in the western Pacific and moved northwestward. They peak the intensity near the Philippines. These typhoons usually continue to move northwestwards and make landfall in the southern China coast. A total of 38 typhoons are plotted in this section (Figs. 2.37–2.74).

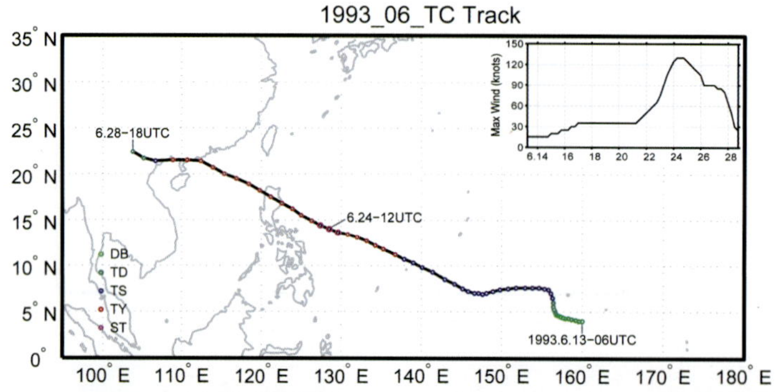

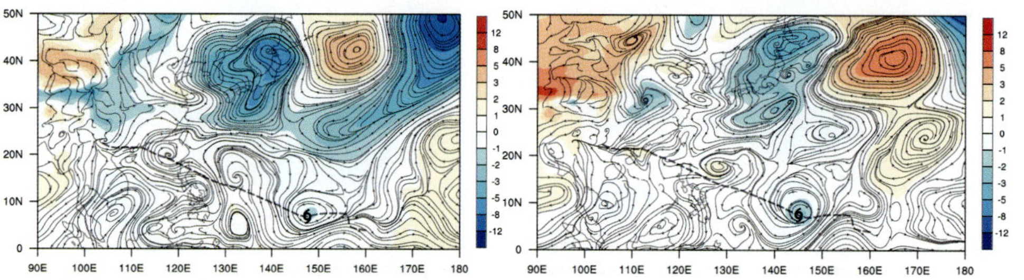

850 hPa flow/height (shading, 10 gpm) anomalies at 0000 UTC 20 and 21 June 1993.

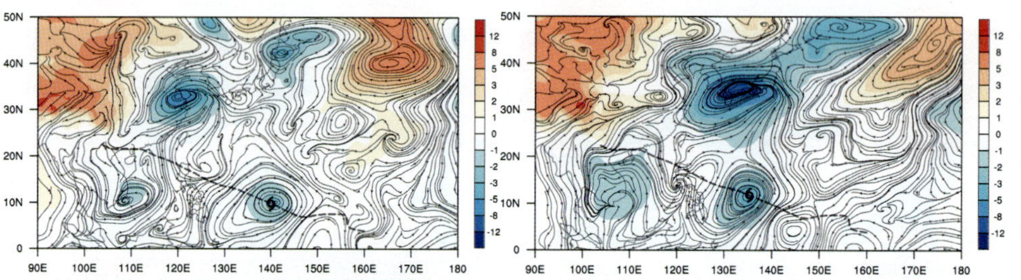

850 hPa flow/height anomalies at 0000 UTC 22 and 23 June 1993.

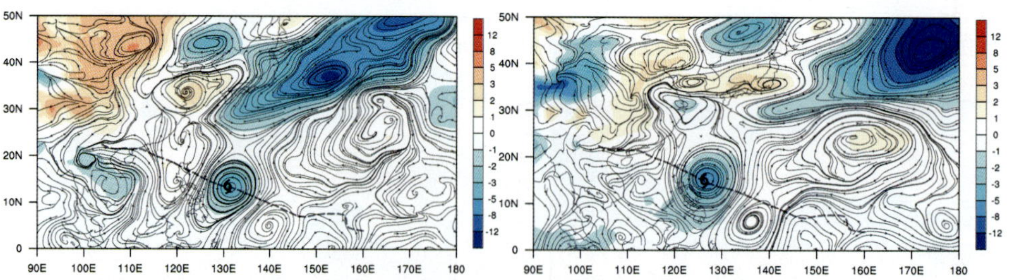

850 hPa flow/height anomalies at 0000 UTC 24 and 25 June 1993.

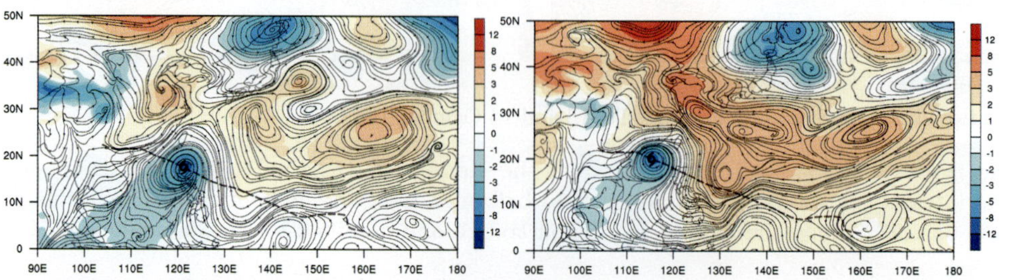

850 hPa flow/height anomalies at 0000 UTC 26 and 27 June 1993.

Figure 2.37. The typhoon (1993_No.06) generated in the equatorial central Pacific on 13 June 1993 and moved west-northwestward along the climatic monsoon trough without the influence of other anomalous vortices at 850 hPa. Its intensity peaked in the east sea of the Philippines on 24 June 1993 before landfall on the island.

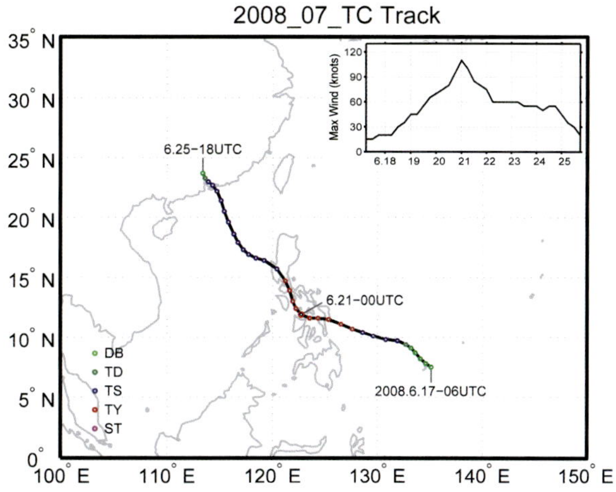

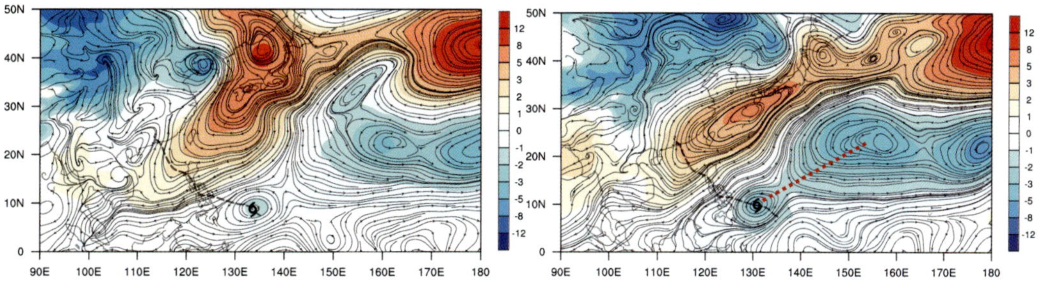

850 hPa flow/height anomalies at 0000 UTC 18 and 19 June 2008.

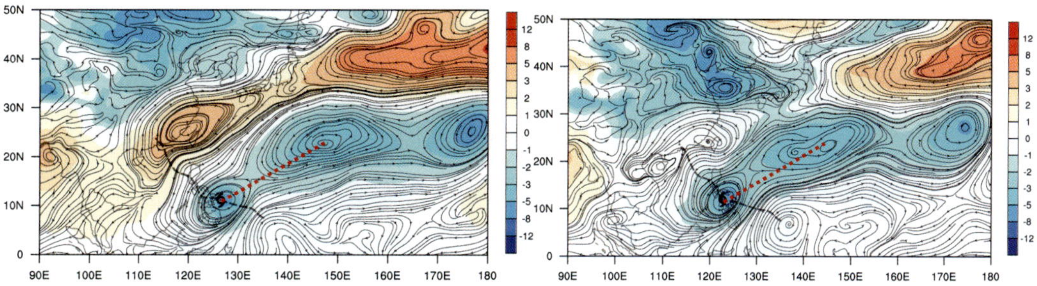

850 hPa flow/height anomalies at 0000 UTC 20 and 21 June 2008.

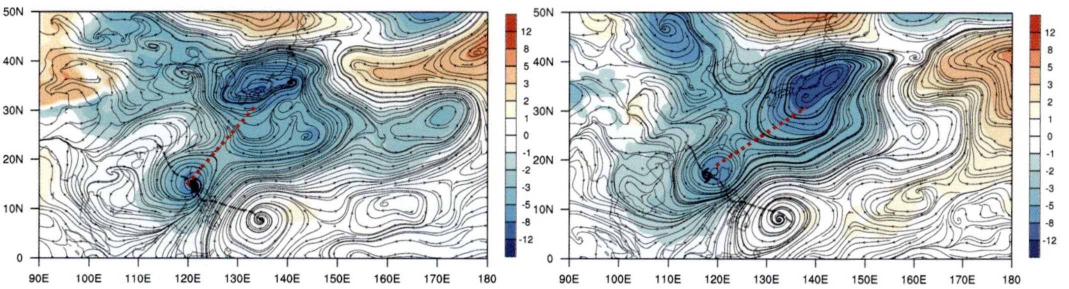

850 hPa flow/height anomalies at 0000 UTC 22 and 23 June 2008.

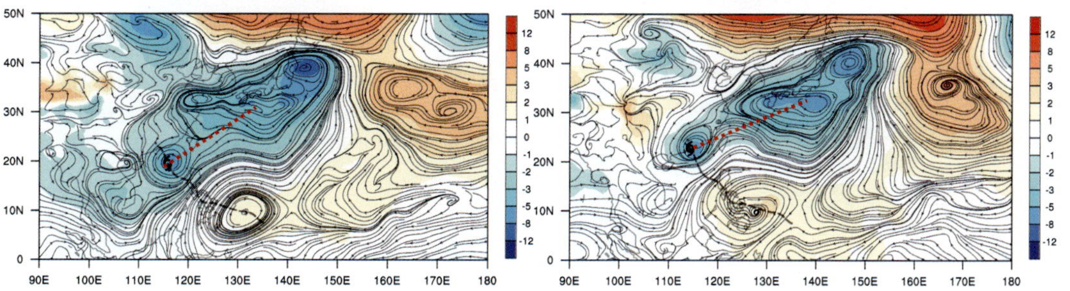

850 hPa flow/height anomalies at 0000 UTC 24 and 25 June 2008.

Figure 2.38. The typhoon (2008_No.07) generated in the equatorial western Pacific on 17 June 2008 and moved northwestward along the climatic monsoon trough. Its intensity peaked in the Philippines on 21 June 2008 and its motion was modified by an anomalous vortex at 850 hPa.

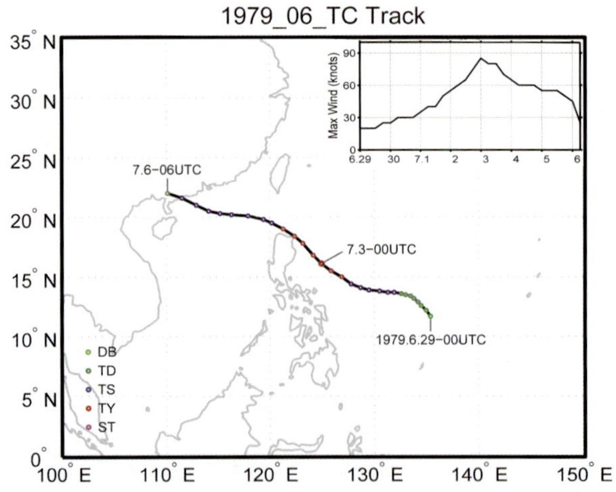

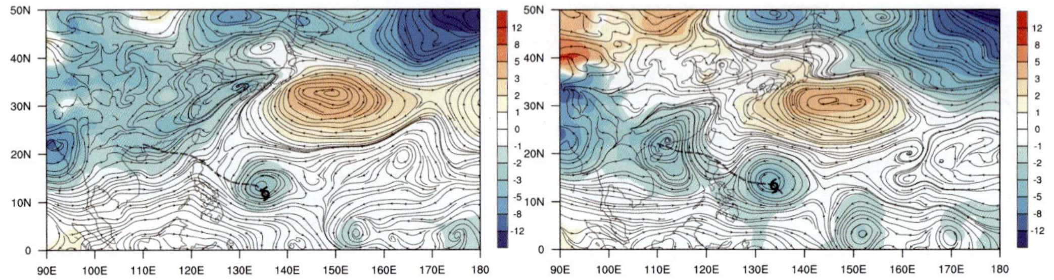

850 hPa flow/height anomalies at 0000 UTC 29 and 30 June 1979.

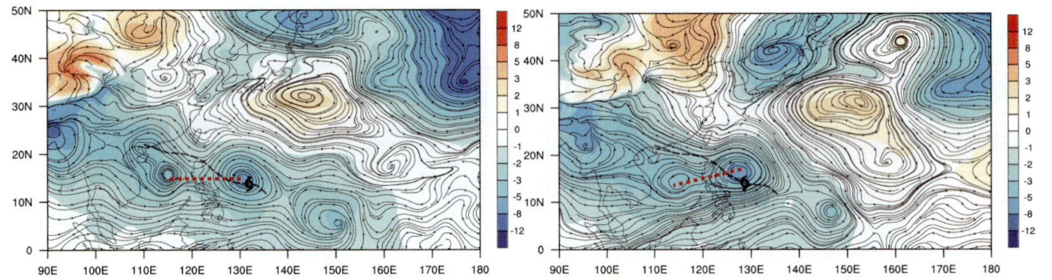

850 hPa flow/height anomalies at 0000 UTC 1 and 2 July 1979.

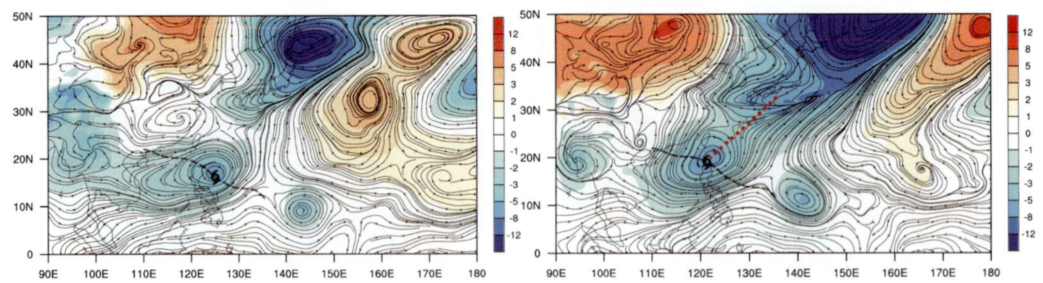

850 hPa flow/height anomalies at 0000 UTC 3 and 4 July 1979.

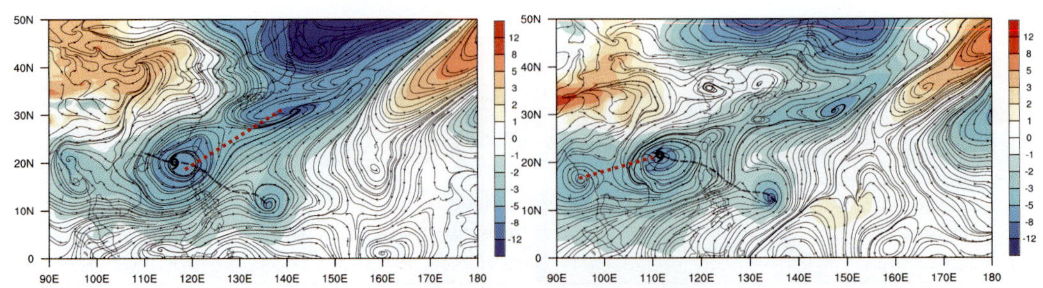

850 hPa flow/height anomalies at 0000 UTC 5 and 6 July 1979.

Figure 2.39. The typhoon (1979_No.06) generated in the tropical western Pacific on 29 June 1979 and moved northwestward along the climatic monsoon trough. Its intensity peaked in the east sea of the Philippines on 3 July 1979 before approaching the island and its motion was modified by anomalous vortices at 850 hPa.

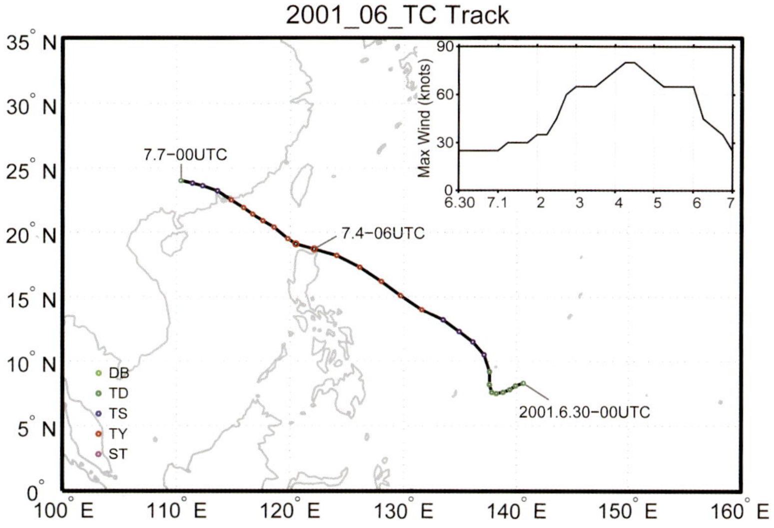

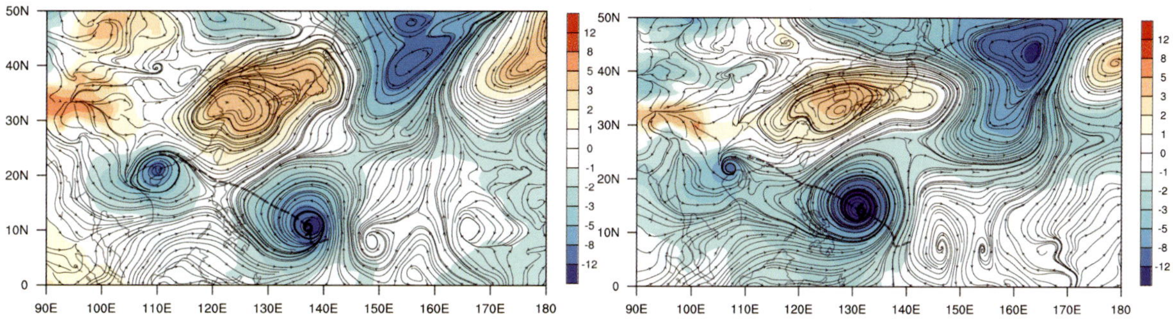

850 hPa flow/height anomalies at 0000 UTC 2 and 3 July 2001.

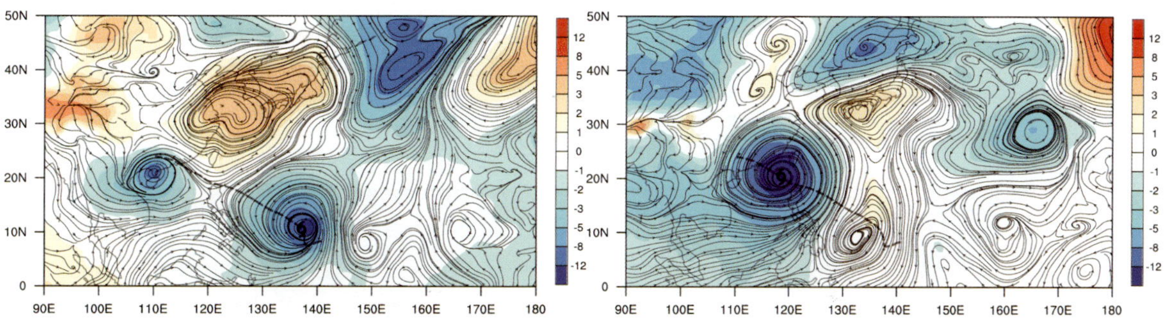

850 hPa flow/height anomalies at 0000 UTC 4 and 5 July 2001.

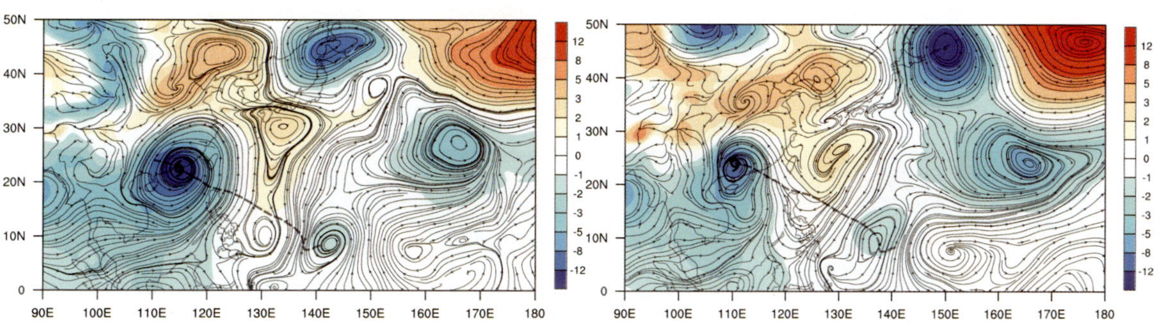

850 hPa flow/height anomalies at 0000 UTC 6 and 7 July 2001.

Figure 2.40. The typhoon (2001_No.06) generated in the equatorial western Pacific on 30 June 2001 and moved northwestward basically along the climatic monsoon trough. Its intensity peaked in the northern Philippines on 4 July 2001 before approaching the island. After a small curvature motion at the beginning stage, its motion was not affected by any anomalous vortices at 850 hPa.

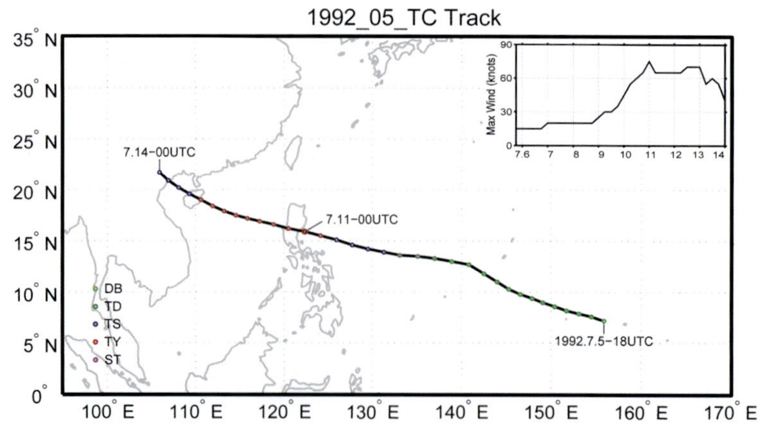

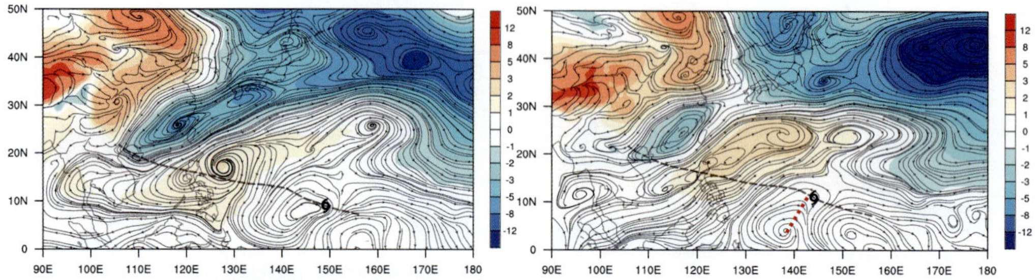

850 hPa flow/height anomalies at 0000 UTC 7 and 8 July 1992.

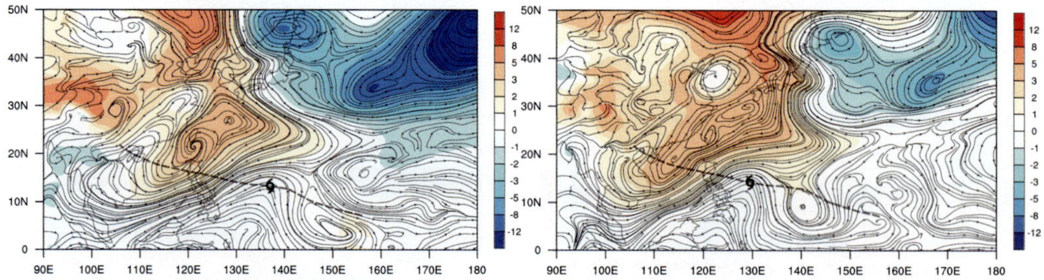

850 hPa flow/height anomalies at 0000 UTC 9 and 10 July 1992.

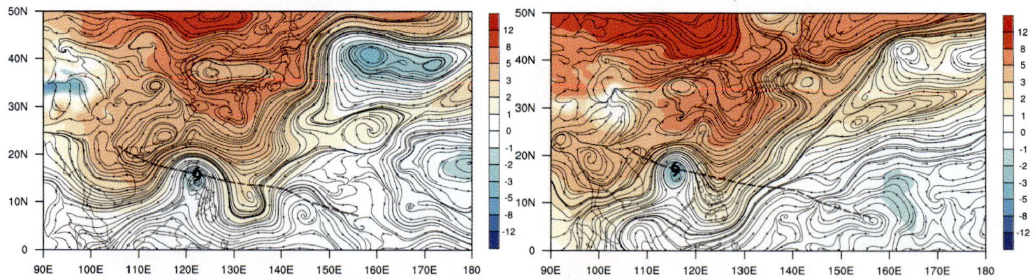

850 hPa flow/height anomalies at 0000 UTC 11 and 12 July 1992.

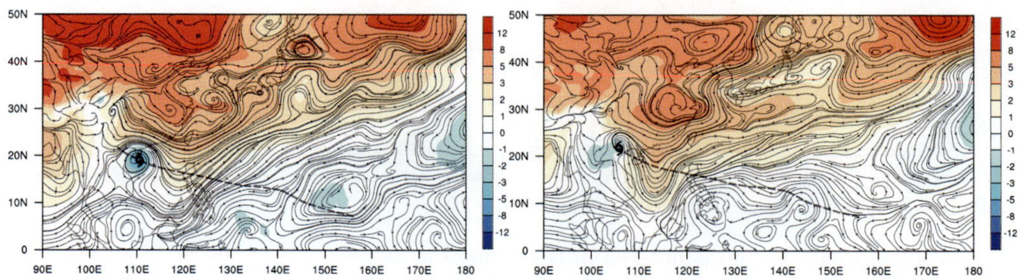

850 hPa flow/height anomalies at 0000 UTC 13 and 14 July 1992.

Figure 2.41. The typhoon (1992_No.05) generated in the equatorial central Pacific on 5 July 1992 and moved west-northwestward near the climatic monsoon trough without interactions with anomalous vortices at 850 hPa. Its intensity peaked in the northern Philippines on 11 July 1992 before landfall on the island.

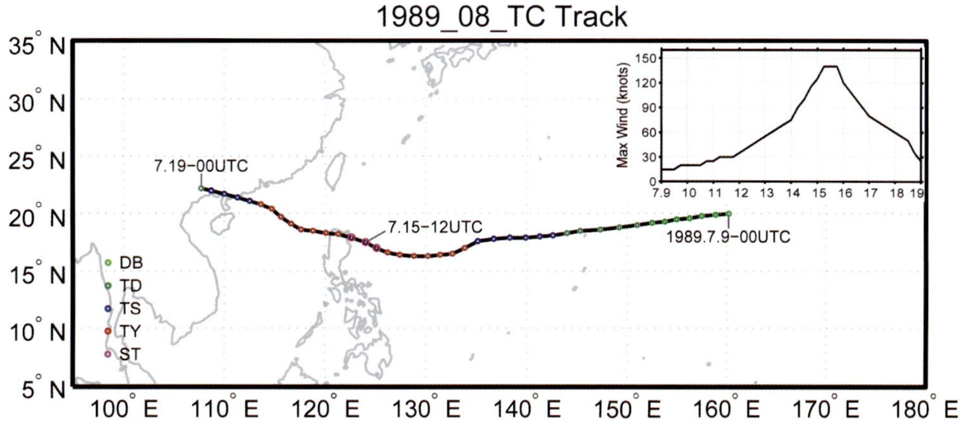

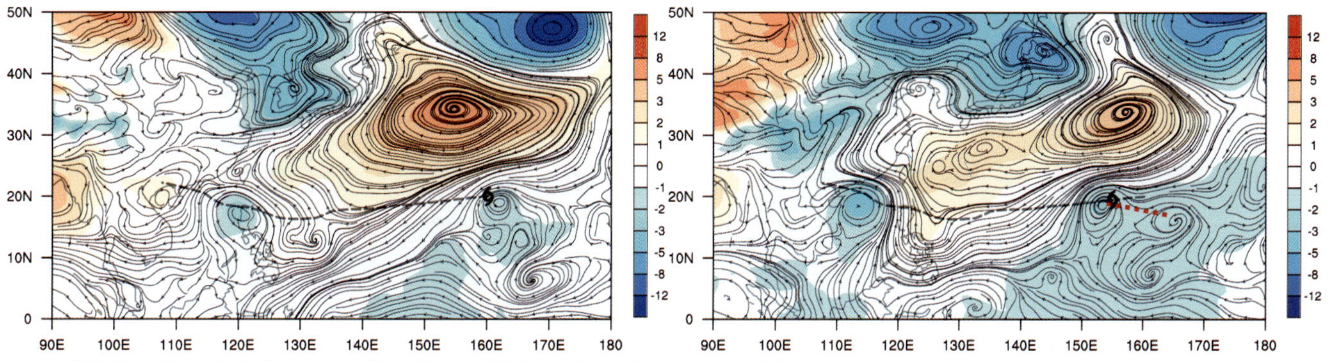

850 hPa flow/height anomalies 0000 UTC on 9 and 10 July 1989.

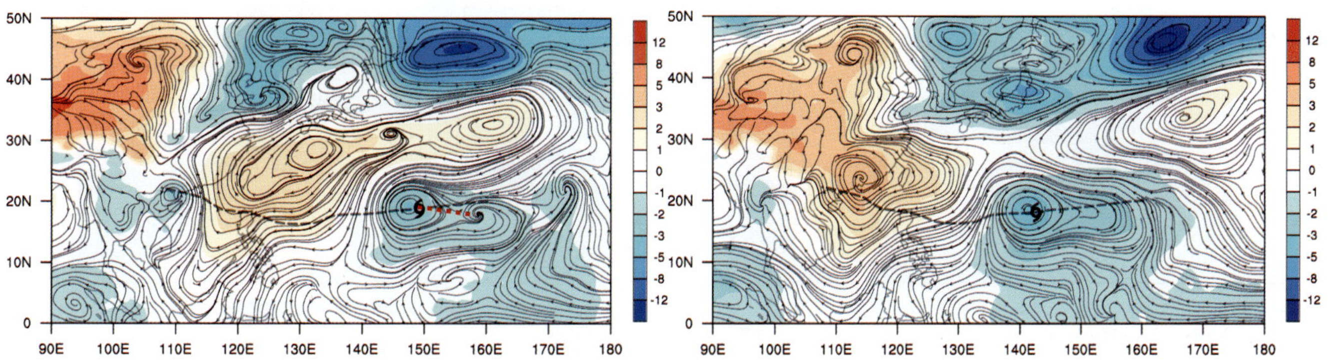

850 hPa flow/height anomalies at 0000 UTC 11 and 12 July 1989.

Figure 2.42. The typhoon (1989_No.08) generated in the subtropical Northwest Pacific on 9 July 1989 and moved westward. Its intensity peaked near the climatic monsoon trough in the east sea of the Philippines on 15 July 1989 before landfall on the island and its motion was modified by another anomalous vortex at 850 hPa.

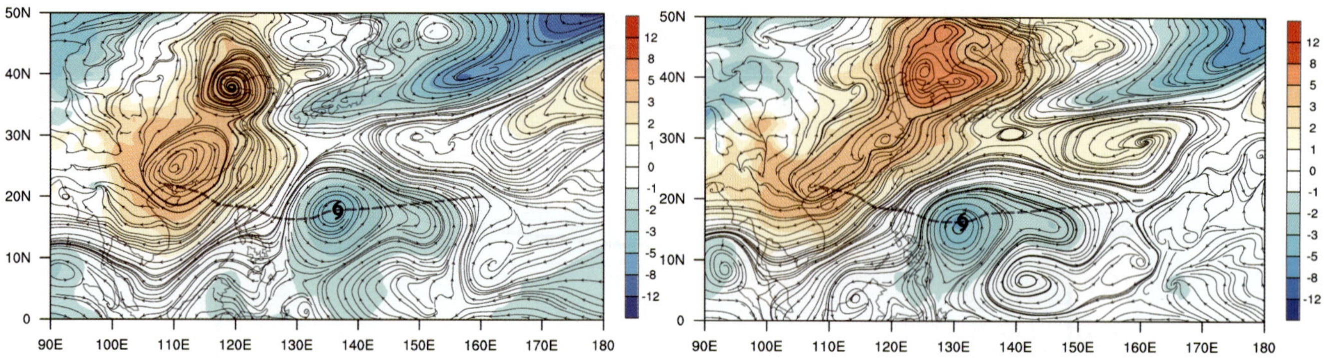

850 hPa flow/height anomalies at 0000 UTC 13 and 14 July 1989.

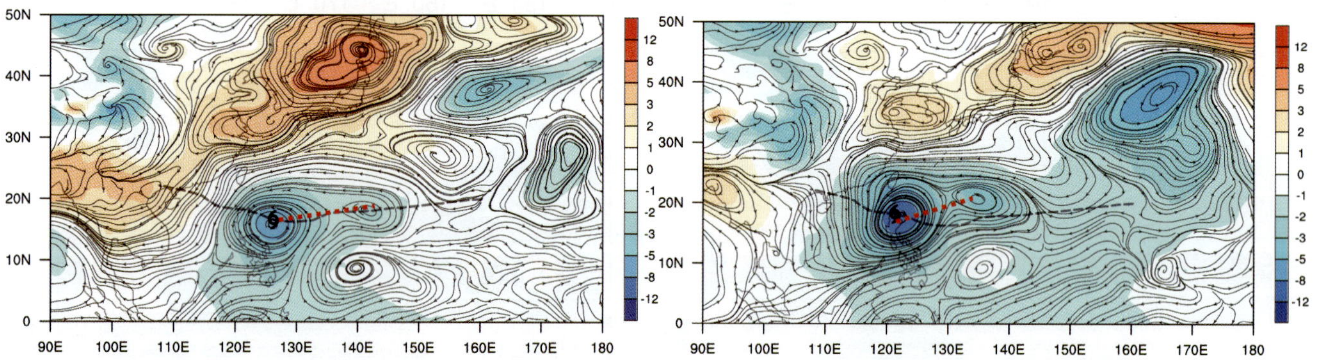

850 hPa flow/height anomalies at 0000 UTC 15 and 16 July 1989.

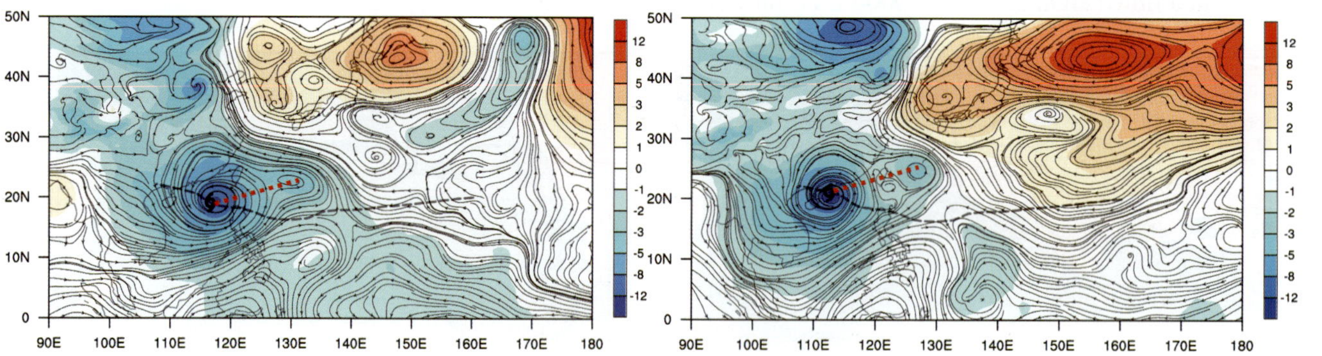

850 hPa flow/height anomalies at 0000 UTC 17 and 18 July 1989.

Figure 2.42. *(Continued)*

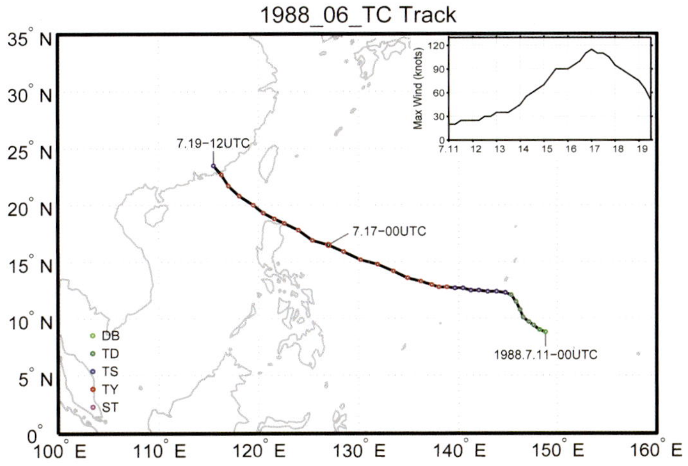

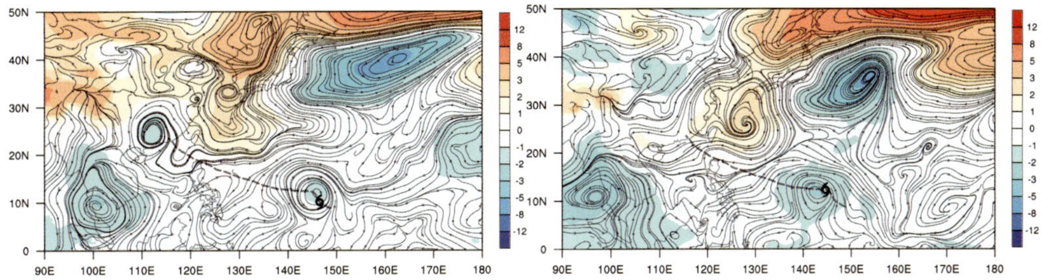

850 hPa flow/height anomalies at 0000 UTC 12 and 13 July 1988.

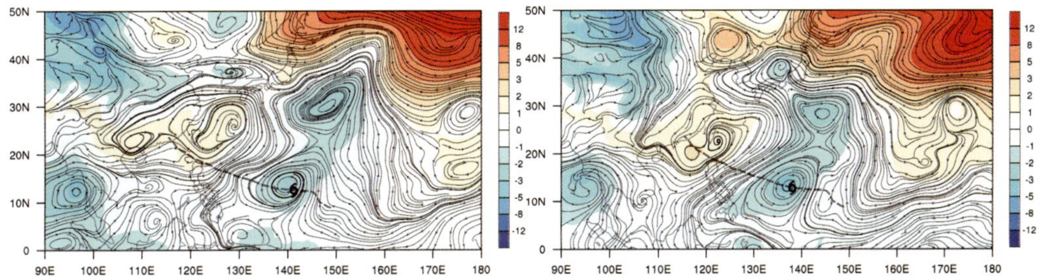

850 hPa flow/height anomalies at 0000 UTC 14 and 15 July 1988.

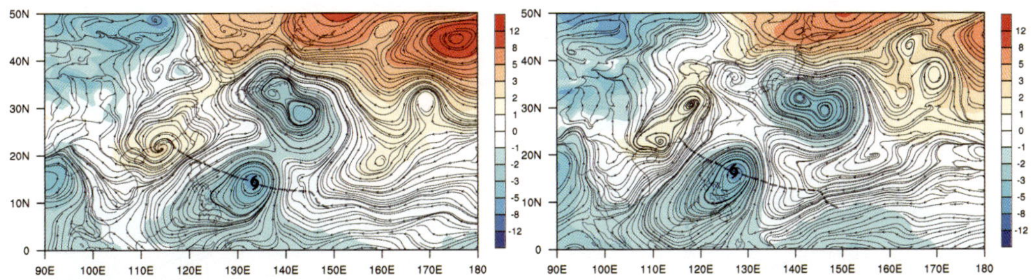

850 hPa flow/height anomalies at 0000 UTC 16 and 17 July 1988.

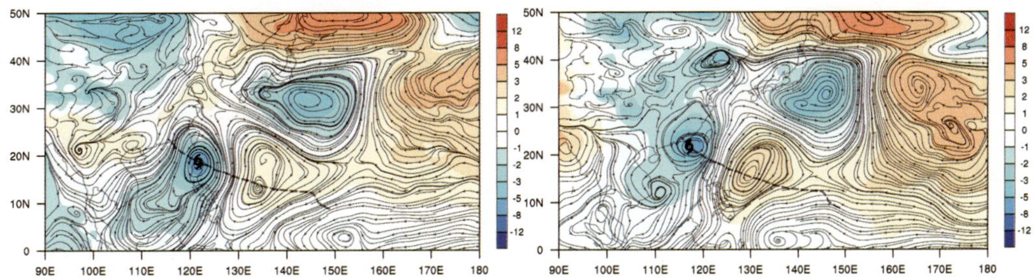

850 hPa flow/height anomalies at 0000 UTC 18 and 19 July 1988.

Figure 2.43. The typhoon (1988_No.06) generated in the equatorial western Pacific on 11 July 1988 and moved northwestward. Its intensity peaked near the climatic monsoon trough in the east sea of the Philippines on 17 July 1988 and weakened when it was approaching the island. After a small curvature motion at the beginning stage, its motion was not affected by anomalous vortices at 850 hPa.

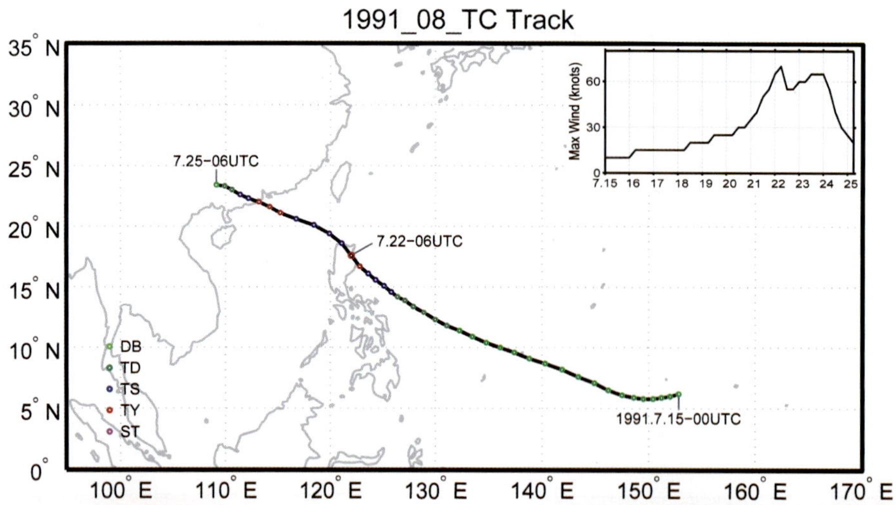

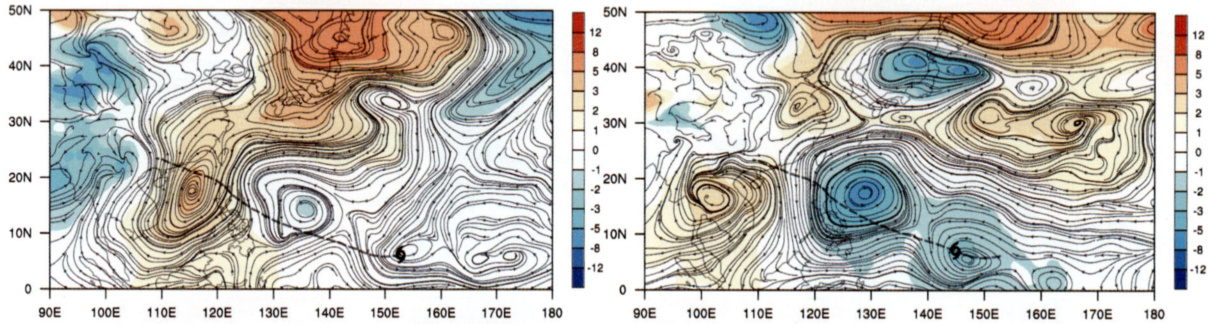

850 hPa flow/height anomalies at 0000 UTC 15 and 17 July 1991.

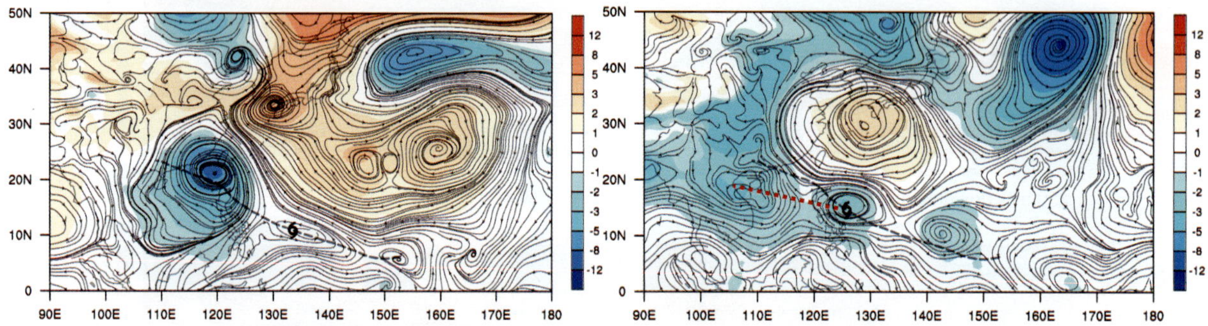

850 hPa flow/height anomalies at 0000 UTC 19 and 21 July 1991.

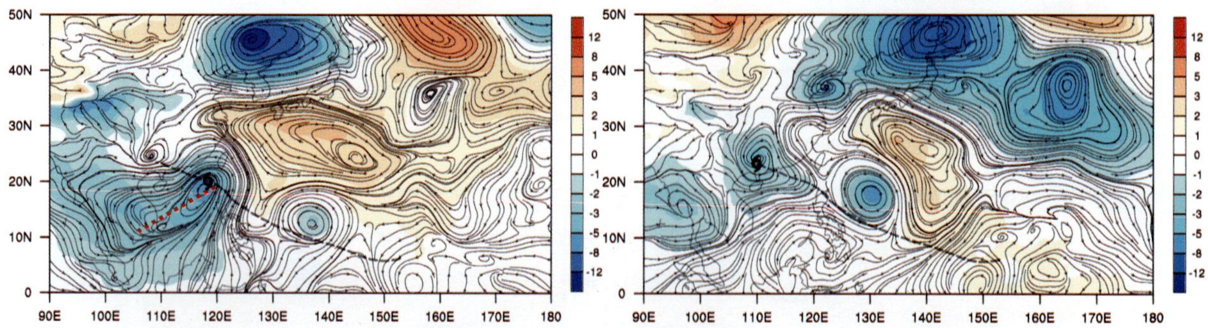

850 hPa flow/height anomalies at 0000 UTC 23 and 25 July 1991.

Figure 2.44. The typhoon (1991_No.08) generated in the equatorial central Pacific on 15 July 1991 and moved northwestward along the climatic monsoon trough. Its intensity peaked in the northern Philippines on 22 July 1991 before landfall on the island and its track had a small change when it was arriving in the Philippines due to the influence of an anomalous vortex at 850 hPa from 21 to 23 July 1991.

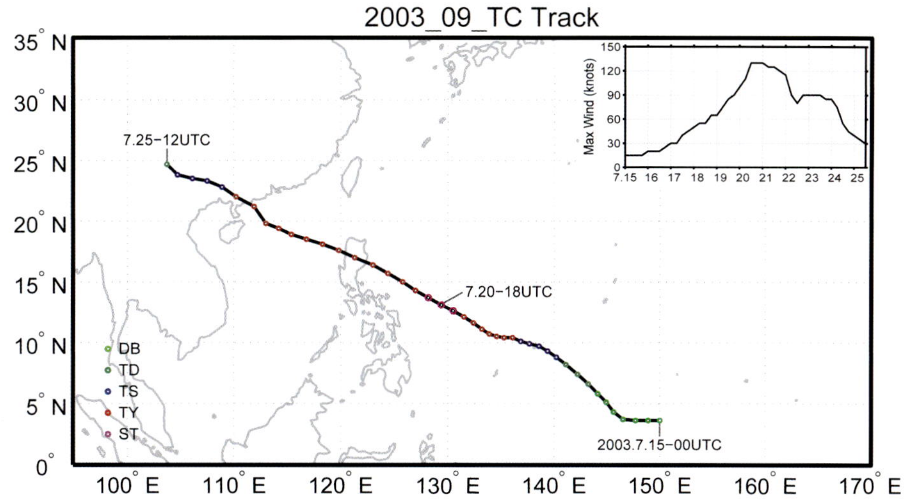

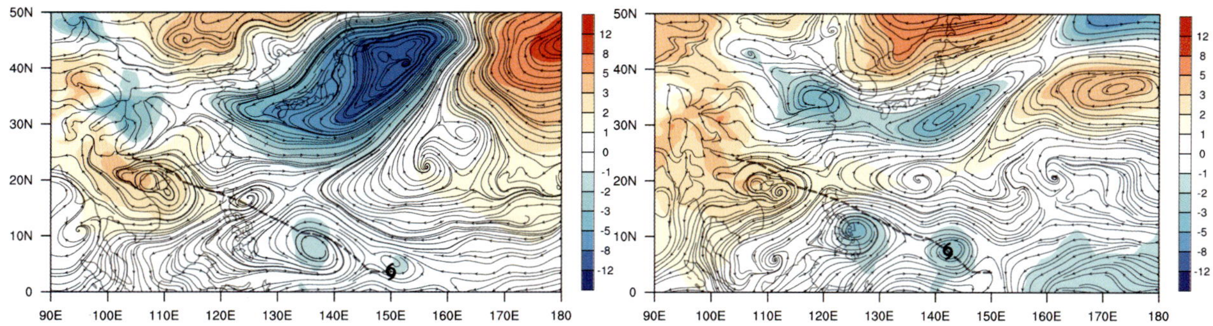

850 hPa flow/height anomalies at 0000 UTC 15 and 17 July 2003.

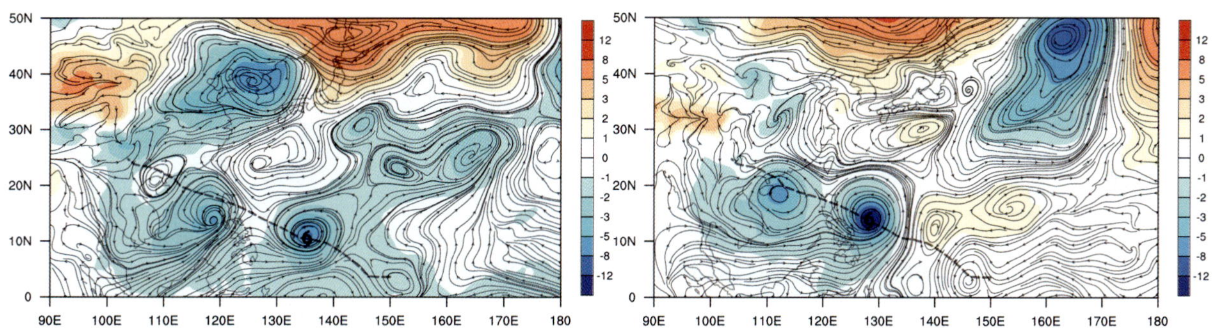

850 hPa flow/height anomalies at 0000 UTC 19 and 21 July 2003.

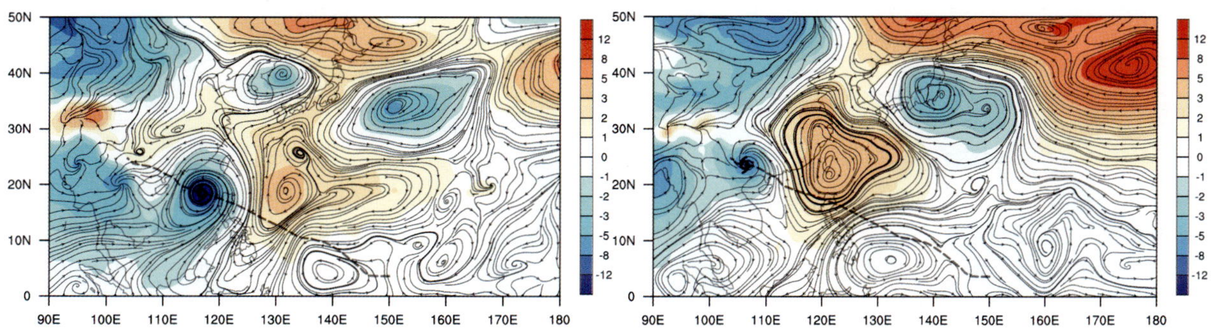

850 hPa flow/height anomalies at 0000 UTC 23 and 25 July 2003.

Figure 2.45. The typhoon (2003_No.09) generated in the equatorial western Pacific on 15 July 2003 and moved northwestward along the climatic monsoon trough. Its intensity peaked in the east sea of the Philippines on 20 July 2003 before landfall on the island and its track was not affected by anomalous vortices at 850 hPa.

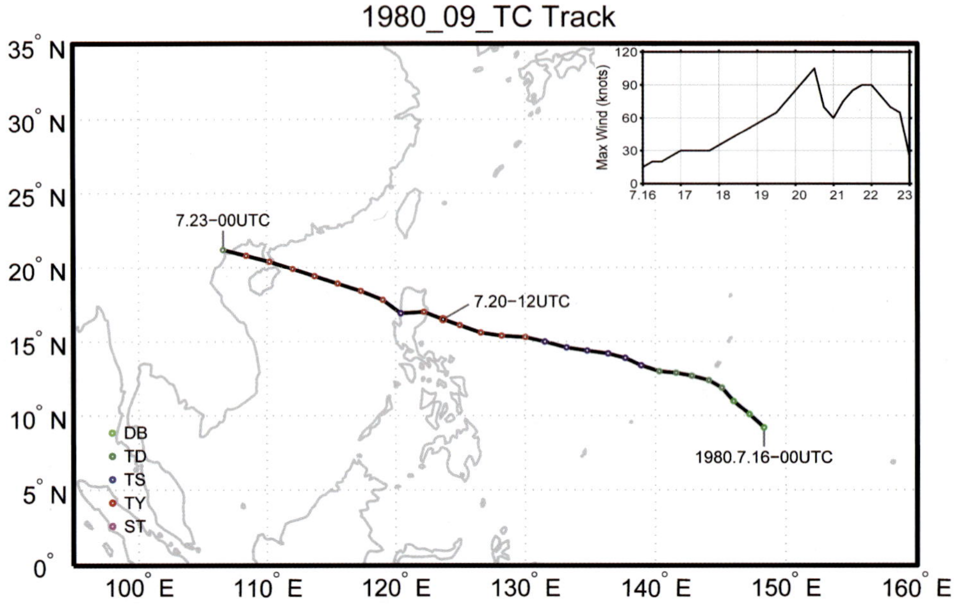

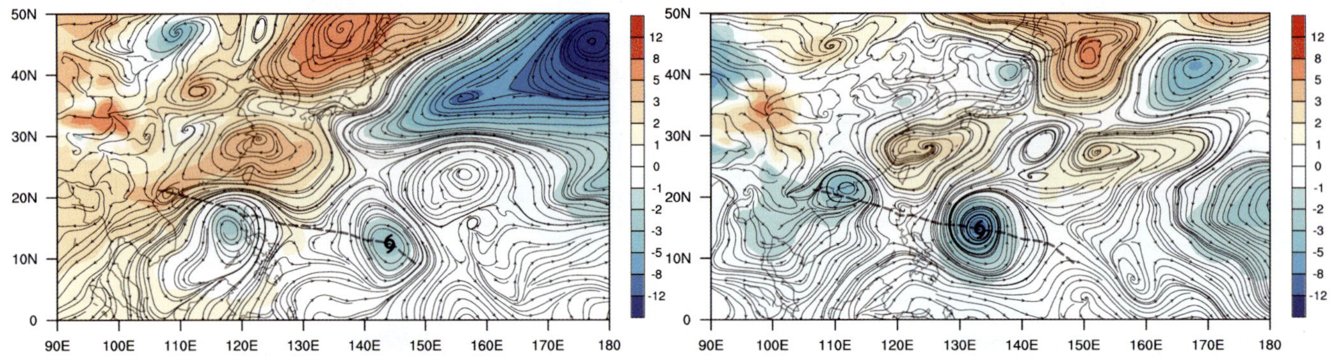

850 hPa flow/height anomalies at 0000 UTC 17 and 19 July 1980.

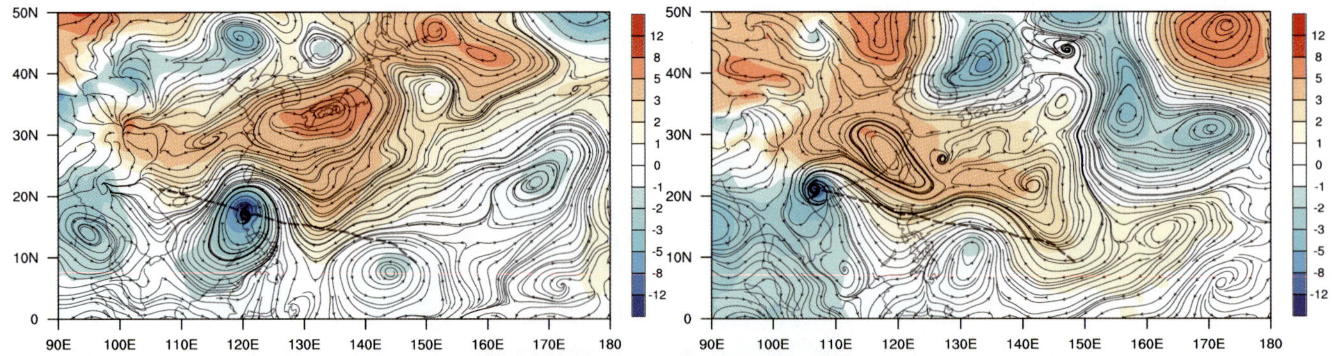

850 hPa flow/height anomalies at 0000 UTC 21 and 23 July 1980.

Figure 2.46. The typhoon (1980_No.09) generated in the equatorial western Pacific on 16 July 1980 and moved west-northwestward without the influence of anomalous vortices at 850 hPa. Its intensity peaked near the climatic monsoon trough in the east sea of the Philippines on 20 July before landfall on the island.

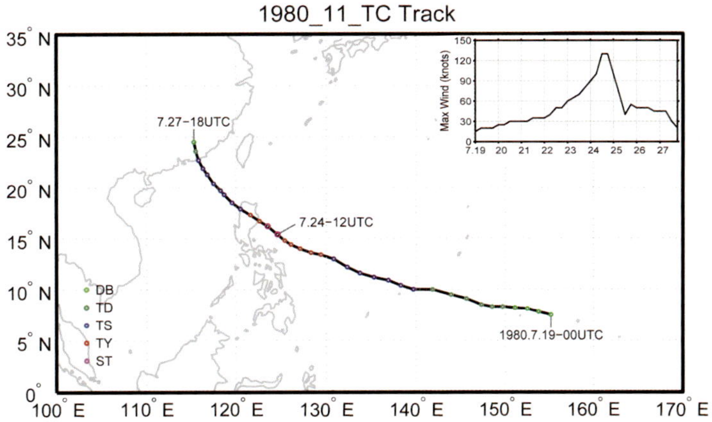

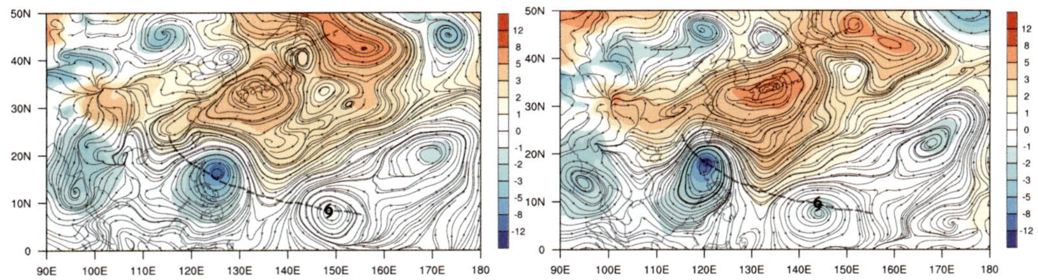

850 hPa flow/height anomalies at 0000 UTC 20 and 21 July 1980.

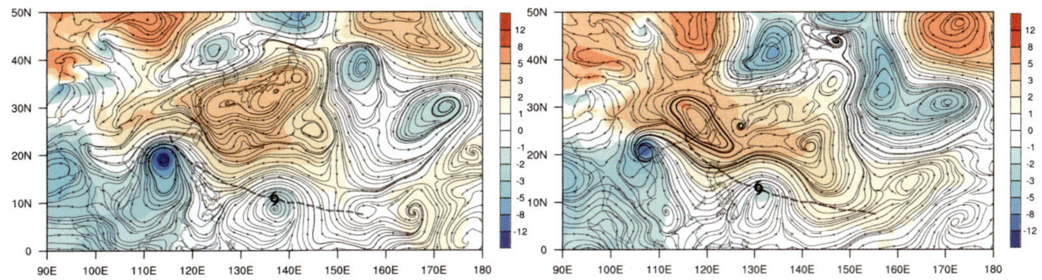

850 hPa flow/height anomalies at 0000 UTC 22 and 23 July 1980.

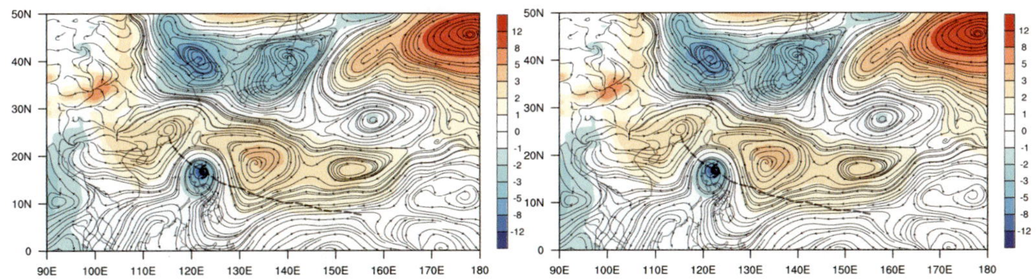

850 hPa flow/height anomalies at 0000 UTC 24 and 25 July 1980.

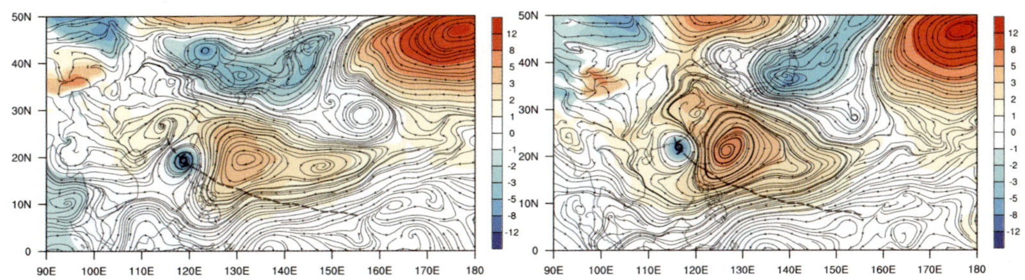

850 hPa flow/height anomalies at 0000 UTC 26 and 27 July 1980.

Figure 2.47. The typhoon (1980_No.11) generated in the equatorial central Pacific on 19 July 1980 and moved northwestward without the influence of other anomalous vortices at 850 hPa. Its intensity peaked near the climatic monsoon trough in the east sea of the Philippines on 24 July 1980 before landfall on the island.

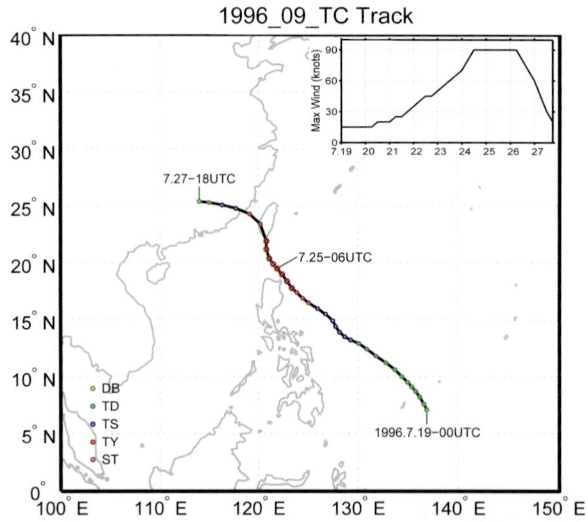

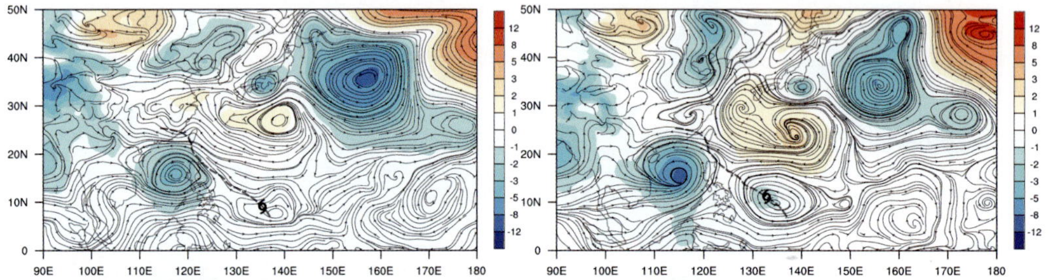

850 hPa flow/height anomalies at 0000 UTC 20 and 21 July 1996.

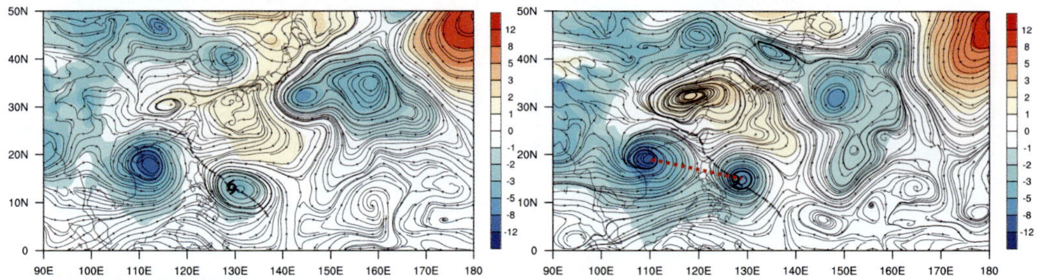

850 hPa flow/height anomalies at 0000 UTC 22 and 23 July 1996.

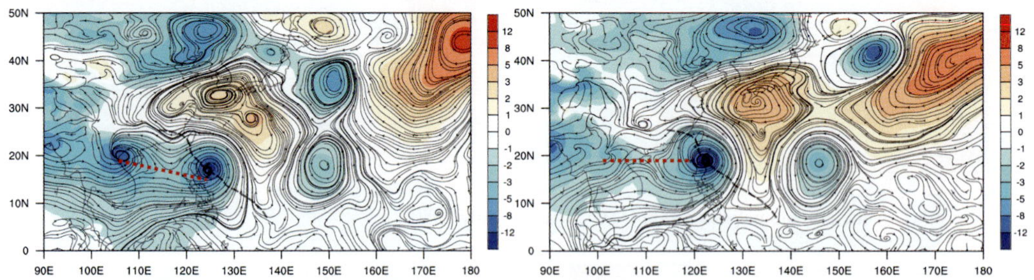

850 hPa flow/height anomalies at 0000 UTC 24 and 25 July 1996.

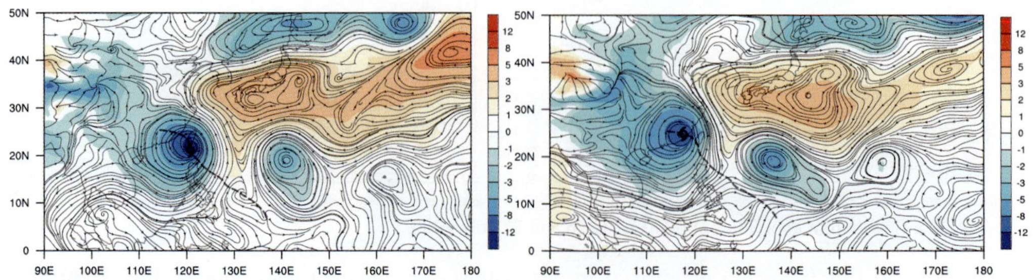

850 hPa flow/height anomalies at 0000 UTC 26 and 27 July 1996.

Figure 2.48. The typhoon (1996_No.09) generated in the equatorial western Pacific on 19 July 1996 and moved northwestward along the climatic monsoon trough. Its intensity peaked in the north sea of the Philippines on 25 July 1996 before landfall on the Taiwan Island and its motion in late stage was modified by an anomalous vortex at 850 hPa.

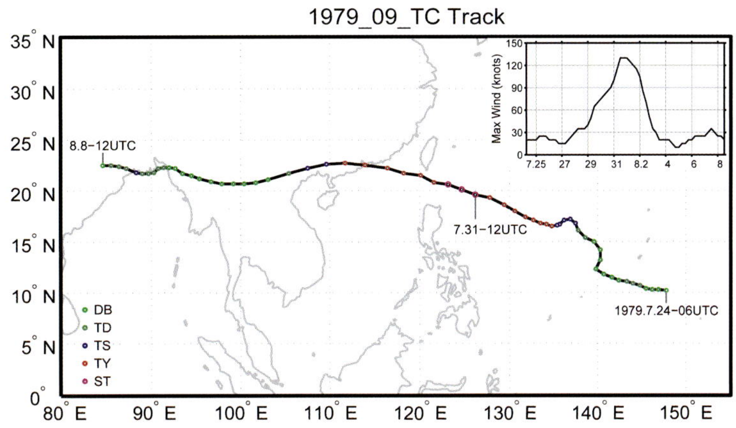

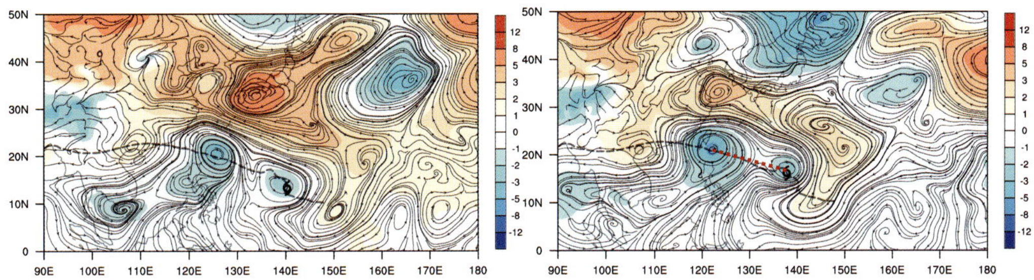

850 hPa flow/height anomalies at 0000 UTC 27 and 28 July 1979.

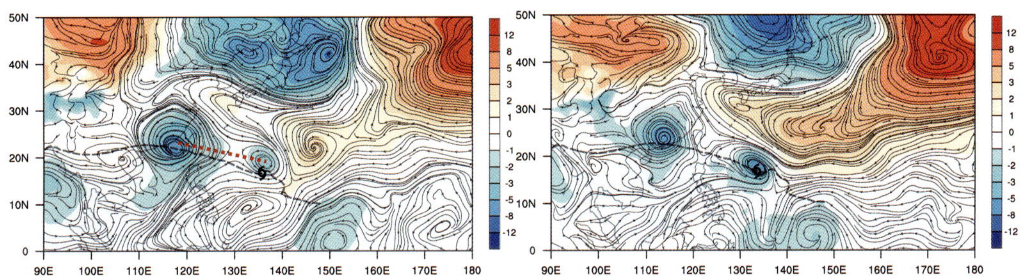

850 hPa flow/height anomalies at 0000 UTC 29 and 30 July 1979.

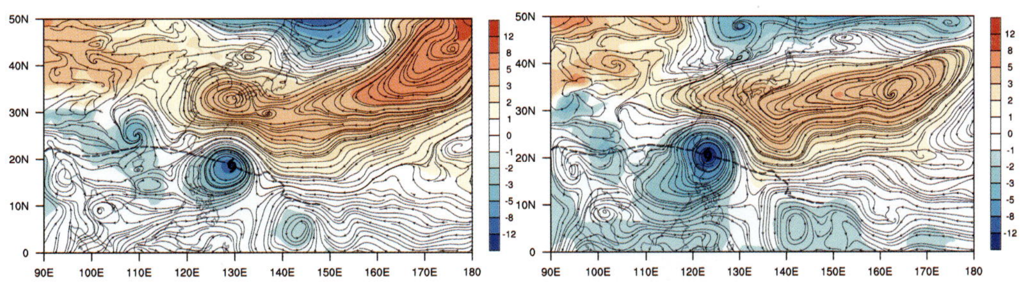

850 hPa flow/height anomalies at 0000 UTC 31 July and 1 August 1979.

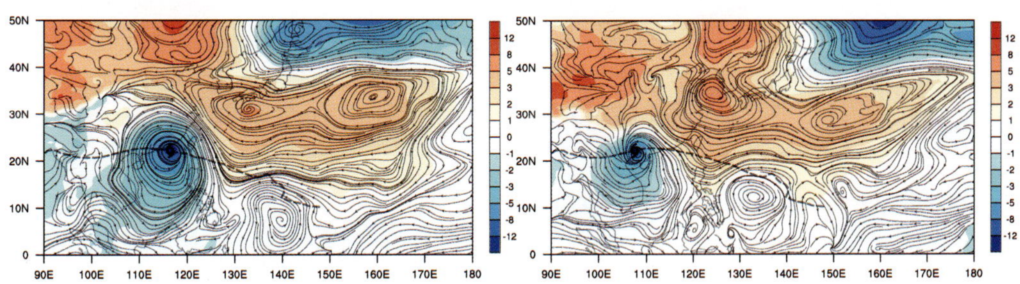

850 hPa flow/height anomalies at 0000 UTC 2 and 3 August 1979.

Figure 2.49. The typhoon (1979_No.09) generated in the tropical western Pacific on 24 July 1979 and moved northwestward along the climatic monsoon trough. Its intensity peaked in the southeast sea of the Taiwan Island on 31 July 1979 before landfall on the South China coast and its motion at the beginning stage was modified by an anomalous vortex at 850 hPa.

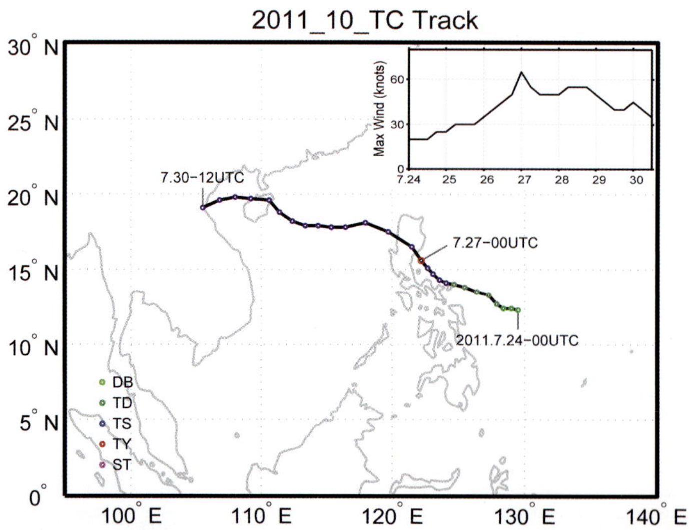

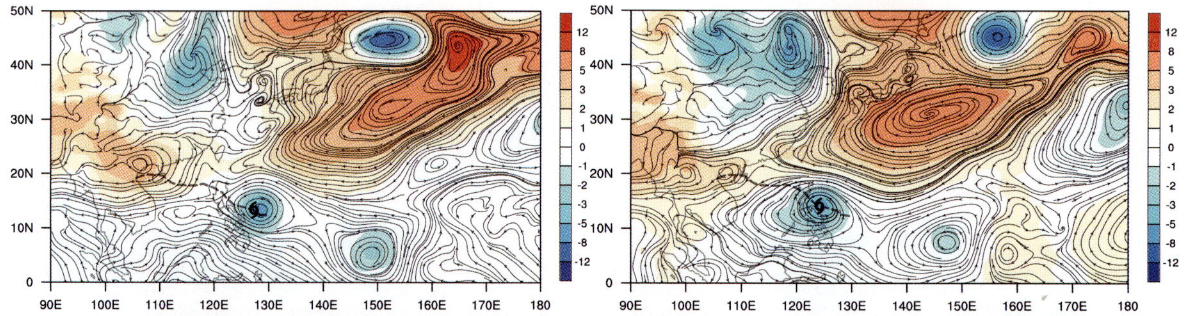

850 hPa flow/height anomalies at 0000 UTC 25 and 26 July 2011.

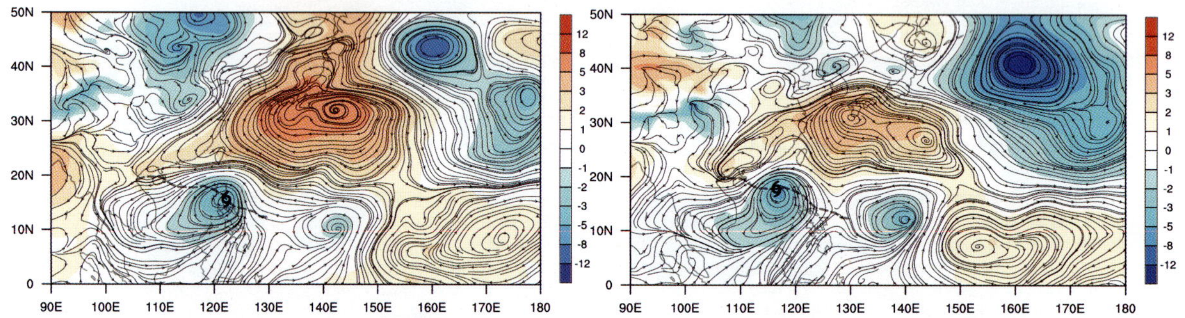

850 hPa flow/height anomalies at 0000 UTC 27 and 28 July 2011.

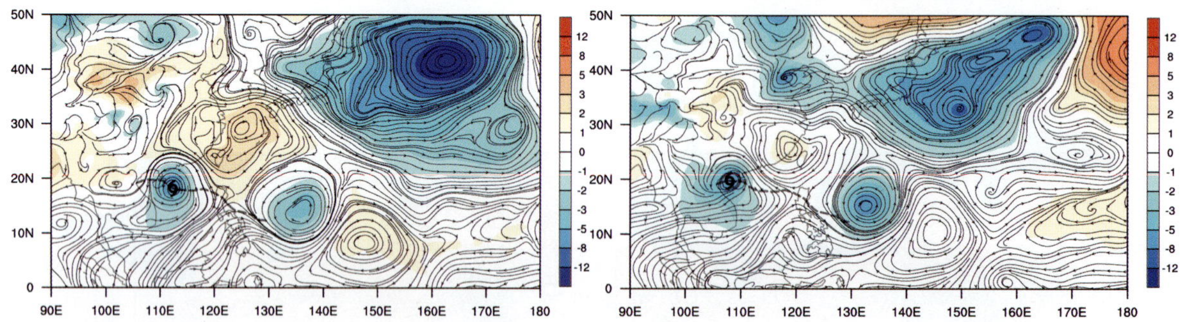

850 hPa flow/height anomalies at 0000 UTC 29 and 30 July 2011.

Figure 2.50. The typhoon (2011_No.10) generated in the east sea of the Philippines on 24 July 2011 and moved northwestward. Its intensity peaked in the east sea of the Philippines on 27 July 2011 before landfall on the island and its motion was modified by anomalous flows at 850 hPa.

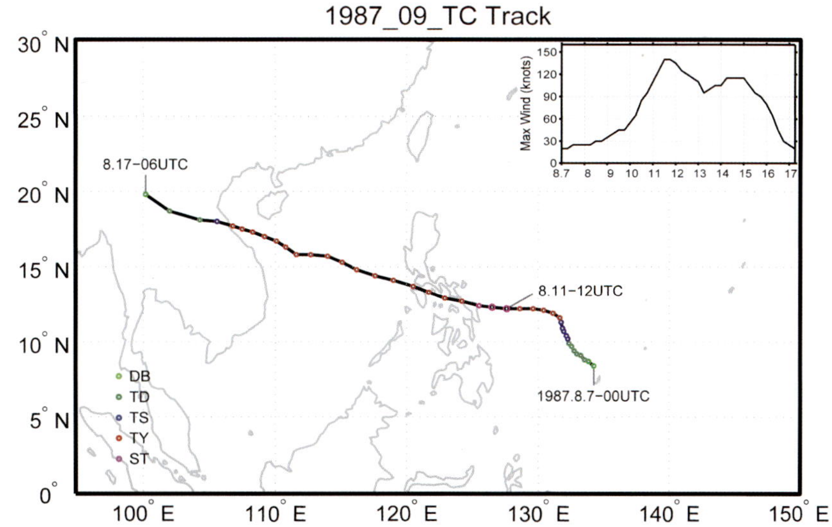

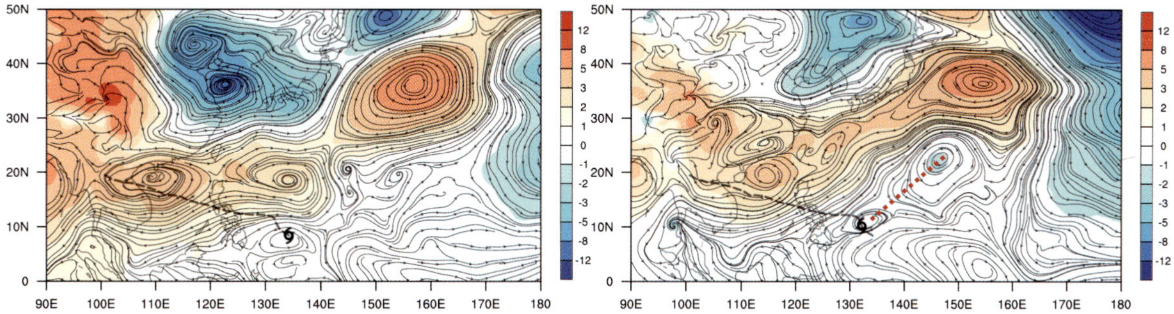

850 hPa flow/height anomalies at 0000 UTC 7 and 9 August 1987.

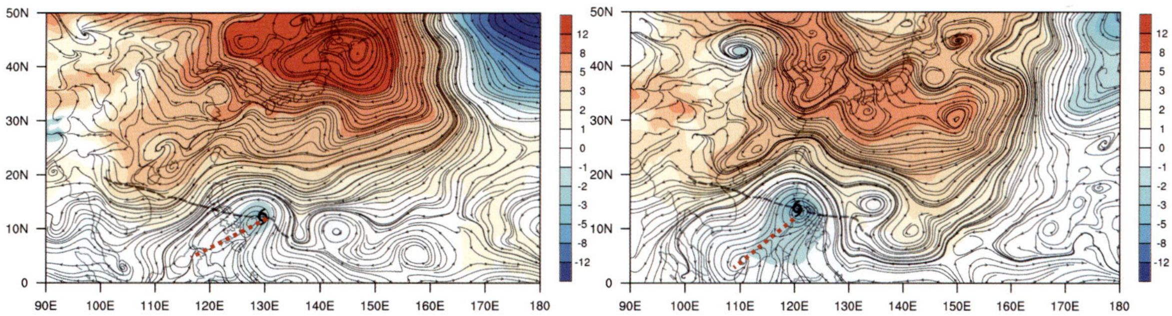

850 hPa flow/height anomalies at 0000 UTC 11 and 13 August 1987.

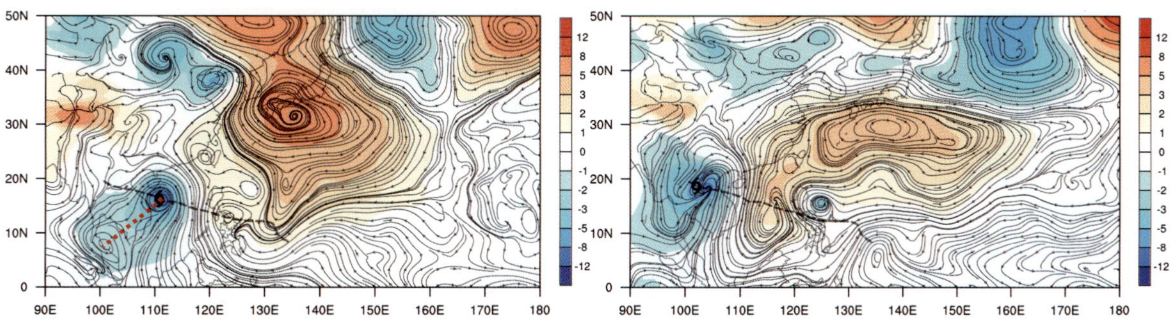

850 hPa flow/height anomalies at 0000 UTC 15 and 17 August 1987.

Figure 2.51. The typhoon (1987_No.09) generated in the equatorial western Pacific on 7 August 1987 and moved northwestward. Its intensity peaked in the east sea of the Philippines on 11 August 1987 before landfall on the central Philippines. At the beginning stage, its motion was influenced by two anomalous vortices at 850 hPa.

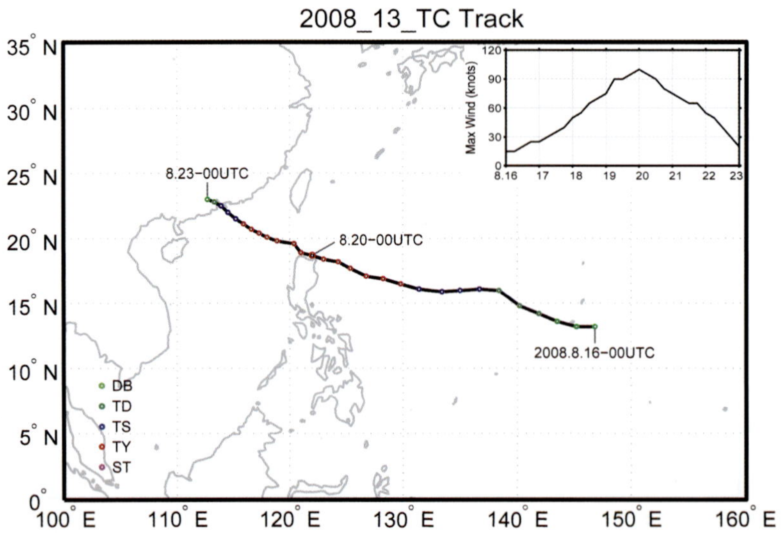

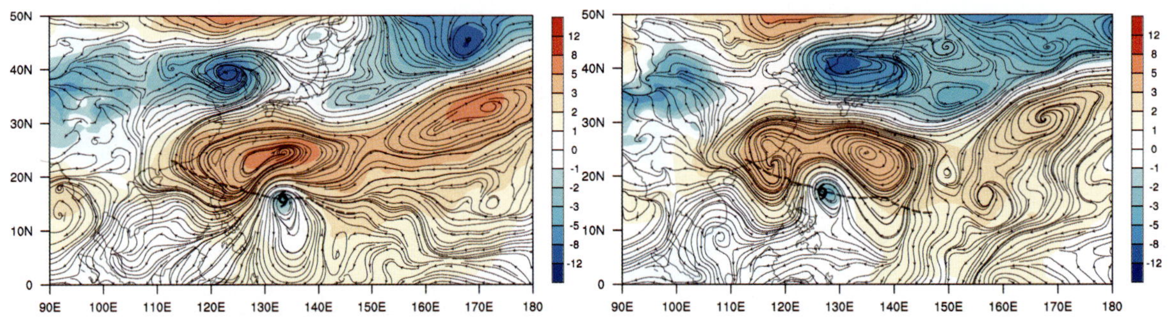

850 hPa flow/height anomalies at 0000 UTC 18 and 19 August 2008.

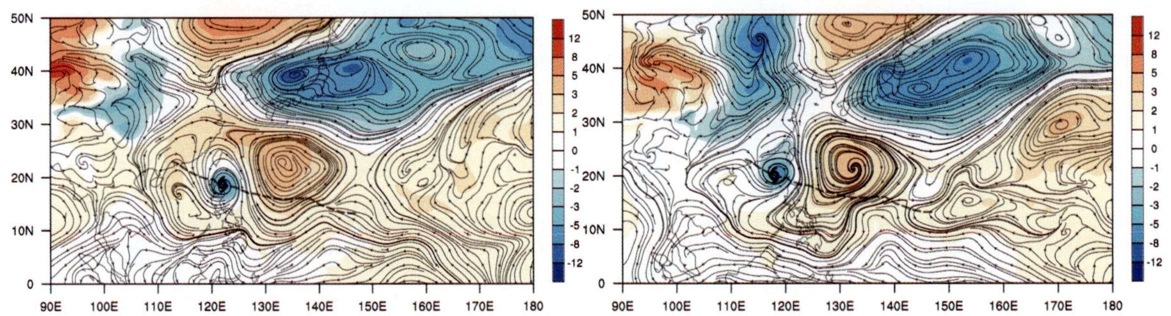

850 hPa flow/height anomalies at 0000 UTC 20 and 21 August 2008.

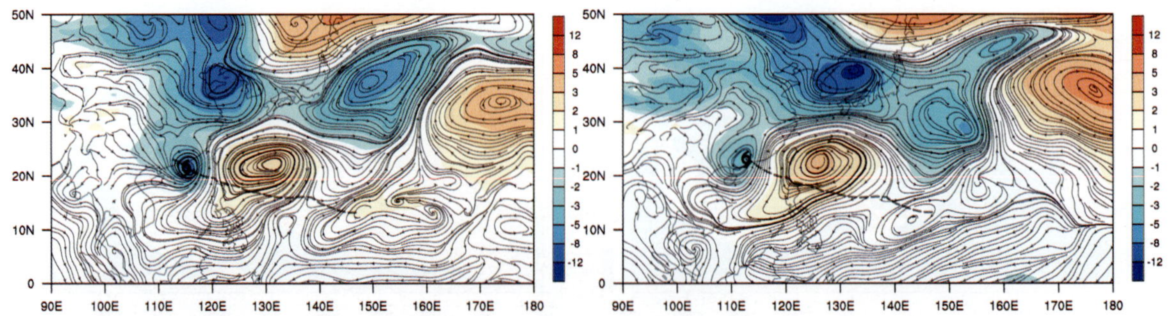

850 hPa flow/height anomalies at 0000 UTC 22 and 23 August 2008.

Figure 2.52. The typhoon (2008_No.13) generated in the tropical western Pacific on 16 August 2008 and moved west-northwestward without the influence of anomalous vortices at 850 hPa. Its intensity peaked near the northern Philippines on 20 August 2008 when approaching the island.

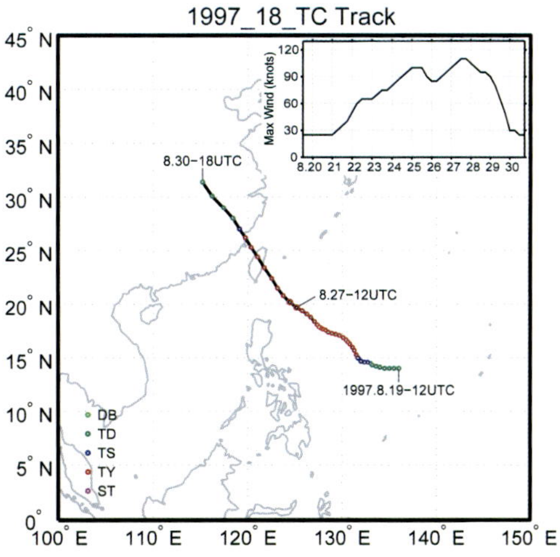

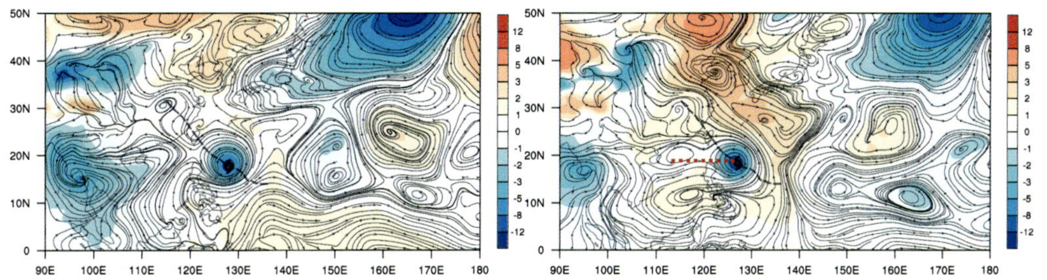

850 hPa flow/height anomalies at 1200 UTC 25 and 26 August 1997.

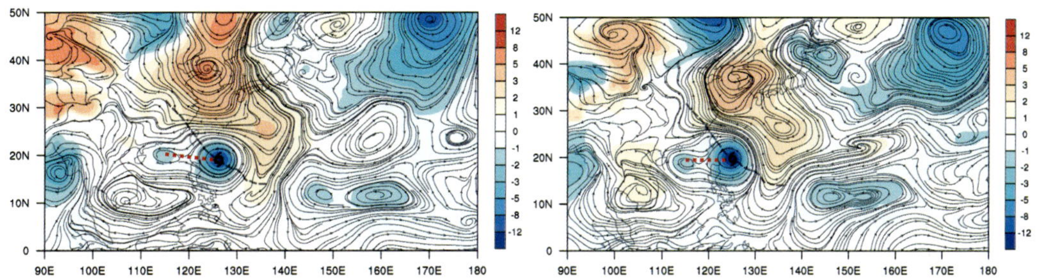

850 hPa flow/height anomalies at 0000 UTC and 1200 UTC 27 August 1997.

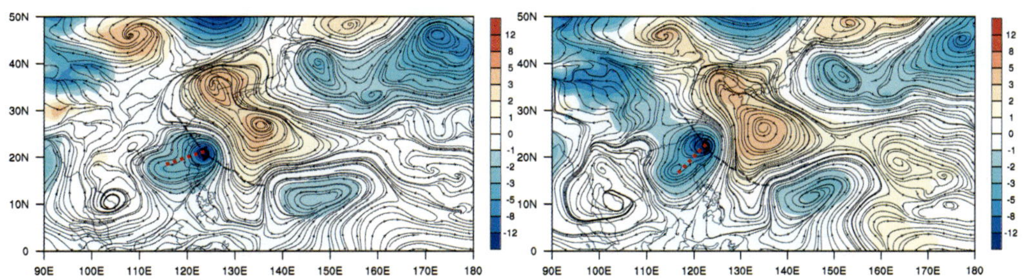

850 hPa flow/height anomalies at 0000 UTC and 1200 UTC 28 August 1997.

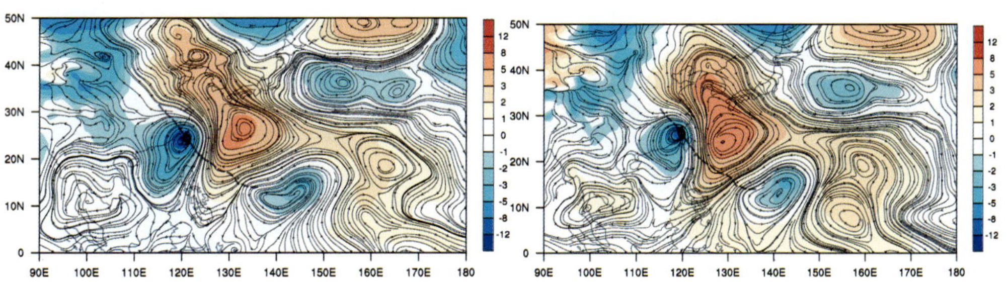

850 hPa flow/height anomalies at 0000 UTC and 1200 UTC 29 August 1997.

Figure 2.53. The typhoon (1997_No.18) generated in the tropical western Pacific on 19 August 1997 and moved northwestward along the climatic monsoon trough. Its intensity peaked in the northeast sea of the Philippines on 27 August 1997 before landfall on the Taiwan Island. Its track and intensity were modified by an anomalous vortex at 850 hPa.

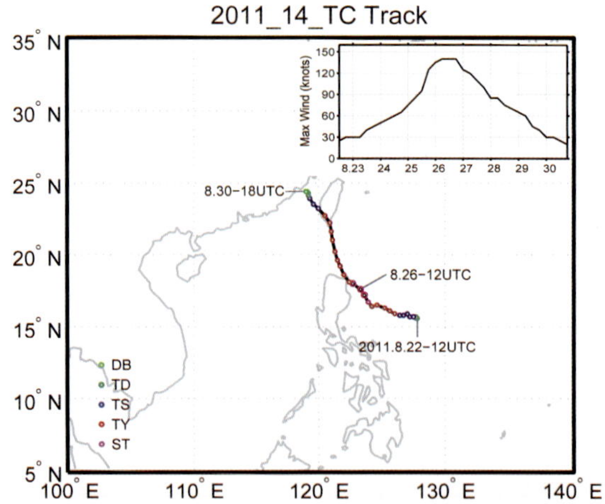

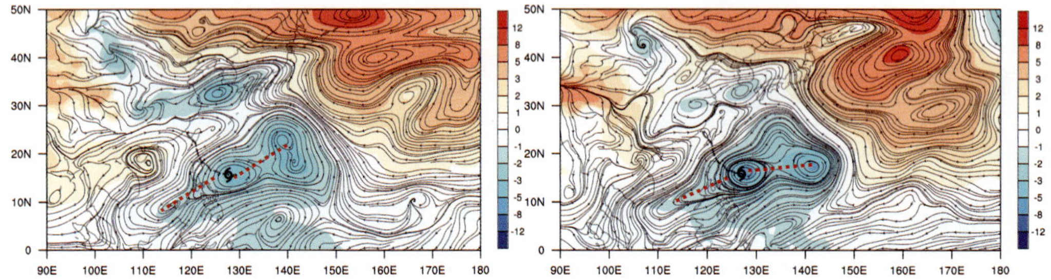

850 hPa flow/height anomalies at 0000 UTC 23 and 24 August 2011.

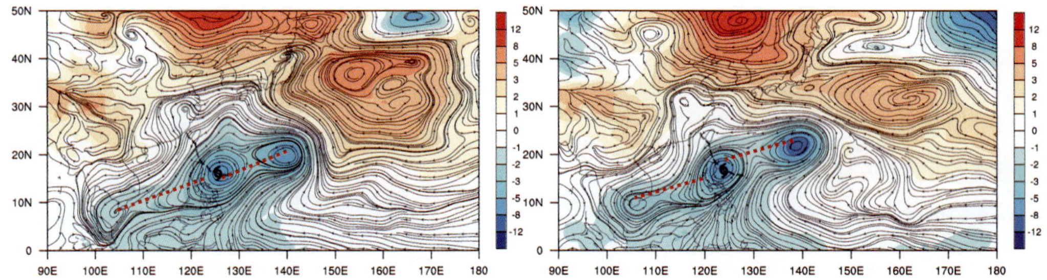

850 hPa flow/height anomalies at 0000 UTC 25 and 26 August 2011.

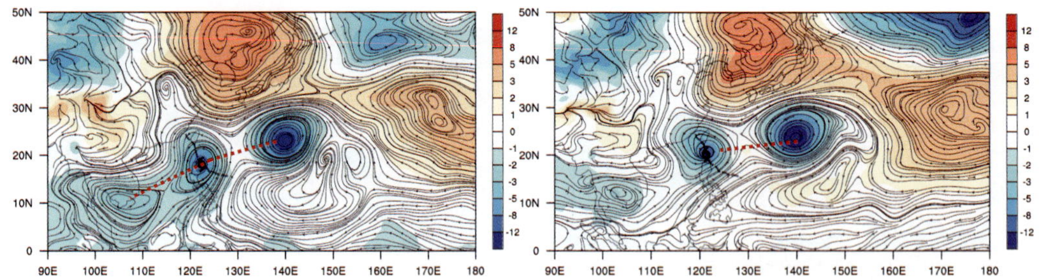

850 hPa flow/height anomalies at 0000 UTC 27 and 28 August 2011.

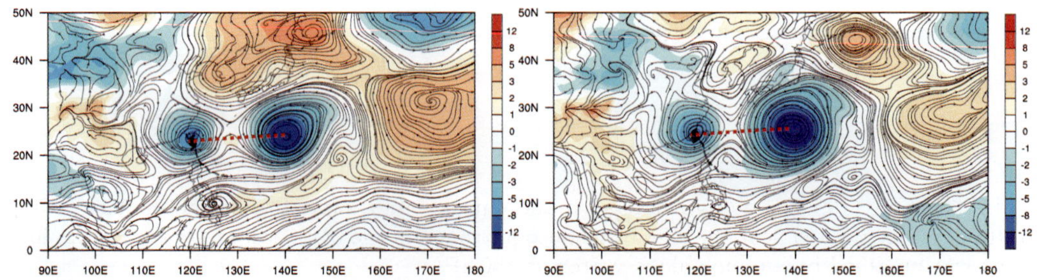

850 hPa flow/height anomalies at 0000 UTC 29 and 30 August 2011.

Figure 2.54. The typhoon (2011_No.14) generated in the northeast sea of the Philippines on 22 August 2011 and moved north-northwestward. Its intensity peaked on 26 August 2011 before approaching the northern Philippines and its track was modified by two anomalous vortices at 850 hPa.

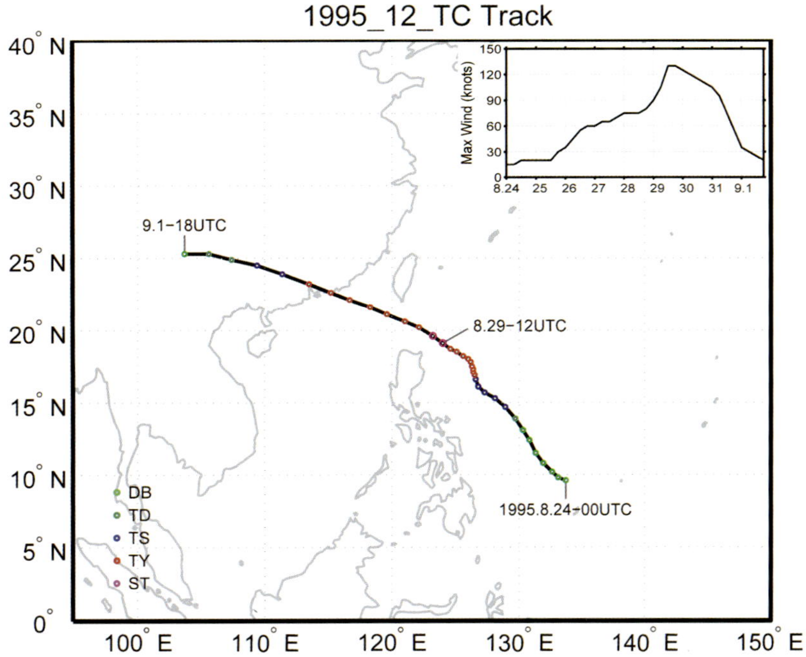

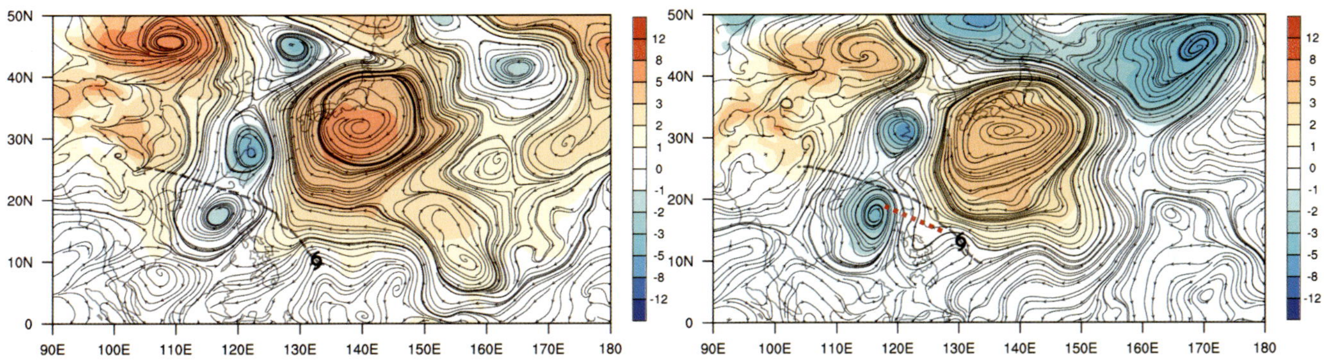

850 hPa flow/height anomalies at 1200 UTC 24 and 25 August 1995.

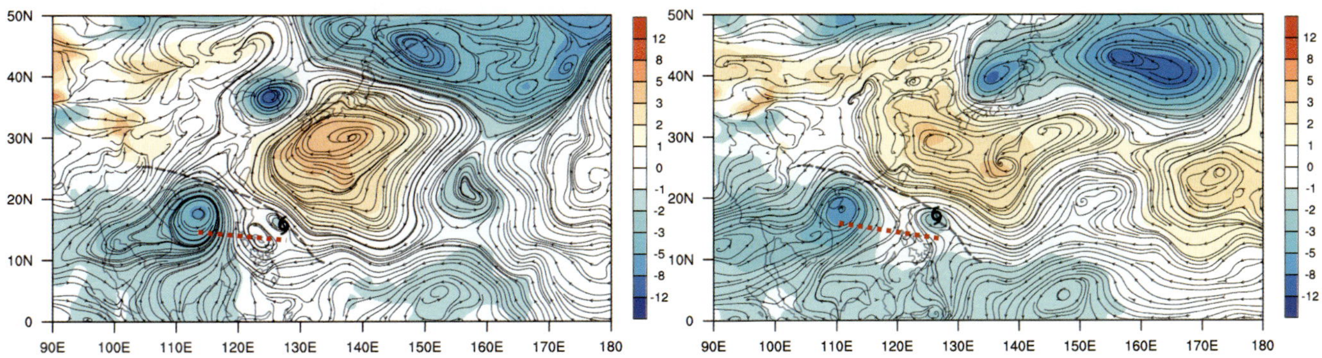

850 hPa flow/height anomalies at 1200 UTC 26 and 27 August 1995.

Figure 2.55. The typhoon (1995_No.12) generated in the equatorial western Pacific on 24 August 1995 and moved northwestward. Its intensity peaked near the climatic monsoon trough in the northeast sea of the Philippines on 29 August 1995 and its track was modified by an anomalous vortex at 850 hPa.

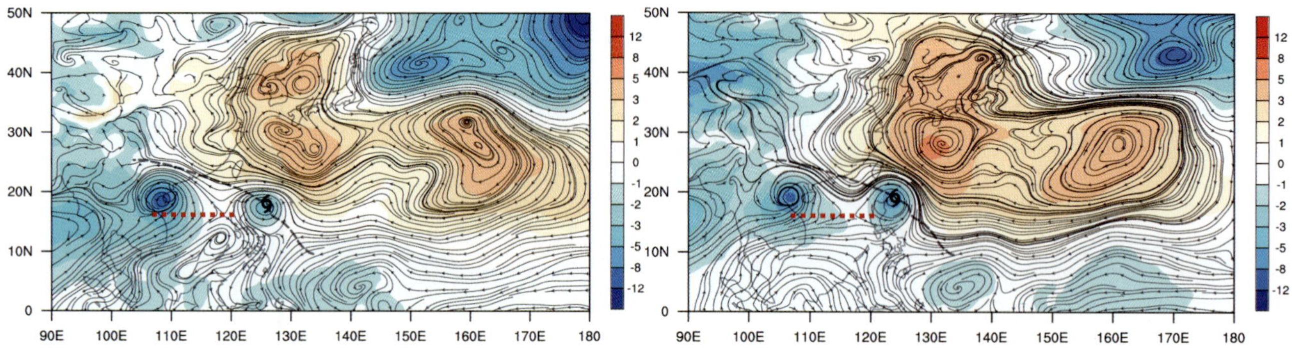

850 hPa flow/height anomalies at 1200 UTC 28 and 29 August 1995.

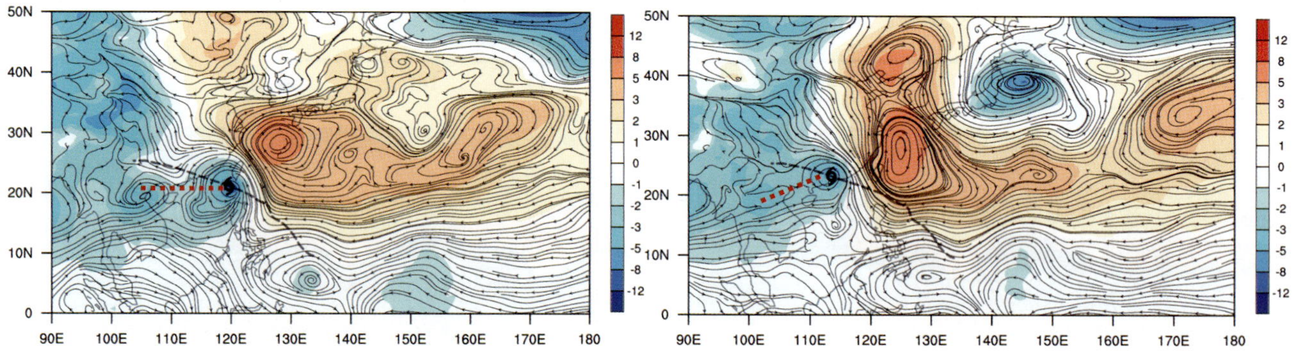

850 hPa flow/height anomalies at 1200 UTC 30 and 31 August 1995.

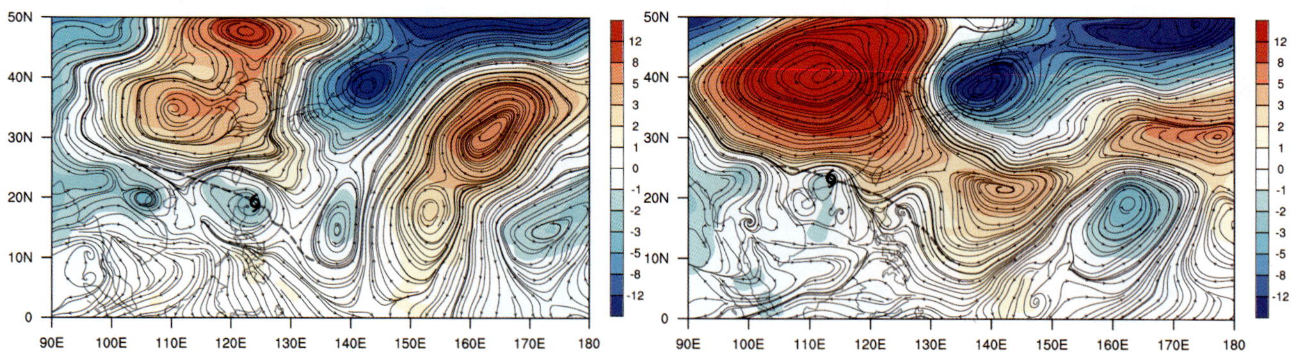

300 hPa flow/height anomalies at 1200 UTC 29 and 31 August 1995.

Figure 2.55. *(Continued)*

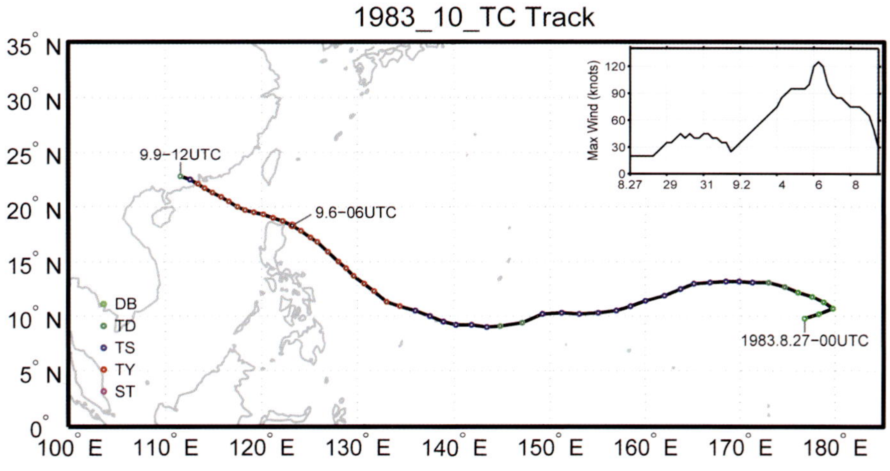

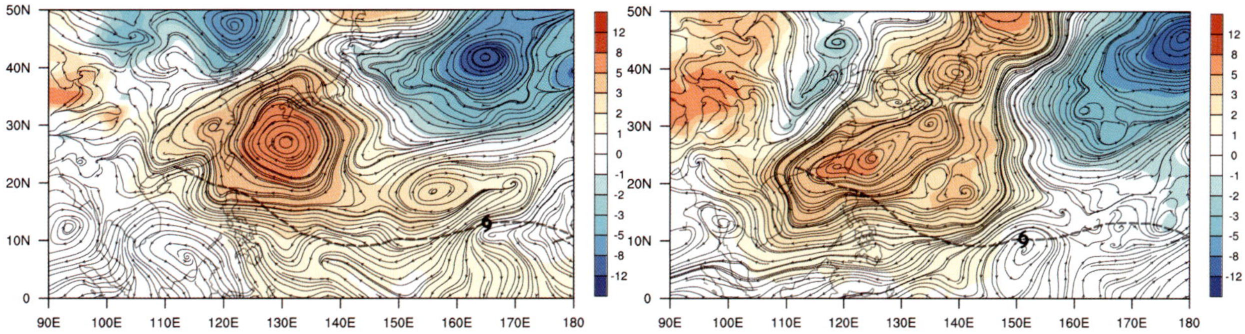

850 hPa flow/height anomalies at 0000 UTC 30 August and 1 September 1983.

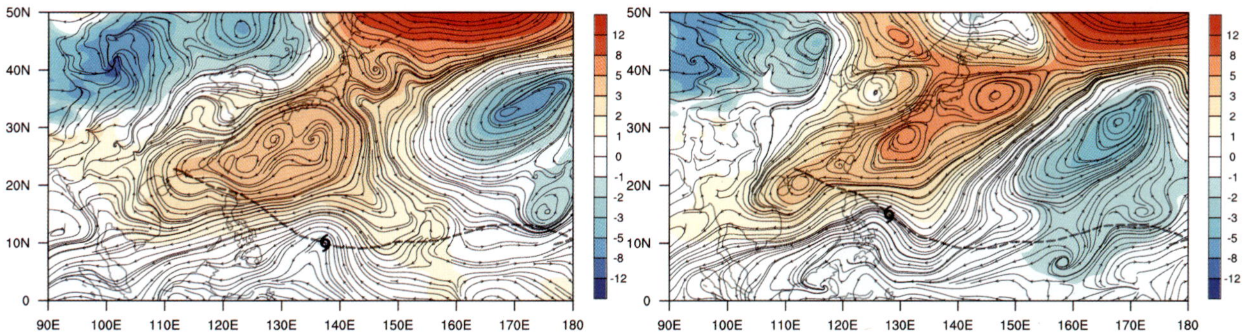

850 hPa flow/height anomalies at 0000 UTC 3 and 5 September 1983.

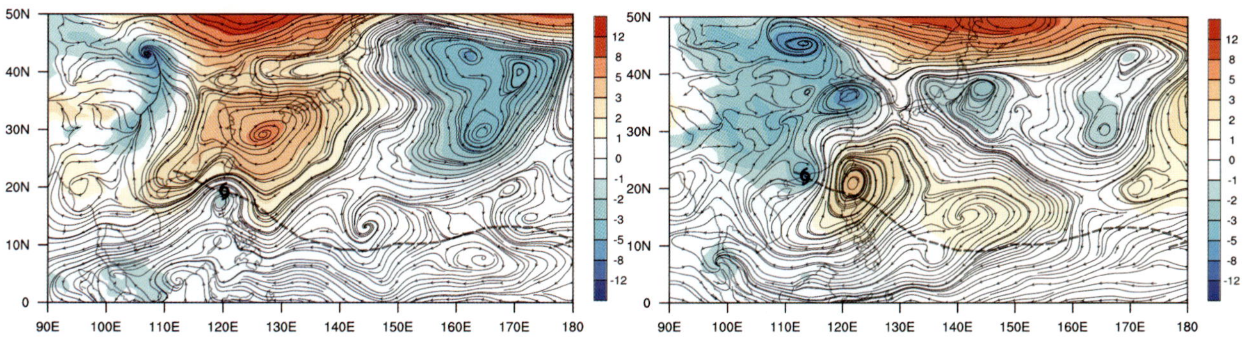

850 hPa flow/height anomalies at 0000 UTC 7 and 9 September 1983.

Figure 2.56. The typhoon (1983_No.10) generated in the equatorial central Pacific on 27 August 1983. After a return at the beginning stage, it moved to westward in the subtropical western Pacific to northwestward in the Philippines and the SCS. In the east sea of the Philippines, its track was along the climatic monsoon trough and its intensity peaked on 6 September 1983 before approaching the northern Philippines. Its path was modified by anomalous flows at 850 hPa.

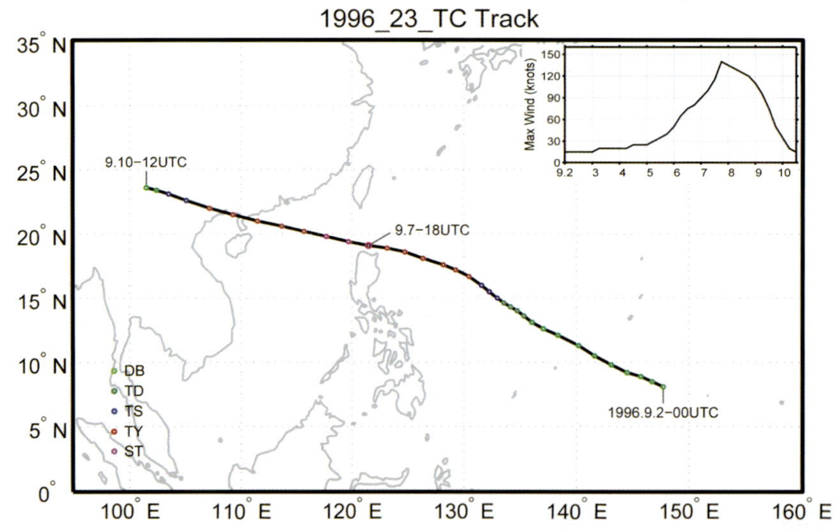

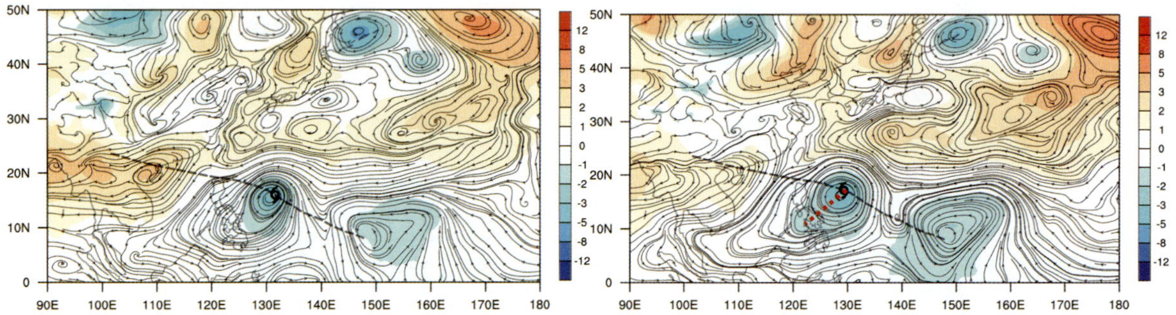

850 hPa flow/height anomalies at 0000 UTC and 1200 UTC 6 September 1996.

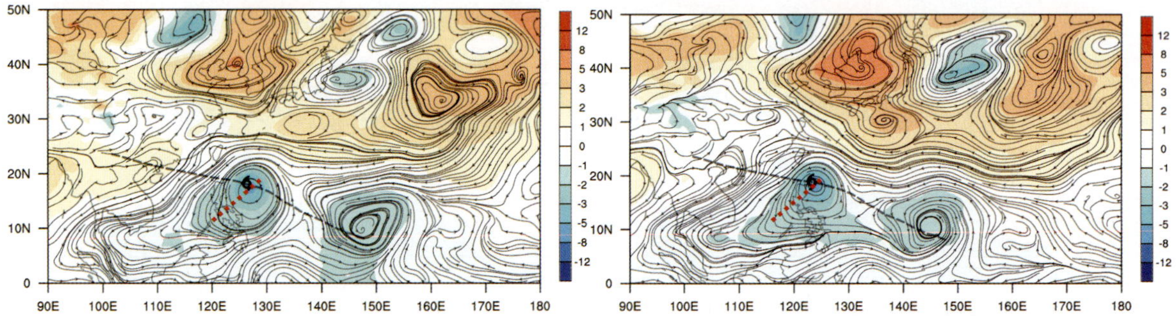

850 hPa flow/height anomalies at 0000 UTC and 1200 UTC 7 September 1996.

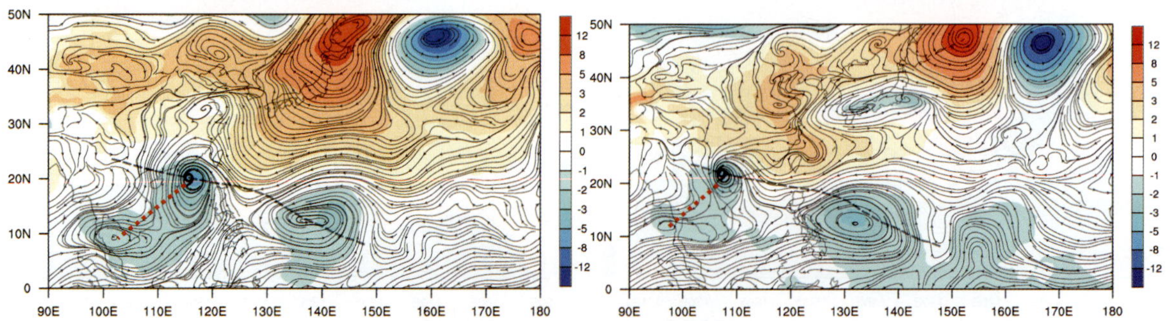

850 hPa flow/height anomalies at 1200 UTC 8 and 9 September 1996.

Figure 2.57. The typhoon (1996_No.23) generated in the equatorial western Pacific on 2 September 1996. It moved northwestward along the climatic monsoon trough so that its intensity peaked on 7 September 1996 when it was approaching the northern Philippines. Its motion near the northern Philippines was modified by an anomalous vortex at 850 hPa.

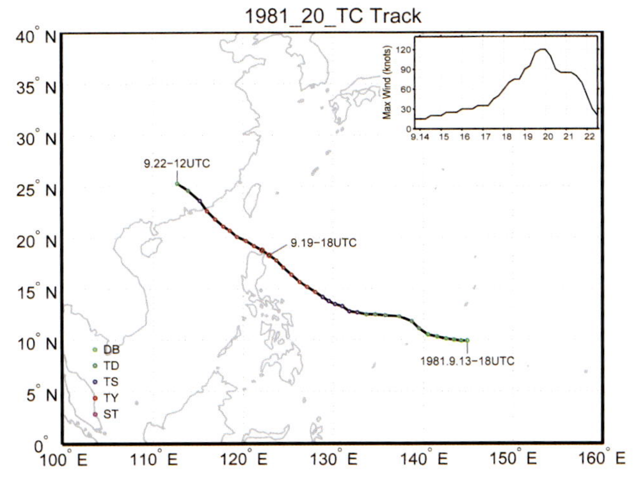

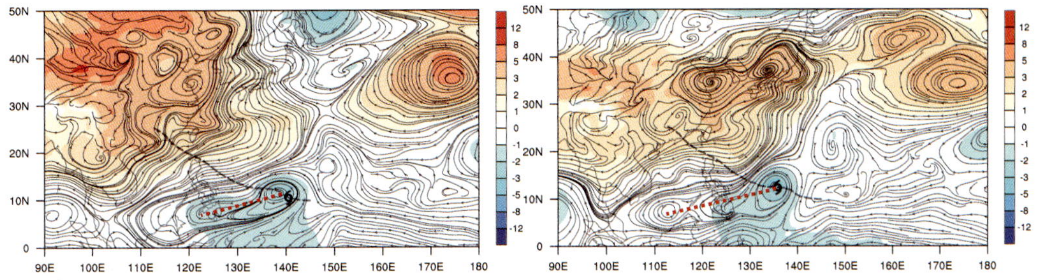

850 hPa flow/height anomalies at 0000 UTC 15 and 16 September 2009.

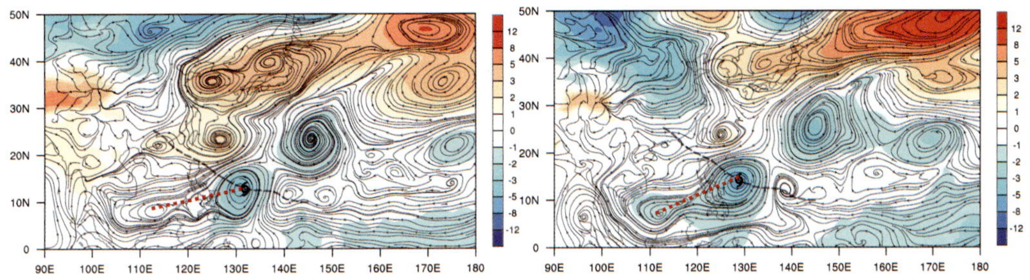

850 hPa flow/height anomalies at 0000 UTC 17 and 18 September 2009.

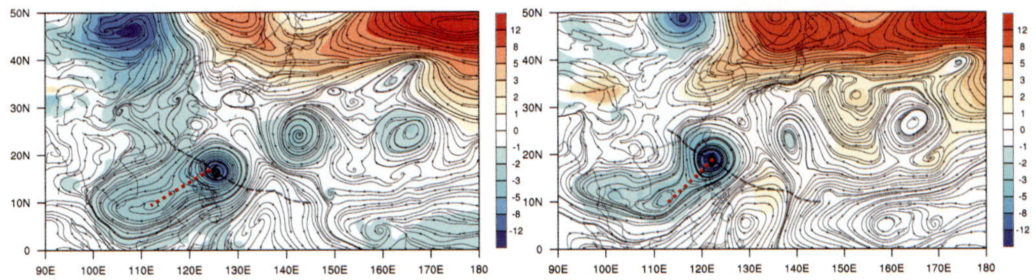

850 hPa flow/height anomalies at 0000 UTC 19 and 20 September 2009.

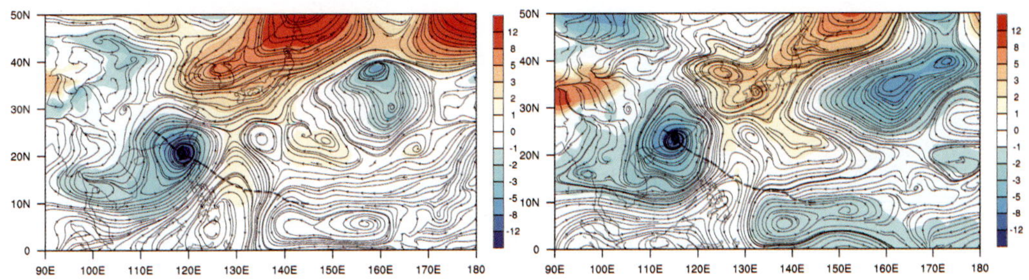

850 hPa flow/height anomalies at 0000 UTC 21 and 22 September 2009.

Figure 2.58. The typhoon (1981_No.20) generated in the tropical western Pacific on 13 September 1981 and moved northwestward. Its intensity peaked near the climatic monsoon trough in the northern Philippines on 19 September 1981 before landfall on the South China coast and its motion was modified by an anomalous vortex at 850 hPa.

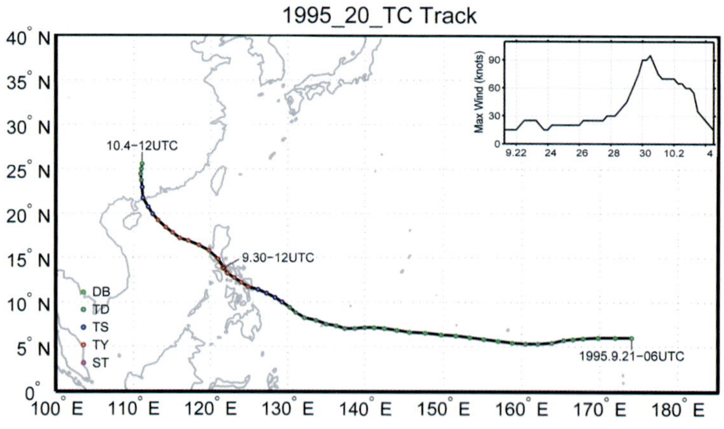

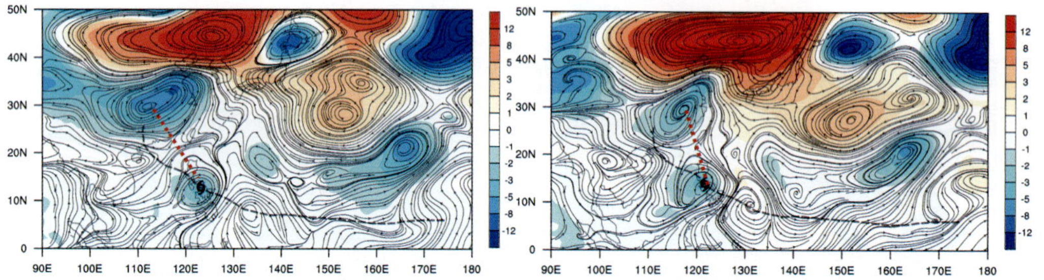

300 hPa flow/height anomalies at 0000 UTC and 1200 UTC 30 September 1995.

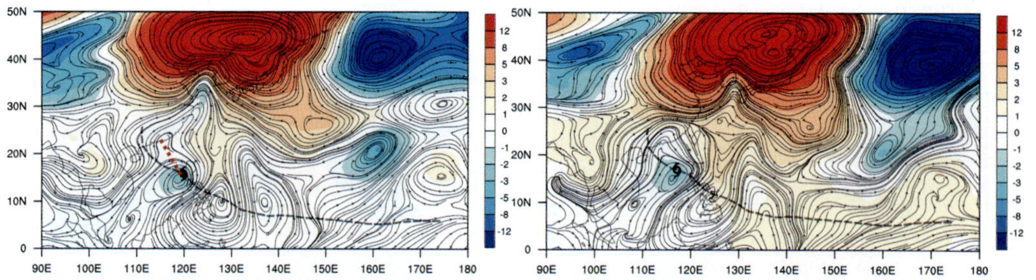

300 hPa flow/height anomalies at 0000 UTC and 1200 UTC 1 October 1995.

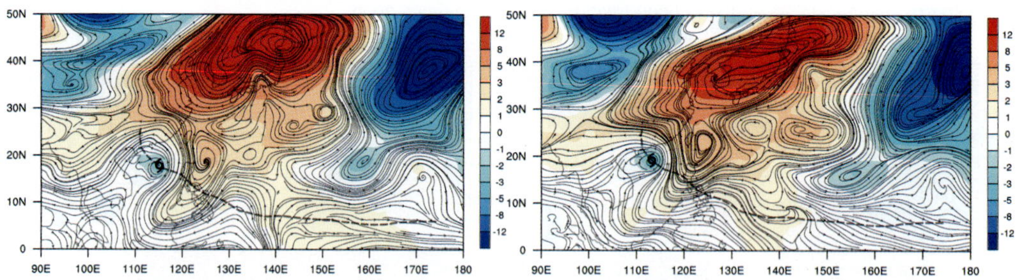

300 hPa flow/height anomalies at 0000 UTC and 1200 UTC 2 October 1995.

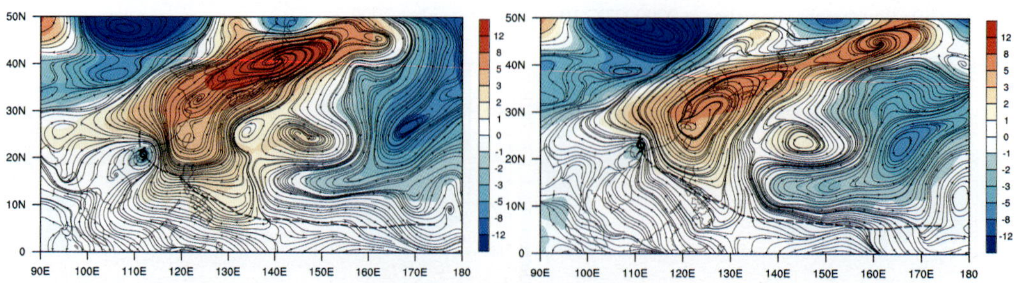

300 hPa flow/height anomalies at 0000 UTC and 1200 UTC 3 October 1995.

Figure 2.59. The typhoon (1995_No.20) generated in the equatorial central Pacific on 21 September 1995 and moved westward in the equatorial western Pacific. Its maximum wind speed reached higher than 60 knots near the climatic monsoon trough in the Philippines and the northern SCS. Its track change in the SCS was due to the influence of an anomalous vortex at 300 hPa.

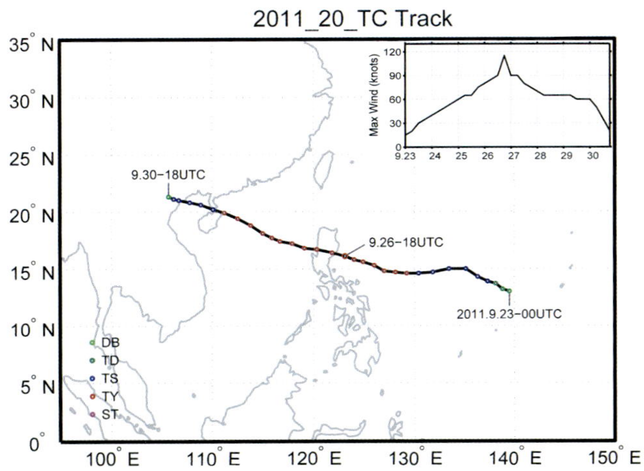

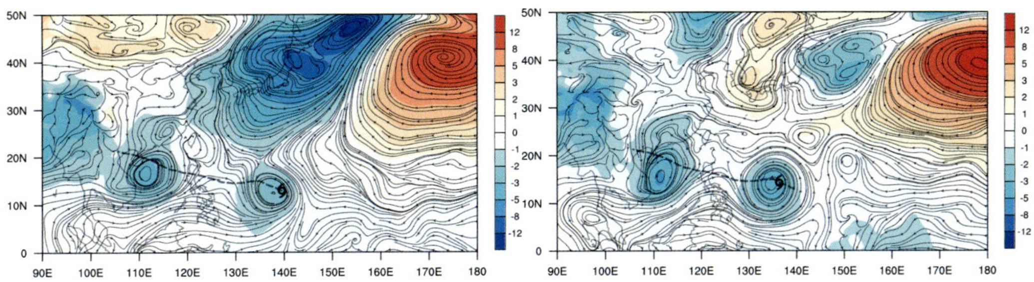

850 hPa flow/height anomalies at 0000 UTC 23 and 24 September 2007.

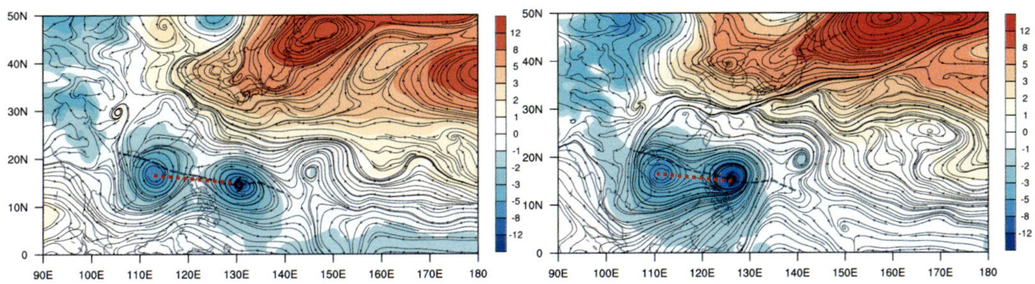

850 hPa flow/height anomalies at 0000 UTC 25 and 26 September 2007.

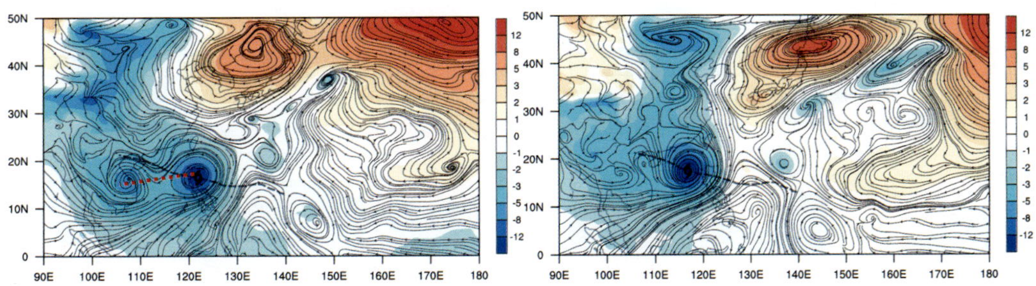

850 hPa flow/height anomalies at 0000 UTC 27 and 28 September 2007.

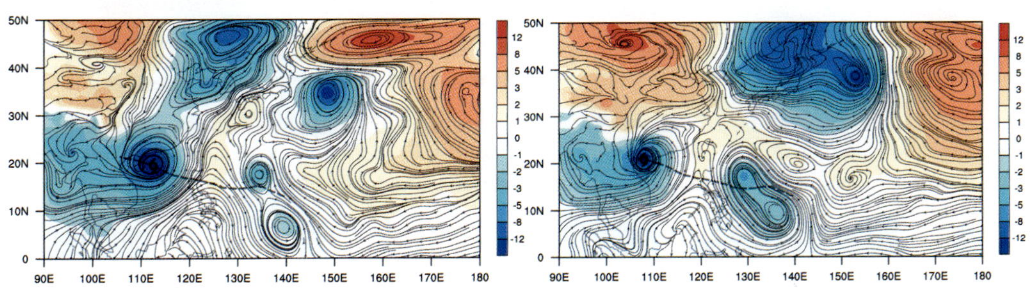

850 hPa flow/height anomalies at 0000 UTC 29 and 30 September 2007.

Figure 2.60. The typhoon (2011_No.20) generated in the tropical western Pacific on 23 September 2011 and moved west-northwestward. Its intensity peaked near the climatic monsoon trough in the northern Philippines on 26 September 2011 before landfall on the island and dissipated when it made landfall on the Southwest China coast. Its motion was modified by an anomalous vortex at 850 hPa.

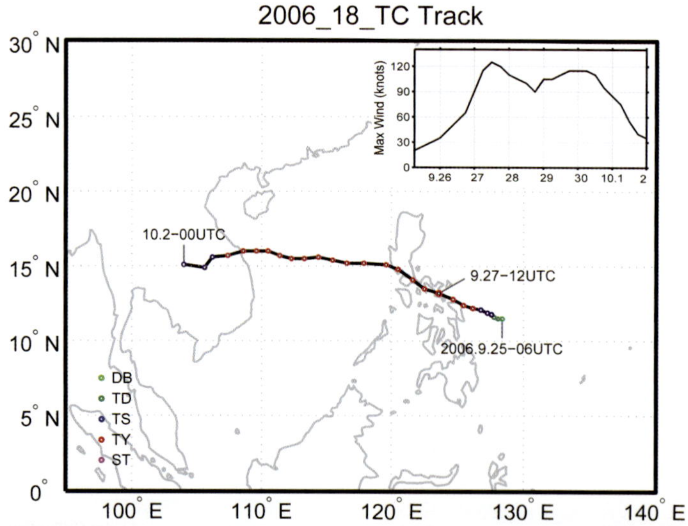

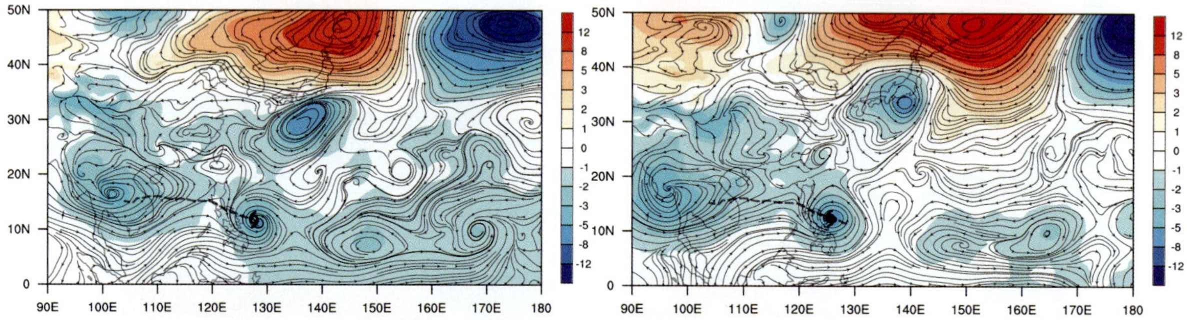

850 hPa flow/height anomalies at 0000 UTC 26 and 27 September 2006.

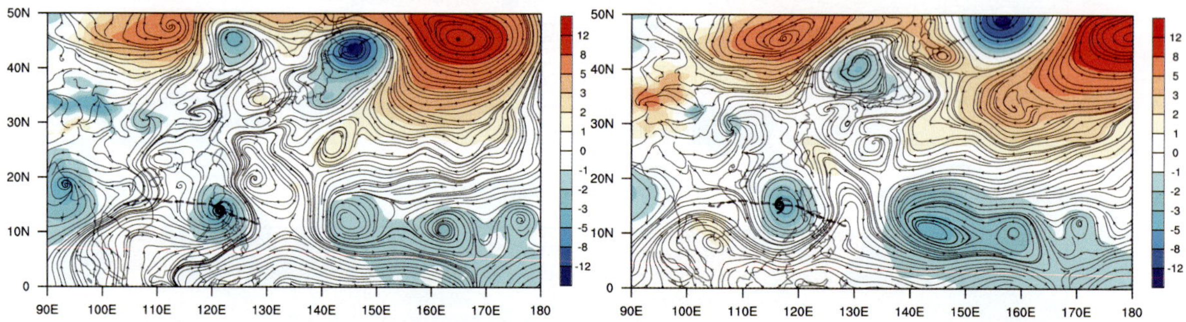

850 hPa flow/height anomalies at 0000 UTC 28 and 29 September 2006.

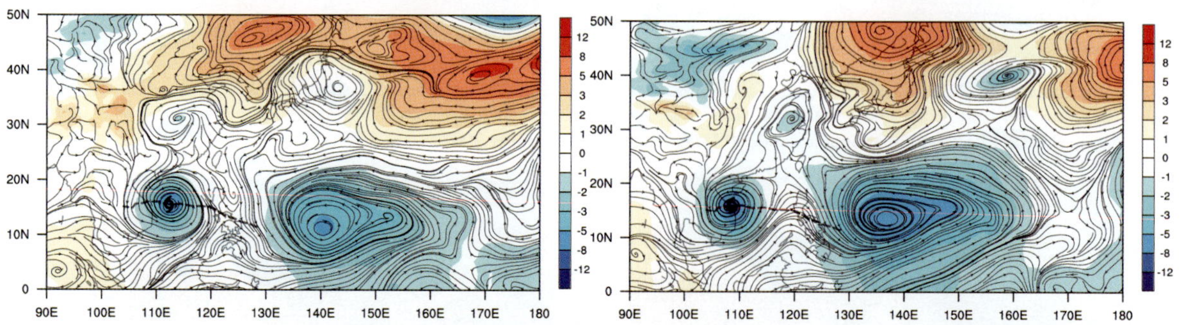

850 hPa flow/height anomalies at 0000 UTC 30 September and 1 October 2006.

Figure 2.61. The typhoon (2006_No.18) generated in the east sea of the Philippines on 25 September 2006 and moved westward without the influence of anomalous vortices at 850 hPa. Its maximum wind speed reached higher than 90 knots near the climatic monsoon trough in the northern SCS.

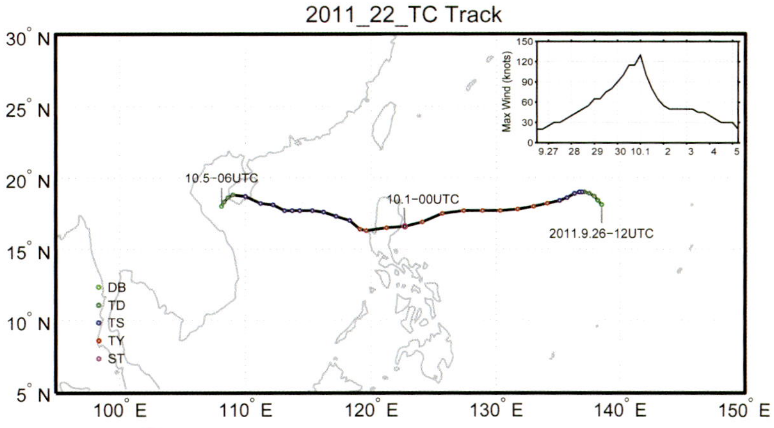

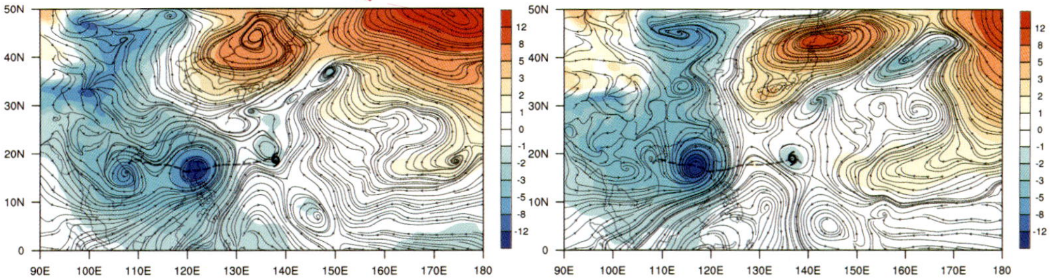

850 hPa flow/height anomalies at 0000 UTC 27 and 28 September 2011.

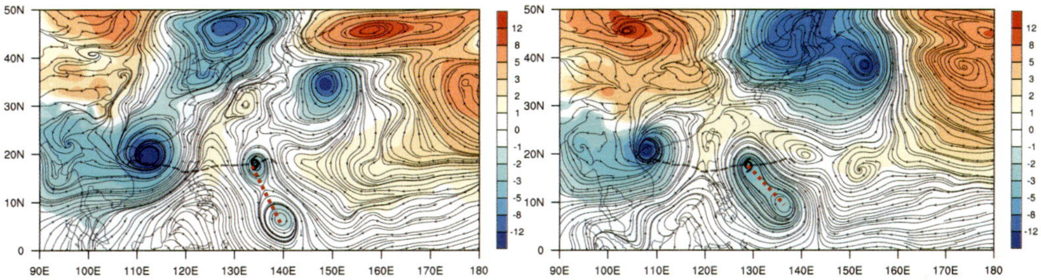

850 hPa flow/height anomalies at 0000 UTC 29 and 30 September 2011.

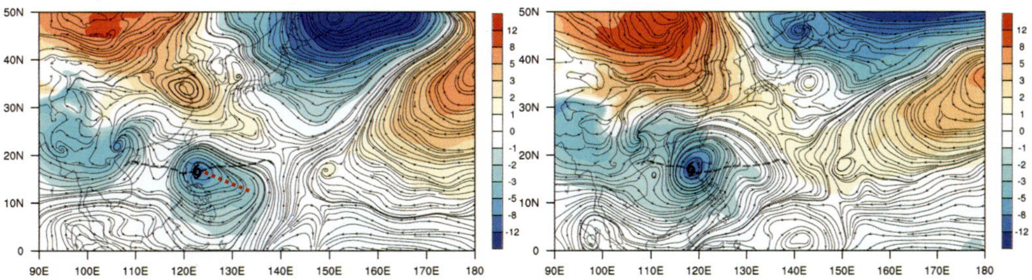

850 hPa flow/height anomalies at 0000 UTC 1 and 2 October 2011.

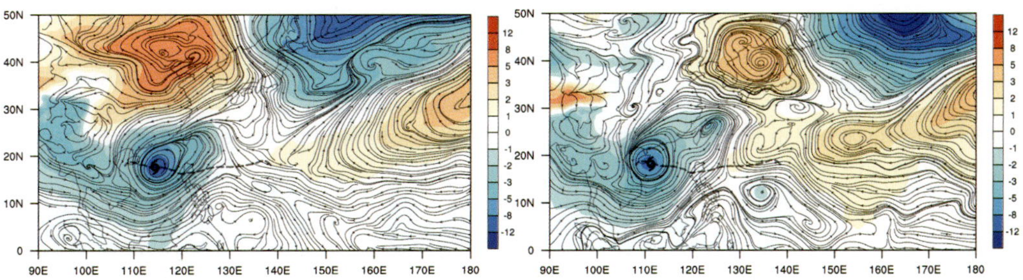

850 hPa flow/height anomalies at 0000 UTC 3 and 4 October 2011.

Figure 2.62. The typhoon (2011_No.22) generated in the tropical western Pacific on 26 September 2011 and moved westward. Its intensity peaked near the climatic monsoon trough in the northern Philippines on 1 October 2011 before landfall on the island and its motion was modified by an anomalous vortex at 850 hPa.

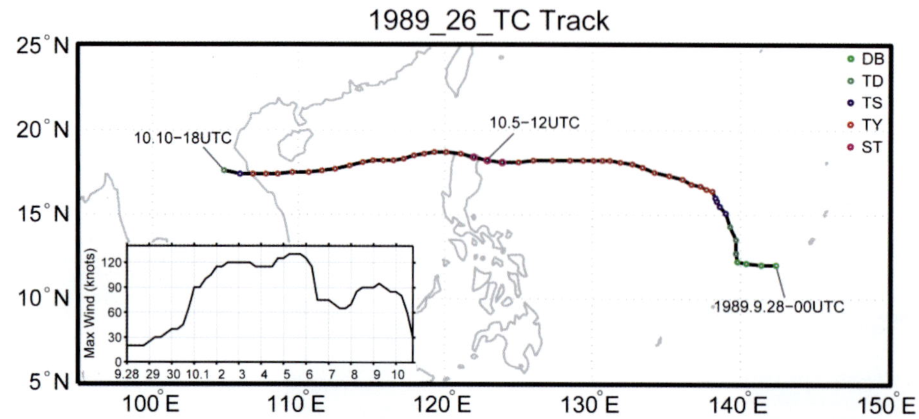

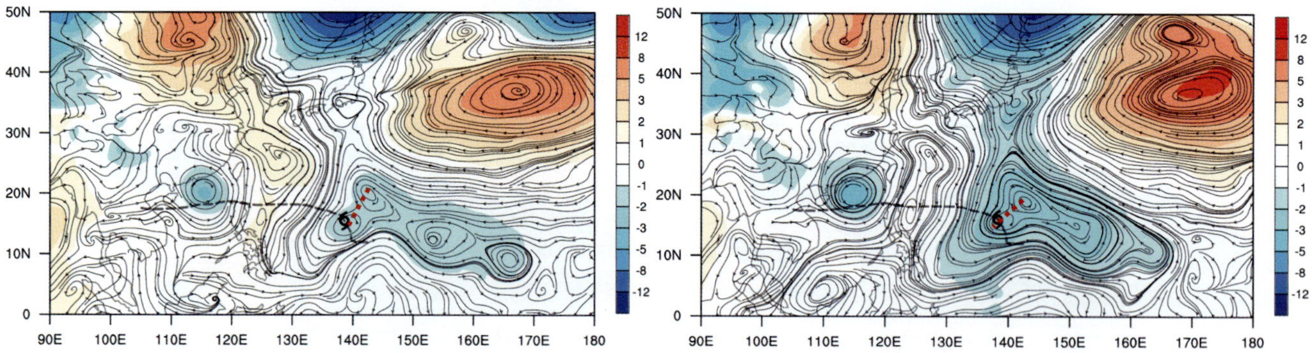

850 hPa flow/height anomalies at 0000 UTC and 1200 UTC 30 September 1989

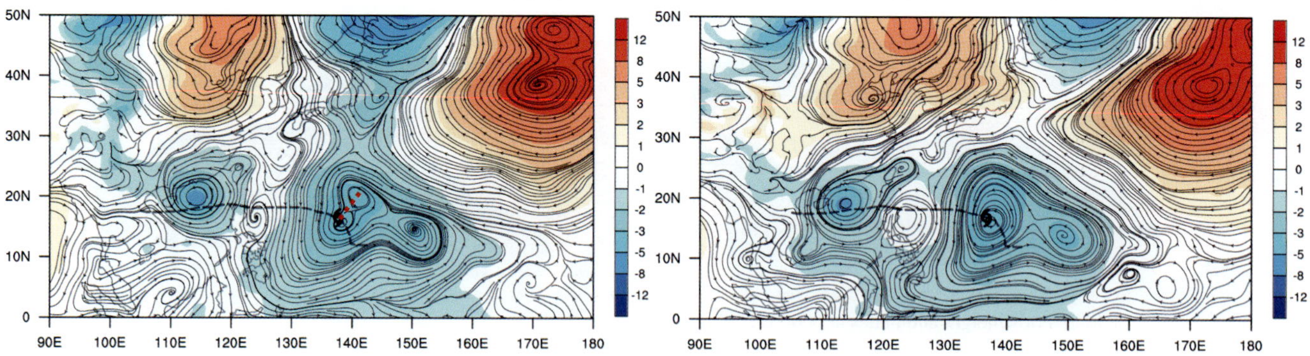

850 hPa flow/height anomalies at 0000 UTC and 1200 UTC 1 October 1989

Figure 2.63. The typhoon (1989_No.26) generated in the tropical western Pacific on 28 September 1989 and its motion was basically along the climatic monsoon trough. Its intensity peaked on 5 October 1989 before landfall on the northern Philippines. The left turning track in the tropical western Pacific was influenced by the interaction and combination with another anomalous vortex at 850 hPa.

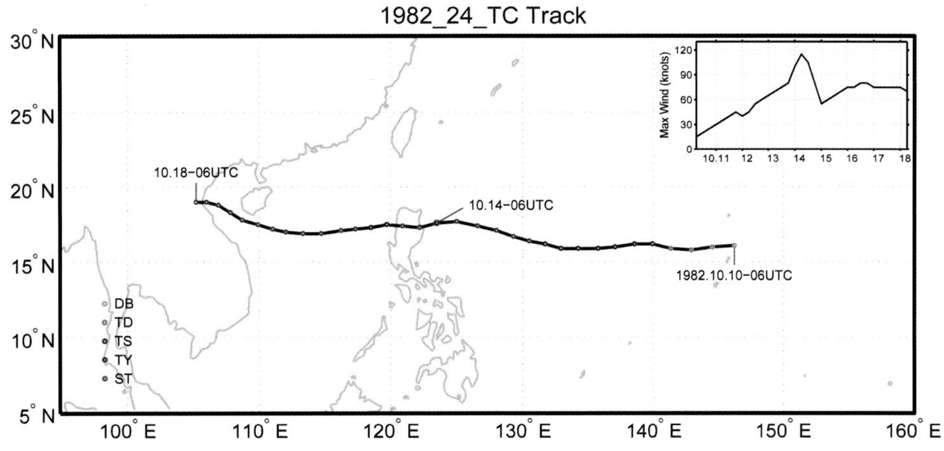

Figure 2.64. The typhoon (1982_No.24) generated in the tropical western Pacific on 10 October 1982 and moved westward basically along the climatic monsoon trough. Its intensity peaked on 14 October 1982 in the northeast sea of the Philippines before landfall on the island.

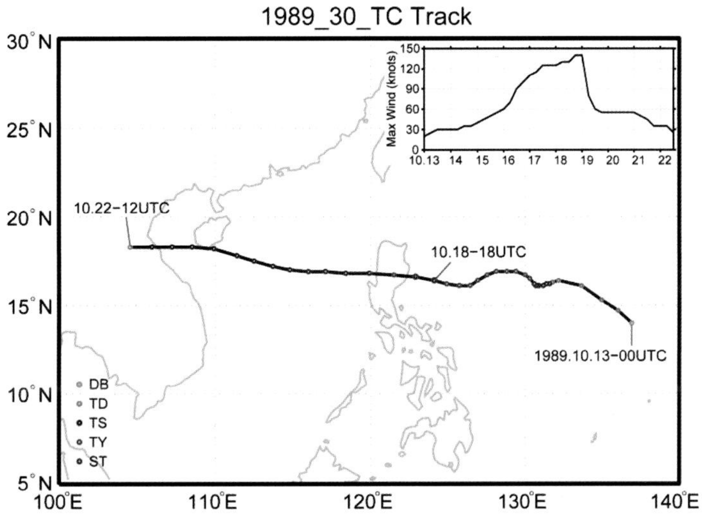

Figure 2.65. The typhoon (1989_No.30) generated in the tropical western Pacific on 13 October 1989 and moved westward near the climatic monsoon trough. Its intensity peaked on 18 October 1989 before landfall on the northern Philippines in the climatic monsoon trough.

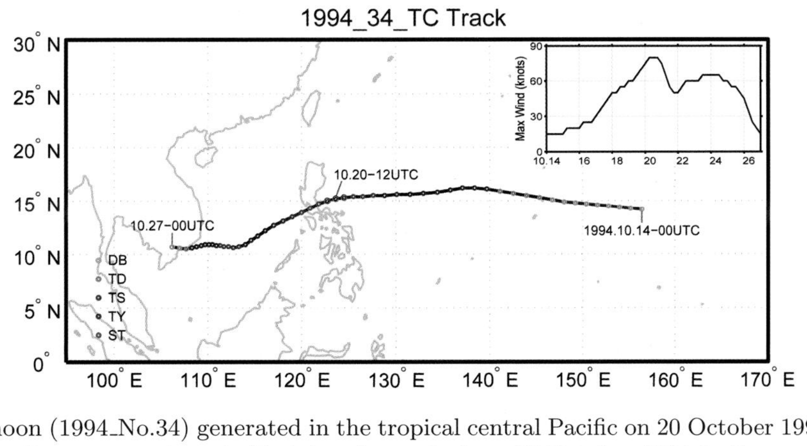

Figure 2.66. The typhoon (1994_No.34) generated in the tropical central Pacific on 20 October 1994 and moved westward basically along the climatic monsoon trough. Its intensity peaked on 20 October 1994 in the east sea of the Philippines before landfall on the island. It re-strengthened in the SCS and dissipated when it made landfall on the south coast of Vietnam.

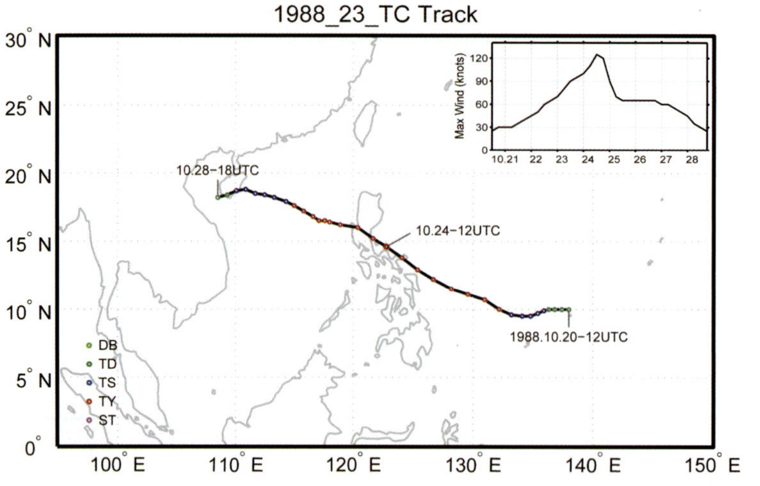

Figure 2.67. The typhoon (1988_No.23) generated in the tropical western Pacific on 20 October 1988 and moved northwestward. Its intensity peaked on 24 October 1988 when it was approaching the climatic monsoon trough before landfall on the northern Philippines.

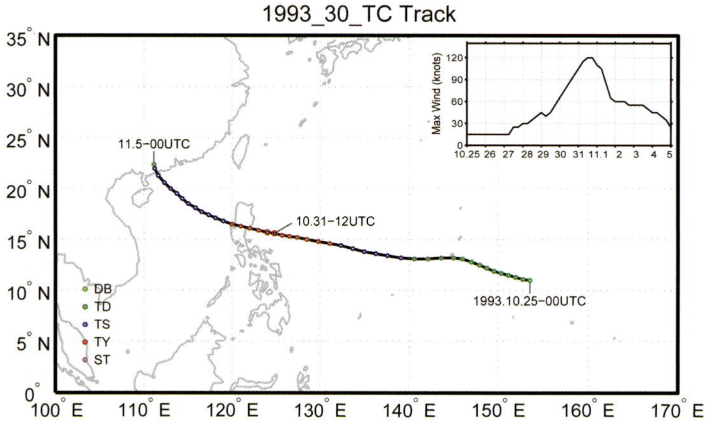

Figure 2.68. The typhoon (1993_No.30) generated in the tropical central Pacific on 25 October 1993 and moved west-northwestward along the climatic monsoon trough. Its intensity peaked on 31 October 1993 in the east sea of the northern Philippines before landfall on the island.

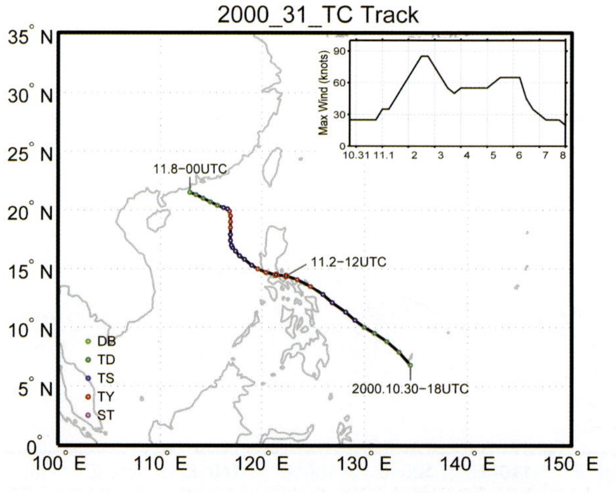

Figure 2.69. The typhoon (2000_No.31) generated in the equatorial western Pacific on 30 October 2000 and moved northwestward. Its intensity peaked near the northern Philippines on 2 November 2000 when it was approaching the climatic monsoon trough before landfall on the island. It re-strengthened in the SCS and dissipated when it was approaching the South China coast.

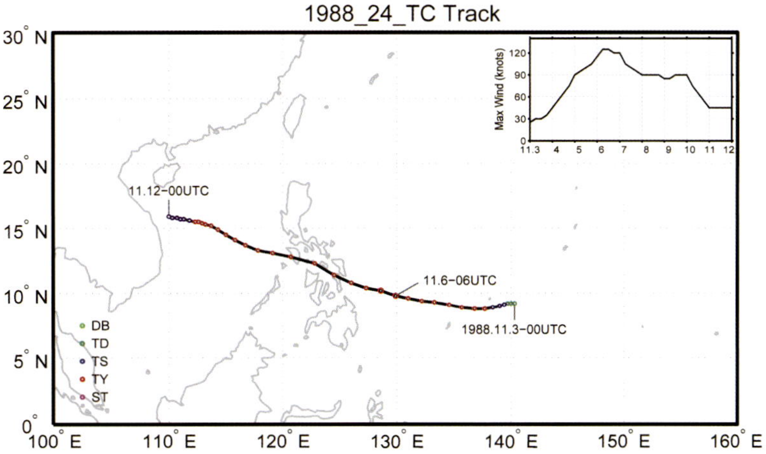

Figure 2.70. The typhoon (1988_No.24) generated in the equatorial western Pacific on 3 November 1988 and moved west-northwestward basically along the climatic monsoon trough. Its intensity peaked on 6 November 1988 in the east sea of the Philippines before landfall on the island.

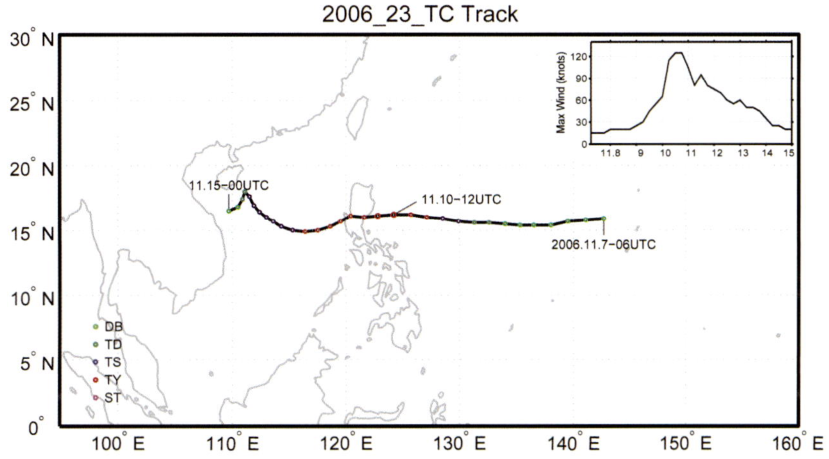

Figure 2.71. The typhoon (2006_No.23) generated in the tropical western Pacific on 7 November 2006 and moved westward basically along the climatic monsoon trough. Its intensity peaked in the northeast sea of the northern Philippines on 10 November 2006 before landfall on the island.

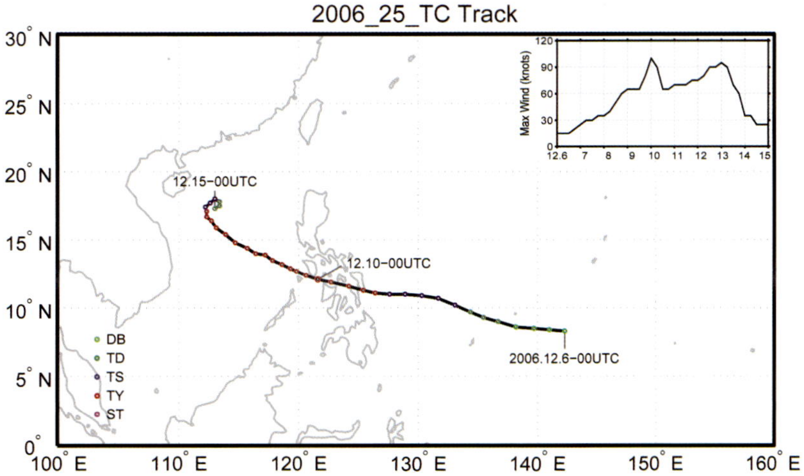

Figure 2.72. The typhoon (2006_No.25) generated in the equatorial western Pacific on 6 December 2006 and moved west-northwestward basically along the climatic monsoon trough. Its intensity peaked on 10 December 2006 in the central Philippines. It then weakened in the central SCS and dissipated after a stagnant state in the northern SCS.

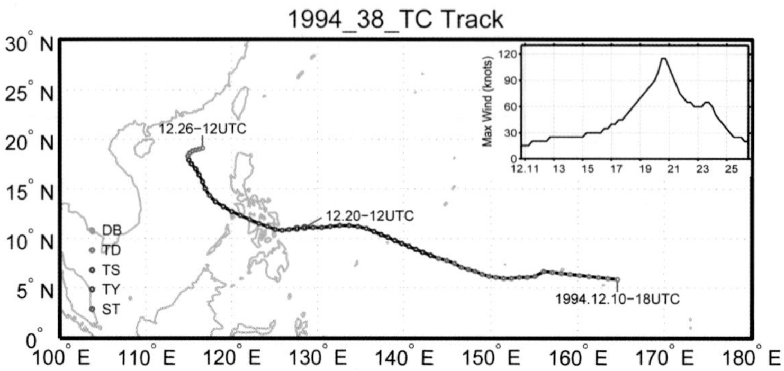

Figure 2.73. The typhoon (1994_No.38) generated in the equatorial central Pacific on 10 December 1994, moving west-northwestward basically along the climatic monsoon trough, and its intensity peaked on 20 December 1994 in the central east sea of the Philippines before landfall on the island. It weakened in the central SCS and dissipated after a stagnant state in the northern SCS.

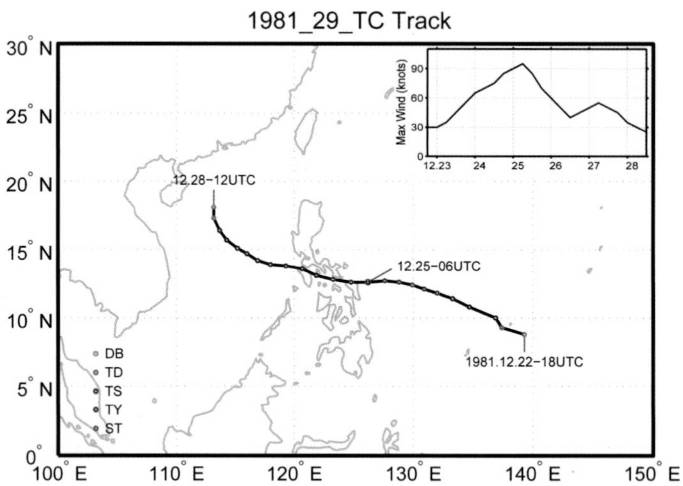

Figure 2.74. The typhoon (1981_No.29) generated in the equatorial western Pacific on 22 December 1981, moving west-northwestward basically along the climatic monsoon trough, and its intensity peaked on 25 December 1981 before landfall on the central Philippines. It weakened in the central SCS and dissipated in the northern SCS.

2.4. Intensity Peaked in the East Seas of China

This section plots 21 typhoons with their intensities peaked in the east seas of China. Their tracks and intensities are related to climatic monsoon troughs and modified by anomalous vortices (Figs. 2.75–2.95).

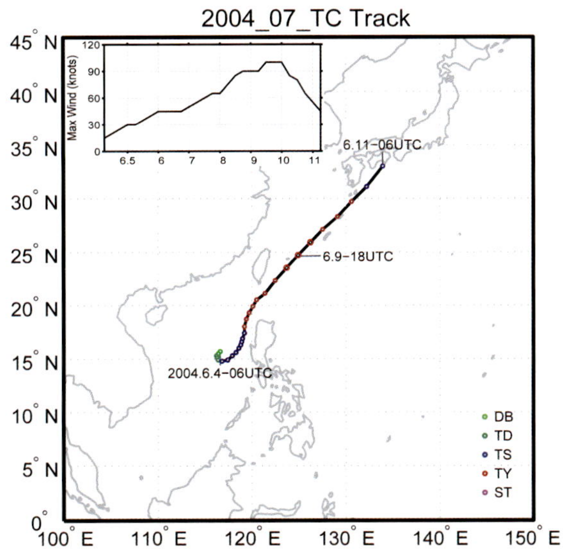

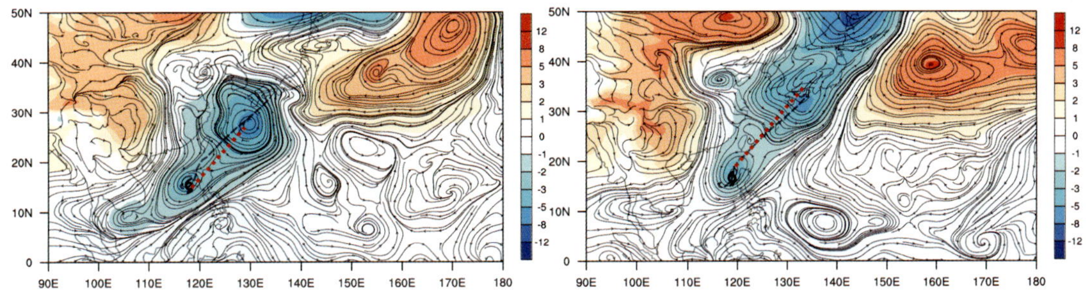

850 hPa flow/height (shading, 10 gpm) anomalies at 0000 UTC 6 and 7 June 2004.

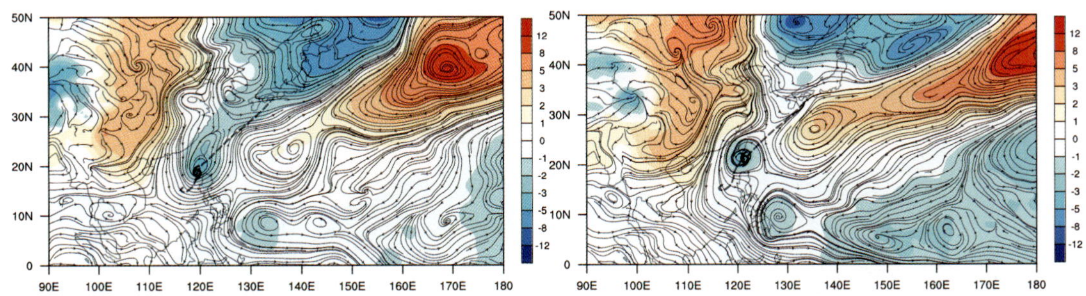

850 hPa flow/height anomalies at 0000 UTC 8 and 9 June 2004.

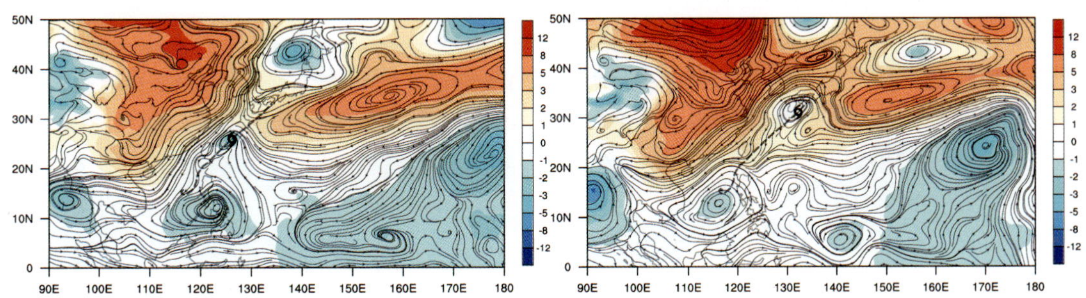

850 hPa flow/height anomalies at 0000 UTC 10 and 11 June 2004.

Figure 2.75. The typhoon (2004_No.07) generated from a tropical disturbance in the SCS near the climatic monsoon trough on 4 June 2004 and moved northeastward along the monsoon flow. Its intensity peaked near the climatic subtropical monsoon trough in the East China Sea on 9 June 2004. A stagnant state at the beginning stage was influenced by an anomalous vortex at 850 hPa. It moved northeastward without the influence of anomalous vortices at 850 hPa.

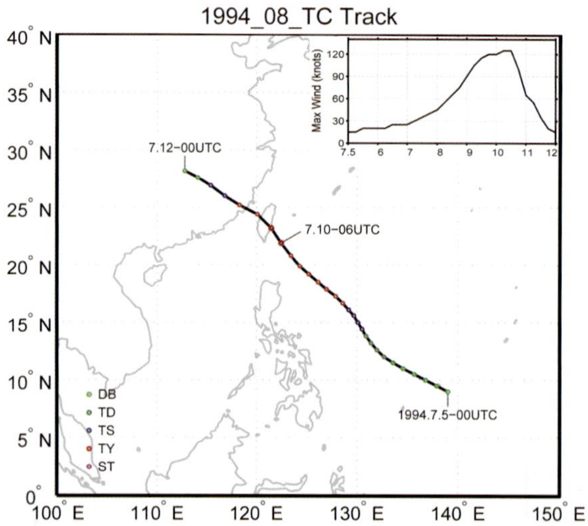

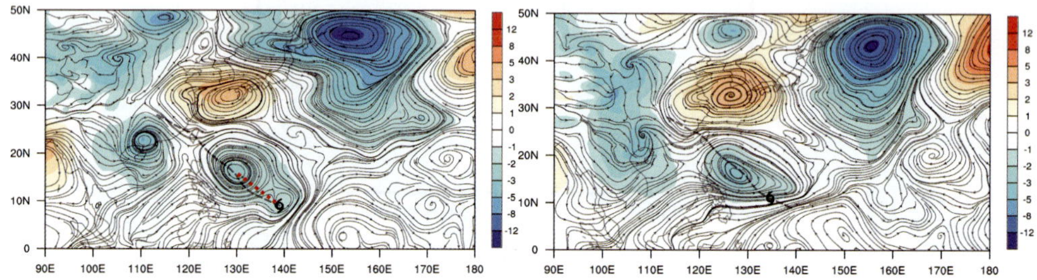

850 hPa flow/height anomalies at 0000 UTC 5 and 6 July 1994.

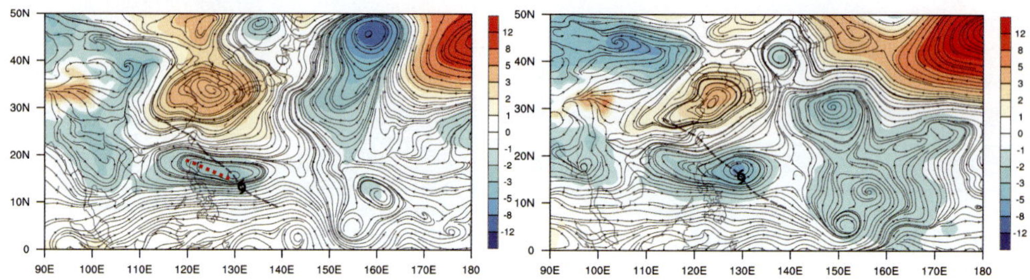

850 hPa flow/height anomalies at 0000 UTC 7 and 8 July 1994.

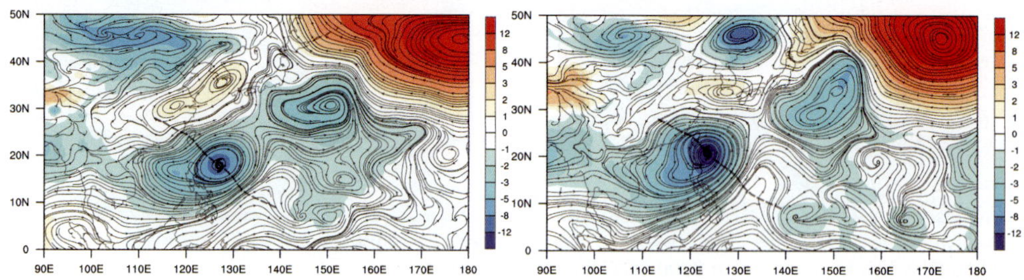

850 hPa flow/height anomalies at 0000 UTC 9 and 10 July 1994.

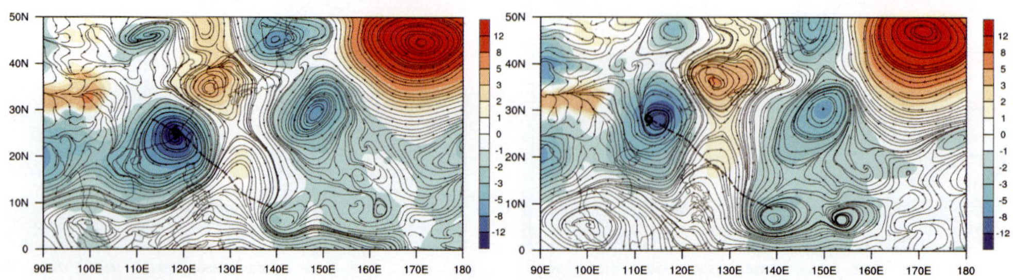

850 hPa flow/height anomalies at 0000 UTC 11 and 12 July 1994.

Figure 2.76. The typhoon (1994_No.08) generated in the equatorial western Pacific on 5 July 1994 and moved northwestward. Its motion was along the climatic monsoon trough so that its intensity peaked on 10 July 1994 before landfall on the Taiwan Island and its motion was modified by an anomalous vortex at 850 hPa.

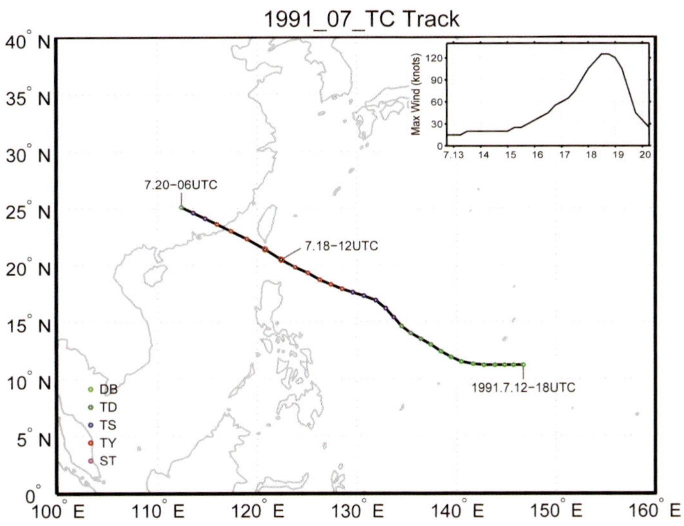

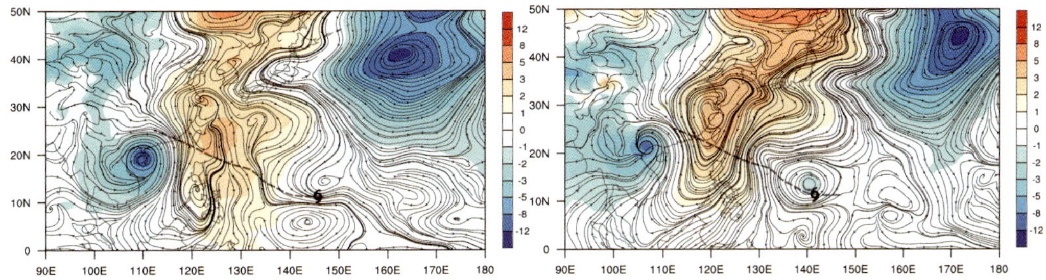

850 hPa flow/height anomalies at 0000 UTC 13 and 14 July 1991.

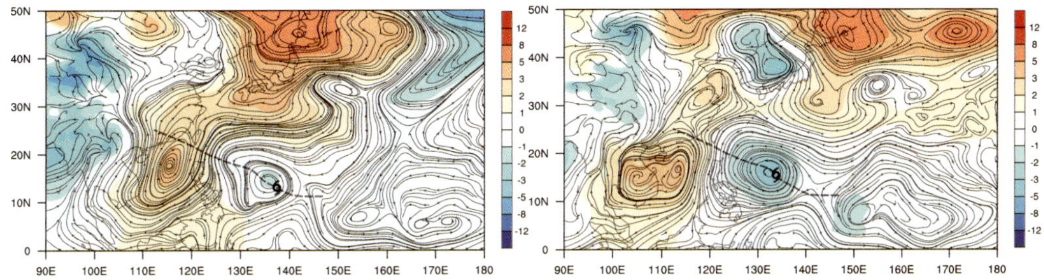

850 hPa flow/height anomalies at 0000 UTC 15 and 16 July 1991.

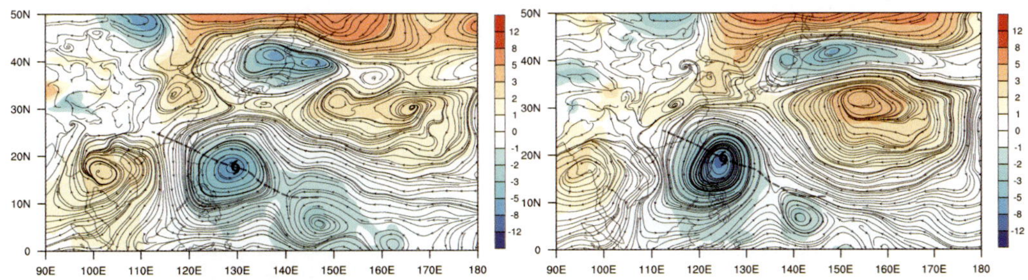

850 hPa flow/height anomalies at 0000 UTC 17 and 18 July 1991.

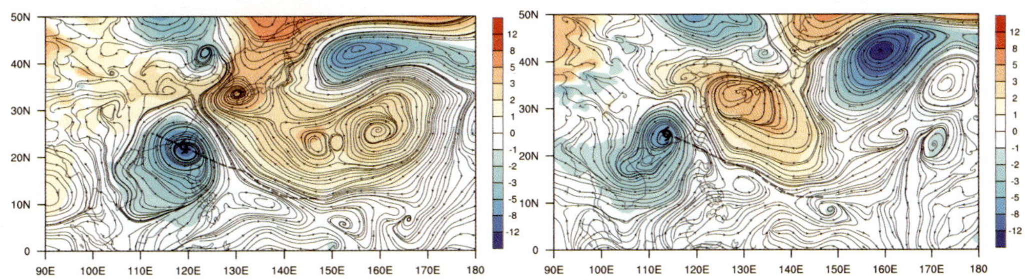

850 hPa flow/height anomalies at 0000 UTC 19 and 20 July 1991.

Figure 2.77. The typhoon (1991_No.07) generated in the tropical western Pacific on 12 July 1991 and moved northwestward along the climatic monsoon trough. Its intensity peaked in the southeast sea of the Taiwan Island on 18 July 1991 before landfall on the South China coast and its motion was not affected by anomalous vortices at 850 hPa.

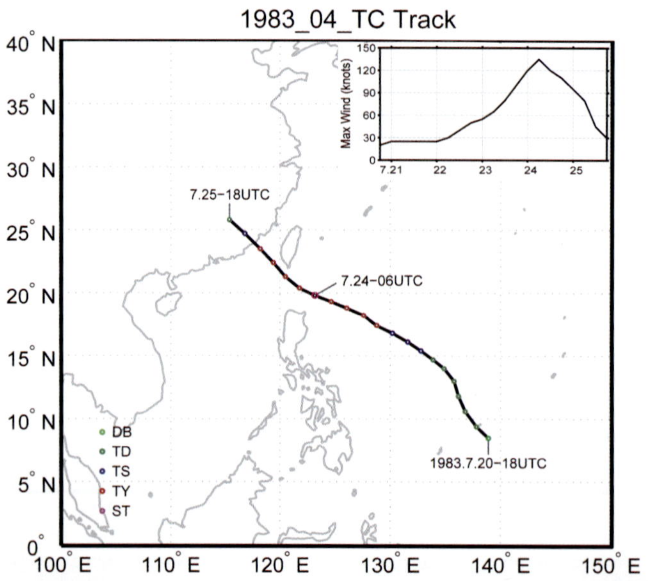

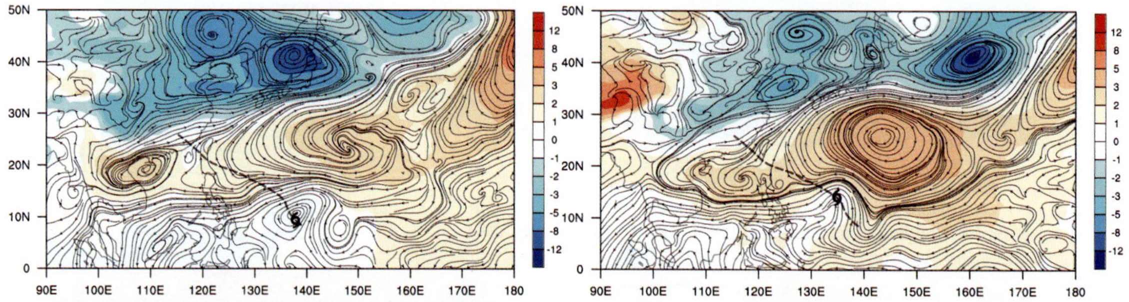

850 hPa flow/height anomalies at 0000 UTC 21 and 22 July 1983.

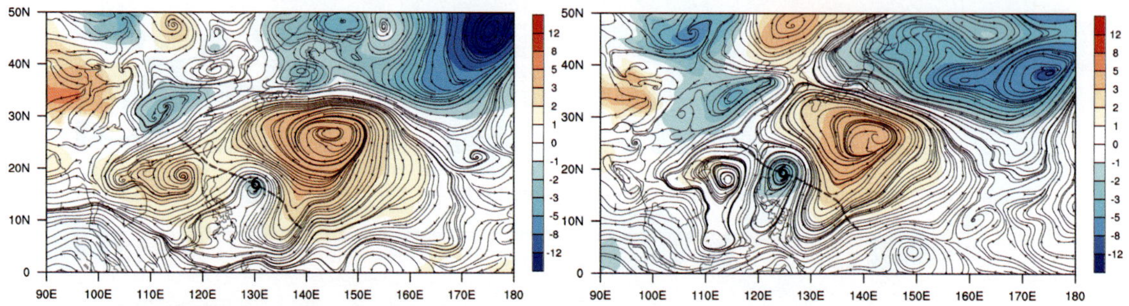

850 hPa flow/height anomalies at 0000 UTC 23 and 24 July 1983.

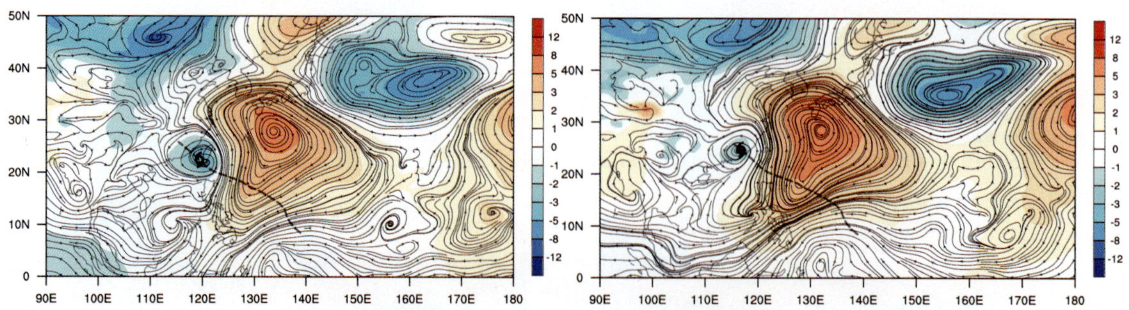

850 hPa flow/height anomalies at 0000 UTC and 1200 UTC 25 July 1983.

Figure 2.78. The typhoon (1983_No.04) generated in the equatorial western Pacific on 20 July 1983 and moved northwestward along the climatic monsoon trough. Its intensity peaked in the southeast sea of the Taiwan Island on 24 July 1983 before landfall on the Southeast China coast and its motion was modified by anomalous flows at 850 hPa.

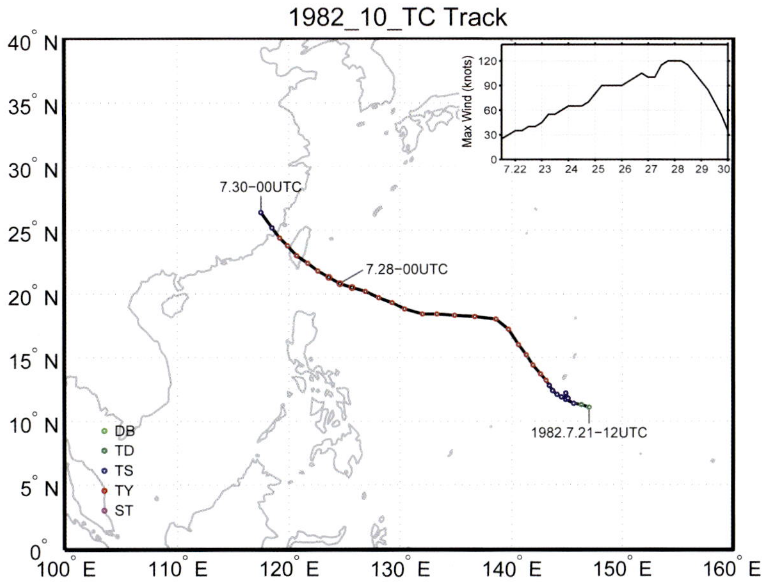

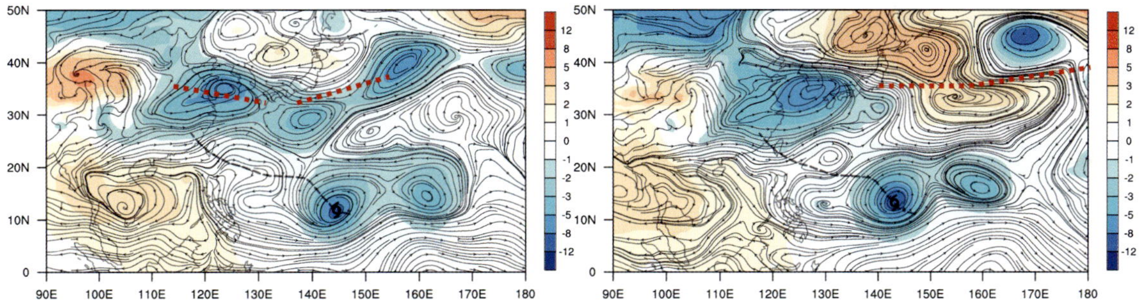

850 hPa flow/height anomalies at 0000 UTC 23 and 24 July 1982.

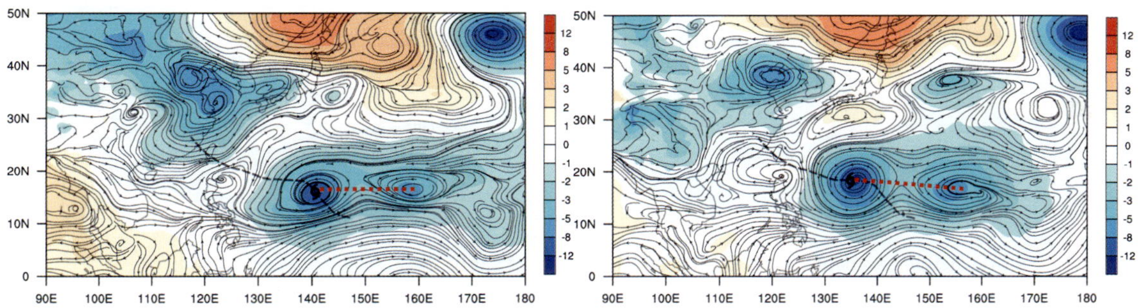

850 hPa flow/height anomalies at 0000 UTC 25 and 26 July 1982.

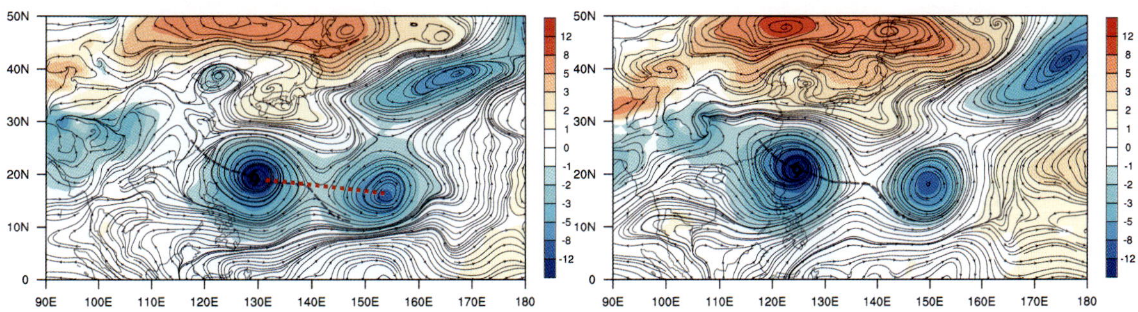

850 hPa flow/height anomalies at 0000 UTC 27 and 28 July 1982.

Figure 2.79. The typhoon (1982_No.10) generated in the tropical western Pacific on 21 July 1982 and moved northwestward along the climatic monsoon trough. Its intensity peaked in the southeast sea of the Taiwan Island on 28 July 1982 before landfall on the Southeast China coast and its track had a small change in the tropical western Pacific due to the influence of anomalous vortices at 850 hPa and 300 hPa.

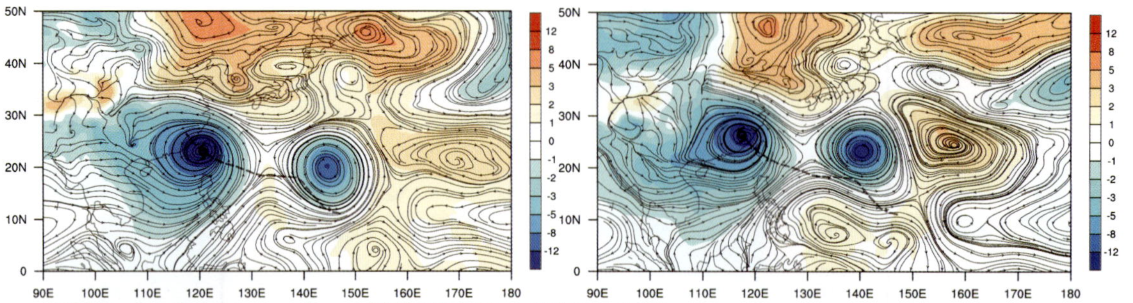

850 hPa flow/height anomalies at 0000 UTC 29 and 30 July 1982.

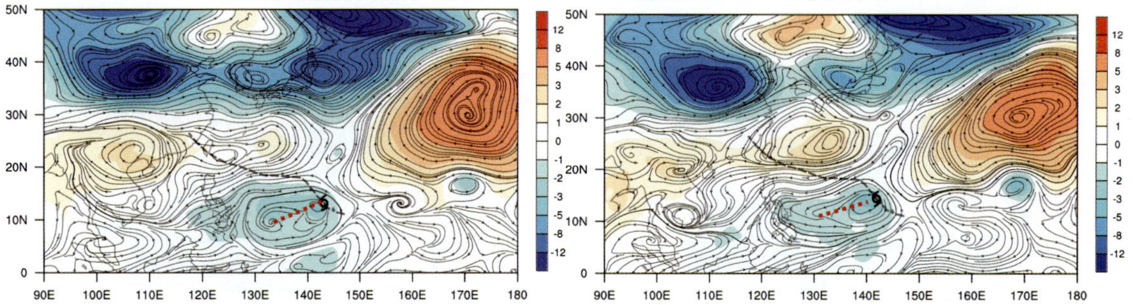

300 hPa flow/height anomalies at 0000 UTC and 1200 UTC 24 July 1982.

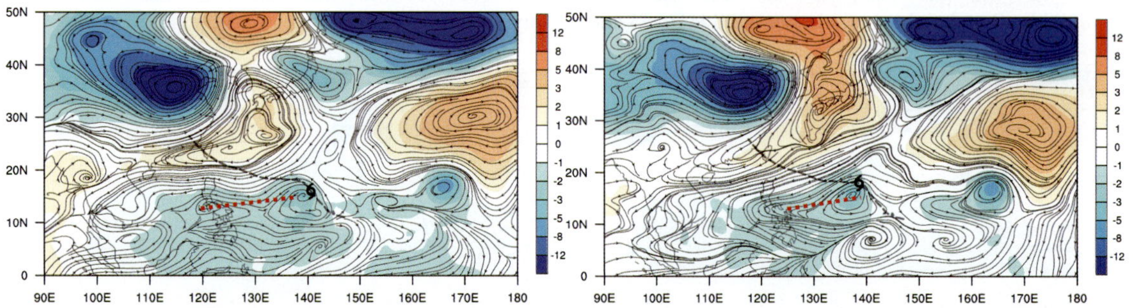

300 hPa flow/height anomalies at 0000 UTC and 1200 UTC 25 July 1982.

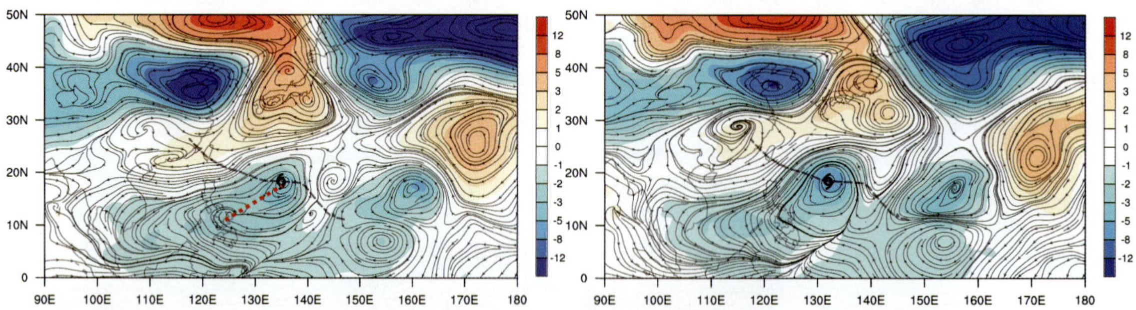

300 hPa flow/height anomalies at 0000 UTC and 1200 UTC 26 July 1982.

Figure 2.79. *(Continued)*

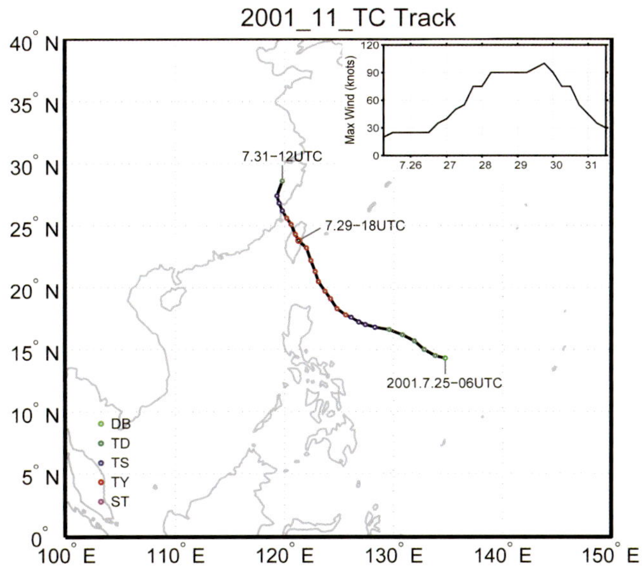

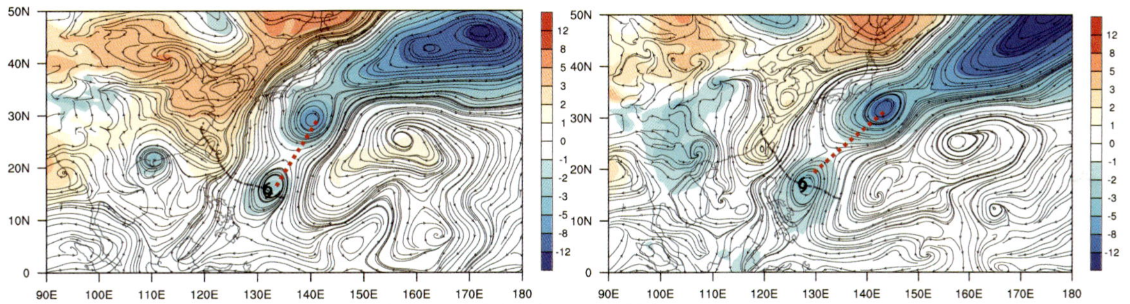

850 hPa flow/height anomalies at 0000 UTC 26 and 27 July 2001.

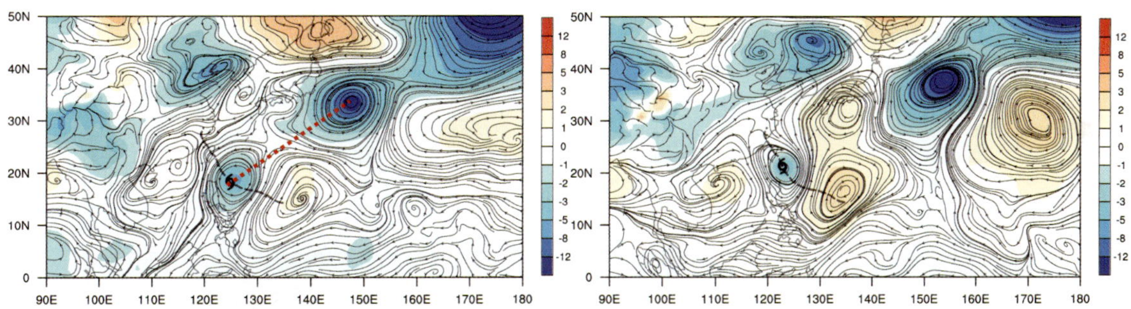

850 hPa flow/height anomalies at 0000 UTC 28 and 29 July 2001.

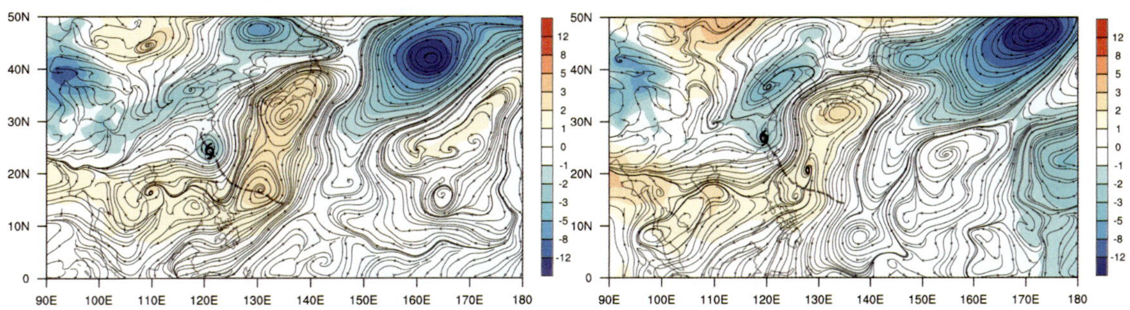

850 hPa flow/height anomalies at 0000 UTC 30 and 31 July 2001.

Figure 2.80. The typhoon (2001_No.11) generated in the tropical western Pacific on 25 July 2001 and moved northwestward along the climatic monsoon trough. Its intensity peaked over the Taiwan Island on 29 July 2001 before landfall on the Southeast China coast and its track was modified by an anomalous vortex at 850 hPa.

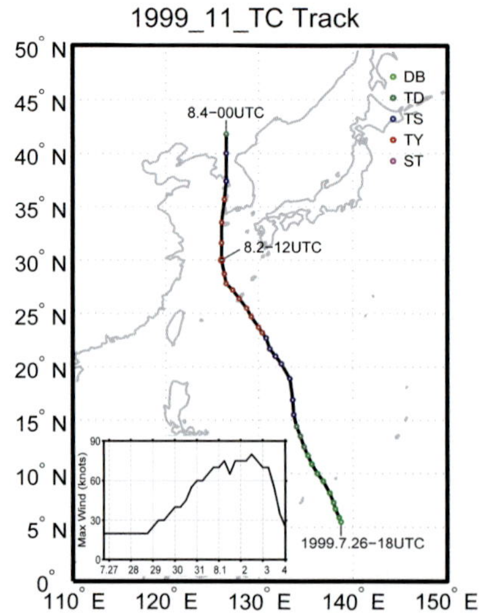

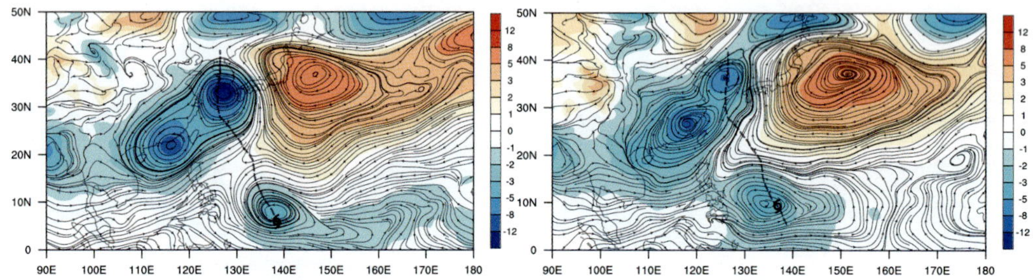

850 hPa flow/height anomalies at 0000 UTC 27 and 28 July 1999.

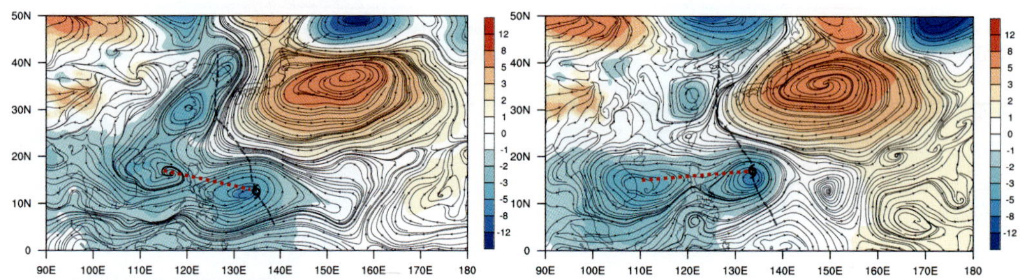

850 hPa flow/height anomalies at 0000 UTC 29 and 30 July 1999.

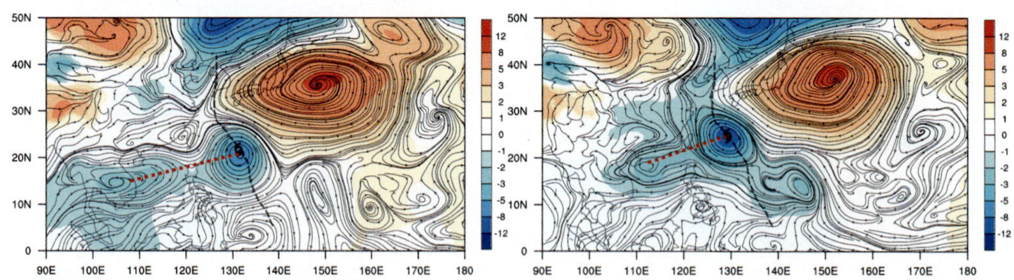

850 hPa flow/height anomalies at 0000 UTC 31 July and 1 August 1999.

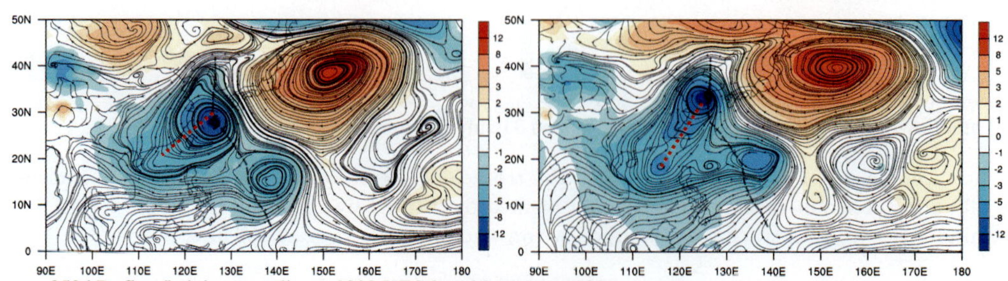

850 hPa flow/height anomalies at 0000 UTC 2 and 3 August 1999.

Figure 2.81. The typhoon (1999_No.11) generated in the equatorial western Pacific on 26 July 1999. Its intensity peaked near the climatic monsoon trough in the East China Sea on 2 August 1999 before landfall on the South Korea coast and its motion was modified by an anomalous vortex at 850 hPa.

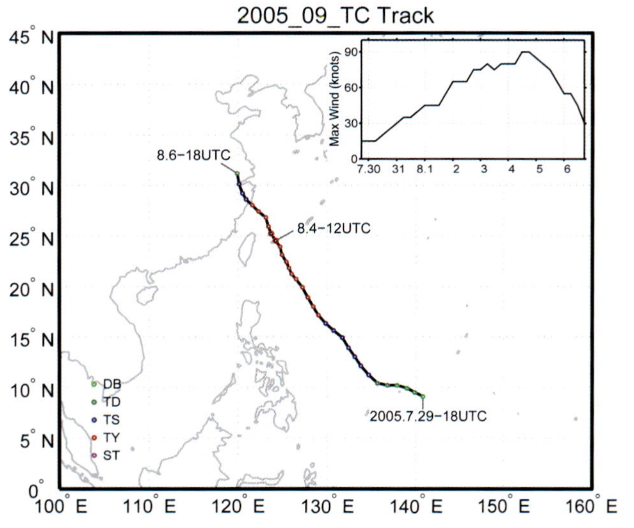

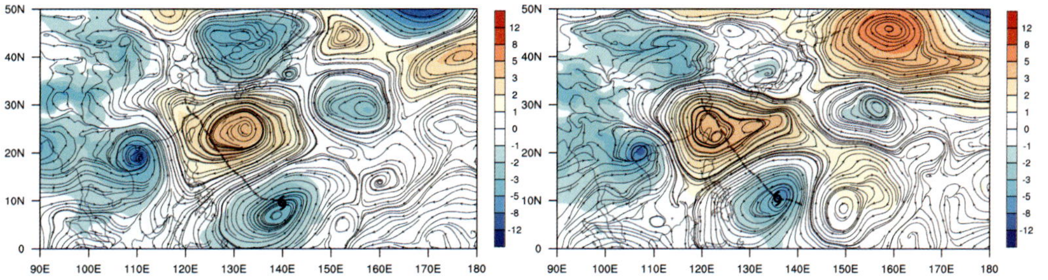

850 hPa flow/height anomalies at 0000 UTC 30 and 31 July 2005.

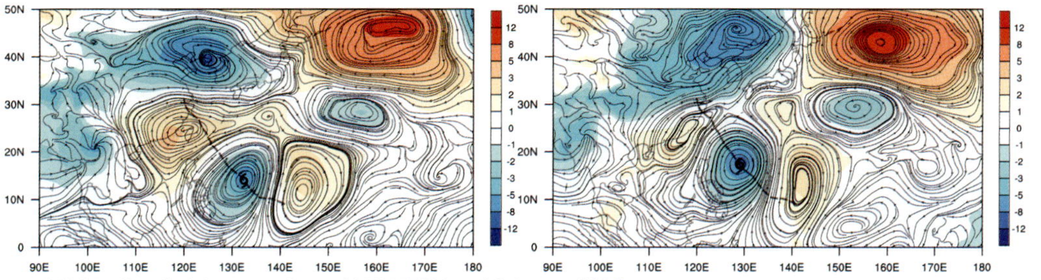

850 hPa flow/height anomalies at 0000 UTC 1 and 2 August 2005.

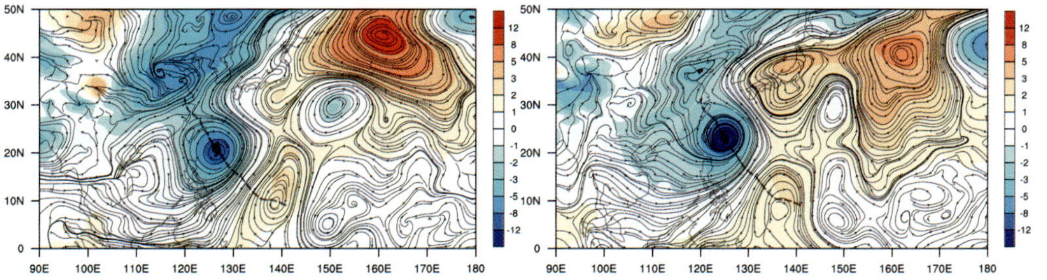

850 hPa flow/height anomalies at 0000 UTC 3 and 4 August 2005.

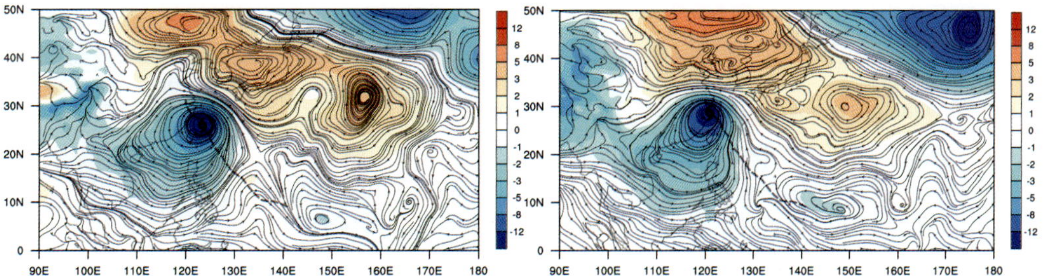

850 hPa flow/height anomalies at 0000 UTC 5 and 6 August 2005.

Figure 2.82. The typhoon (2005_No.09) generated in the equatorial western Pacific on 29 July 2006 and moved north-northwestward along the climatic monsoon trough. Its intensity peaked in the east sea of the Taiwan Island on 4 August 2005 before landfall on the East China coast and its motion was not affected by anomalous vortices at 850 hPa.

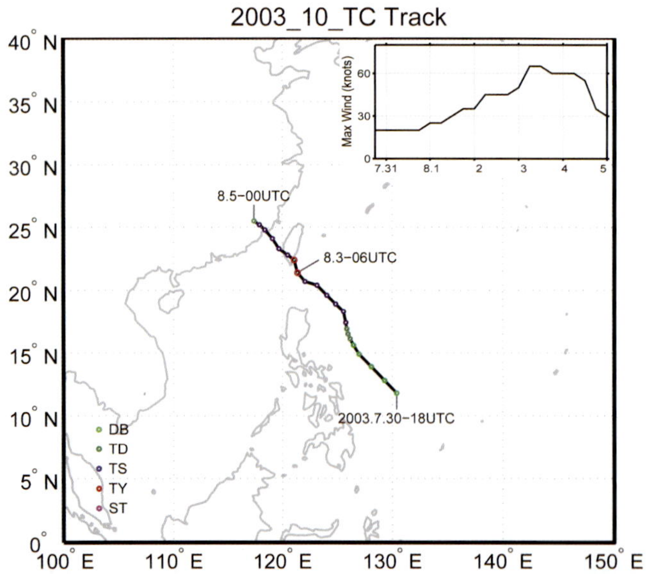

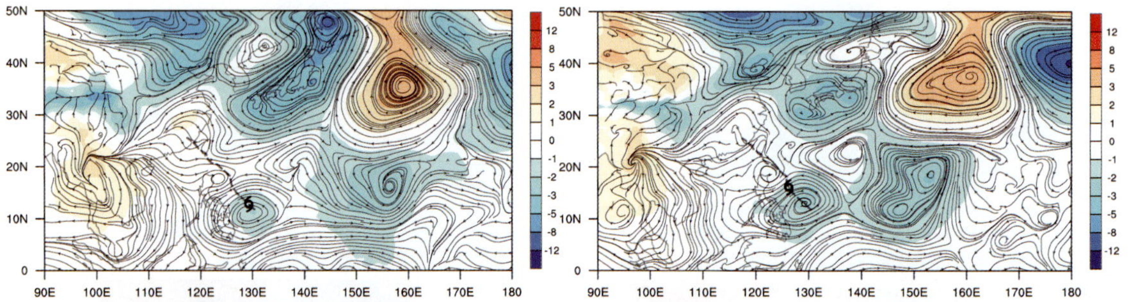

850 hPa flow/height anomalies at 0000 UTC 31 July and 1 August 2005.

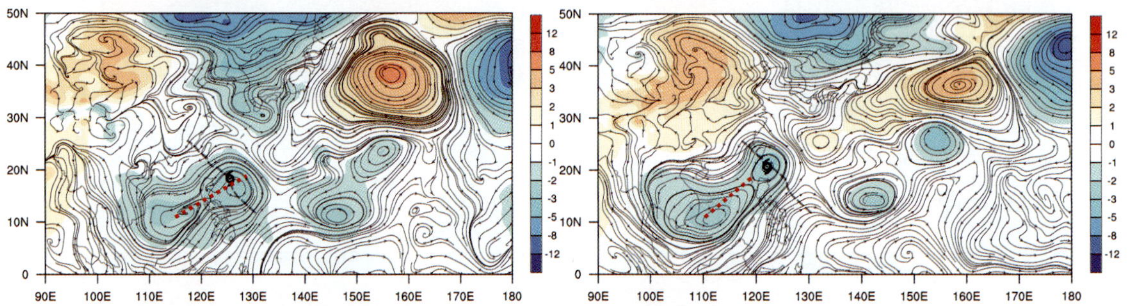

850 hPa flow/height anomalies at 0000 UTC 2 and 3 August 2005.

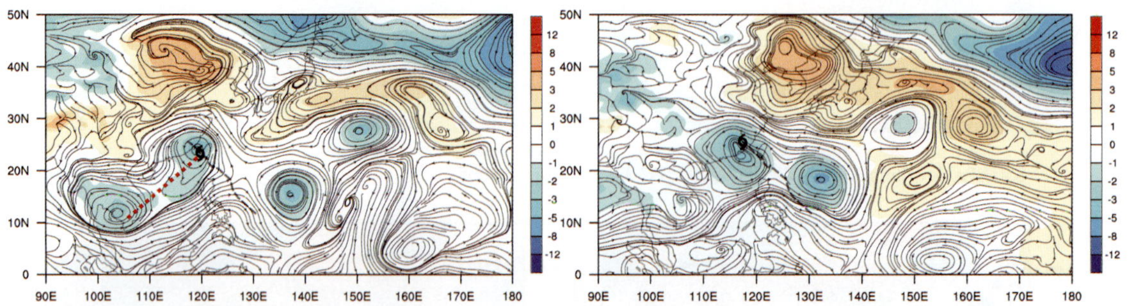

850 hPa flow/height anomalies at 0000 UTC 4 and 5 August 2005.

Figure 2.83. The typhoon (2003_No.10) generated in the east sea of the Philippines on 30 July 2003 and moved north-northwestward. Its intensity peaked in the northern Balintang Channel when it was approaching the climatic monsoon trough and weakened when it made landfall on the Taiwan Island. Its motion was modified by an anomalous vortex at 850 hPa.

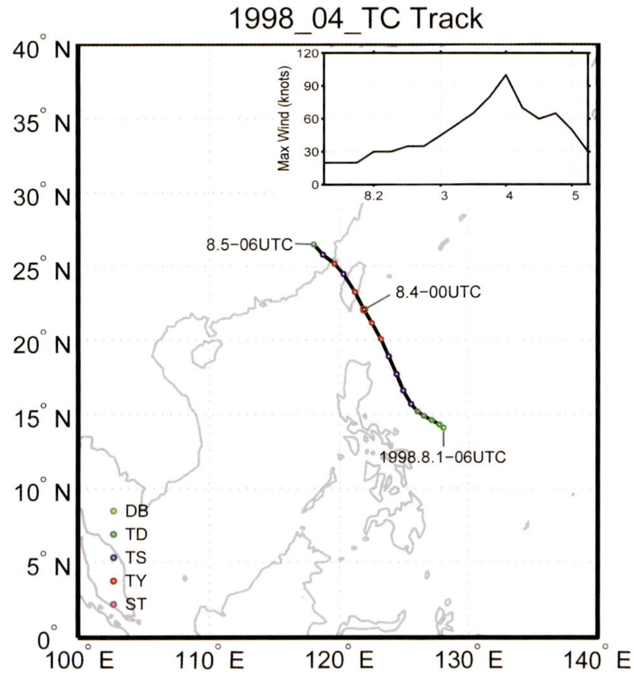

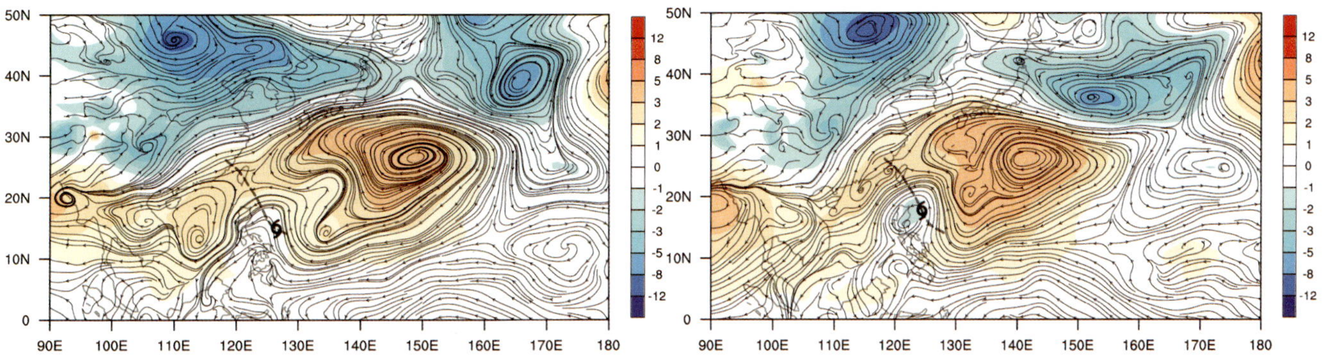

850 hPa flow/height anomalies at 0000 UTC 2 and 3 August 1998.

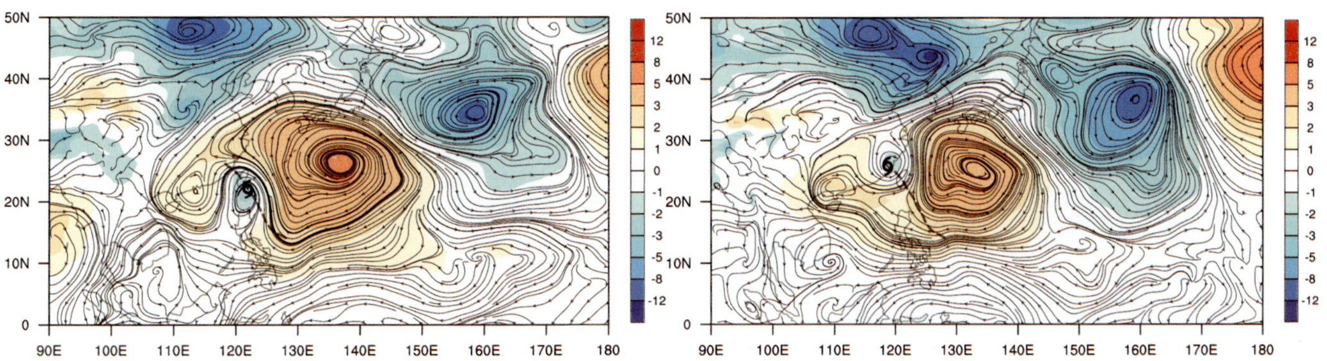

850 hPa flow/height anomalies at 0000 UTC 4 and 5 August 1998.

Figure 2.84. The typhoon (1998_No.04) generated in the east sea of the Philippines on 1 August 1998 and moved north-northwestward without the influence of anomalous vortices at 850 hPa. Its intensity peaked on 4 August 1998 when it was approaching the climatic monsoon trough before landfall on the Taiwan Island and the Southeast China coast.

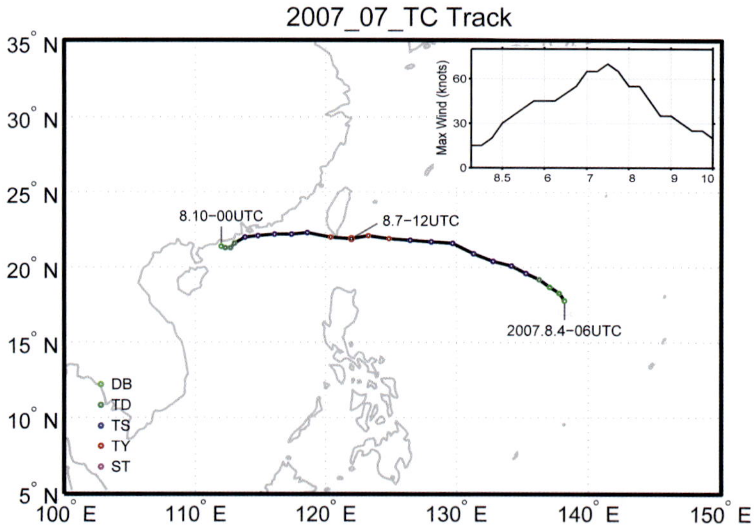

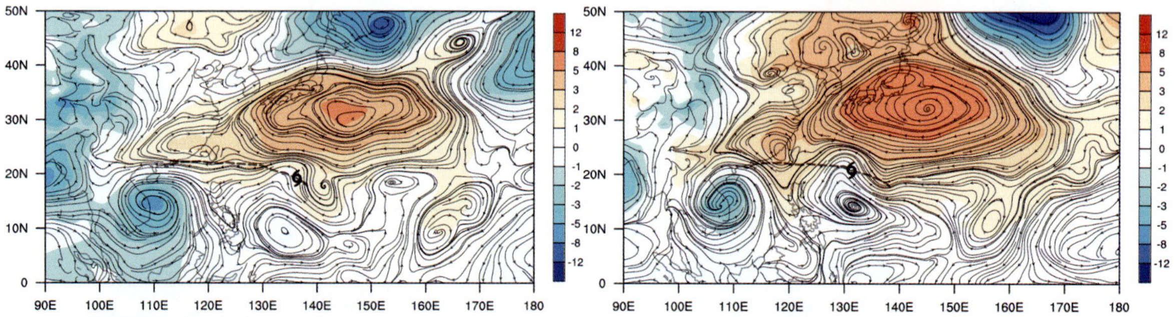

850 hPa flow/height anomalies at 0000 UTC 5 and 6 August 2007.

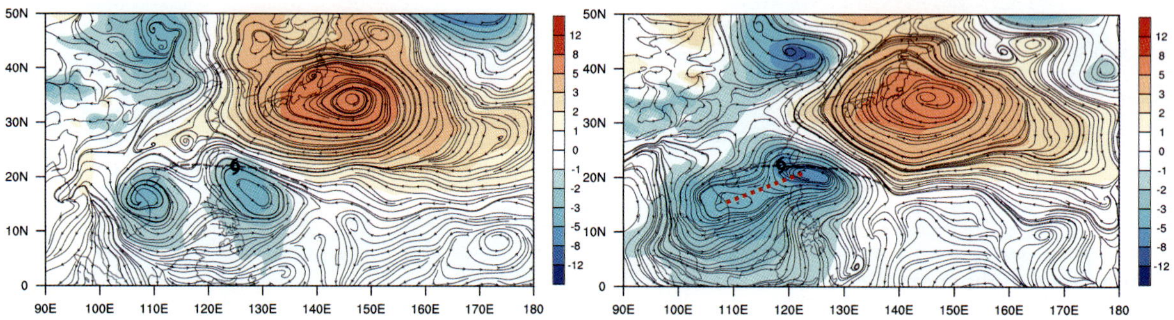

850 hPa flow/height anomalies at 0000 UTC 7 and 8 August 2007.

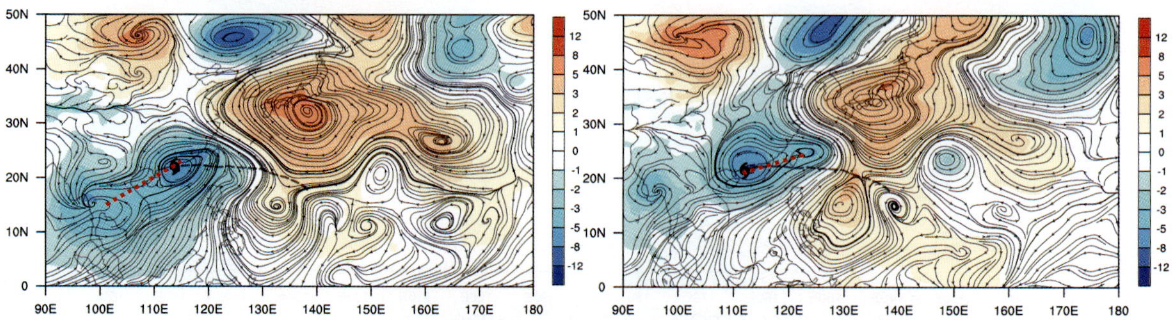

850 hPa flow/height anomalies at 0000 UTC 9 and 10 August 2007.

Figure 2.85. The typhoon (2007_No.07) generated in the tropical western Pacific on 4 August 2007 and moved westward. Its intensity peaked in the south sea of the Taiwan Island on 7 August 2007 when it was approaching the climatic monsoon trough and its motion was modified by anomalous vortices at 850 hPa.

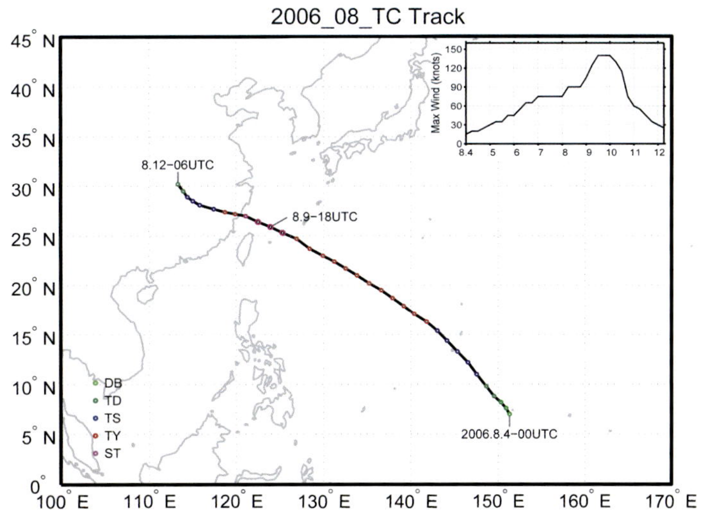

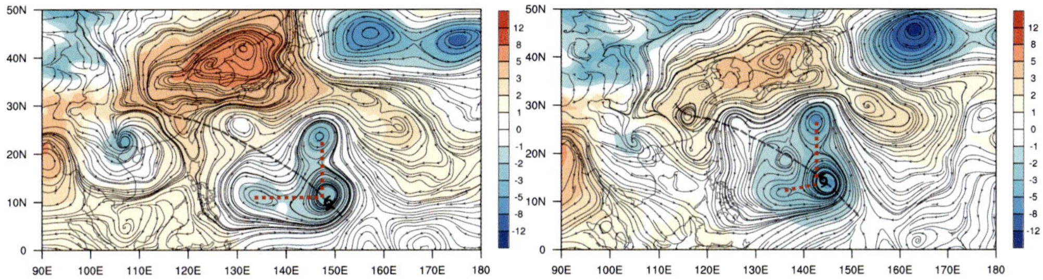

850 hPa flow/height anomalies at 0000 UTC 5 and 6 August 2006.

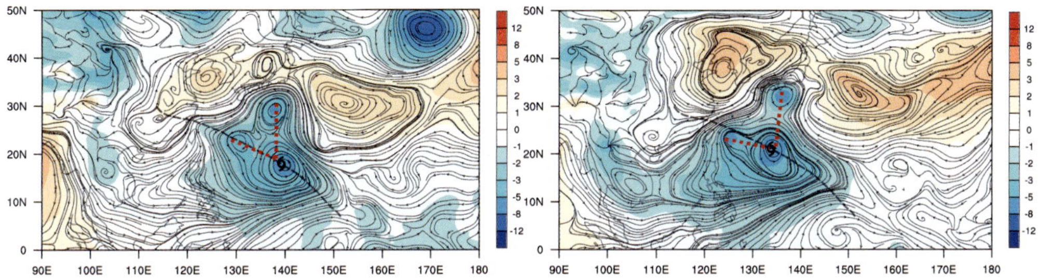

850 hPa flow/height anomalies at 0000 UTC 7 and 8 August 2006.

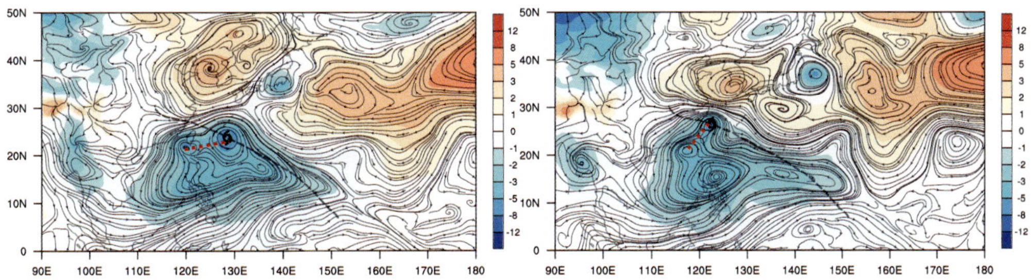

850 hPa flow/height anomalies at 0000 UTC 9 and 10 August 2006.

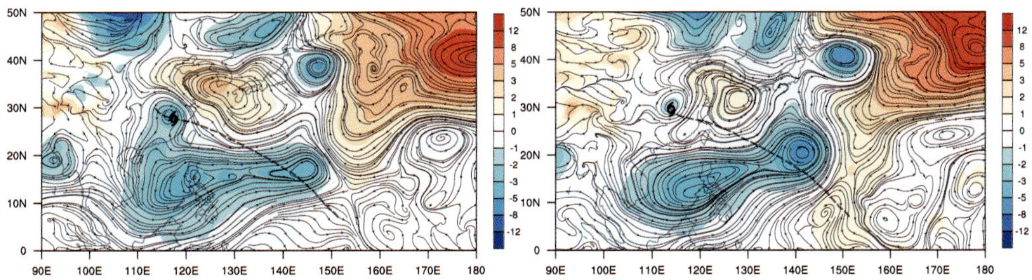

850 hPa flow/height anomalies at 0000 UTC 11 and 12 August 2006.

Figure 2.86. The typhoon (2006_No.08) generated in the equatorial western Pacific on 4 August 2006 and moved northwestward along the climatic monsoon trough. Its intensity peaked in the East China Sea on 9 August 2006 before landfall on the Southeast China coast and its motion was modified by two anomalous vortices at 850 hPa.

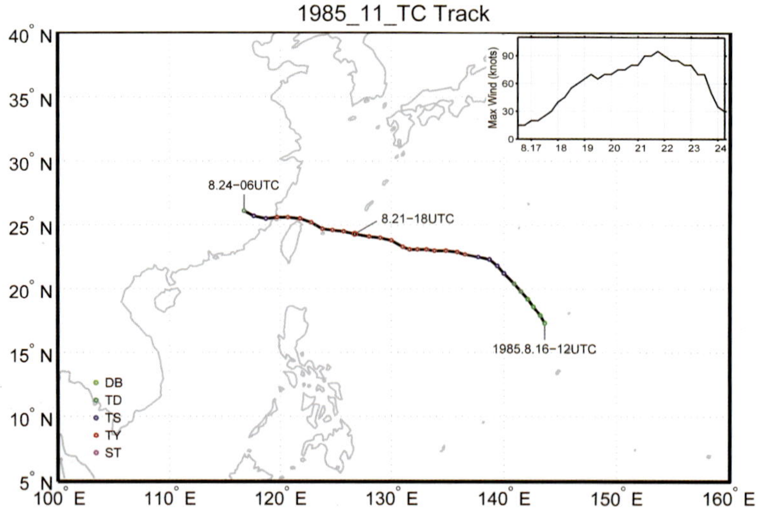

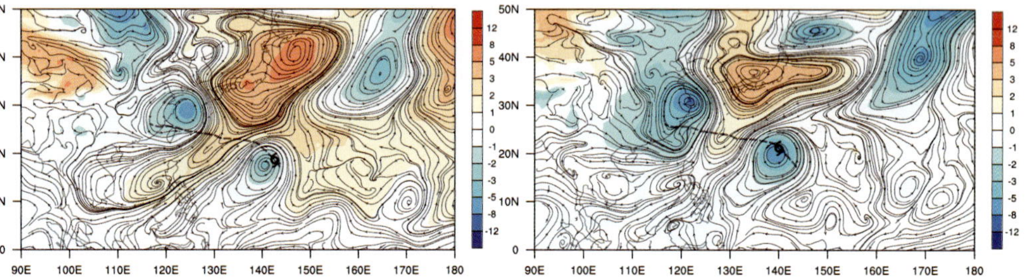

850 hPa flow/height anomalies at 0000 UTC 17 and 18 August 1985.

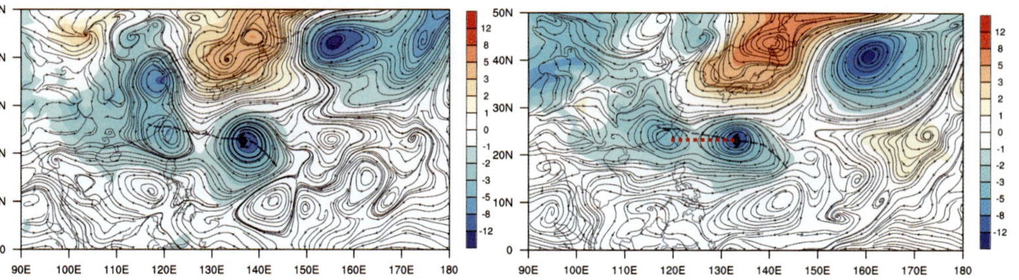

850 hPa flow/height anomalies at 0000 UTC 19 and 20 August 1985.

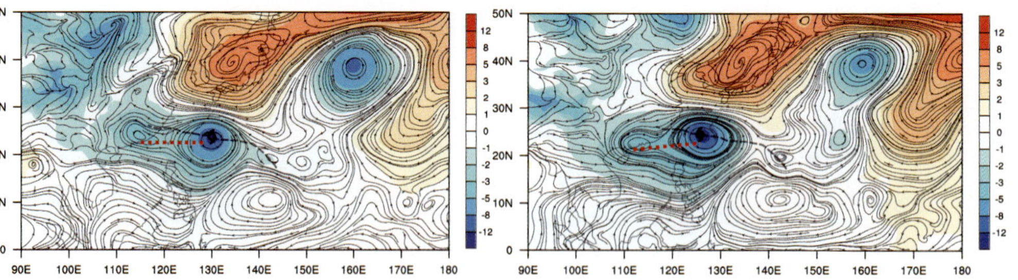

850 hPa flow/height anomalies at 0000 UTC 21 and 22 August 1985.

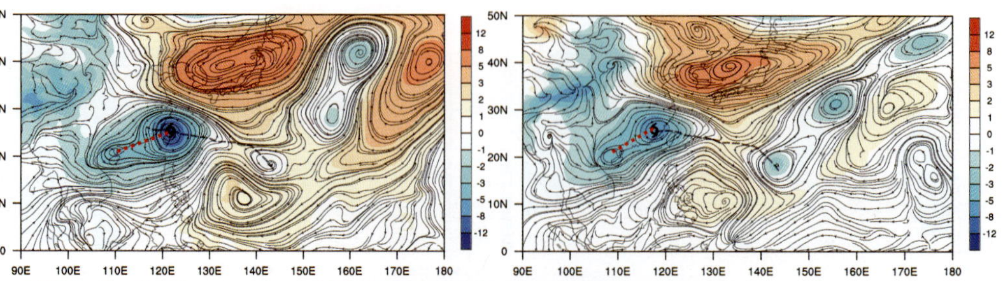

850 hPa flow/height anomalies at 0000 UTC 23 and 24 August 1985.

Figure 2.87. The typhoon (1985_No.11) generated in the tropical western Pacific on 16 August 1985 and moved west-northwestward along the climatic monsoon trough. Its motion was modified by an anomalous vortex at 850 hPa and its intensity peaked in the east sea of the Taiwan Island on 21 August 1985 before landfall on the Southeast China coast.

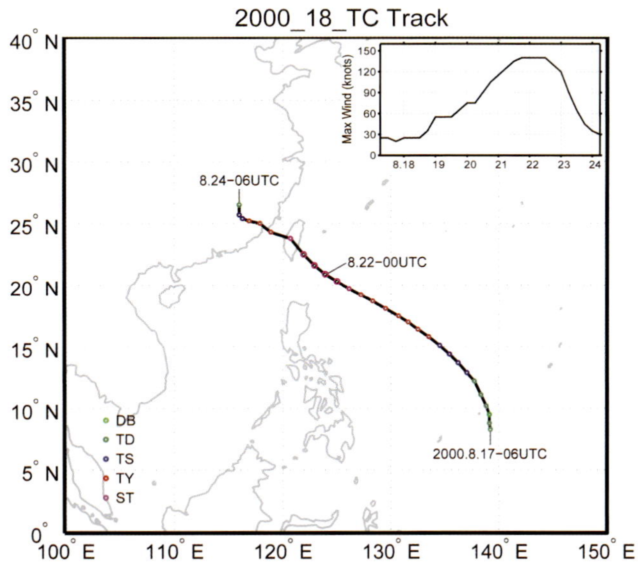

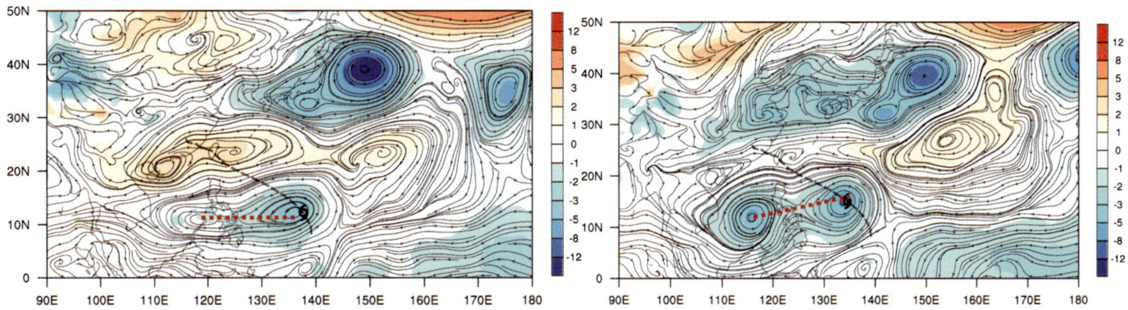

850 hPa flow/height anomalies at 1200 UTC 18 and 19 August 2000.

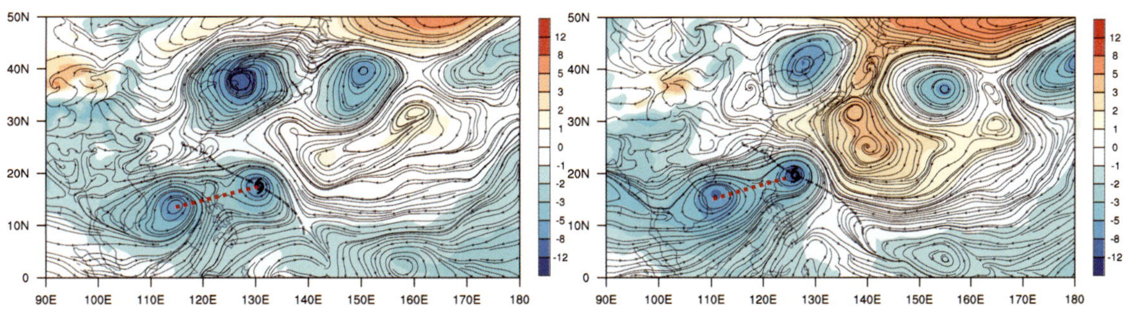

850 hPa flow/height anomalies at 1200 UTC 20 and 21 August 2000.

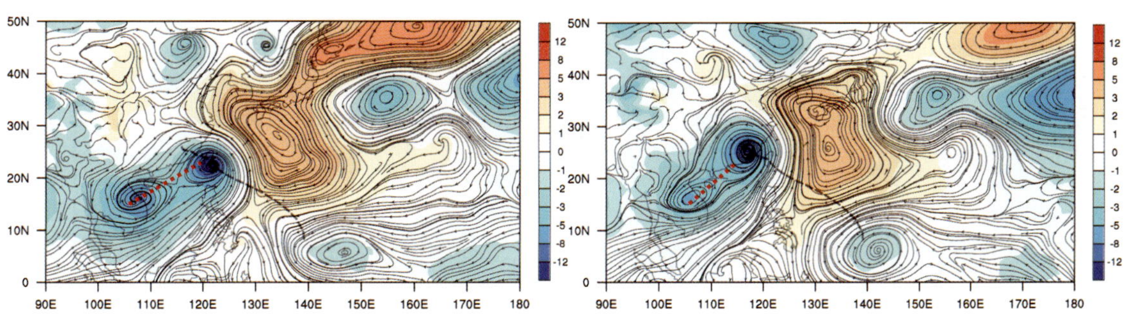

850 hPa flow/height anomalies at 1200 UTC 22 and 23 August 2000.

Figure 2.88. The typhoon (2000_No.18) generated in the equatorial western Pacific on 17 August 2000 and moved northwestward basically along the climatic monsoon trough. Its intensity peaked on 22 August 2008 before landfall on the Taiwan Island. Its track with small change was modified by an anomalous vortex at 850 hPa.

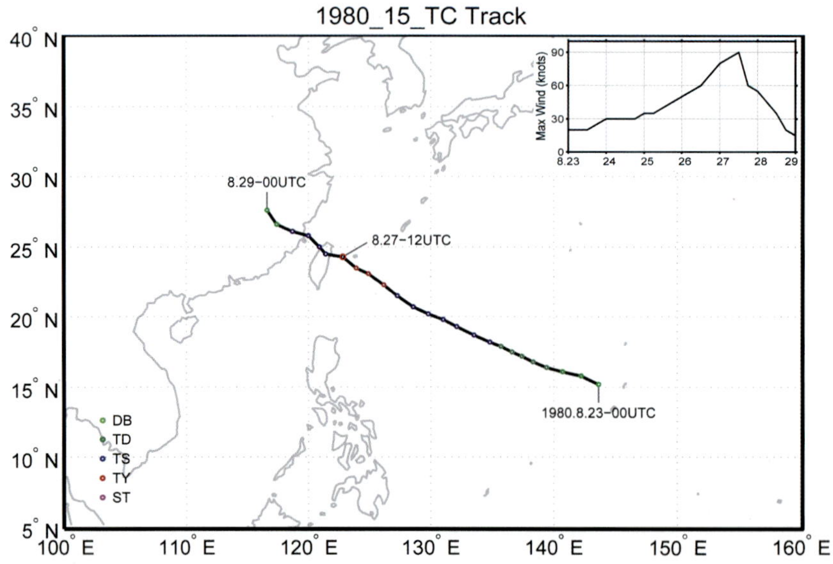

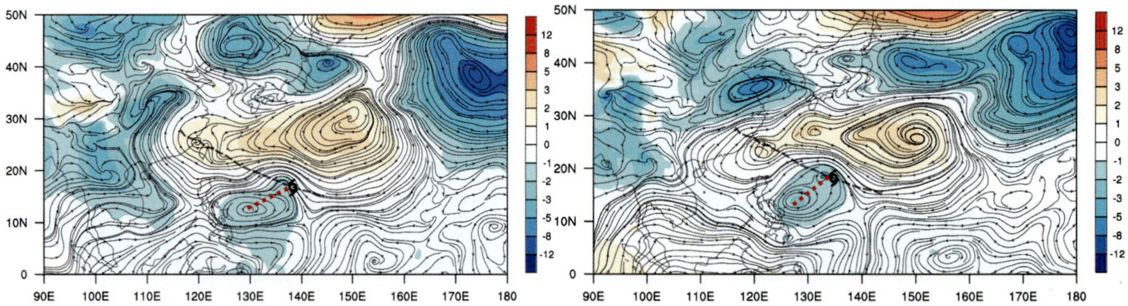

850 hPa flow/height anomalies at 0000 UTC 24 and 25 August 1980.

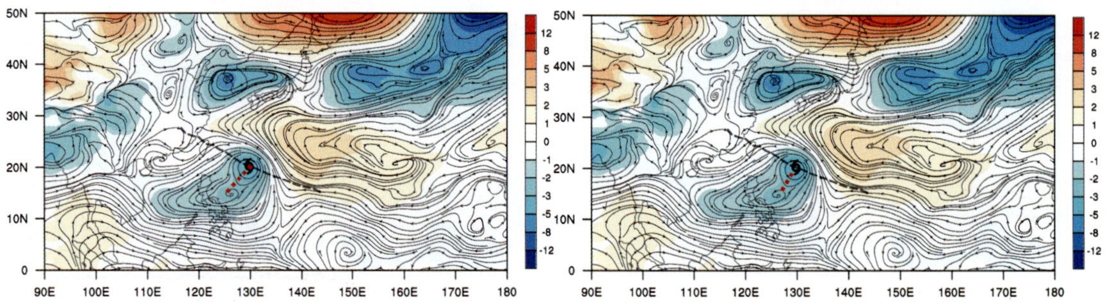

850 hPa flow/height anomalies at 0000 UTC 26 and 27 August 1980.

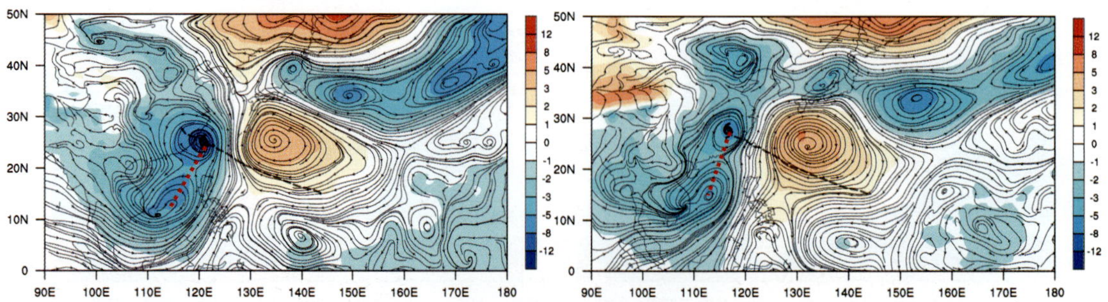

850 hPa flow/height anomalies at 0000 UTC 28 and 29 August 1980.

Figure 2.89. The typhoon (1980_No.15) generated in the tropical western Pacific on 23 August 1980 and moved northwestward basically along the climatic monsoon trough. Its intensity peaked near the Taiwan Island on 27 August 1980 before landfall on the island and its motion was modified by an anomalous vortex at 850 hPa.

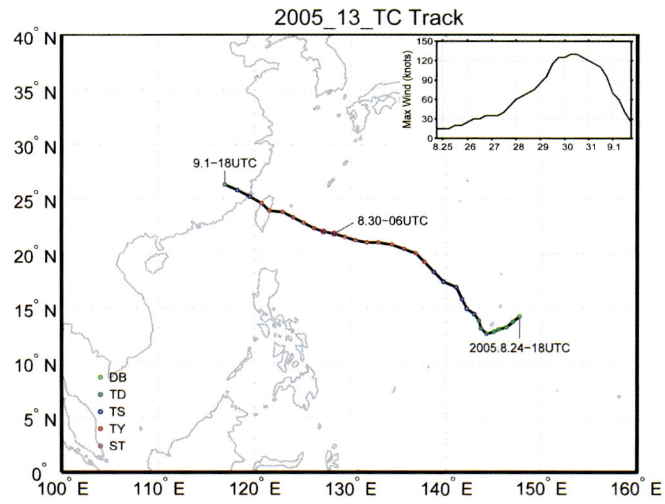

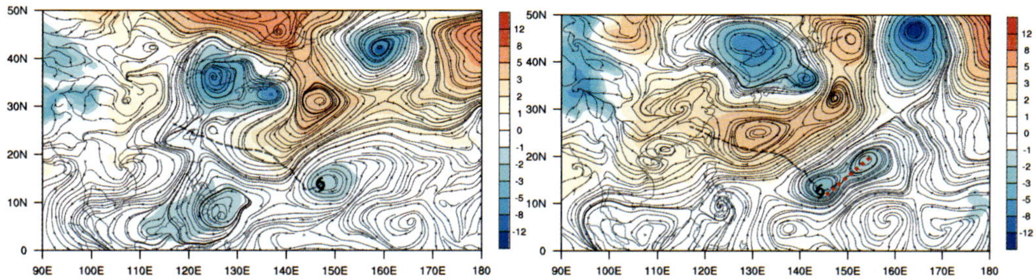

850 hPa flow/height anomalies at 0000 UTC 25 and 26 August 2005.

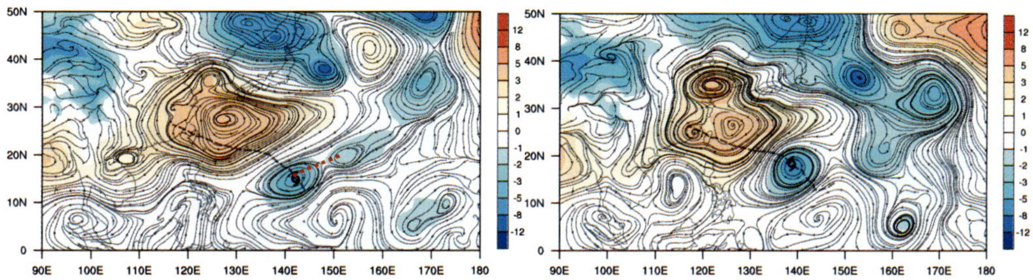

850 hPa flow/height anomalies at 0000 UTC 27 and 28 August 2005.

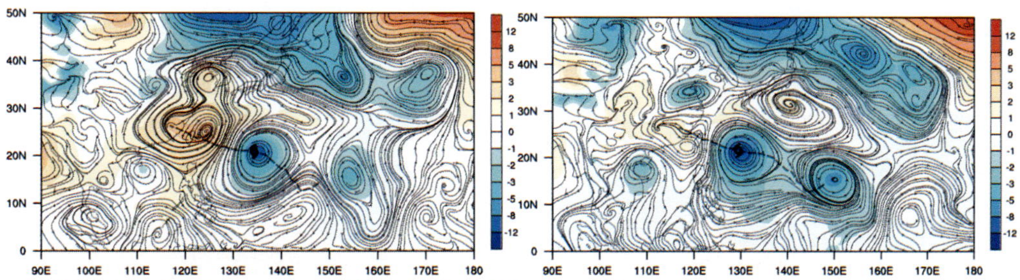

850 hPa flow/height anomalies at 0000 UTC 29 and 30 August 2005.

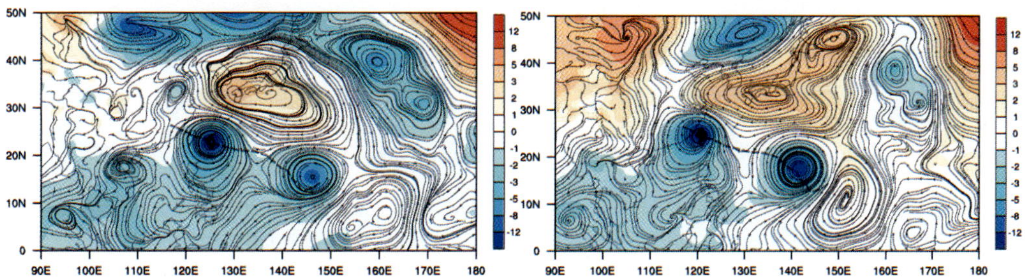

850 hPa flow/height anomalies at 0000 UTC 31 August and 1 September 2005.

Figure 2.90. The typhoon (2005_No.13) generated in the tropical western Pacific on 24 August 2005 and moved northwestward along the climatic monsoon trough. Its intensity peaked in the southeast sea of the Taiwan Island on 30 August 2005 before landfall on the island and the Southeast China coast and its motion was modified by an anomalous vortex at 850 hPa.

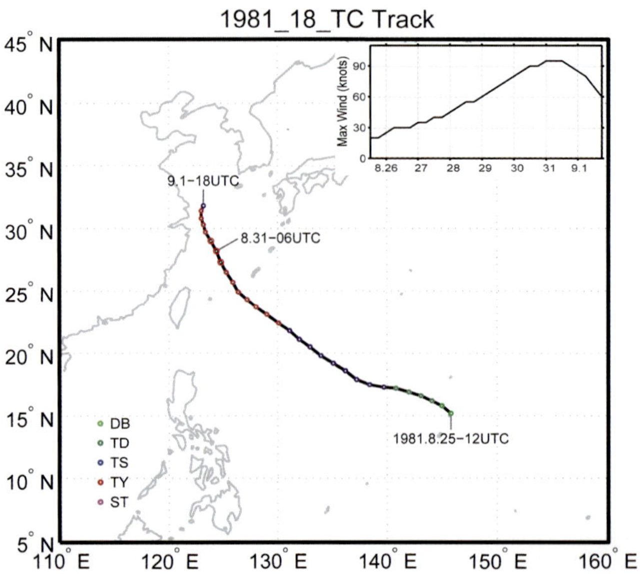

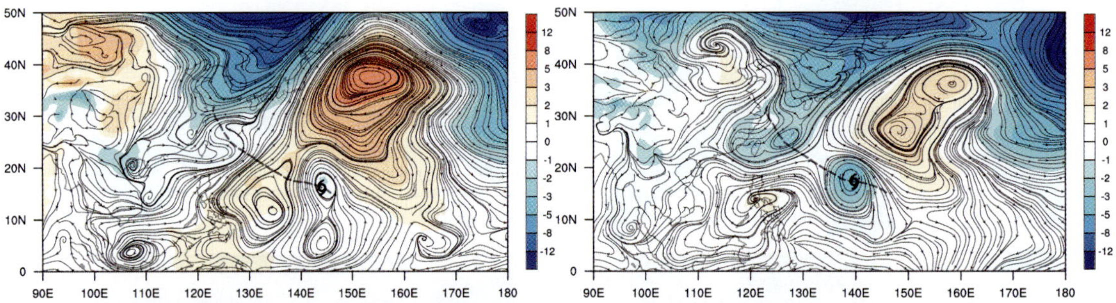

850 hPa flow/height anomalies at 0000 UTC 26 and 27 August 1981.

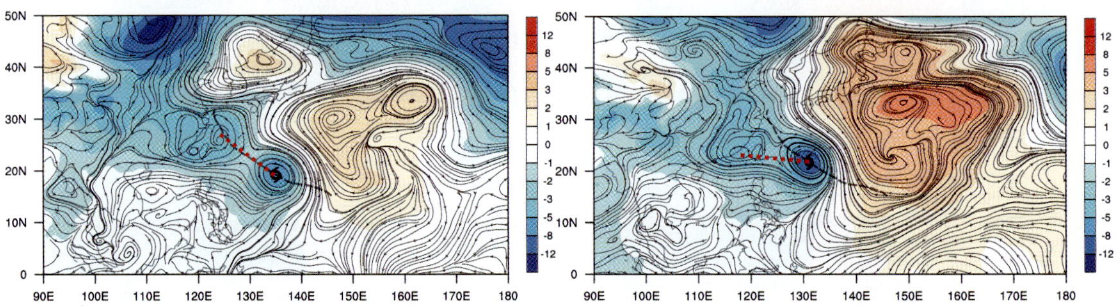

850 hPa flow/height anomalies at 0000 UTC 28 and 29 August 1981.

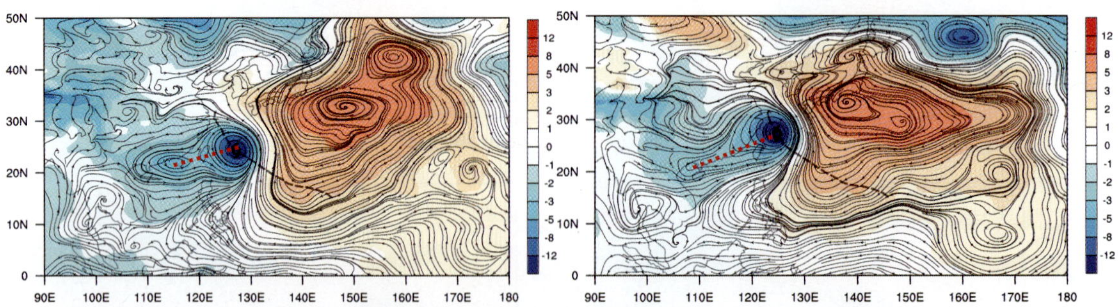

850 hPa flow/height anomalies at 0000 UTC 30 and 31 August 1981.

Figure 2.91. The typhoon (1981_No.18) generated in the tropical western Pacific on 25 August 1981 and moved northwestward basically along the climatic monsoon trough. Its intensity peaked in the East China Sea on 31 August 1981 and its motion in the East China Sea was modified by an anomalous vortex at 850 hPa.

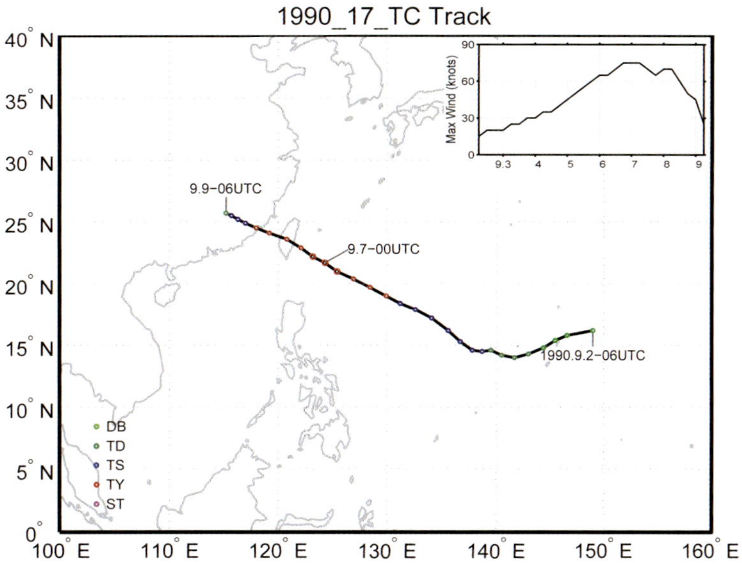

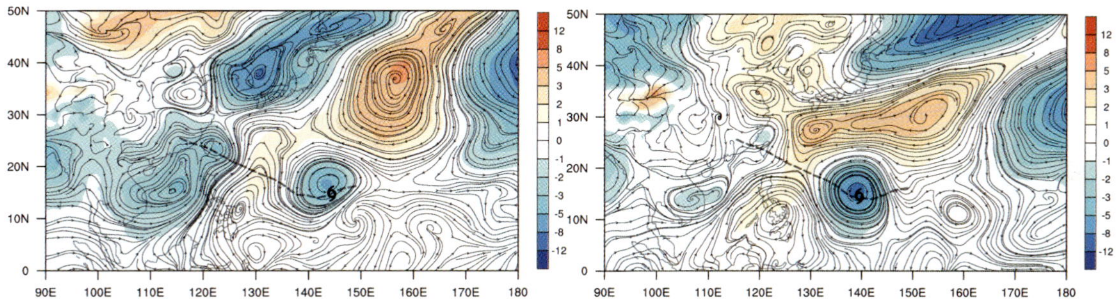

850 hPa flow/height anomalies at 0000 UTC 3 and 4 September 1990.

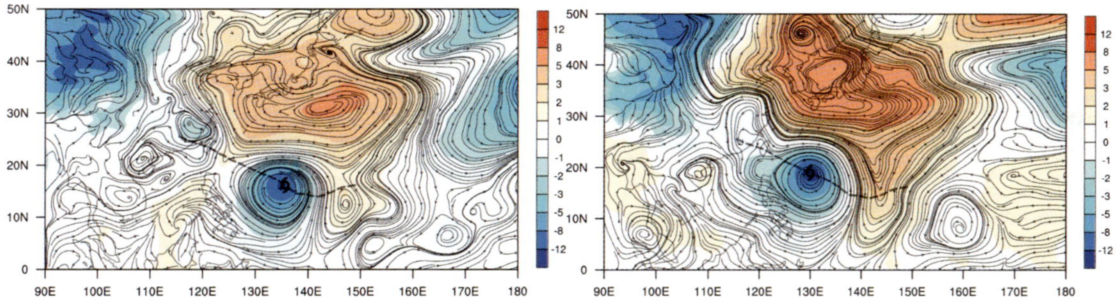

850 hPa flow/height anomalies at 0000 UTC 5 and 6 September 1990.

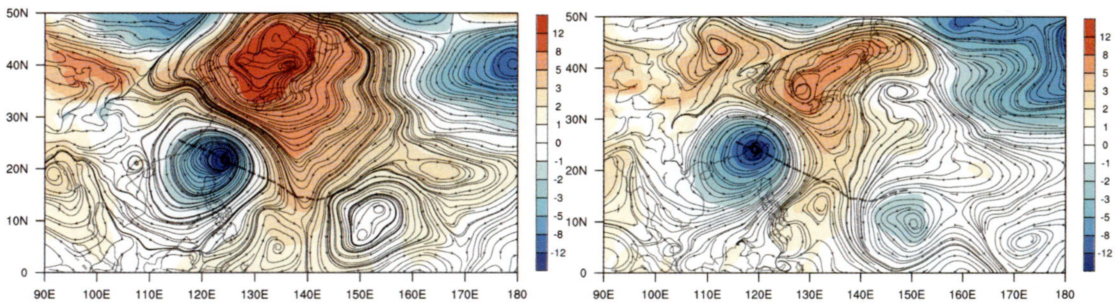

850 hPa flow/height anomalies at 0000 UTC 7 and 8 September 1990.

Figure 2.92. The typhoon (1990_No.17) generated in the tropical western Pacific on 2 September 1990 and moved west-northwestward along the climatic monsoon trough. Its intensity peaked in the southeast sea of Taiwan Island on 7 September 1990 before landfall on the island and its motion was influenced by anomalous flows at 850 hPa.

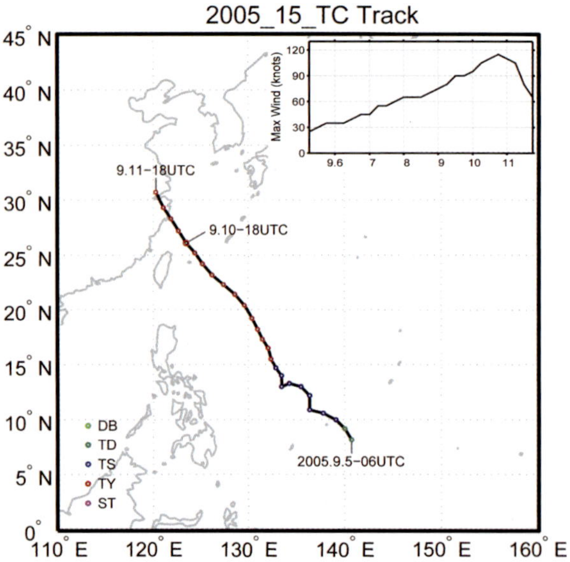

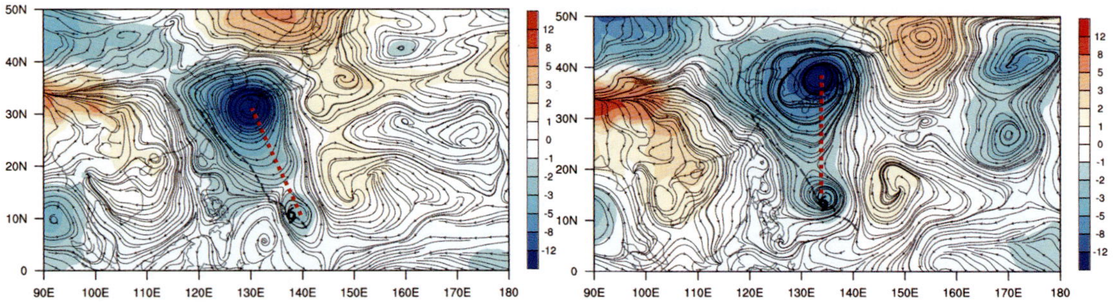

850 hPa flow/height anomalies at 0000 UTC 6 and 7 September 2005.

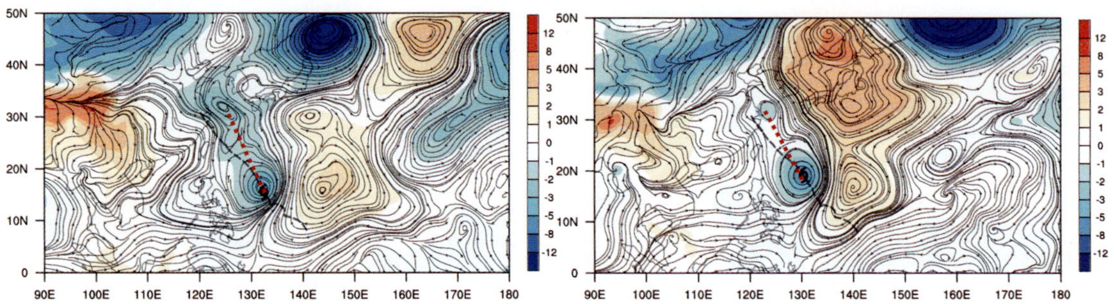

850 hPa flow/height anomalies at 0000 UTC 8 and 9 September 2005.

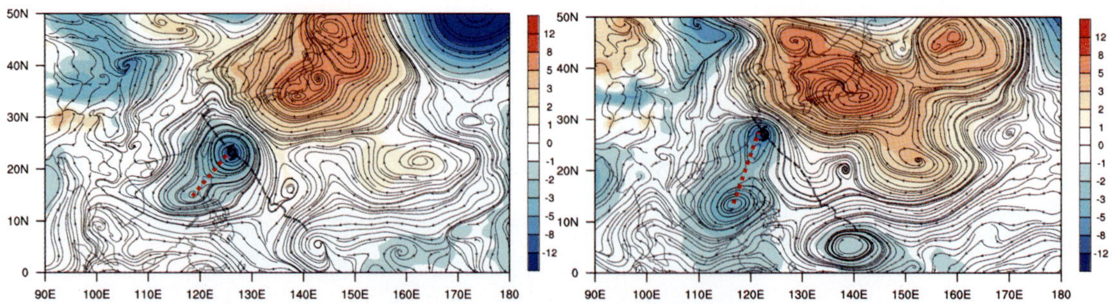

850 hPa flow/height anomalies at 0000 UTC 10 and 11 September 2005.

Figure 2.93. The typhoon (2005_No.15) generated in the equatorial western Pacific on 5 September 2005 and moved northwestward along the climatic monsoon trough. Its intensity peaked in the East China Sea on 10 September 2005 before landfall on the East China coast and its motion was modified by anomalous vortices at 850 hPa.

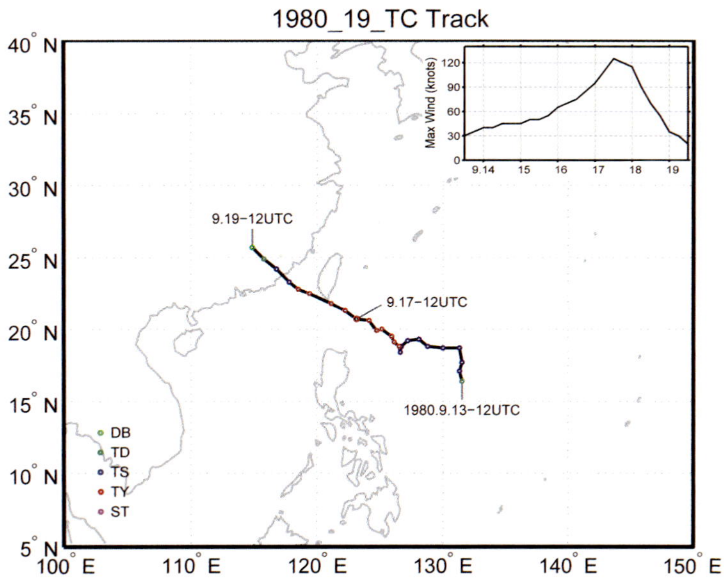

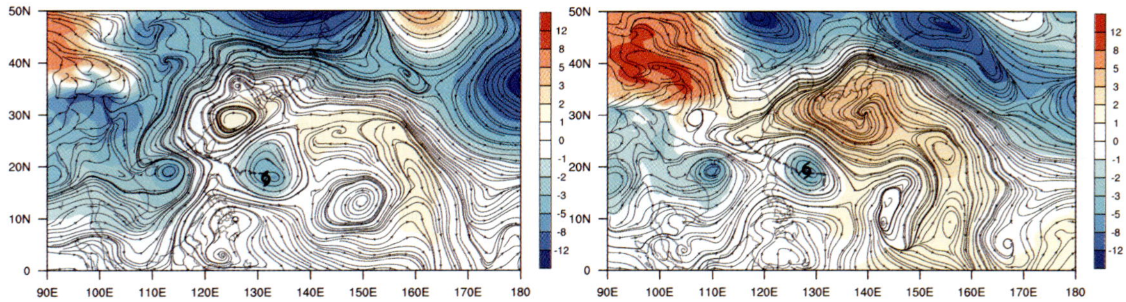

850 hPa flow/height anomalies at 0000 UTC 14 and 15 September 1980.

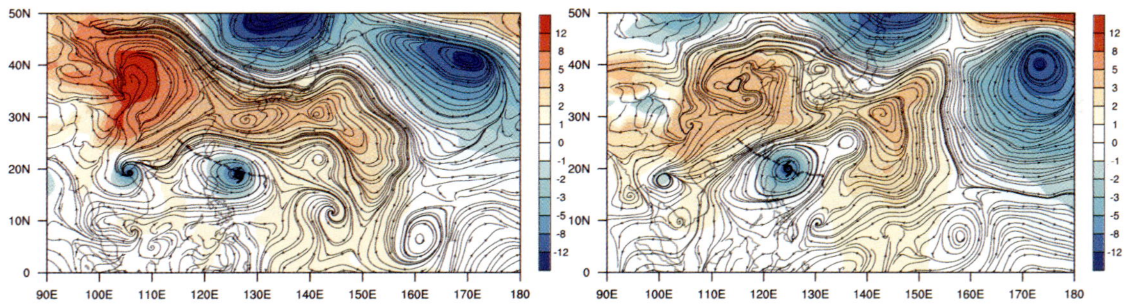

850 hPa flow/height anomalies at 0000 UTC 16 and 17 September 1980.

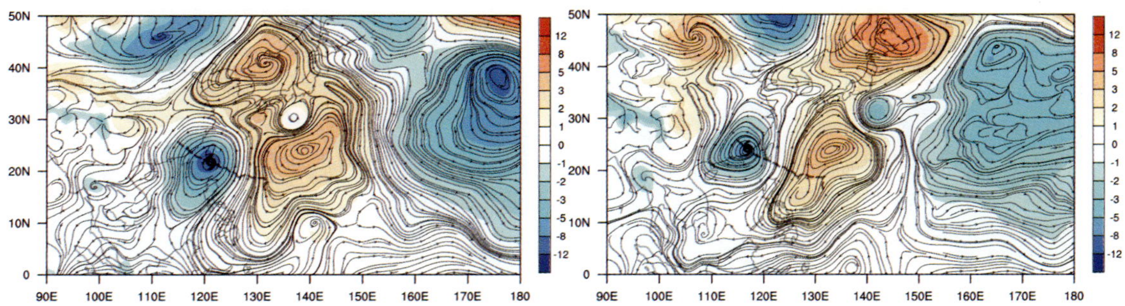

850 hPa flow/height anomalies at 0000 UTC 18 and 19 September 1980.

Figure 2.94. The typhoon (1980_No.19) generated in the tropical western Pacific on 13 September 1980 and had some small curvatures at the beginning stage. Its intensity peaked near the climatic monsoon trough in the southeast sea of Taiwan Island on 17 September 1980 before landfall on the Southeast China coast and its motion was not affected by anomalous vortices at 850 hPa.

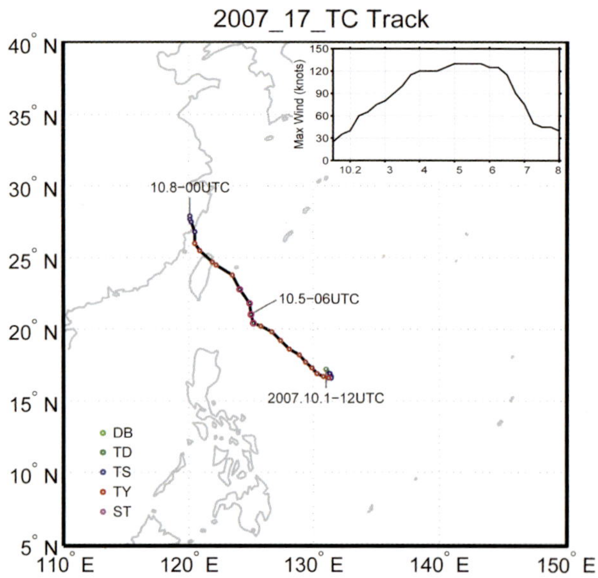

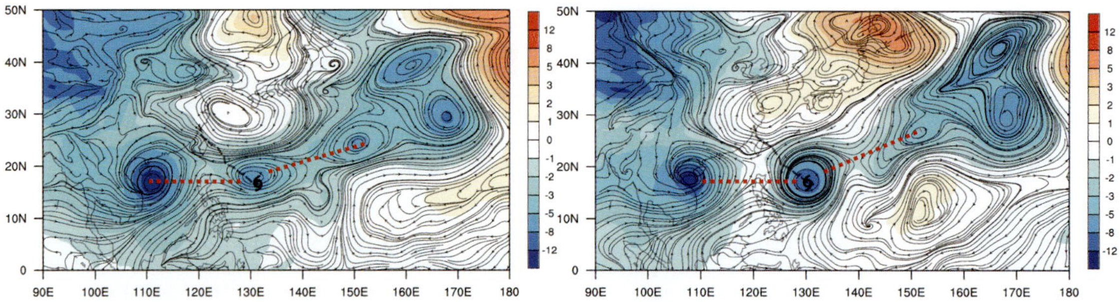

850 hPa flow/height anomalies at 0000 UTC 2 and 3 October 2007.

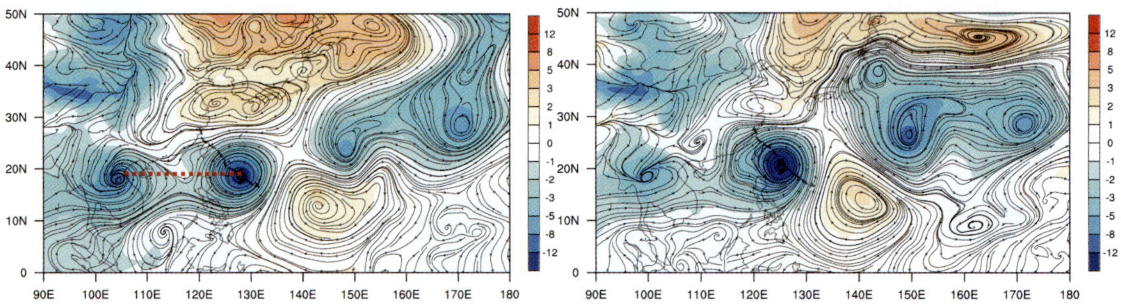

850 hPa flow/height anomalies at 0000 UTC 4 and 5 October 2007.

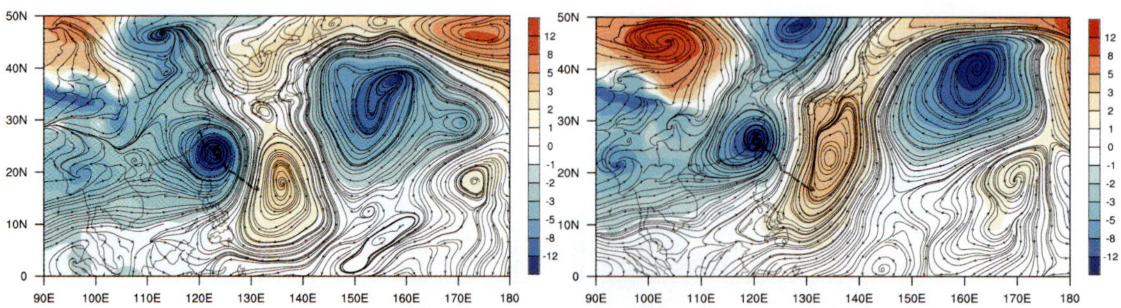

850 hPa flow/height anomalies at 0000 UTC 6 and 7 October 2007.

Figure 2.95. The typhoon (2007_No.17) generated in the tropical western Pacific on 1 October 2007 and moved northwestward along the climatic monsoon trough. Its intensity peaked in the southeast sea of Taiwan Island on 5 October 2007 before landfall on the island and its motion was modified by two anomalous vortices at 850 hPa.

2.5. Intensity Peaked in the Western Pacific

This section plots 16 typhoons, which intensities peaked in the western Pacific, without obvious turning motions. Their tracks with small changes were modified by anomalous vortices or anomalous circulations (Figs. 2.96–2.111).

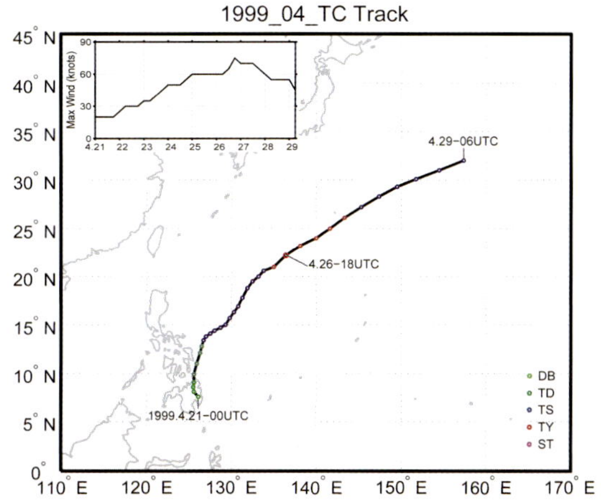

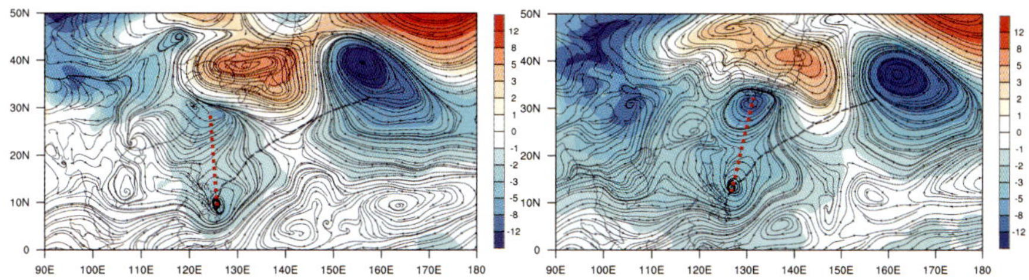

850 hPa flow/height (shading, 10 gpm) anomalies at 0000 UTC 22 and 23 April 1999.

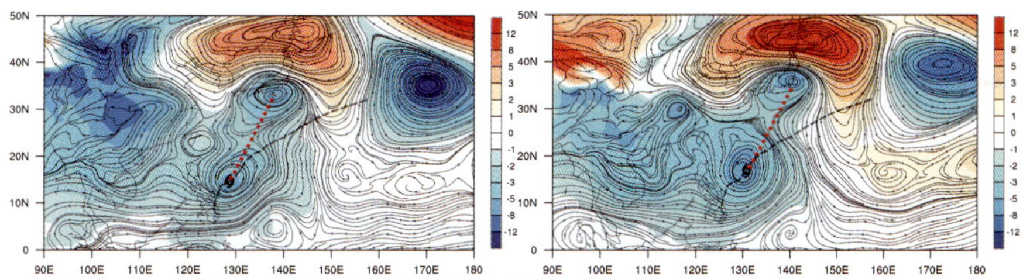

850 hPa flow/height anomalies at 0000 UTC 24 and 25 April 1999.

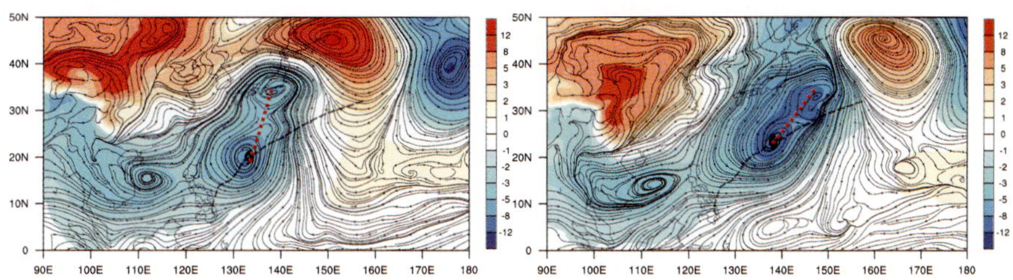

850 hPa flow/height anomalies at 0000 UTC 26 and 27 April 1999.

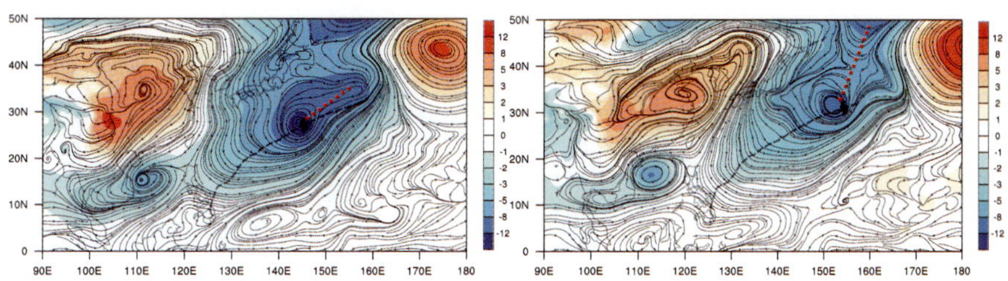

850 hPa flow/height anomalies at 0000 UTC 28 and 29 April 1999.

Figure 2.96. The typhoon (1999_No.04) generated in the central Philippines on 21 April 1999 and moved northeast. Its intensity peaked in the subtropical Northwest Pacific on 26 April 1999 and its track was modified by an anomalous vortex at 850 hPa.

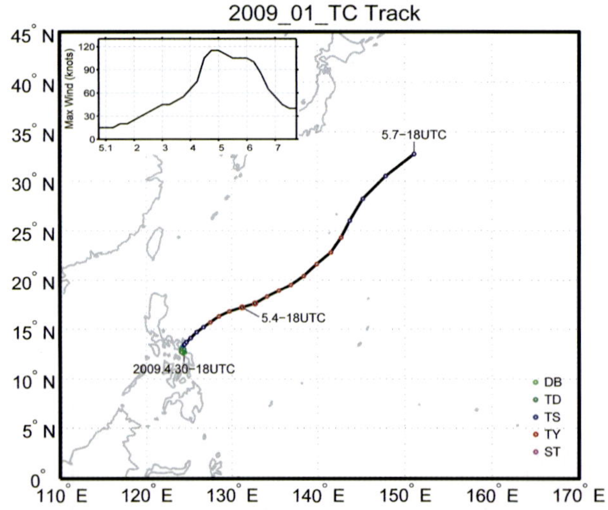

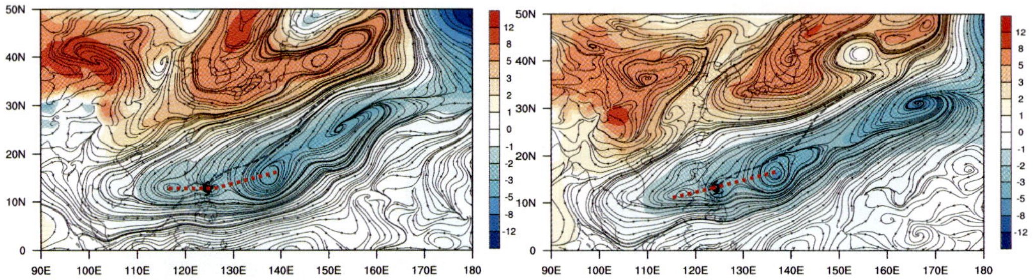

850 hPa flow/height anomalies at 0000 UTC 1 and 2 May 2009.

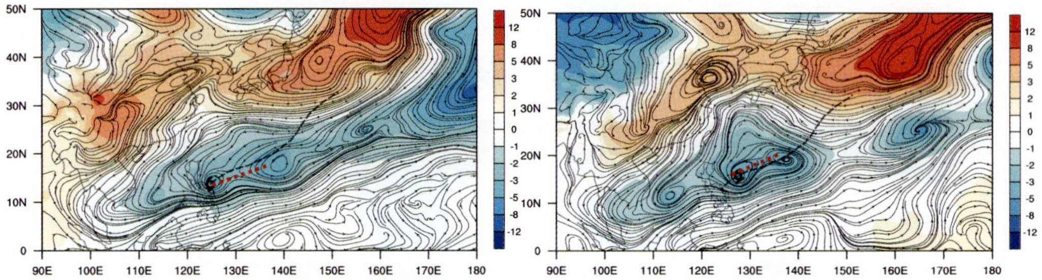

850 hPa flow/height anomalies at 0000 UTC 3 and 4 May 2009.

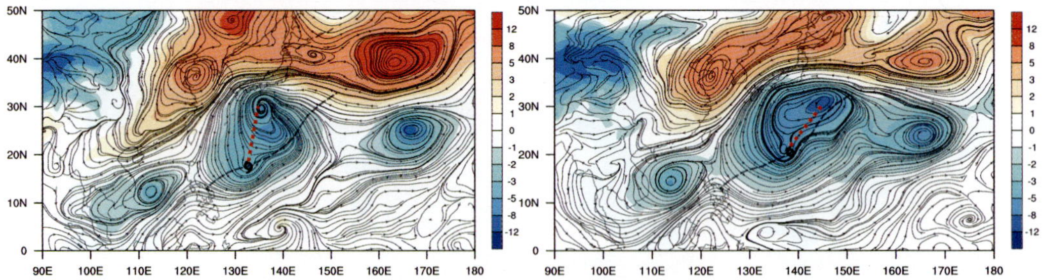

850 hPa flow/height anomalies at 0000 UTC 5 and 6 May 2009.

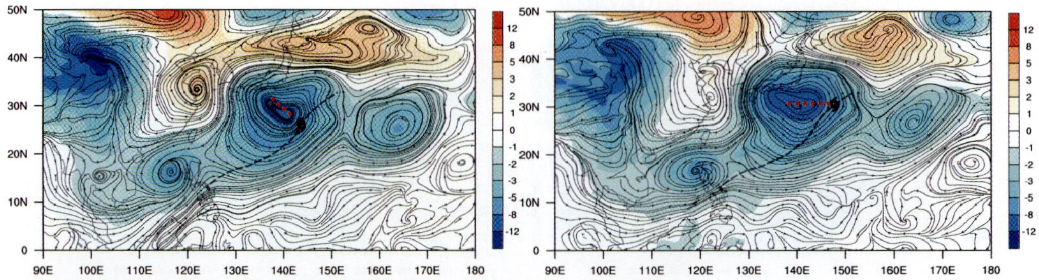

850 hPa flow/height anomalies at 0000 UTC and 1200 UTC 7 May 2009.

Figure 2.97. The typhoon (2009_No.01) generated in the central Philippines on 30 April 2009 and moved northeastward. Its intensity peaked in the tropical western Pacific on 4 May 2009 and its track was modified by anomalous vortices at 850 hPa.

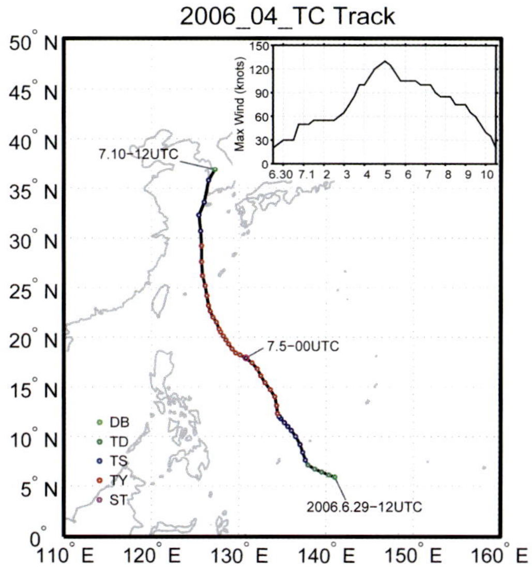

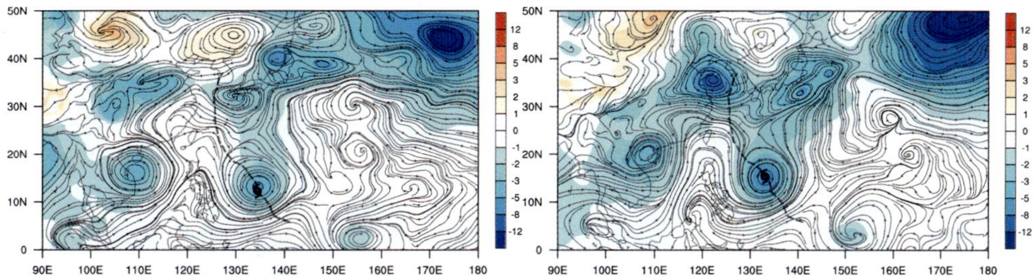
850 hPa flow/height anomalies at 0000 UTC 3 and 4 July 2006.

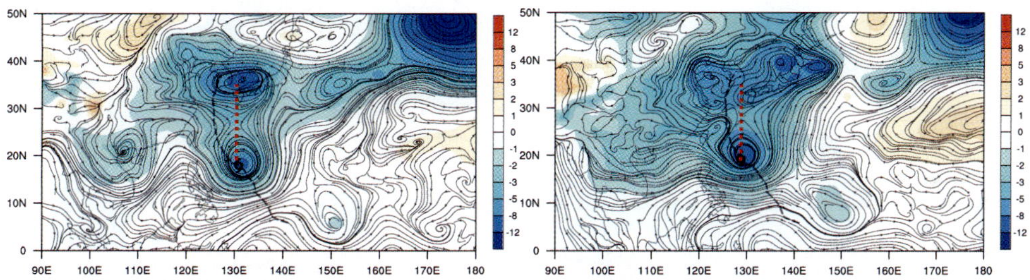
850 hPa flow/height anomalies at 0000 UTC 5 and 6 July 2006.

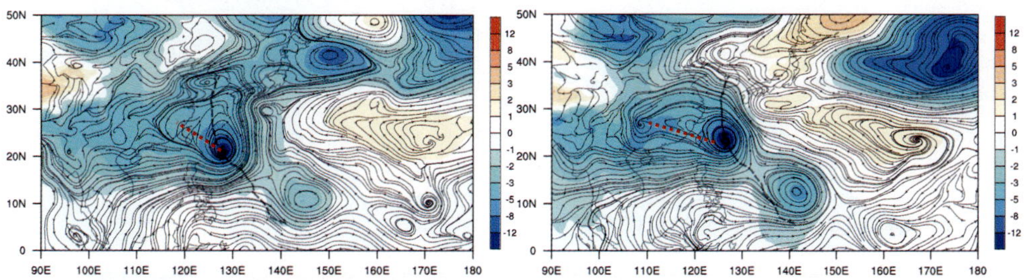
850 hPa flow/height anomalies at 0000 UTC 7 and 8 July 2006.

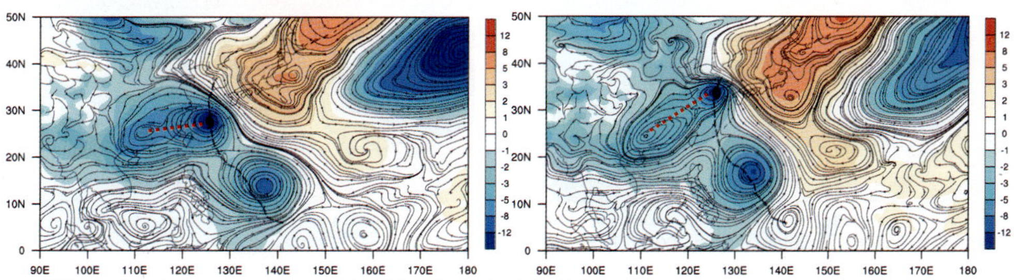
850 hPa flow/height anomalies at 0000 UTC 9 and 10 July 2006.

Figure 2.98. The typhoon (2006_No.04) generated in the equatorial western Pacific on 29 June 2006 and moved north-northwestward. Its intensity peaked in the northeast sea of the Philippines on 5 July 2006 when it was approaching the climatic monsoon trough and its motion with a small change in the east seas of China was modified by anomalous vortices at 850 hPa.

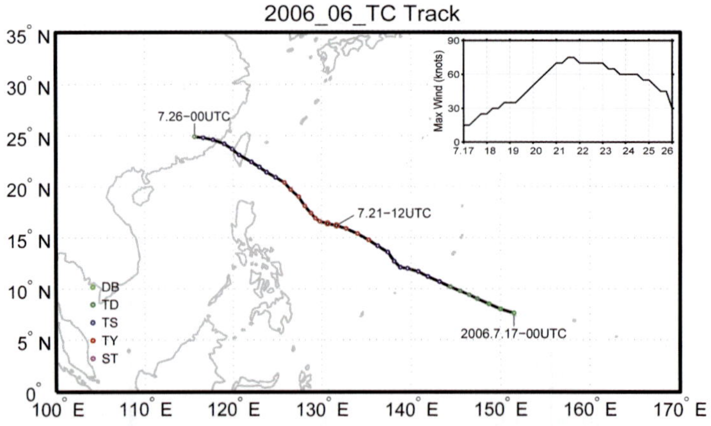

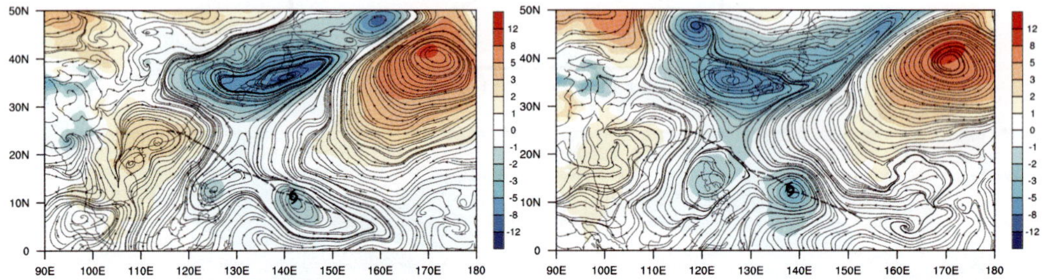

850 hPa flow/height anomalies at 0000 UTC 19 and 20 July 2006.

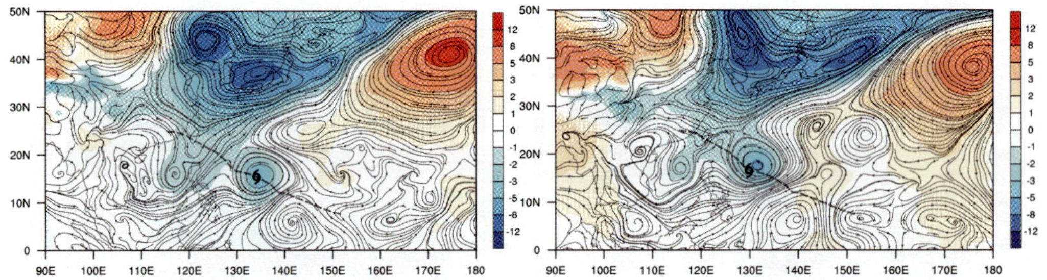

850 hPa flow/height anomalies at 0000 UTC 21 and 22 July 2006.

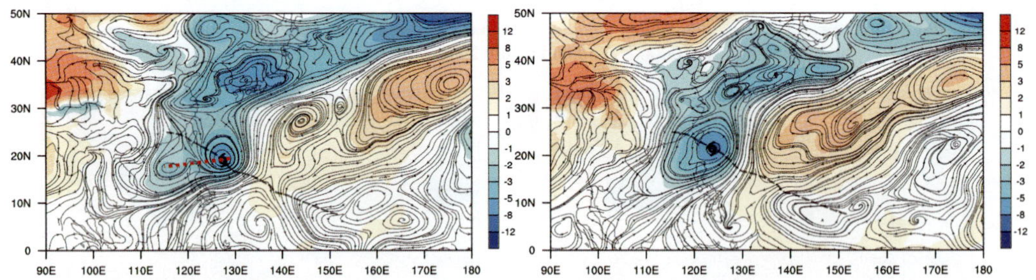

850 hPa flow/height anomalies at 0000 UTC 23 and 24 July 2006.

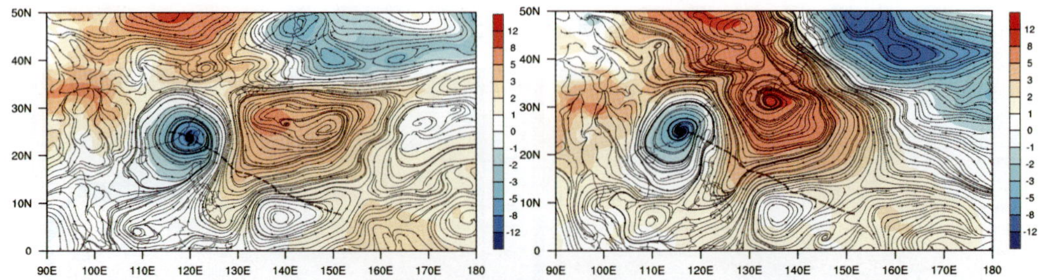

850 hPa flow/height anomalies at 0000 UTC 25 and 26 July 2006.

Figure 2.99. The typhoon (2006_No.06) generated in the equatorial central Pacific on 17 July 2006 and moved northwestward along the climatic monsoon trough. Its intensity peaked in the northeast sea of the Philippines on 21 July 2006 and its motion with a small change was influenced by an anomalous vortex at 850 hPa.

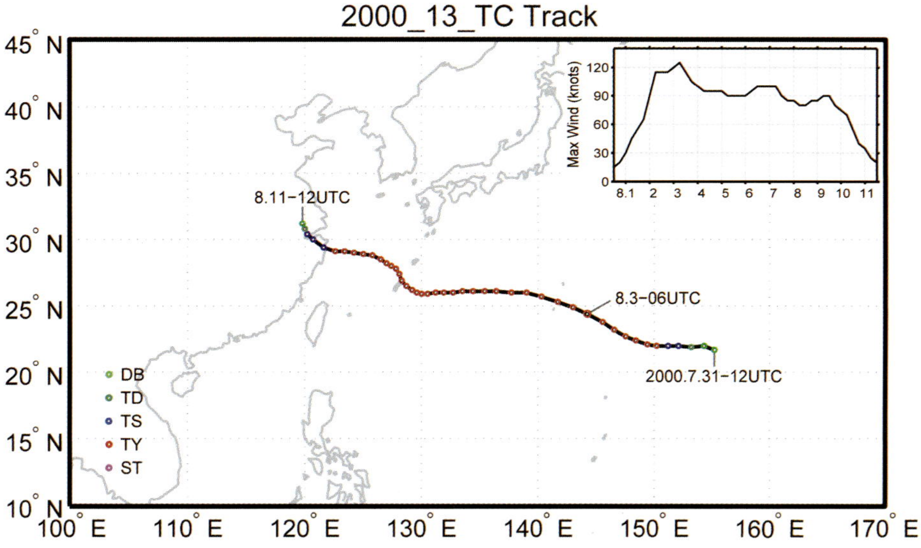

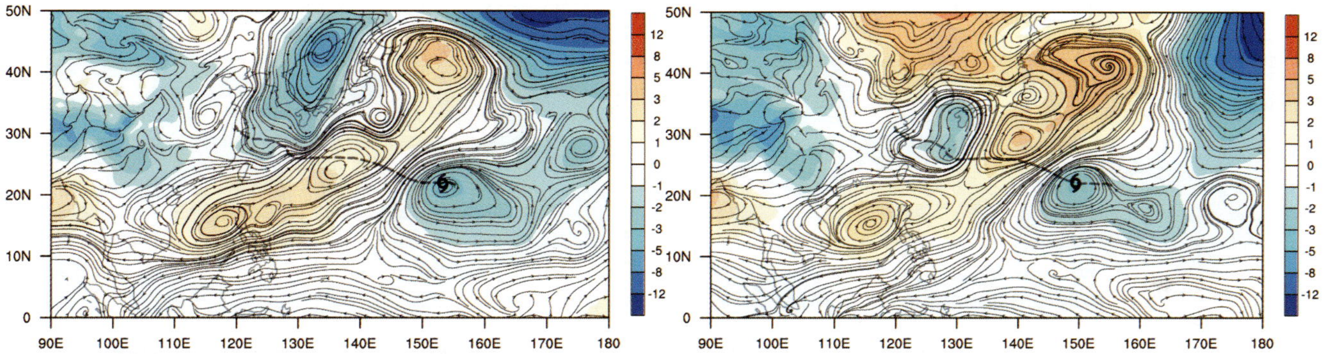

850 hPa flow/height anomalies at 0000 UTC 1 and 2 August 2000.

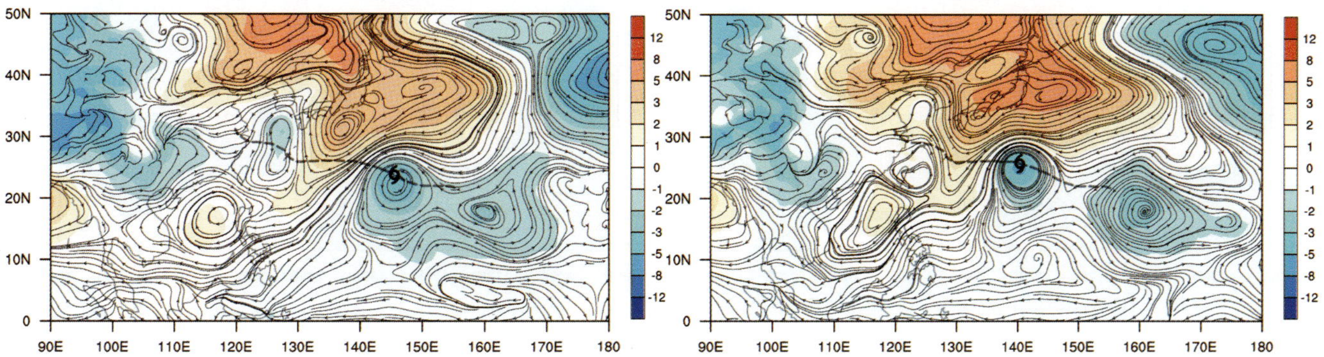

850 hPa flow/height anomalies at 0000 UTC 3 and 4 August 2000.

Figure 2.100. The typhoon (2000_No.13) generated in the tropical central Pacific on 31 July 2000 and moved west-northwestward without the influence of other anomalous vortices at 850 hPa.

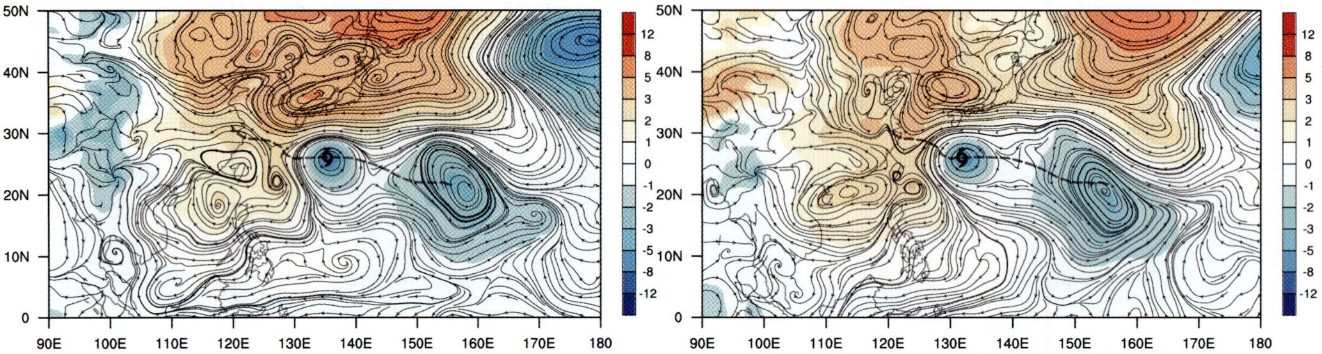

850 hPa flow/height anomalies at 0000 UTC 5 and 6 August 2000.

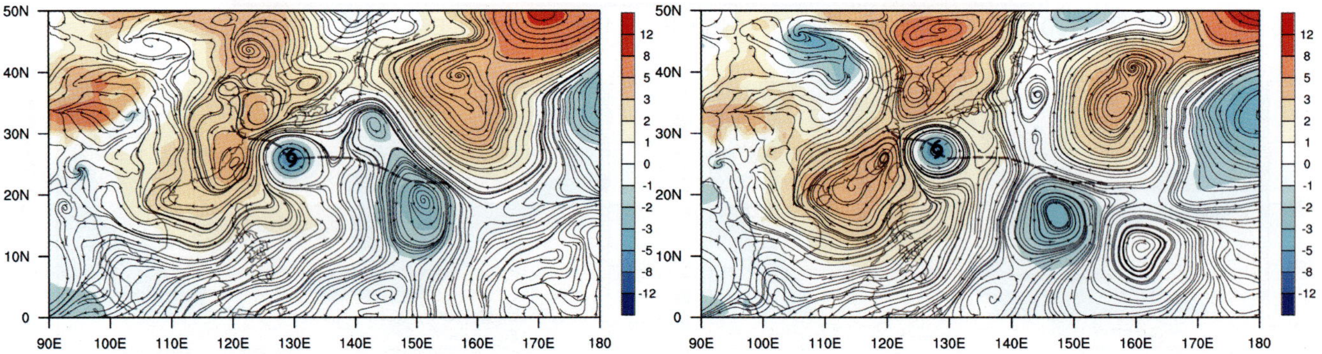

850 hPa flow/height anomalies at 0000 UTC 7 and 8 August 2000.

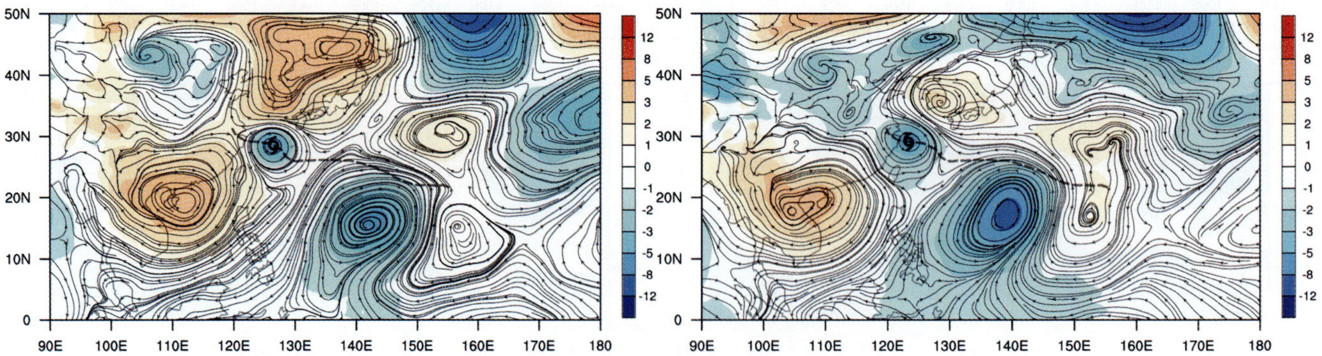

850 hPa flow/height anomalies at 0000 UTC 9 and 10 August 2000.

Figure 2.100. *(Continued)*

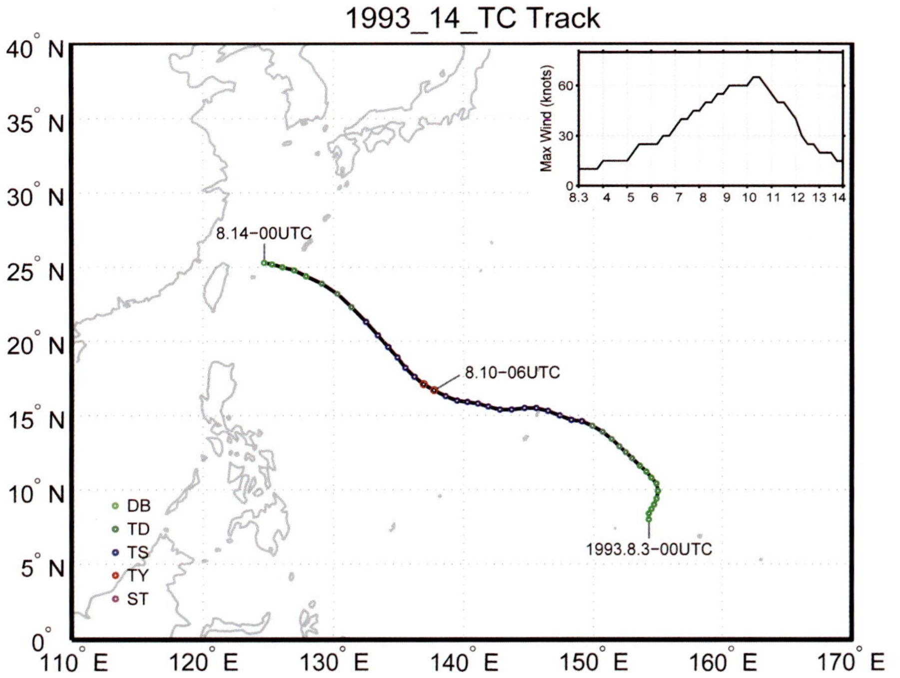

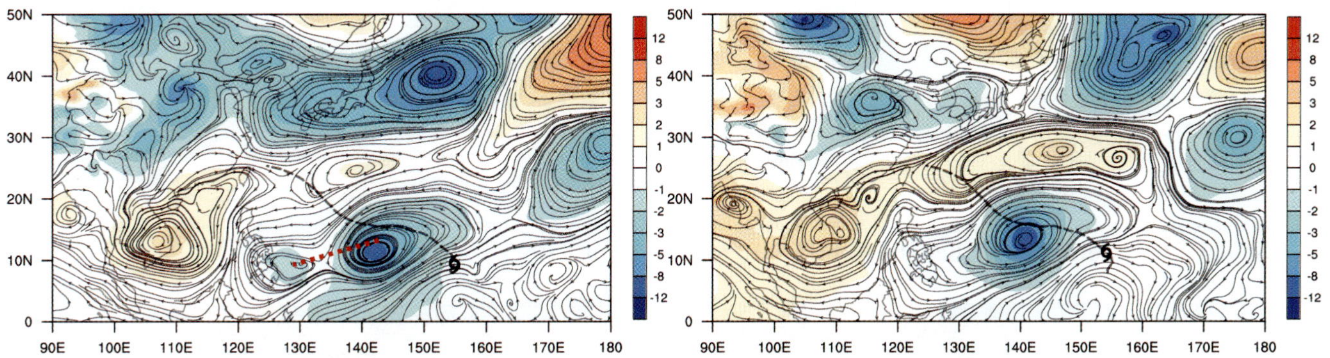

850 hPa flow/height anomalies at 0000 UTC 4 and 5 August 1993.

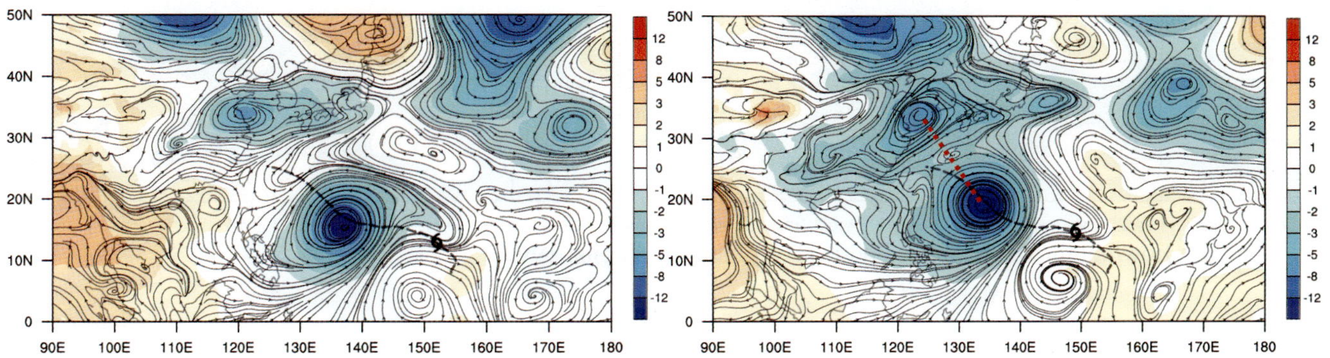

850 hPa flow/height anomalies at 0000 UTC 6 and 7 August 1993.

Figure 2.101. The typhoon (1993_No.14) generated in the equatorial central Pacific on 3 August 1993 and moved northwestward along the climatic monsoon trough. Its intensity peaked in the tropical western Pacific on 10 August 1993 and its motion with a small change was modified by anomalous vortices at 850 hPa.

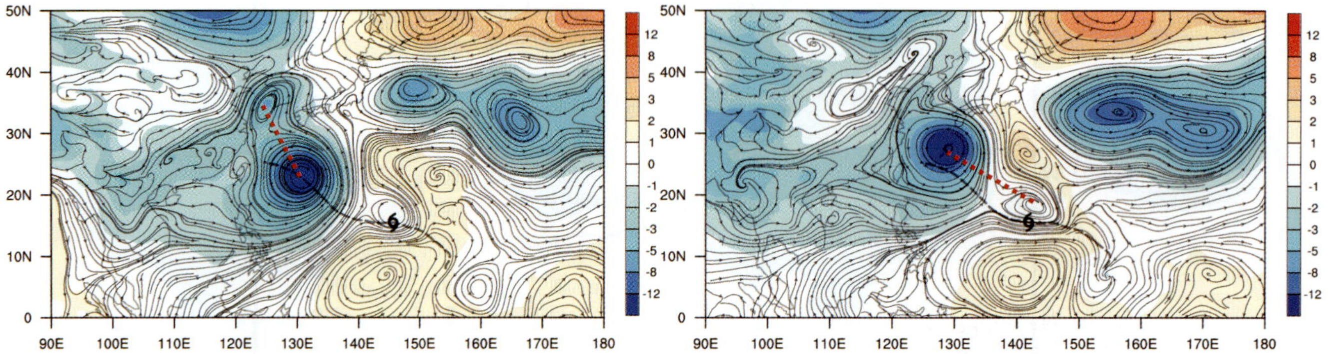

850 hPa flow/height anomalies at 0000 UTC 8 and 9 August 1993.

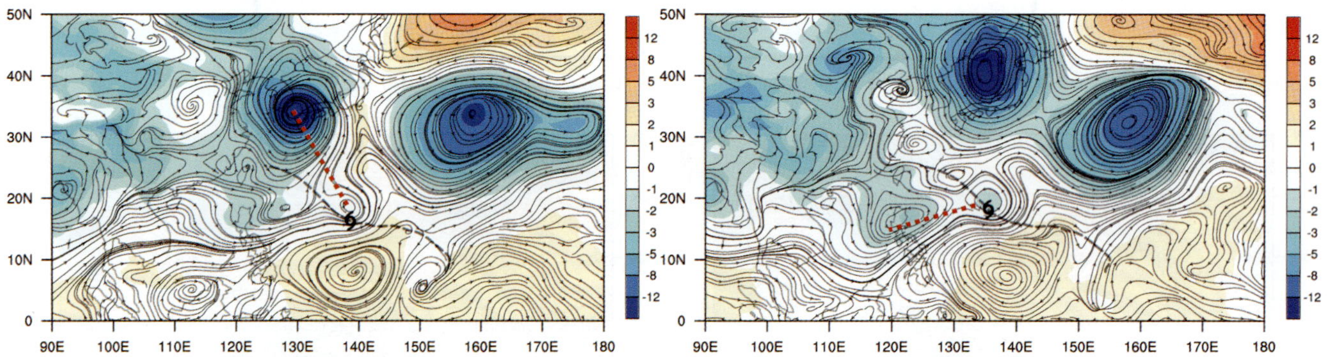

850 hPa flow/height anomalies at 0000 UTC 10 and 11 August 1993.

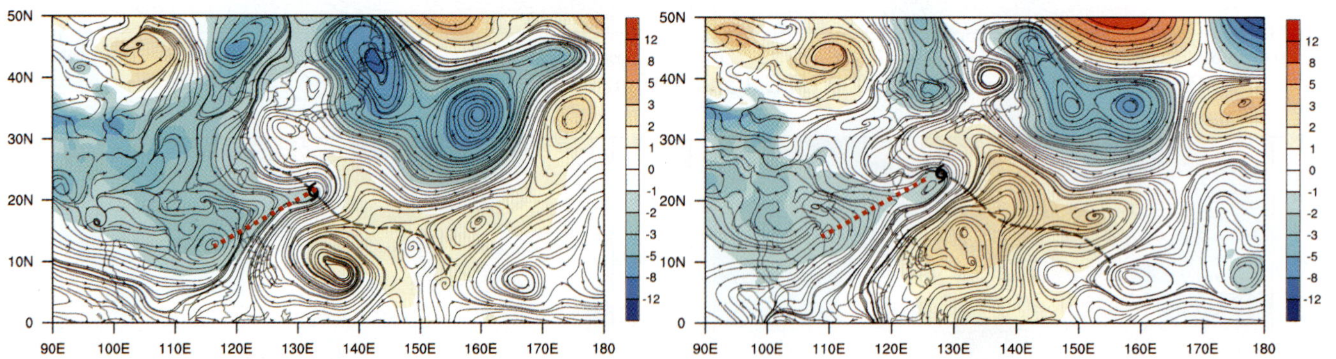

850 hPa flow/height anomalies at 0000 UTC 12 and 13 August 1993.

Figure 2.101. *(Continued)*

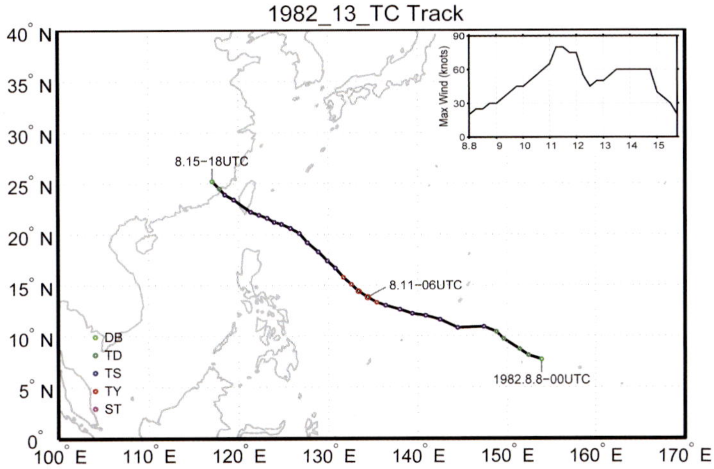

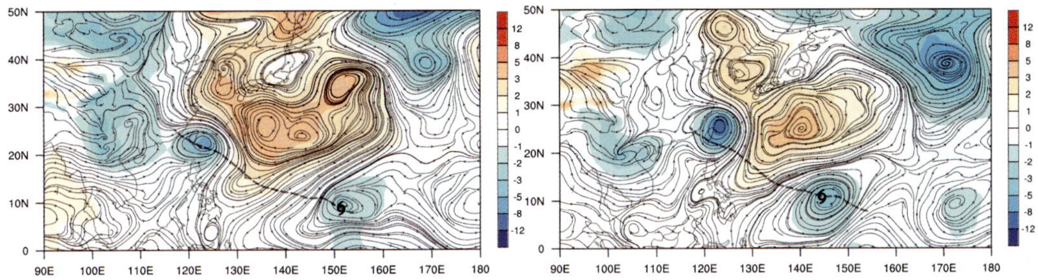

850 hPa flow/height anomalies at 1200 UTC 8 and 9 August 1982.

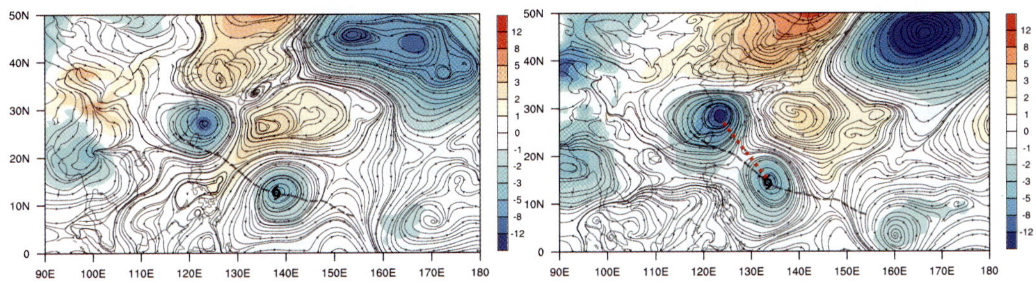

850 hPa flow/height anomalies at 1200 UTC 10 and 11 August 1982.

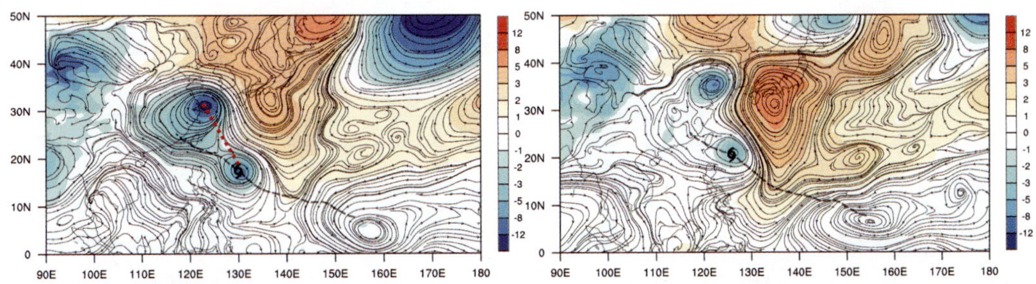

850 hPa flow/height anomalies at 1200 UTC 12 and 13 August 1982.

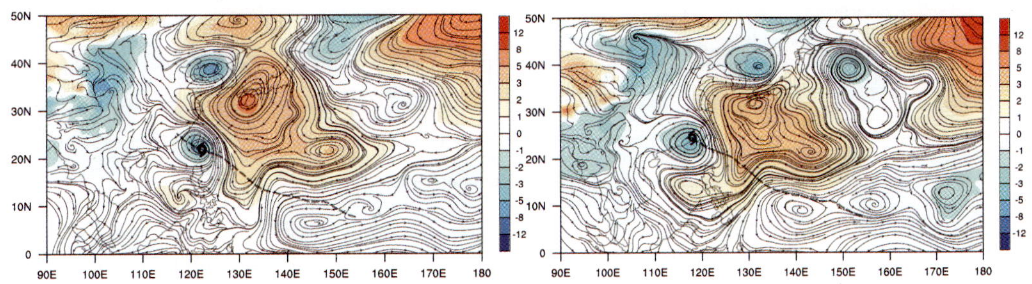

850 hPa flow/height anomalies at 1200 UTC 14 and 15 August 1982.

Figure 2.102. The typhoon (1982_No.13) generated in the equatorial central Pacific on 8 August 1982 and moved northwestward along the climatic monsoon trough. Its intensity peaked in the tropical western Pacific on 11 August 1982 and its motion with a small change was modified by an anomalous vortex at 850 hPa.

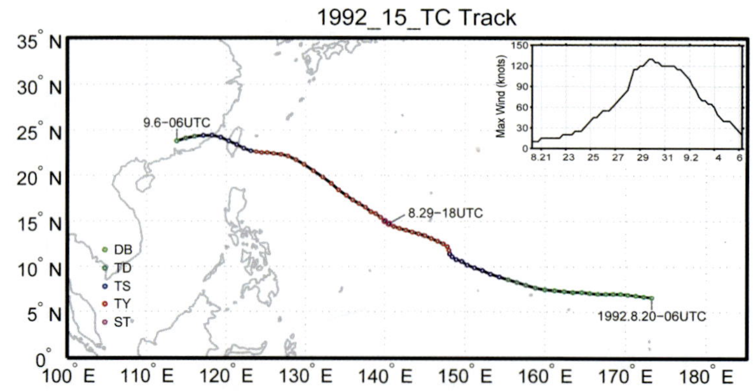

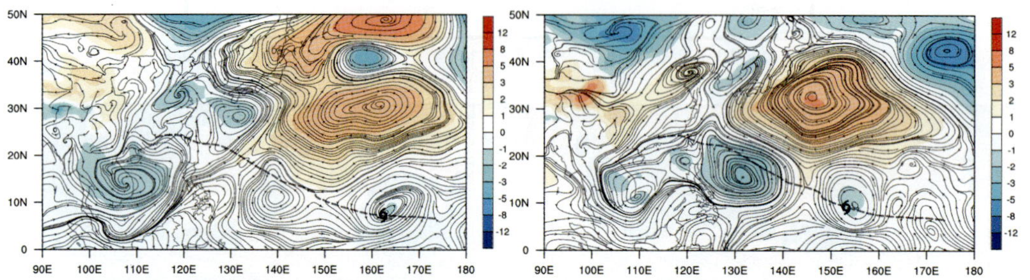

850 hPa flow/height anomalies at 0000 UTC 23 and 25 August 1992.

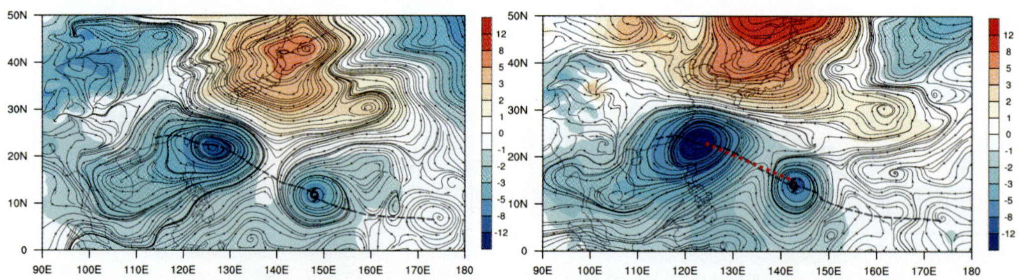

850 hPa flow/height anomalies at 0000 UTC 27 and 29 August 1992.

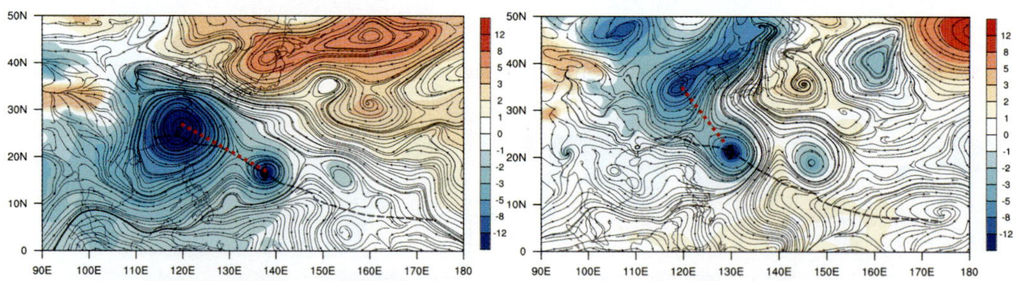

850 hPa flow/height anomalies at 0000 UTC 31 August and 2 September 1992.

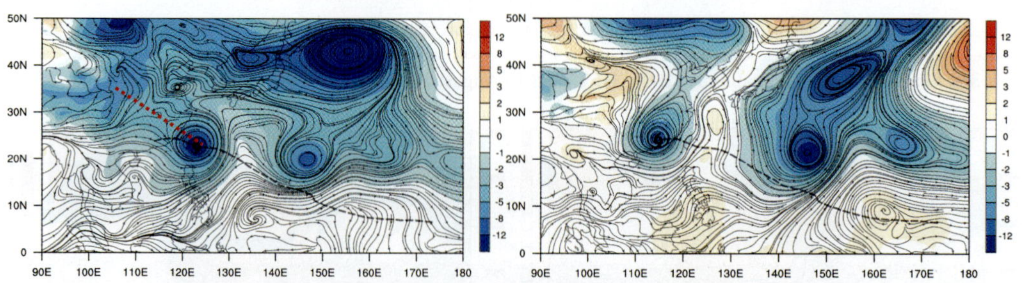

850 hPa flow/height anomalies at 0000 UTC 4 and 6 September 1992.

Figure 2.103. The typhoon (1992_No.15) generated in the equatorial central Pacific on 20 August 1992 and moved west-northwestward along the climatic monsoon trough. Its intensity peaked in the tropical western Pacific on 29 August 1992 and its motion with a small change was modified by an anomalous vortex at 850 hPa.

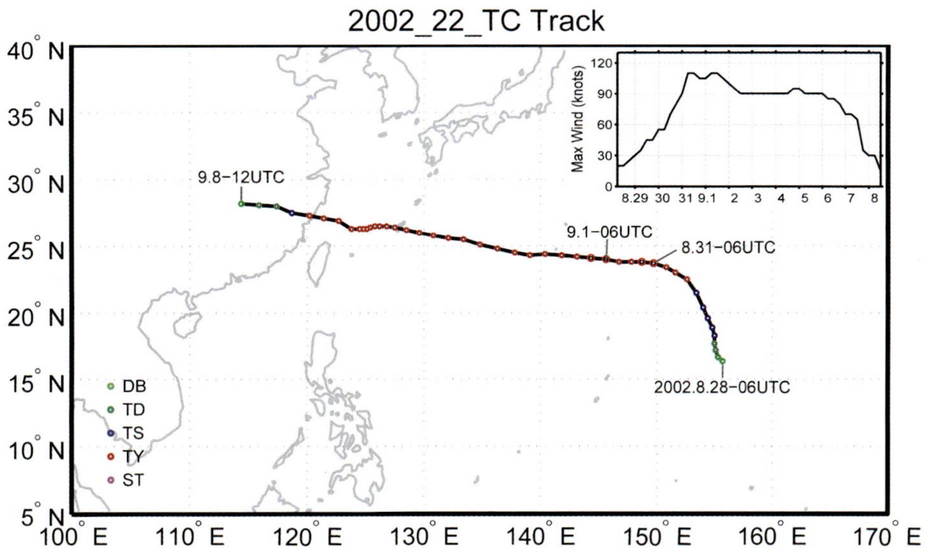

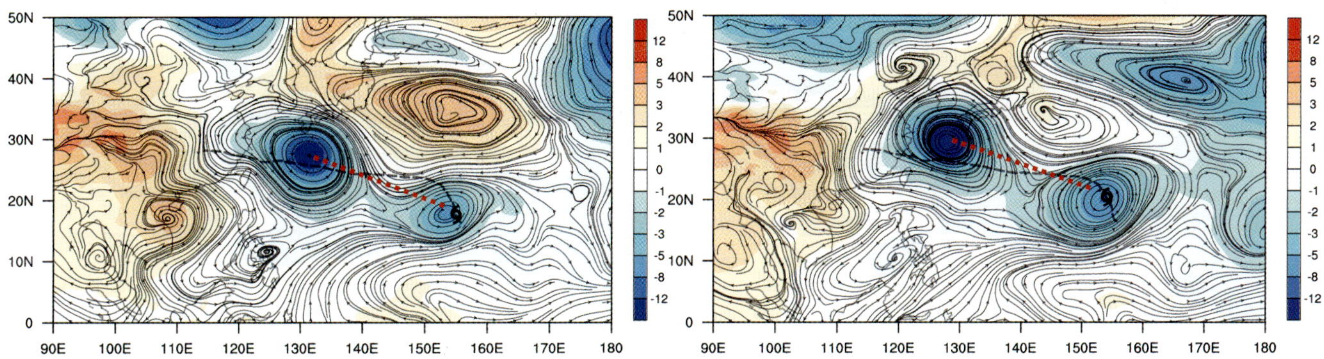

850 hPa flow/height anomalies at 0000 UTC 29 and 30 August 2002.

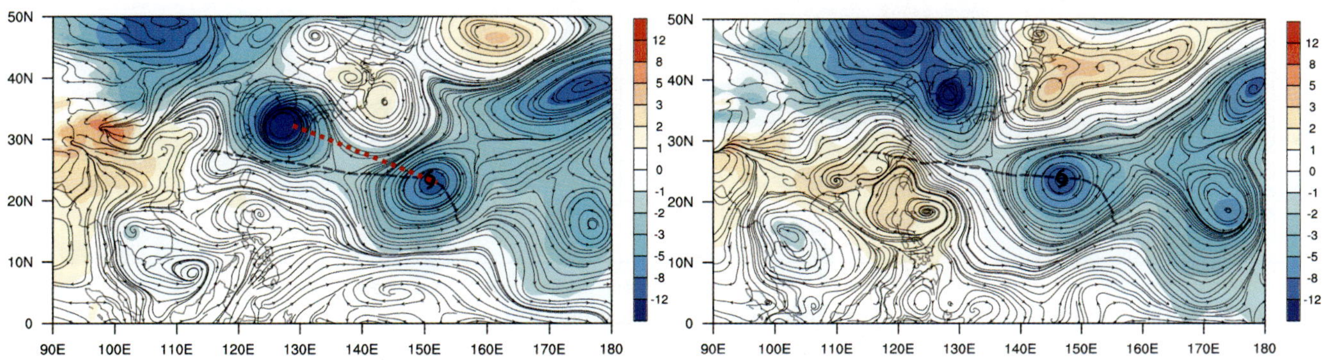

850 hPa flow/height anomalies at 0000 UTC 31 August and 1 September 2002.

Figure 2.104. The typhoon (2002_No.22) generated in the tropical central Pacific on 28 August 2002. It turned left after genesis due to the influence of an anomalous vortex at 850 hPa. After the intensity peaked on 31 August 2002, its motion was not affected by anomalous vortices at 850 hPa.

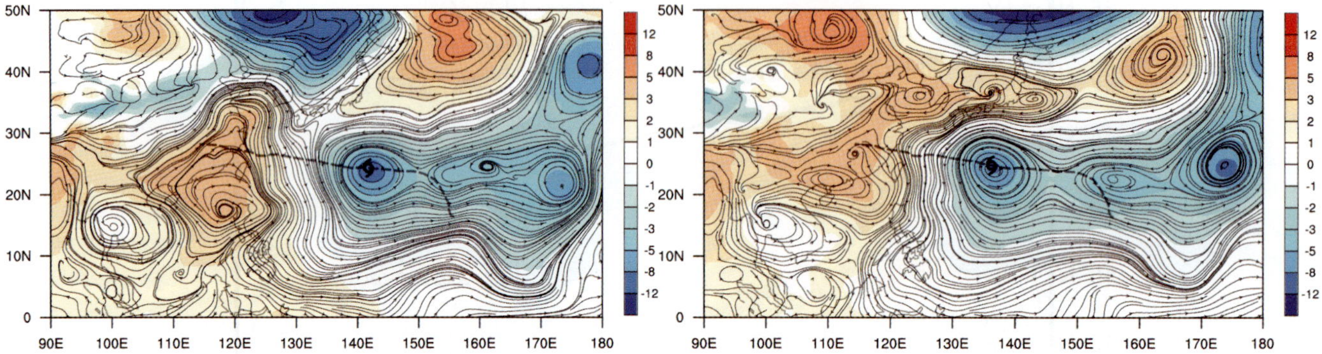

850 hPa flow/height anomalies at 0000 UTC 2 and 3 September 2002.

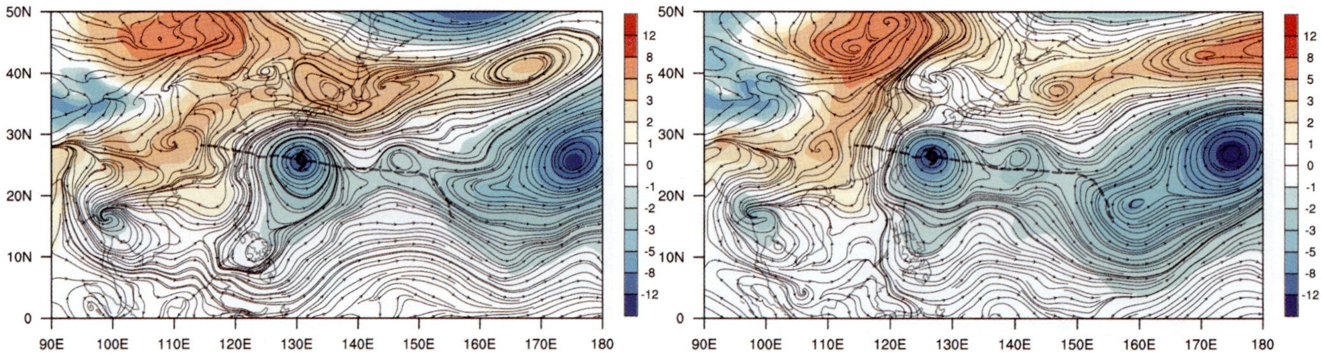

850 hPa flow/height anomalies at 0000 UTC 4 and 5 September 2002.

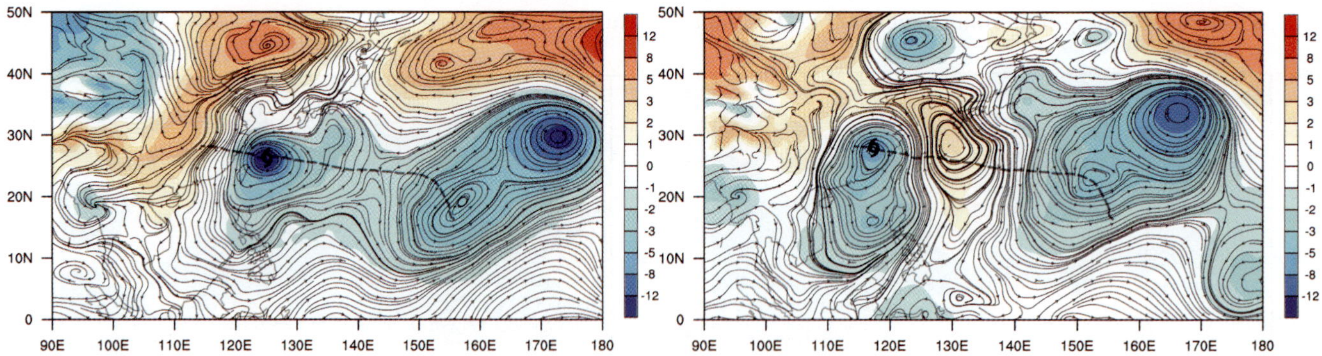

850 hPa flow/height anomalies at 0000 UTC 6 and 8 September 2002.

Figure 2.104. *(Continued)*

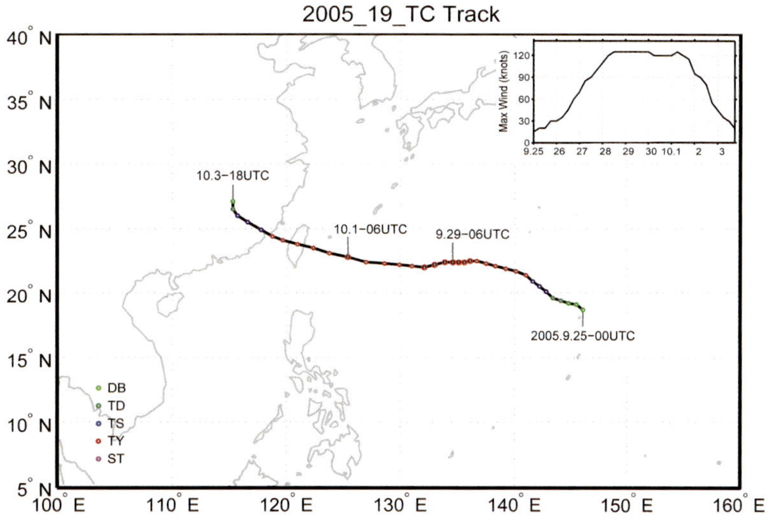

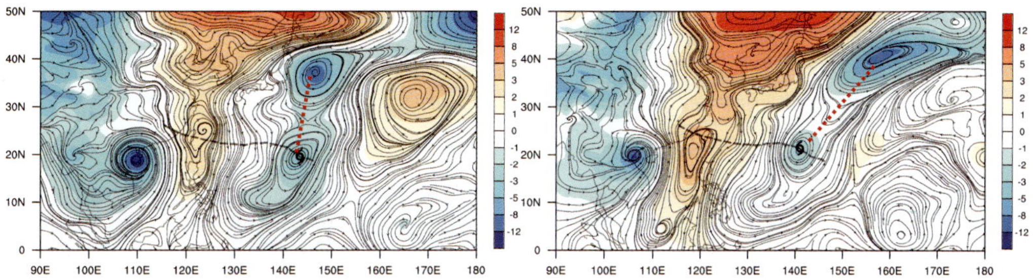

850 hPa flow/height anomalies at 0000 UTC 26 and 27 September 2005.

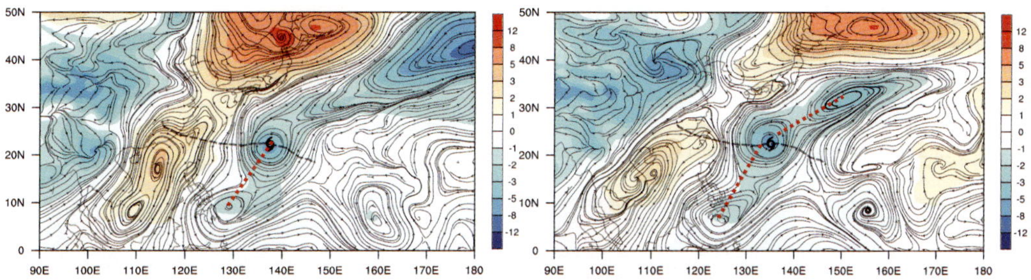

850 hPa flow/height anomalies at 0000 UTC 28 and 29 September 2005.

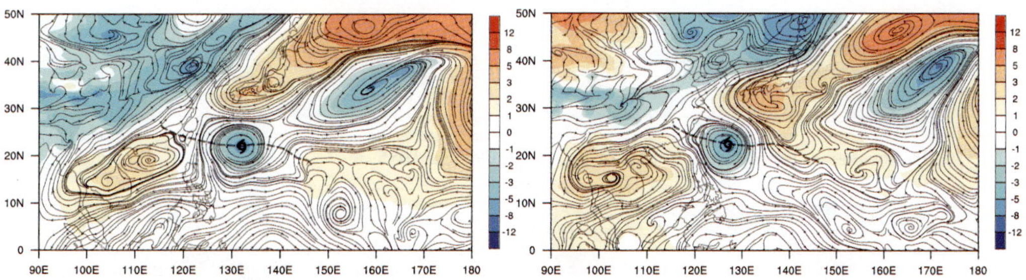

850 hPa flow/height anomalies at 0000 UTC 30 September and 1 October 2005.

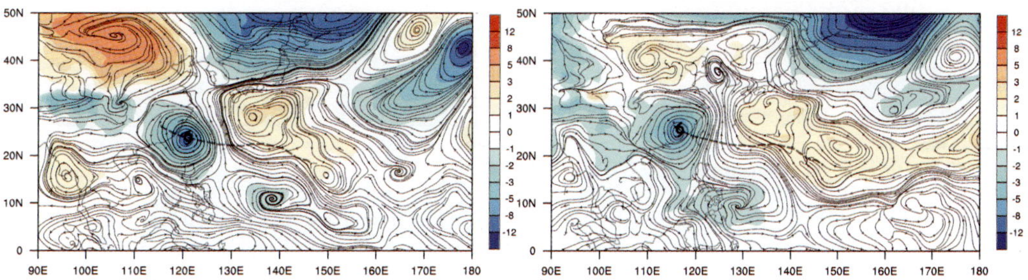

850 hPa flow/height anomalies at 0000 UTC 2 and 3 October 2005.

Figure 2.105. The typhoon (2005_No.19) generated in the tropical western Pacific on 25 September 2005 and moved westward near the climatic monsoon trough. Its intensity peaked in the subtropical Northwest Pacific on 29 September 2005 and its motion was modified by two anomalous vortices at 850 hPa.

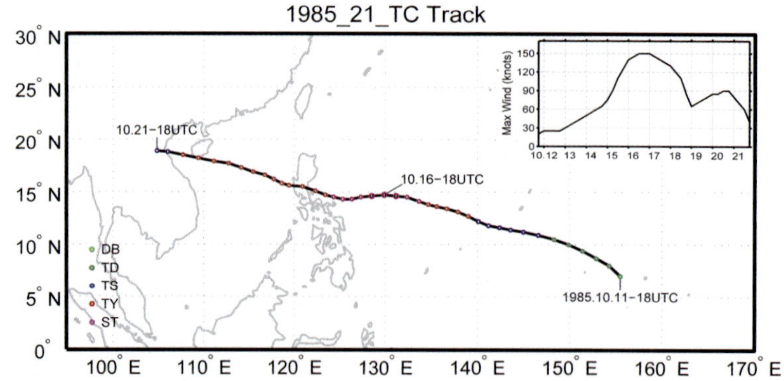

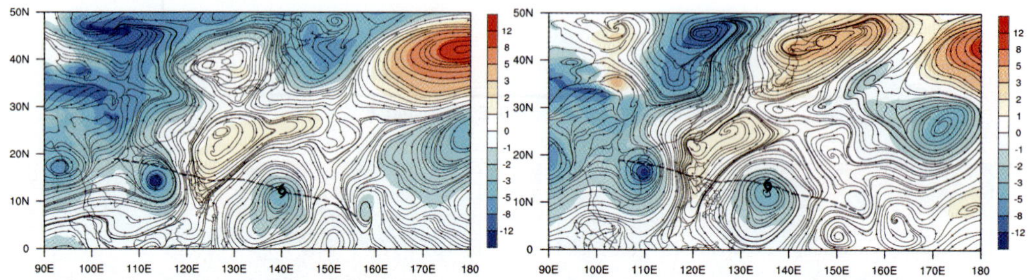

850 hPa flow/height anomalies at 1200 UTC 14 and 15 October 1985.

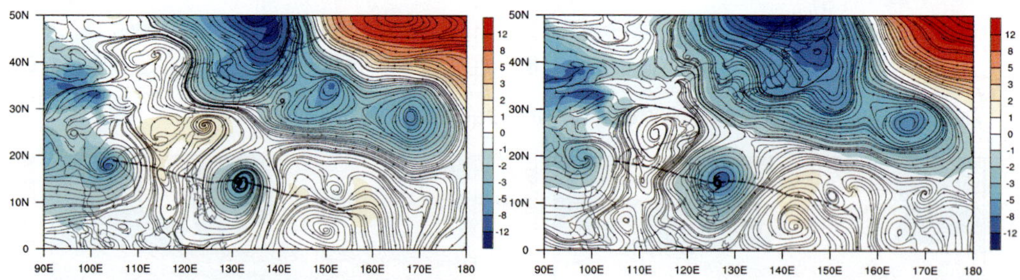

850 hPa flow/height anomalies at 1200 UTC 16 and 17 October 1985.

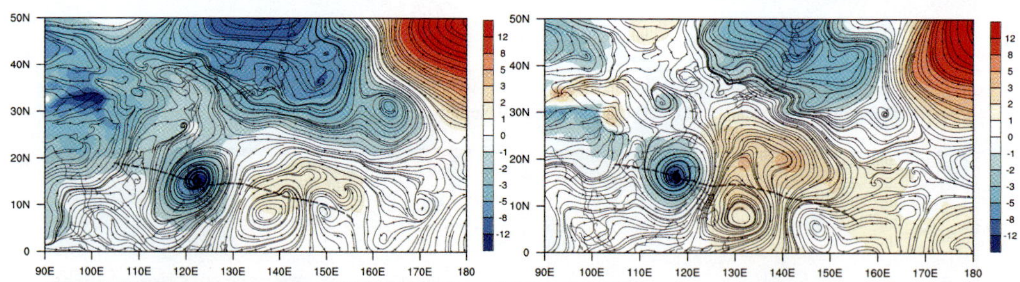

850 hPa flow/height anomalies at 1200 UTC 18 and 19 October 1985.

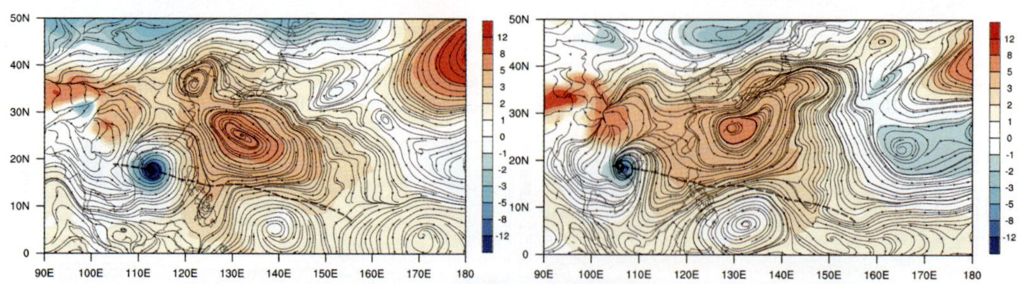

850 hPa flow/height anomalies at 1200 UTC 20 and 21 October 1985.

Figure 2.106. The typhoon (1985_No.21) generated in the equatorial central Pacific on 11 October 1985 and moved westward along the climatic monsoon trough. Its intensity peaked in the east sea of the northern Philippines on 16 October 1985 before landfall on the island and its track was not affected by anomalous vortices at 850 hPa.

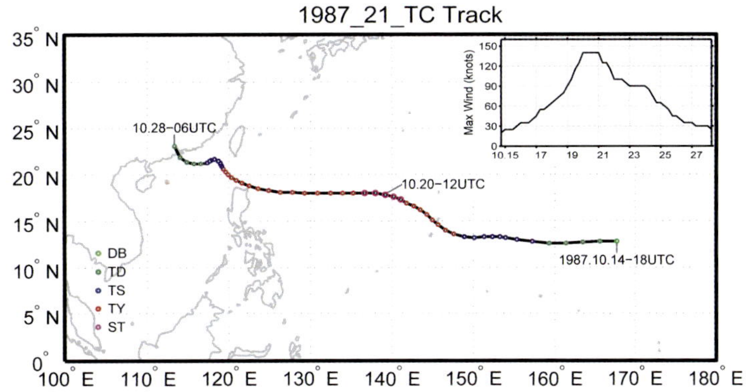

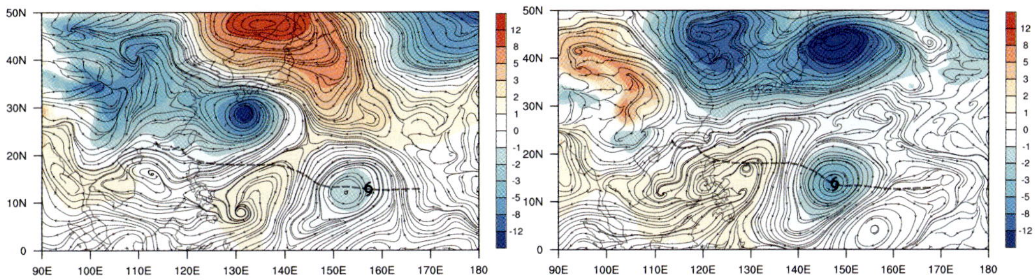
850 hPa flow/height anomalies at 0000 UTC 16 and 18 October 1987.

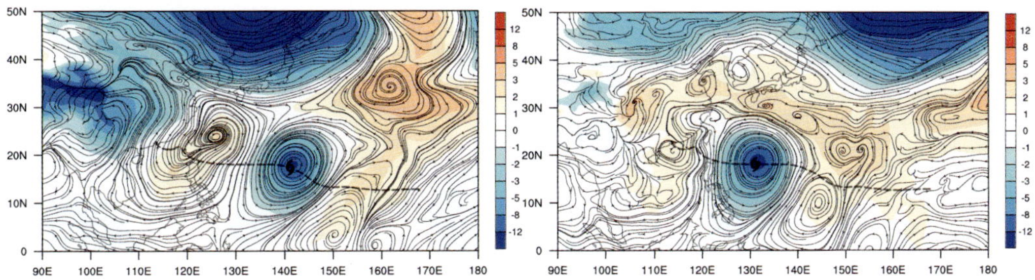
850 hPa flow/height anomalies at 0000 UTC 20 and 22 October 1987.

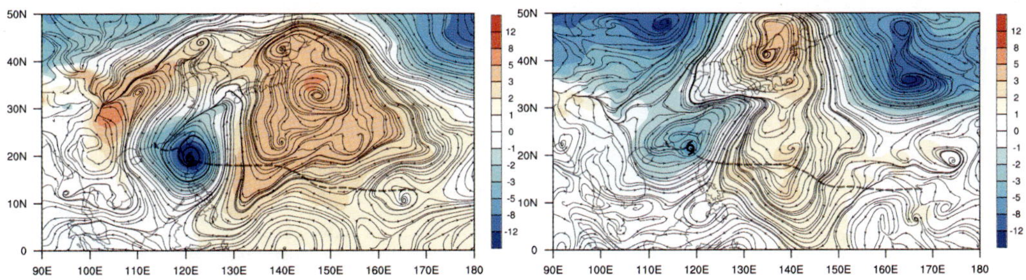
850 hPa flow/height anomalies at 0000 UTC 24 and 26 October 1987.

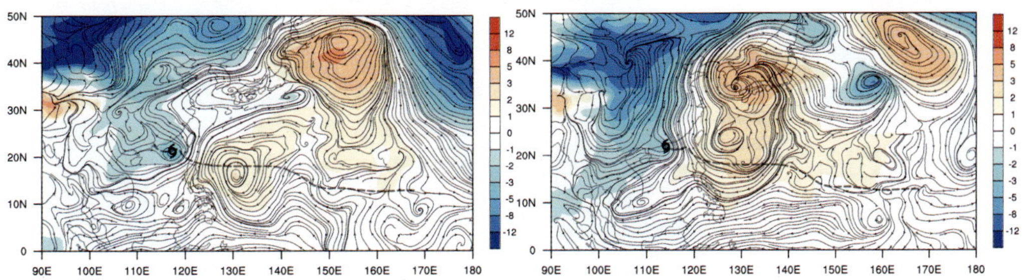
850 hPa flow/height anomalies at 0000 UTC 27 and 28 October 1987.

Figure 2.107. The typhoon (1987_No.21) generated in the tropical central Pacific on 14 October 1987 and moved westward along the climatic monsoon trough. Its intensity peaked in the tropical western Pacific on 20 October 1987 and its track with a small change was influenced by anomalous circulations at 850 hPa.

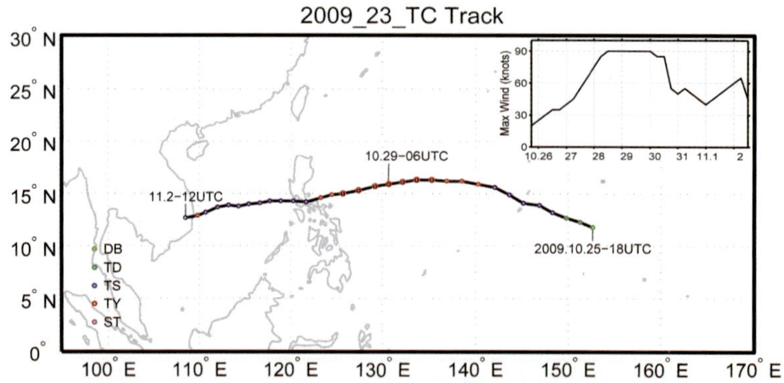

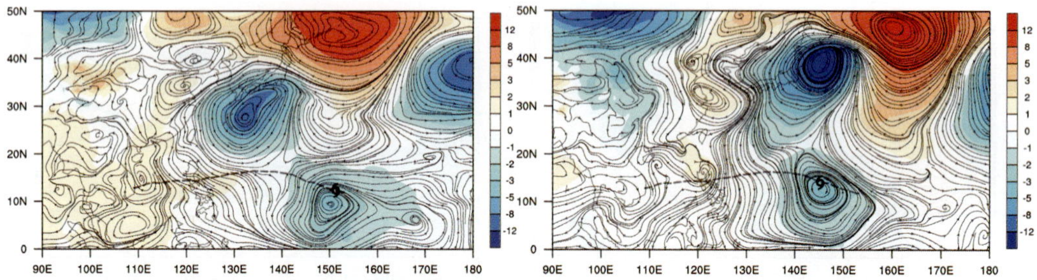

850 hPa flow/height anomalies at 0000 UTC 26 and 27 October 2009.

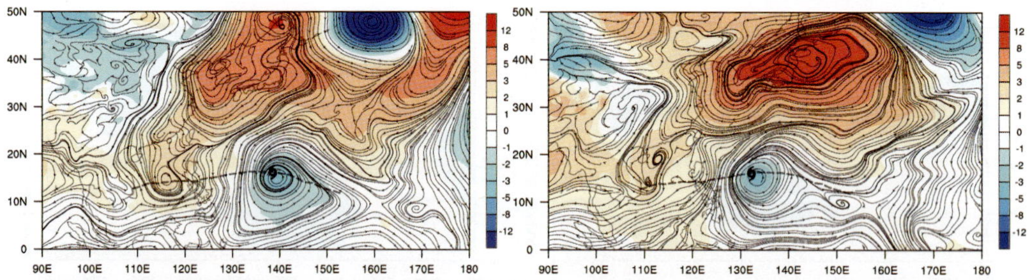

850 hPa flow/height anomalies at 0000 UTC 28 and 29 October 2009.

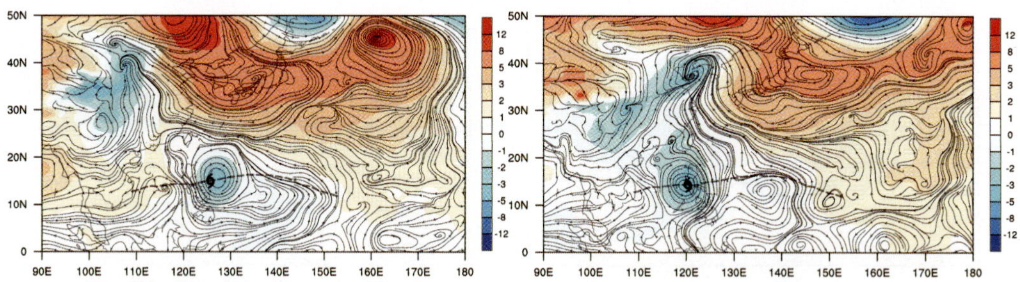

850 hPa flow/height anomalies at 0000 UTC 30 and 31 October 2009.

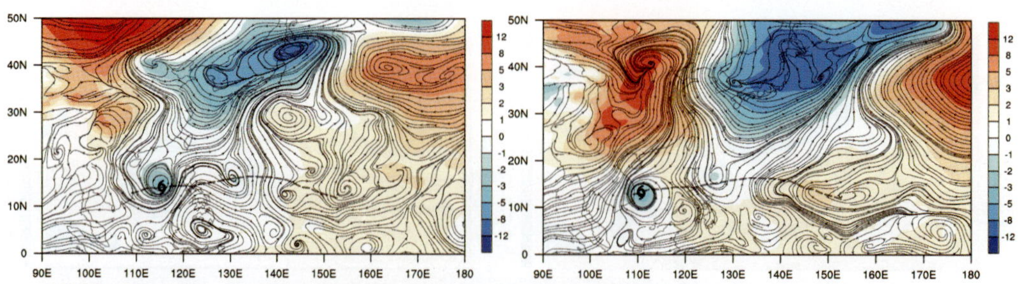

850 hPa flow/height anomalies at 0000 UTC 1 and 2 November 2009.

Figure 2.108. The typhoon (2009_No.23) generated in the tropical central Pacific on 25 October 2009 and moved westward along the climatic monsoon trough. Its intensity peaked in the east sea of the northern Philippines on 29 October 2009 before landfall on the island and its track with a small change was influenced by anomalous circulations at 850 hPa.

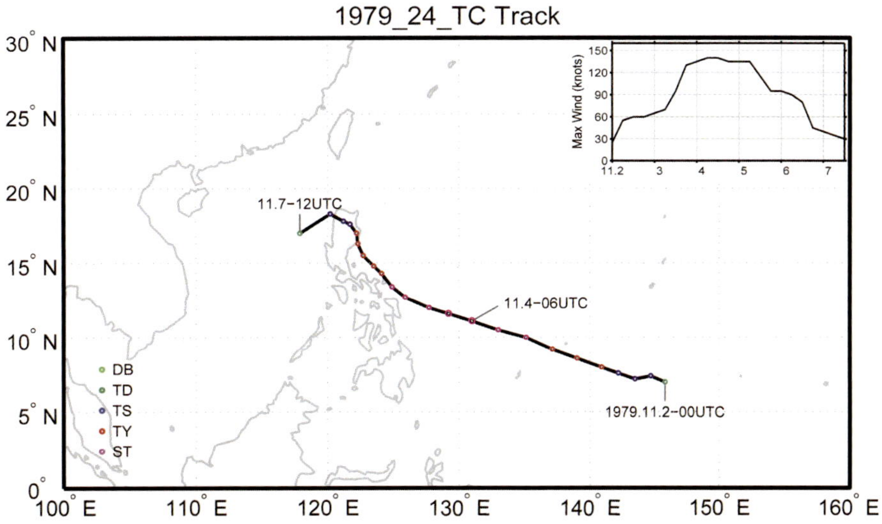

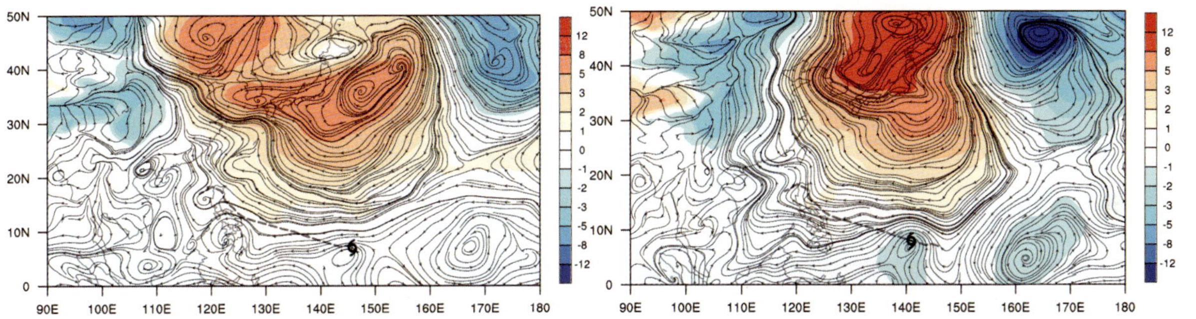

850 hPa flow/height anomalies at 0000 UTC 2 and 3 November 1979.

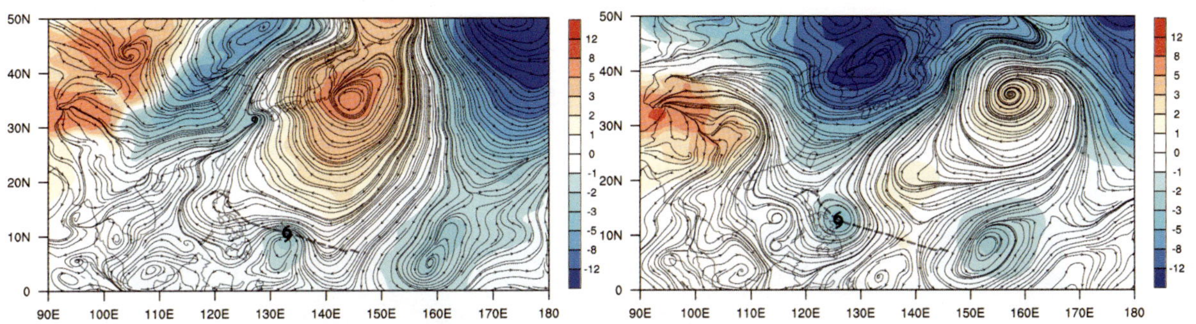

850 hPa flow/height anomalies at 0000 UTC 4 and 5 November 1979.

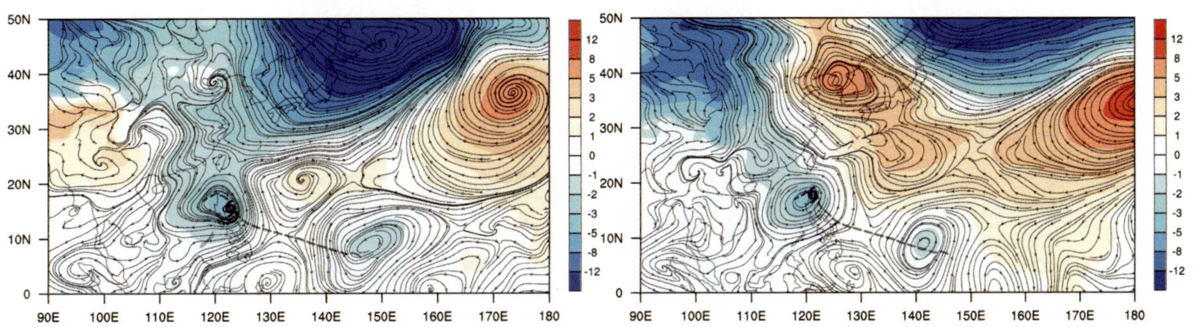

850 hPa flow/height anomalies at 0000 UTC 6 and 7 November 1979.

Figure 2.109. The typhoon (1979_No.24) generated in the equatorial western Pacific on 2 November 1979 and moved northwestward along the climatic monsoon trough. Its intensity peaked in the east sea of the Philippines on 4 November 1979 before landfall on the island and its track with a small change was modified by anomalous circulations at 850 hPa.

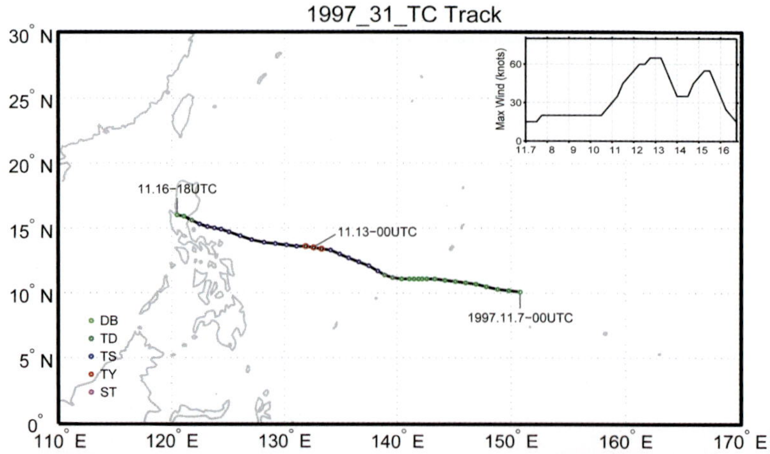

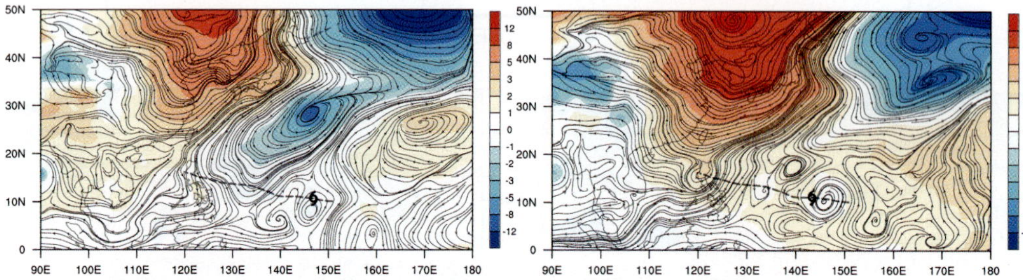

850 hPa flow/height anomalies at 0000 UTC 8 and 9 November 1997.

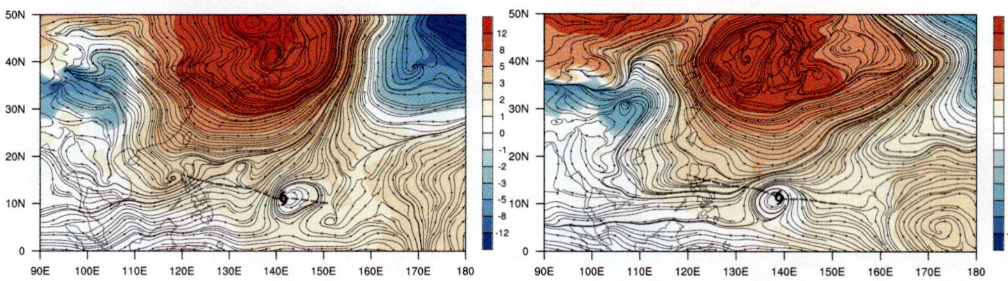

850 hPa flow/height anomalies at 0000 UTC 10 and 11 November 1997.

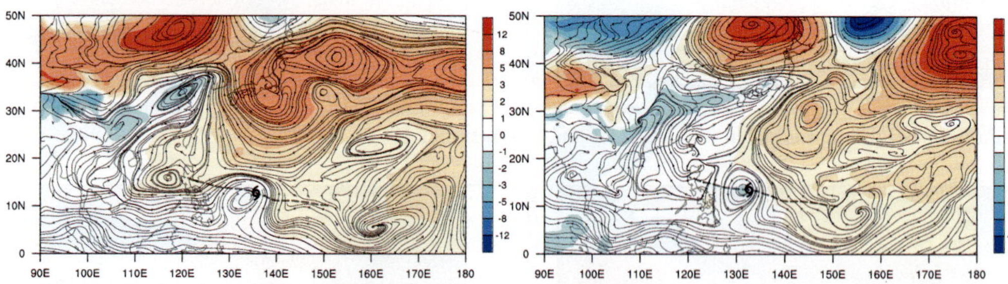

850 hPa flow/height anomalies at 0000 UTC 12 and 13 November 1997.

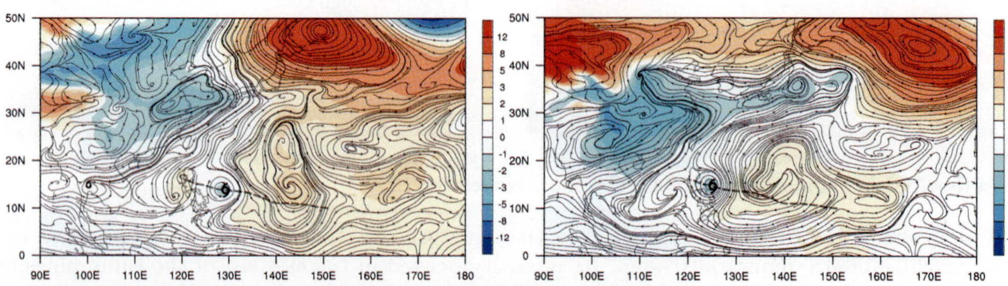

850 hPa flow/height anomalies at 0000 UTC 14 and 15 November 1997.

Figure 2.110. The typhoon (1997_No.31) generated in the tropical central Pacific on 7 November 1997 and moved west-northwestward along the climatic monsoon trough. Its intensity peaked in the tropical western Pacific on 13 November 1997 and its track was not affected by anomalous circulations at 850 hPa.

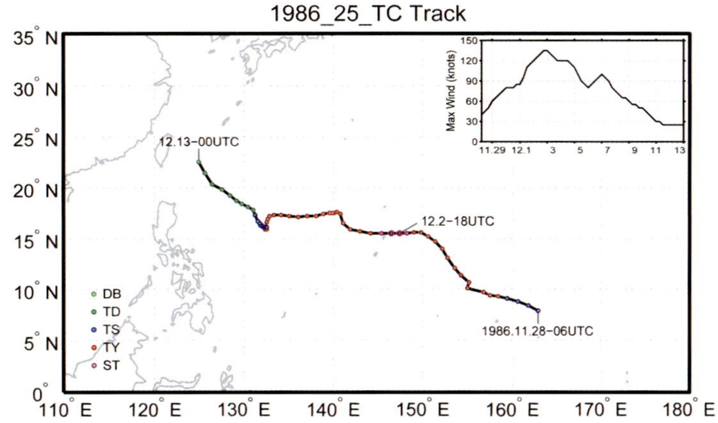

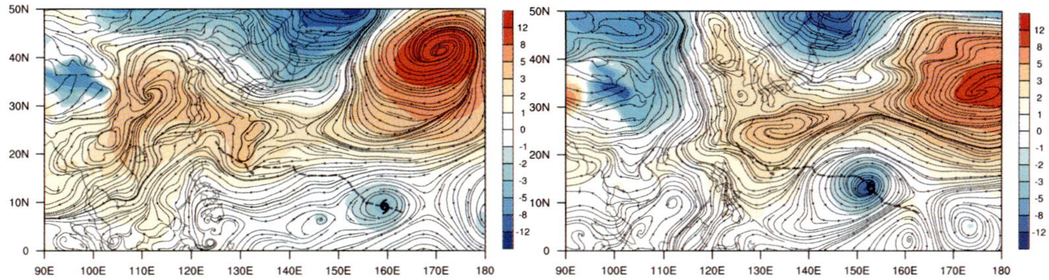

850 hPa flow/height anomalies at 0000 UTC 29 November and 1 December 1986.

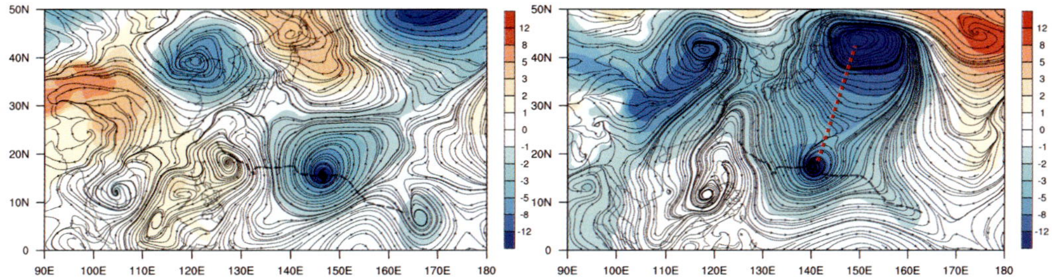

850 hPa flow/height anomalies at 0000 UTC 3 and 5 December 1986.

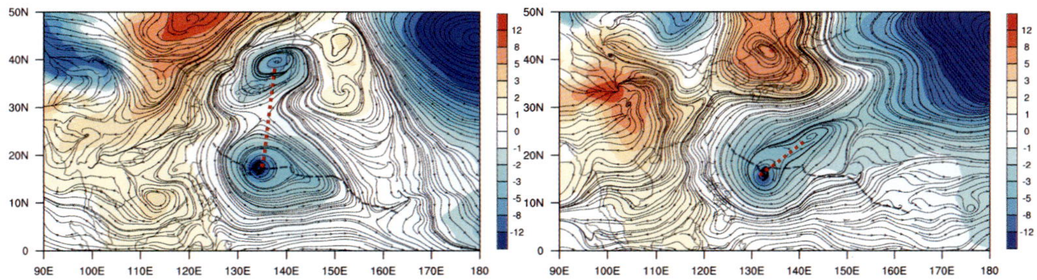

850 hPa flow/height anomalies at 0000 UTC 7 and 9 December 1986.

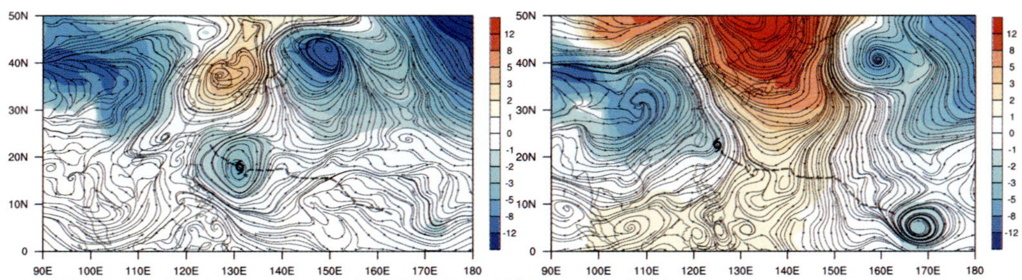

850 hPa flow/height anomalies at 0000 UTC 11 and 13 December 1986.

Figure 2.111. The typhoon (1986_No.25) generated in the equatorial central Pacific on 28 November 1986 and moved west-northwestward along the climatic monsoon trough. Its intensity peaked in the tropical western Pacific on 2 December 1986 and its track was modified by anomalous circulations and vortices at 850 hPa.

Chapter 3

Intensity Peaked in the South China Sea

Chapter 3 describes typhoons, which intensities peaked in the South China Sea (SCS) but they generated in the SCS (Sec. 3.1), in the Philippines area (Sec. 3.2), in the equatorial Pacific (Sec. 3.3), and in the tropical Pacific (Sec. 3.4), respectively. These typhoons experienced right turns or left turns of motions due to the influence of other anomalous vortices.

3.1. Geneses in the South China Sea

There were 10 typhoons that generated from tropical disturbances, then reached their peak intensity in the SCS, and moved to other regions or made landfall over Southeast Asian continents. The tracks of these typhoons were influenced by anomalous vortices at 300 hPa and/or 850 hPa (Figs. 3.1–3.10).

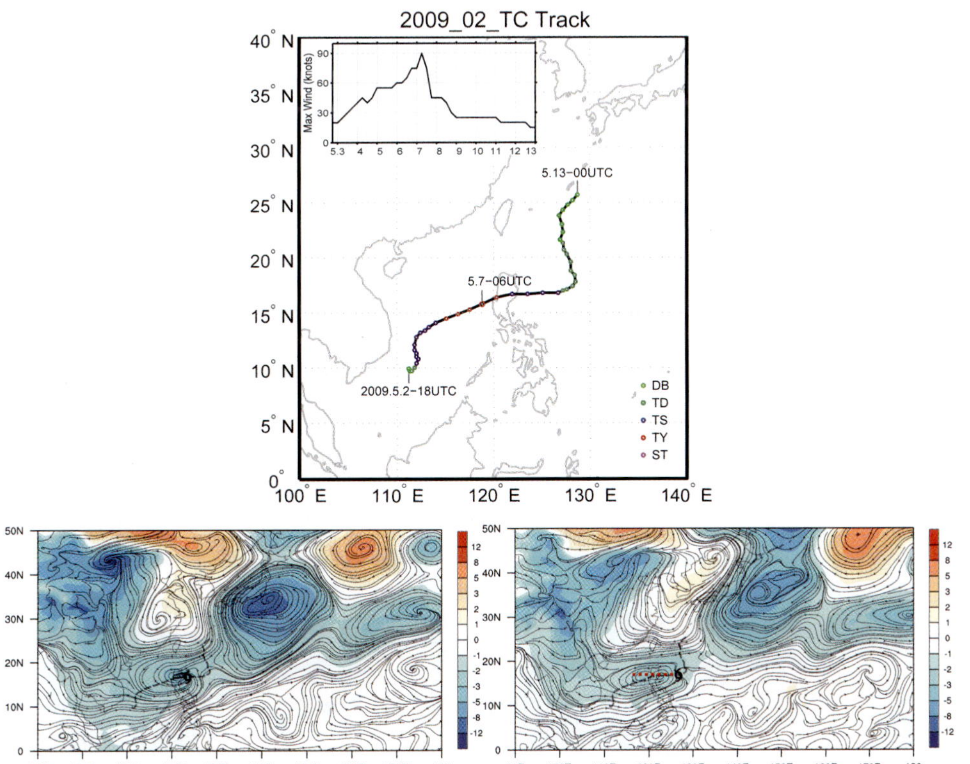

850 hPa flow/height (shading, 10 gpm) anomalies at 0000 UTC and 1200 UTC 8 May 2009.

Figure 3.1. The typhoon (2009_No.02) generated in the southern SCS on 2 May 2009 and moved north-northeastward. Its intensity peaked in the central eastern SCS on 7 May 2009 before landfall on the northern Philippines. It turned left after passing the island. This track was due to the attraction and interaction with anomalous vortices at 850 hPa and 300 hPa in the east sea of the Philippines. Symbols used are same as in Fig. 2.1.

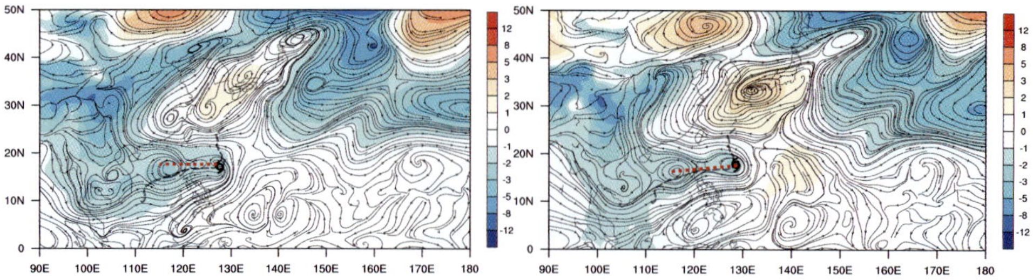

850 hPa flow/height anomalies at 0000 UTC and 1200 UTC 9 May 2009.

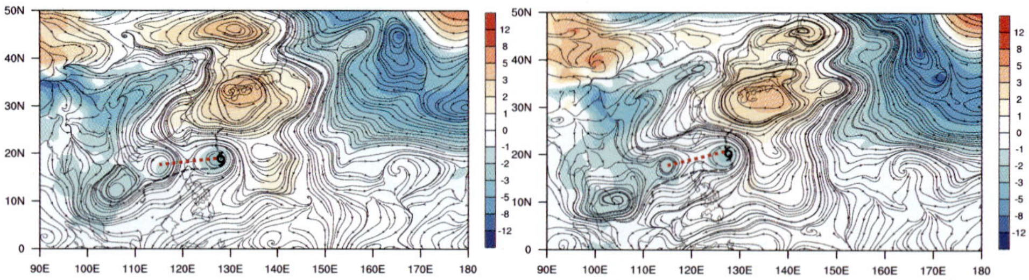

850 hPa flow/height anomalies at 0000 UTC and 1200 UTC 10 May 2009.

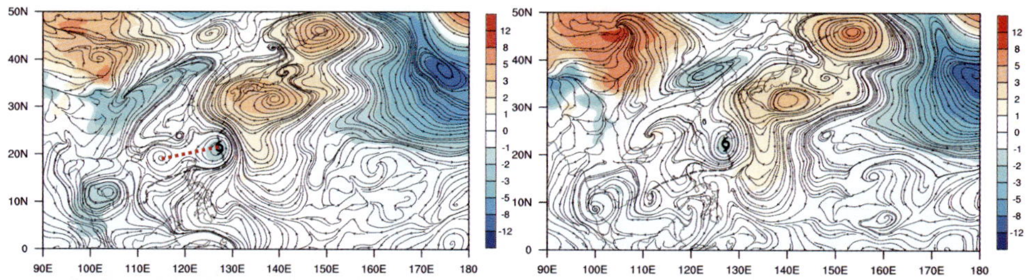

850 hPa flow/height anomalies at 0000 UTC and 1200 UTC 11 May 2009.

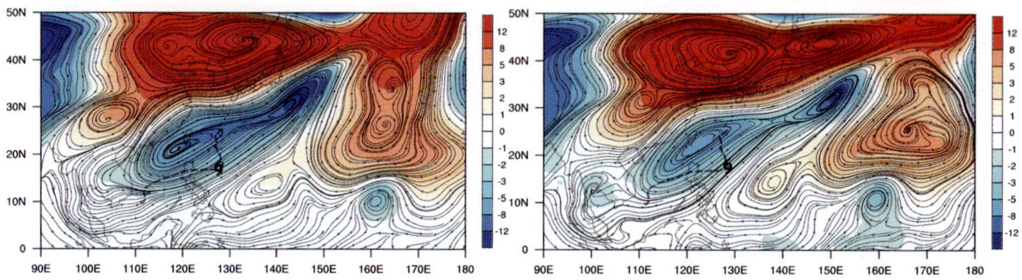

300 hPa flow/height anomalies at 0000 UTC and 1200 UTC 9 May 2009.

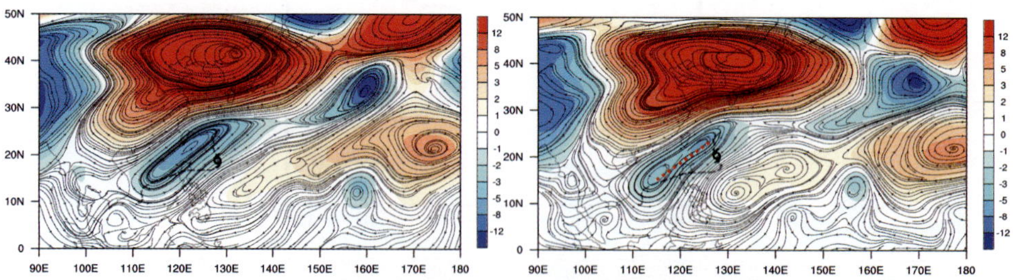

300 hPa flow/height anomalies at 0000 UTC and 1200 UTC 10 May 2009.

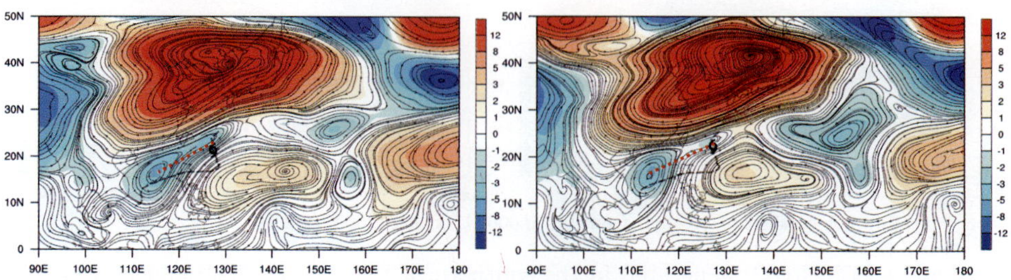

300 hPa flow/height anomalies at 0000 UTC and 1200 UTC 11 May 2009.

Figure 3.1. *(Continued)*

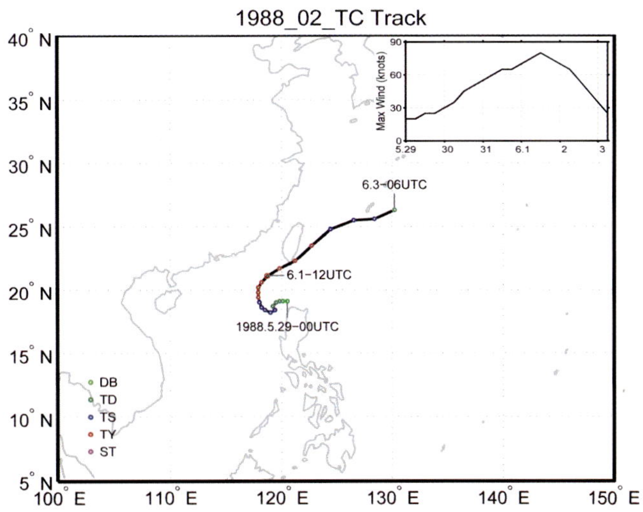

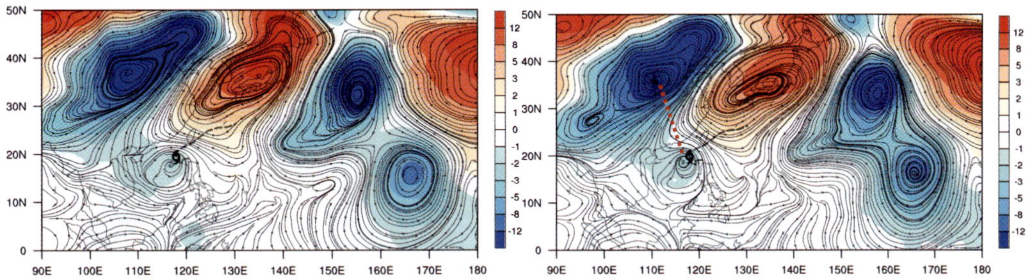

300 hPa flow/height anomalies at 1200 UTC and 1800 UTC 31 May 1988.

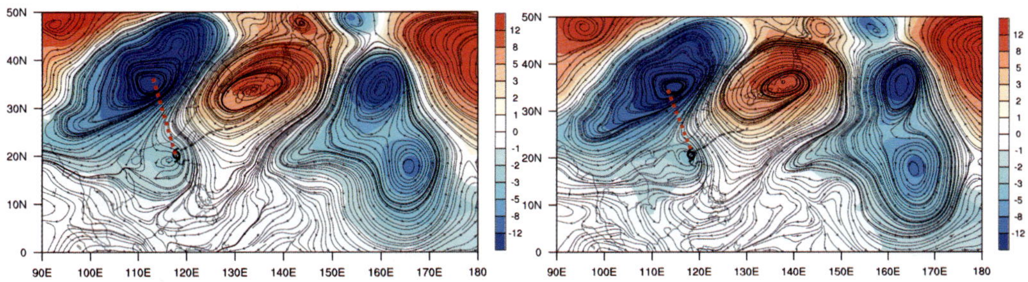

300 hPa flow/height anomalies at 0000 UTC and 0600 UTC 1 June 1988.

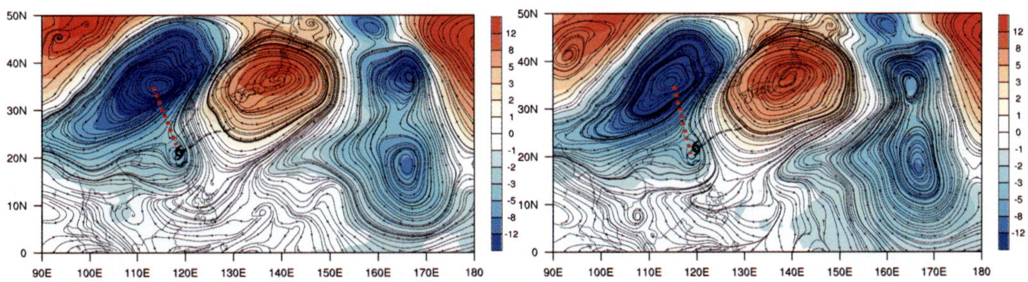

300 hPa flow/height anomalies at 1200 UTC and 1800 UTC 1 June 1988.

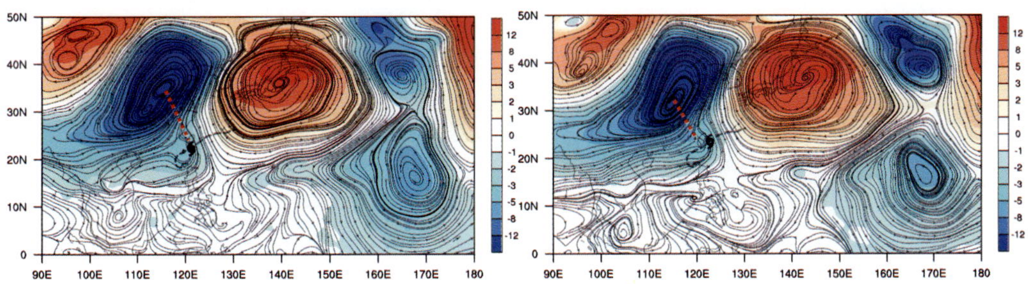

300 hPa flow/height anomalies at 0000 UTC and 0600 UTC 2 June 1988.

Figure 3.2. The typhoon (1988_No.02) generated near the northern Philippines on 29 May 1988 and had a curvature motion in the northeastern SCS. This motion was influenced by an anomalous vortex in China at 300 hPa. Its intensity peaked on 1 June 1988 before it was approaching the Taiwan Island.

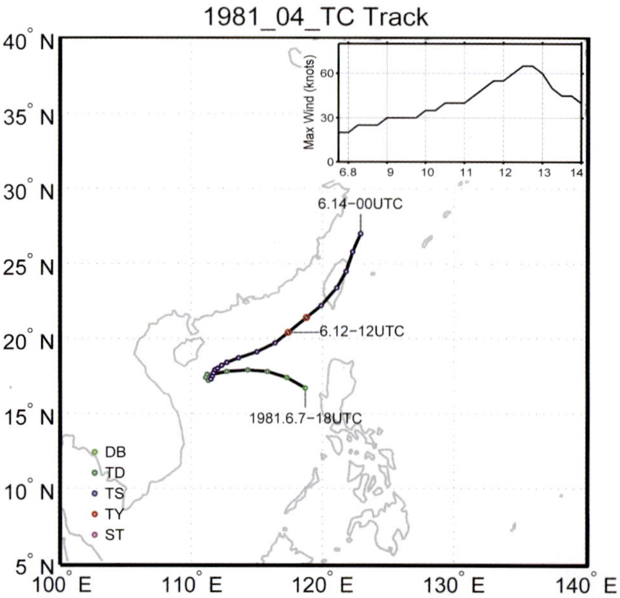

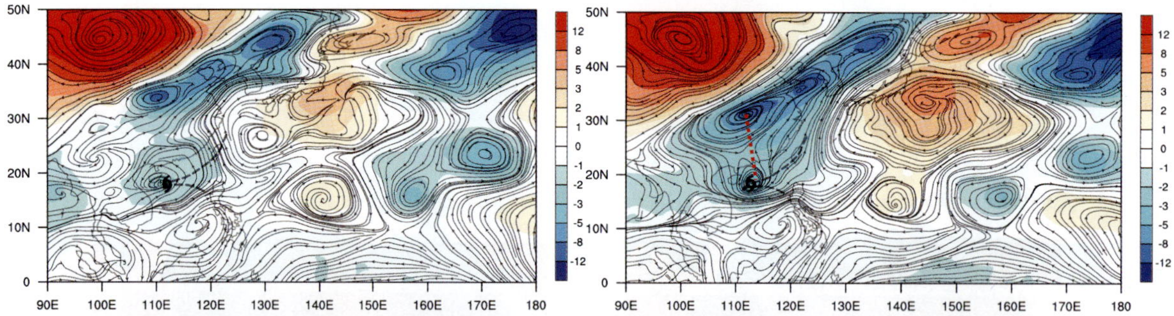

300 hPa flow/height anomalies at 0000 UTC and 1200 UTC 11 June 1981.

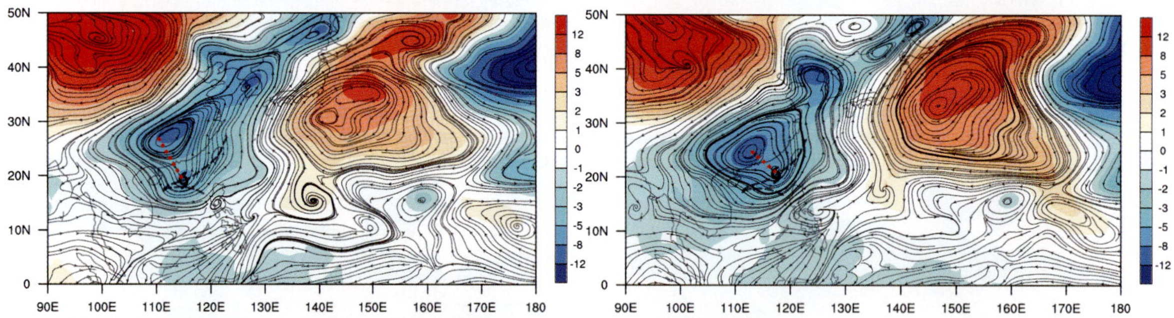

300 hPa flow/height anomalies at 0000 UTC and 1200 UTC 12 June 1981.

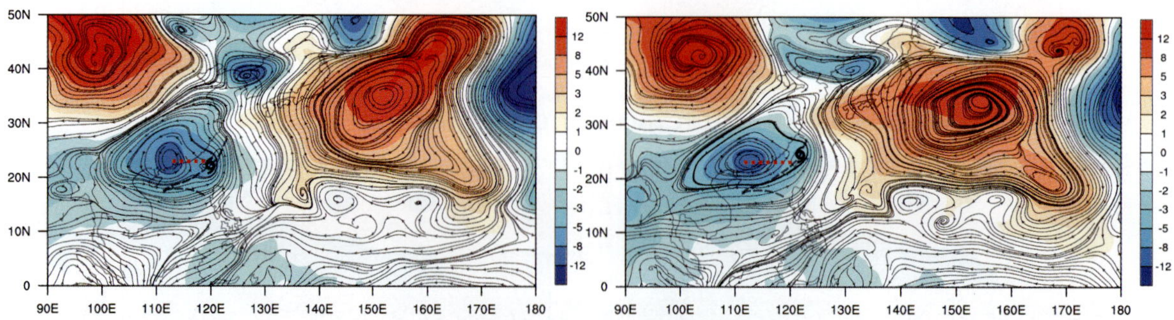

300 hPa flow/height anomalies at 0000 UTC and 1200 UTC 13 June 1981.

Figure 3.3. The typhoon (1981_No.04) generated in the central eastern SCS on 7 June 1981. Its intensity peaked on 12 June 1981 before landfall on the Taiwan Island. It had a returning motion that was influenced by the attraction and interaction with an anomalous vortex at 300 hPa.

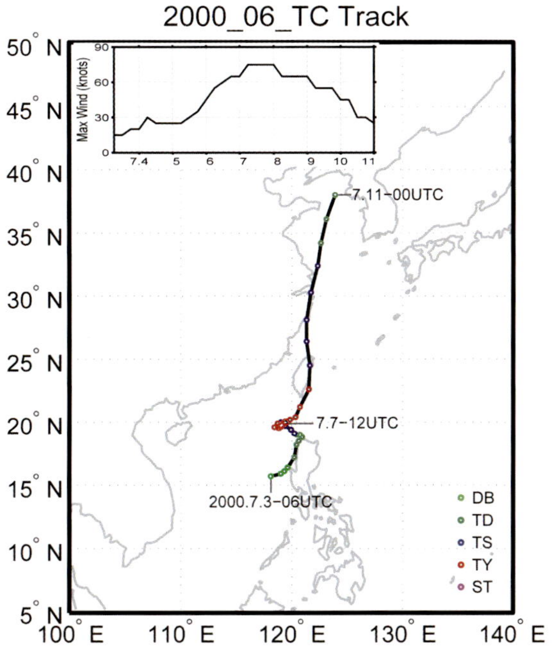

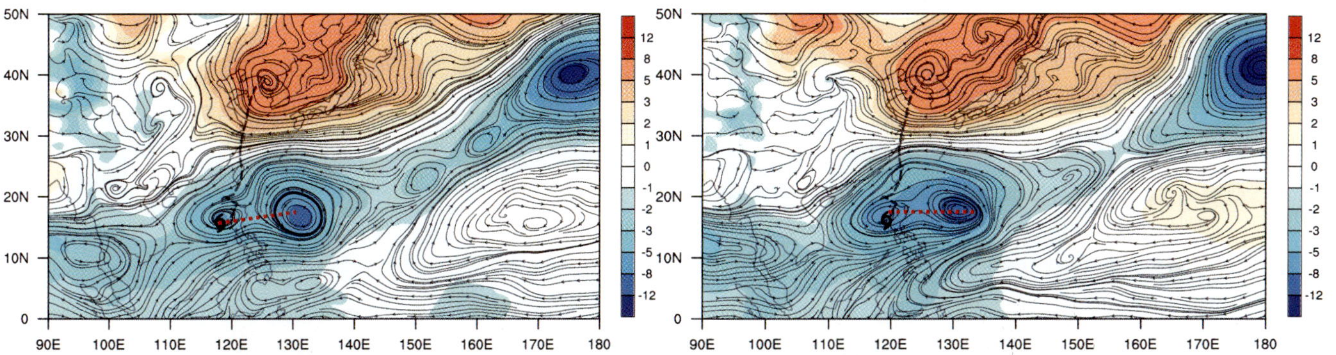

850 hPa flow/height anomalies at 0600 UTC and 1800 UTC 3 July 2007.

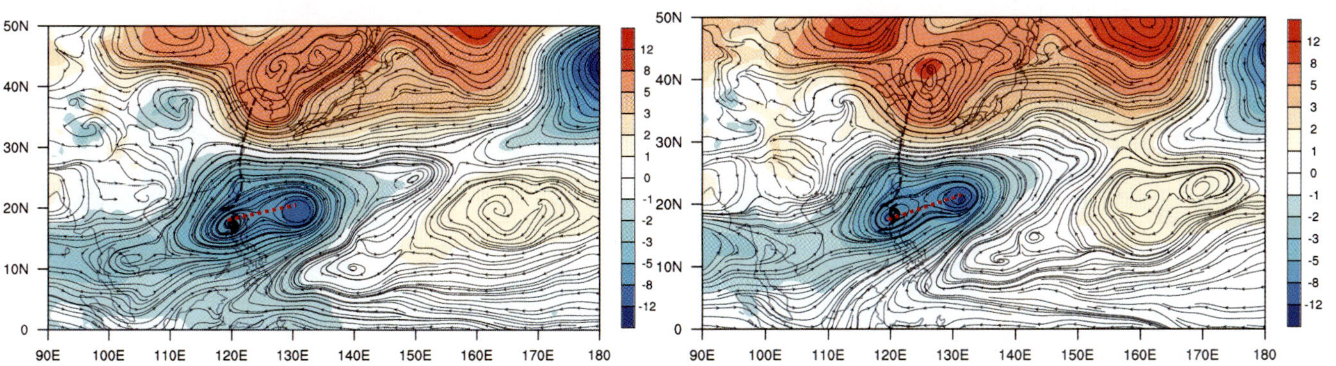

850 hPa flow/height anomalies at 0600 UTC and 1800 UTC 4 July 2007.

Figure 3.4. The typhoon (2000_No.06) generated in the central eastern SCS on 3 July 2000 and had a stagnant state for a few days between the northern Philippines and the Taiwan Island. This motion was influenced by the interaction with an anomalous vortex at 850 hPa. Its intensity peaked between the two islands on 7 July 2000.

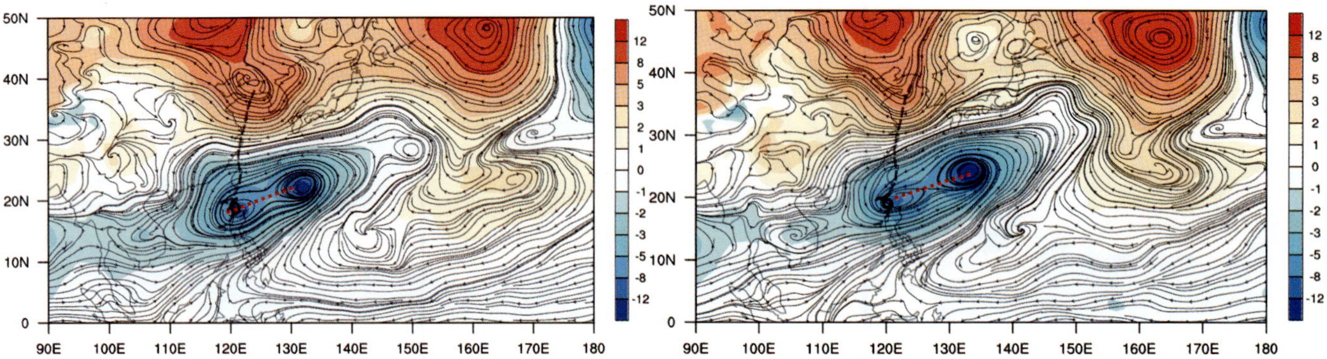

850 hPa flow/height anomalies at 0600 UTC and 1800 UTC 5 July 2007.

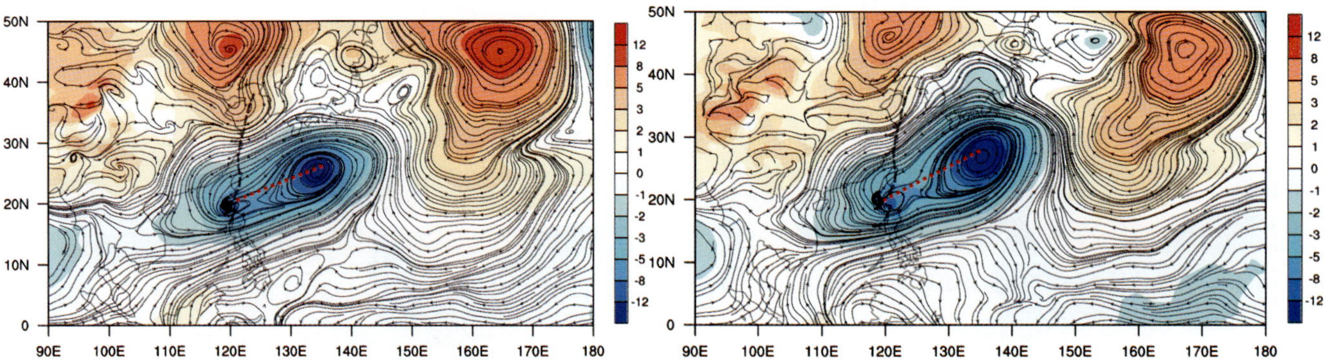

850 hPa flow/height anomalies at 0600 UTC and 1800 UTC 6 July 2007.

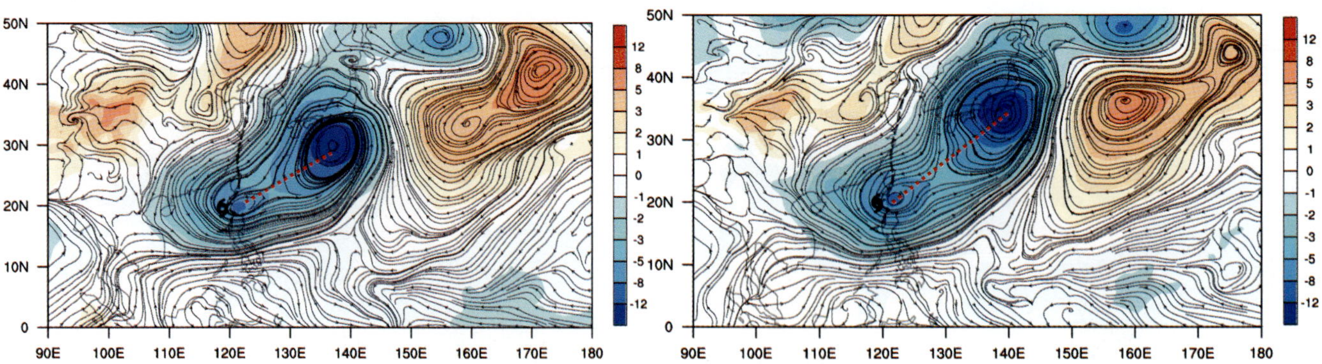

850 hPa flow/height anomalies at 0600 UTC and 1800 UTC 7 July 2007.

Figure 3.4. (*Continued*)

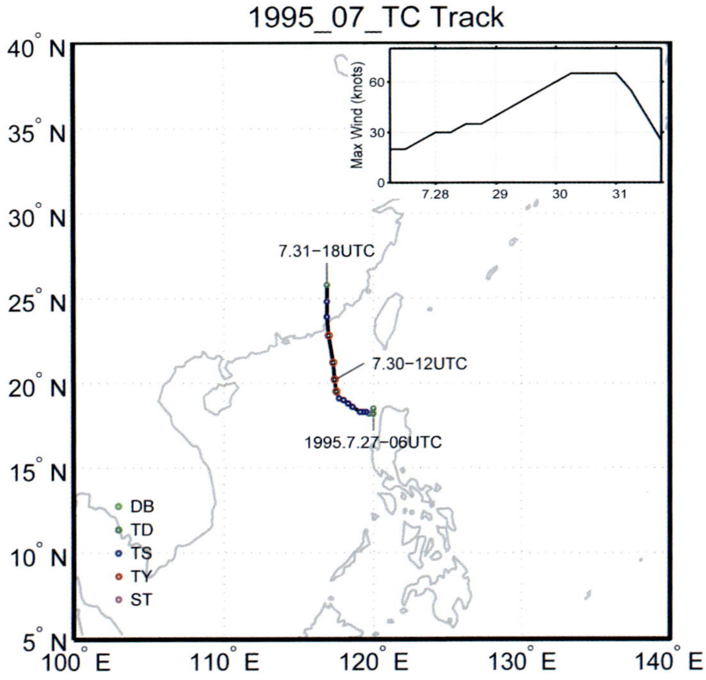

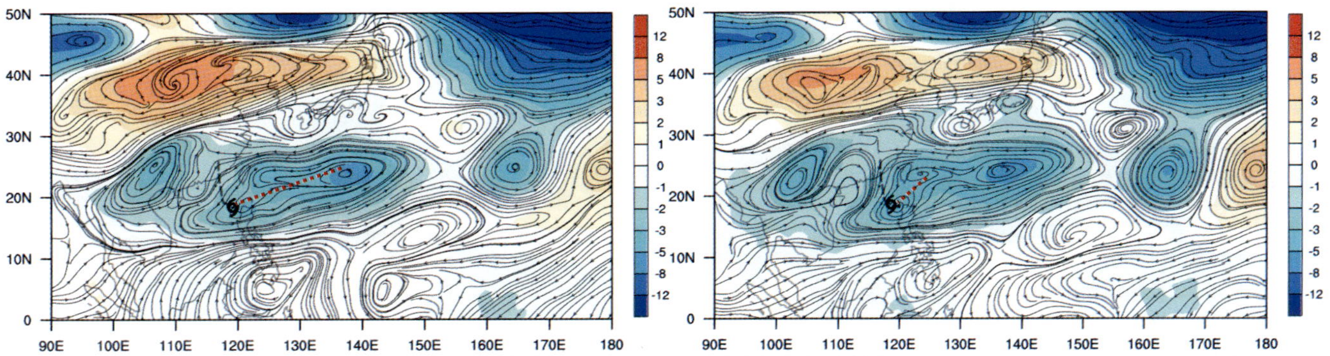

300 hPa flow/height anomalies at 0000 UTC and 0600 UTC 29 July 1995.

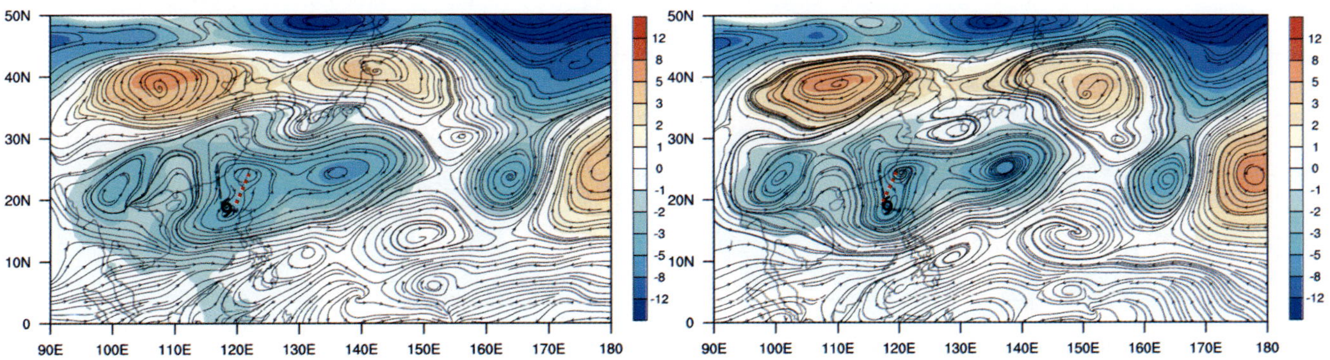

300 hPa flow/height anomalies at 1200 UTC and 1800 UTC 29 July 1995.

Figure 3.5. The typhoon (1995_No.07) generated near the northern Philippines on 27 July 1995 and had a motion of turning right in the northeastern SCS. This motion was influenced by the interaction with anomalous vortices at 300 hPa. Its intensity peaked in the northeastern SCS on 30 July 1995 before landfall on the South China coast.

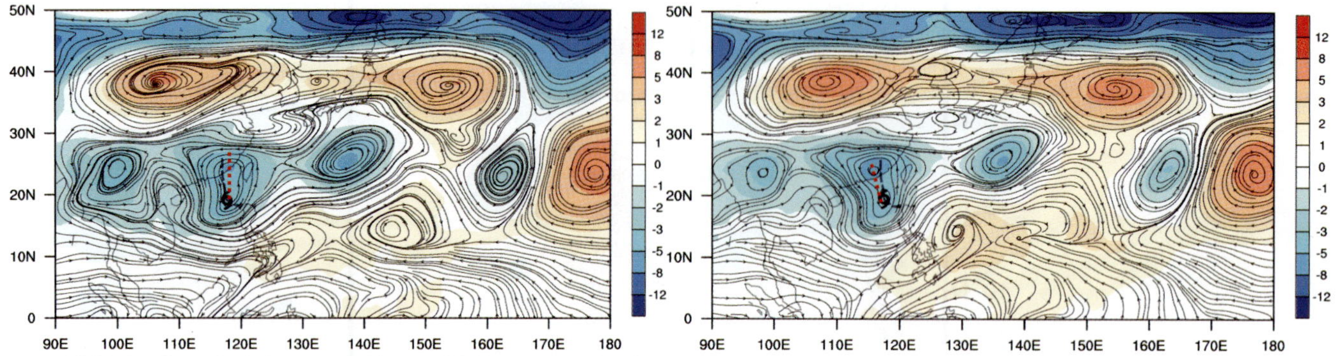

300 hPa flow/height anomalies at 0000 UTC and 0600 UTC 30 July 1995.

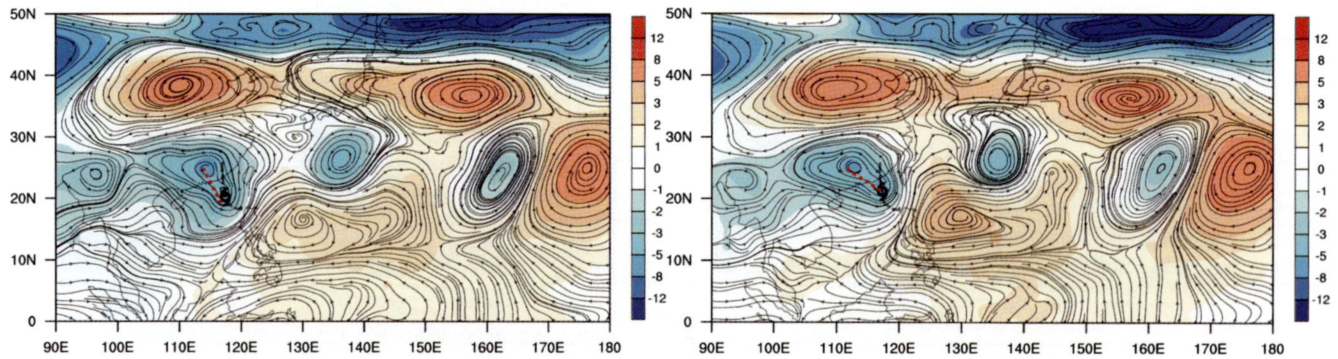

300 hPa flow/height anomalies at 1200 UTC and 1800 UTC 30 July 1995.

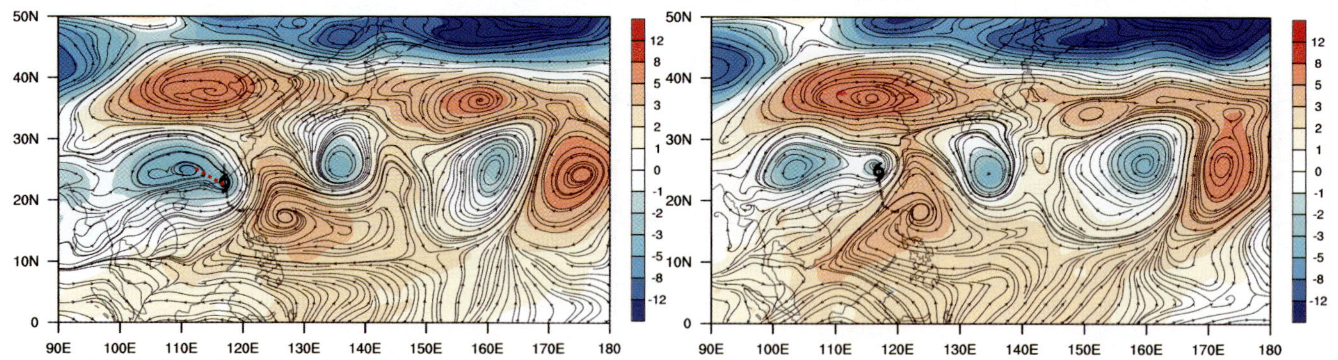

300 hPa flow/height anomalies at 0000 UTC and 1200 UTC 31 July 1995.

Figure 3.5. (*Continued*)

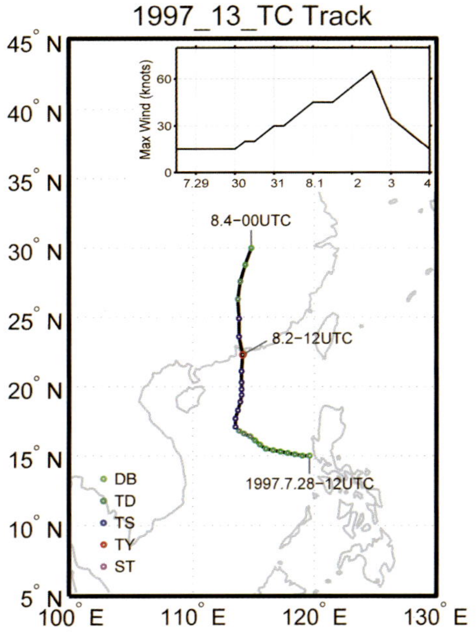

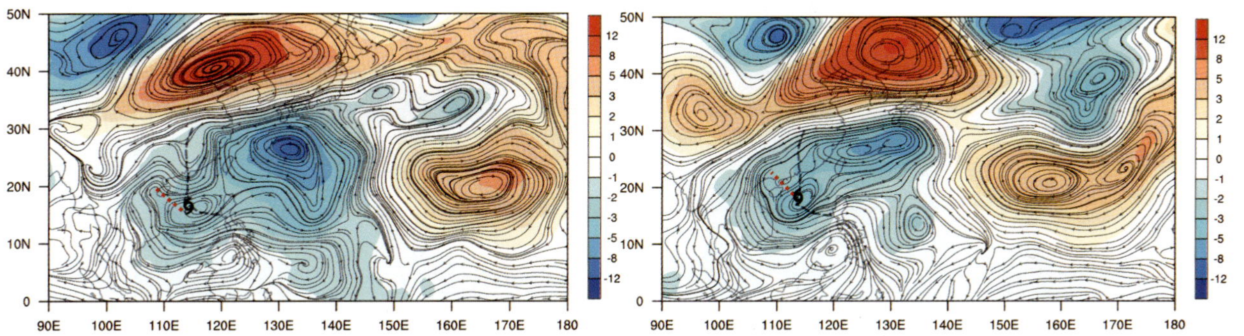

300 hPa flow/height anomalies at 0000 UTC 31 July and 1 August 1997.

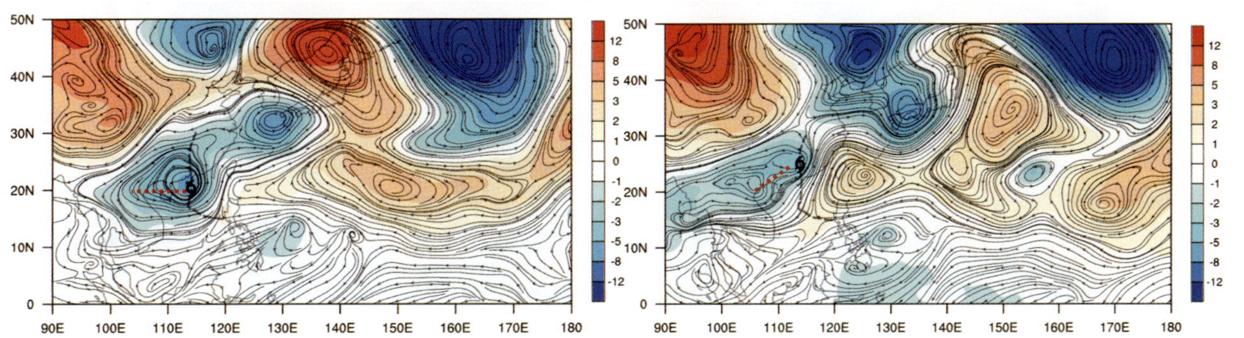

300 hPa flow/height anomalies at 0000 UTC 2 and 3 August 1997.

Figure 3.6. The typhoon (1997_No.13) generated in the west sea of the Philippines on 28 July 1997 and turned right in the northern SCS. This motion was influenced by an anomalous vortex at 300 hPa. Its intensity peaked near the climatic monsoon trough in the northernmost SCS on 2 August 1997 before landfall on the South China coast.

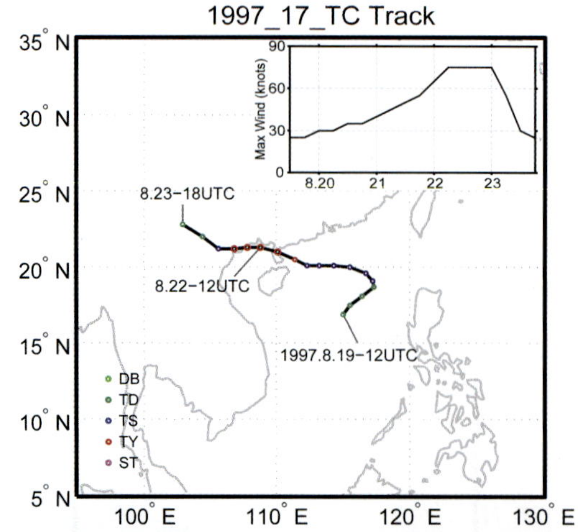

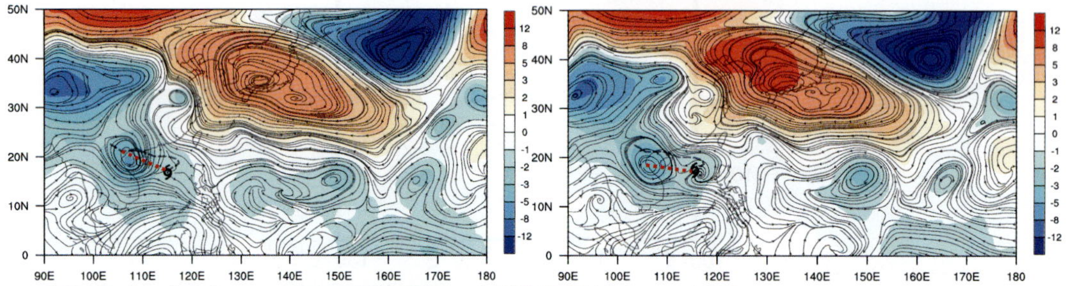

300 hPa flow/height anomalies at 1200 UTC and 1800 UTC 19 August 1997.

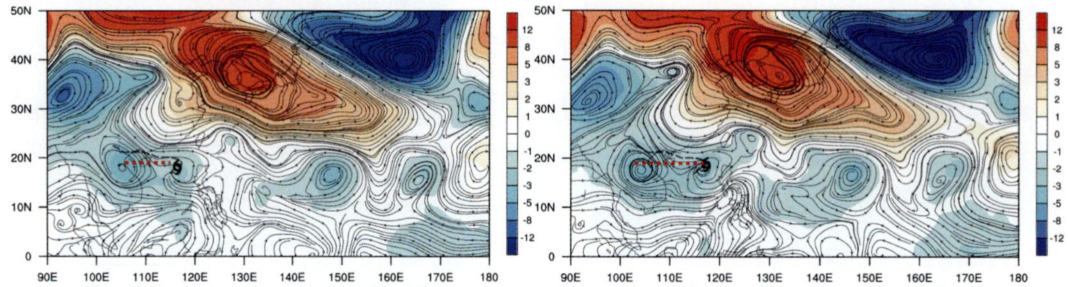

300 hPa flow/height anomalies at 0000 UTC and 0600 UTC 20 August 1997.

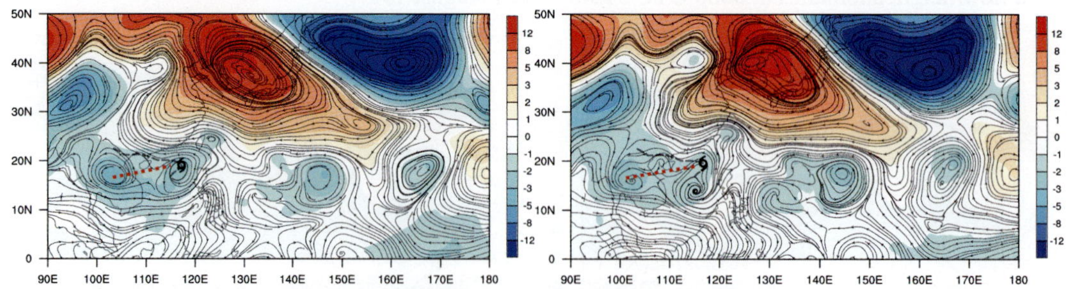

300 hPa flow/height anomalies at 1200 UTC and 1800 UTC 20 August 1997.

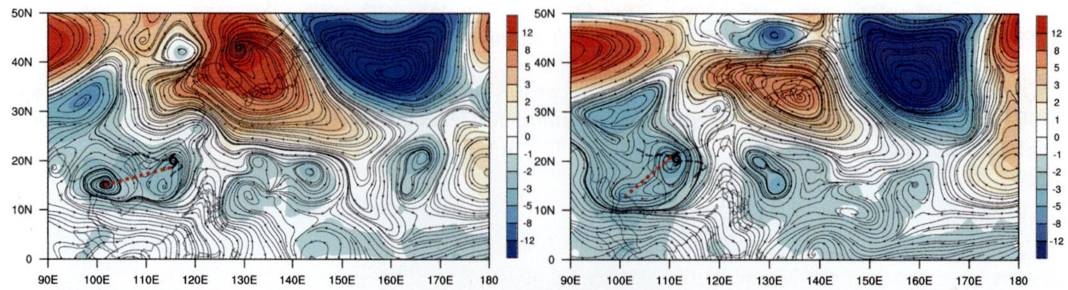

300 hPa flow/height anomalies at 0000 UTC 21 and 22 August 1997.

Figure 3.7. The typhoon (1997_No.17) generated in the central northern SCS on 19 August 1995 and turned left in the northern SCS. This motion was influenced by the interaction with an anomalous vortex at 300 hPa. Its intensity peaked in the South China offshore region on 22 August 1997 when it was approaching the climatic monsoon trough before landfall on the Southwest China coast.

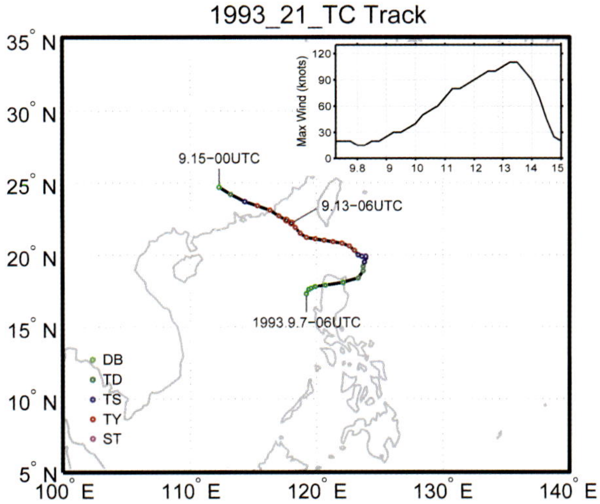

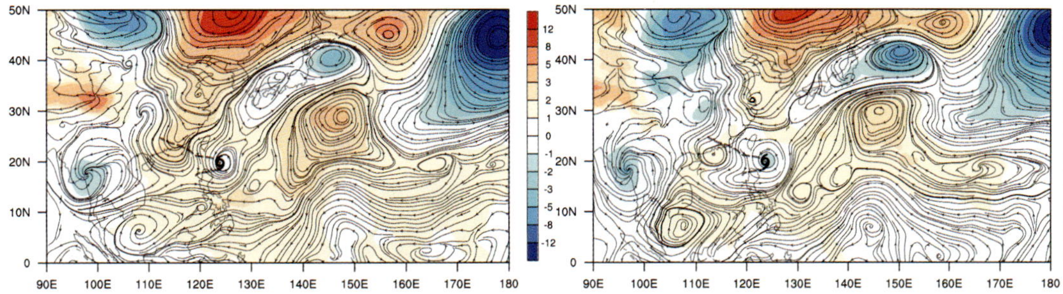

850 hPa flow/height anomalies at 0000 UTC and 1200 UTC 10 September 1993.

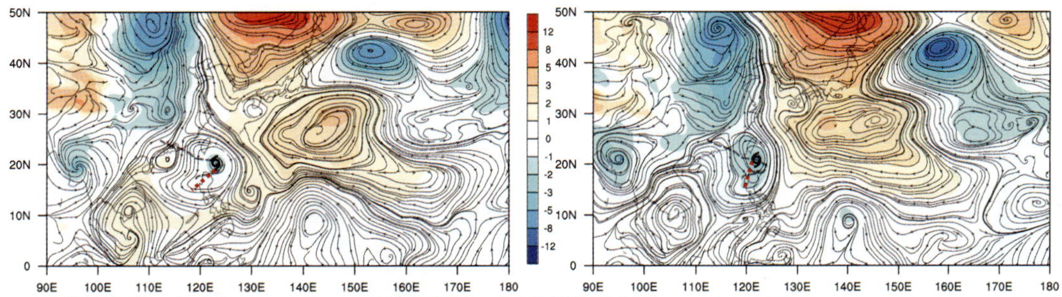

850 hPa flow/height anomalies at 0000 UTC and 1200 UTC 11 September 1993.

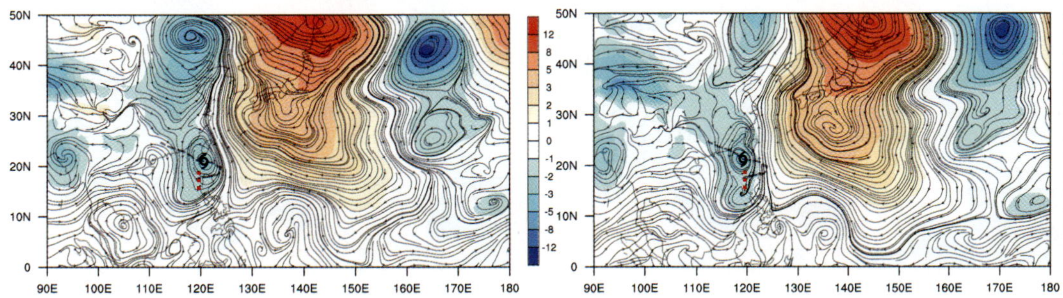

850 hPa flow/height anomalies at 0000 UTC and 1200 UTC 12 September 1993.

Figure 3.8. The typhoon (1993_No.21) generated in the northwest sea of the Philippines on 7 September 1993 and turned left. Its intensity peaked in the northernmost SCS on 13 September 1993 when it was approaching the climatic monsoon trough before landfall on the South China coast. This motion was due to the attraction and interaction with anomalous vortices at 850 hPa and 300 hPa.

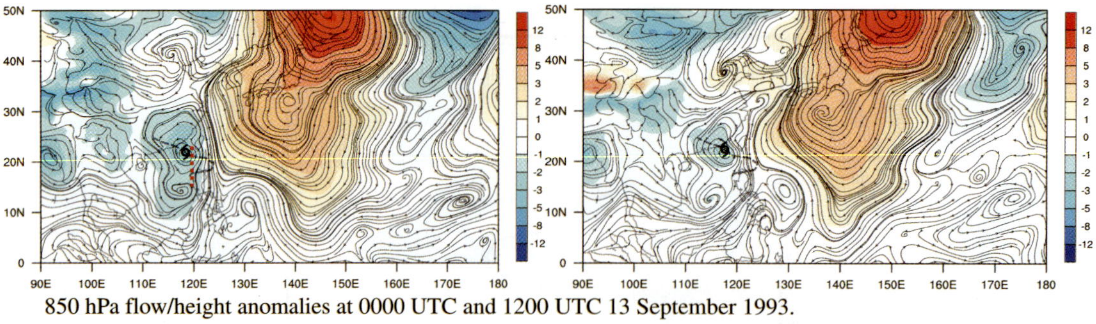

850 hPa flow/height anomalies at 0000 UTC and 1200 UTC 13 September 1993.

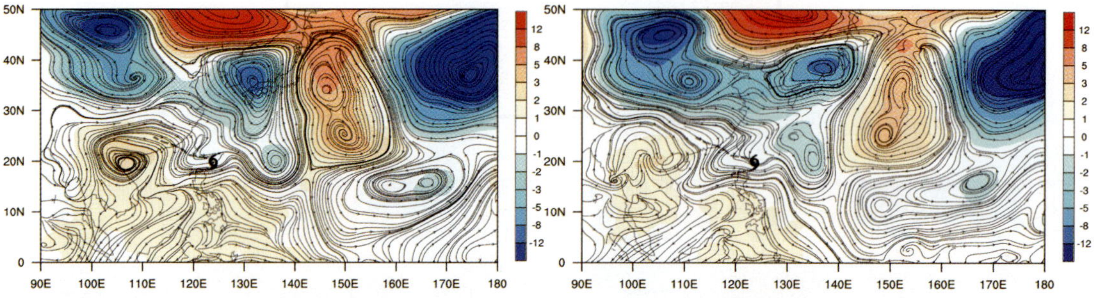

300 hPa flow/height anomalies at 0000 UTC and 1200 UTC 10 September 1993.

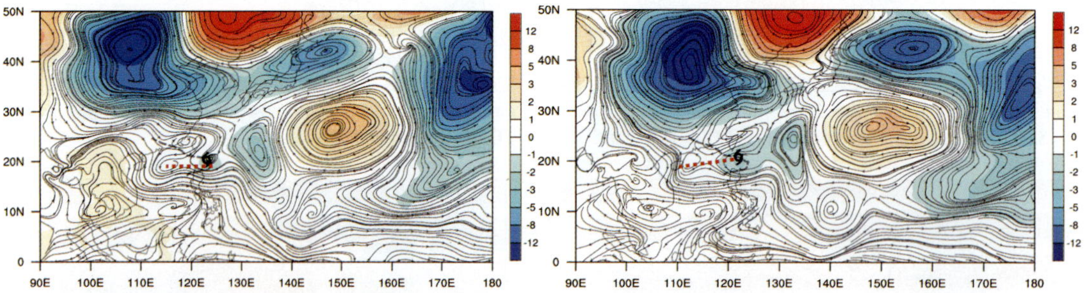

300 hPa flow/height anomalies at 0000 UTC and 1200 UTC 11 September 1993.

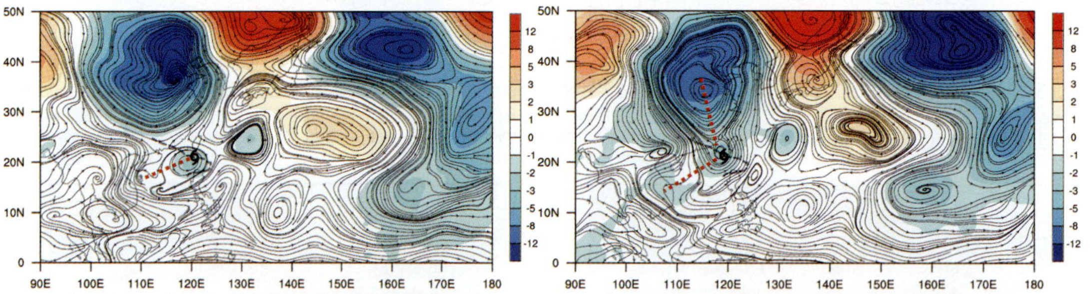

300 hPa flow/height anomalies at 0000 UTC and 1200 UTC 12 September 1993.

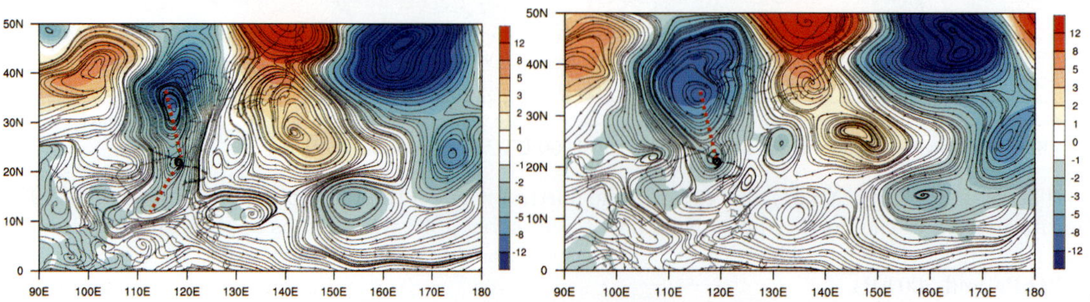

300 hPa flow/height anomalies at 0000 UTC and 1200 UTC 13 September 1993.

Figure 3.8. *(Continued)*

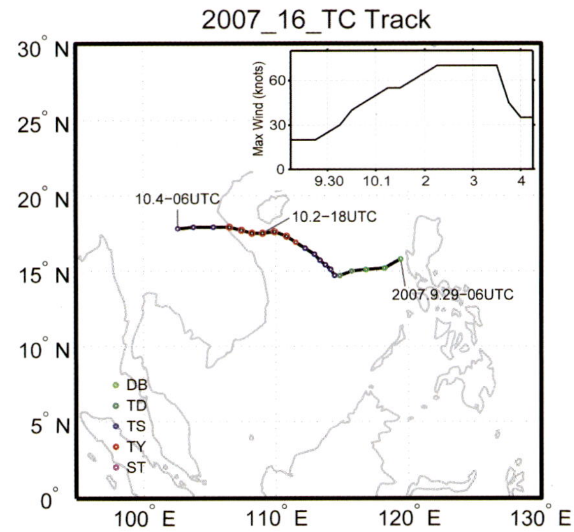

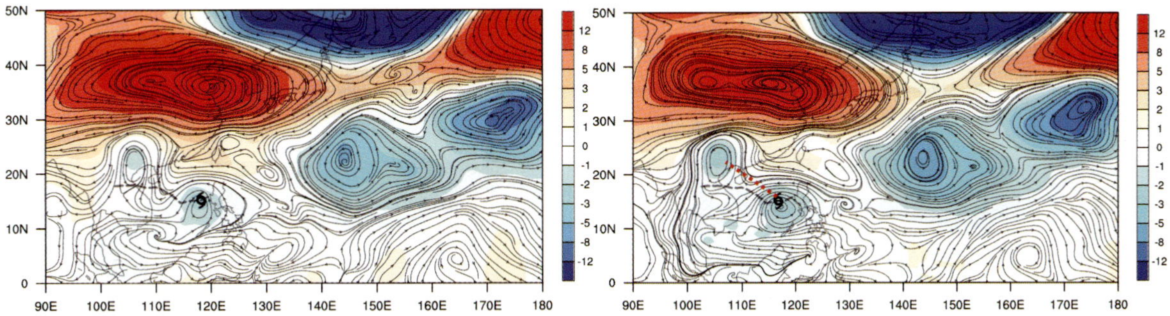

300 hPa flow/height anomalies at 1200 UTC and 1800 UTC 29 September 2007.

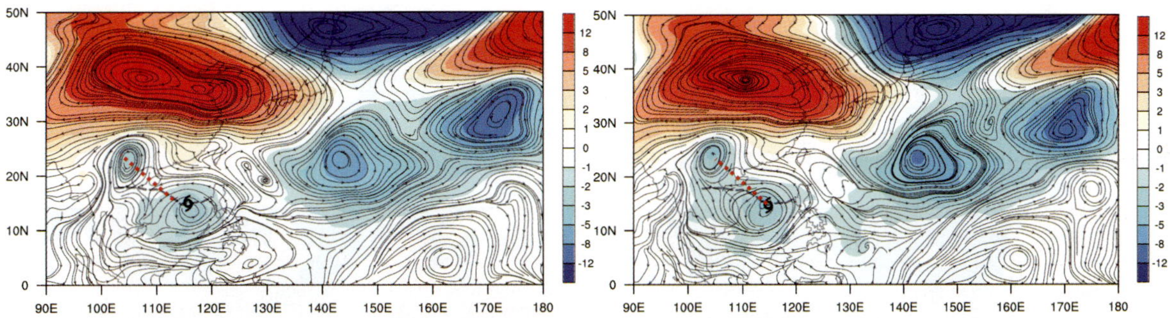

850 hPa flow/height anomalies at 0000 UTC and 0600 UTC 30 September 1997.

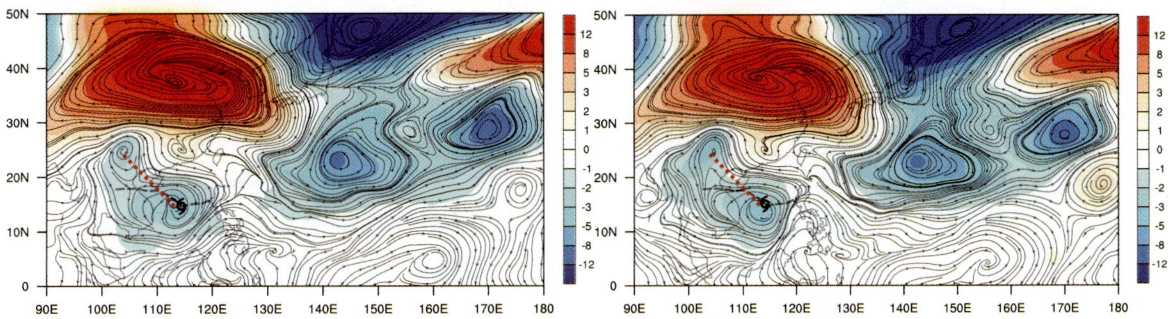

850 hPa flow/height anomalies at 1200 UTC and 1800 UTC 30 September 1997.

Figure 3.9. The typhoon (2007_No.16) generated in the northeastern SCS on 29 September 2007 and had a motion of small curvature that was influenced by an anomalous vortex at 300 hPa. Its intensity peaked in the south sea of the Hainan Island on 2 October 2007 when it was approaching the climatic monsoon trough before landfall on the northeast coast of Vietnam.

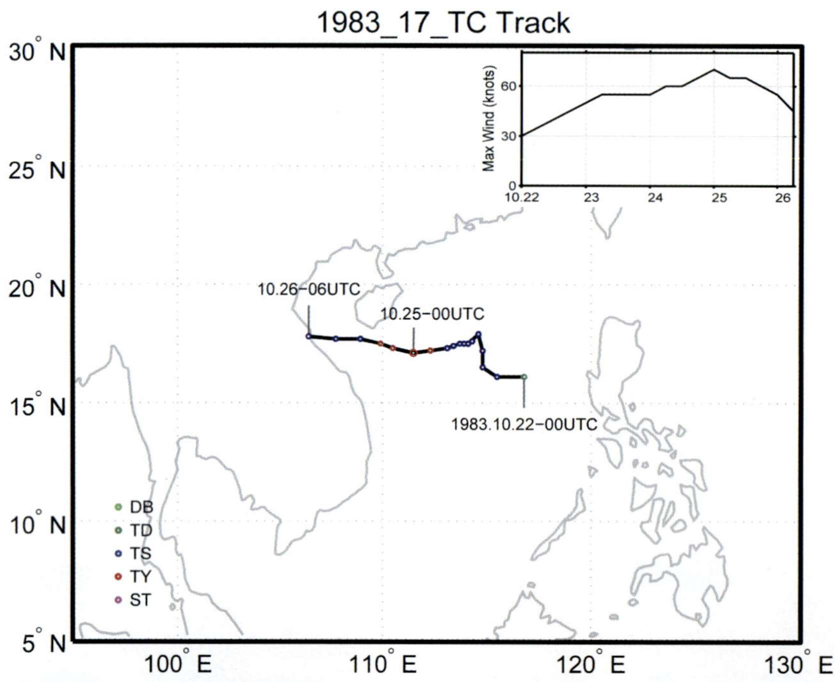

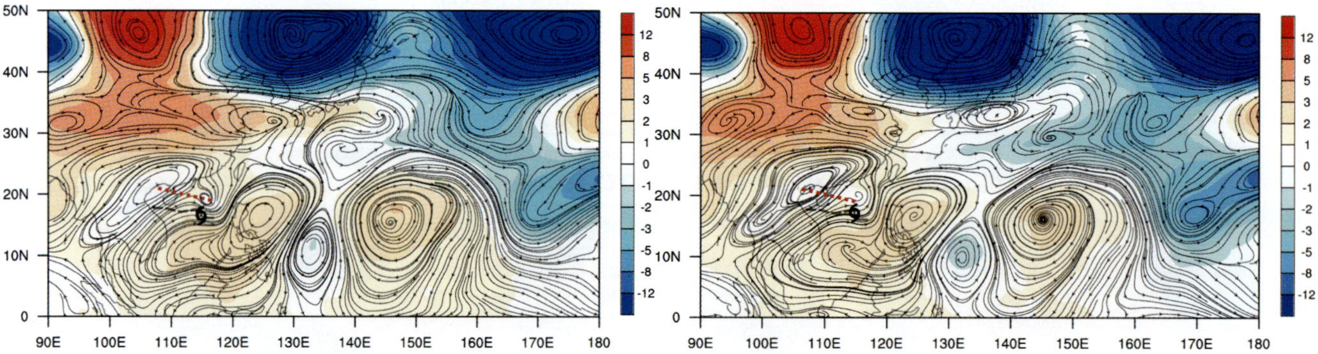

300 hPa flow/height anomalies at 1200 UTC and 1800 UTC 22 October 1983.

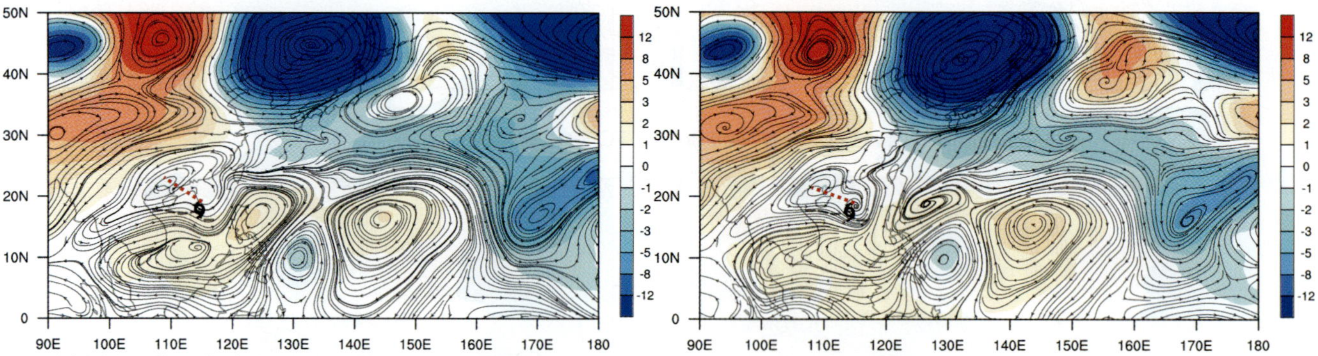

300 hPa flow/height anomalies at 0000 UTC and 0600 UTC 23 October 1983.

Figure 3.10. The typhoon (1983_No.17) generated in the northeastern SCS on 22 October 1983 and had a motion of small curvature in the central northern SCS. Its motion was influenced by an anomalous vortex at 300 hPa and the climatic monsoon trough, so the intensity peaked on 25 October 1983 before landfall on the northeast coast of Vietnam.

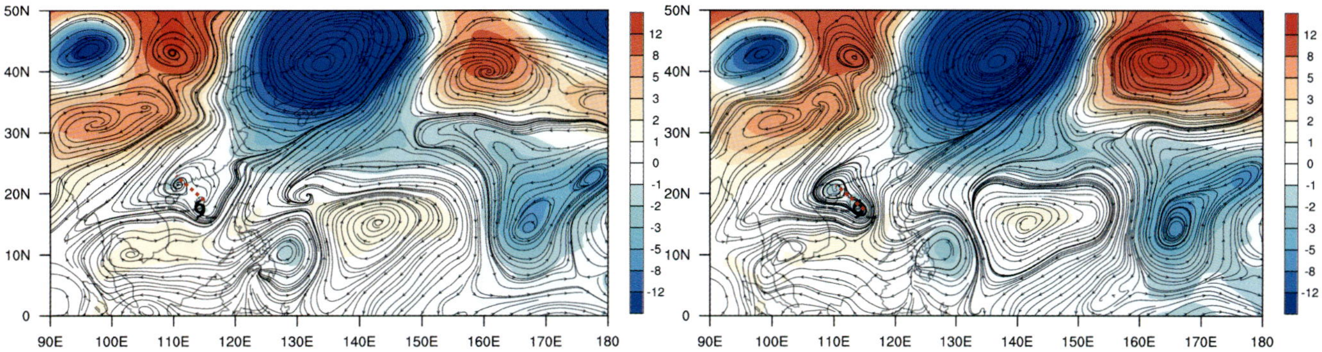

300 hPa flow/height anomalies at 1200 UTC and 1800 UTC 23 October 1983.

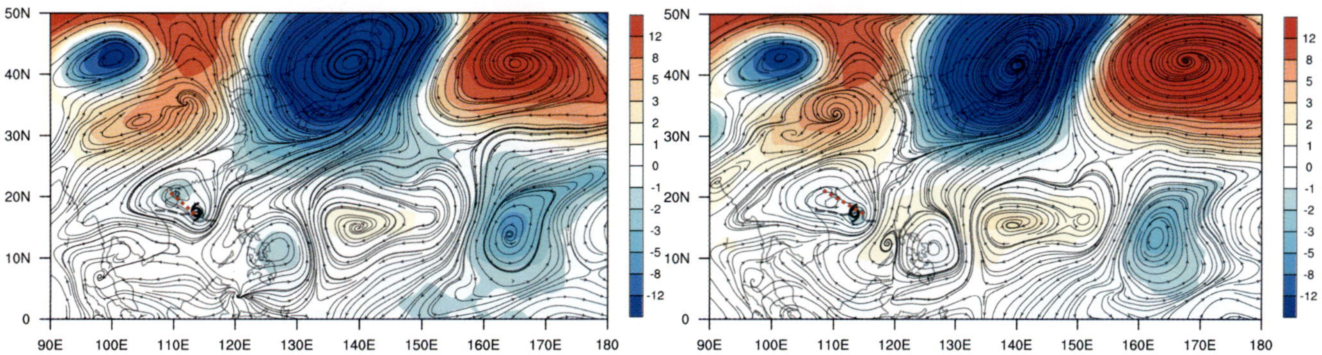

300 hPa flow/height anomalies at 0000 UTC and 0600 UTC 24 October 1983.

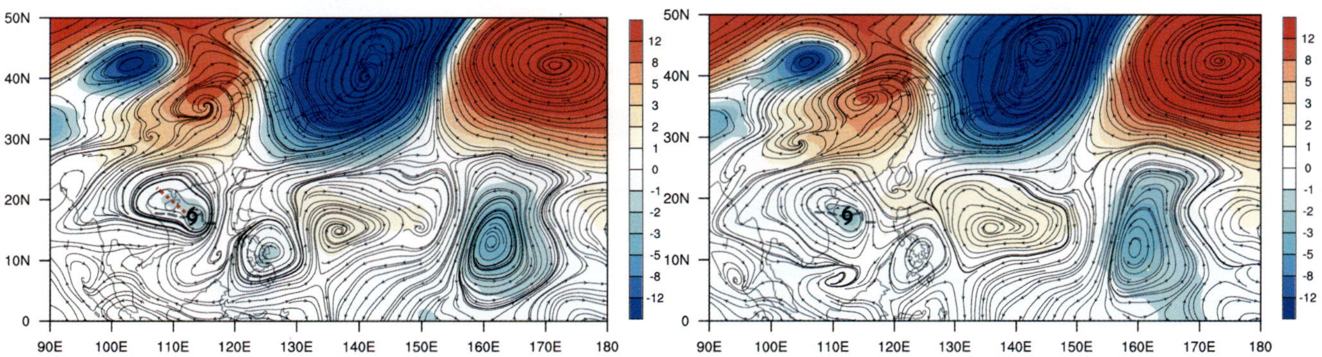

300 hPa flow/height anomalies at 1200 UTC and 1800 UTC 24 October 1983.

Figure 3.10. *(Continued)*

3.2. Geneses in the Philippines

Four typhoons generated near the Philippines and their intensities peaked in the SCS. The unusual motions of these typhoons were influenced by anomalous vortices and their intensities peaked near the climatic monsoon troughs (Figs. 3.11–3.14).

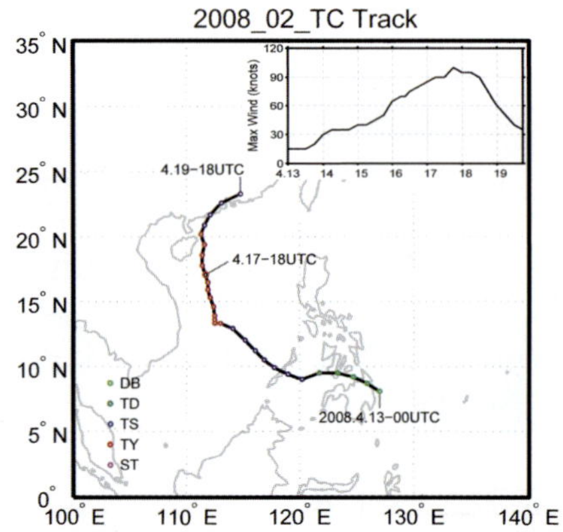

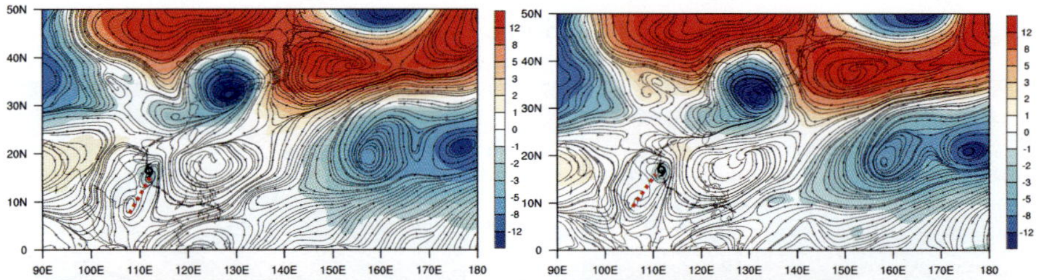

300 hPa flow/height (shading, 10 gpm) anomalies at 1200 UTC and 1800 UTC 17 April 2008.

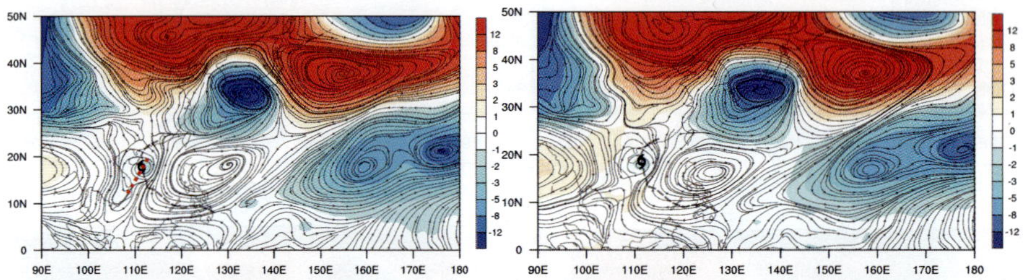

300 hPa flow/height anomalies at 0000 UTC and 0600 UTC 18 April 2008.

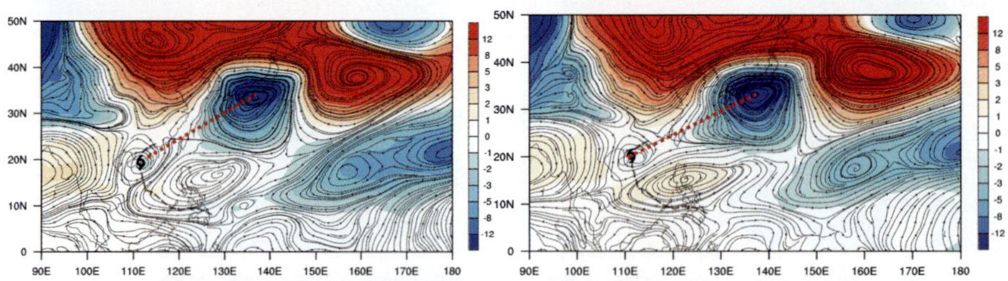

300 hPa flow/height anomalies at 1200 UTC and 1800 UTC 18 April 2008.

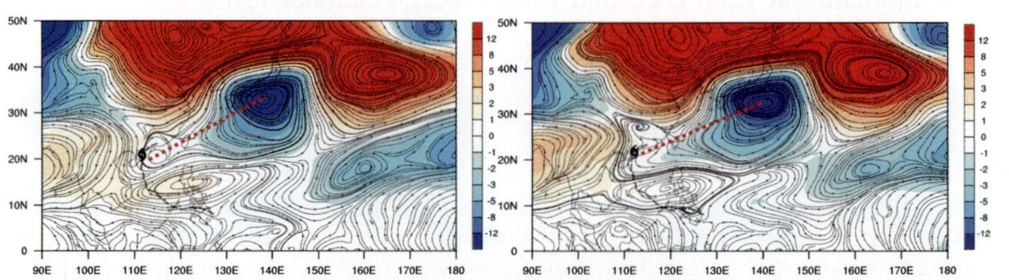

300 hPa flow/height anomalies at 0000 UTC and 0600 UTC 19 April 2008.

Figure 3.11. The typhoon (2008_No.02) generated in the eastern Philippines on 13 April 2008 and moved westward. Its intensity peaked on 17 April 2008 in the northern SCS due to the combination of two anomalous vortices at 300 hPa before landfall on the South China coast. In the SCS, its right turn was influenced by the interaction and attraction with two anomalous vortices at 300 hPa.

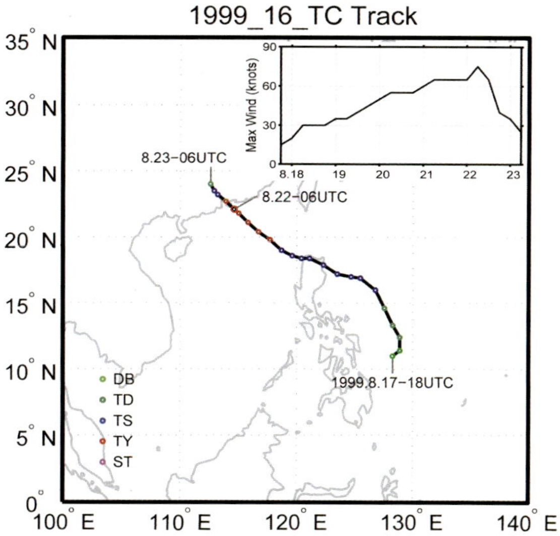

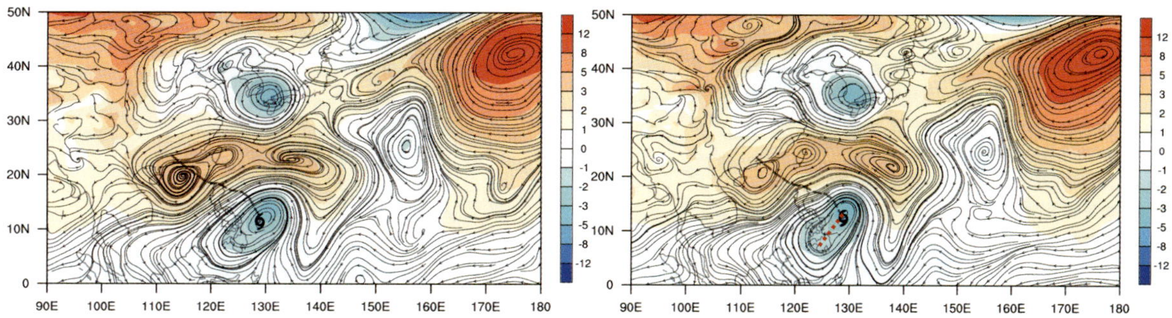

850 hPa flow/height anomalies at 0000 UTC and 0600 UTC 18 August 1999.

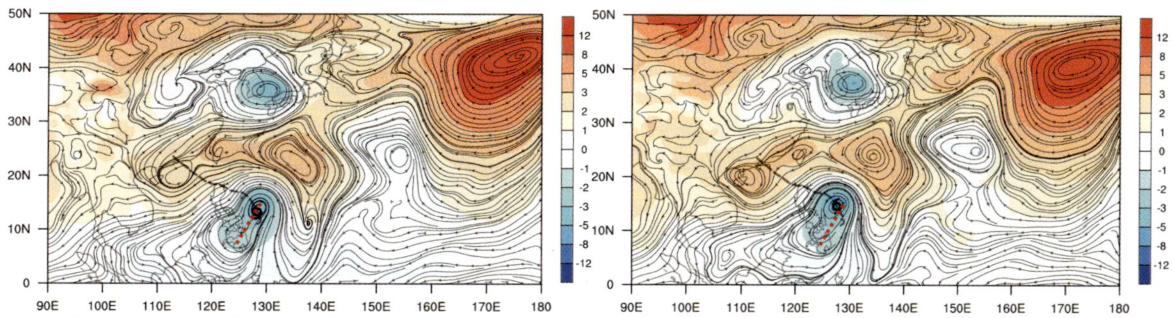

850 hPa flow/height anomalies at 1200 UTC and 1800 UTC 18 August 1999.

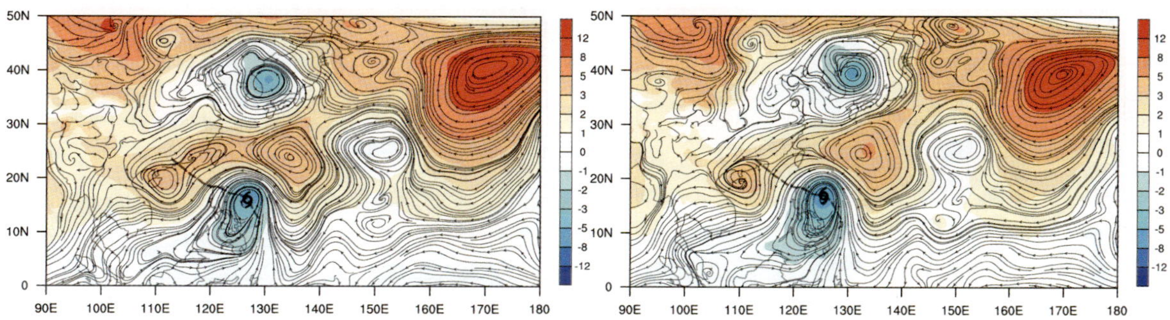

850 hPa flow/height anomalies at 0000 UTC and 0600 UTC 19 August 1999.

Figure 3.12. The typhoon (1999_No.16) generated in the east sea of the Philippines on 17 August 1999 and turned left because of attraction and interaction with another anomalous vortex at 850 hPa. Its intensity peaked in the northernmost SCS on 22 August 1999 when it was approaching the climatic monsoon trough before landfall on the South China coast.

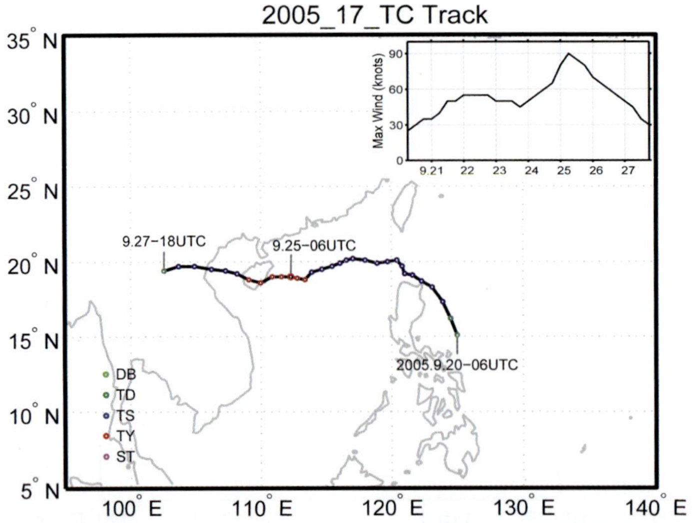

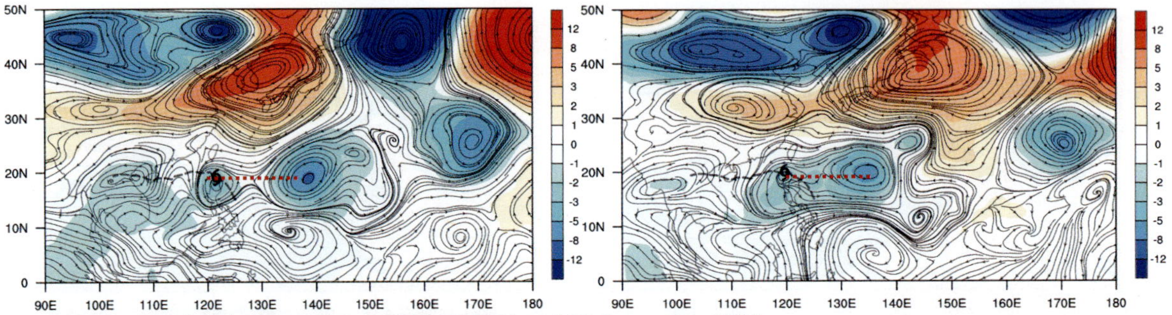

300 hPa flow/height anomalies at 1200 UTC 21 and 22 September 2005.

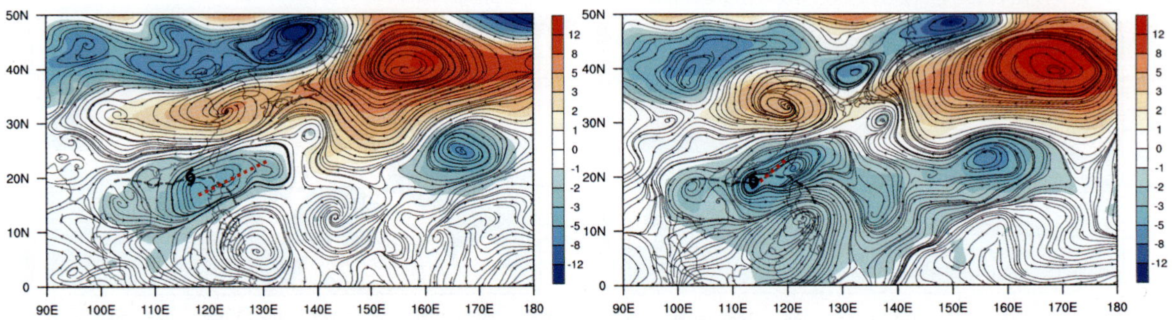

300 hPa flow/height anomalies at 1200 UTC 23 and 24 September 2005.

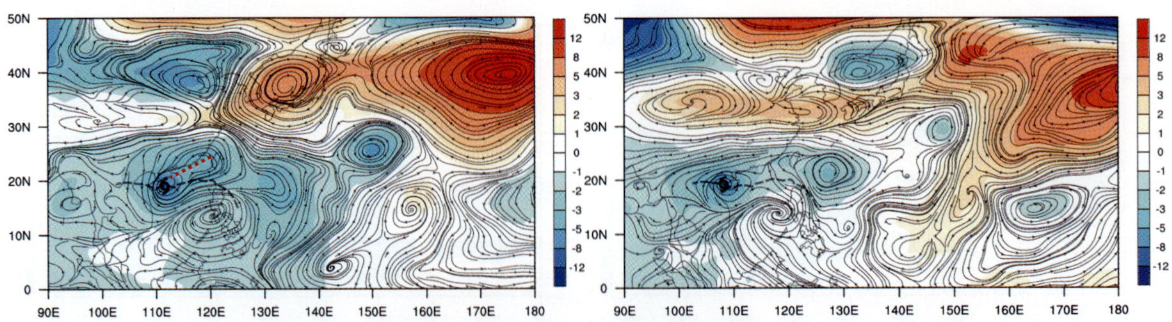

300 hPa flow/height anomalies at 1200 UTC 25 and 26 September 2005.

Figure 3.13. The typhoon (2005_No.17) generated in the east sea of the Philippines on 20 September 2005 and moved northwestward with a left turn basically along the climatic monsoon trough. Its intensity peaked on 25 September 2005 before landfall on the Hainan Island and its motion was influenced by an anomalous vortex at 300 hPa.

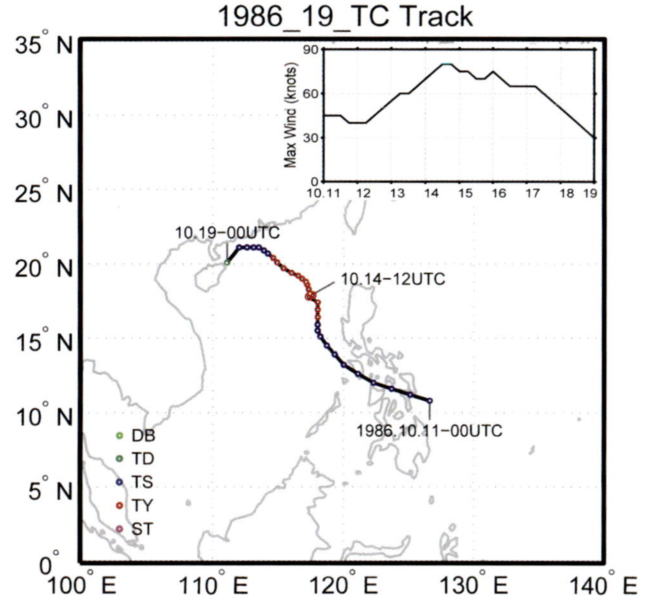

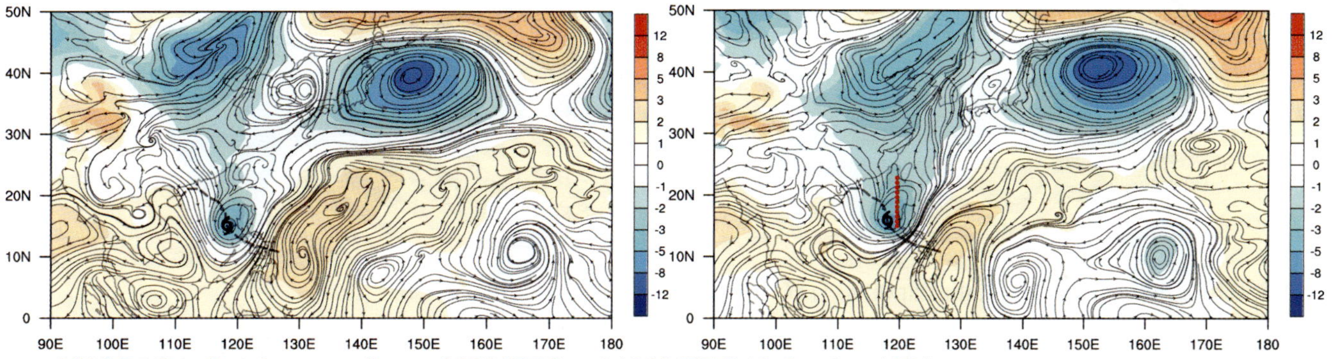

850 hPa flow/height anomalies at 0000 UTC and 1200 UTC 13 October 1986.

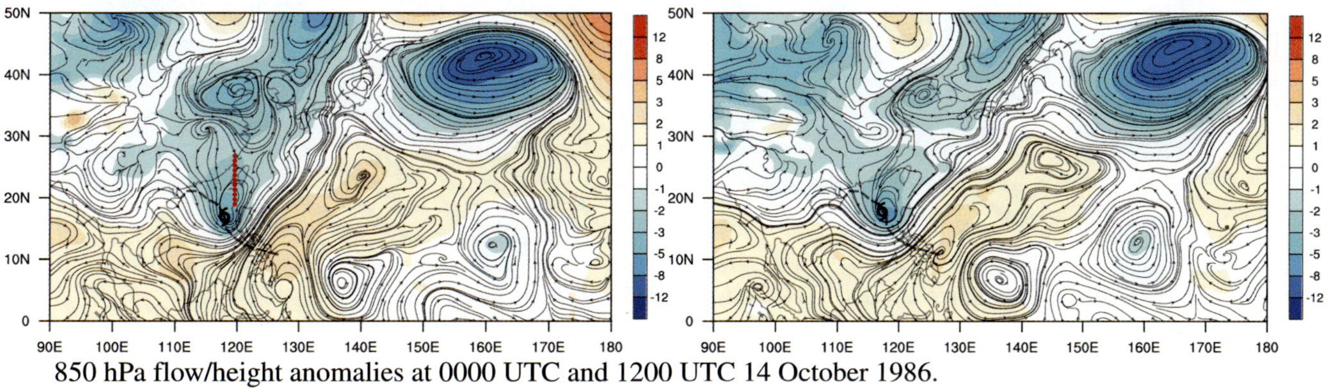

850 hPa flow/height anomalies at 0000 UTC and 1200 UTC 14 October 1986.

Figure 3.14. The typhoon (1986_No.19) generated in the east sea of the Philippines on 11 October 1986 and moved northwestward crossing the central Philippines. Its intensity peaked in the northeastern SCS on 14 October 1986 when it was approaching the climatic monsoon trough. Its right turn and left turn were influenced by two anomalous vortices at 850 hPa.

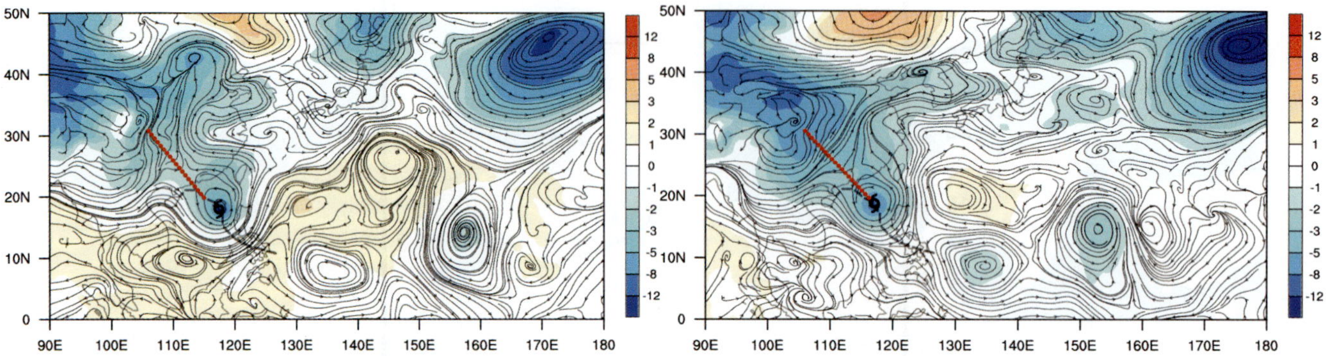

850 hPa flow/height anomalies at 0000 UTC and 1200 UTC 15 October 1986.

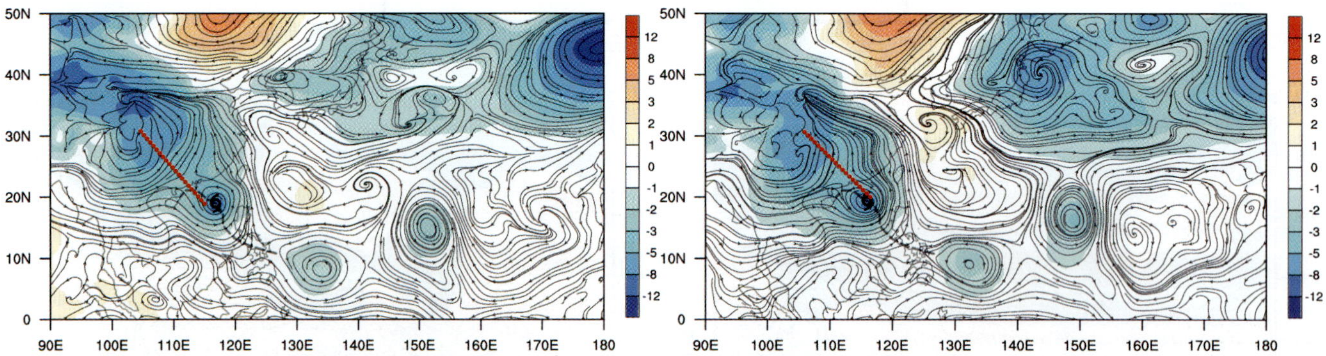

850 hPa flow/height anomalies at 0000 UTC and 1200 UTC 16 October 1986.

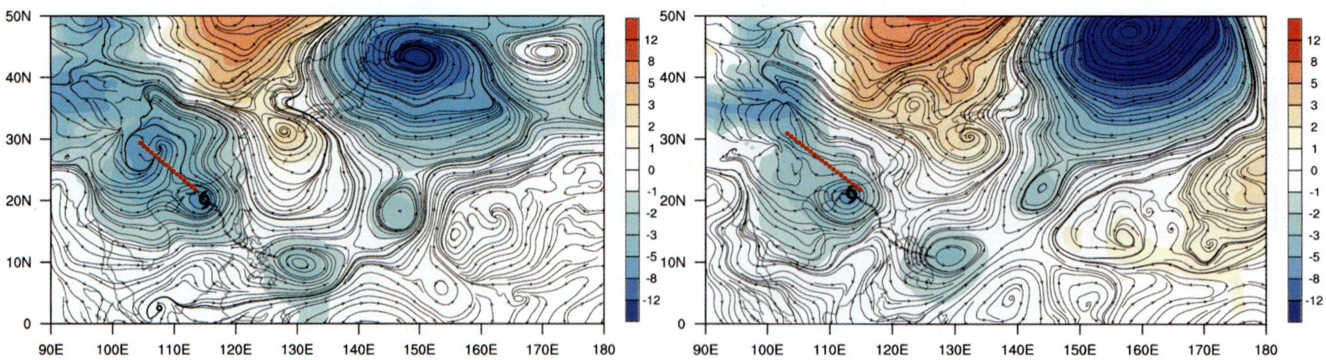

850 hPa flow/height anomalies at 0000 UTC and 1200 UTC 17 October 1986.

Figure 3.14. (*Continued*)

3.3. Geneses in the Equatorial Pacific

Seven typhoons generated in the equatorial Pacific and their intensities peaked in the SCS. The turning motions of these typhoons were due to the interaction with anomalous vortices. In this section, five typhoons are plotted (Figs. 3.15–3.19) and the other two typhoons (2006_No.02 and 2009_No.03), which also generated in the equatorial Pacific and peaked in the SCS, have been plotted in Chapter 1 as two examples.

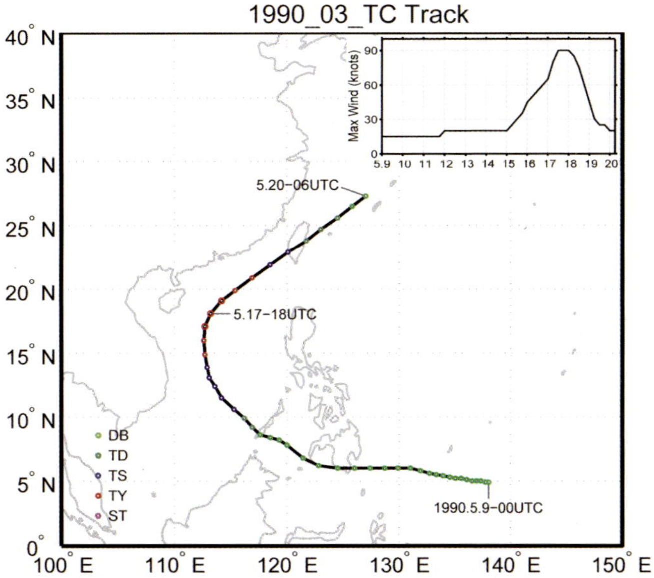

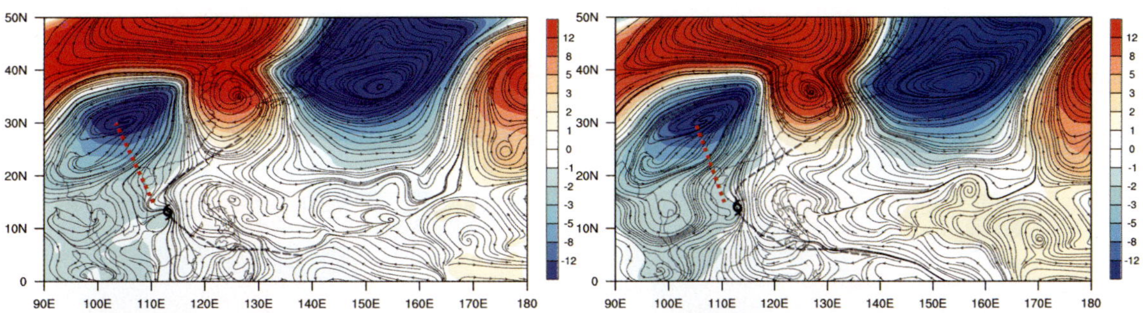

300 hPa flow/height (shading, 10 gpm) anomalies at 1200 UTC and 1800 UTC 16 May 1990.

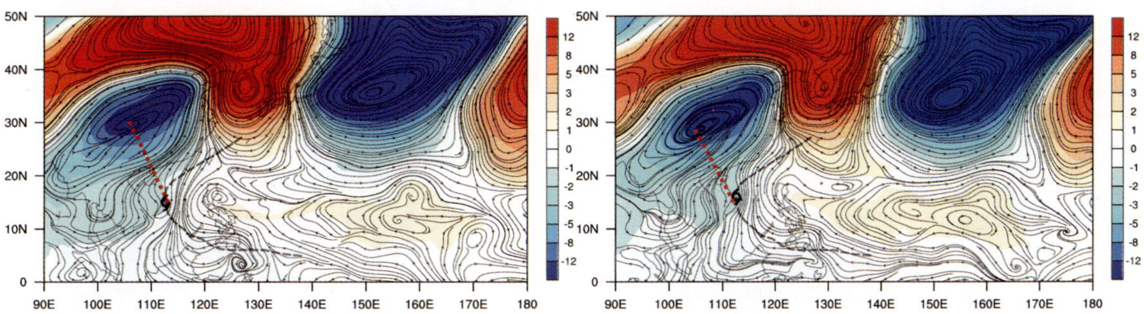

300 hPa flow/height anomalies at 0000 UTC and 0600 UTC 17 May 1990.

Figure 3.15. The typhoon (1990_No.03) generated in the equatorial western Pacific on 9 May 1990 and moved westward to the central SCS. Its intensity peaked in the southeast sea of the Hainan Island on 17 May 1990 when it was approaching the climatic monsoon trough before landfall on the Taiwan Island. Its right turn in the SCS was influenced by another anomalous vortex from the China continent at 300 hPa.

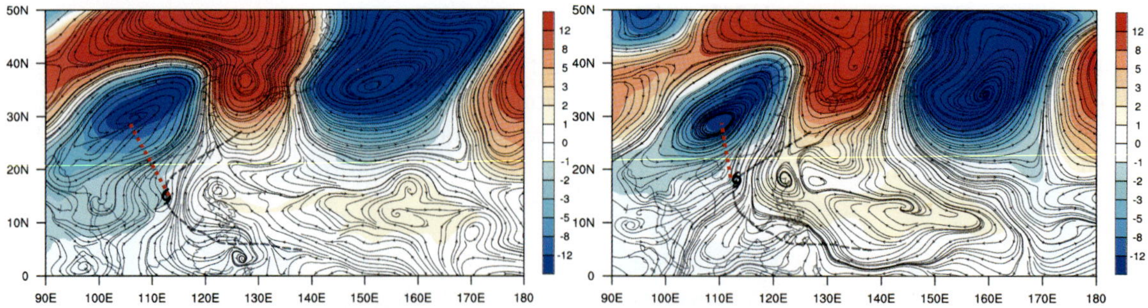

300 hPa flow/height anomalies at 1200 UTC and 1800 UTC 17 May 1990.

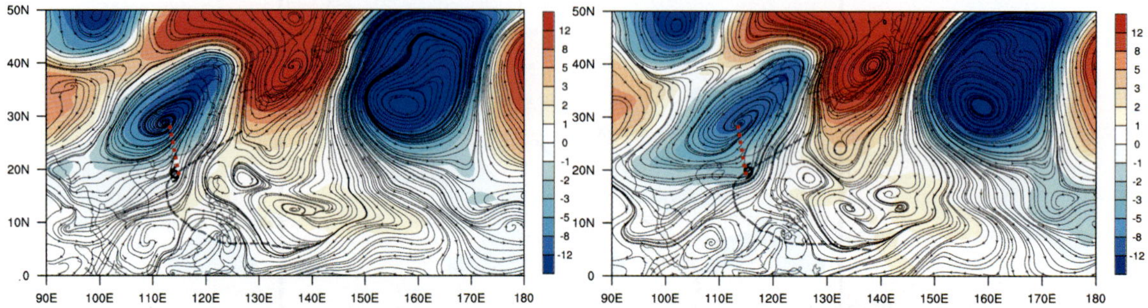

300 hPa flow/height anomalies at 0000 UTC and 0600 UTC 18 May 1990.

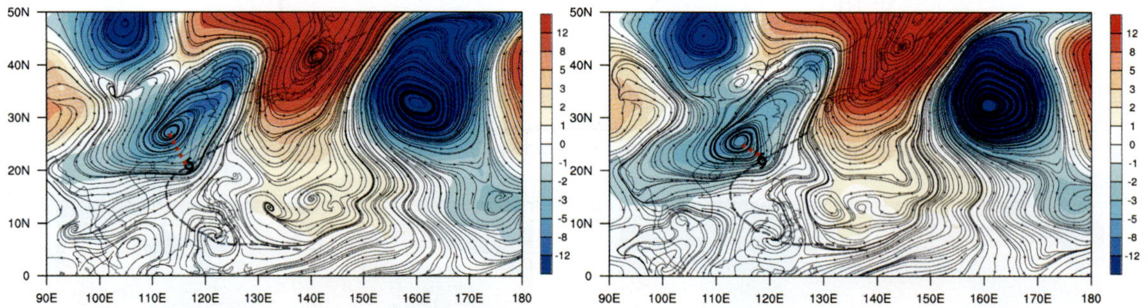

300 hPa flow/height anomalies at 1200 UTC and 1800 UTC 18 May 1990.

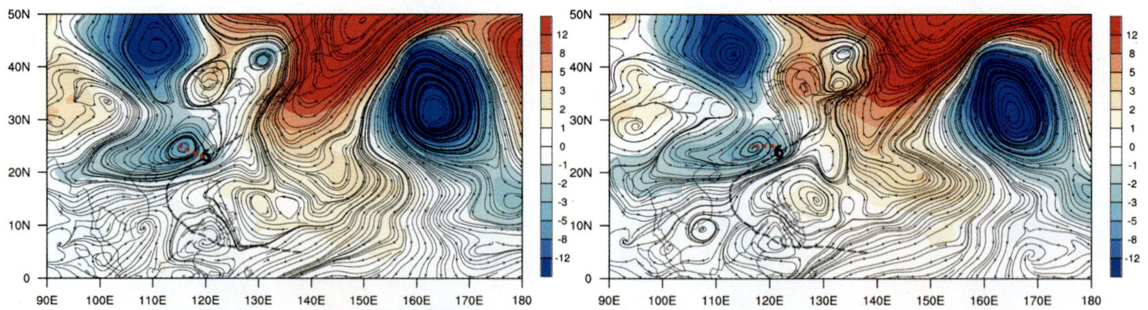

300 hPa flow/height anomalies at 0000 UTC and 0600 UTC 19 May 1990.

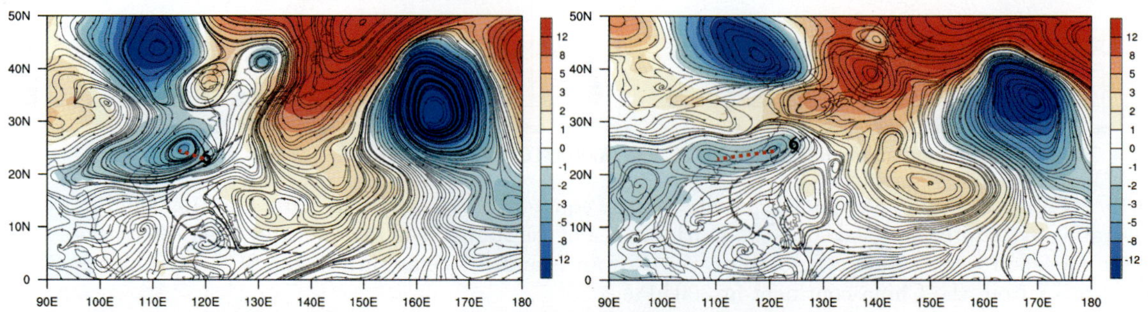

300 hPa flow/height anomalies at 1200 UTC and 1800 UTC 19 May 1990.

Figure 3.15. *(Continued)*

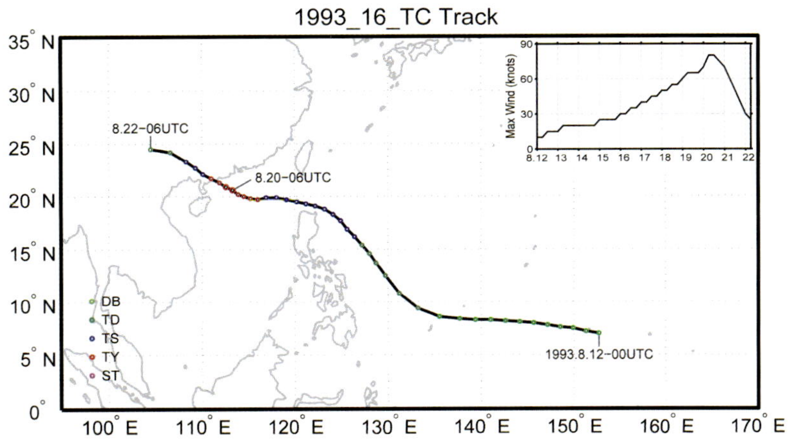

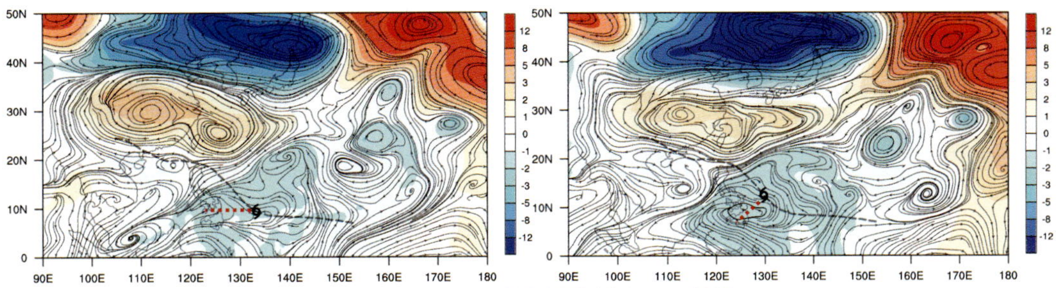

300 hPa flow/height anomalies at 0000 UTC and 1200 UTC 15 August 1993.

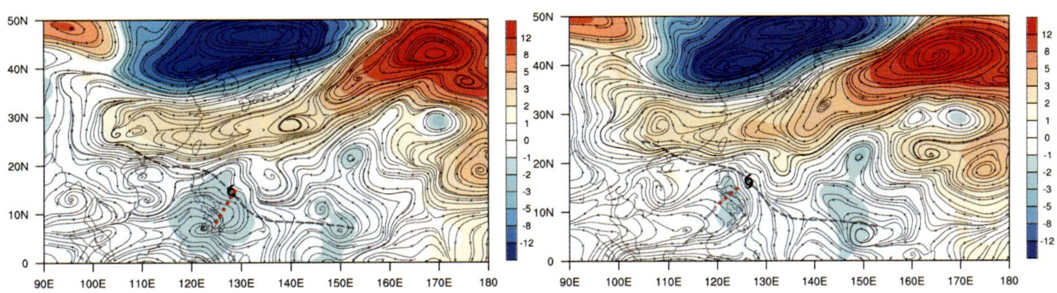

300 hPa flow/height anomalies at 0000 UTC and 1200 UTC 16 August 1993.

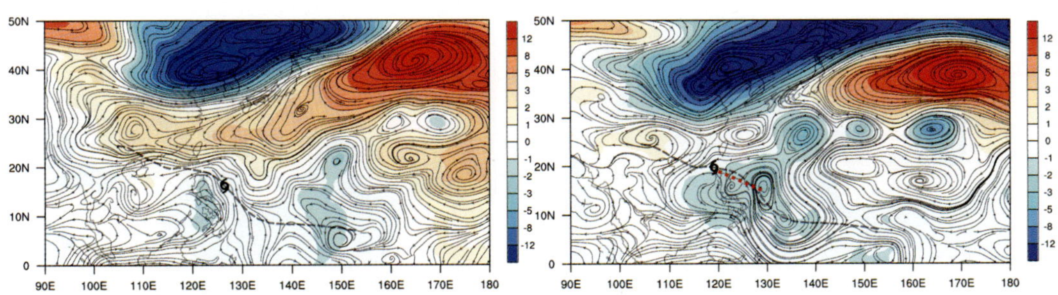

300 hPa flow/height anomalies at 1200 UTC 17 and 18 August 1993.

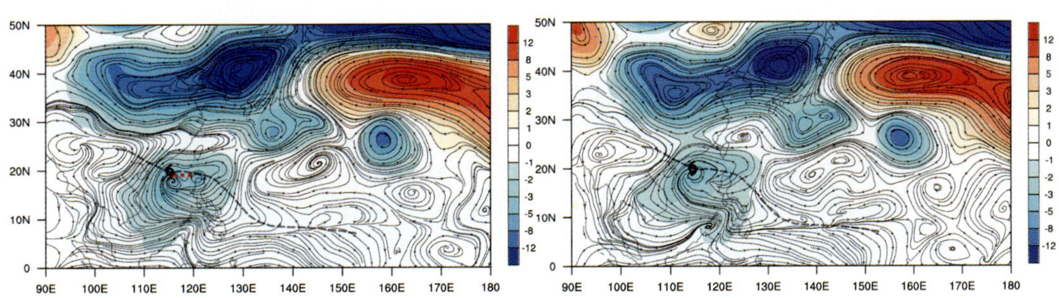

300 hPa flow/height anomalies at 1200 UTC and 1800 UTC 19 August 1993.

Figure 3.16. The typhoon (1993_No.16) generated in the equatorial central Pacific on 12 August 1993 and moved westward. Its intensity peaked in the northernmost SCS on 20 August 1993 when it was approaching the climatic monsoon trough before landfall on the South China coast. Its left turn near the Philippines was influenced by the interaction and attraction with two anomalous vortices at 300 hPa.

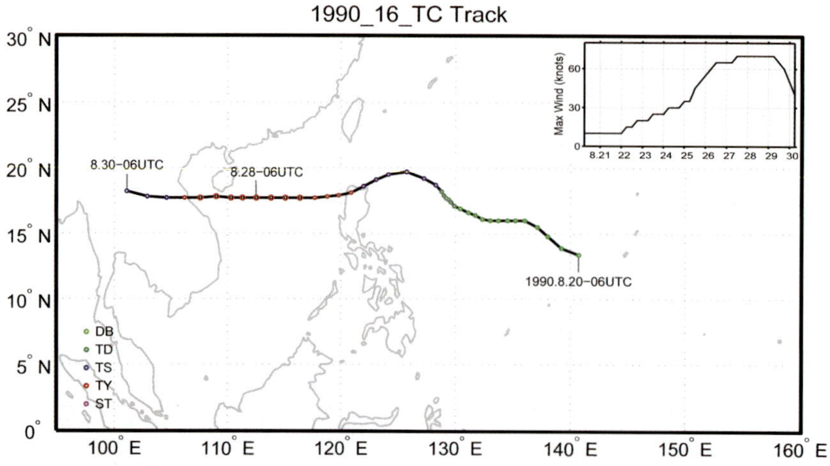

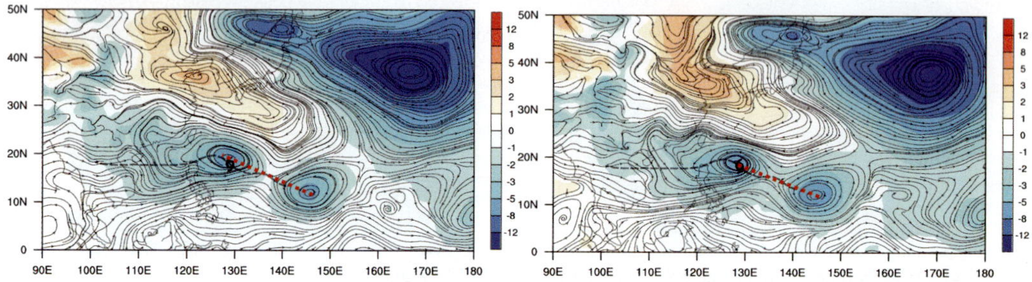

850 hPa flow/height anomalies at 1200 UTC and 1800 UTC 24 August 1990.

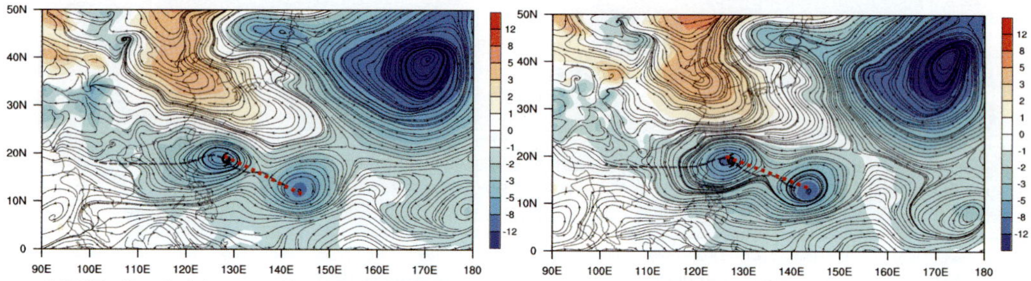

850 hPa flow/height anomalies at 0000 UTC and 0600 UTC 25 August 1990.

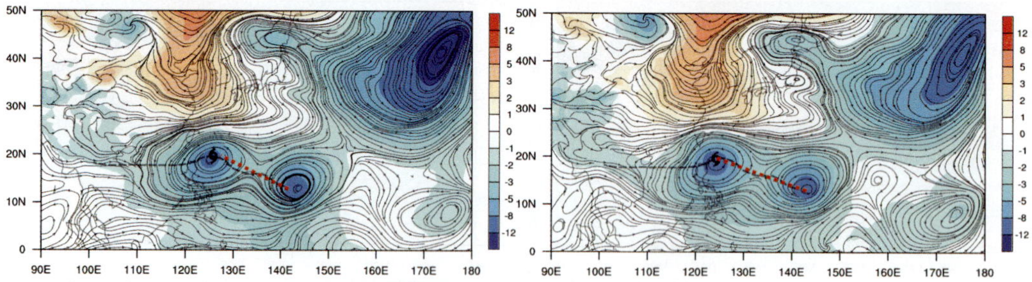

850 hPa flow/height anomalies at 1200 UTC and 1800 UTC 25 August 1990.

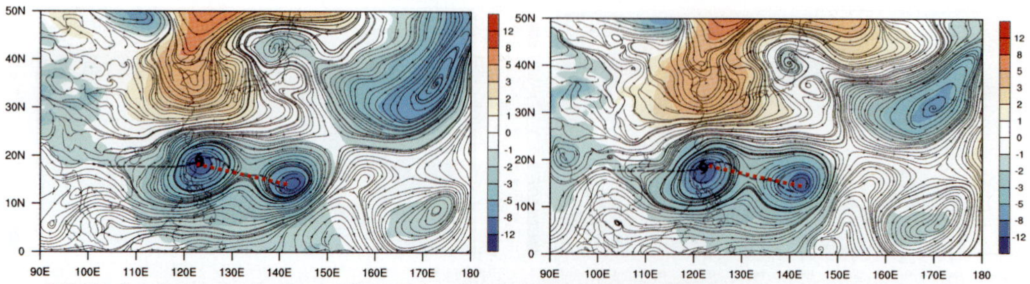

850 hPa flow/height anomalies at 0000 UTC and 0600 UTC 26 August 1990.

Figure 3.17. The typhoon (1990_No.16) generated in the equatorial western Pacific on 20 August 1990 and moved westward. Its intensity peaked in the southeast sea of the Hainan Island on 28 August 1990 when it was approaching the climatic monsoon trough before landfall on the east coast of Vietnam. Its left turn in the northeast sea of the Philippines was influenced by another anomalous vortex at 850 hPa.

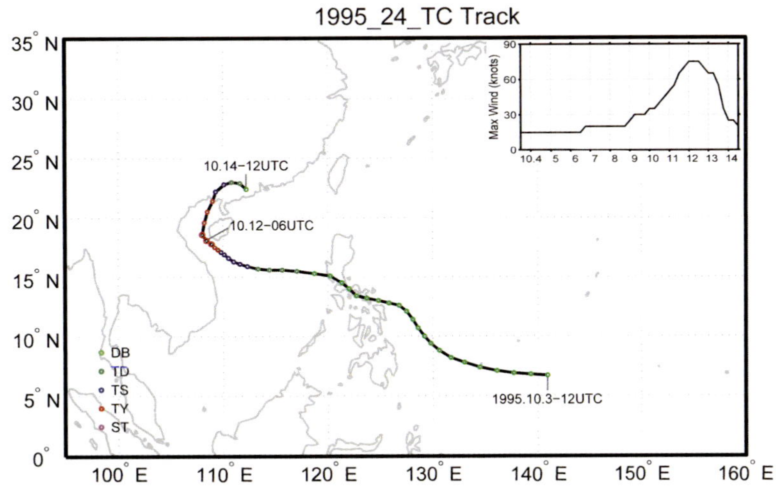

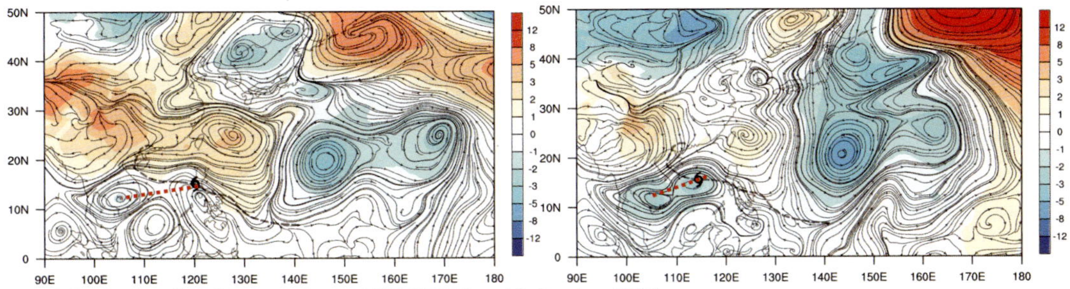

850 hPa flow/height anomalies at 1200 UTC 8 and 9 October 1995.

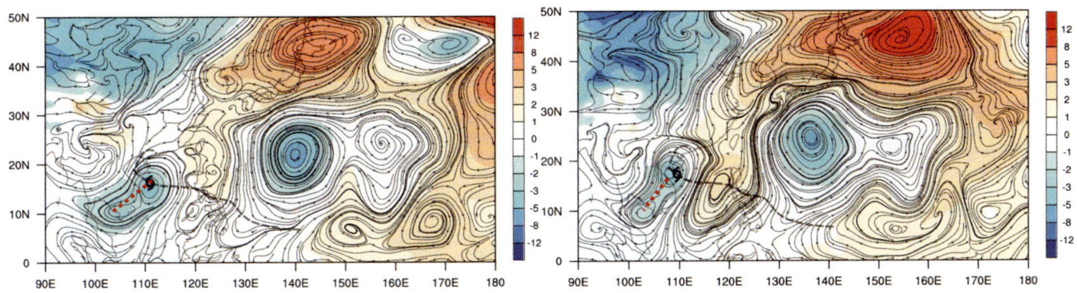

850 hPa flow/height anomalies at 1200 UTC 10 and 11 October 1995.

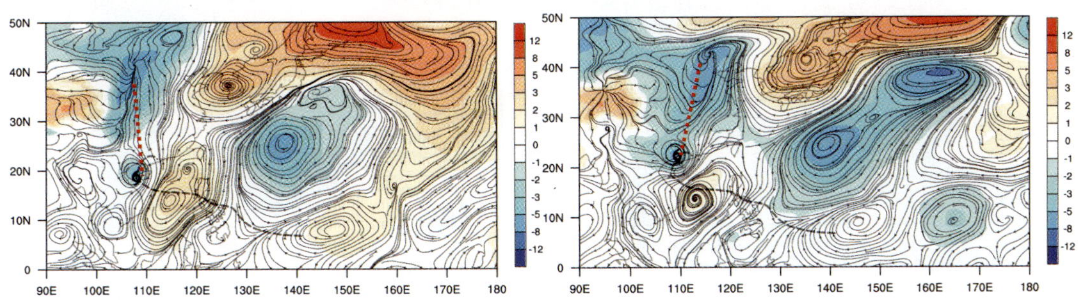

850 hPa flow/height anomalies at 1200 UTC 12 and 13 October 1995.

Figure 3.18. The typhoon (1995_No.24) generated in the equatorial western Pacific on 3 October 1995 and moved northwestward. Its intensity peaked in the west sea of the Hainan Island on 12 October 1995 when it was approaching the climatic monsoon trough before landfall on the Southwest China coast. Its right turn in the northern SCS was influenced by two anomalous vortices at 850 hPa.

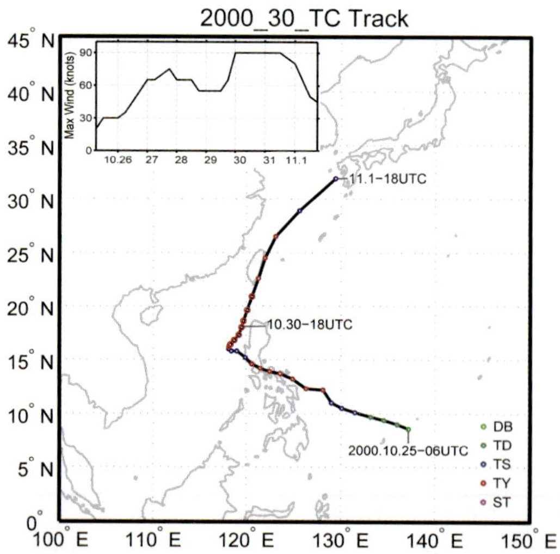

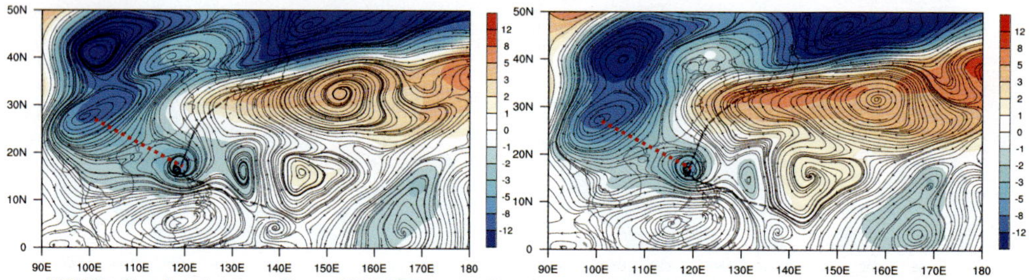

300 hPa flow/height anomalies at 0000 UTC and 0600 UTC 30 October 2000.

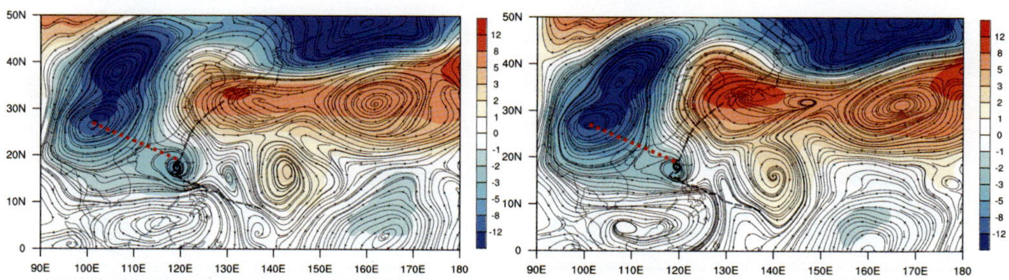

300 hPa flow/height anomalies at 1200 UTC and 1800 UTC 30 October 2000.

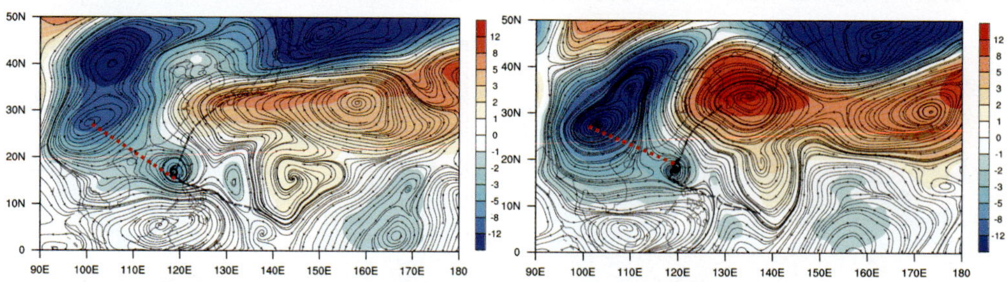

300 hPa flow/height anomalies at 0000 UTC and 0600 UTC 31 October 2000.

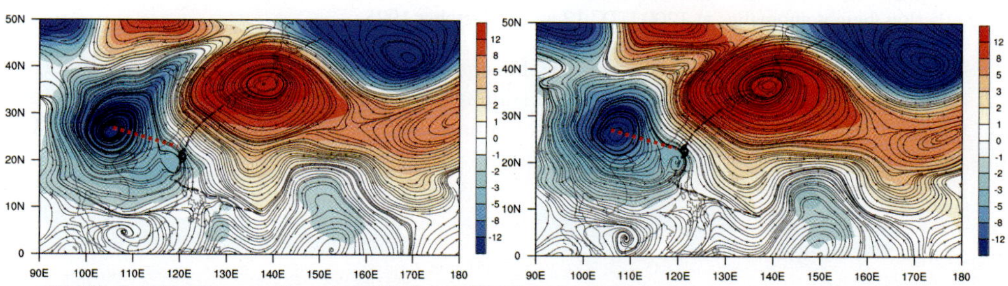

300 hPa flow/height anomalies at 1200 UTC and 1800 UTC 31 October 2000.

Figure 3.19. The typhoon (2000_No.30) generated in the equatorial western Pacific on 25 October 2000 and moved northwestward. Its intensity peaked near the northern Philippines on 30 October 2000 when it was approaching the climatic monsoon trough. Its right turn was influenced by another anomalous vortex in the China continent at 300 hPa.

3.4. Geneses in the Tropical Pacific

Ten typhoons generated in the tropical Pacific and their intensities peaked in the SCS. The right turns and left turns of these typhoons were due to the interaction of anomalous vortices (Figs. 3.20–3.29).

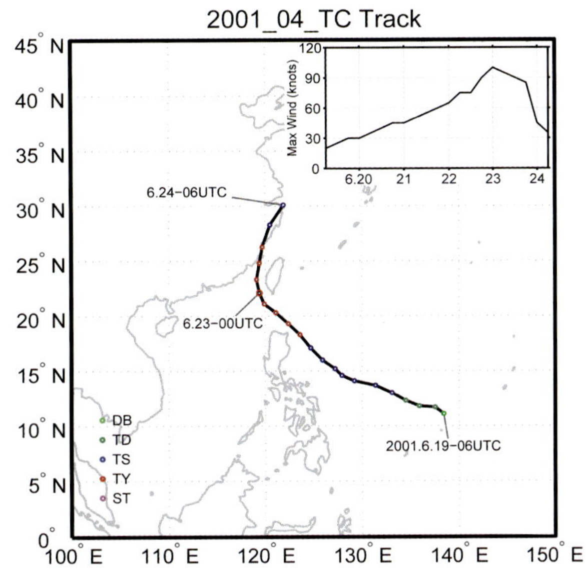

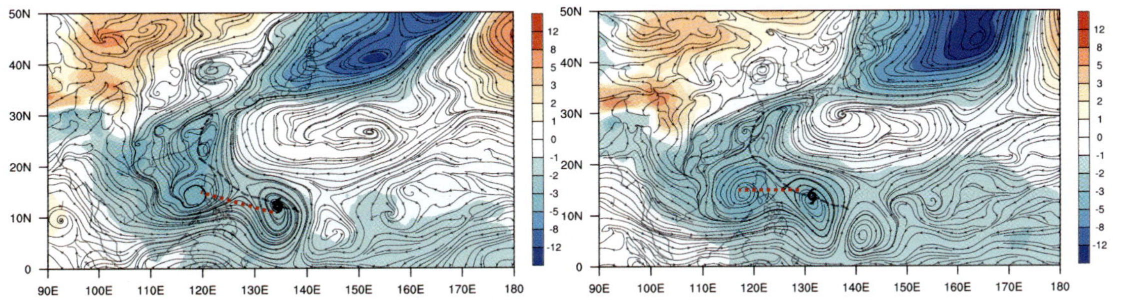

850 hPa flow/height (shading, 10 gpm) anomalies at 0000 UTC and 1200 UTC 20 June 2001.

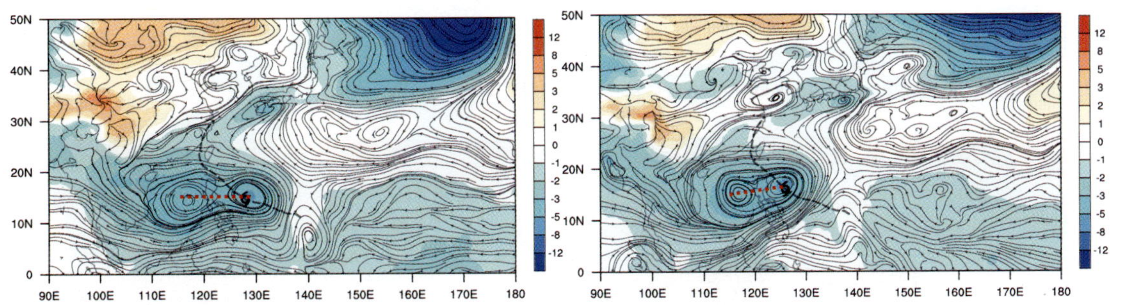

850 hPa flow/height anomalies at 0000 UTC and 1200 UTC 21 June 2001.

Figure 3.20. The typhoon (2001_No.04) generated in the tropical western Pacific on 19 June 2001 and moved northwestward along the climatic monsoon trough. Its intensity peaked near the Taiwan Strait on 23 June 2001 before landfall on the Southeast China coast. Its right turn near the Taiwan Island was influenced by anomalous vortices at 850 hPa and 300 hPa.

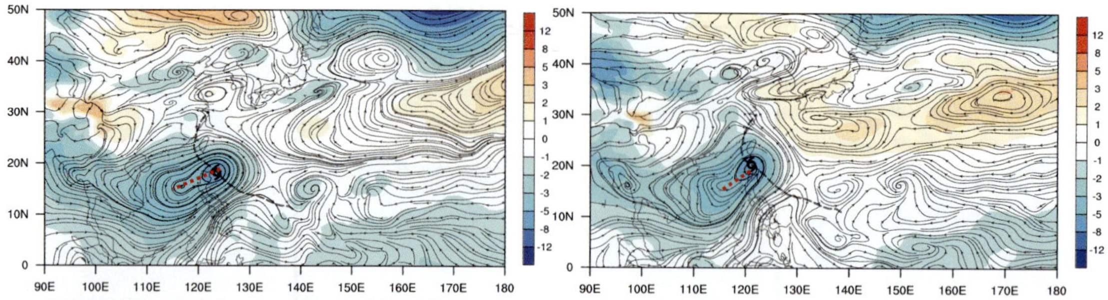

850 hPa flow/height anomalies at 0000 UTC and 1200 UTC 22 June 2001.

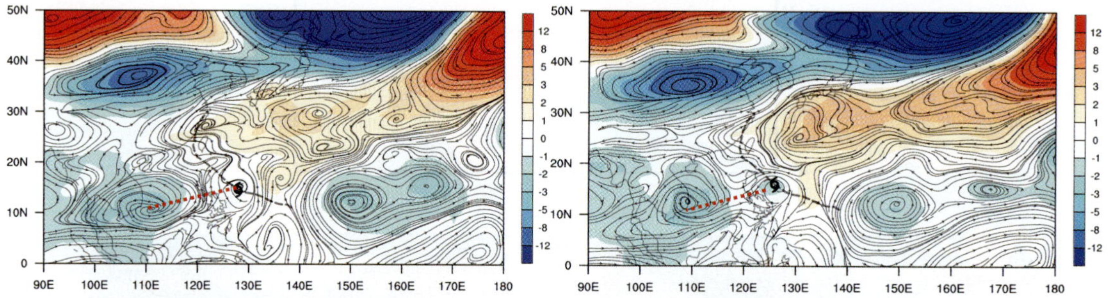

300 hPa flow/height anomalies at 0000 UTC and 1200 UTC 21 June 2001.

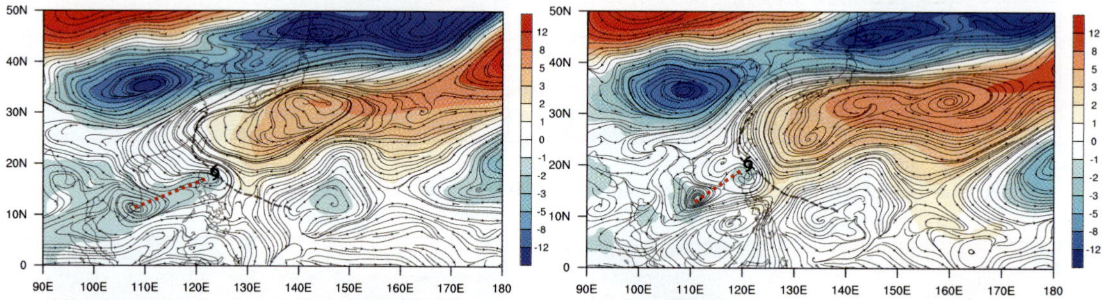

300 hPa flow/height anomalies at 0000 UTC and 1200 UTC 22 June 2001.

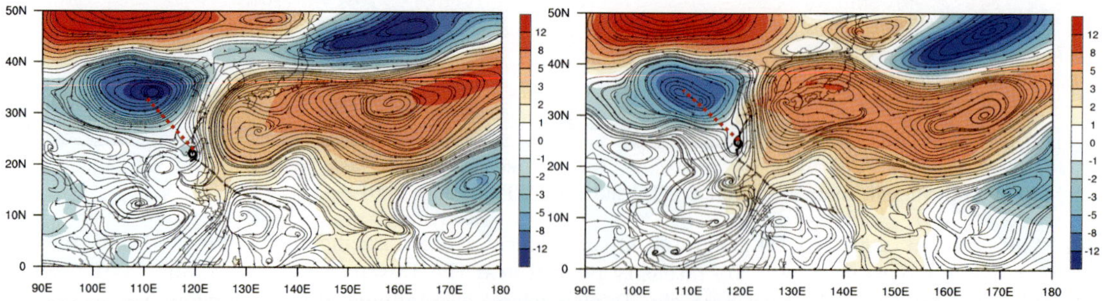

300 hPa flow/height anomalies at 0000 UTC and 1200 UTC 23 June 2001.

Figure 3.20. *(Continued)*

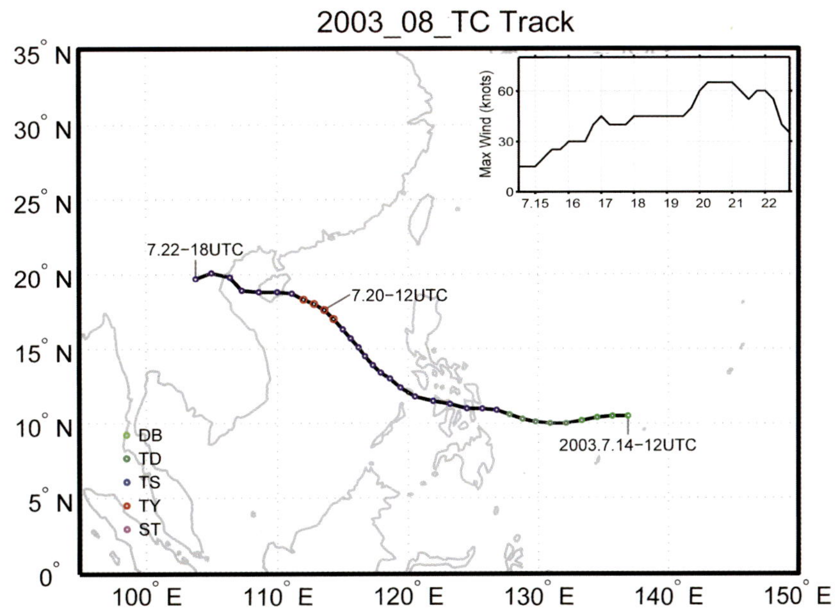

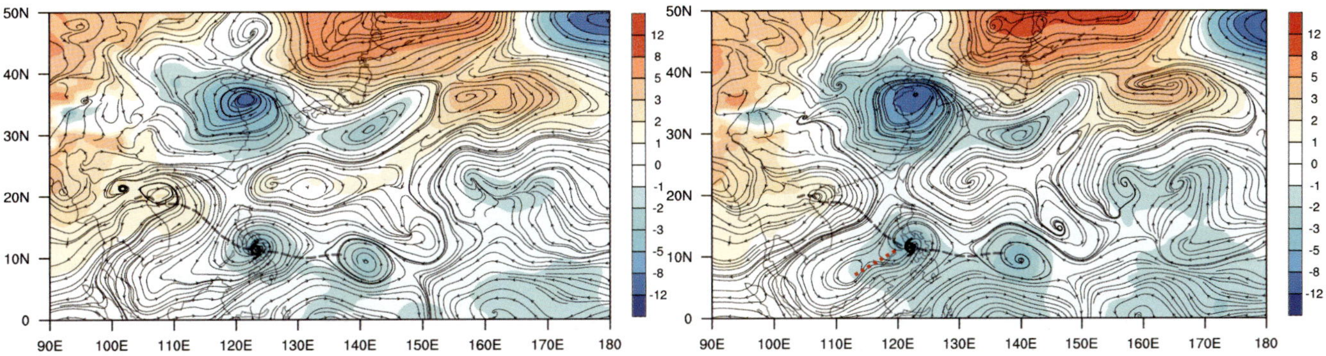

850 hPa flow/height anomalies at 1200 UTC and 1800 UTC 17 July 2003.

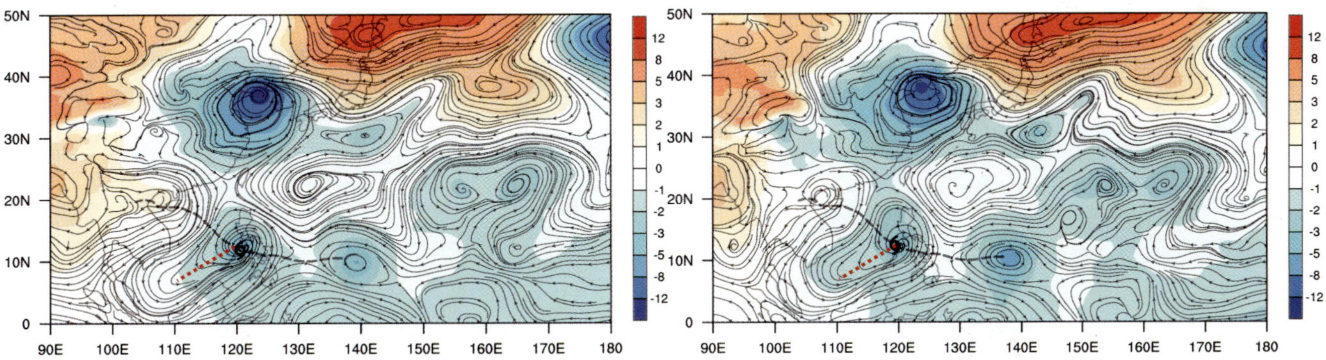

850 hPa flow/height anomalies at 0000 UTC and 0600 UTC 18 July 2003.

Figure 3.21. The typhoon (2003_No.08) generated in the tropical western Pacific on 14 July 2003 and moved westward. Its intensity peaked in the southeast sea of the Hainan Island on 20 July 2003 when it was approaching the climatic monsoon trough before landfall on the island. Its left turn in the northern SCS was influenced by another anomalous vortex at 850 hPa.

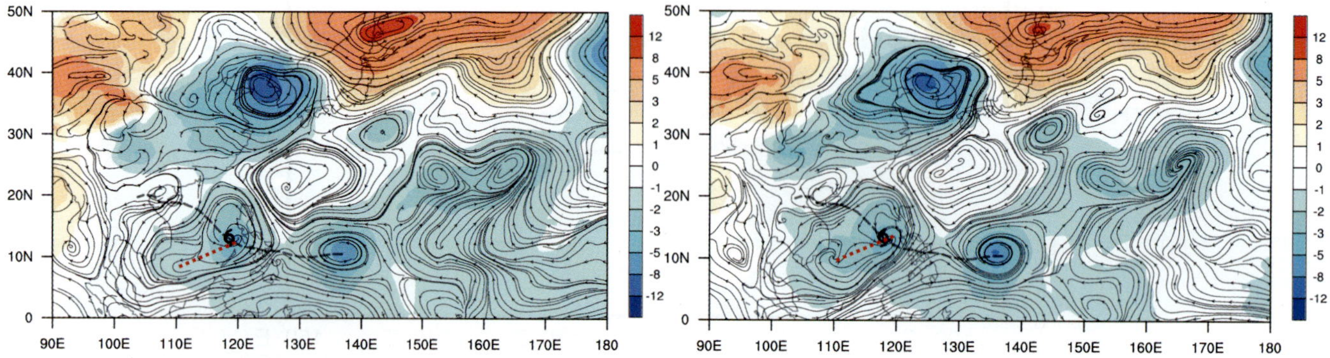

850 hPa flow/height anomalies at 1200 UTC and 1800 UTC 18 July 2003.

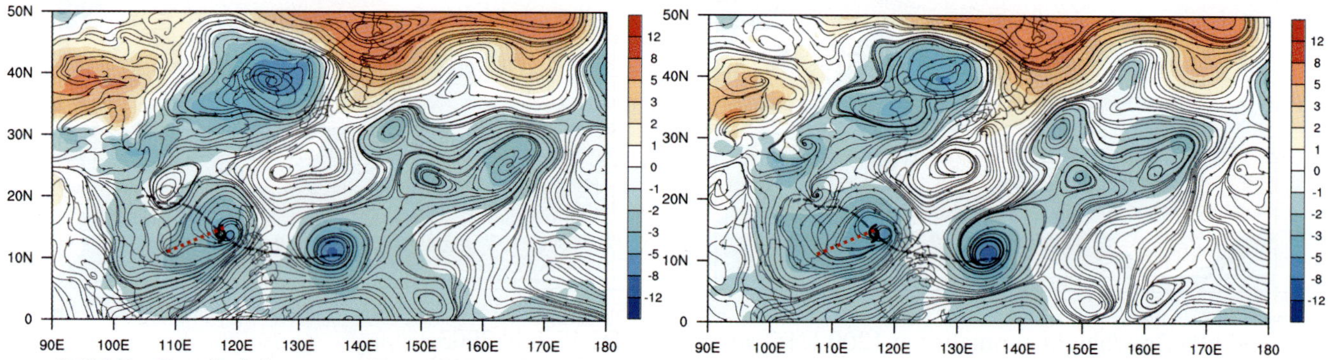

850 hPa flow/height anomalies at 0000 UTC and 0600 UTC 19 July 2003.

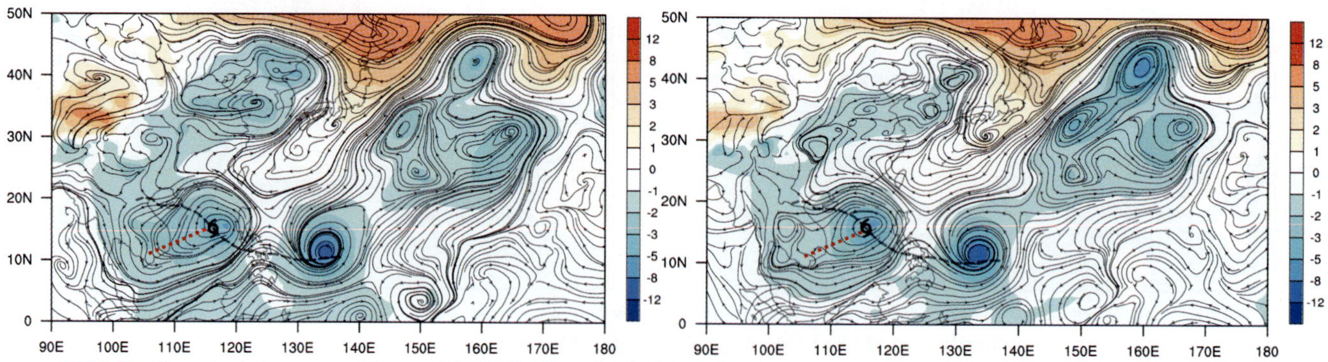

850 hPa flow/height anomalies at 1200 UTC and 1800 UTC 19 July 2003.

Figure 3.21. *(Continued)*

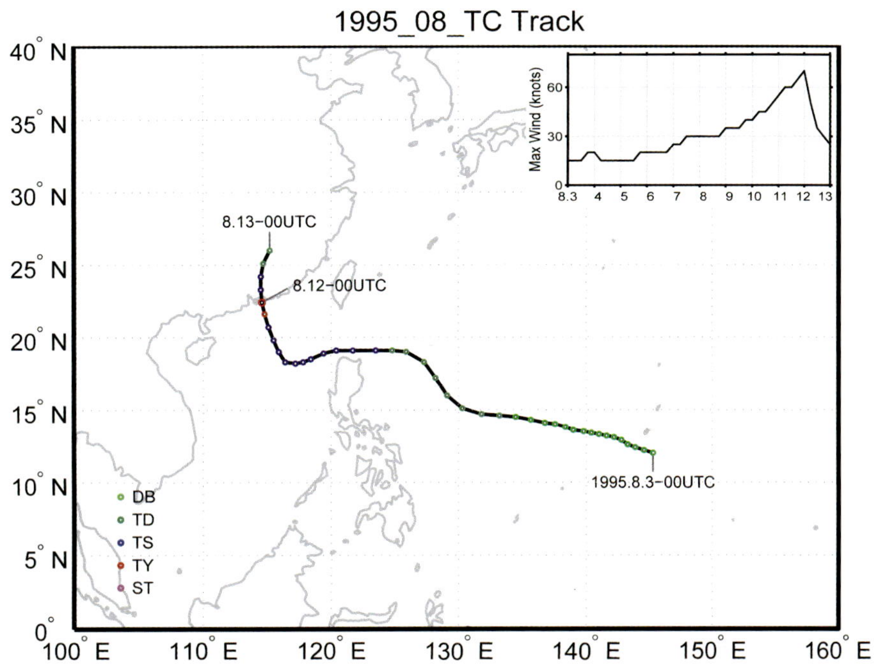

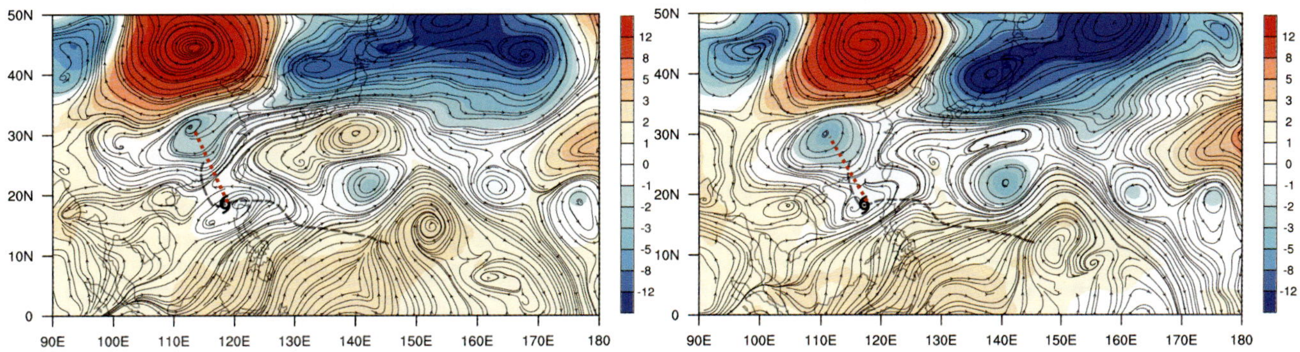

300 hPa flow/height anomalies at 0000 UTC and 1200 UTC 10 August 1995.

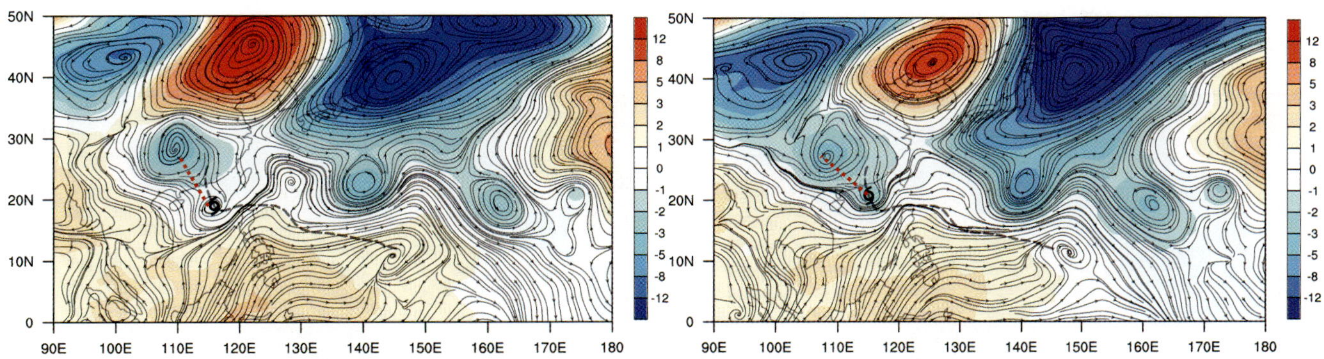

300 hPa flow/height anomalies at 0000 UTC and 1200 UTC on 11 August 1995.

Figure 3.22. The typhoon (1995_No.08) generated in the tropical western Pacific on 3 August 1995 and moved northwestward along the climatic monsoon trough. Its intensity peaked on 12 August 1995 before landfall on the South China coast. Its right turn in the northern SCS was influenced by another anomalous vortex at 300 hPa.

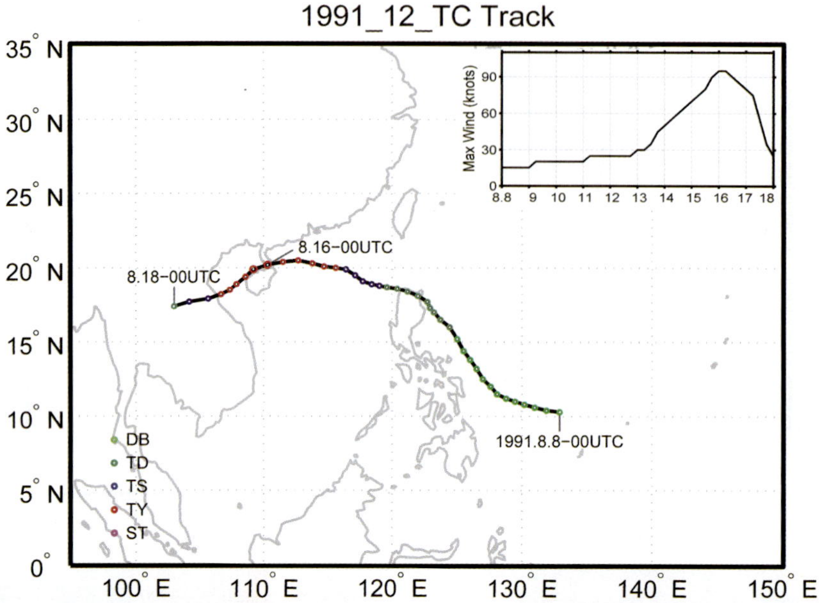

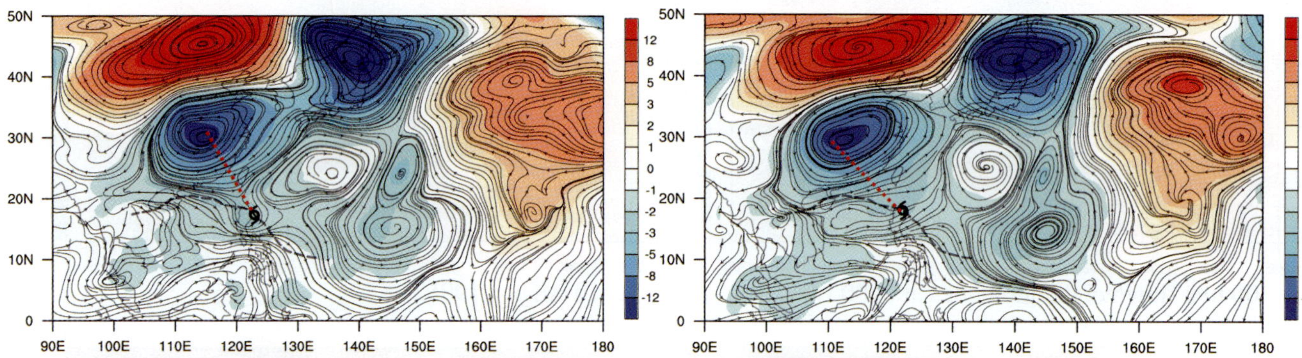

300 hPa flow/height anomalies at 0000 UTC and 1200 UTC 12 August 1991.

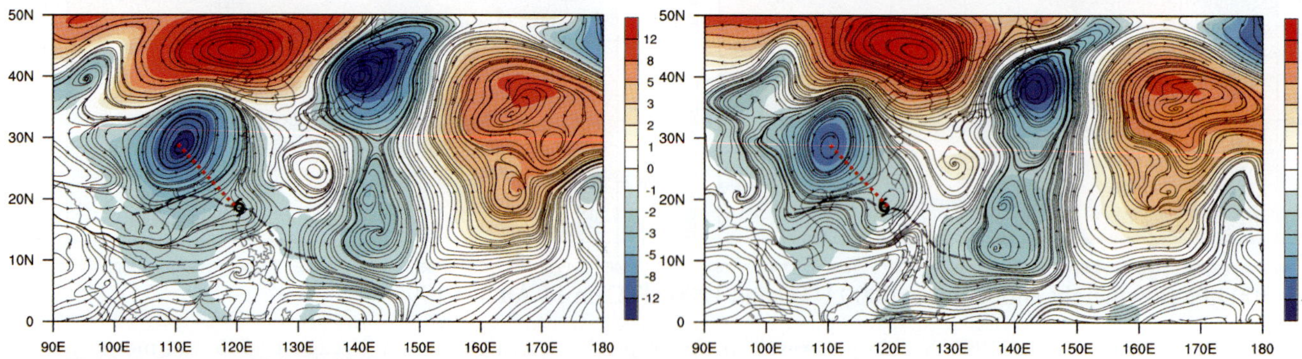

300 hPa flow/height anomalies at 0000 UTC and 1200 UTC 13 August 1991.

Figure 3.23. The typhoon (1991_No.12) generated in the tropical western Pacific on 8 August 1991 and moved northwestward. Its intensity peaked near the Hainan Island on 16 August 1991 when it was approaching the climatic monsoon trough before landfall on the northeast coast of Vietnam. Its left turn in the northern SCS was influenced by another anomalous vortex at 300 hPa.

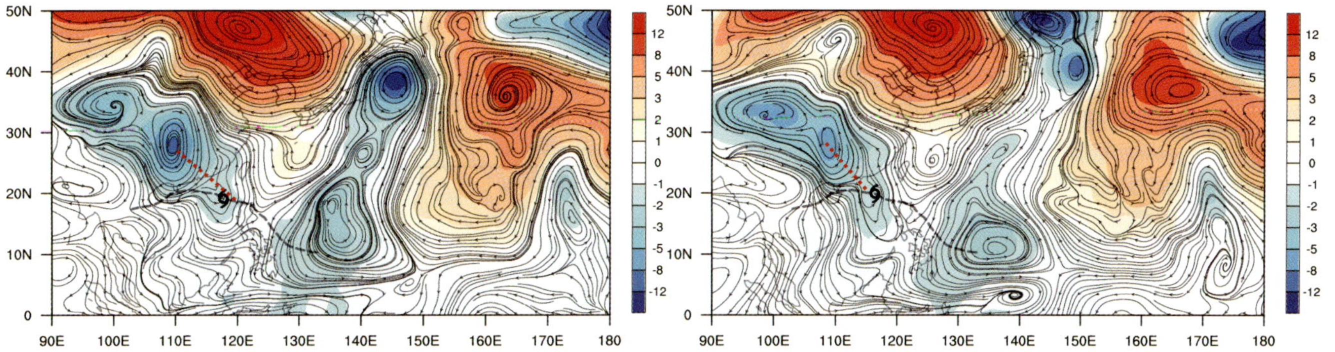

300 hPa flow/height anomalies at 0000 UTC and 1200 UTC 14 August 1991.

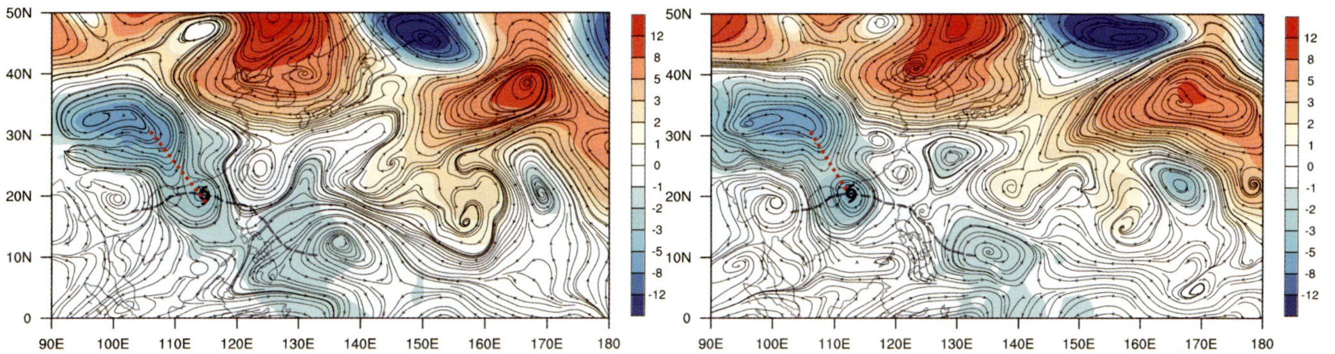

300 hPa flow/height anomalies at 0000 UTC and 1200 UTC 15 August 1991.

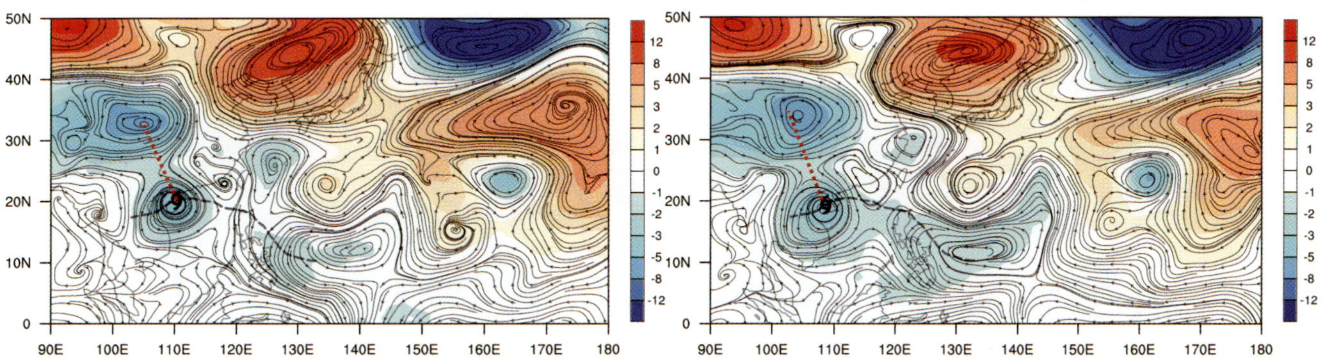

300 hPa flow/height anomalies at 0000 UTC and 1200 UTC 16 August 1991.

Figure 3.23. *(Continued)*

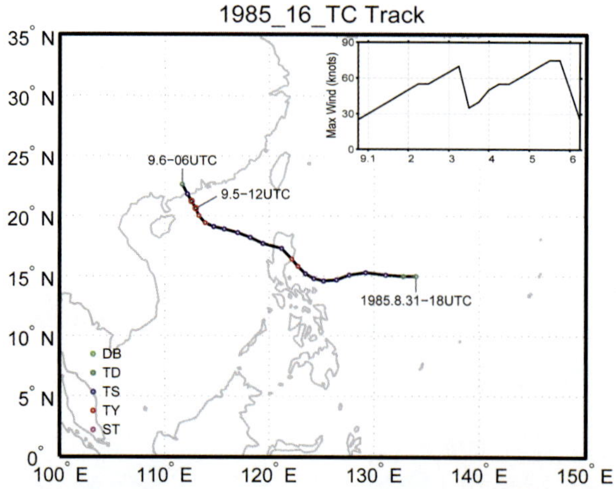

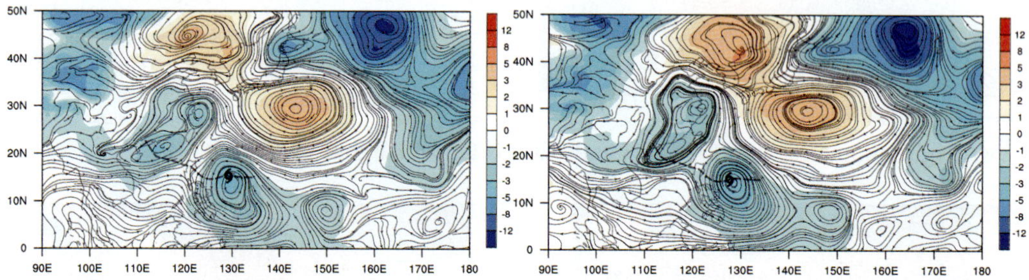

850 hPaflow/height anomalies at 1200 UTC and 1800 UTC 1 September 1985.

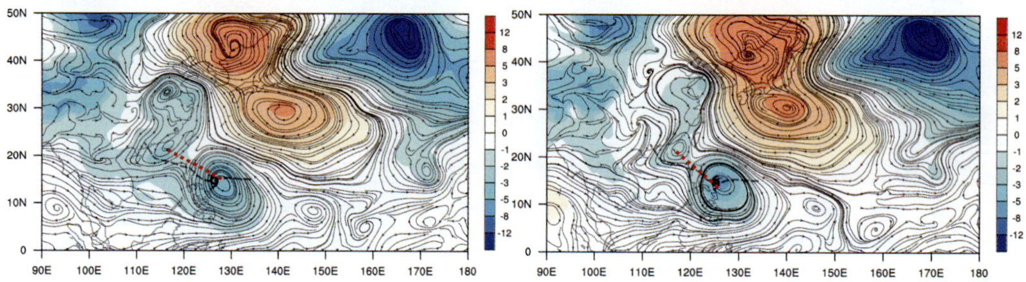

850 hPa flow/height anomalies at 0000 UTC and 0600 UTC 2 September 1985.

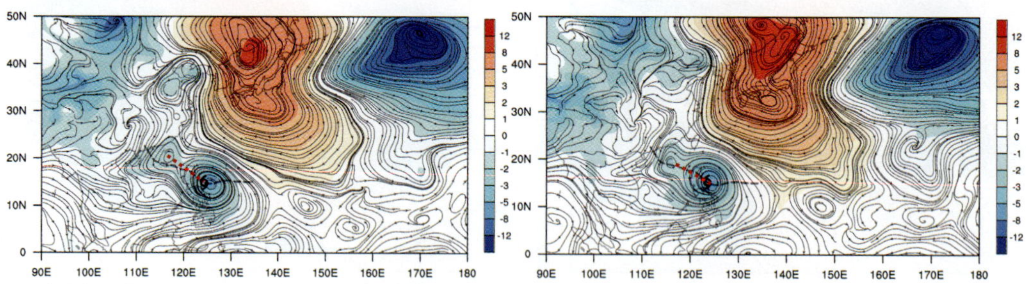

850 hPa flow/height anomalies at 1200 UTC and 1800 UTC 2 September 1985.

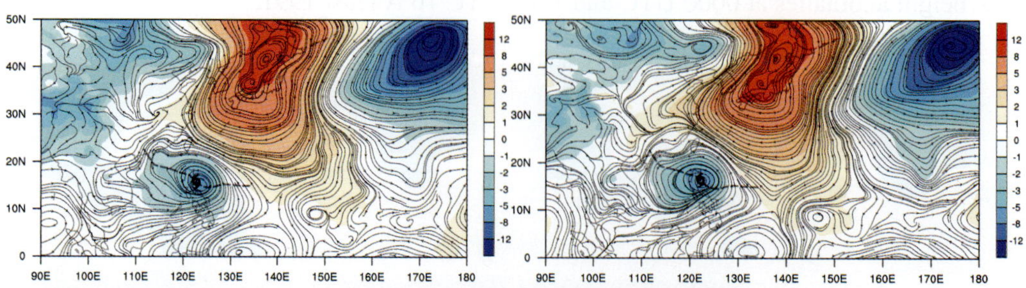

850 hPa flow/height anomalies at 0000 UTC and 0600 UTC 3 September 1985.

Figure 3.24. The typhoon (1985_No.16) generated in the tropical western Pacific on 31 August 1985 and moved northwestward. Its intensity peaked in the northernmost SCS on 5 September 1985 when it was approaching the climatic monsoon trough before landfall on the South China coast. A small curvature in its track was influenced by another anomalous vortex at 850 hPa.

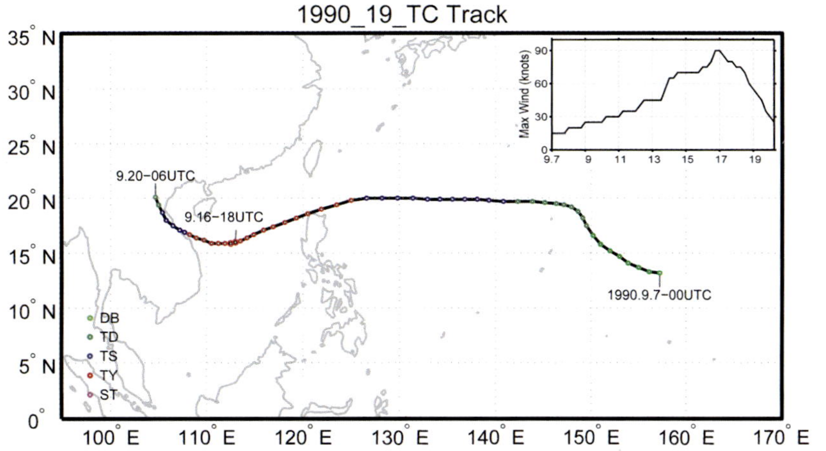

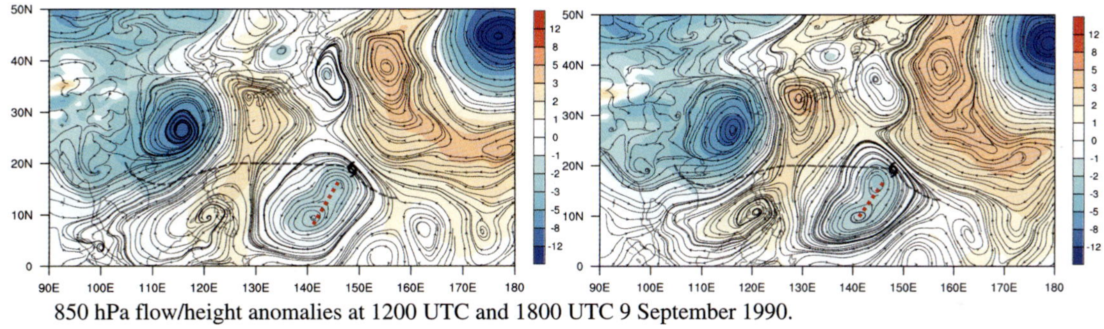

850 hPa flow/height anomalies at 1200 UTC and 1800 UTC 9 September 1990.

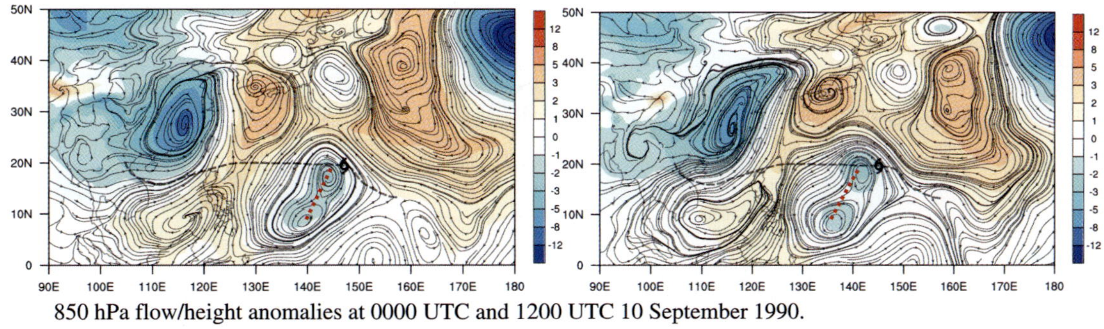

850 hPa flow/height anomalies at 0000 UTC and 1200 UTC 10 September 1990.

Figure 3.25. The typhoon (1990_No.19) generated in the tropical central Pacific on 7 September 1990 and moved westward for a long distance. Its intensity peaked in the southeast sea of the Hainan Island on 16 September 1990 when it was approaching the climatic monsoon trough before landfall on the northeast coast of Vietnam. Its turning was influenced by anomalous vortices at 850 hPa.

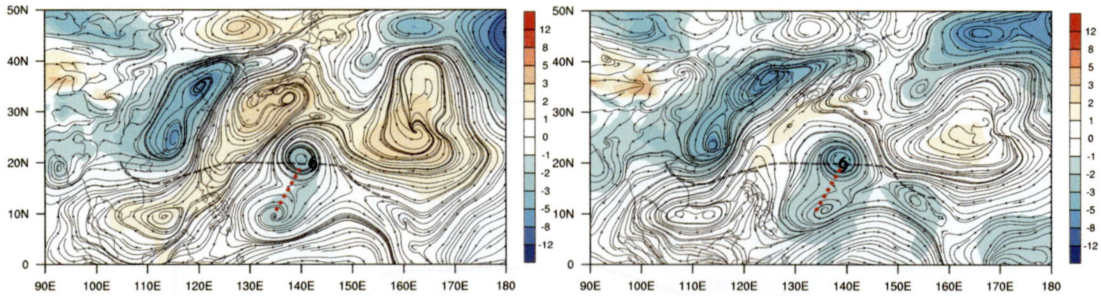

850 hPa flow/height anomalies at 0000 UTC and 1200 UTC 11 September 1990.

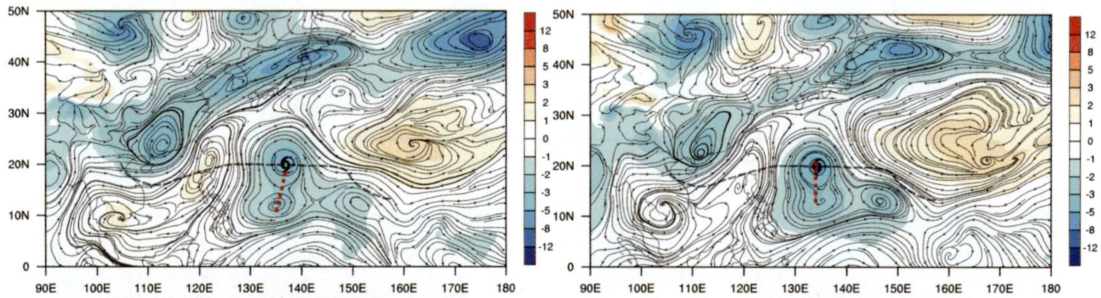

850 hPa flow/height anomalies at 0000 UTC and 1200 UTC 12 September 1990.

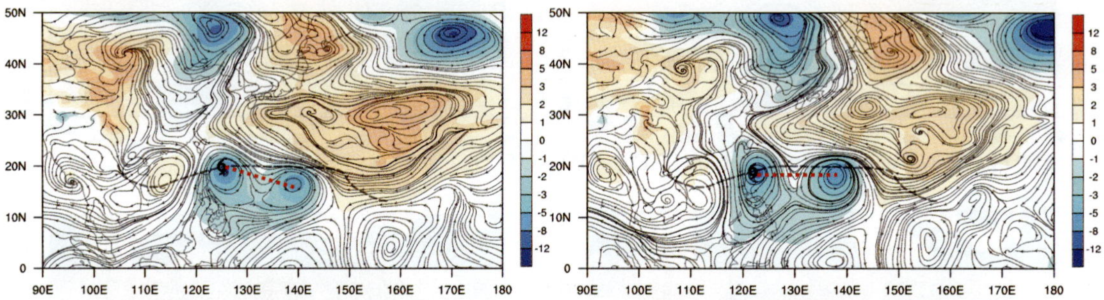

850 hPa flow/height anomalies at 0000 UTC and 1200 UTC 14 September 1990.

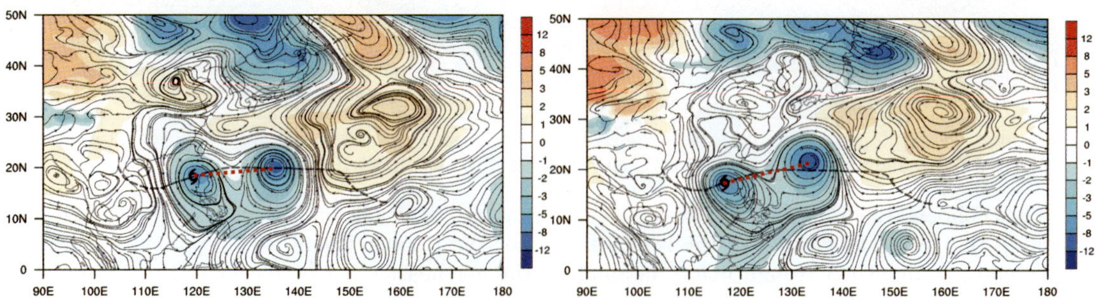

850 hPa flow/height anomalies at 0000 UTC and 1200 UTC 15 September 1990.

Figure 3.25. *(Continued)*

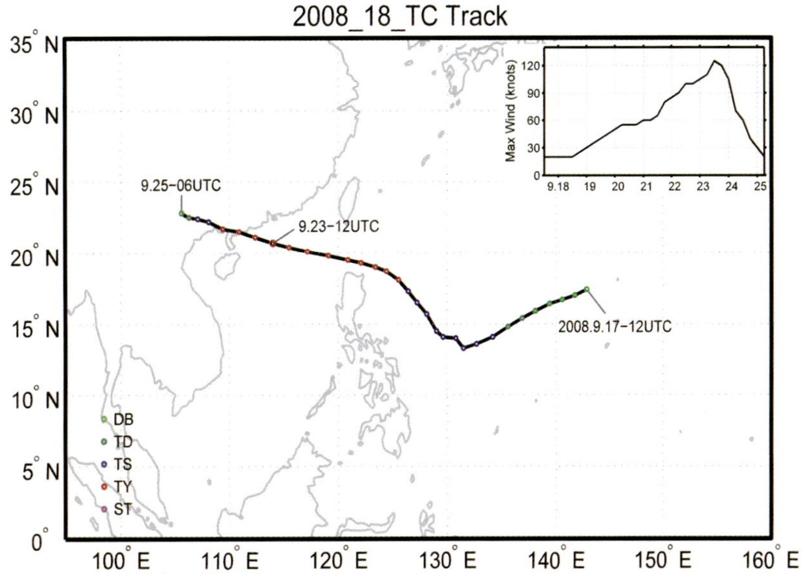

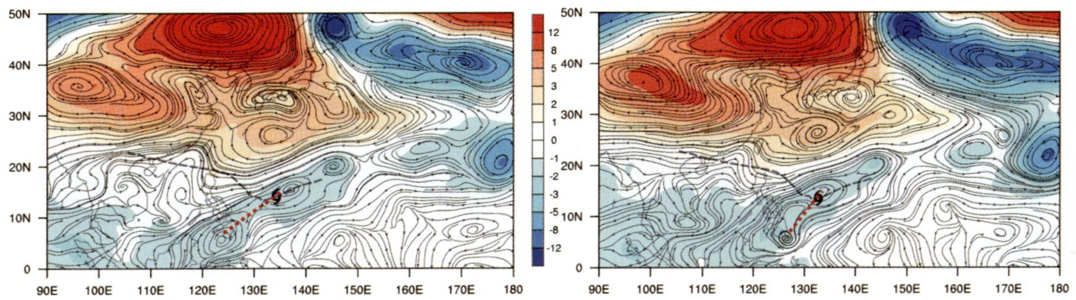

300 hPa flow/height anomalies at 0600 UTC and 1200 UTC 19 September 2008.

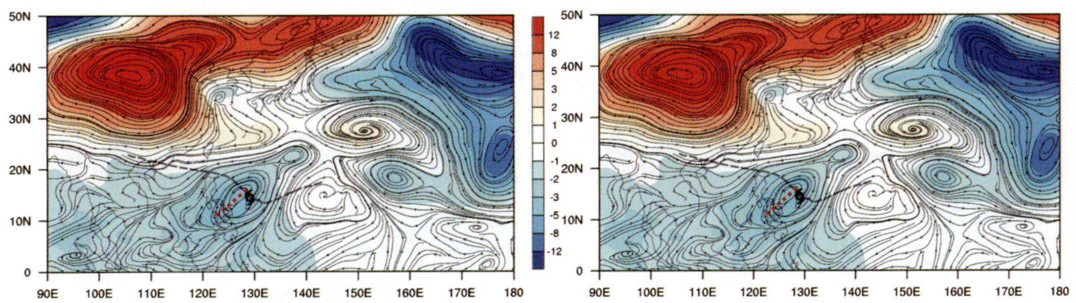

300 hPa flow/height anomalies at 0000 UTC and 1200 UTC 20 September 2008.

Figure 3.26. The typhoon (2008_No.18) generated in the tropical western Pacific on 17 September 2008 and moved southwestward at the beginning stage. Its intensity peaked in the northernmost SCS on 23 September 2008 when it was approaching the climatic monsoon trough before landfall on the South China coast. Its right turn in the east sea of the Philippines was influenced by anomalous vortices at 300 hPa and 850 hPa.

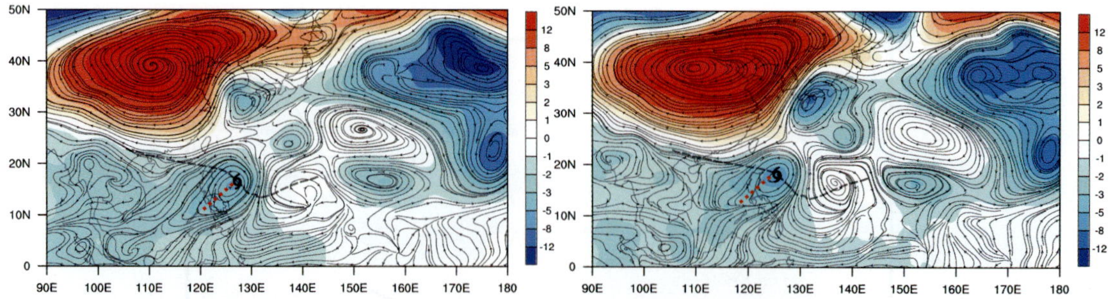

300 hPa flow/height anomalies at 0000 UTC and 1200 UTC 21 September 2008.

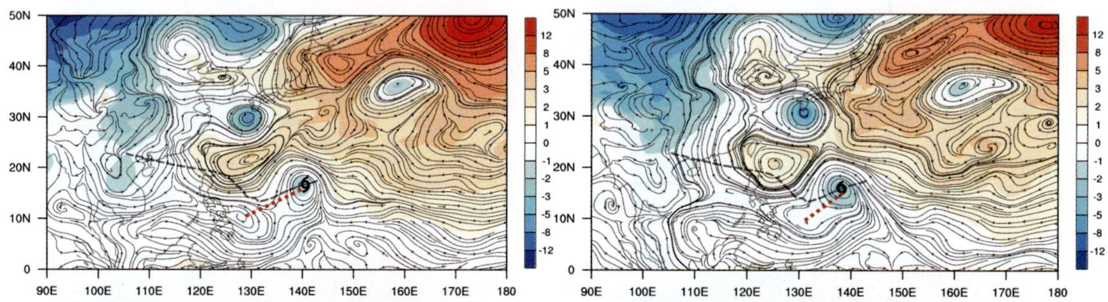

850 hPa flow/height anomalies at 0000 UTC and 1200 UTC 18 September 2008.

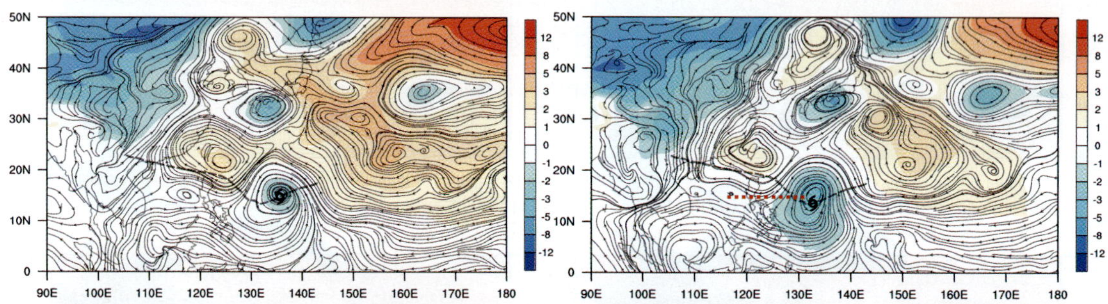

850 hPa flow/height anomalies at 0000 UTC and 1200 UTC 19 September 2008.

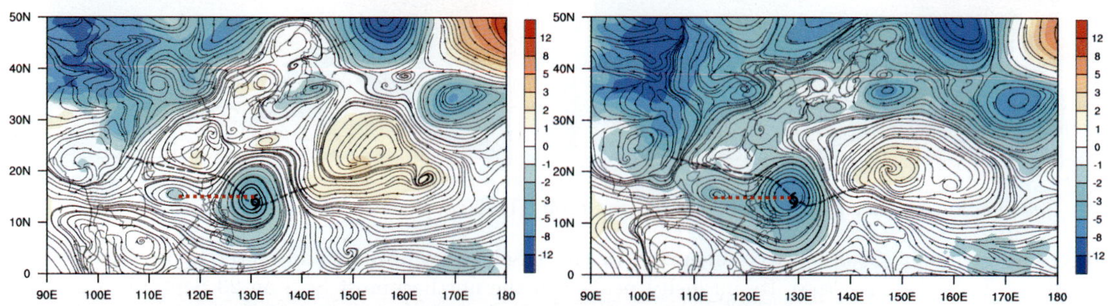

850 hPa flow/height anomalies at 0000 UTC and 1200 UTC 20 September 2008.

Figure 3.26. *(Continued)*

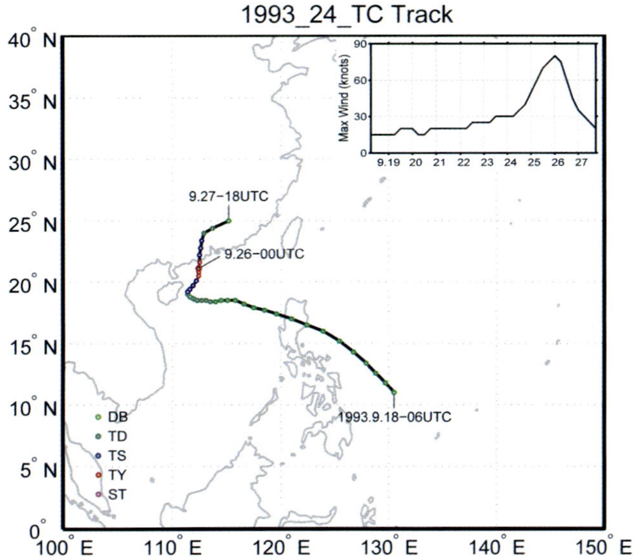

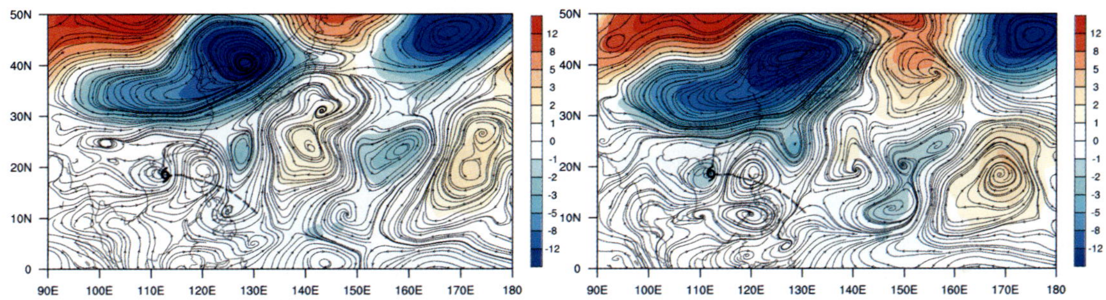

300 hPa flow/height anomalies at 0000 UTC and 1200 UTC 23 September 1993.

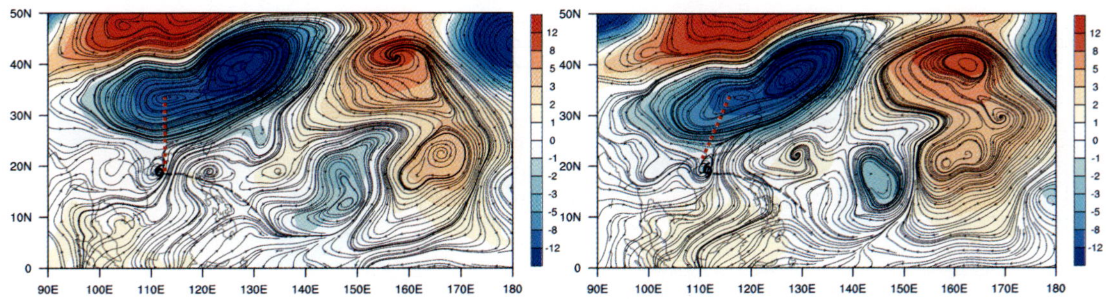

300 hPa flow/height anomalies at 0000 UTC and 1200 UTC 24 September 1993.

Figure 3.27. The typhoon (1993_No.24) generated in the east sea of the Philippines on 18 September 1993 and moved northwestward. Its intensity peaked in the northernmost SCS on 26 September 1993 when it was in the climatic monsoon trough before landfall on the South China coast. Its rapid right turn in the northern SCS was influenced by anomalous vortices at 300 hPa.

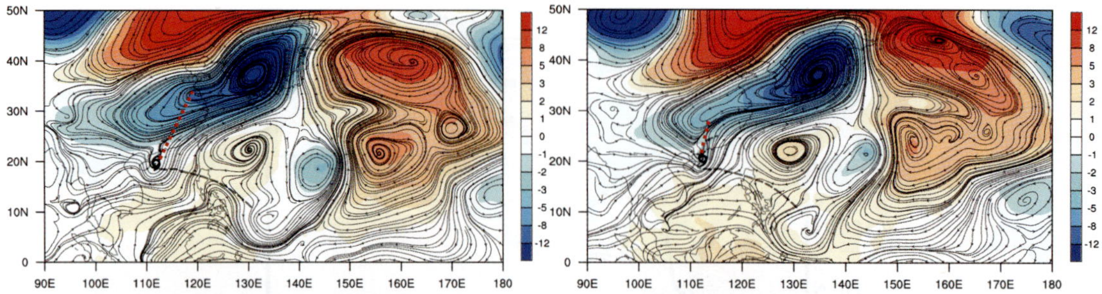

300 hPa flow/height anomalies at 0000 UTC and 1200 UTC 25 September 1993.

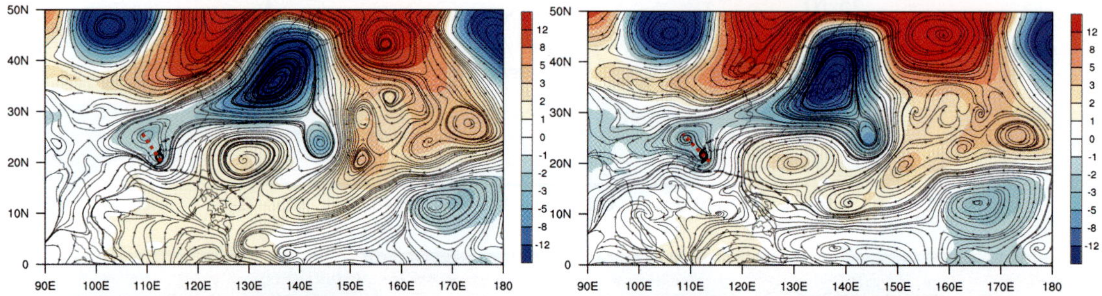

300 hPa flow/height anomalies at 0000 UTC and 0600 UTC 26 September 1993.

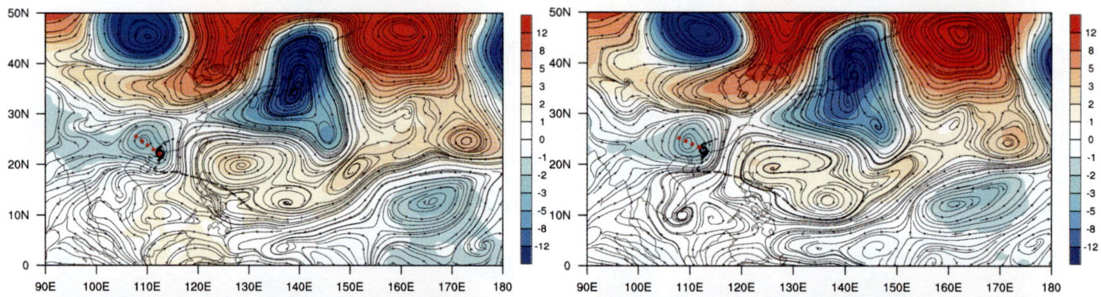

300 hPa flow/height anomalies at 1200 UTC and 1800 UTC 26 September 1993.

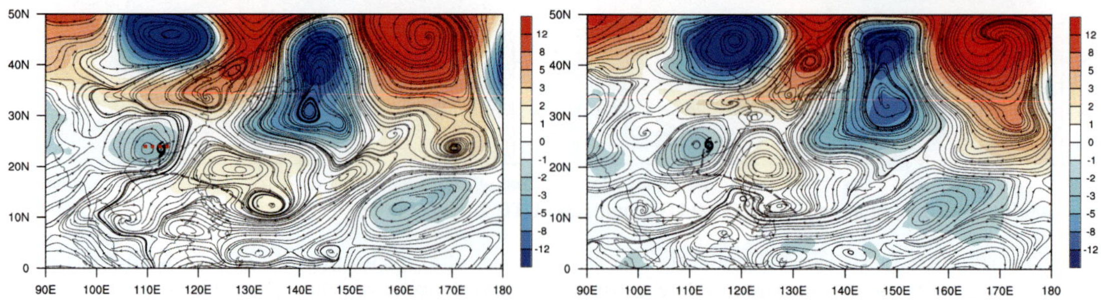

300 hPa flow/height anomalies at 0000 UTC and 1200 UTC 27 September 1993.

Figure 3.27. *(Continued)*

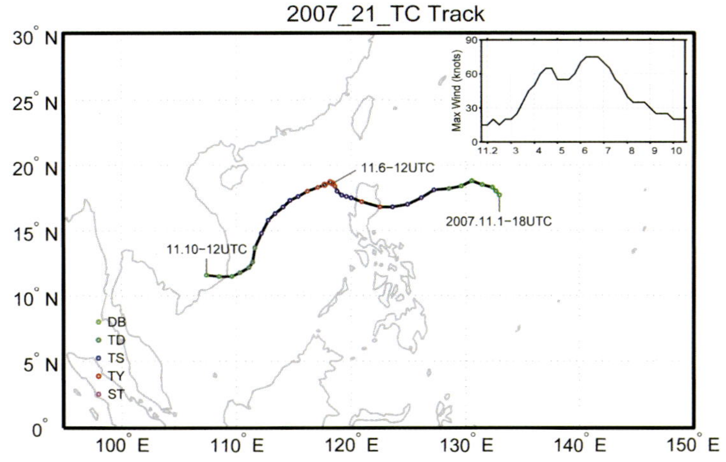

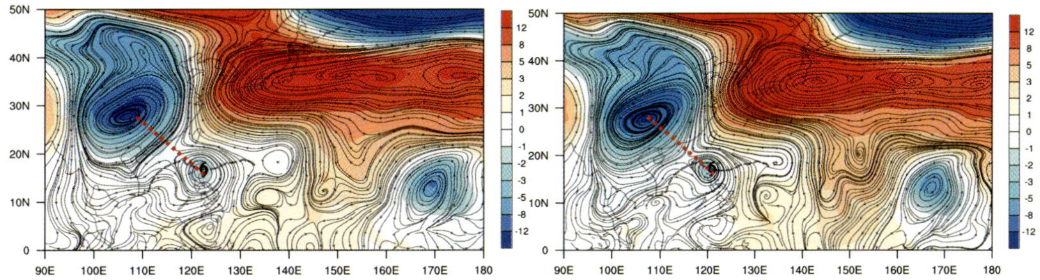

300 hPa flow/height anomalies at 1200 UTC and 1800 UTC 4 November 2007.

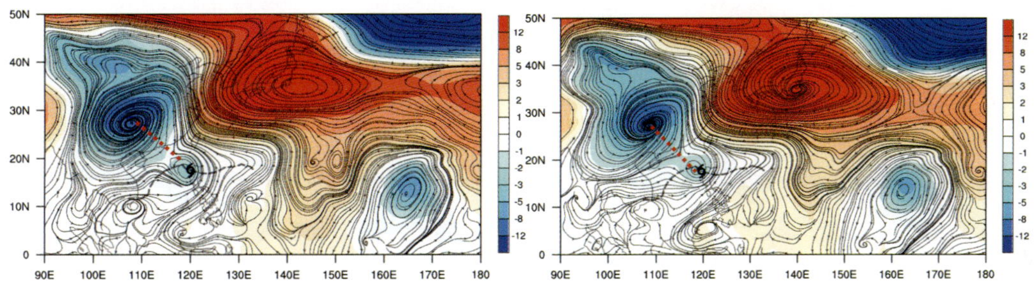

300 hPa flow/height anomalies at 0000 UTC and 0600 UTC 5 November 2007.

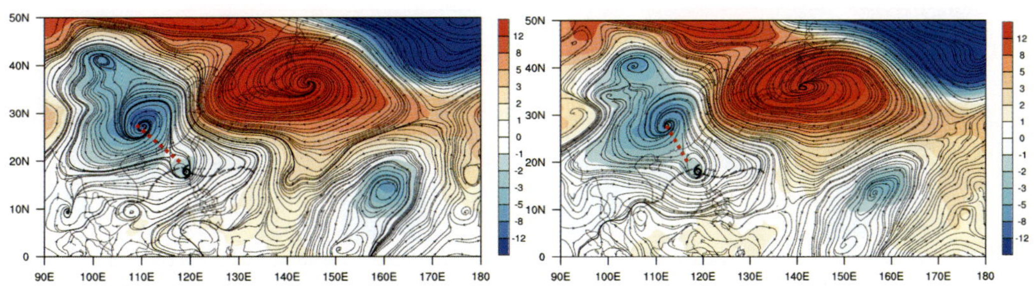

300 hPa flow/height anomalies at 1200 UTC and 1800 UTC 5 November 2007.

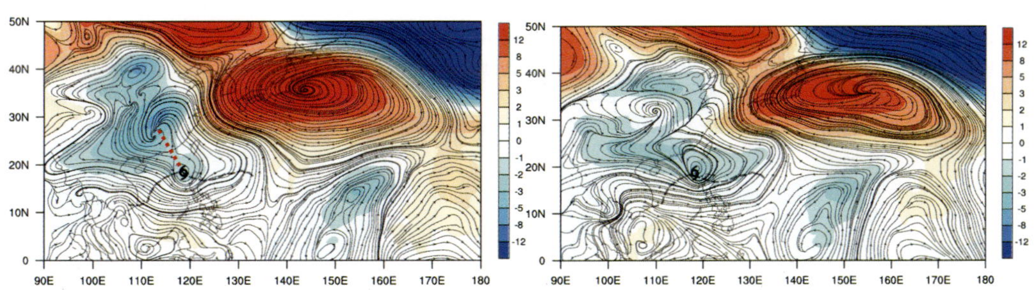

300 hPa flow/height anomalies at 0000 UTC and 1200 UTC 6 November 2007.

Figure 3.28. The typhoon (2007_No.21) generated in the tropical western Pacific on 1 November 2007 and moved westward along the climatic monsoon trough. Its intensity peaked in the northeastern SCS on 6 November 2007 at the moment from right turn to left turn. This right turn in the northern Philippines was influenced by an anomalous vortex at 300 hPa.

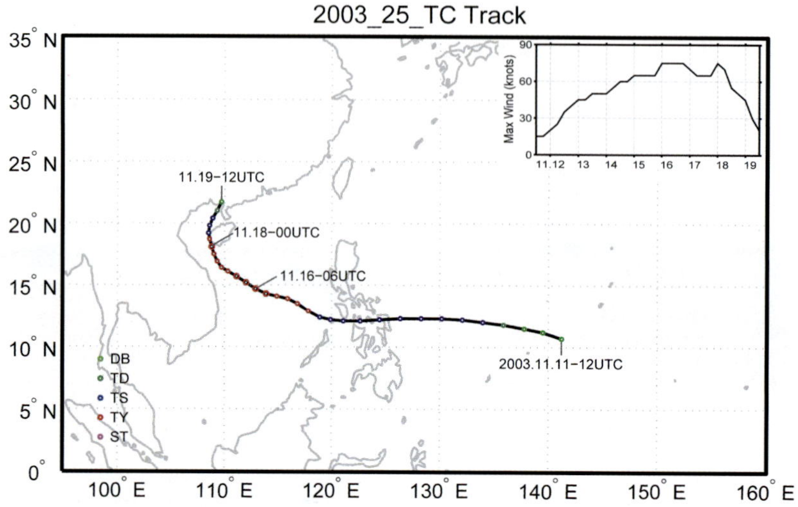

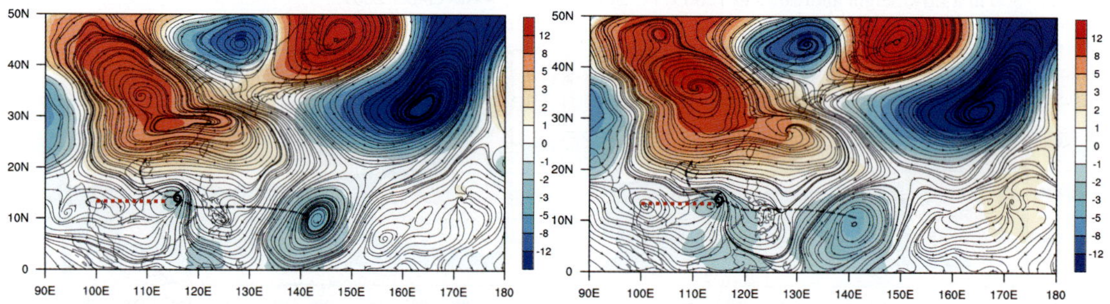

300 hPa flow/height anomalies at 1200 UTC and 1800 UTC 15 November 2003.

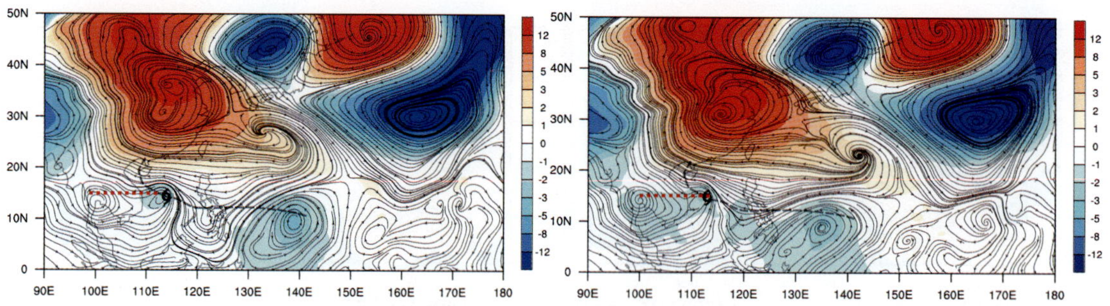

300 hPa flow/height anomalies at 0000 UTC and 0600 UTC 16 November 2003.

Figure 3.29. The typhoon (2003_No.25) generated on 11 November 2003 in the tropical western Pacific and moved westward along the Pacific tropical monsoon trough. Its intensity peaked two times in the central SCS on 16 November 2003 and near the Hainan Island on 18 November 2003 before landfall on the South China coast, respectively. Its right turn in the northern SCS was influenced by anomalous vortex at 300 hPa and anomalous southwesterly flow at 850 hPa.

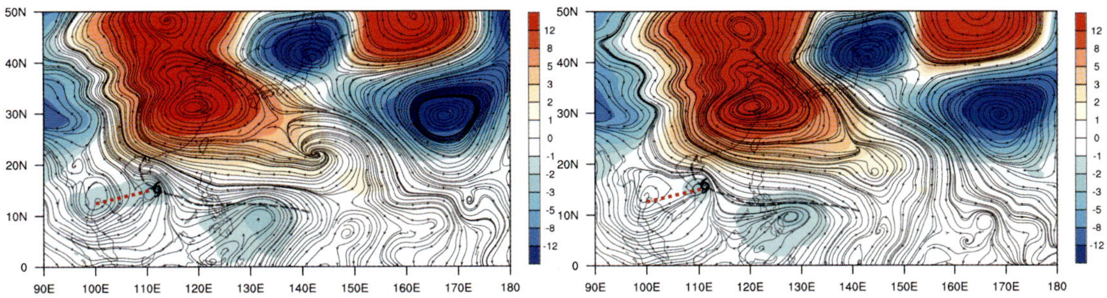

300 hPa flow/height anomalies at 1200 UTC and 1800 UTC 16 November 2003.

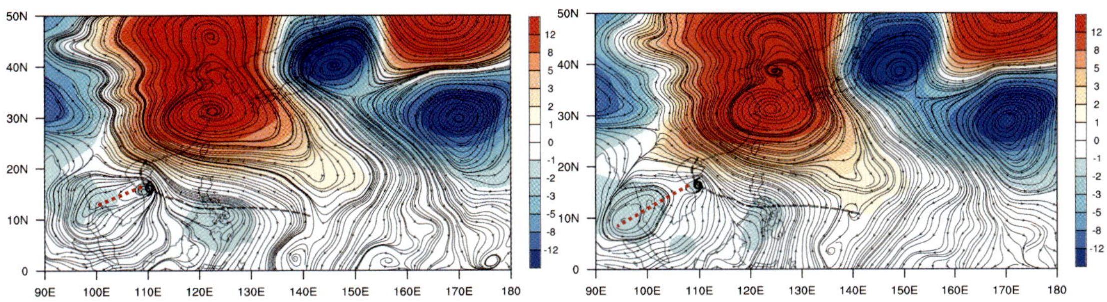

300 hPa flow/height anomalies at 0000 UTC and 0600 UTC 17 November 2003.

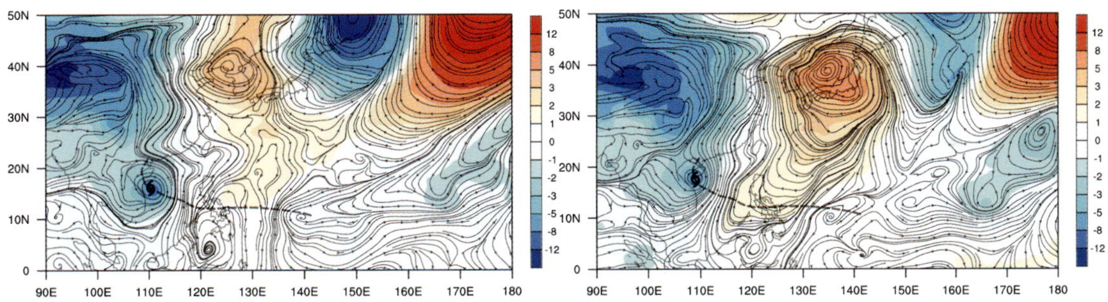

850 hPa flow/height anomalies at 1200 UTC and 1800 UTC 17 November 2003.

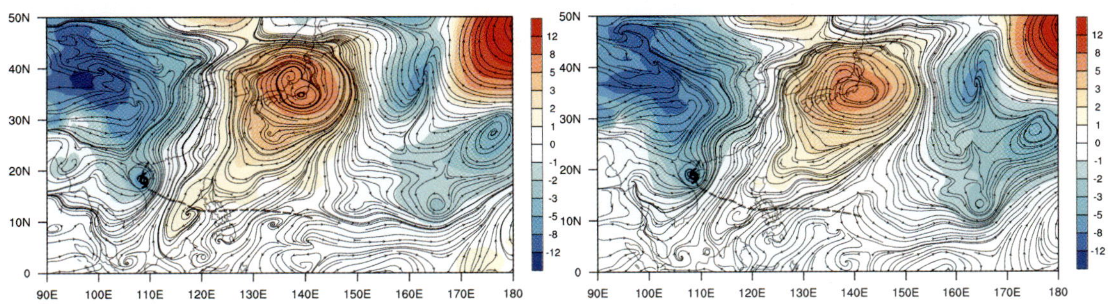

850 hPa flow/height anomalies at 0000 UTC and 0600 UTC 18 November 2003.

Figure 3.29. *(Continued)*

Chapter 4

Intensity Peaked in the Philippines

Chapter 4 describes typhoons with their intensities peaked in the Philippines and generated, respectively, in the Philippines (Sec. 4.1), in the equatorial western Pacific (Sec. 4.2), in the equatorial central Pacific (Sec. 4.3), and in the tropical Pacific (Sec. 4.4).

4.1. Geneses in the Philippines

Two typhoons generated from tropical disturbances and their intensities peaked in the Philippines. They occurred in June 1991 and November 1986, respectively (Figs. 4.1 and 4.2).

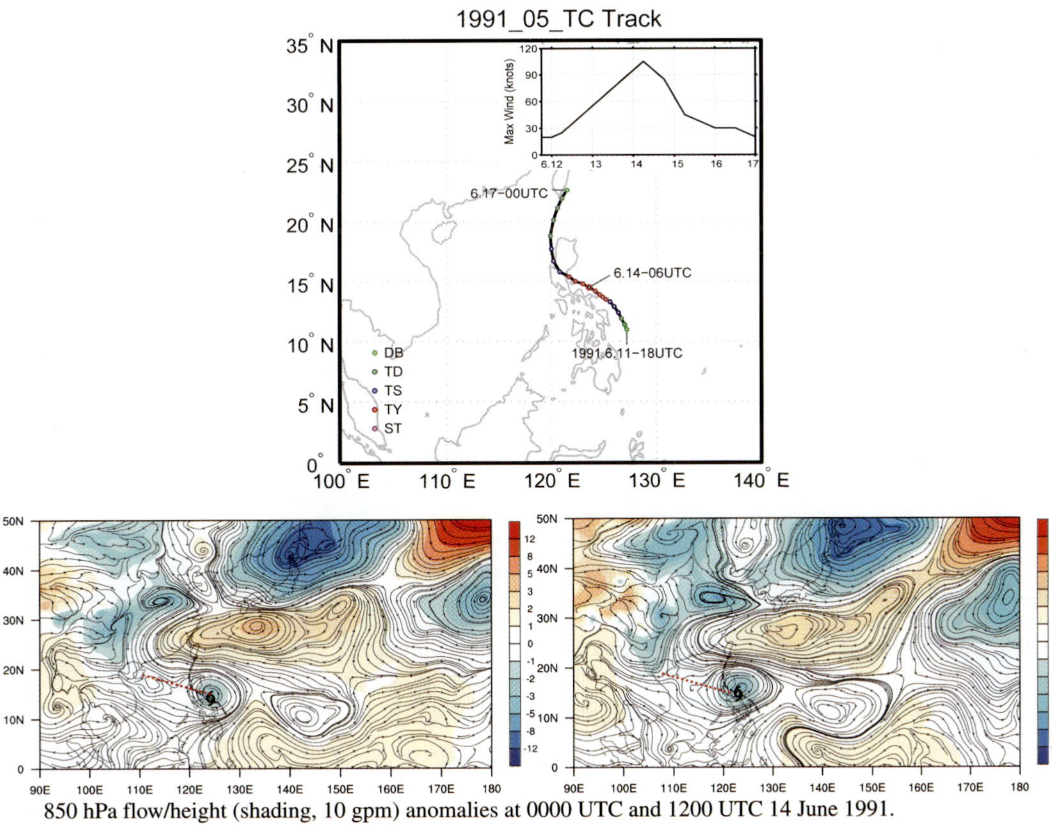

850 hPa flow/height (shading, 10 gpm) anomalies at 0000 UTC and 1200 UTC 14 June 1991.

Figure 4.1. The typhoon (1991_No.05) generated in the east sea of Philippines on 11 June 1991 and moved northwestward along the climatic monsoon trough. Its intensity peaked in the central Philippines on 14 June 1991 before landfall on the island. Its right turn in the northern Philippines was influenced by another anomalous vortex at 850 hPa. Symbols used are same as in Fig. 2.1.

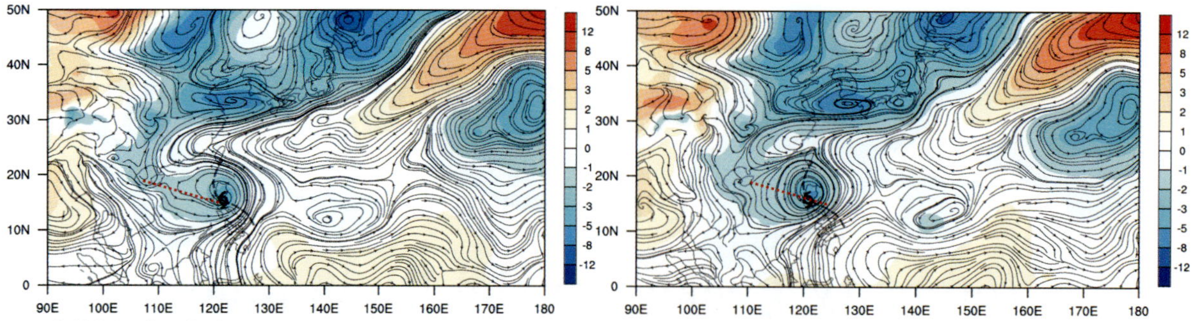

850 hPa flow/height anomalies at 0000 UTC and 0600 UTC 15 June 1991.

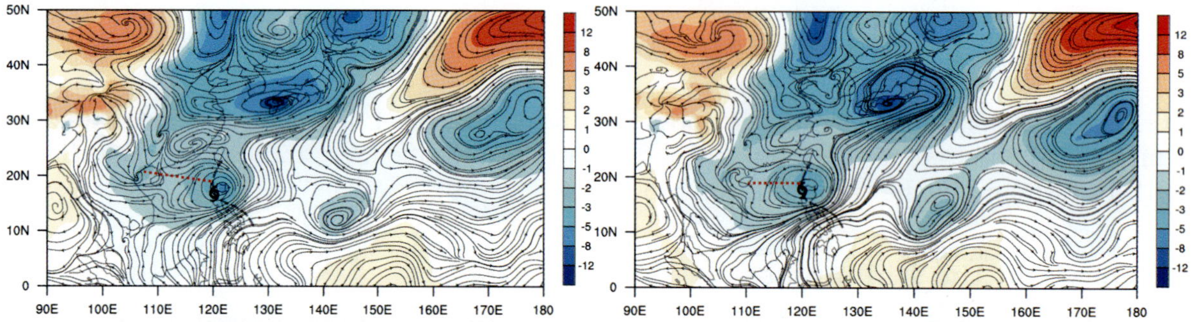

850 hPa flow/height anomalies at 1200 UTC and 1800 UTC 15 June 1991.

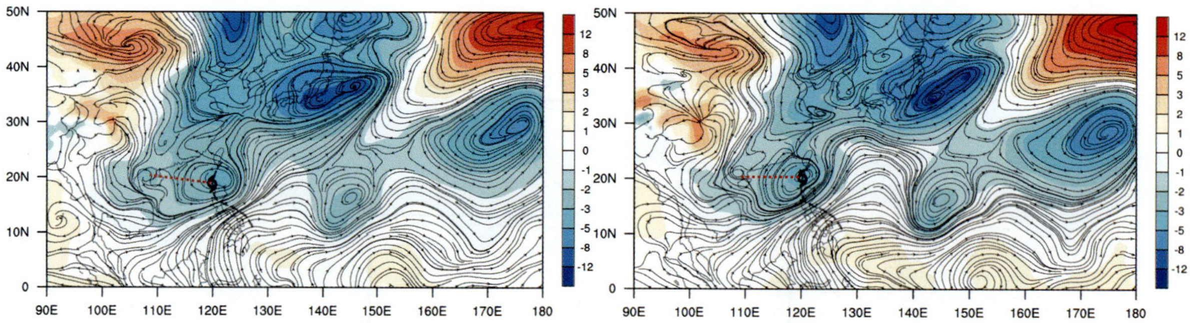

850 hPa flow/height anomalies at 0000 UTC and 0600 UTC 16 June 1991.

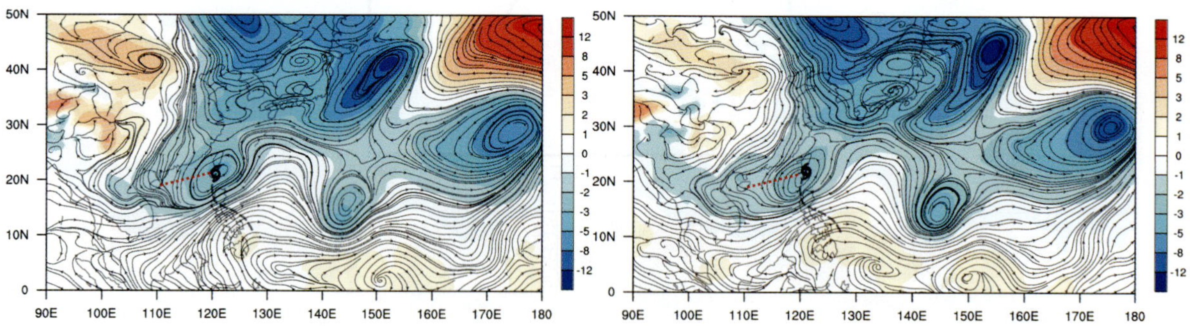

850 hPa flow/height anomalies at 1200 UTC and 1800 UTC 16 June 1991.

Figure 4.1. *(Continued)*

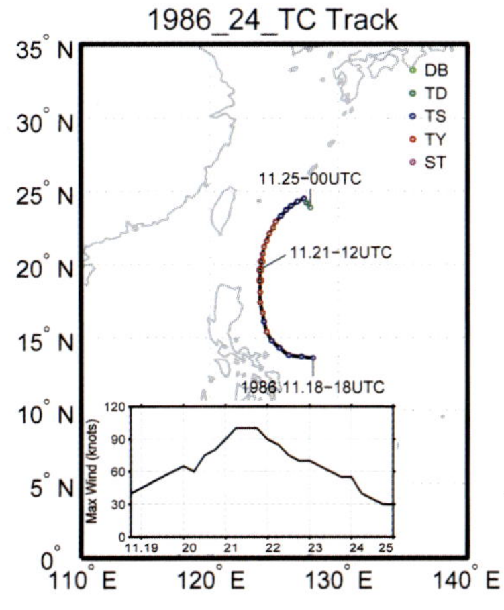

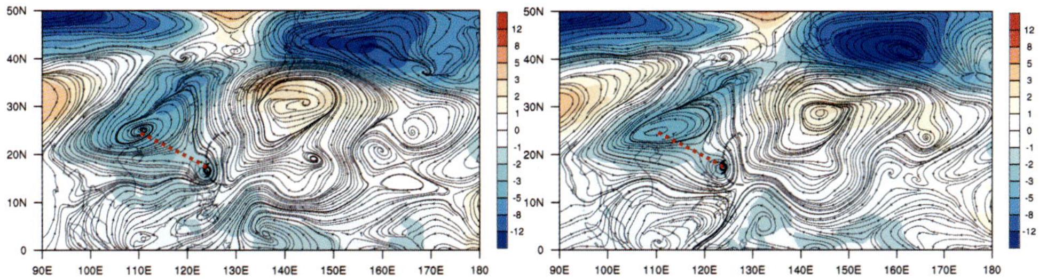

300 hPa flow/height anomalies at 1200 UTC and 1800 UTC 20 November 1986.

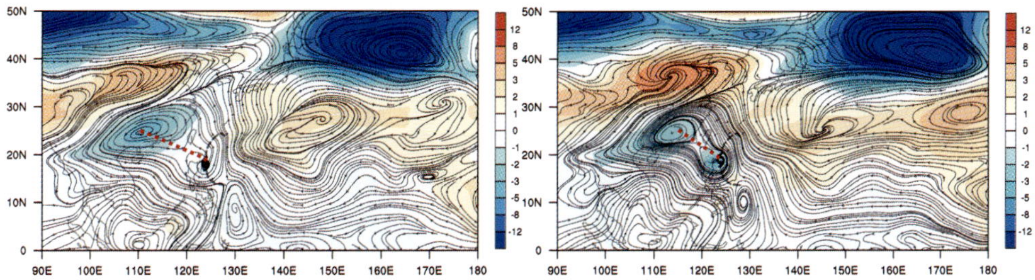

300 hPa flow/height anomalies at 0000 UTC and 0600 UTC 21 November 1986.

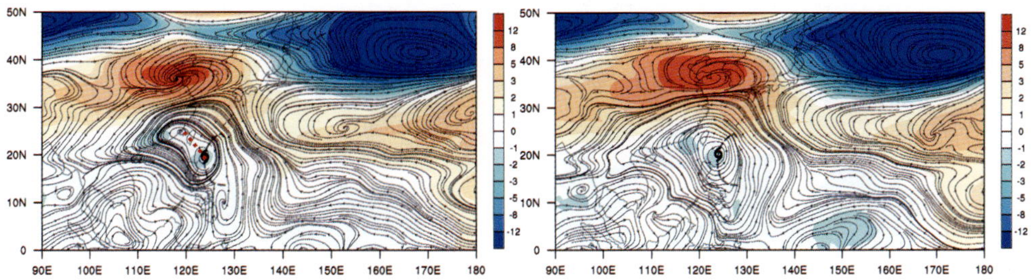

300 hPa flow/height anomalies at 1200 UTC and 1800 UTC 21 November 1986.

Figure 4.2. The typhoon (1986_No.24) generated in the tropical east sea of the Philippines on 18 November 1986. Its intensity peaked in the northeast sea of the Philippines on 21 November 1986. Its right turn in the east sea of the Philippines was influenced by another anomalous vortex at 300 hPa.

4.2. Geneses in the Equatorial Western Pacific

There were 13 typhoons that generated in the equatorial western Pacific and their intensities peaked in the Philippines. The right turns or left turns of these typhoons were due to the influences of anomalous vortices (Figs. 4.3–4.15).

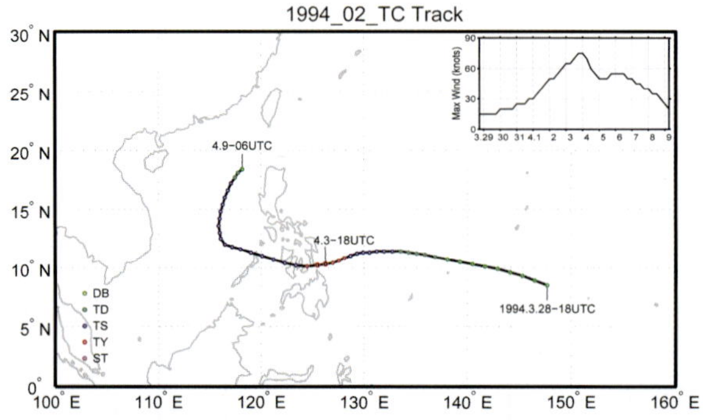

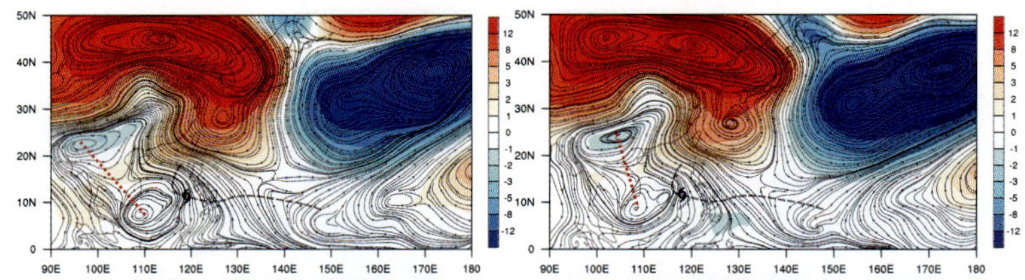

300 hPa flow/height (shading, 10 gpm) anomalies at 1200 UTC and 1800 UTC 5 April 1994.

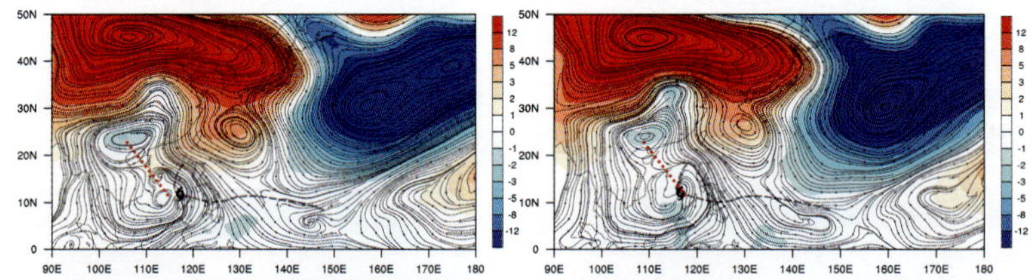

300 hPa flow/height anomalies at 0000 UTC and 0600 UTC 6 April 1994.

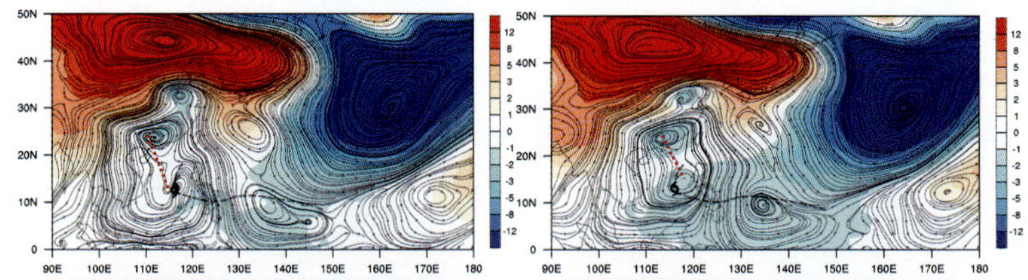

300 hPa flow/height anomalies at 1200 UTC and 1800 UTC 6 April 1994.

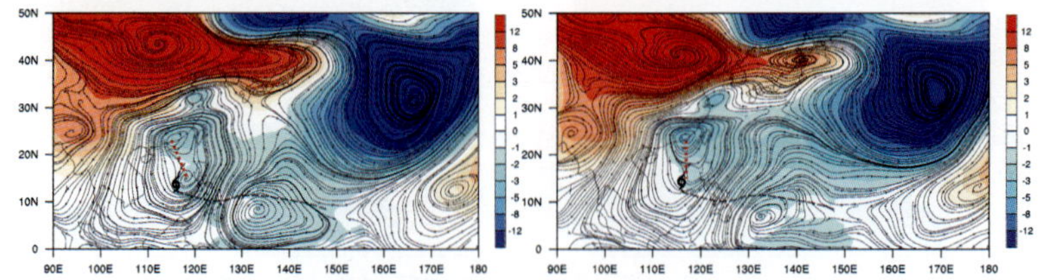

300 hPa flow/height anomalies at 0000 UTC and 0600 UTC 7 April 1994.

Figure 4.3. The typhoon (1994_No.02) generated in the equatorial western Pacific on 28 March 1994 and moved westward in the tropical western Pacific. Its intensity peaked in the east sea of the Philippines on 3 April 1994 before landfall on the island. Its right turn in the South China Sea (SCS) was influenced by another anomalous vortex at 300 hPa.

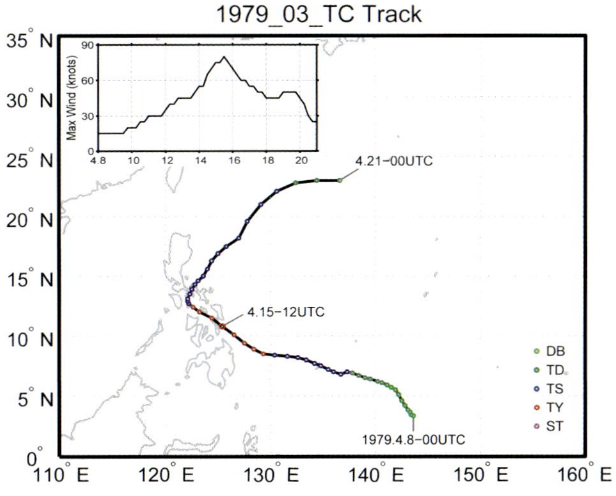

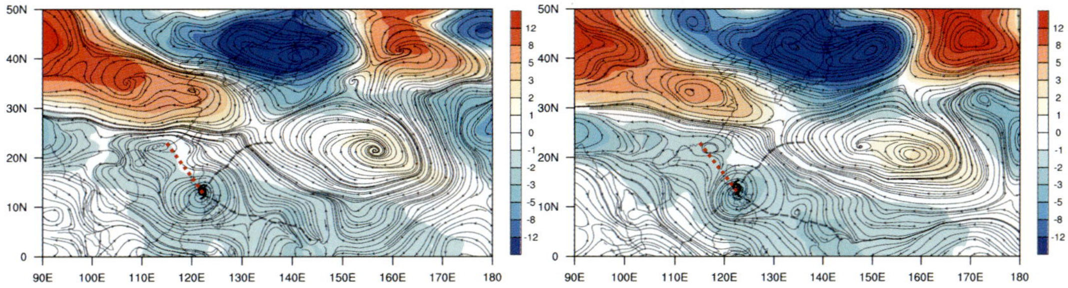

300 hPa flow/height anomalies at 0000 UTC and 1200 UTC 17 April 1979.

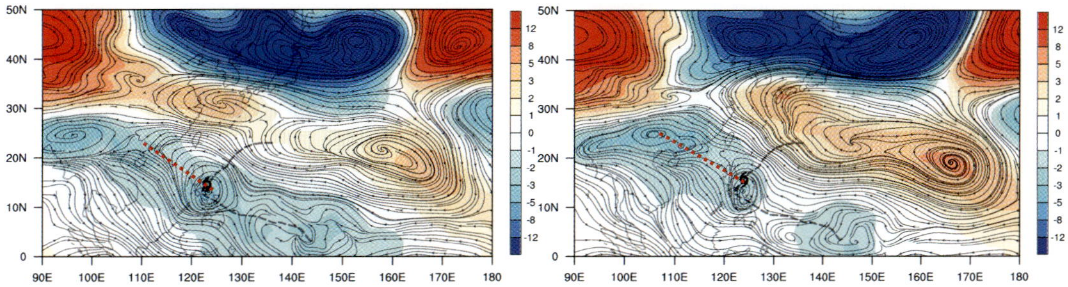

300 hPa flow/height anomalies at 0000 UTC and 1200 UTC 18 April 1979.

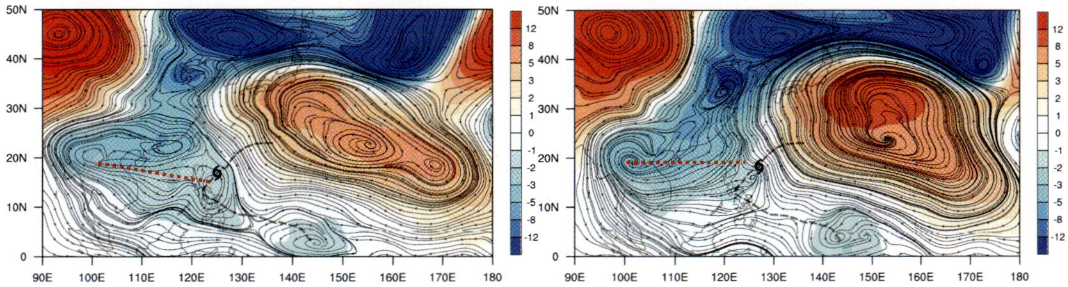

300 hPa flow/height anomalies at 0000 UTC and 1200 UTC 19 April 1979.

Figure 4.4. The typhoon (1979_No.03) generated in the equatorial western Pacific on 8 April 1979 and its intensity peaked in the east sea of the Philippines on 15 April 1979 when it was approaching the climatic monsoon trough before landfall on the island. Its right turn was influenced by an anomalous vortex at 300 hPa.

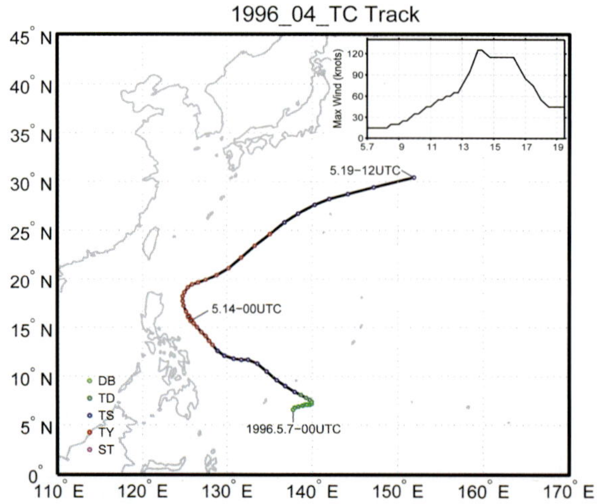

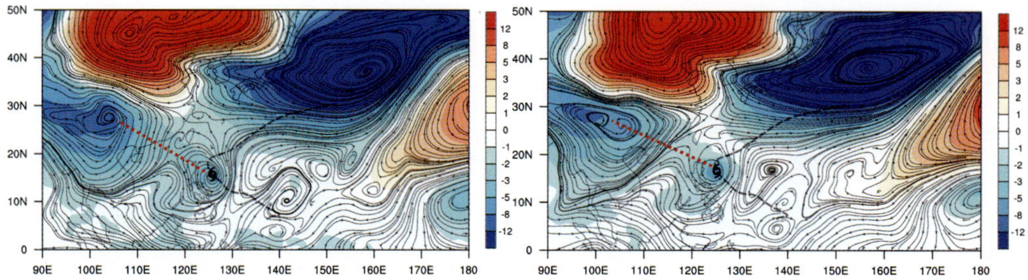

300 hPa flow/height anomalies at 0000 UTC and 1200 UTC 14 May 1996.

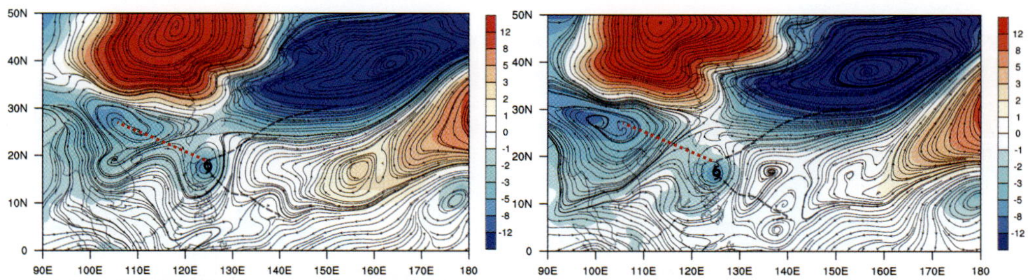

300 hPa flow/height anomalies at 0000 UTC and 1200 UTC 15 May 1996.

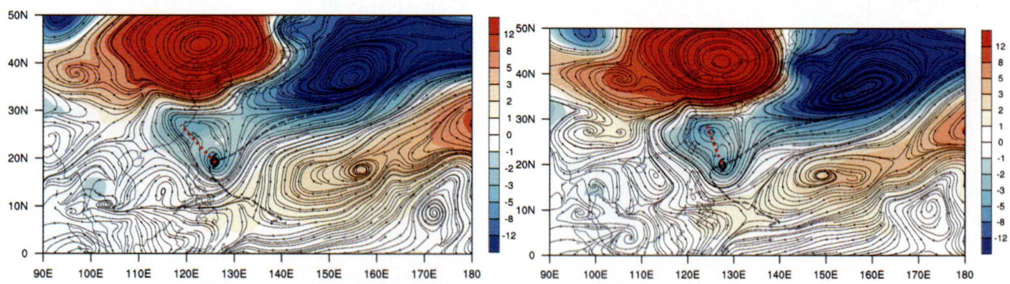

300 hPa flow/height anomalies at 0000 UTC and 1200 UTC 16 May 1996.

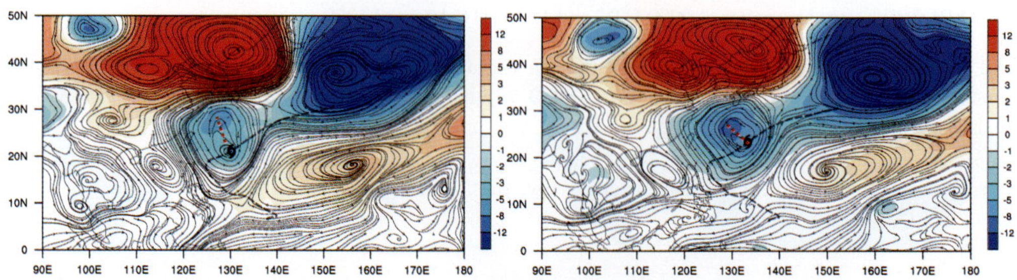

300 hPa flow/height anomalies at 0000 UTC and 1200 UTC 17 May 1996.

Figure 4.5: The typhoon (1996_No.04) generated in the equatorial western Pacific on 7 May 1996 and moved northwestward along the climatic monsoon trough. Its intensity peaked in the east sea of the Philippines on 14 May 1996. Its right turn was influenced by an anomalous vortex at 300 hPa.

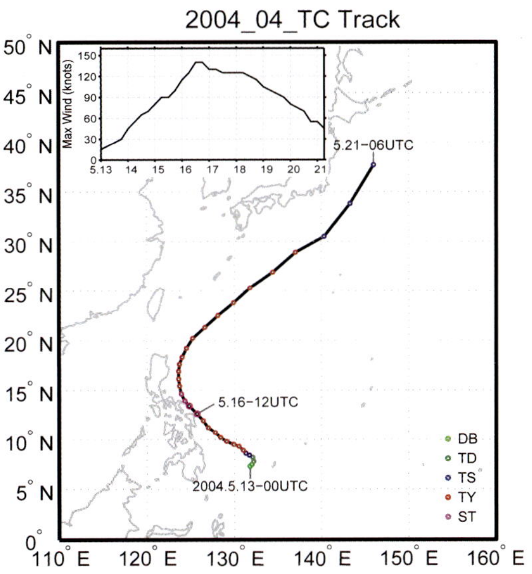

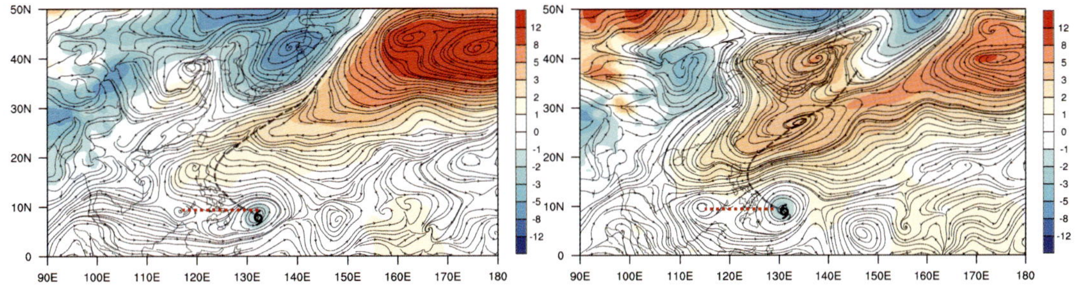

850 hPa flow/height anomalies at 1200 UTC 13 and 14 May 2004.

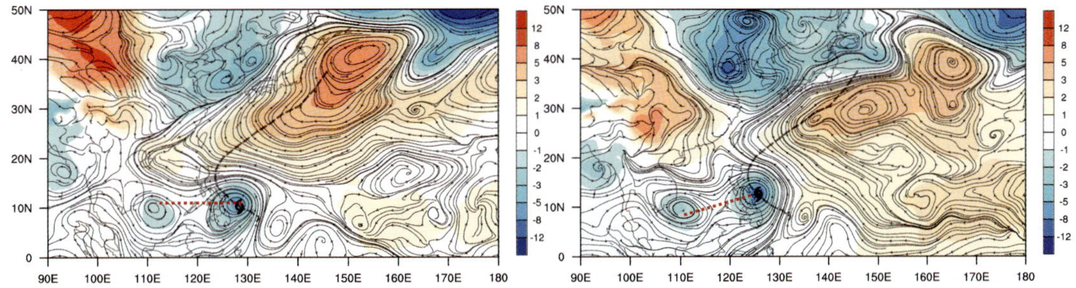

850 hPa flow/height anomalies at 1200 UTC 15 and 16 May 2004.

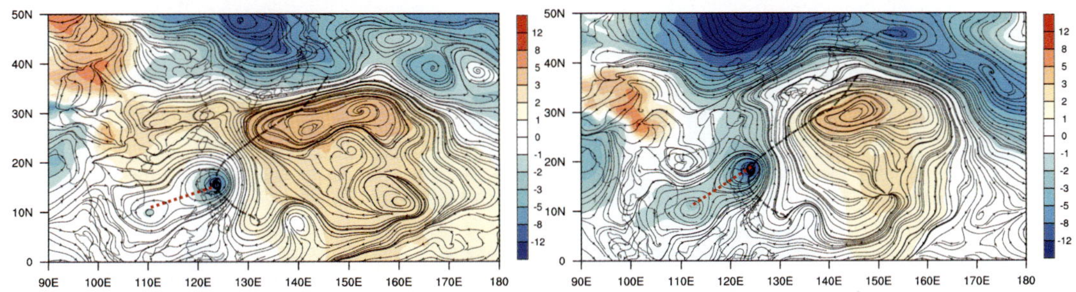

850 hPa flow/height anomalies at 1200 UTC 17 and 18 May 2004.

Figure 4.6. The typhoon (2004_No.04) generated in the equatorial western Pacific on 13 May 2004 and moved northwestward along the climatic monsoon trough. Its intensity peaked in the east coast of the central Philippines on 16 May 2004. Its right turn near the Philippines was influenced by another anomalous vortex at 850 hPa.

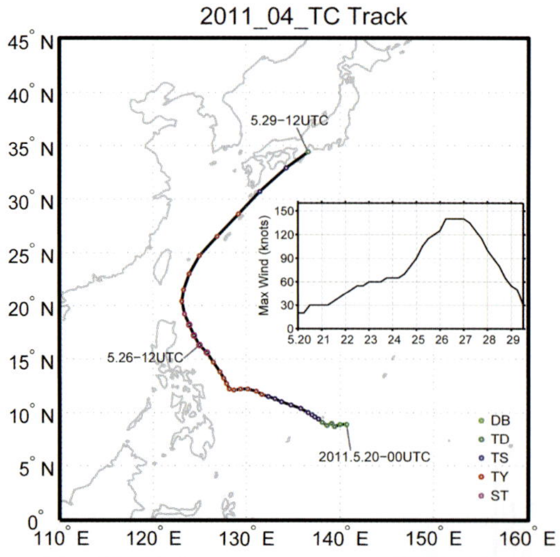

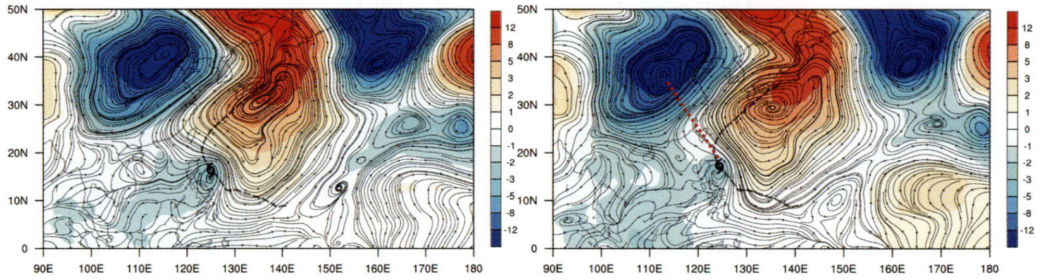

300 hPa flow/height anomalies at 1200 UTC and 1800 UTC 26 May 2011.

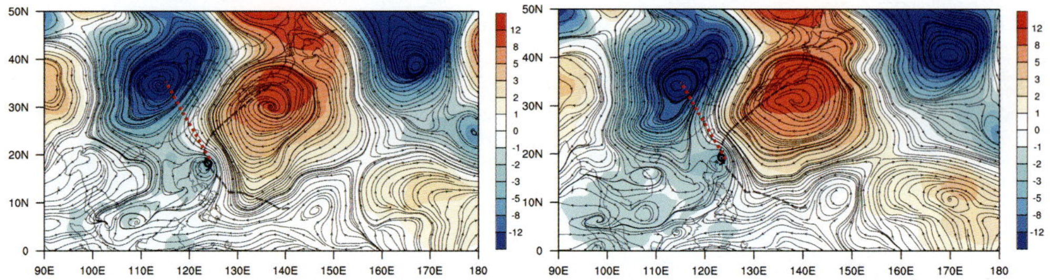

300 hPa flow/height anomalies at 0000 UTC and 0600 UTC 27 May 2011.

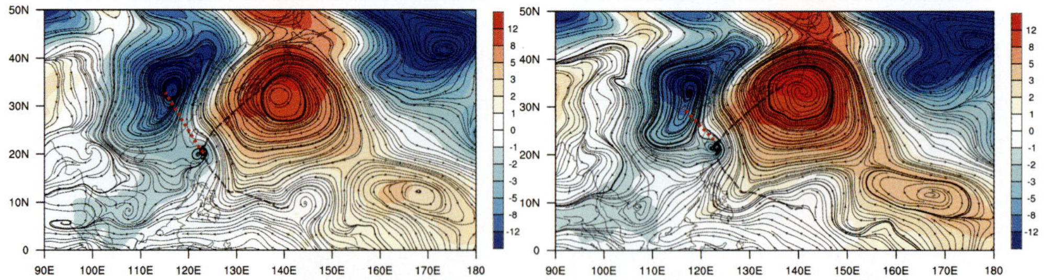

300 hPa flow/height anomalies at 1200 UTC and 1800 UTC 27 May 2011.

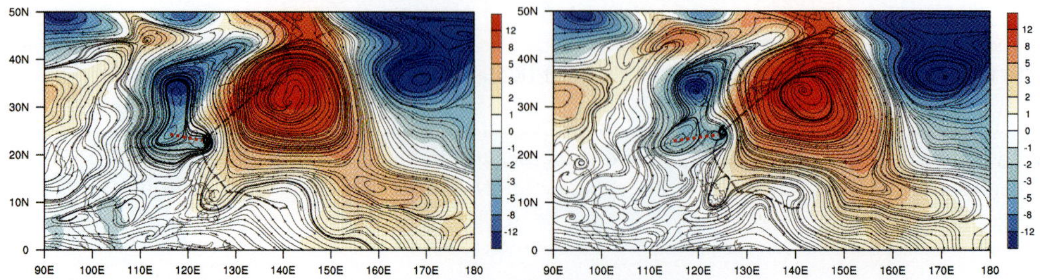

300 hPa flow/height anomalies at 0000 UTC and 0600 UTC 28 May 2011.

Figure 4.7. The typhoon (2011_No.04) generated in the equatorial western Pacific on 20 May 2011 and moved northwestward along the climatic monsoon trough. Its intensity peaked in the northeast sea of the Philippines on 26 May 2011. Its right turn during its lifetime was influenced by another anomalous vortex at 300 hPa.

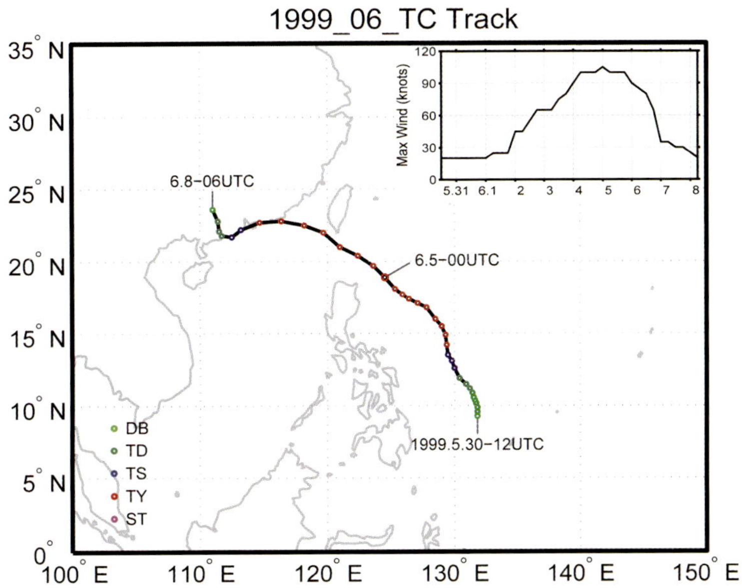

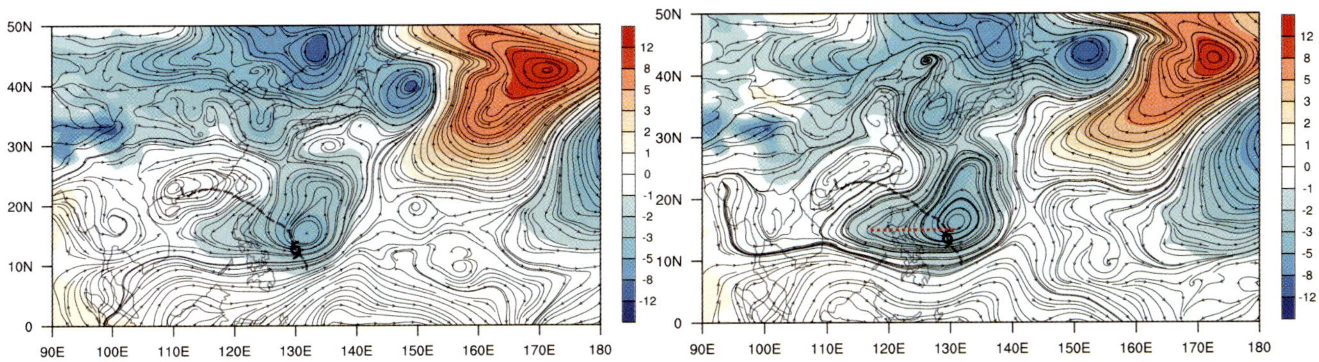

850 hPa flow/height anomalies at 0000 UTC and 1200 UTC 2 June 1999.

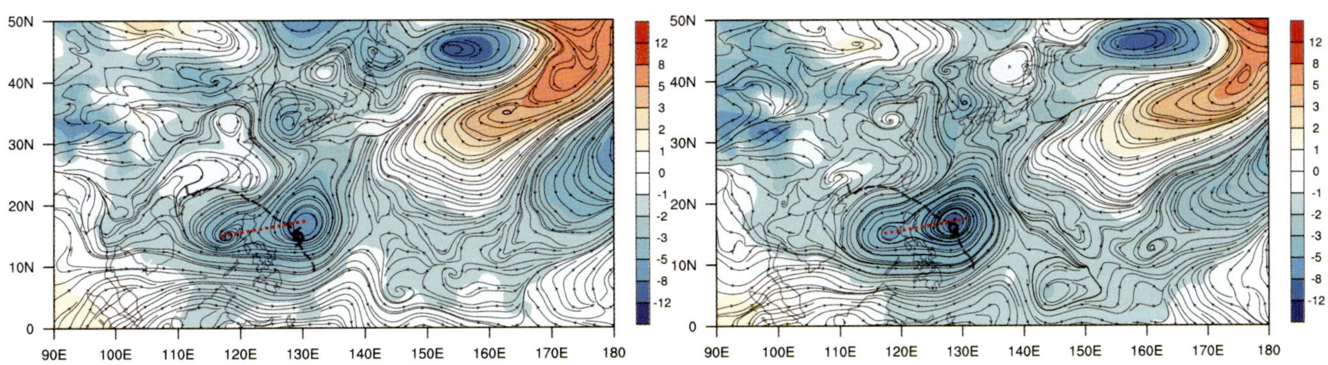

850 hPa flow/height anomalies at 0000 UTC and 1200 UTC 3 June 1999.

Figure 4.8. The typhoon (1999_No.06) generated in the equatorial western Pacific on 30 May 1999 and moved northwestward. Its intensity peaked in the northeast sea of the Philippines on 5 June 1999. Its left turn in the northeast sea of the Philippines and the northern SCS was influenced by another anomalous vortex at 850 hPa.

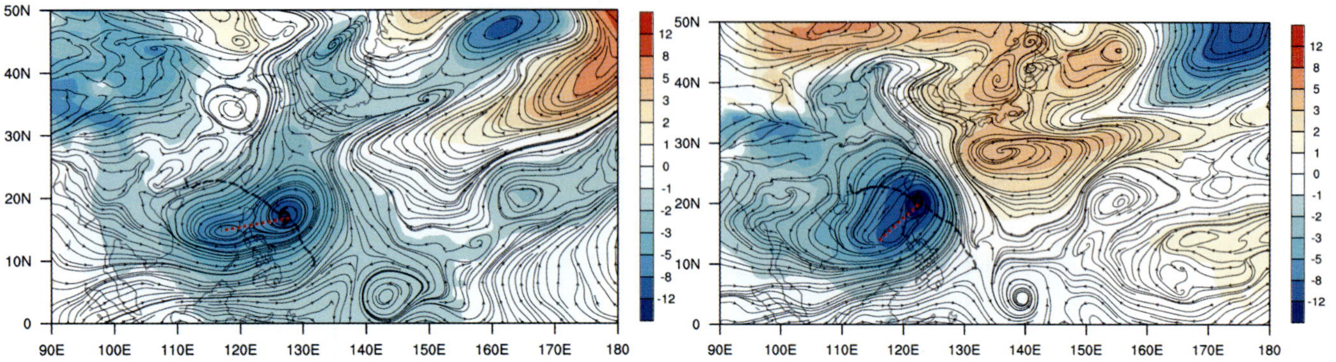

850 hPa flow/height anomalies at 0000 UTC and 1200 UTC 4 June 1999.

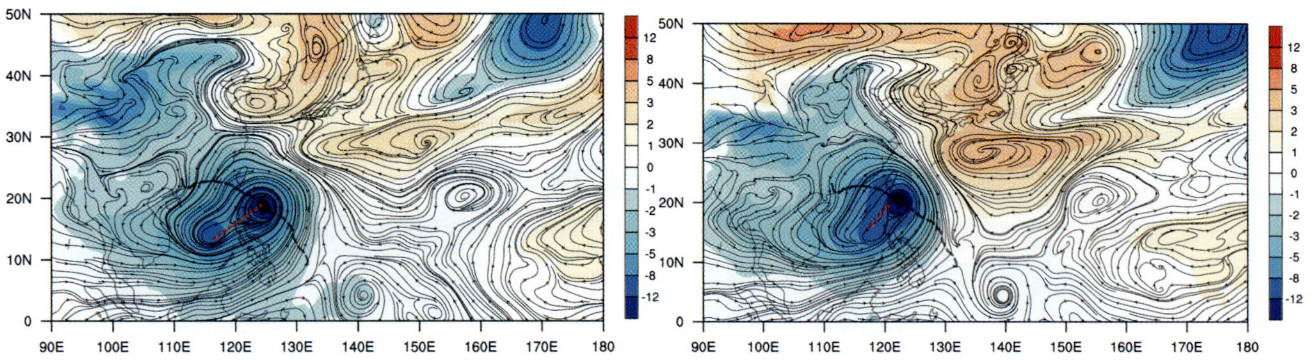

850 hPa flow/height anomalies at 0000 UTC and 1200 UTC 5 June 1999.

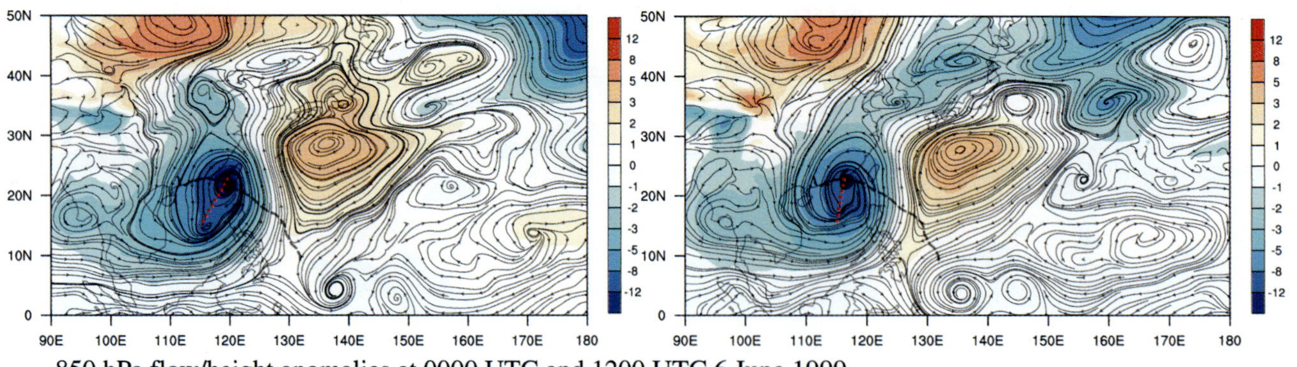

850 hPa flow/height anomalies at 0000 UTC and 1200 UTC 6 June 1999.

Figure 4.8. *(Continued)*

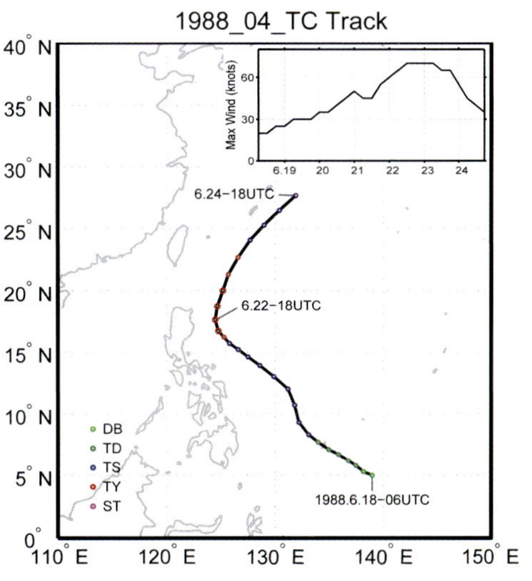

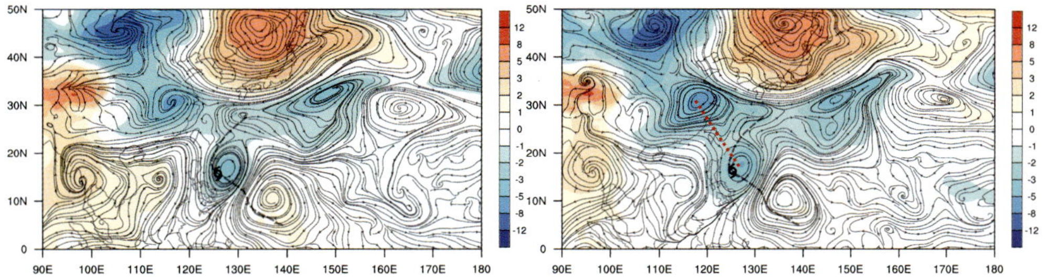

850 hPa flow/height anomalies at 0000 UTC and 0600 UTC 22 June 1988.

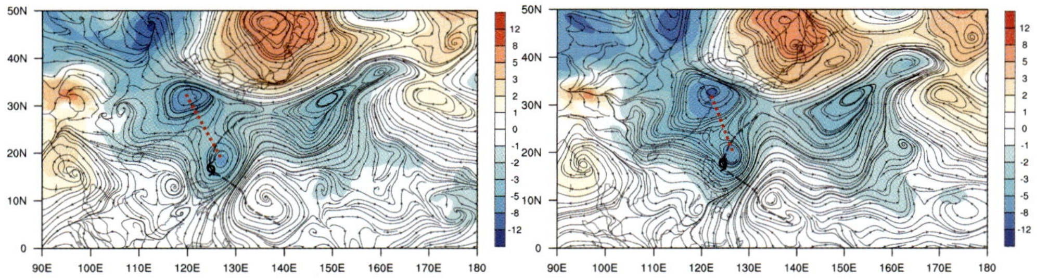

850 hPa flow/height anomalies at 1200 UTC and 1800 UTC 22 June 1988.

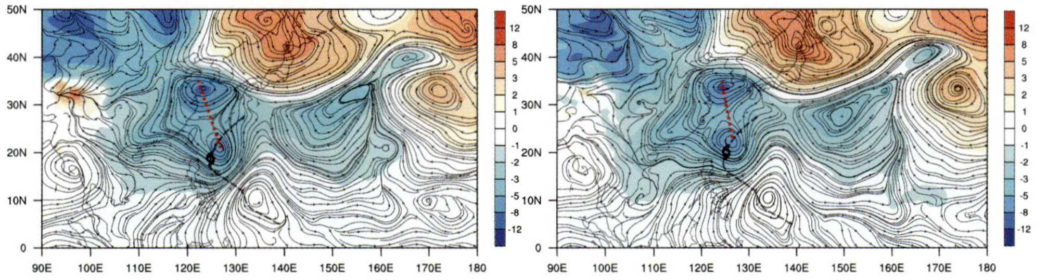

850 hPa flow/height anomalies at 0000 UTC and 0600 UTC 23 June 1988.

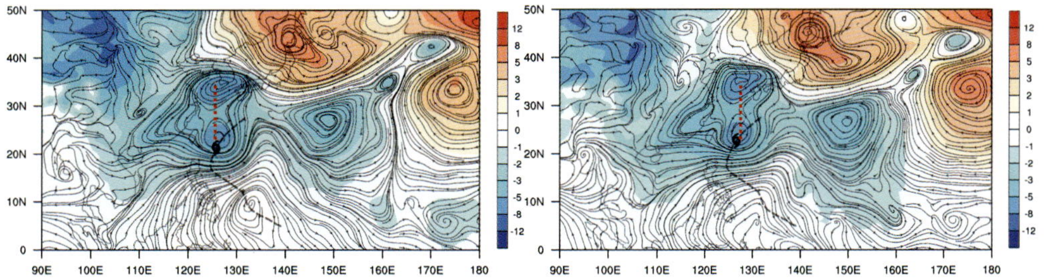

850 hPa flow/height anomalies at 1200 UTC and 1800 UTC 23 June 1988.

Figure 4.9. The typhoon (1988_No.04) generated in the equatorial western Pacific on 18 June 1988 and moved northwestward along the climatic monsoon trough. Its intensity peaked in the northeast sea of the Philippines on 22 June 1988. Its right turn in the east sea of the Philippines was influenced by another anomalous vortex at 850 hPa.

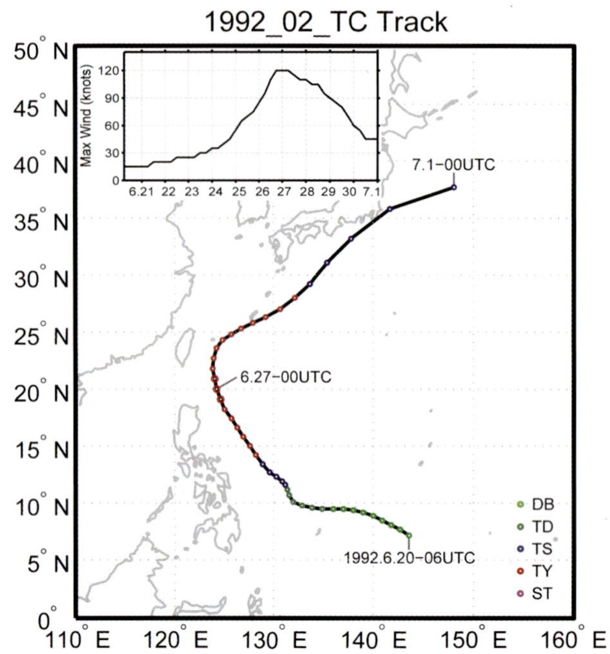

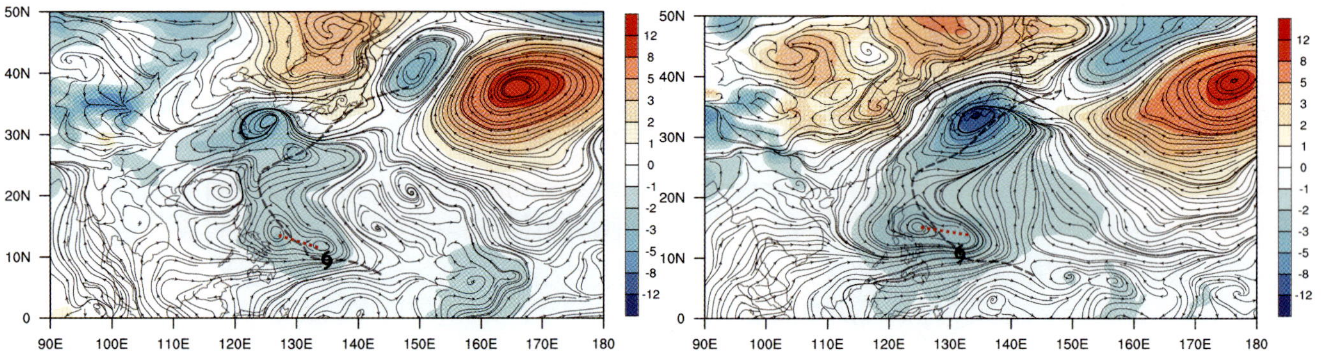

850 hPa flow/height anomalies at 1200 UTC 22 and 23 June 1992.

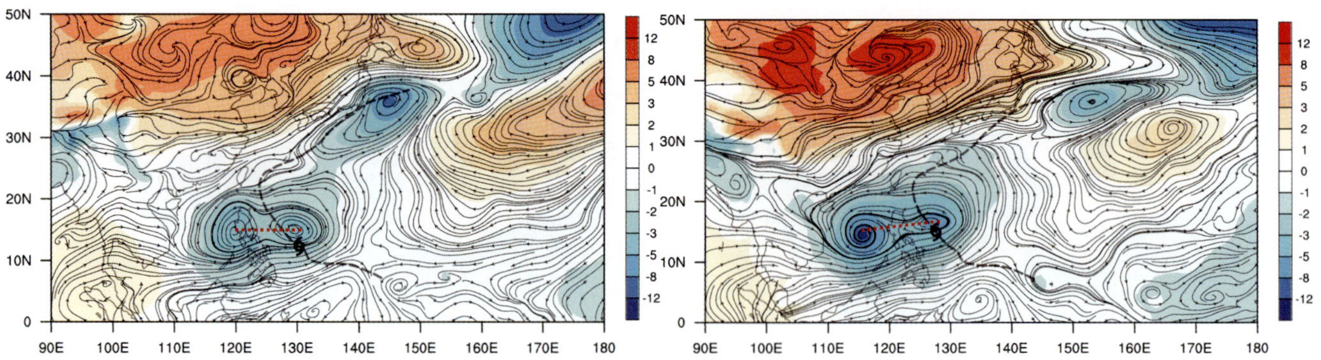

850 hPa flow/height anomalies at 1200 UTC 24 and 25 June 1992.

Figure 4.10. The typhoon (1992_No.02) generated in the equatorial western Pacific on 20 June 1992 and moved northwestward along the climatic monsoon trough. Its intensity peaked in the northeast sea of the Philippines on 27 June 1992. The right turn of its motion in the southeast sea of the Taiwan Island was influenced by anomalous vortices at 850 hPa and 300 hPa.

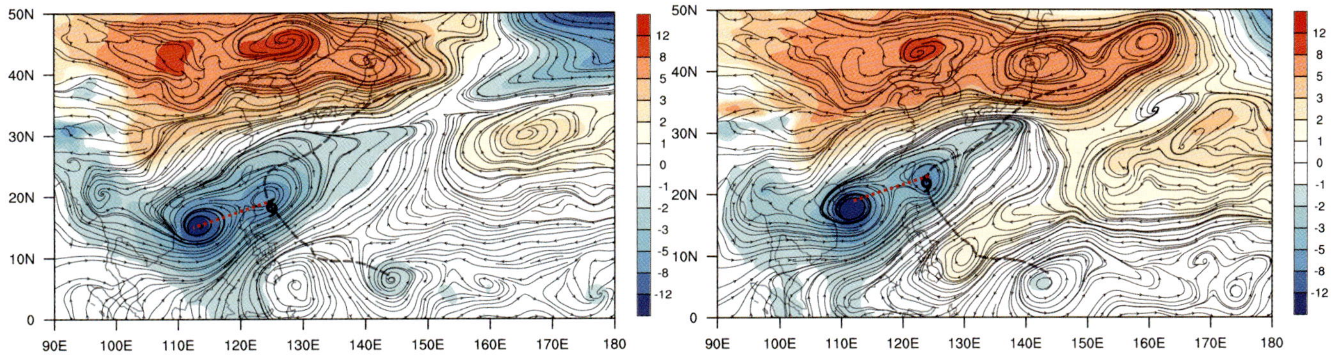

850 hPa flow/height anomalies at 1200 UTC 26 and 27 June 1992.

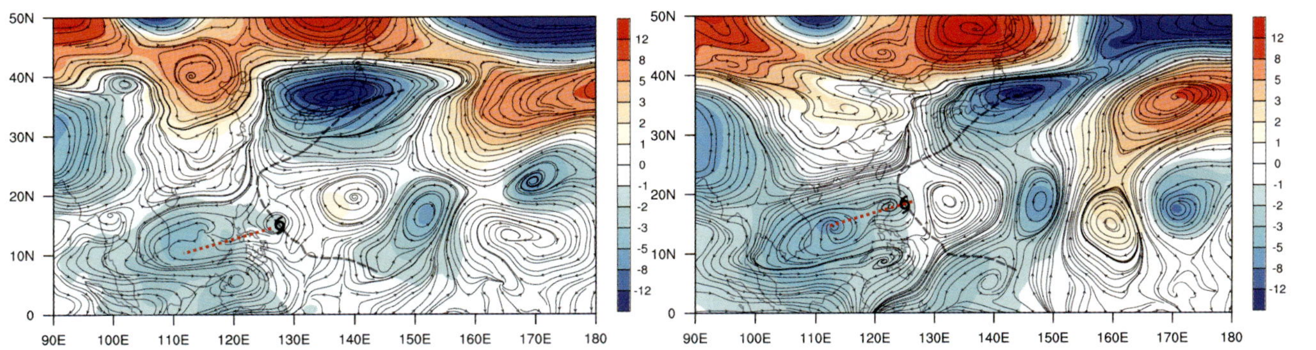

300 hPa flow/height anomalies at 1200 UTC 25 and 26 June 1992.

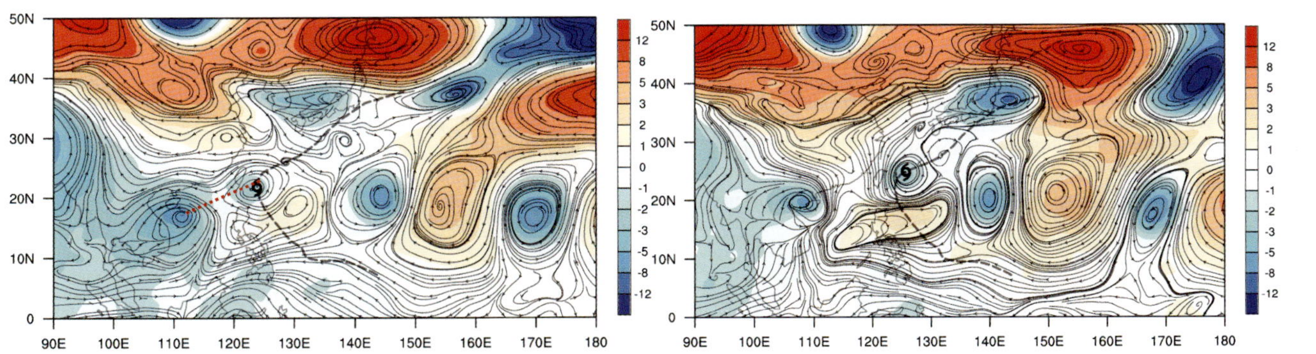

300 hPa flow/height anomalies at 1200 UTC 27 and 28 June 1992.

Figure 4.10. *(Continued)*

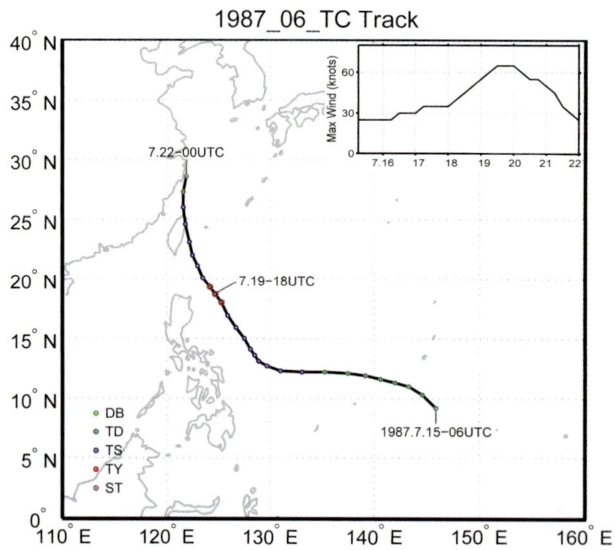

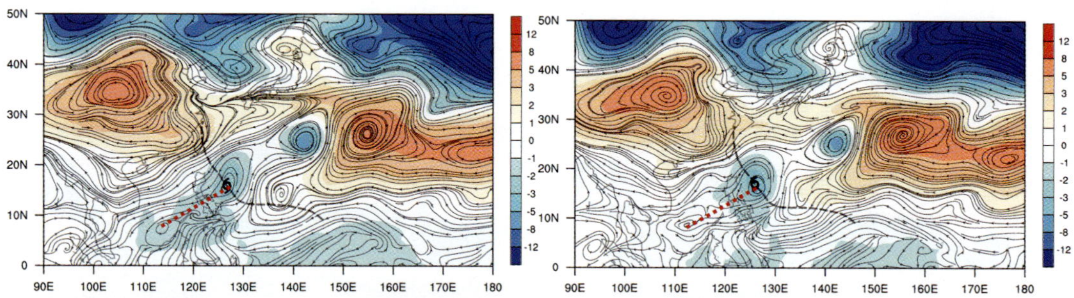

300 hPa flow/height anomalies at 0000 UTC and 0600 UTC 19 July 1987.

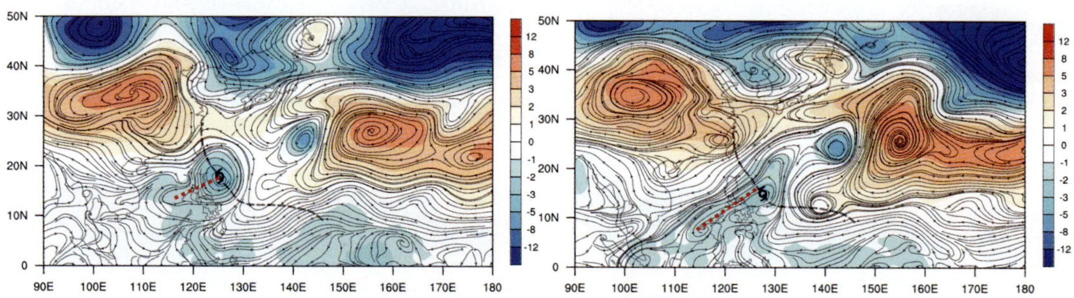

300 hPa flow/height anomalies at 1200 UTC and 1800 UTC 19 July 1987.

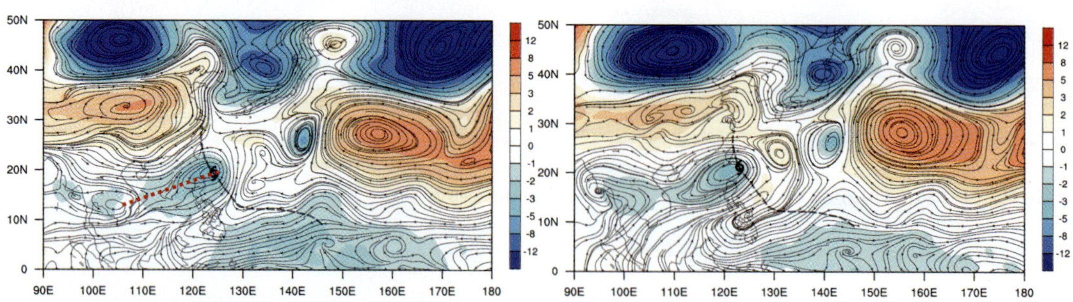

300 hPa flow/height anomalies at 0000 UTC and 1200 UTC 20 July 1987.

Figure 4.11. The typhoon (1987_No.06) generated in the equatorial western Pacific on 15 July 1987 and moved northwestward along the climatic monsoon trough. Its intensity peaked in the northeast sea of the Philippines on 19 July 1987 before landfall on the Taiwan Island. The right turn of its motion in the east sea of the Philippines was influenced by another anomalous vortex at 300 hPa.

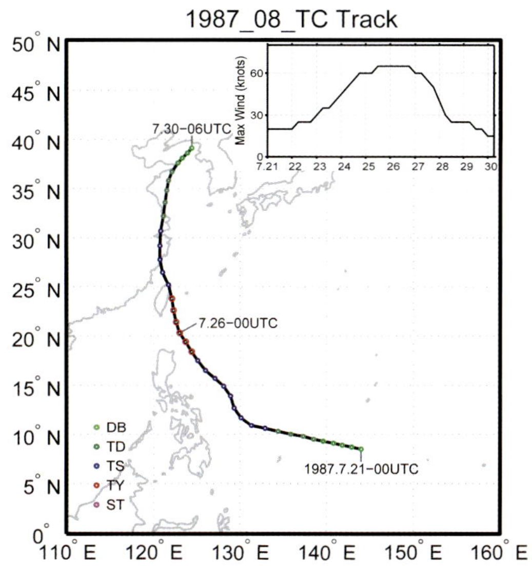

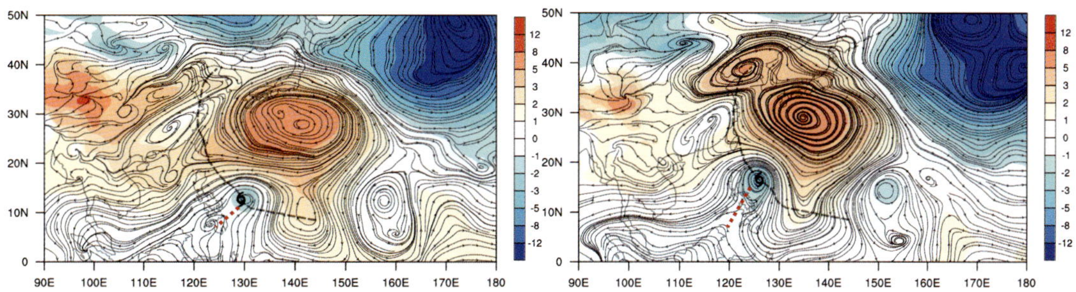

850 hPa flow/height anomalies at 0000 UTC 24 and 25 July 1987.

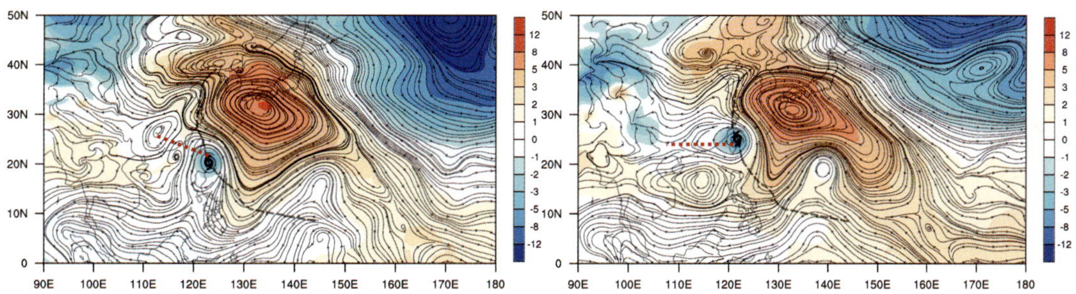

850 hPa flow/height anomalies at 0000 UTC 26 and 27 July 1987.

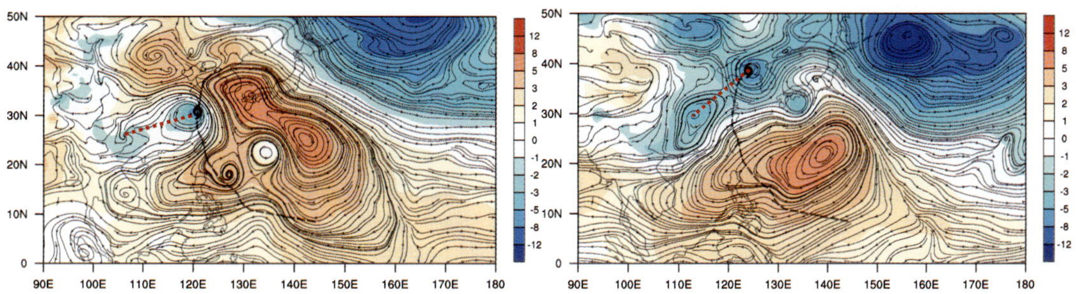

850 hPa flow/height anomalies at 0000 UTC 28 and 29 July 1987.

Figure 4.12. The typhoon (1987_No.08) generated in the equatorial western Pacific on 21 July 1987 and moved northwestward along the climatic monsoon trough. Its intensity peaked in the northeast sea of the Philippines on 26 July 1987 before landfall on the Taiwan Island. The right turn of its motion in the east sea of the Philippines was influenced by anomalous vortices at 850 hPa.

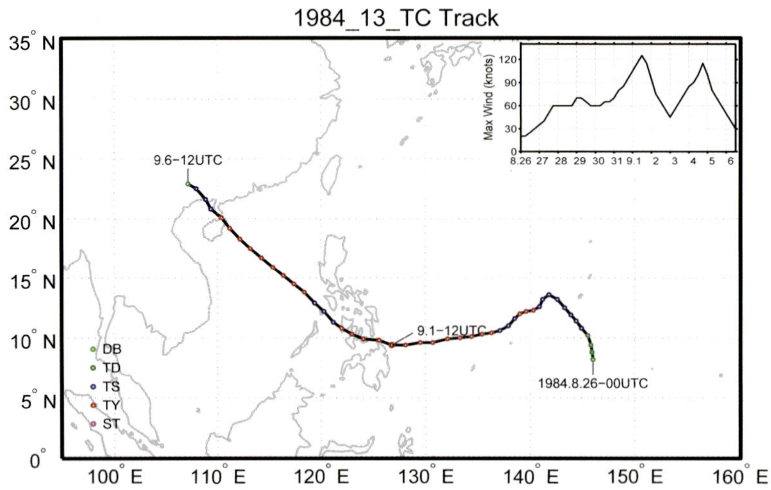

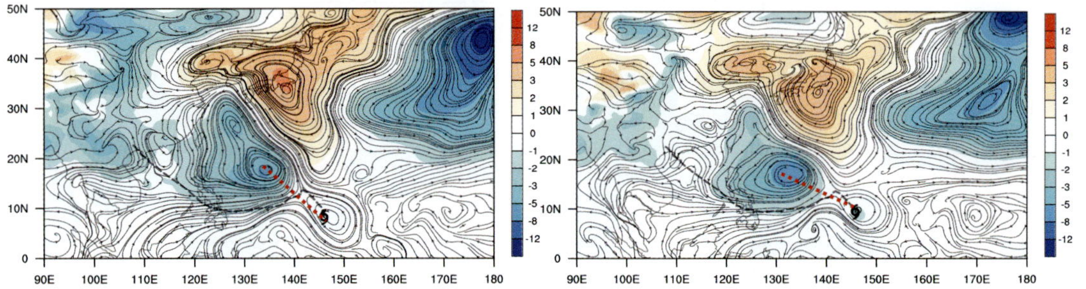

850 hPa flow/height anomalies at 0000 UTC and 1200 UTC 26 August 1984.

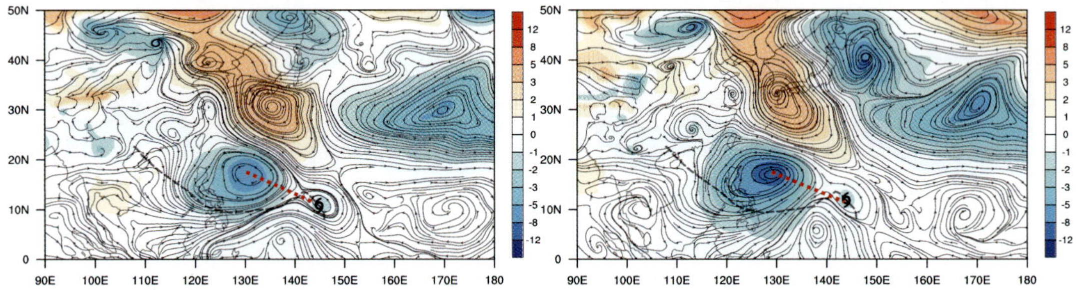

850 hPa flow/height anomalies at 0000 UTC and 1200 UTC 27 August 1984.

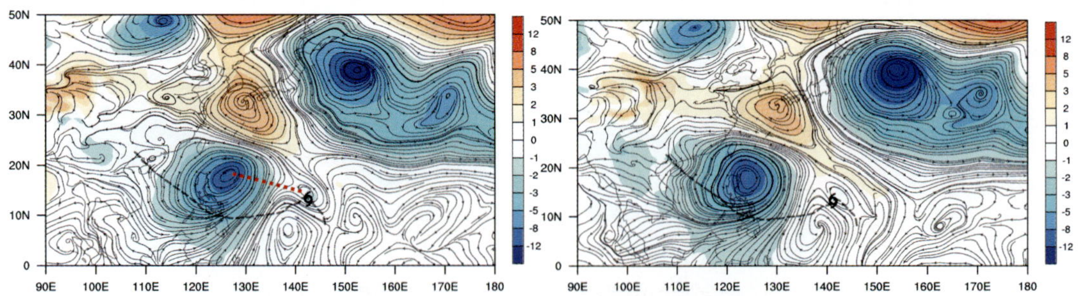

850 hPa flow/height anomalies at 0000 UTC and 1200 UTC 28 August 1984.

Figure 4.13. The typhoon (1984_No.13) generated in the equatorial western Pacific on 26 August 1984 and its intensity peaked on 1 September 1984 before landfall on the central Philippines. The left turn of its motion at the beginning stage was influenced by another anomalous vortex at 850 hPa.

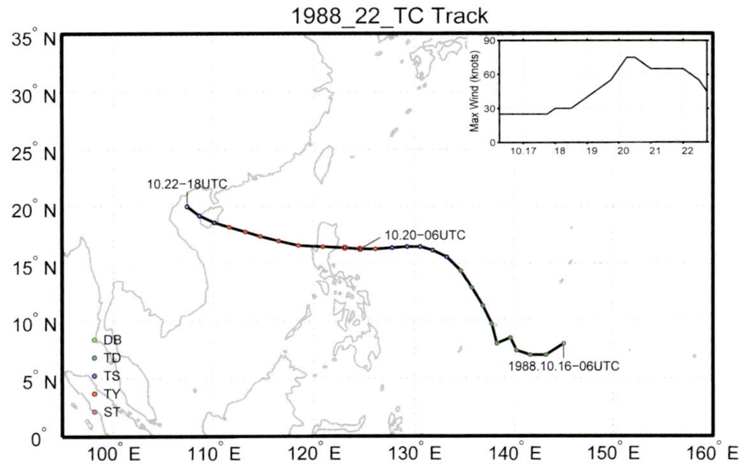

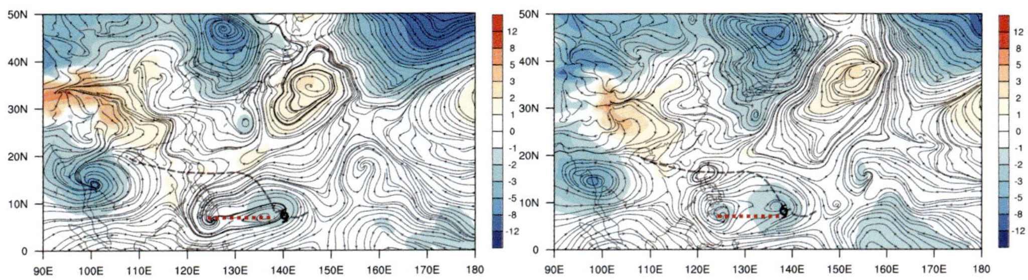

850 hPa flow/height anomalies at 0000 UTC and 1200 UTC 17 October 1988.

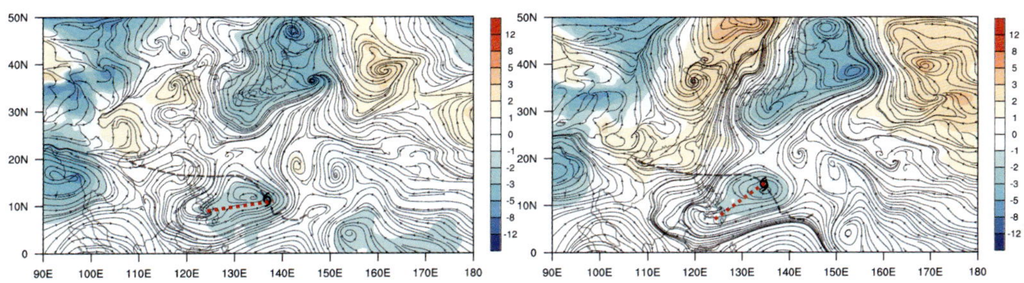

850 hPa flow/height anomalies at 0000 UTC and 1200 UTC 18 October 1988.

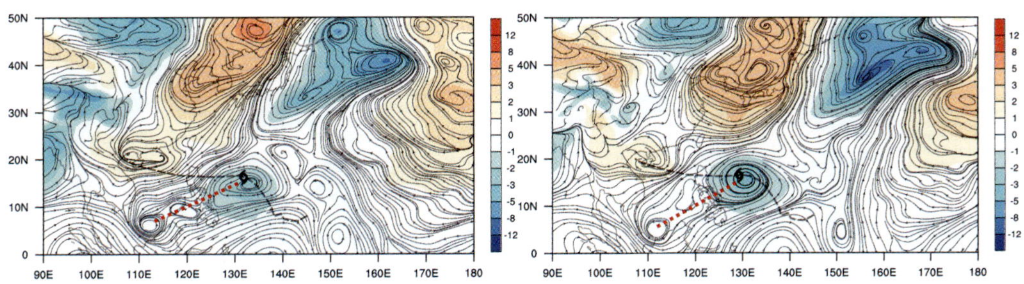

850 hPa flow/height anomalies at 0000 UTC and 1200 UTC 19 October 1988.

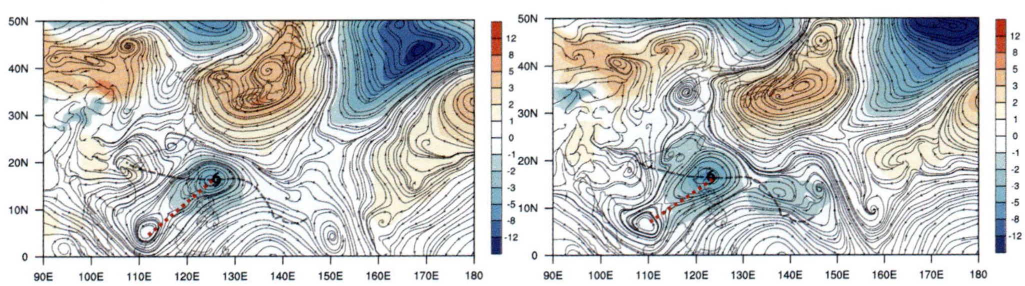

850 hPa flow/height anomalies at 0000 UTC and 1200 UTC 20 October 1988.

Figure 4.14. The typhoon (1988_No.22) generated in the equatorial western Pacific on 16 October 1988 and moved northwestward along the climatic monsoon trough. Its intensity peaked in the northeast sea of the Philippines on 20 October 1988 before landfall on the island. The left turn of its motion in the tropical western Pacific was influenced by another anomalous vortex at 850 hPa.

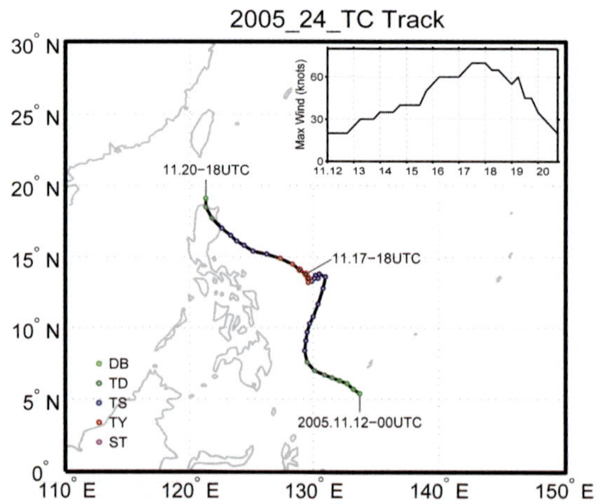

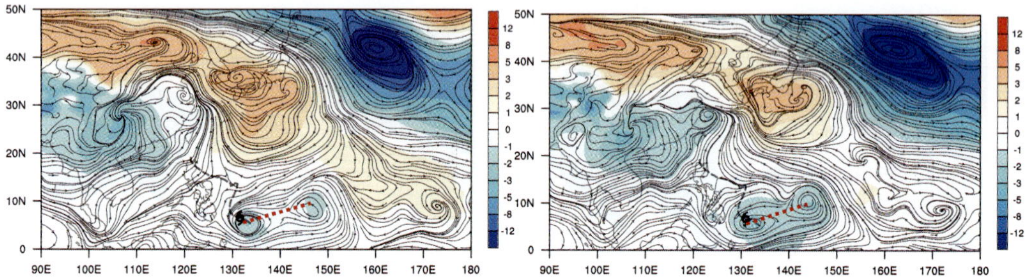
850 hPa flow/height anomalies at 0000 UTC and 0600 UTC 13 November 2005.

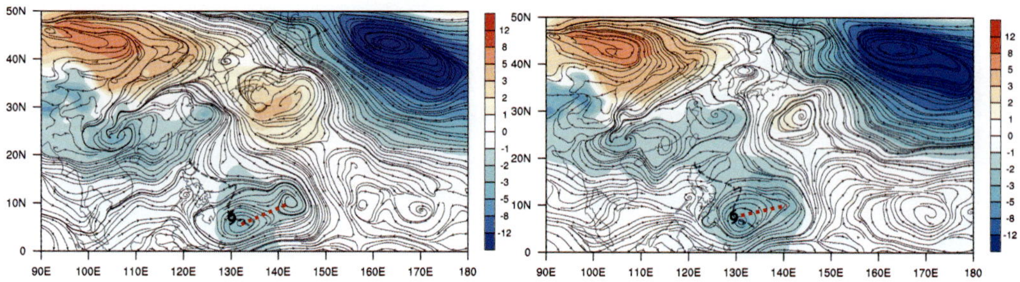
850 hPa flow/height anomalies at 1200 UTC and 1800 UTC 13 November 2005.

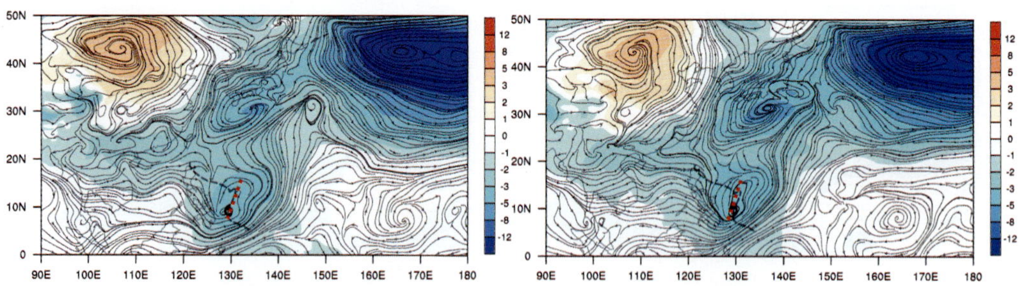
850 hPa flow/height anomalies at 0600 UTC and 1200 UTC 14 November 2005.

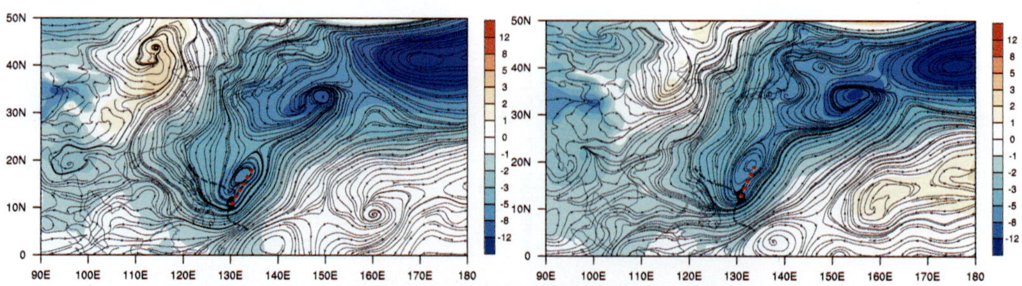
850 hPa flow/height anomalies at 0000 UTC and 1200 UTC 15 November 2005.

Figure 4.15. The typhoon (2005_No.24) generated in the equatorial western Pacific on 12 November 2005 and moved northwestward. Its intensity peaked in the central east sea of the Philippines on 17 November 2005 when it was approaching the climatic monsoon trough. The right turn and left turn of its motion were influenced by another anomalous vortex at 850 hPa.

4.3. Geneses in the Equatorial Central Pacific

There were 14 typhoons that generated in the equatorial central Pacific and their intensities peaked in the Philippines. The right turns and left turns of their motions were due to the influences of anomalous vortices (Figs. 4.16–4.28).

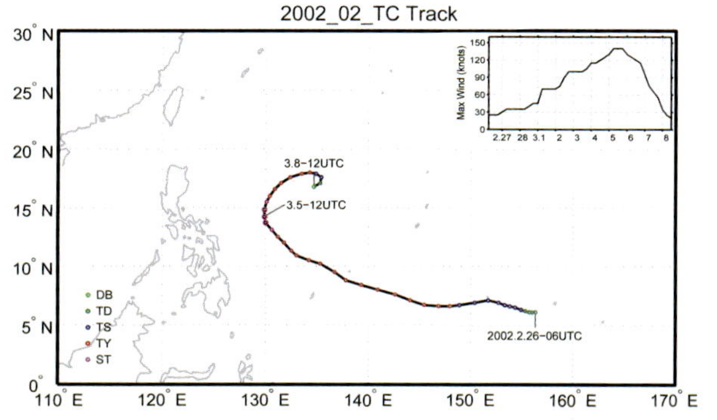

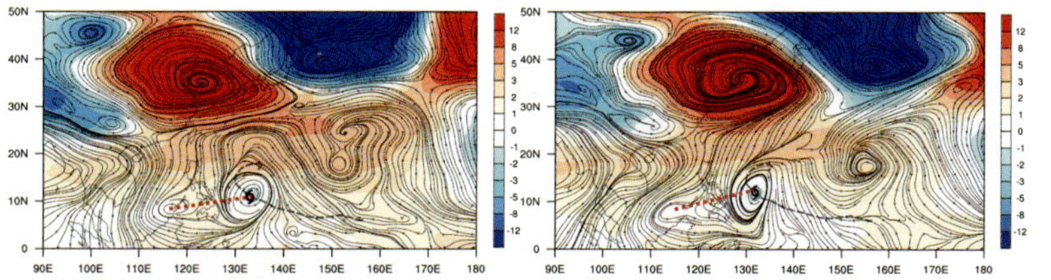

300 hPa flow/height (shading, 10 gpm) anomalie sat 0000 UTC and 1200 UTC 4 March 2002.

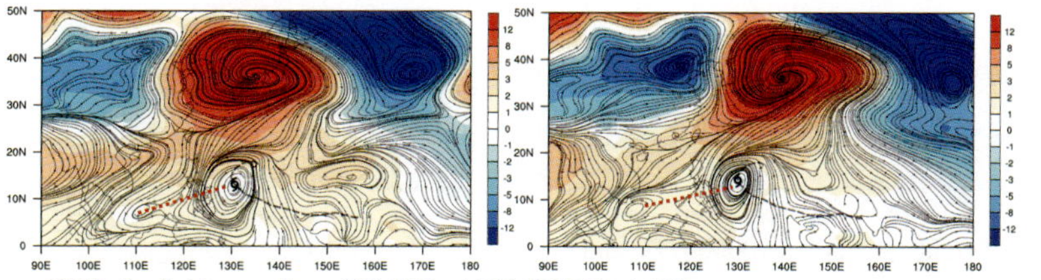

300 hPa flow/height anomalies at 0000 UTC and 1200 UTC 5 March 2002.

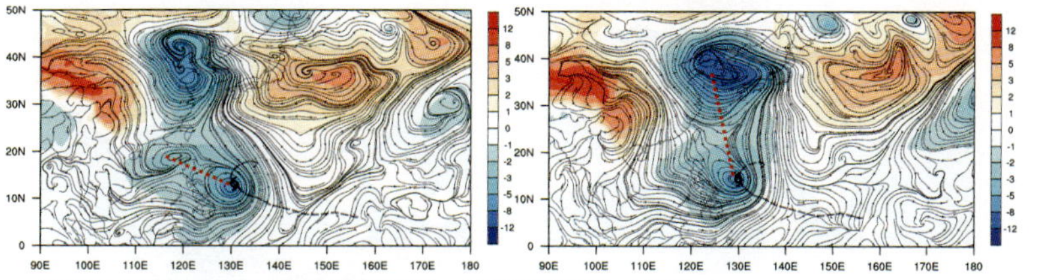

850 hPa flow/height anomalies at 0000 UTC and 1200 UTC 5 March 2002.

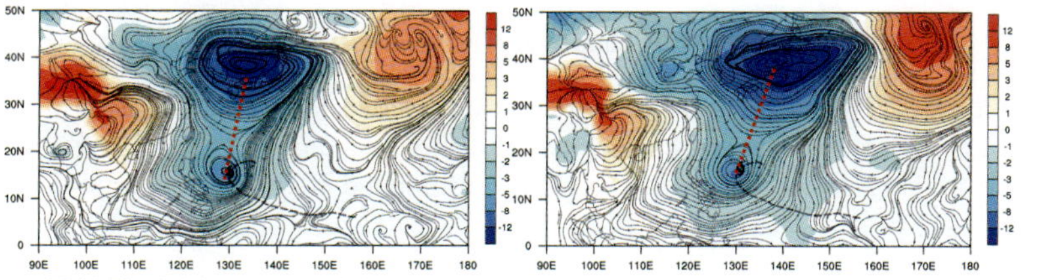

850 hPa flow/height anomalies at 0000 UTC and 1200 UTC 6 March 2002.

Figure 4.16. The typhoon (2002_No.02) generated in the equatorial central Pacific on 26 February 2002 and moved westward. Its intensity peaked on 5 March 2002 at the moment when it was turning right. The right turn of its motion was influenced by anomalous vortices at 300 hPa and 850 hPa.

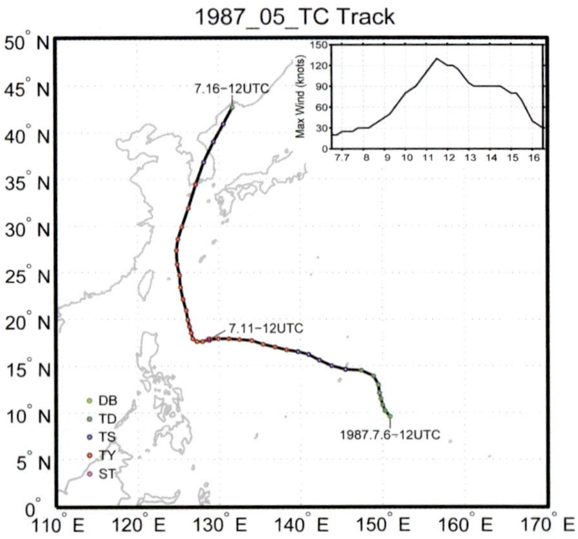

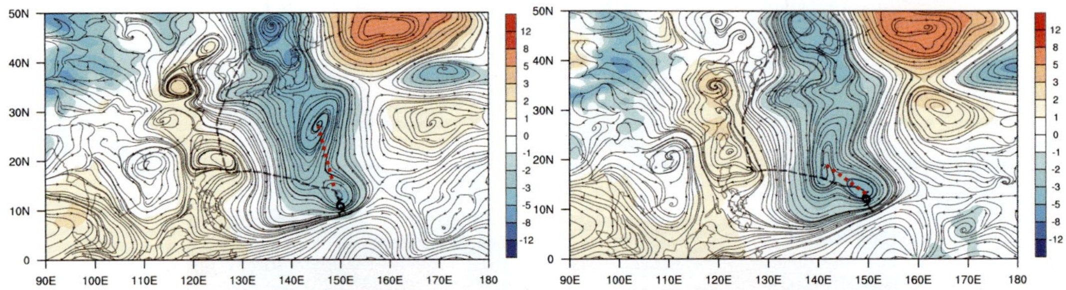

850 hPa flow/height anomalies at 0000 UTC and 1200 UTC 7 July 1987.

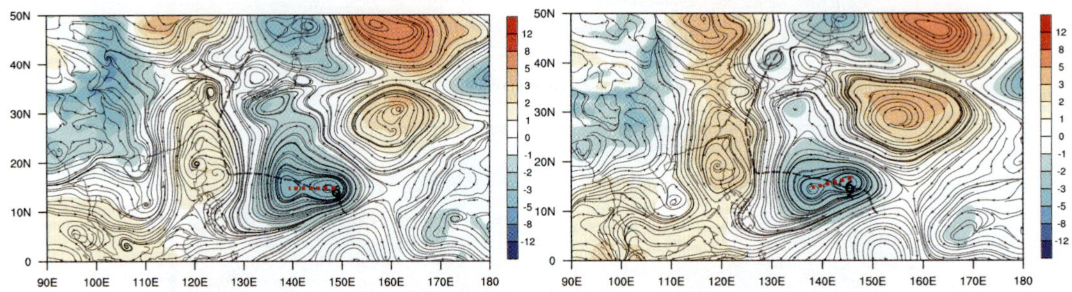

850 hPa flow/height anomalies at 0000 UTC and 1200 UTC 8 July 1987.

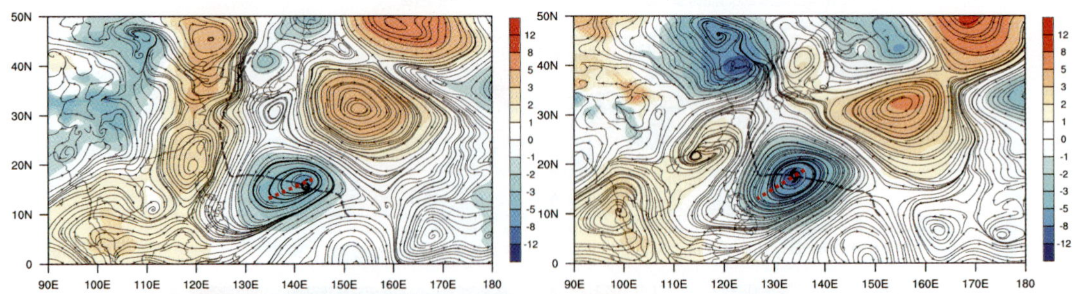

850 hPa flow/height anomalies at 0000 UTC and 1200 UTC 9 July 1987.

Figure 4.17. The typhoon (1987_No.05) generated in the equatorial central Pacific on 6 July 1987 and moved northwestward basically along the climatic monsoon trough. Its intensity peaked on 11 July 1987 at the moment when it was turning right. The left turn and right turn of its motion were influenced by anomalous vortices at 850 hPa and 300 hPa.

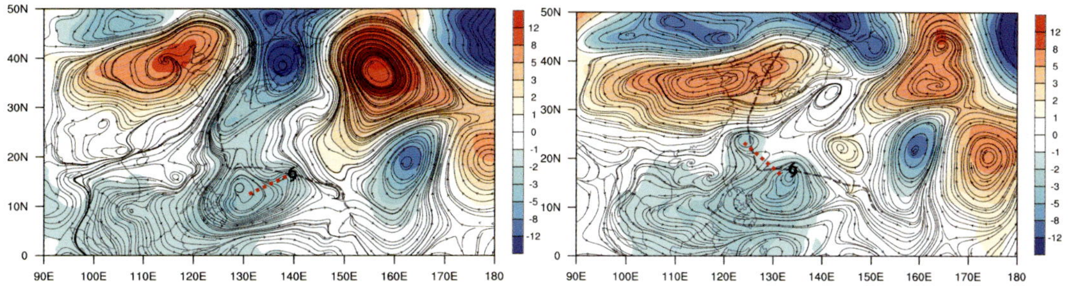

300 hPa flow/height anomalies at 1200 UTC 9 and 10 July 1987.

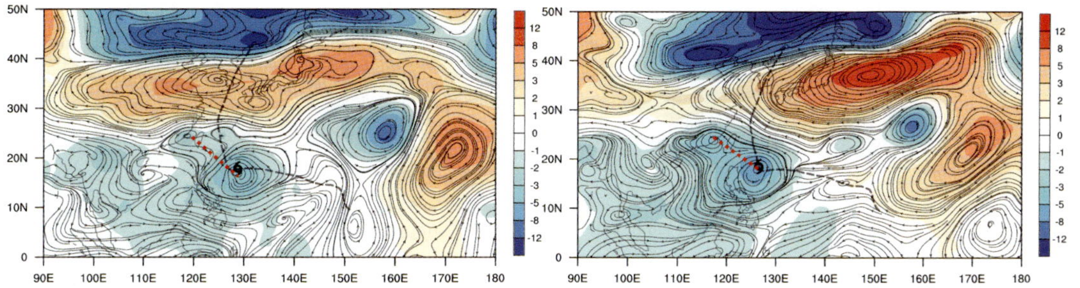

300 hPa flow/height anomalies at 1200 UTC 11 and 12 July 1987.

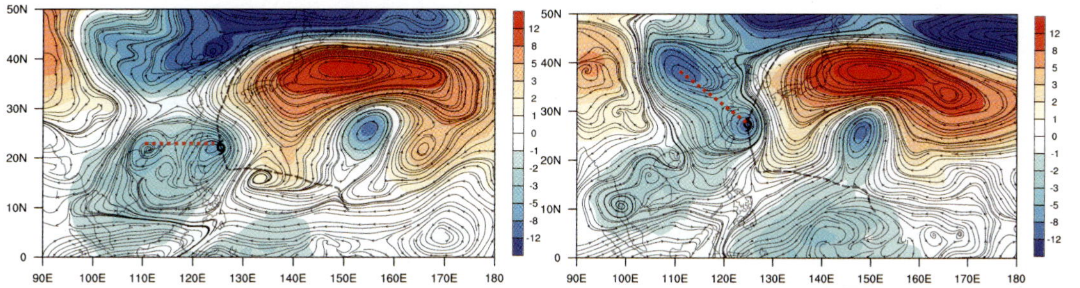

300 hPa flow/height anomalies at 1200 UTC 13 and 14 July 1987.

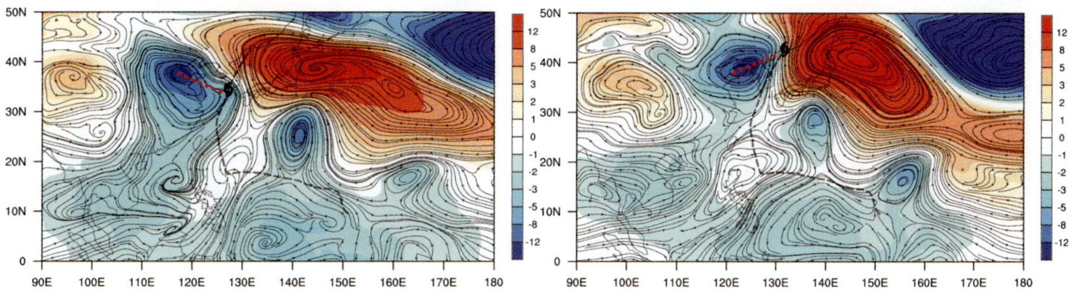

300 hPa flow/height anomalies at 1200 UTC 15 and 16 July 1987.

Figure 4.17. *(Continued)*

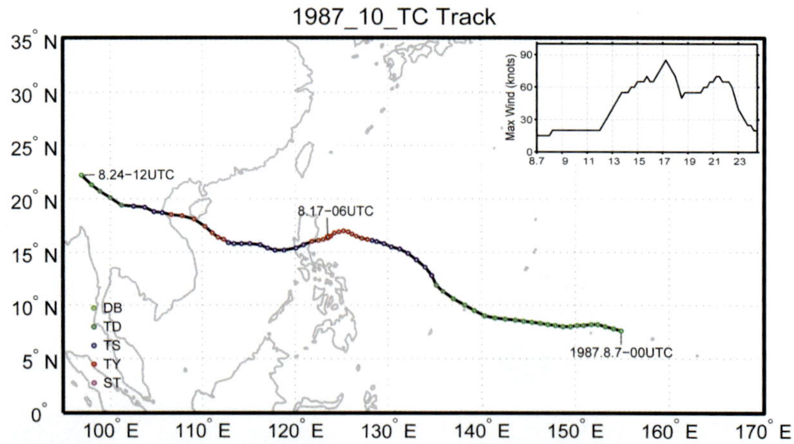

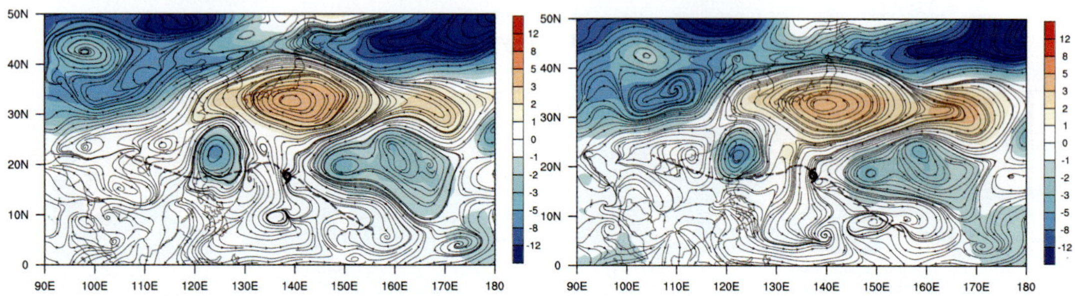

300 hPa flow/height anomalies at 1200 UTC and 1800 UTC 18 August 2003.

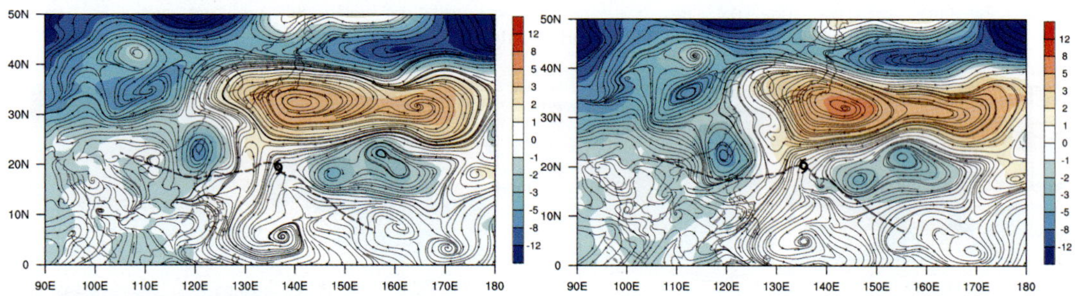

300 hPa flow/height anomalies at 0000 UTC and 0600 UTC 19 August 2003.

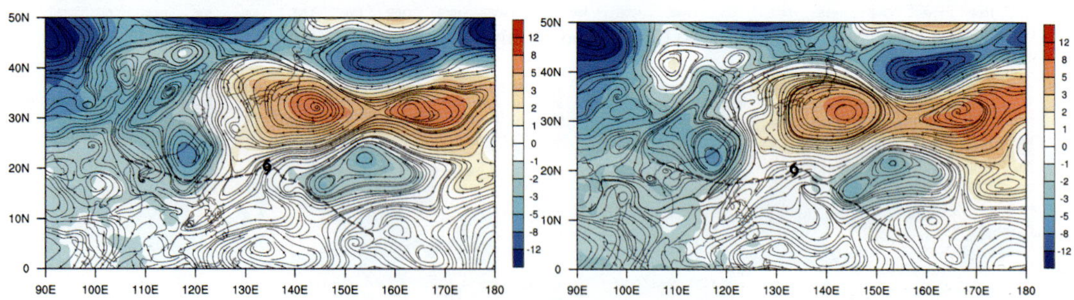

300 hPa flow/height anomalies at 1200 UTC and 1800 UTC 19 August 2003.

Figure 4.18. The typhoon (1987_No.10) generated in the equatorial central Pacific on 7 August 1987 and moved west-northwestward basically along the climatic monsoon trough. Its intensity peaked in the northeast sea of the Philippines on 17 August 1987 before landfall on the island. The left turn of its motion was influenced by anomalous anti-cyclonic circulation at 300 hPa.

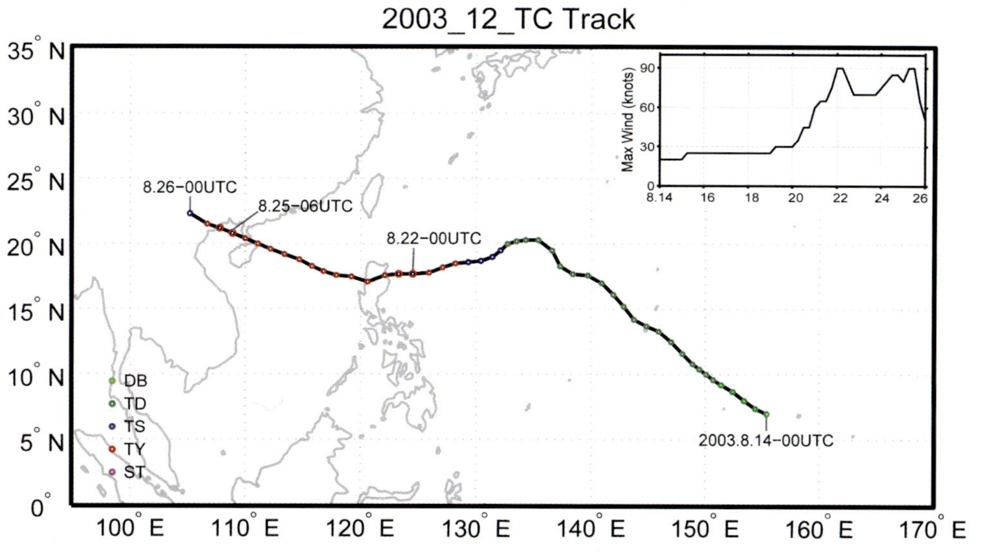

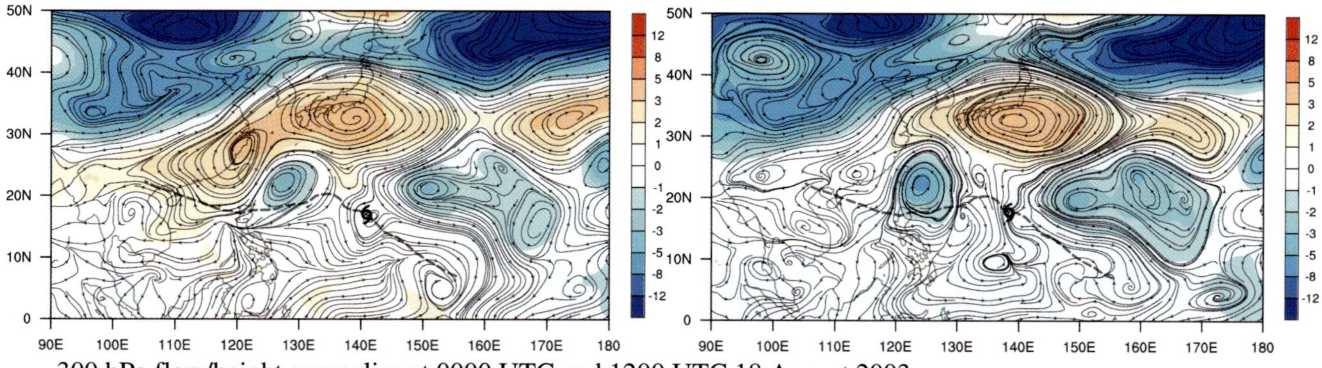

300 hPa flow/height anomalies at 0000 UTC and 1200 UTC 18 August 2003.

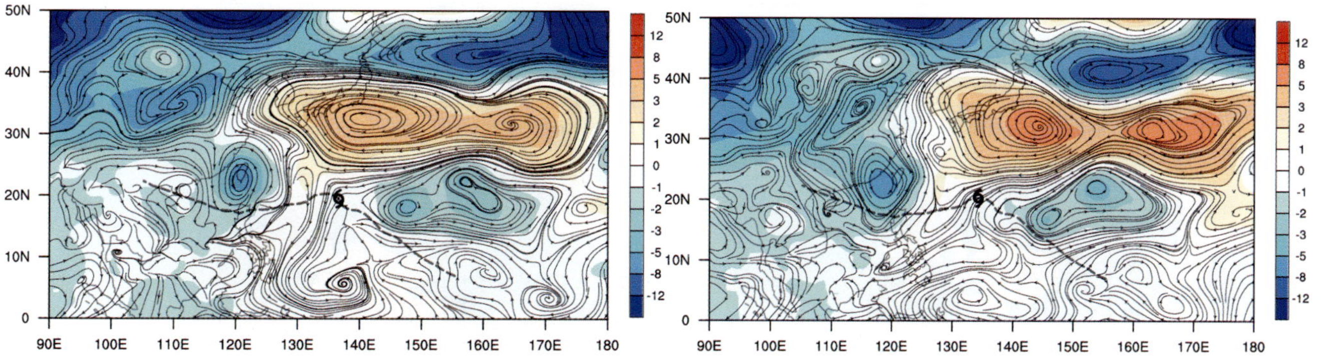

300 hPa flow/height anomalies at 0000 UTC and 1200 UTC 19 August 2003.

Figure 4.19. The typhoon (2003_No.12) generated in the equatorial central Pacific on 14 August 2003 and moved northwestward. Its intensity peaked in the northeast sea of the Philippines on 25 August 2003 before landfall on the island. The second peak of its intensity occurred when it is was near the SCS climatic monsoon trough on 25 August 2003 before landfall on the South China coast. The left turn of its motion in the subtropical western Pacific was influenced by anomalous anti-cyclonic circulation at 300 hPa.

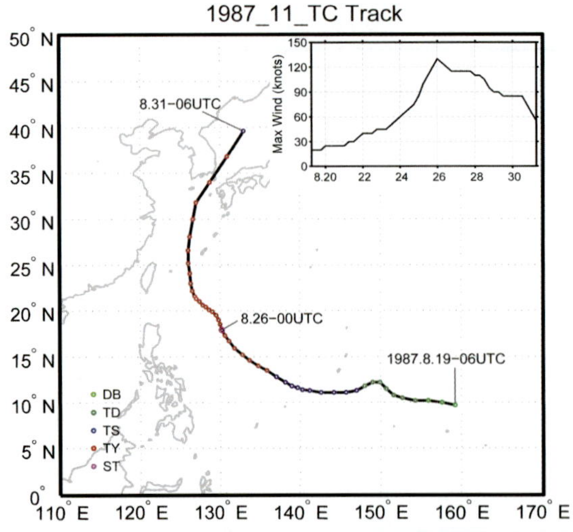

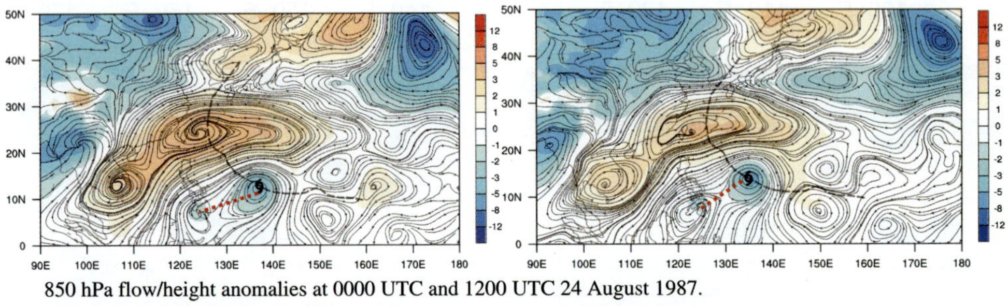

850 hPa flow/height anomalies at 0000 UTC and 1200 UTC 24 August 1987.

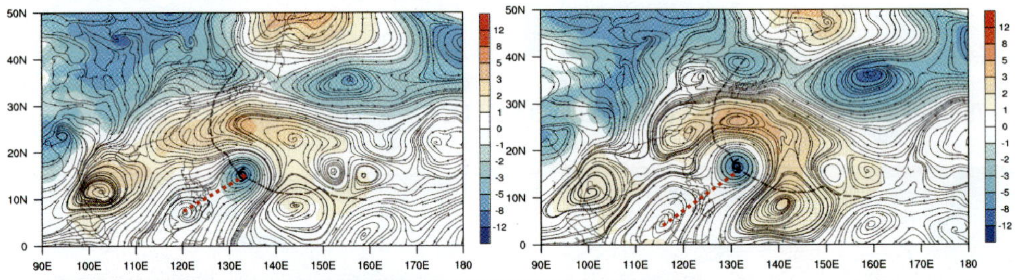

850 hPa flow/height anomalies at 0000 UTC and 1200 UTC 25 August 1987.

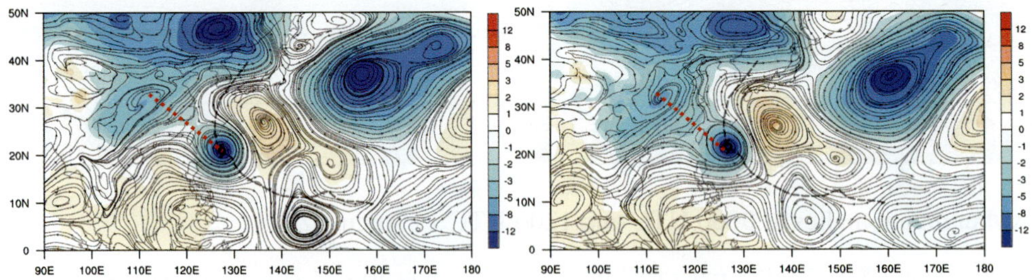

850 hPa flow/height anomalies at 0000 UTC and 1200 UTC 28 August 1987.

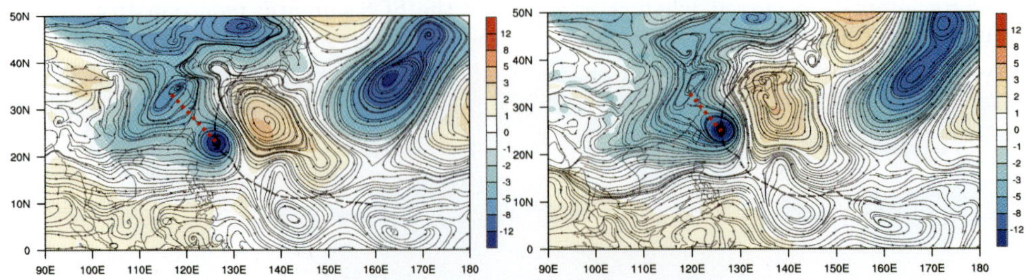

850 hPa flow/height anomalies at 0000 UTC and 1200 UTC 29 August 1987.

Figure 4.20. The typhoon (1987_No.11) generated in the equatorial central Pacific on 19 August 1987 and moved westward along the climatic monsoon trough. Its intensity peaked in the northeast sea of the Philippines on 26 August 1987. The right turn of its motion in the east seas of China was influenced by anomalous vortices at 850 hPa and 300 hPa.

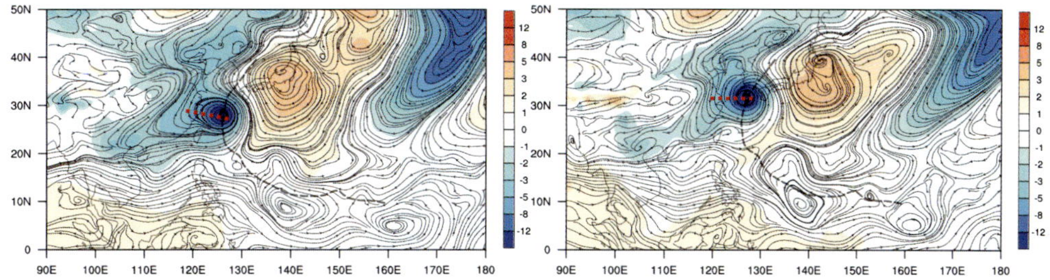

850 hPa flow/height anomalies at 0000 UTC and 1200 UTC 30 August 1987.

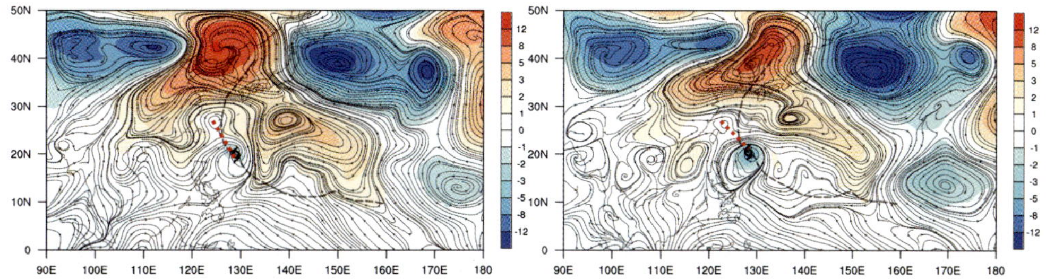

300 hPa flow/height anomalies at 0000 UTC and 1200 UTC 27 August 1987.

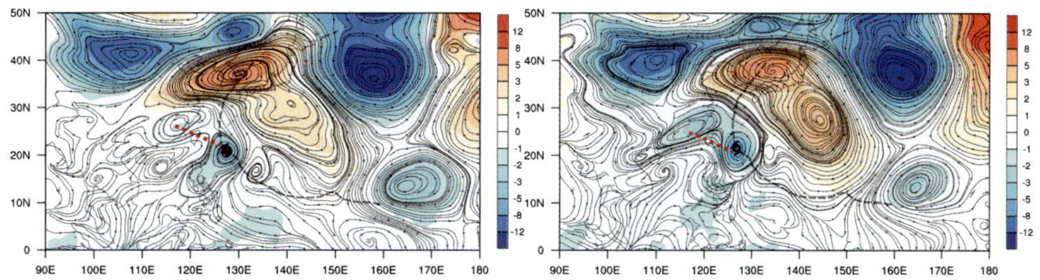

300 hPa flow/height anomalies at 0000 UTC and 1200 UTC 28 August 1987.

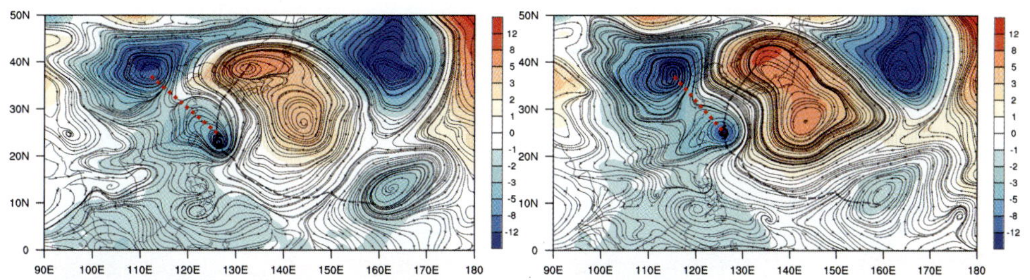

300 hPa flow/height anomalies at 0000 UTC and 1200 UTC 29 August 1987.

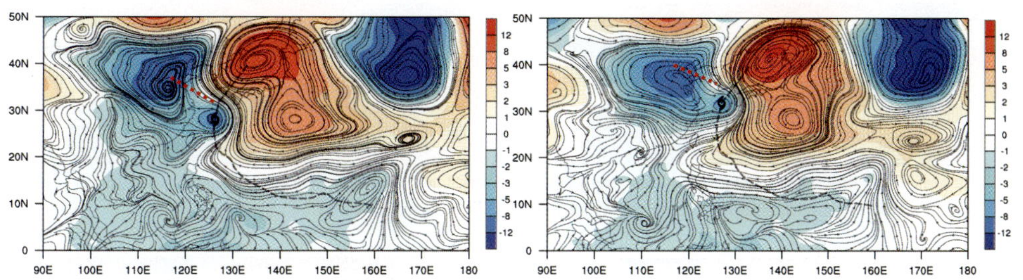

300 hPa flow/height anomalies at 0000 UTC and 1200 UTC 30 August 1987.

Figure 4.20. *(Continued)*

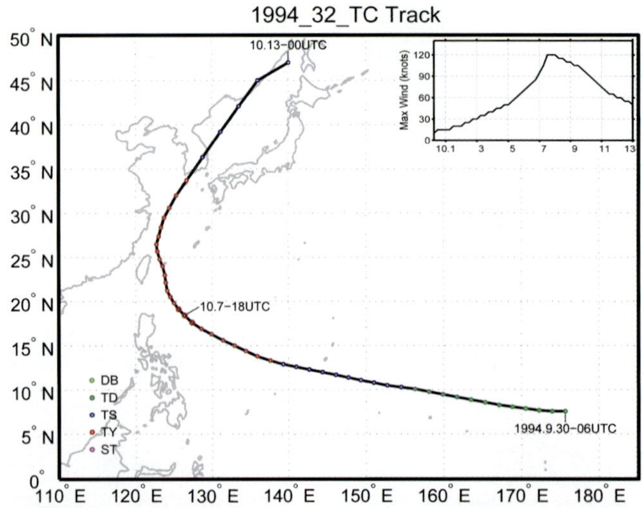

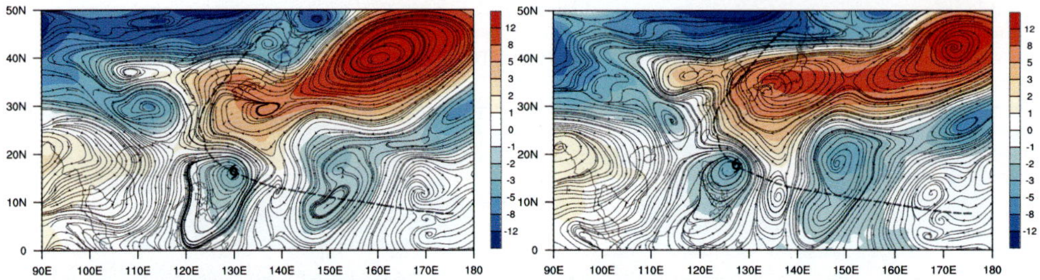

300 hPa flow/height anomalies at 0000 UTC and 1200 UTC 7 October 1994.

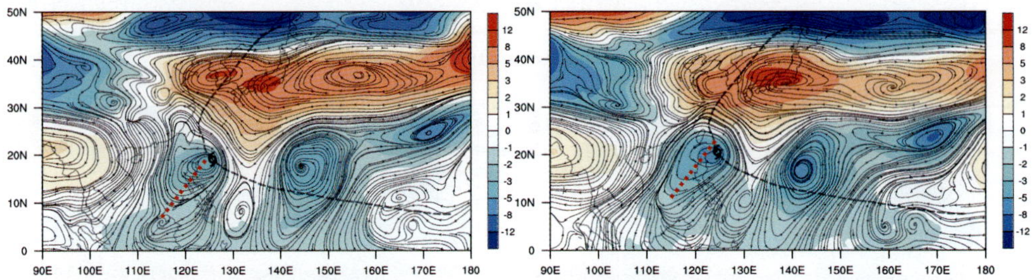

300 hPa flow/height anomalies at 0000 UTC and 1200 UTC 8 October 1994.

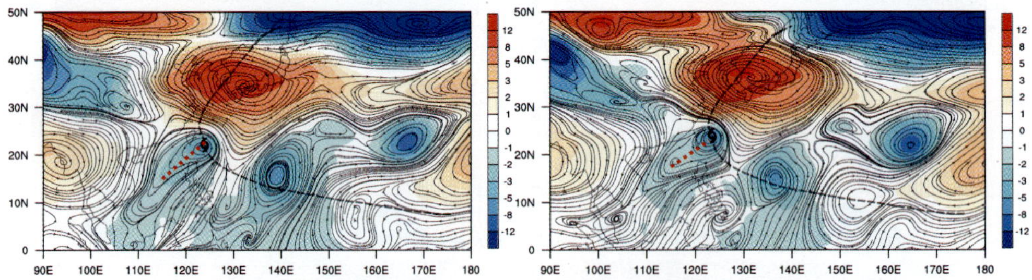

300 hPa flow/height anomalies at 0000 UTC and 1200 UTC 9 October 1994.

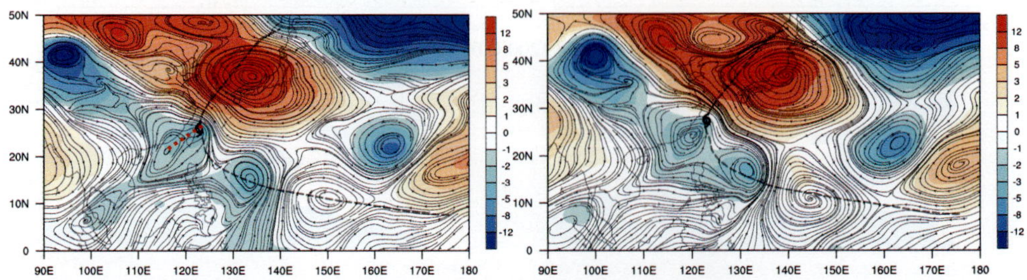

300 hPa flow/height anomalies at 0000 UTC and 1200 UTC 10 October 1994.

Figure 4.21. The typhoon (1994_No.32) generated in the equatorial central Pacific on 30 September 1994 and moved northwestward along the climatic monsoon trough. Its intensity peaked in the northeast sea of the Philippines on 7 October 1994. The right turn of its motion in the east seas of China was influenced by another anomalous vortex at 300 hPa.

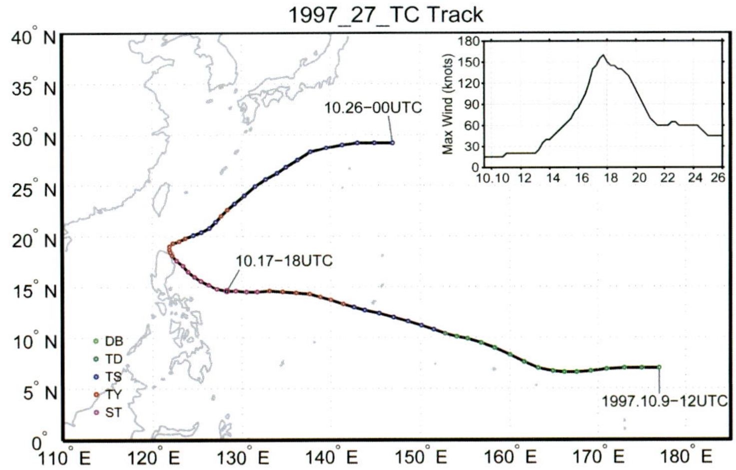

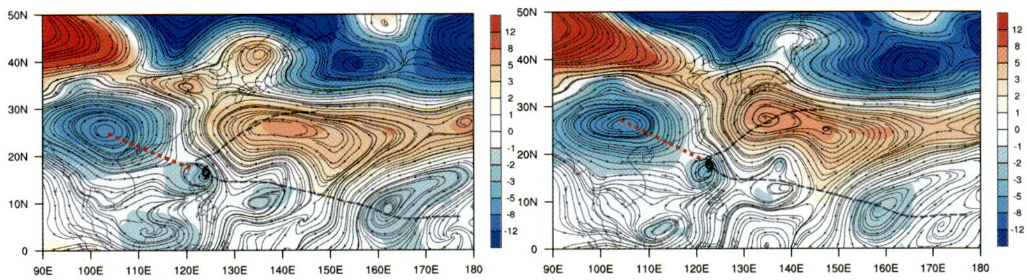

300 hPa flow/height anomalies at 0000 UTC and 1200 UTC 19 October 1979.

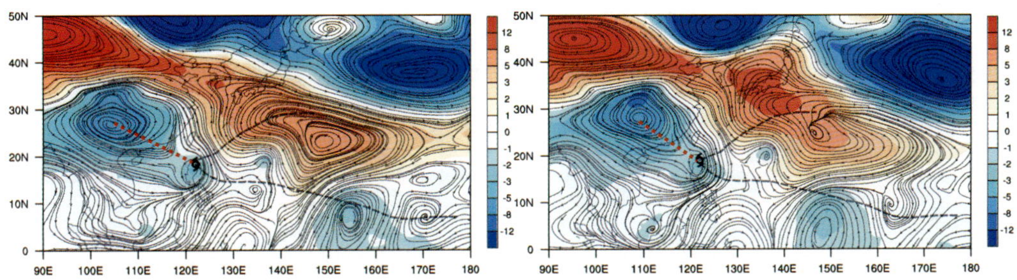

300 hPa flow/height anomalies at 0000 UTC and 1200 UTC 20 October 1979.

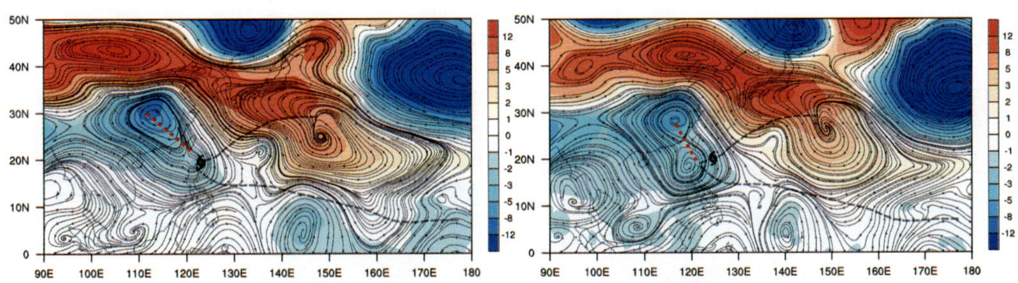

300 hPa flow/height anomalies at 0000 UTC and 1200 UTC 21 October 1979.

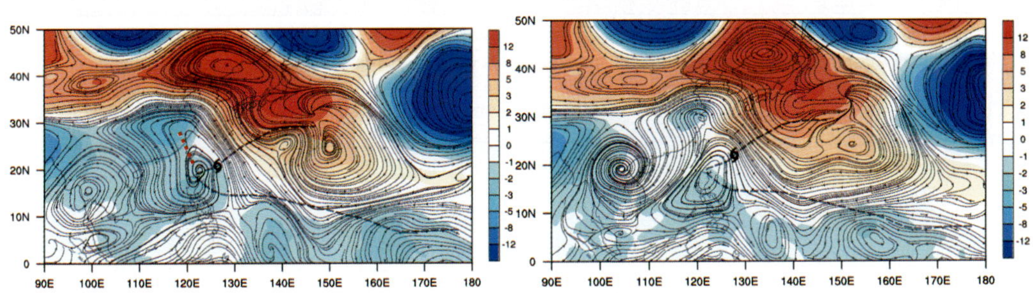

300 hPa flow/height anomalies at 0000 UTC and 1200 UTC 22 October 1979.

Figure 4.22. The typhoon (1997_No.27) generated in the equatorial central Pacific on 9 October 1997 and moved northwestward along the climatic monsoon trough. Its intensity peaked in the east sea of the Philippines on 17 October 1997 before landfall on the island. The right turn of its motion near the Philippines was influenced by another anomalous vortex at 300 hPa.

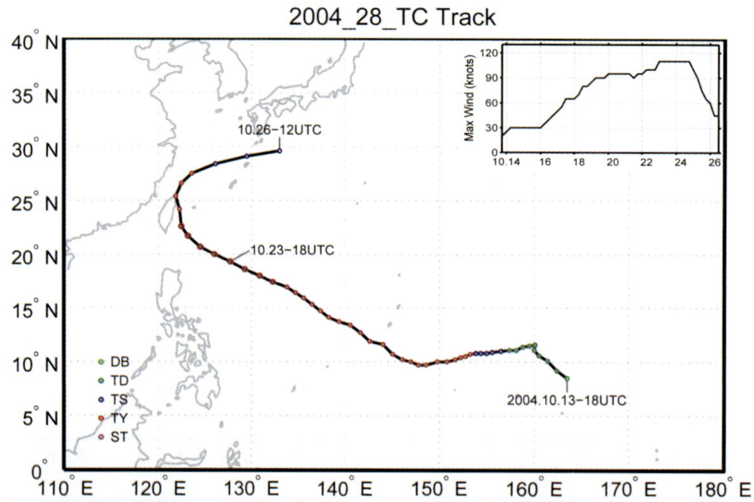

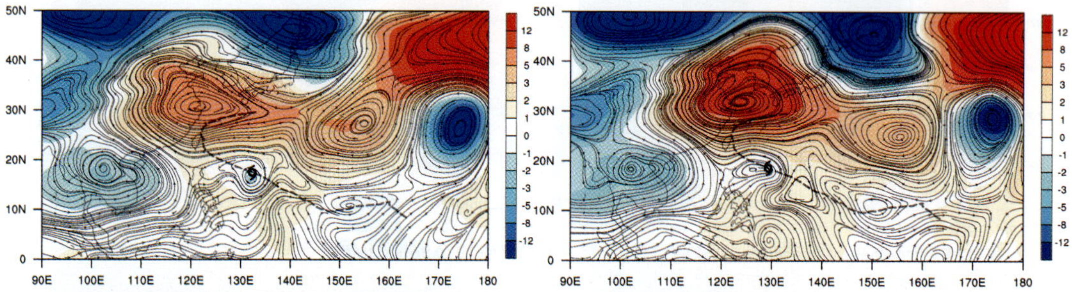

300 hPa flow/height anomalies at 0000 UTC and 1200 UTC 23 October 2004.

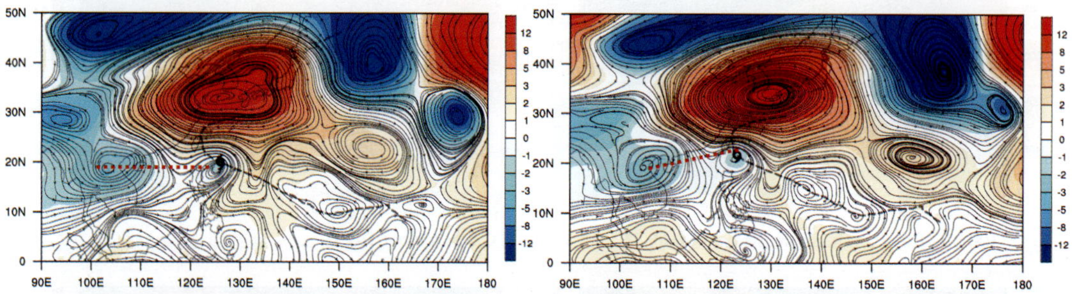

300 hPa flow/height anomalies at 0000 UTC and 1200 UTC 24 October 2004.

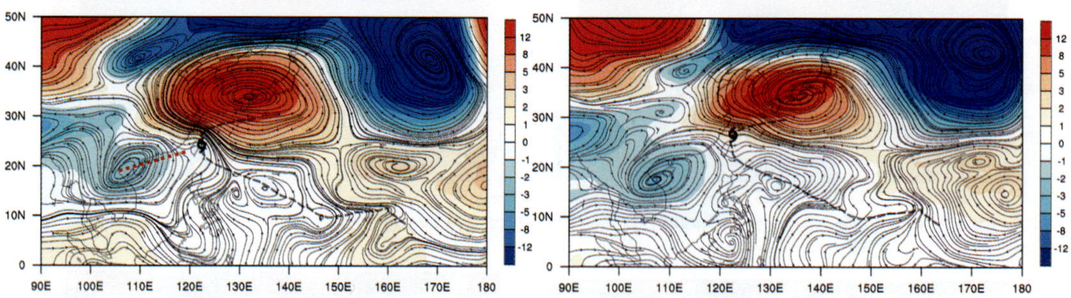

300 hPa flow/height anomalies at 0000 UTC and 1200 UTC 25 October 2004.

Figure 4.23. The typhoon (2004_No.28) generated in the equatorial central Pacific on 13 October 2004 and moved northwestward along the climatic monsoon trough. Its intensity peaked in the southeast sea of the Taiwan Island on 23 October 2004 before landfall on the island. The right turn of its motion near the Taiwan Island was influenced by another anomalous vortex at 300 hPa.

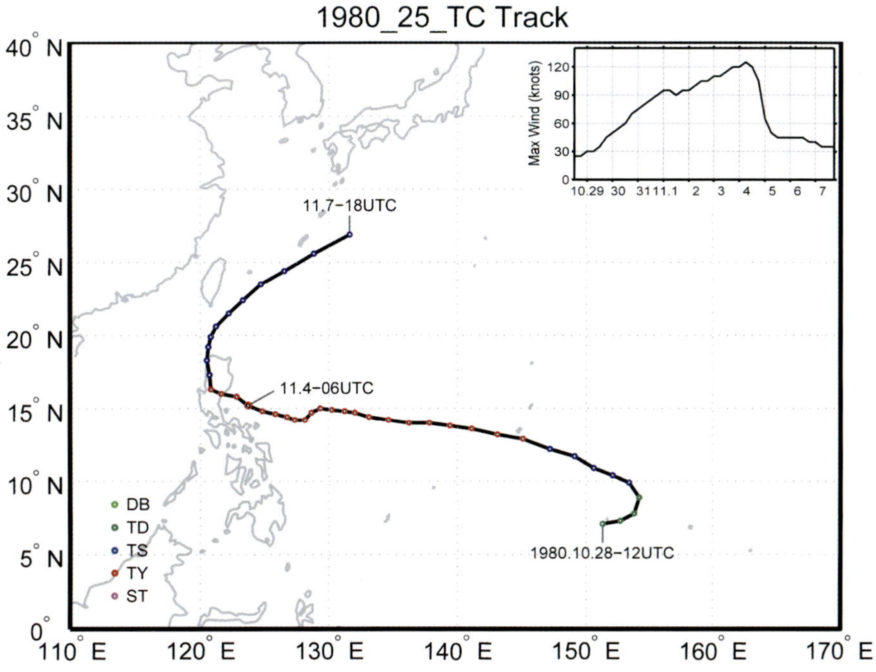

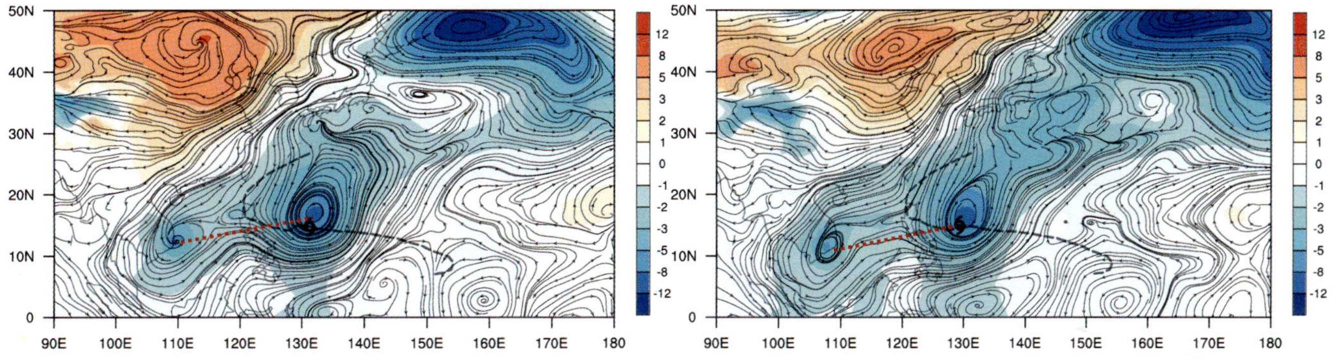

850 hPa flow/height anomalies at 0000 UTC and 1200 UTC 2 November 1980.

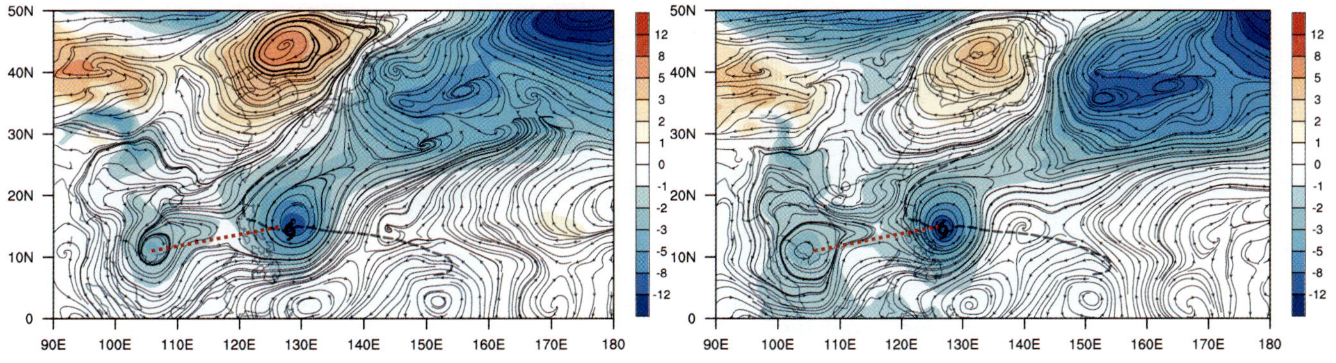

850 hPa flow/height anomalies at 0000 UTC and 1200 UTC 3 November 1980.

Figure 4.24. The typhoon (1980_No.25) generated in the equatorial central Pacific on 28 October 1980 and moved northwestward along the climatic monsoon trough. Its intensity peaked in the northeast sea of the Philippines on 4 November 1980 before landfall on the island. The right turn of its motion near the Philippines was influenced by anomalous vortices at 850 hPa.

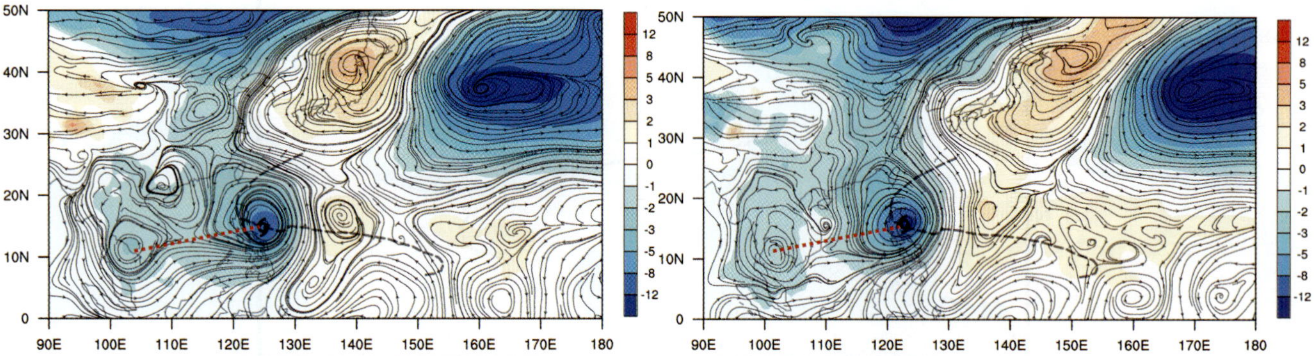

850 hPa flow/height anomalies at 0000 UTC and 1200 UTC 4 November 1980.

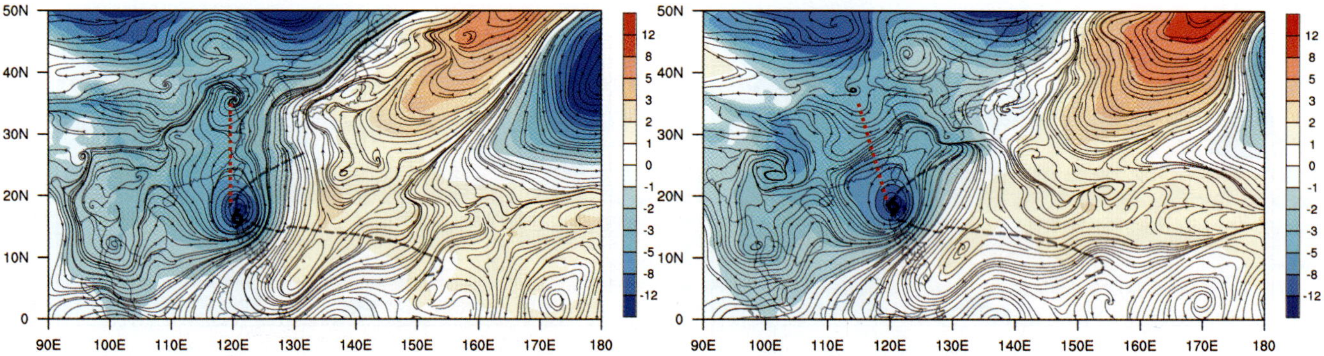

850 hPa flow/height anomalies at 0000 UTC and 1200 UTC 5 November 1980.

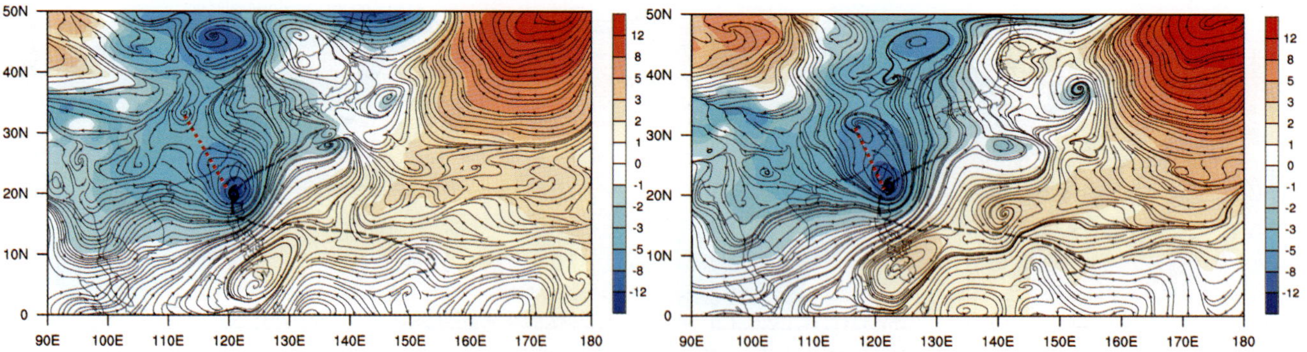

850 hPa flow/height anomalies at 0000 UTC and 1200 UTC 6 November 1980.

Figure 4.24. *(Continued)*

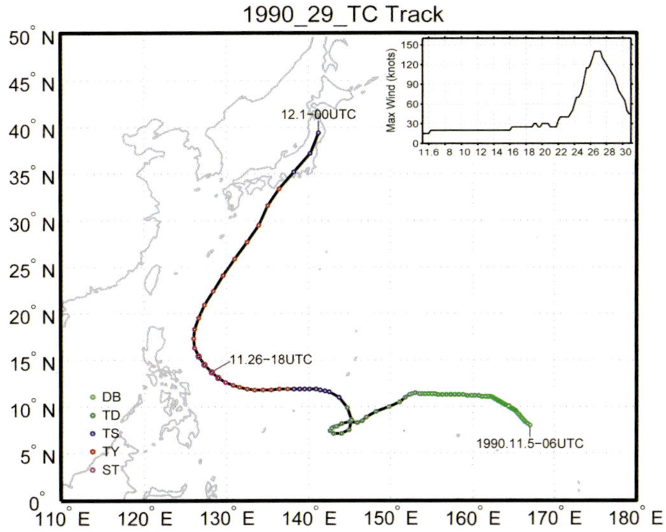

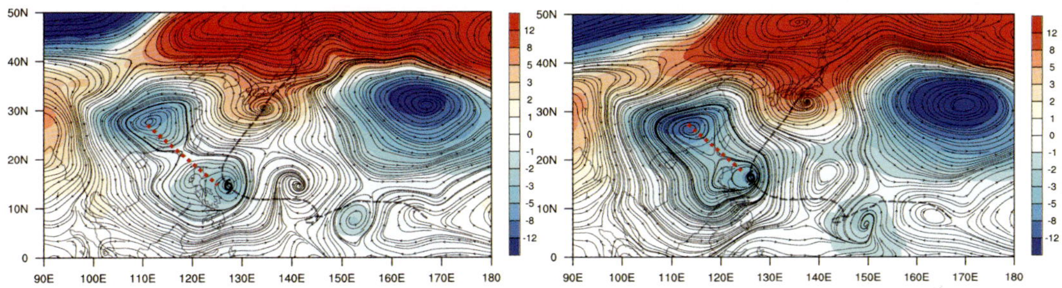

300 hPa flow/height anomalies at 0000 UTC and 1200 UTC 27 November 1990.

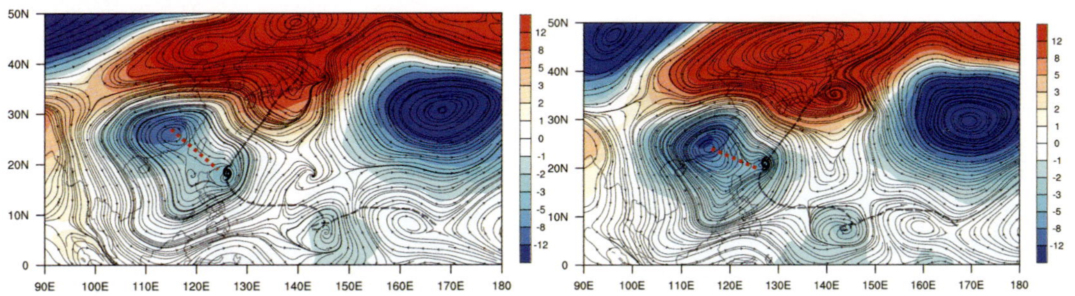

300 hPa flow/height anomalies at 0000 UTC and 1200 UTC 28 November 1990.

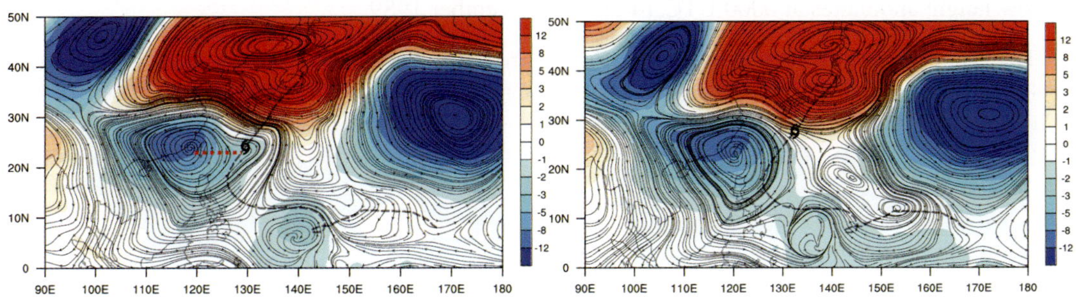

300 hPa flow/height anomalies at 0000 UTC and 1200 UTC 29 November 1990.

Figure 4.25. The typhoon (1990_No.29) generated in the equatorial central Pacific on 5 November 1990 and moved northwestward along the climatic monsoon trough. Its intensity peaked in the east sea of the Philippines on 26 November 1990. The right turn of its motion in the east sea of the Philippines was influenced by another anomalous vortex at 300 hPa.

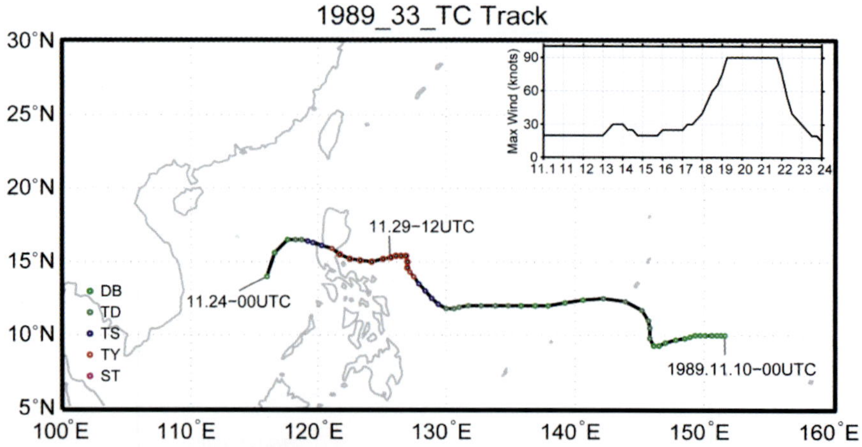

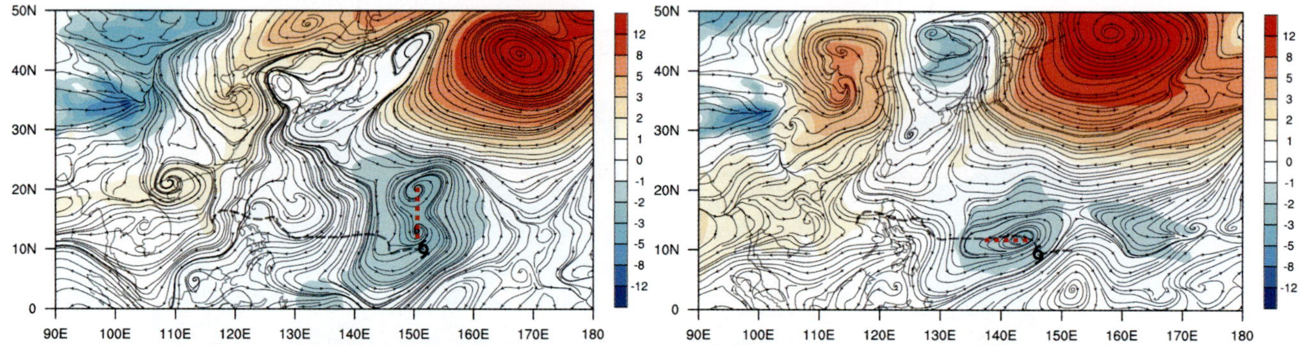

850 hPa flow/height anomalies at 0000 UTC 10 and 12 November 1989

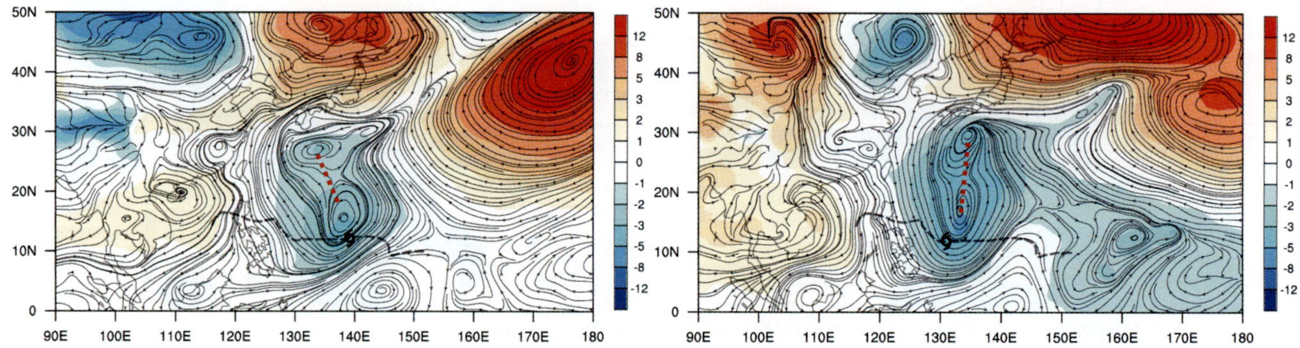

850 hPa flow/height anomalies at 0000 UTC 14 and 17 November 1989

Figure 4.26. The typhoon (1989_No.33) generated in the equatorial central Pacific on 10 November 1989 and experienced a complex motion near the climatic monsoon trough. Its intensity peaked on 29 November 1989 before landfall on the northern Philippines. Its turning motion was influenced by other vortices at 850 hPa.

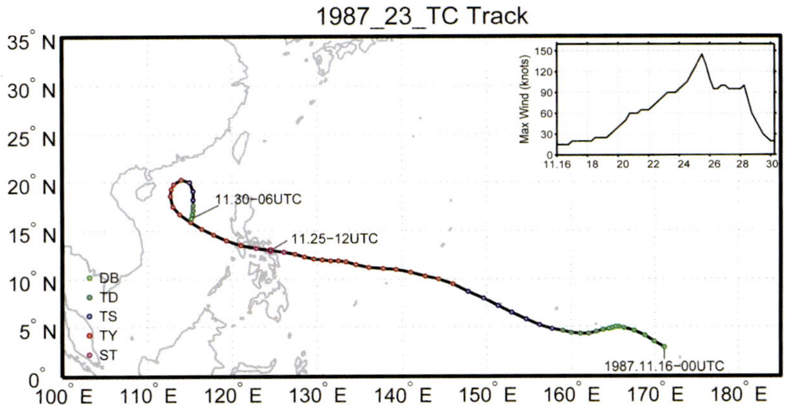

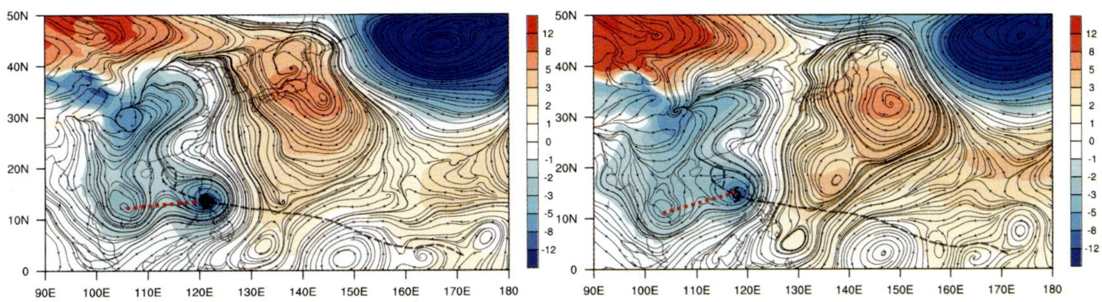

850 hPa flow/height anomalies at 0000 UTC and 1200 UTC 26 November 1987.

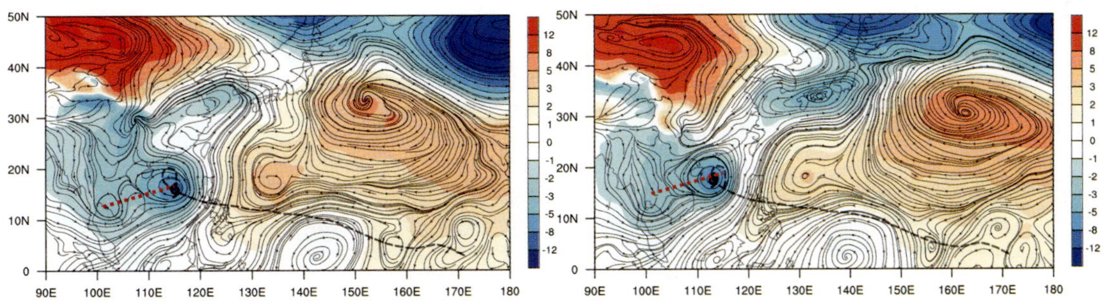

850 hPa flow/height anomalies at 0000 UTC and 1200 UTC 27 November 1987.

Figure 4.27. The typhoon (1987_No.23) generated in the equatorial central Pacific on 16 November 1987 and moved west-northwestward along the climatic monsoon trough. Its intensity peaked over the central eastern Philippines on 25 November 1987 when it made landfall on the island. Its circular motion in the northern SCS was influenced by anomalous vortices at 850 hPa.

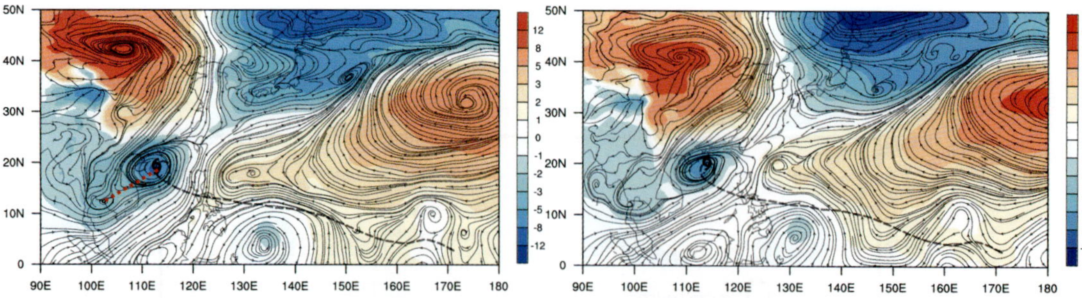

850 hPa flow/height anomalies at 0000 UTC and 1200 UTC 28 November 1987.

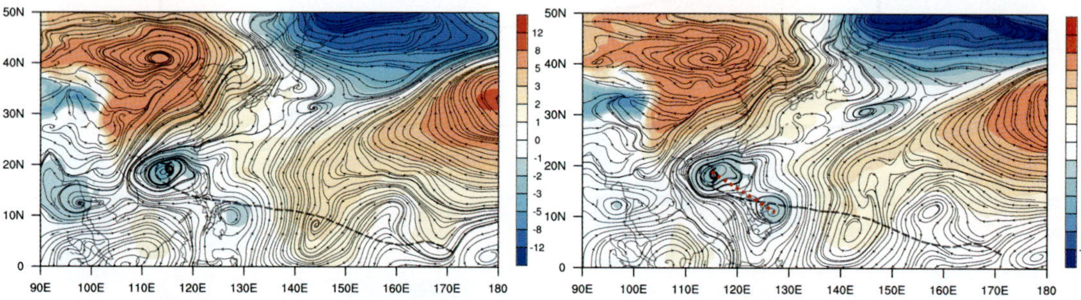

850 hPa flow/height anomalies at 0000 UTC and 0600 UTC 29 November 1987.

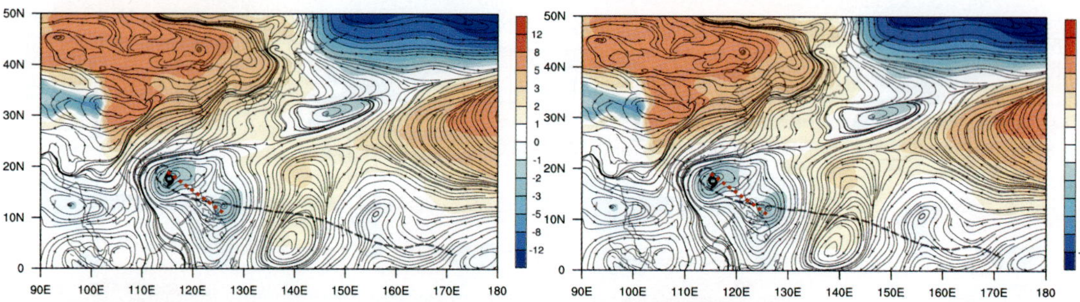

850 hPa flow/height anomalies at 1200 UTC and 1800 UTC 29 November 1987.

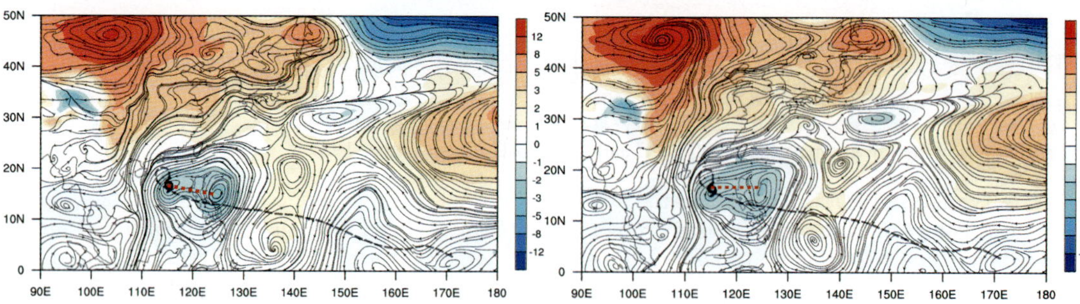

850 hPa flow/height anomalies at 0000 UTC and 0600 UTC 30 November 1987.

Figure 4.27. *(Continued)*

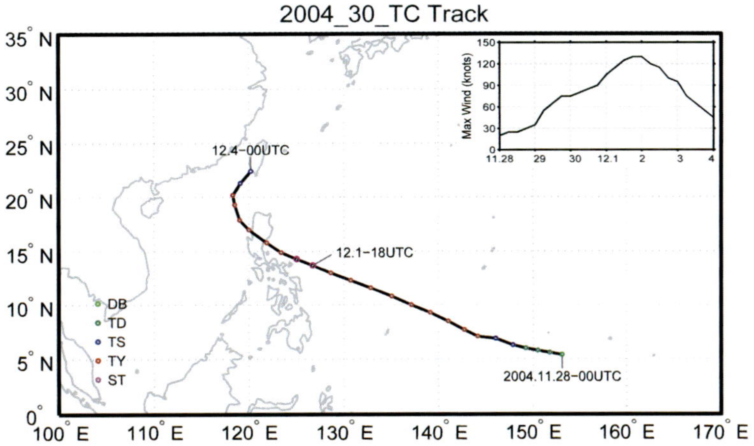

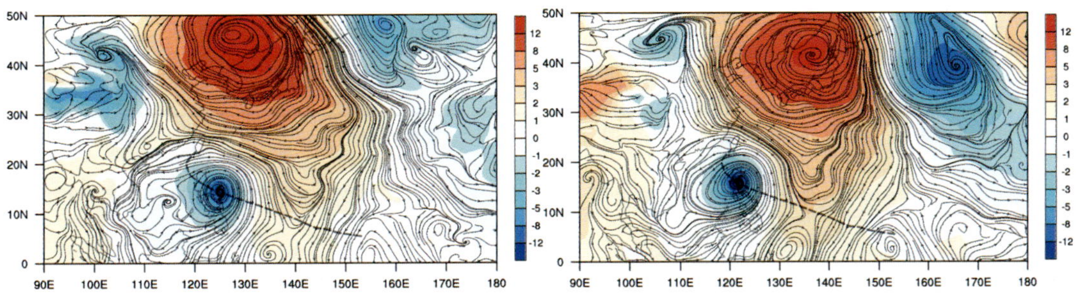

850 hPa flow/height anomalies at 0000 UTC and 1200 UTC 2 December 2004.

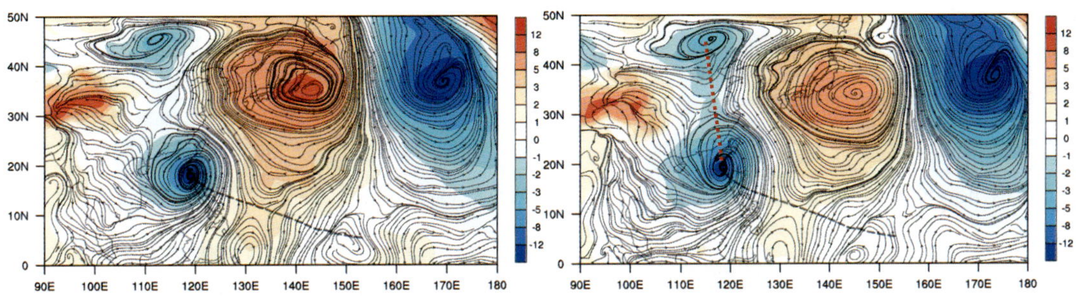

850 hPa flow/height anomalies at 0000 UTC and 0600 UTC 3 December 2004.

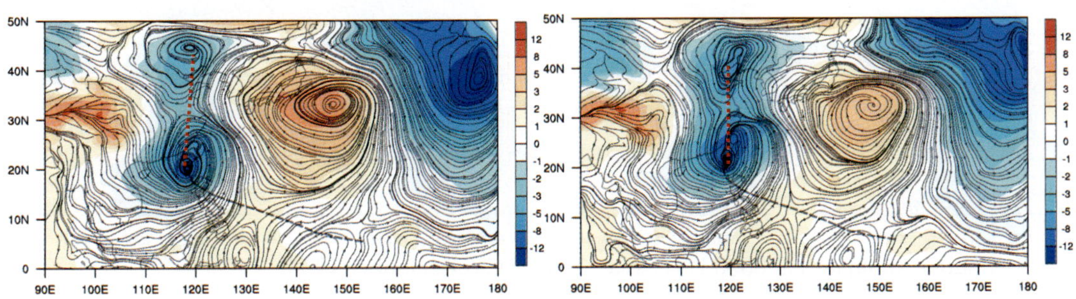

850 hPa flow/height anomalies at 1200 UTC and 1800 UTC 3 December 2004.

Figure 4.28. The typhoon (2004_No.30) generated in the equatorial central Pacific on 28 November 2004 and moved northwestward along the climatic monsoon trough. Its intensity peaked in the east sea of the Philippines on 1 December 2004 before landfall on the island. The right turn of its motion in the northeastern SCS was influenced by another anomalous vortex at 850 hPa.

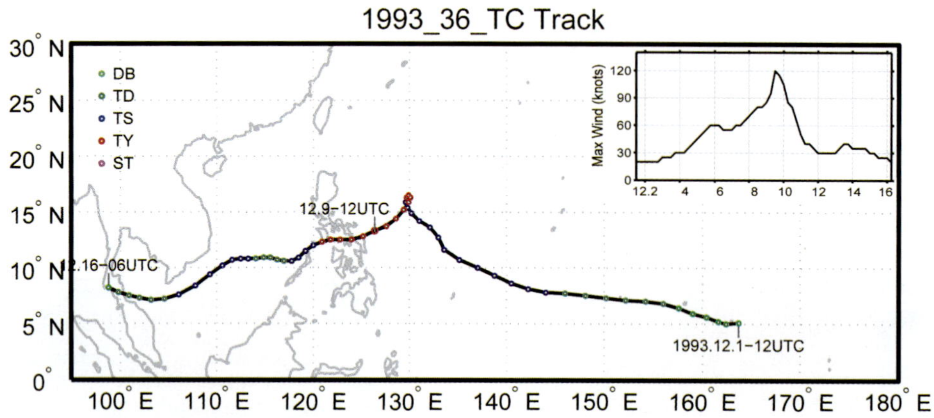

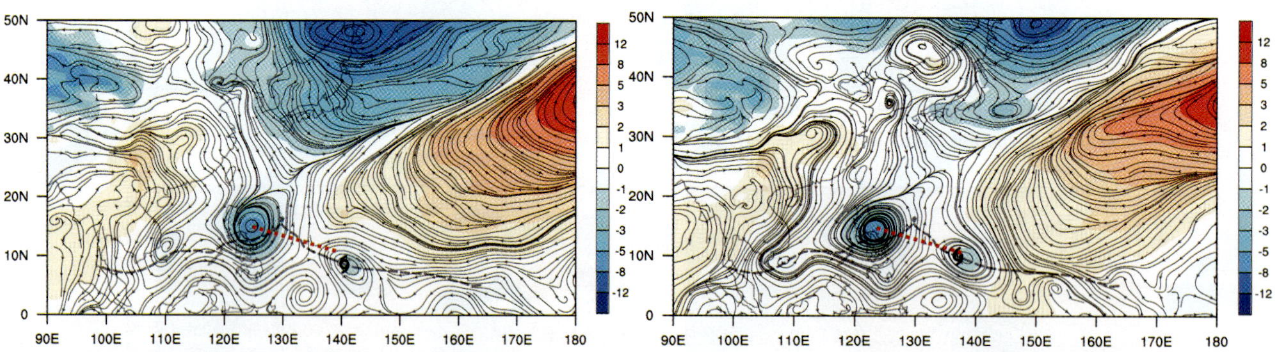

850 hPa flow/height anomalies at 0000 UTC and 1200 UTC 5 December 1993.

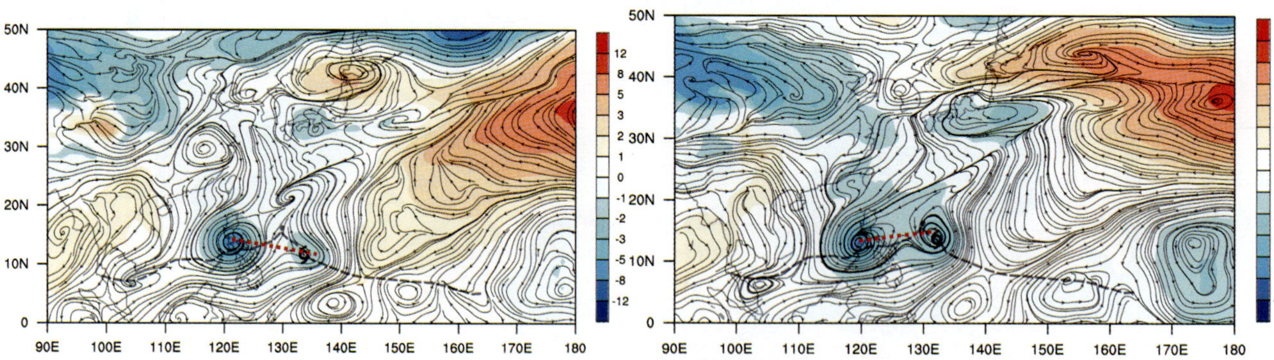

850 hPa flow/height anomalies at 0000 UTC and 1200 UTC 6 December 1993.

Figure 4.29. The typhoon (1993_No.36) generated in the equatorial central Pacific on 1 December 1993 and moved northwestward along the climatic monsoon trough. Its intensity peaked in the east sea of the Philippines on 12 December 1993 before landfall on the island. The left turn of its motion with a stagnant state in the east sea of the Philippines was influenced by another anomalous vortex at 850 hPa.

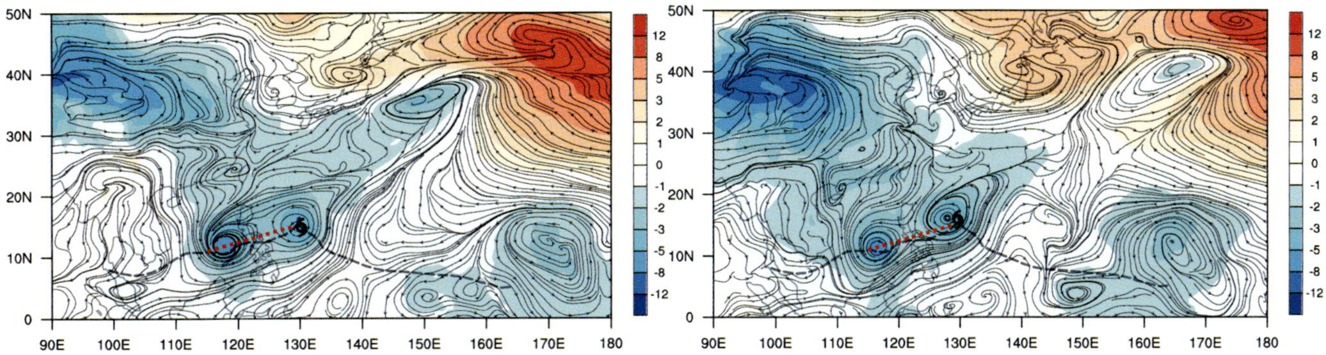

850 hPa flow/height anomalies at 0000 UTC and 1200 UTC 7 December 1993.

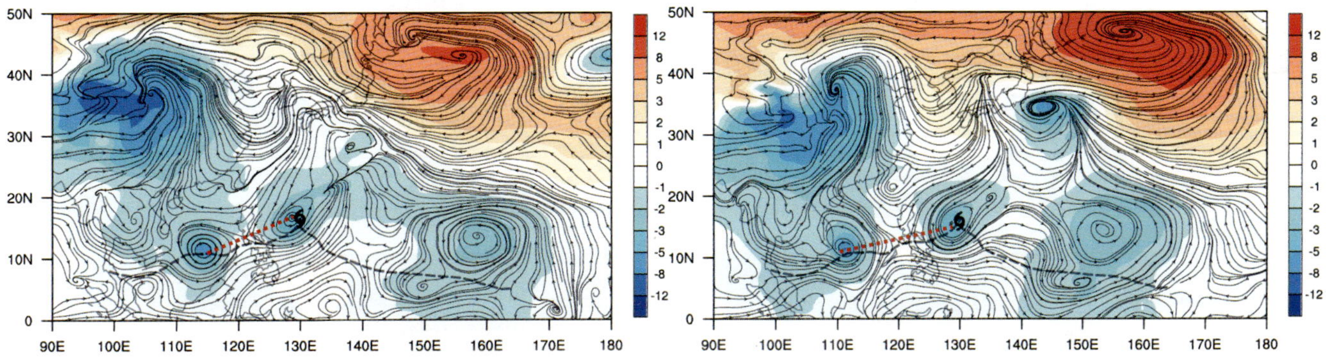

850 hPa flow/height anomalies at 0000 UTC and 1200 UTC 8 December 1993.

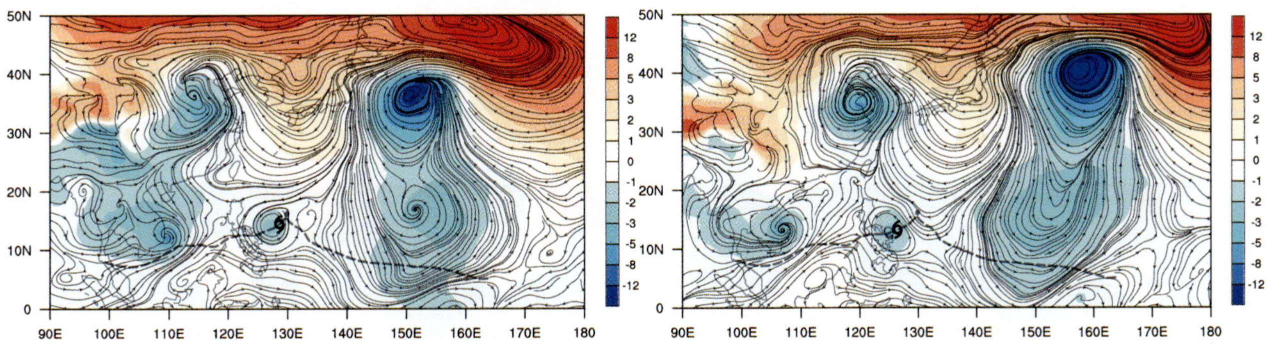

850 hPa flow/height anomalies at 0000 UTC and 1200 UTC 9 December 1993.

Figure 4.29. *(Continued)*

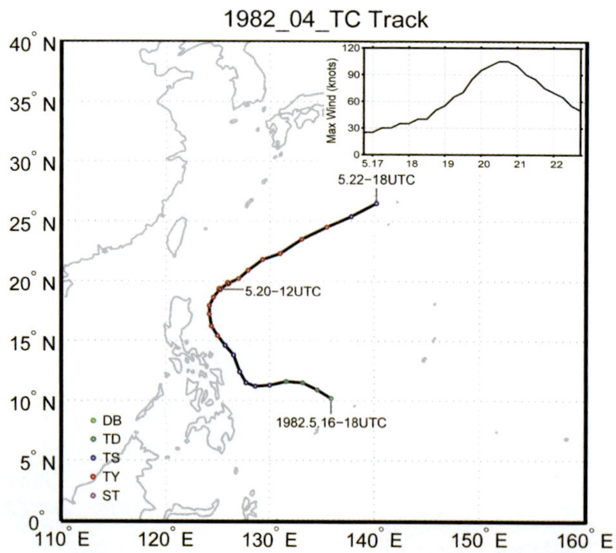

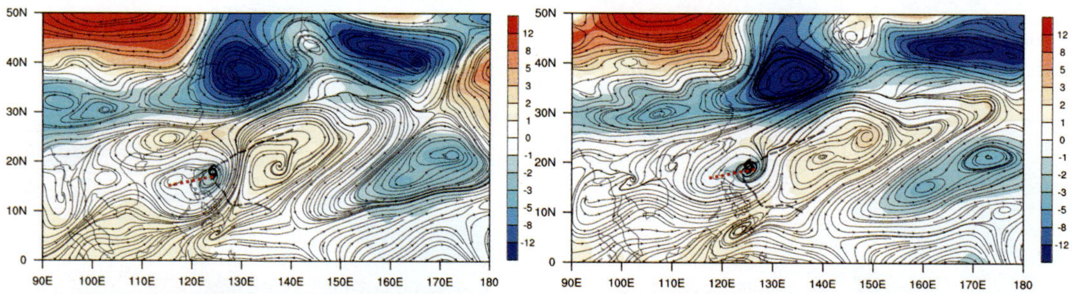

300 hPa flow/height anomalies at 0000 UTC and 1200 UTC 20 May 1982.

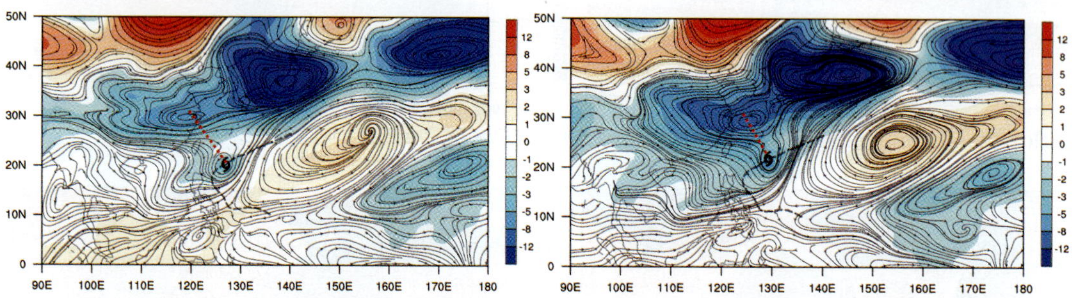

300 hPa flow/height anomalies at 0000 UTC and 1200 UTC 21 May 1982.

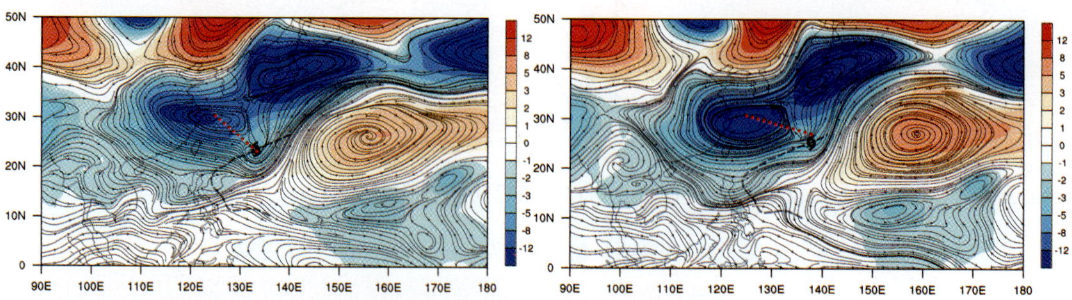

300 hPa flow/height anomalies at 0000 UTC and 1200 UTC 22 May 1982.

Figure 4.30. The typhoon (1982_No.04) generated in the tropical western Pacific on 16 May 1982 and moved northwestward along the climatic monsoon trough. Its intensity peaked in the northeast sea of the Philippines on 20 May 1982. Its right turn in the east sea of the Philippines was influenced by anomalous vortices at 300 hPa.

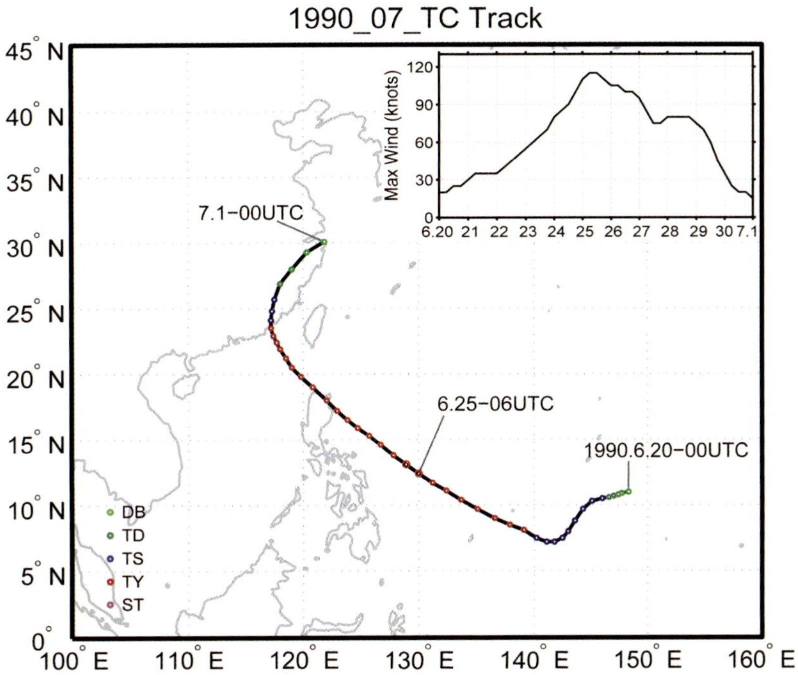

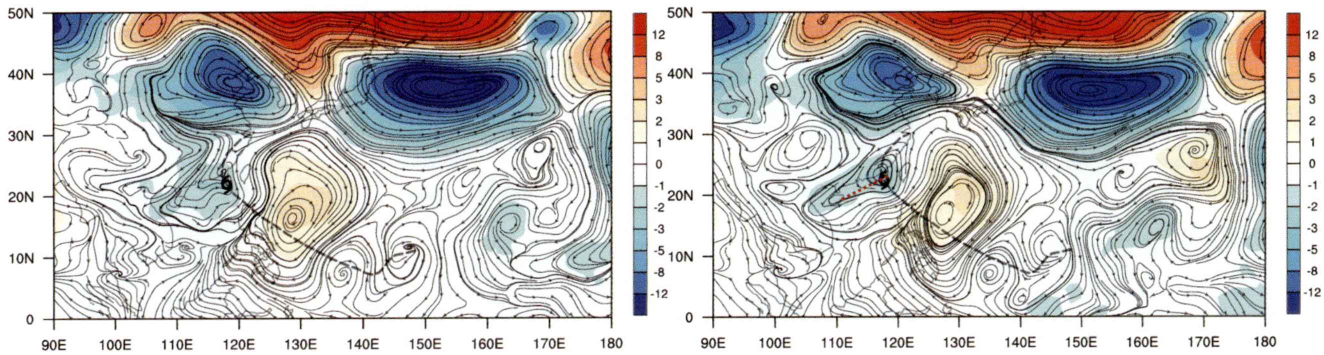

300 hPa flow/height anomalies at 1200 UTC and 1800 UTC 28 June 1990.

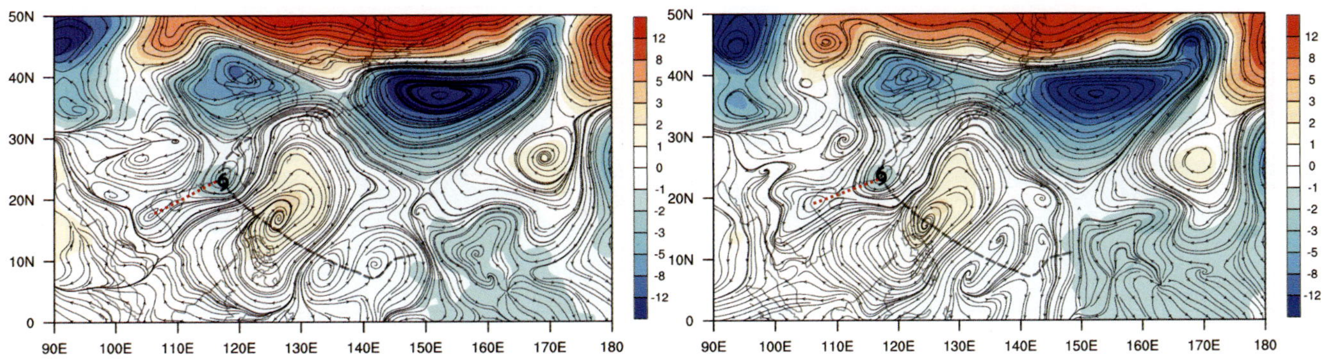

300 hPa flow/height anomalies at 0000 UTC and 0600 UTC 29 June 1990.

Figure 4.31. The typhoon (1990_No.07) generated in the tropical western Pacific on 20 June 1990 and moved southeastward at the beginning stage. It moved northwestward along the climatic monsoon trough and its intensity peaked in the east sea of the Philippines on 25 June 1990. The right turn of its motion in the northeastern SCS was influenced by anomalous vortices at 850 hPa and 300 hPa.

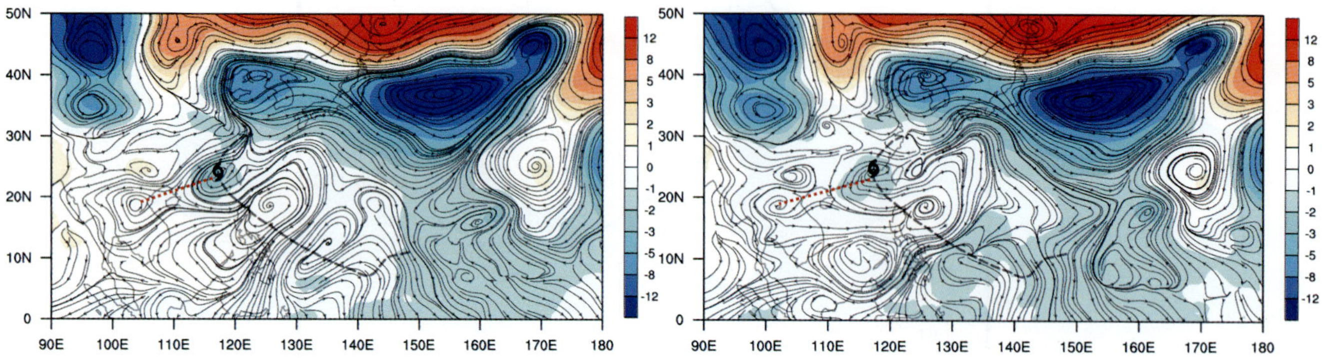

300 hPa flow/height anomalies at 1200 UTC and 1800 UTC 29 June 1990.

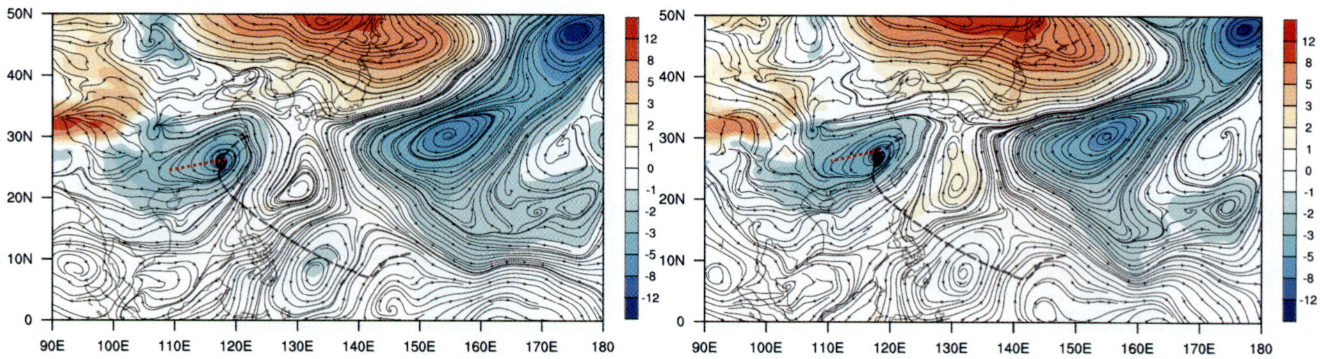

850 hPa flow/height anomalies at 0000 UTC and 0600 UTC 30 June 1990.

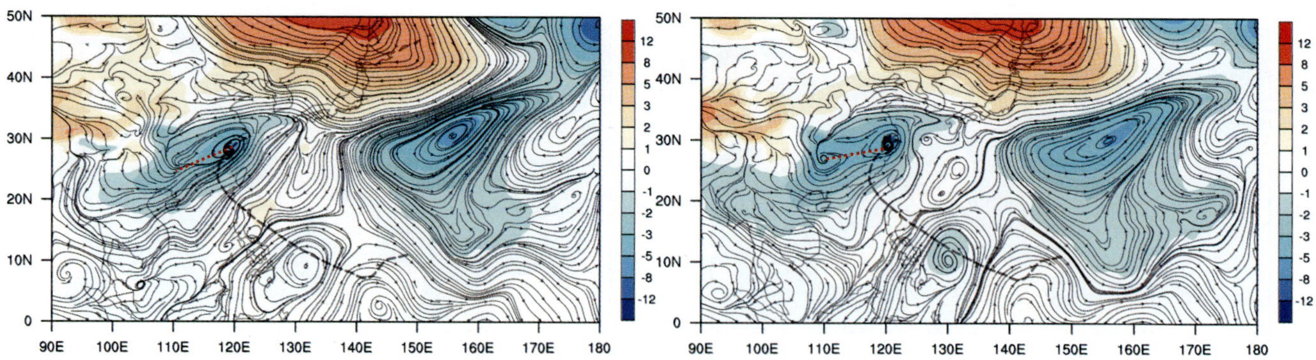

850 hPa flow/height anomalies at 1200 UTC and 1800 UTC 30 June 1990.

Figure 4.31. *(Continued)*

4.4. Geneses in the Tropical Pacific

There were 16 typhoons that generated in the tropical Pacific and their intensities peaked in the Philippines, including 15 typhoons shown in this section (Figs. 4.30–4.44) and the typhoon Megi (2010_No.15) plotted in Chapter 1 as an example. Their right turns or left turns during their lifetime were due to the influence of anomalous vortices.

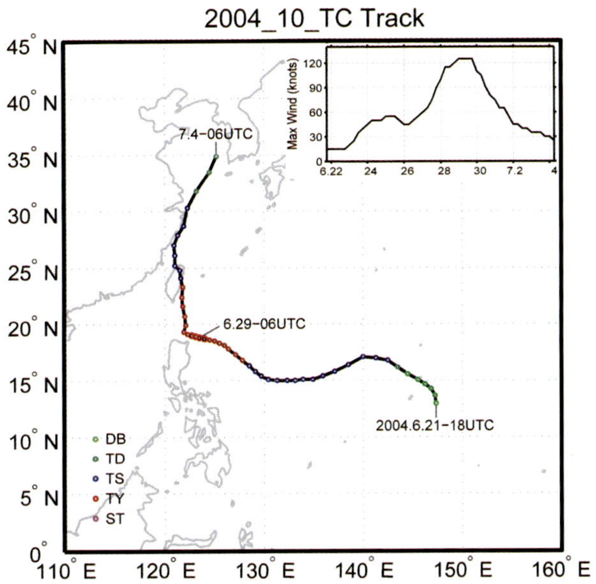

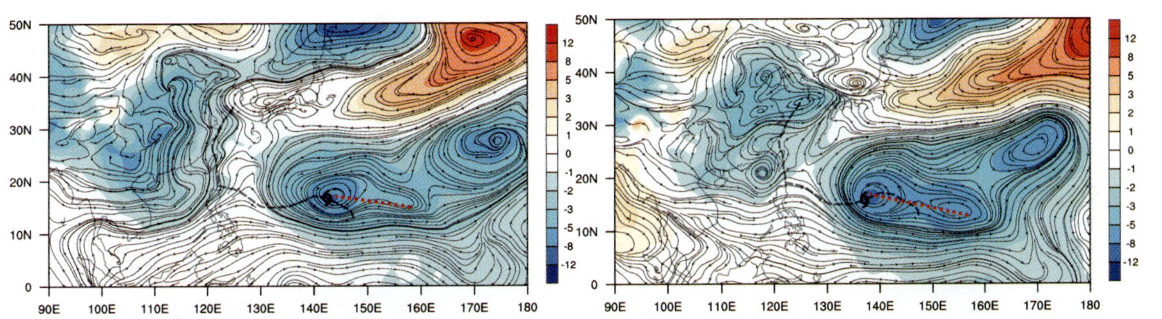

850 hPa flow/height (shading, 10 gpm) anomalies at 1200 UTC 23 and 24 June 2004.

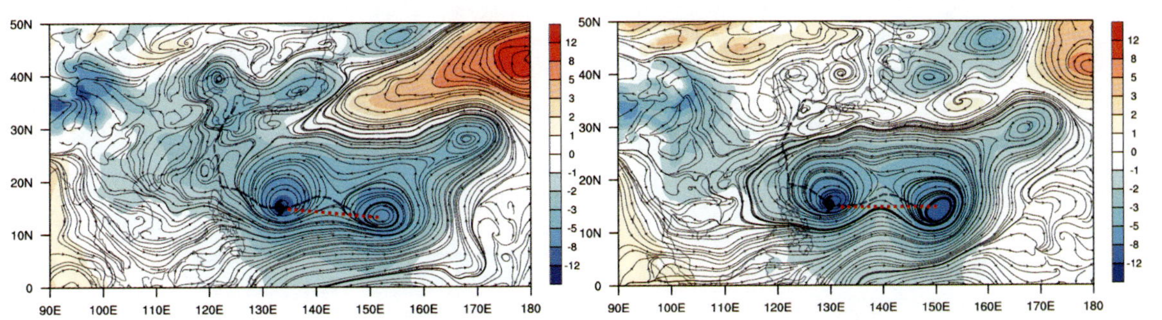

850 hPa flow/height anomalies at 1200 UTC 25 and 26 June 2004.

Figure 4.32. The typhoon (2004_No.10) generated in the tropical western Pacific on 21 June 2004 and moved westward. Its intensity peaked in the northeast sea of the Philippines on 29 June 2004 when it was approaching the climatic monsoon trough before landfall on the Taiwan Island. The right turn of its motion was influenced by anomalous vortices at 850 hPa and 300 hPa.

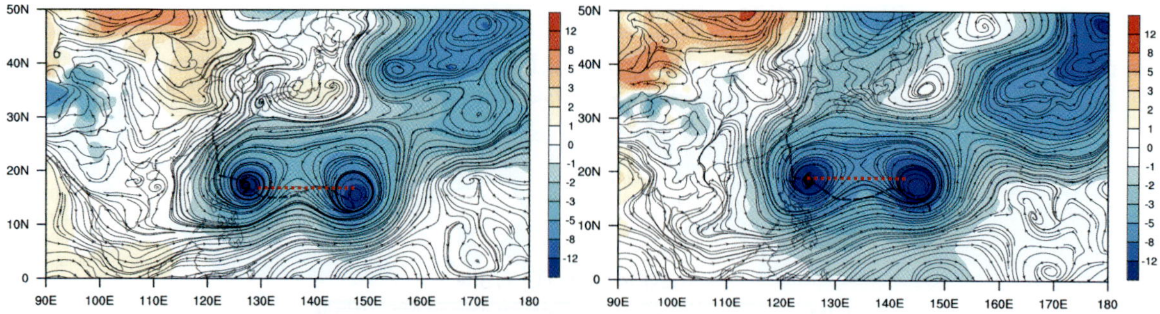

850 hPa flow/height anomalies at 1200 UTC 27 and 28 June 2004.

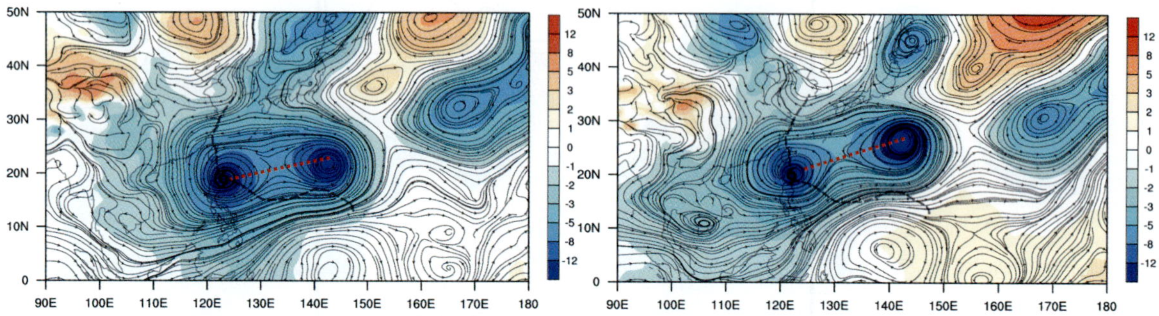

850 hPa flow/height anomalies at 1200 UTC 29 and 30 June 2004.

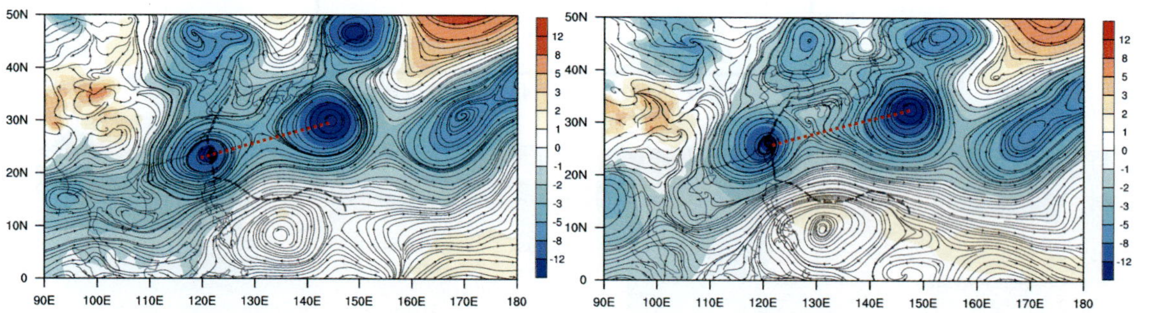

850 hPa flow/height anomalies at 1200 UTC 1 and 2 July 2004.

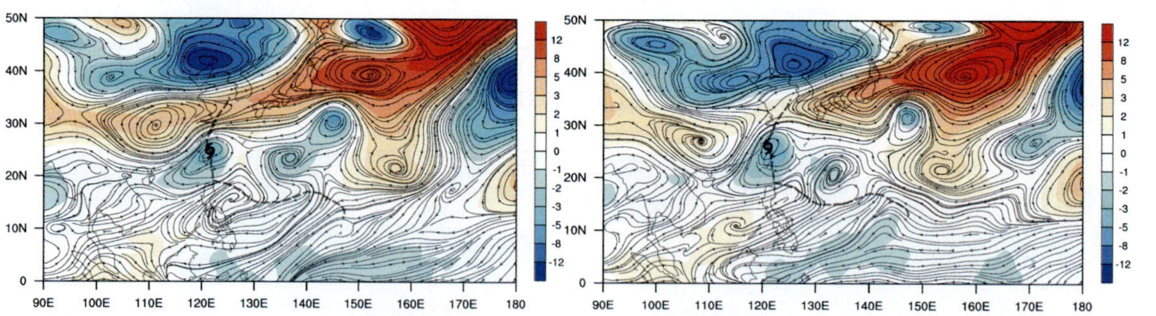

300 hPa flow/height anomalies at 0000 UTC and 1200 UTC 2 July 2004.

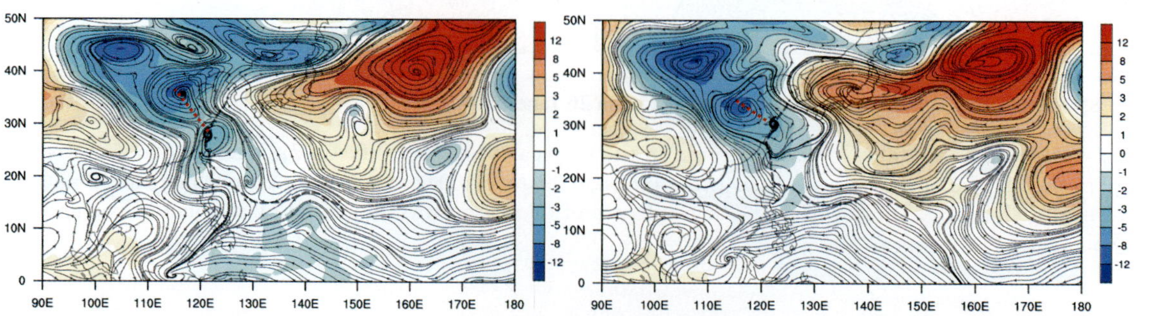

300 hPa flow/height anomalies at 0000 UTC and 1200 UTC 3 July 2004.

Figure 4.32. (*Continued*)

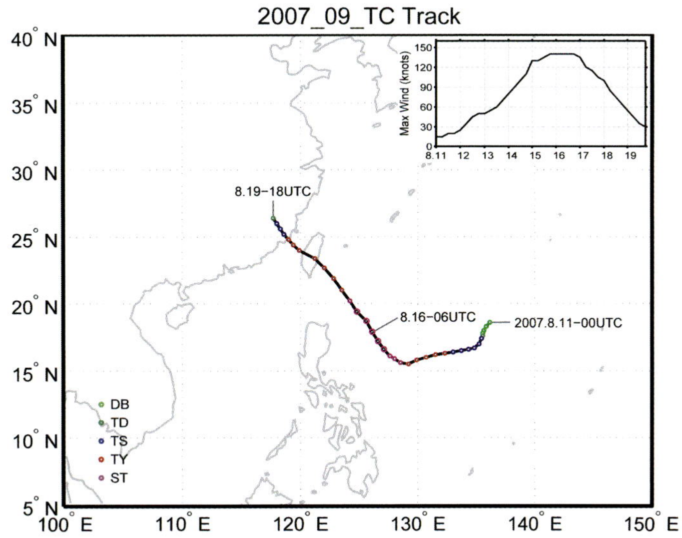

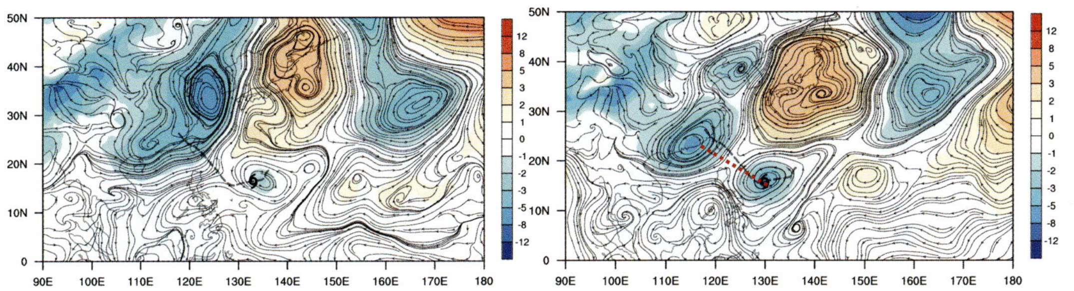

850 hPa flow/height anomalies at 1200 UTC 13 and 14 August 2007.

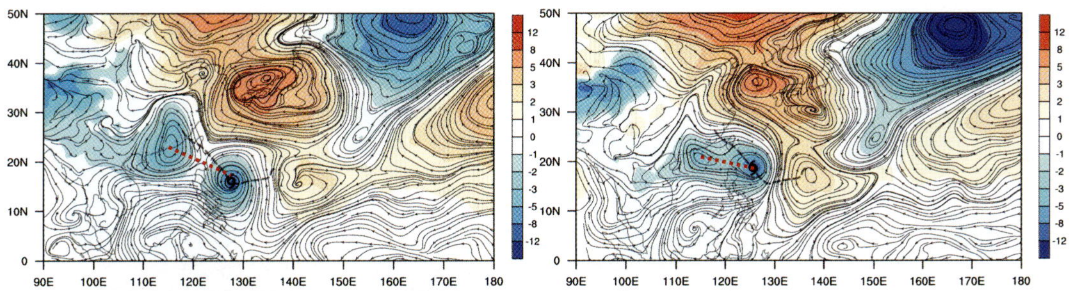

850 hPa flow/height anomalies at 1200 UTC 15 and 16 August 2007.

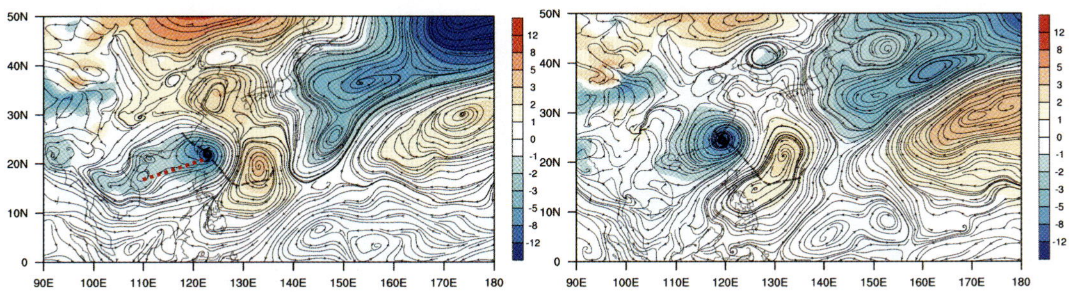

850 hPa flow/height anomalies at 1200 UTC 17 and 18 August 2007.

Figure 4.33. The typhoon (2007_No.09) generated in the tropical western Pacific on 11 August 2007 and moved southwestward at the beginning stage. Before its intensity peaked on 16 August 2007, the typhoon turned right due to the influence by another anomalous vortex at 850 hPa.

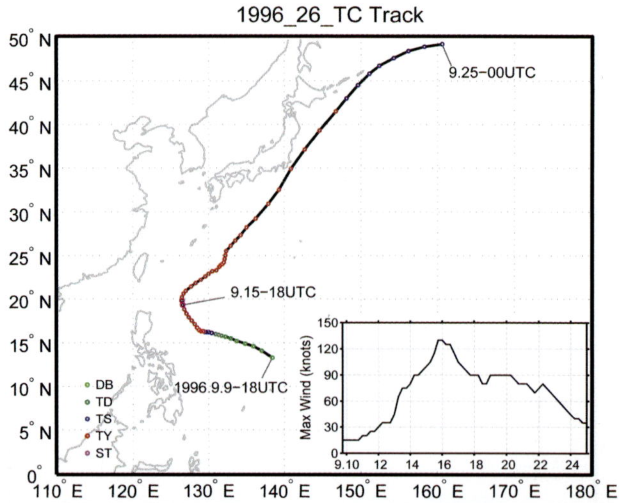

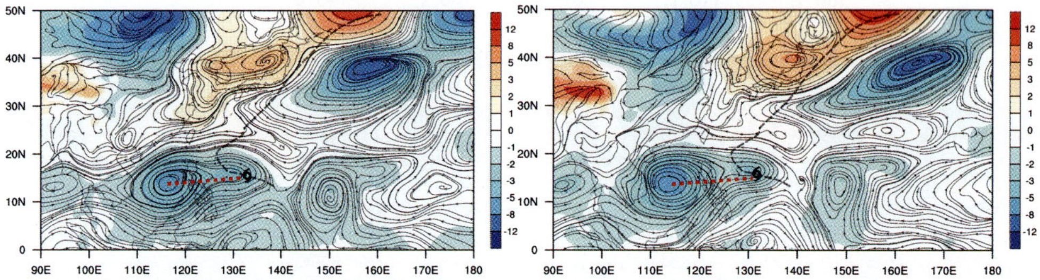

850 hPa flow/height anomalies at 0000 UTC and 1200 UTC 11 September 1996.

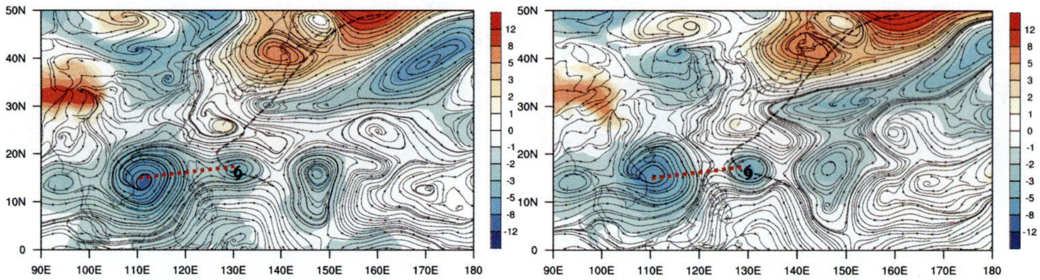

850 hPa flow/height anomalies at 0000 UTC and 1200 UTC 12 September 1996.

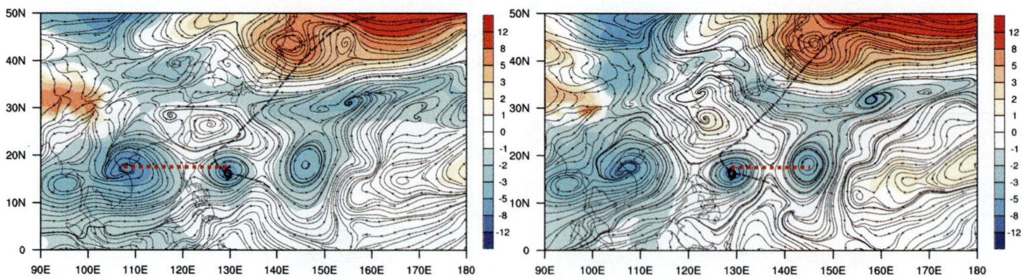

850 hPa flow/height anomalies at 0000 UTC and 1200 UTC 13 September 1996.

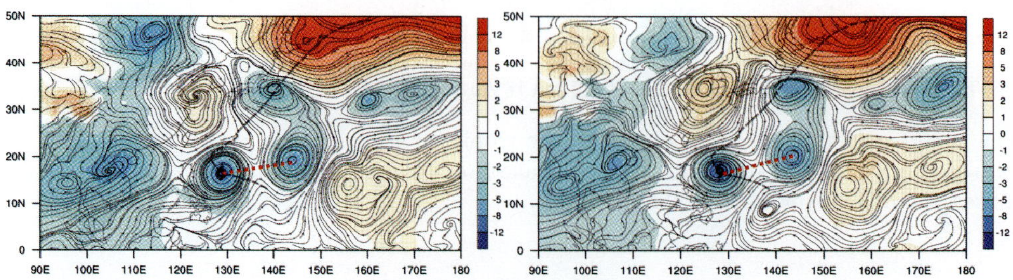

850 hPa flow/height anomalies at 0000 UTC and 1200 UTC 14 September 1996.

Figure 4.34. The typhoon (1996_No.26) generated in the tropical western Pacific on 9 September 1996 and moved northwestward along the climatic monsoon trough. Its intensity peaked in the northeast sea of the Philippines on 15 September 1996 at the moment when it was turning right. This right turn near the Philippines was influenced by anomalous vortices at 850 hPa.

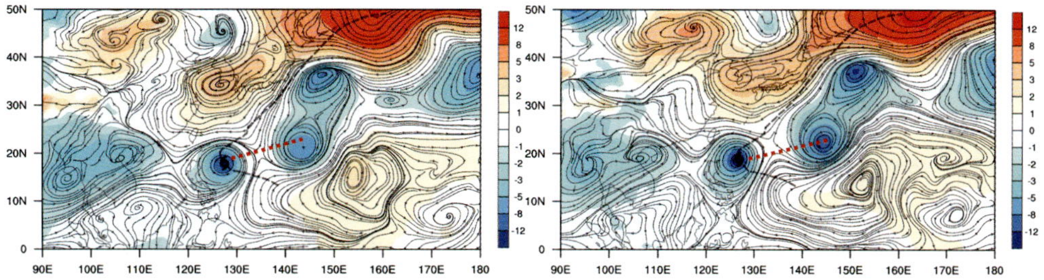

850 hPa flow/height anomalies at 0000 UTC and 1200 UTC 15 September 1996.

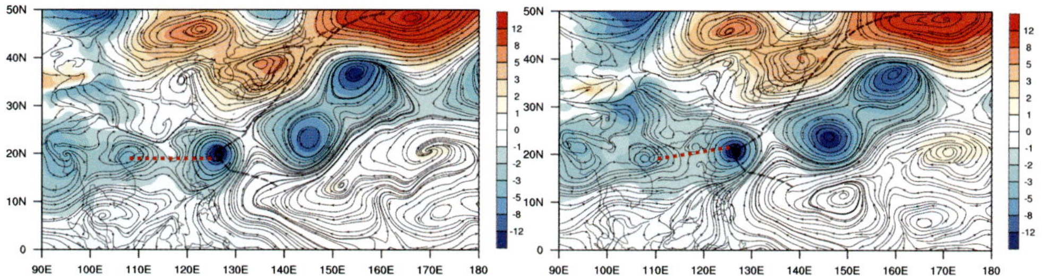

850 hPa flow/height anomalies at 0000 UTC and 1200 UTC 16 September 1996.

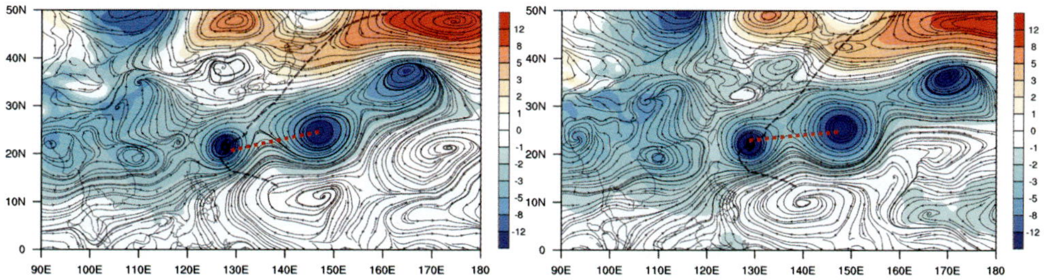

850 hPa flow/height anomalies at 0000 UTC and 1200 UTC 17 September 1996.

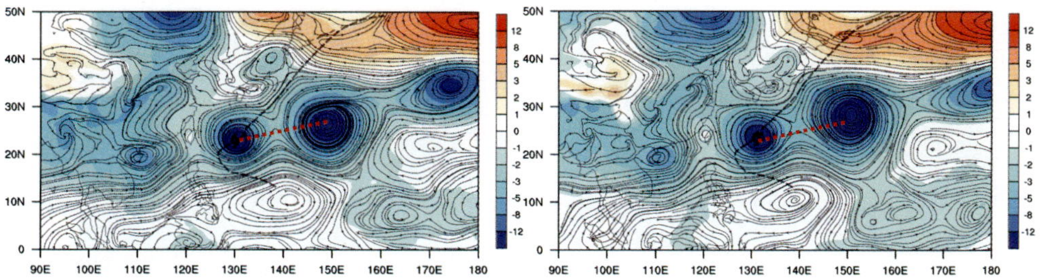

850 hPa flow/height anomalies at 0000 UTC and 1200 UTC 18 September 1996.

Figure 4.34. *(Continued)*

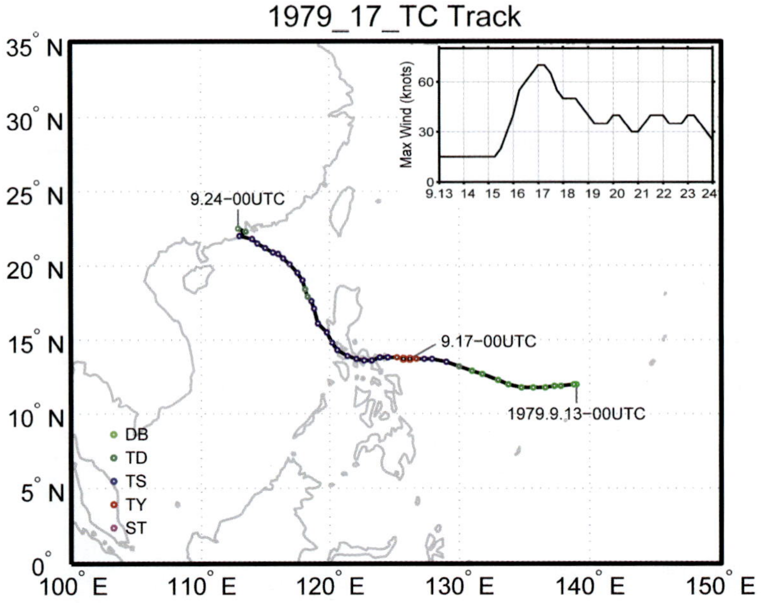

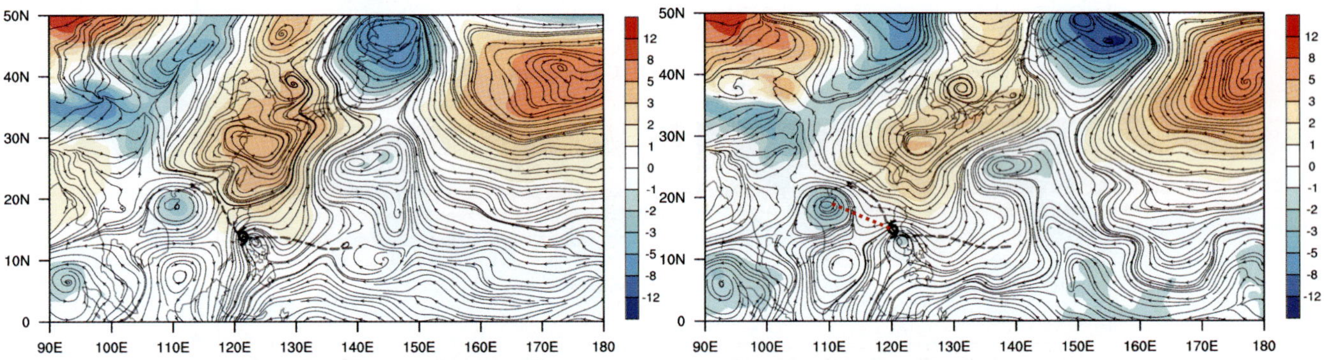

850 hPa flow/height anomalies at 0000 UTC and 1200 UTC 19 September 1979.

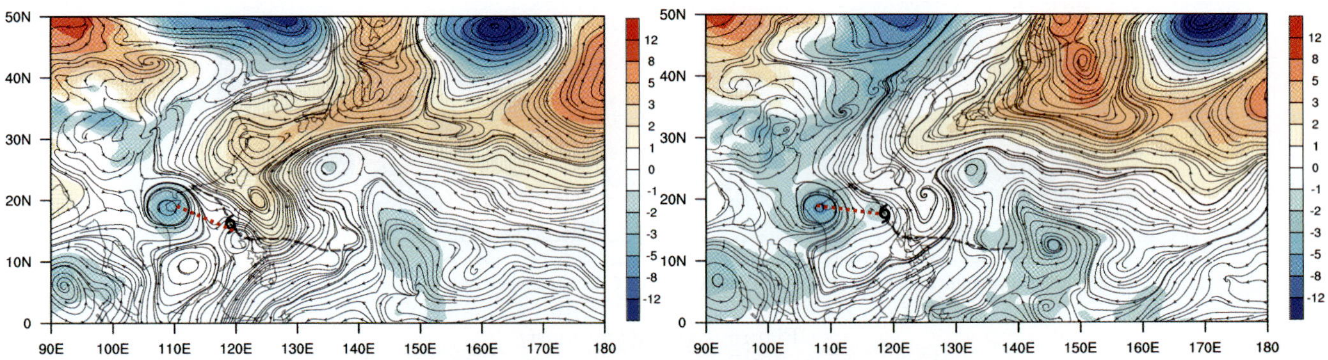

850 hPa flow/height anomalies at 0000 UTC and 1200 UTC 20 September 1979.

Figure 4.35. The typhoon (1979_No.17) generated in the tropical western Pacific on 13 September 1997 and moved westward. Its intensity peaked on 17 September 1979 before landfall on the central Philippines. The right turn of its motion near the Philippines was influenced by another anomalous vortex at 850 hPa.

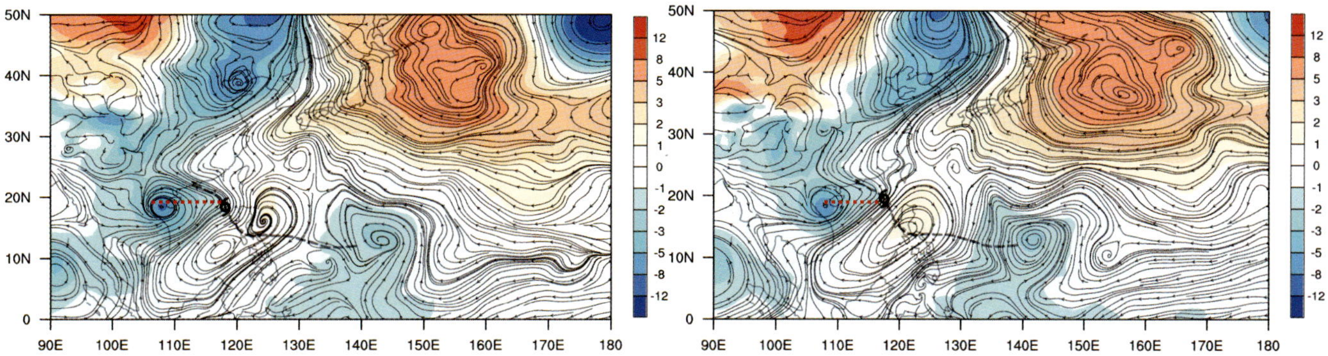

850 hPa flow/height anomalies at 0000 UTC and 1200 UTC 21 September 1979.

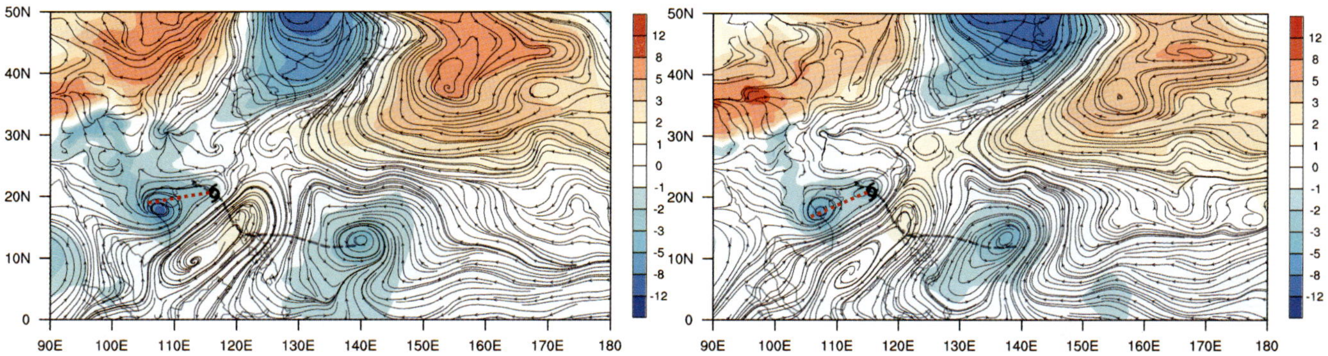

850 hPa flow/height anomalies at 0000 UTC and 1200 UTC 22 September 1979.

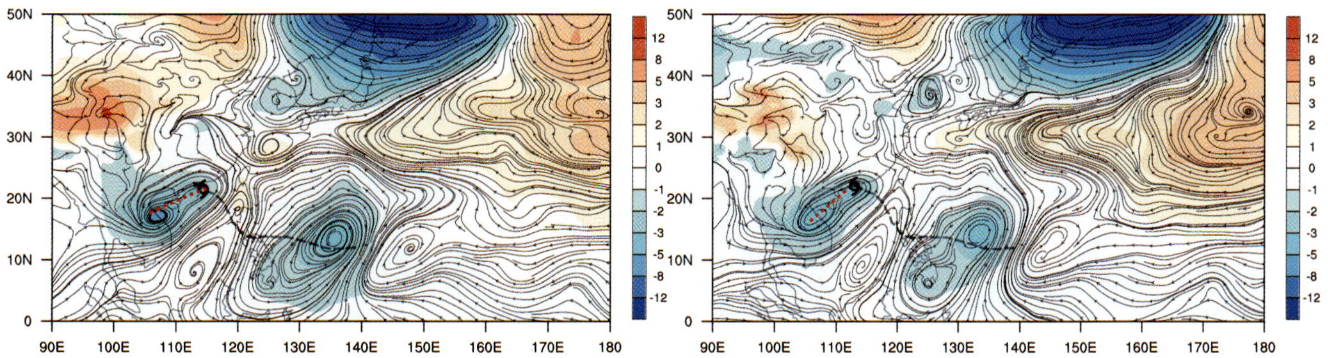

850 hPa flow/height anomalies at 0000 UTC and 1200 UTC 23 September 1979.

Figure 4.35. *(Continued)*

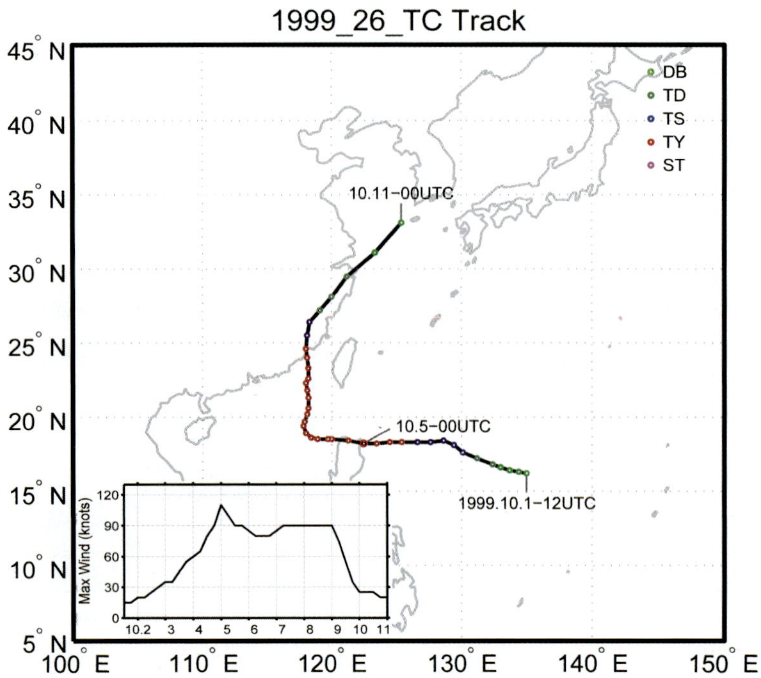

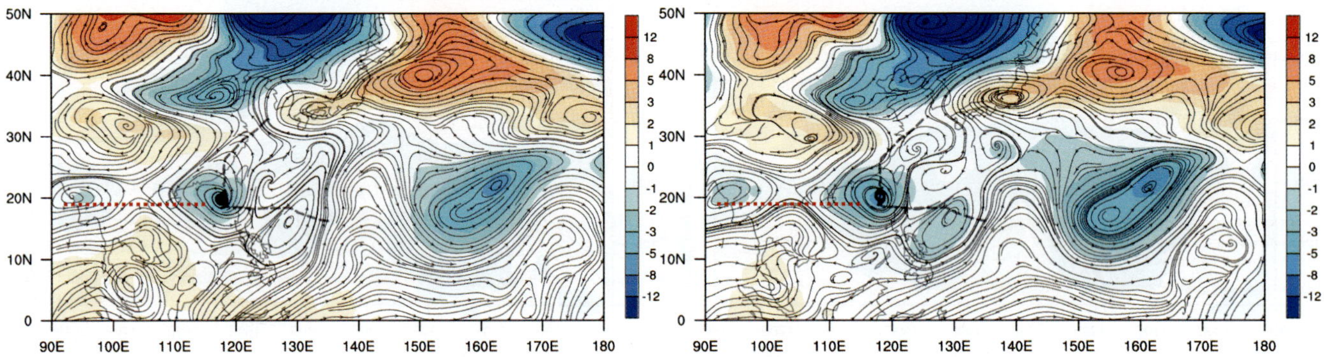

300 hPa flow/height anomalies at 0000 UTC and 0600 UTC 7 October 1999.

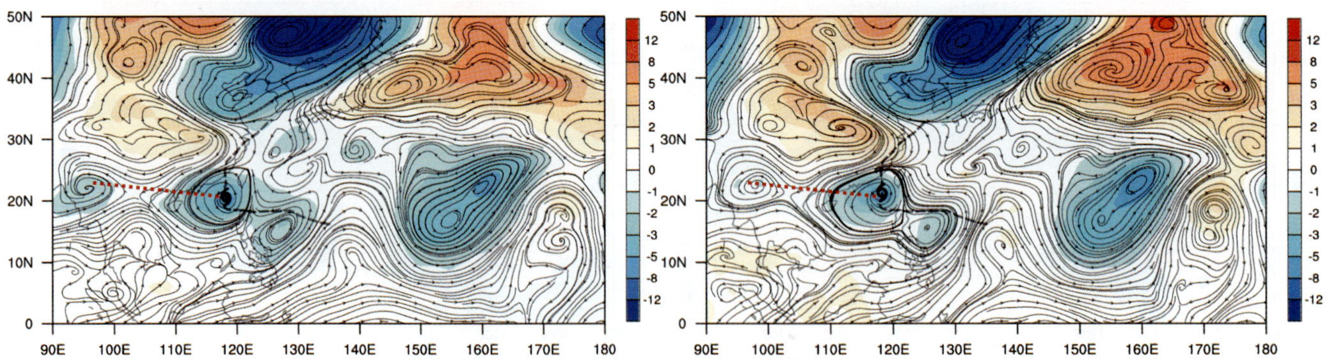

300 hPa flow/height anomalies at 1200 UTC and 1800 UTC 7 October 1999.

Figure 4.36. The typhoon (1999_No.26) generated in the tropical western Pacific on 1 October 1999 and moved westward along the climatic monsoon trough so that its intensity peaked in the northern Philippines on 5 October 1999 before landfall on the island. After passing the Philippines, it turned northward rapidly on 6–7 October 1999. This right turn was influenced by another anomalous vortex at 300 hPa.

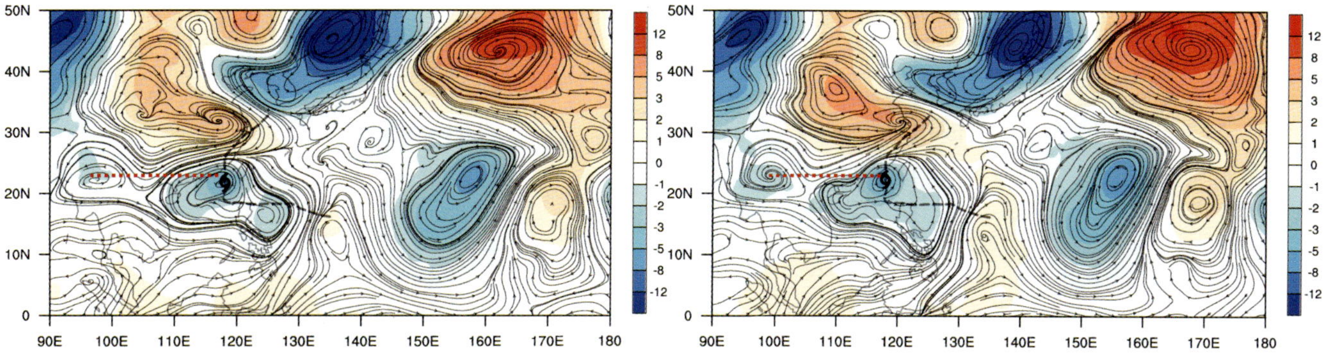

300 hPa flow/height anomalies at 0000 UTC and 0600 UTC 8 October 1999.

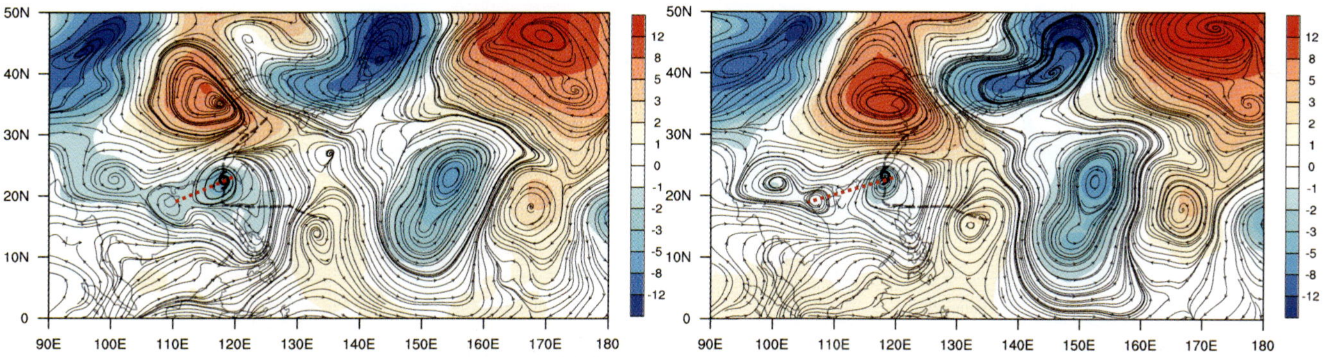

300 hPa flow/height anomalies at 1200 UTC and 1800 UTC 8 October 1999.

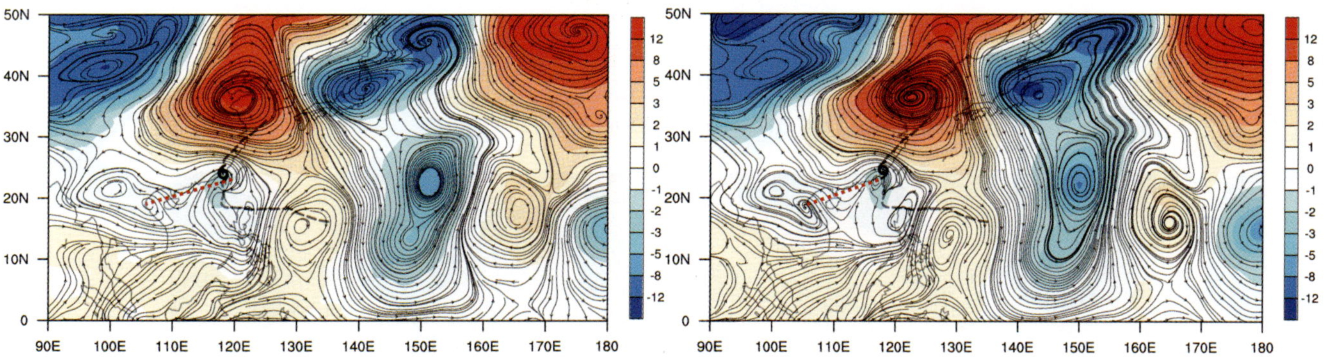

300 hPa flow/height anomalies at 0000 UTC and 0600 UTC 9 October 1999.

Figure 4.36. *(Continued)*

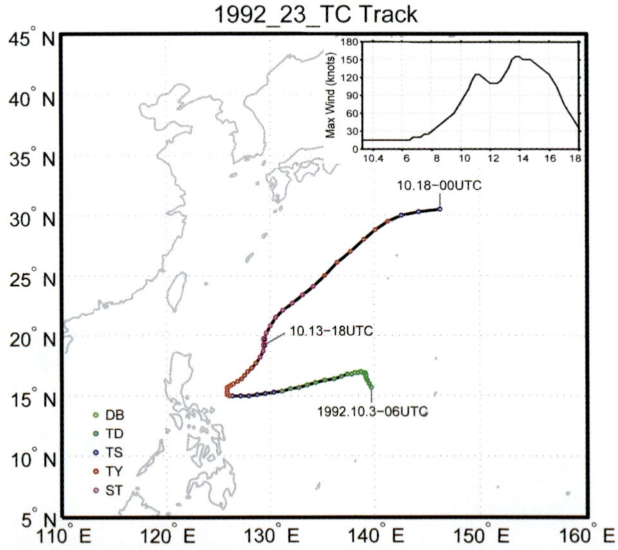

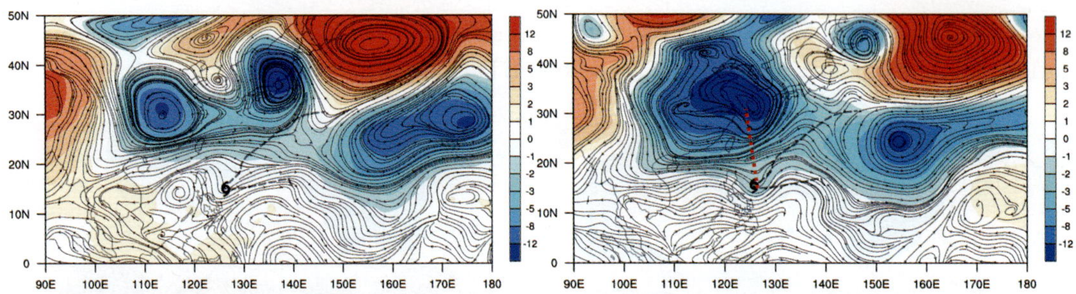

300 hPa flow/height anomalies at 1200 UTC 9 and 10 October 1992.

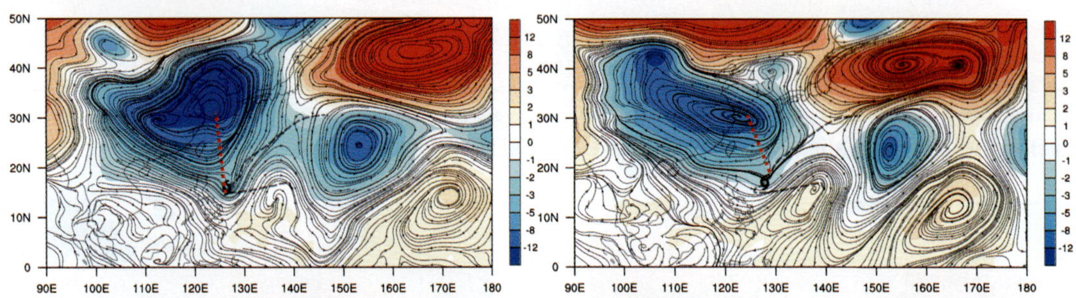

300 hPa flow/height anomalies at 1200 UTC 11 and 12 October 1992.

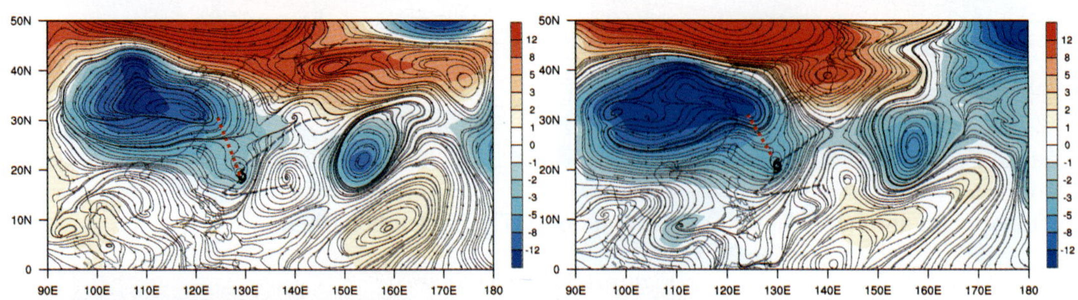

300 hPa flow/height anomalies at 1200 UTC 13 and 14 October 1992.

Figure 4.37. The typhoon (1992_No.23) generated in the tropical western Pacific on 3 October 1992. Its intensity peaked in the northeast sea of the Philippines on 13 October 1992 when it was approaching the climatic monsoon trough. The right turn of its motion in the east sea of the Philippines was influenced by another anomalous vortex at 300 hPa.

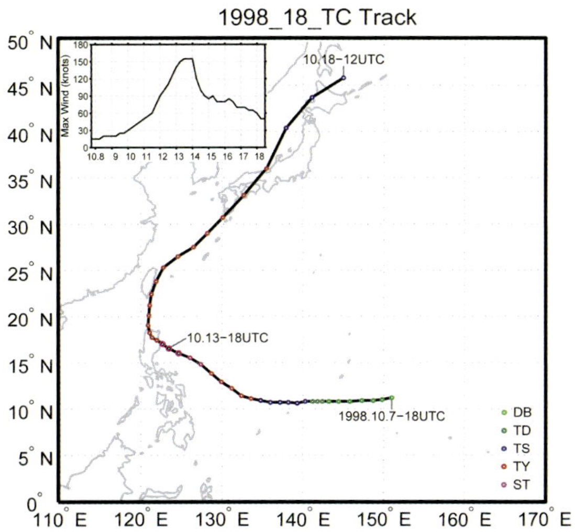

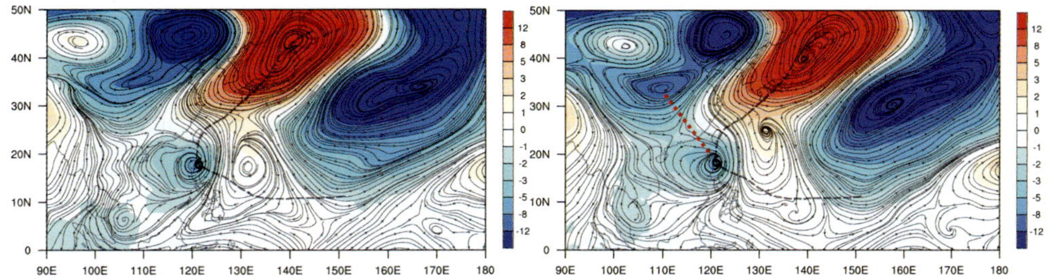

300 hPa flow/height anomalies at 0012 UTC and 1800 UTC 14 October 1998.

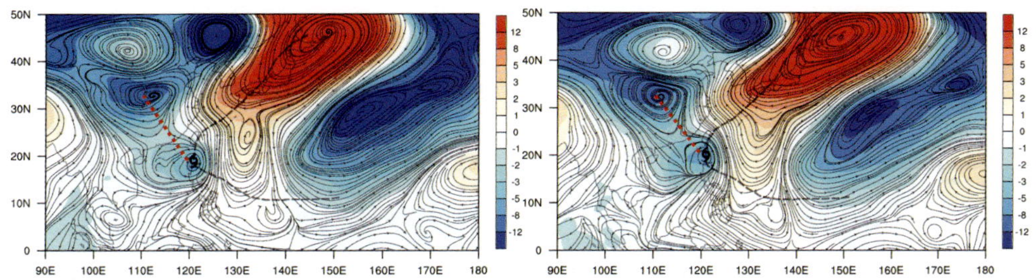

300 hPa flow/height anomalies at 0000 UTC and 0600 UTC 15 October 1998.

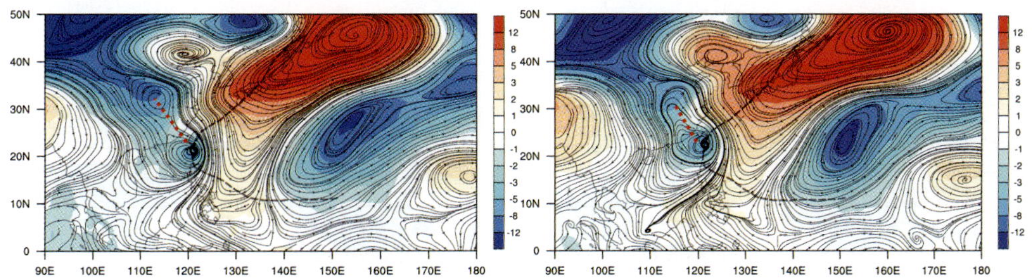

300 hPa flow/height anomalies at 1200 UTC and 1800 UTC 15 October 1998.

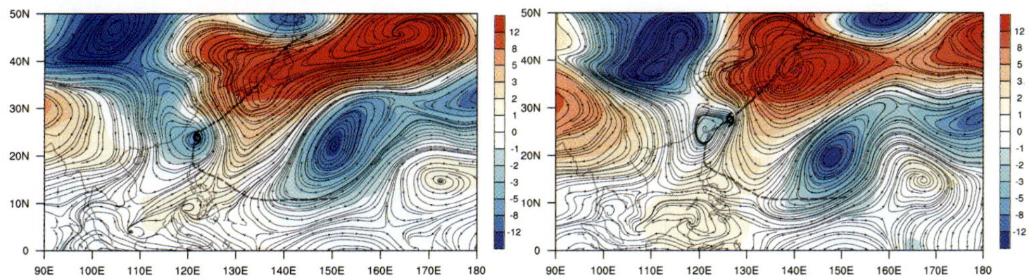

300 hPa flow/height anomalies at 1200 UTC and 1800 UTC 16 October 1998.

Figure 4.38. The typhoon (1998_No.18) generated in the tropical western Pacific on 7 October 1998 and moved west-northwestward along the climatic monsoon trough. Its intensity peaked near the northern Philippines on 13 October 1998 before landfall on the island. The right turn of its motion from the Philippines to the Taiwan Island and Japan was influenced by an anomalous vortex at 300 hPa.

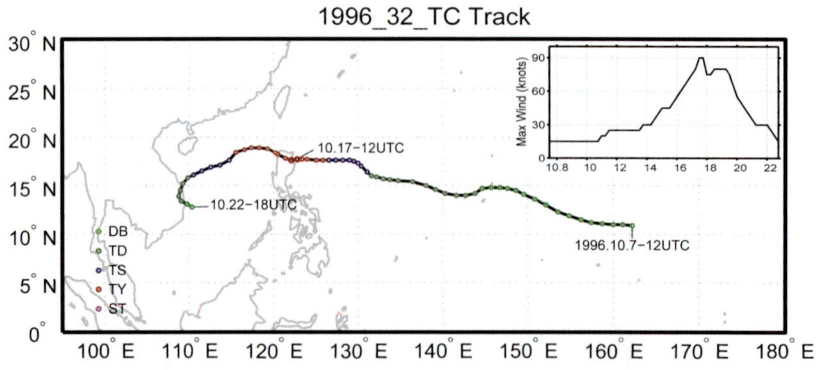

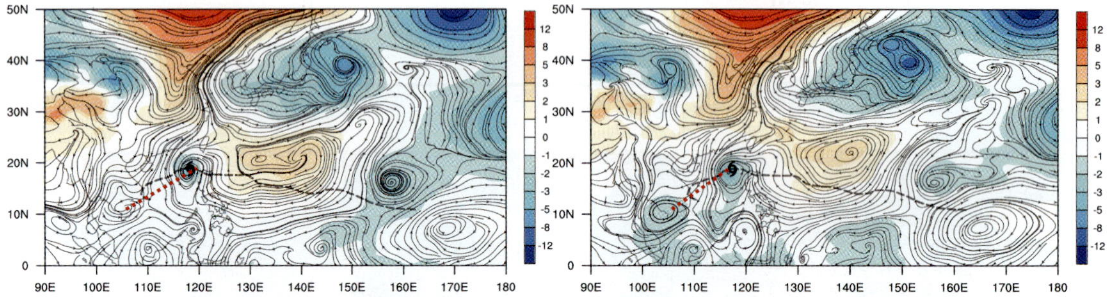

850 hPa flow/height anomalies at 0000 UTC and 0600 UTC 19 October 1996.

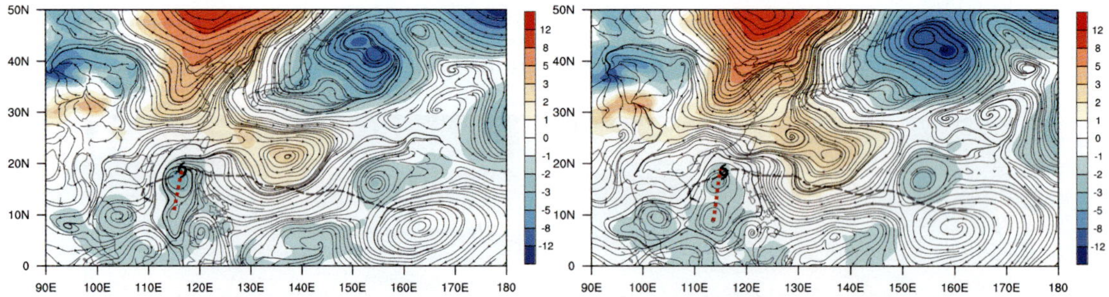

850 hPa flow/height anomalies at 1200 UTC and 1800 UTC 19 October 1996.

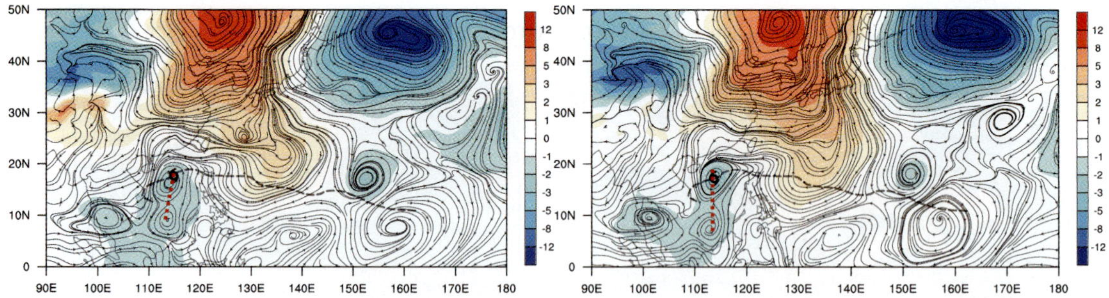

850 hPa flow/height anomalies at 0000 UTC and 0600 UTC 20 October 1996.

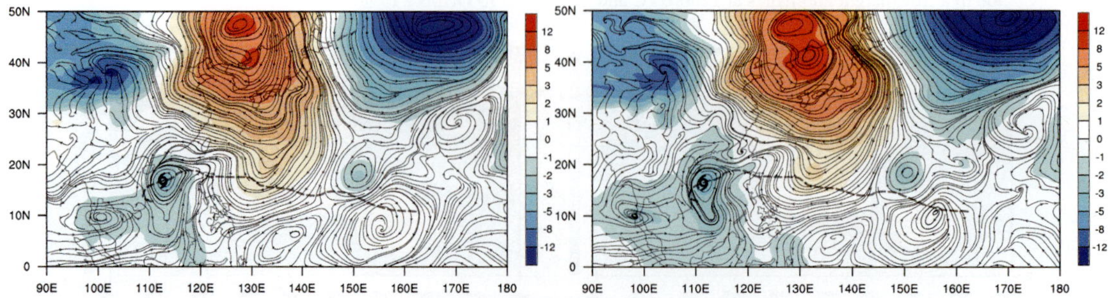

850 hPaflow/height anomalies at 1200 UTC and 1800 UTC 20 October 1996.

Figure 4.39. The typhoon (1996_No.32) generated in the tropical central Pacific on 7 October 1996 and moved west-northwestward along the Pacific tropical climatic monsoon trough. Its intensity peaked on 17 October 1996 before landfall on the northern Philippines. The left turn of its motion in the SCS was influenced by another anomalous vortex at 850 hPa.

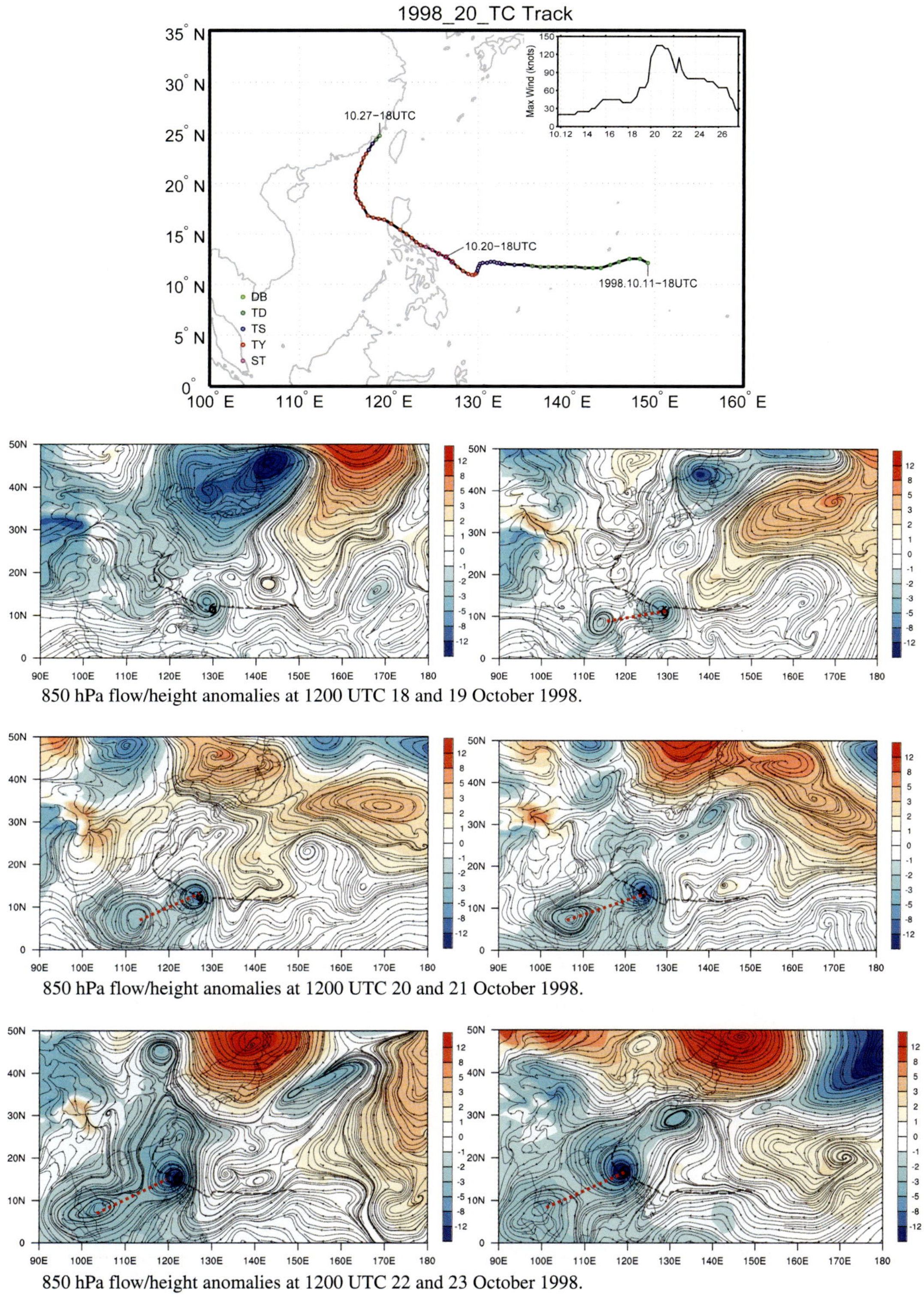

Figure 4.40. The typhoon (1998_No.20) generated in the tropical western Pacific on 11 October 1998 and moved westward along the climatic monsoon trough. Its intensity peaked in the east sea of the Philippines on 20 October 1998 before landfall on the island. The right turn of its motion in the SCS was influenced by anomalous vortices at 850 hPa and 300 hPa.

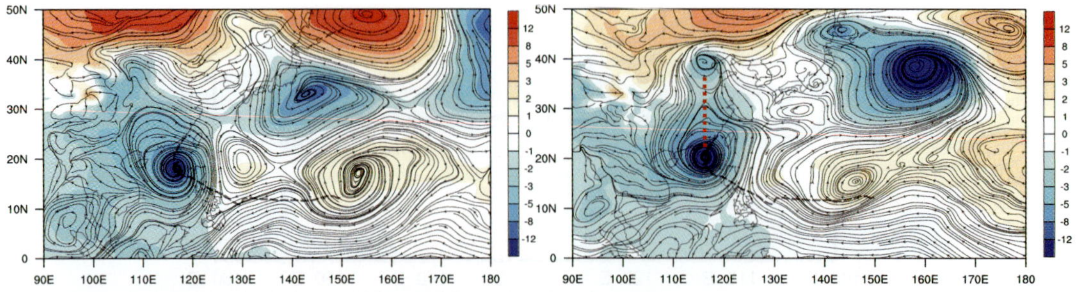

850 hPa flow/height anomalies at 1200 UTC 24 and 25 October 1998.

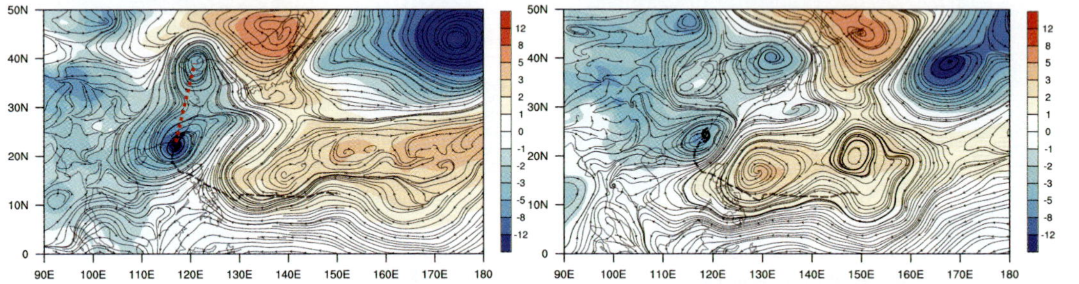

850 hPa flow/height anomalies at 0000 UTC and 1200 UTC 26 October 1998.

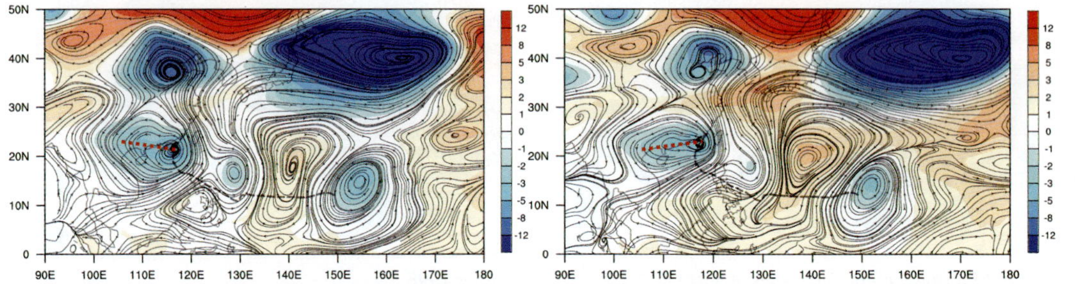

300 hPa flow/height anomalies at 0000 UTC and 1200 UTC 26 October 1998.

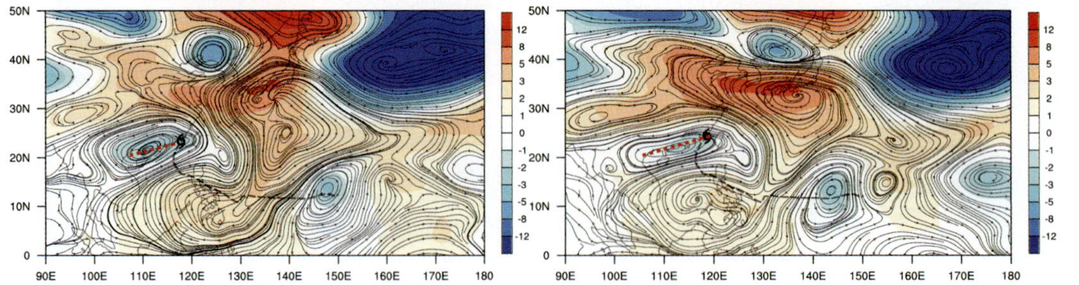

300 hPa flow/height anomalies at 0000 UTC and 1200 UTC 27 October 1998.

Figure 4.40. *(Continued)*

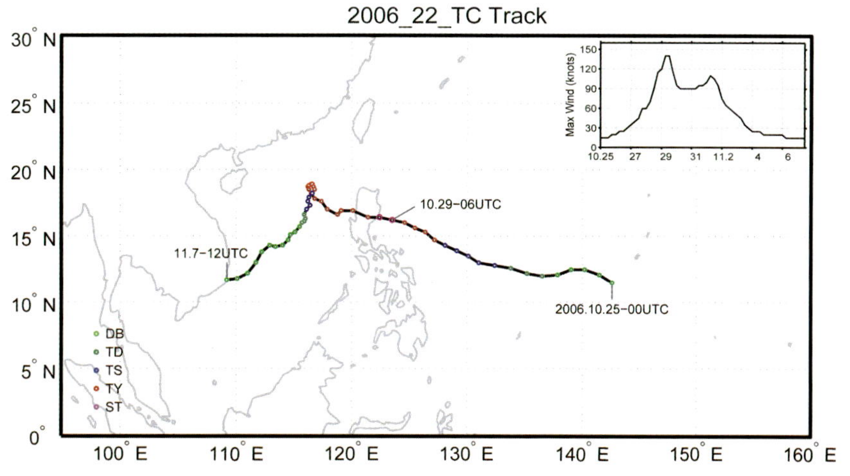

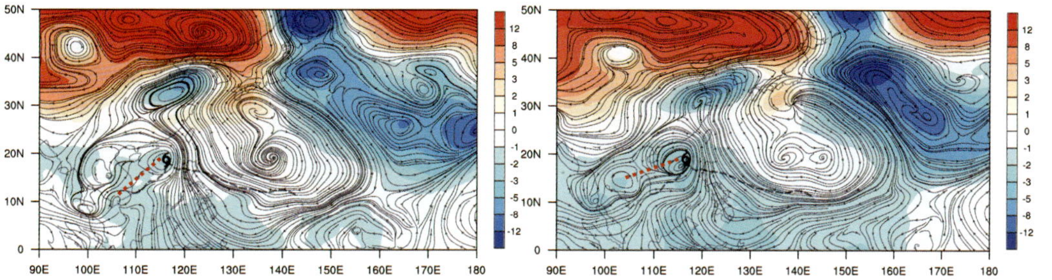

300 hPa flow/height anomalies at 0000 UTC and 1200 UTC 1 November 2006.

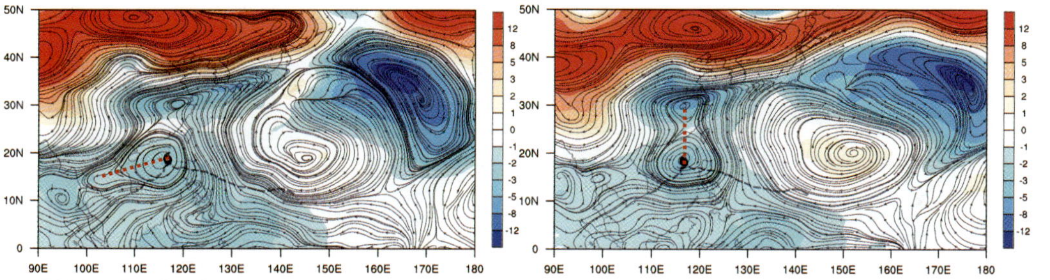

300 hPa flow/height anomalies at 0000 UTC and 1200 UTC 2 November 2006.

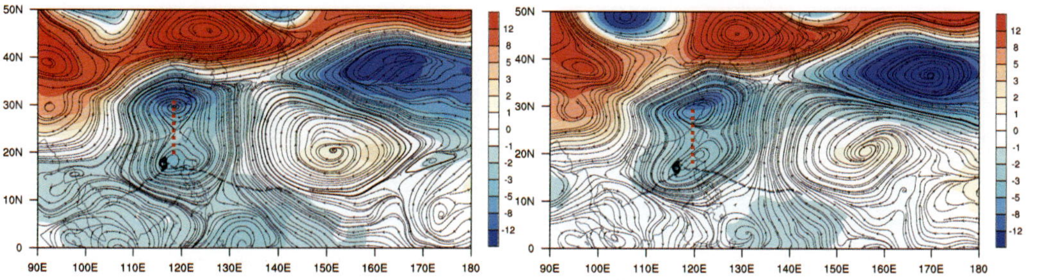

300 hPa flow/height anomalies at 0000 UTC and 1200 UTC 3 November 2006.

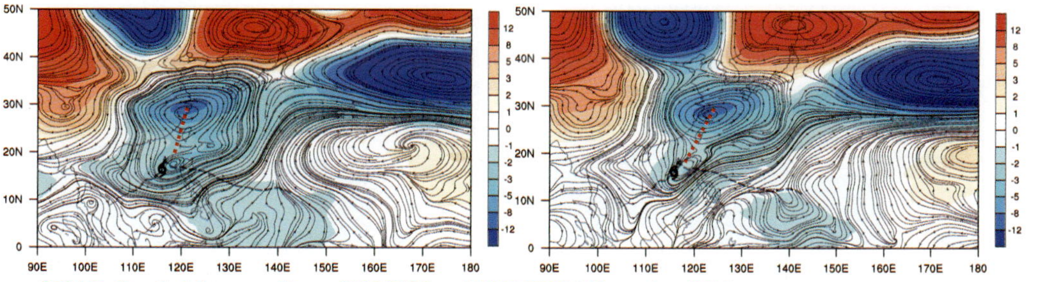

300 hPa flow/height anomalies at 0000 UTC and 1200 UTC 4 November 2006.

Figure 4.41. The typhoon (2006_No.22) generated in the tropical western Pacific on 25 October 2006 and moved west-northwestward along the climatic monsoon trough. Its intensity peaked on 29 October 2006 before landfall on the northern Philippines. The left turn of its motion with a stagnant state in the northern SCS was influenced by anomalous vortices at 300 hPa.

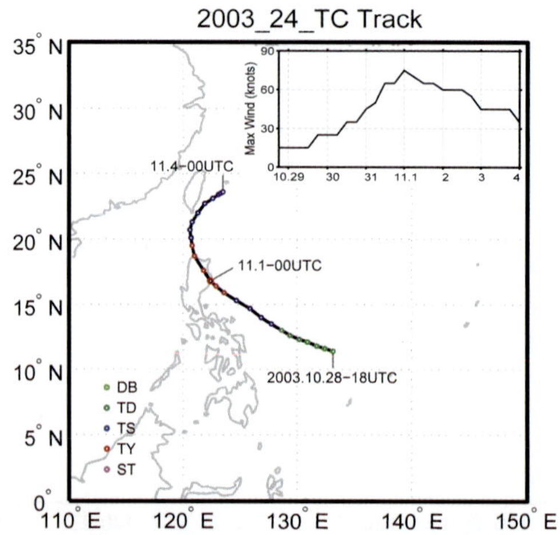

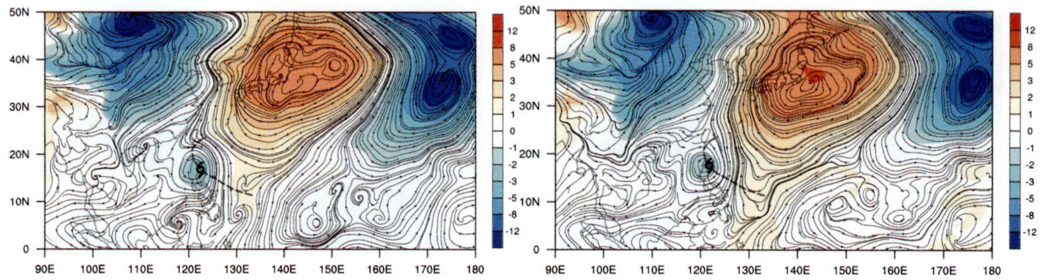

300 hPa flow/height anomalies at 0000 UTC and 0600 UTC 1 November 2003.

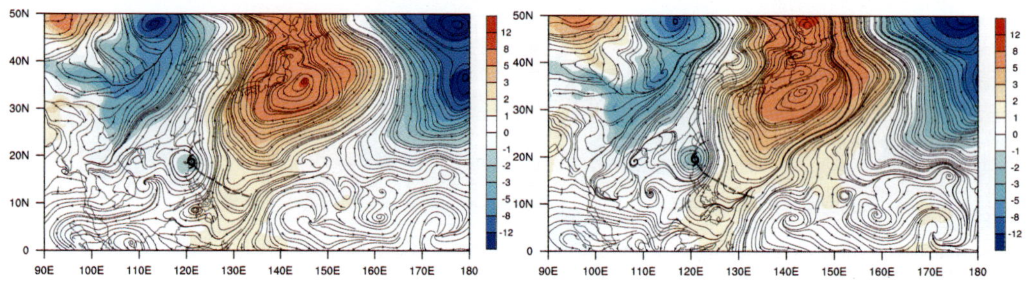

300 hPa flow/height anomalies at 1200 UTC and 1800 UTC 1 November 2003.

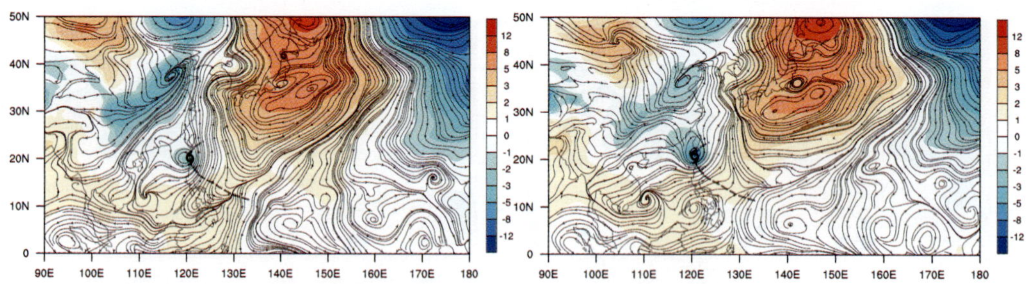

300 hPa flow/height anomalies at 0000 UTC and 0600 UTC 2 November 2003.

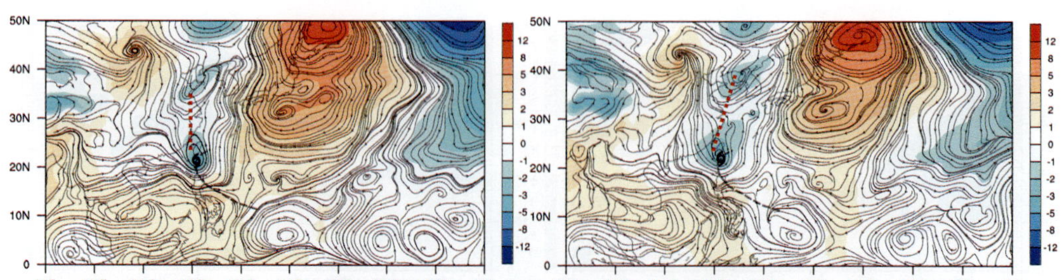

300 hPa flow/height anomalies at 1200 UTC and 1800 UTC 2 November 2003.

Figure 4.42. The typhoon (2003_No.24) generated in the tropical western Pacific on 28 October 2003 and moved northwestward along the climatic monsoon trough. Its intensity peaked on 1 November 2003 before landfall on the northeastern Philippines. The right turn of its motion was influenced by anomalous anti-cyclonic circulation at 300 hPa.

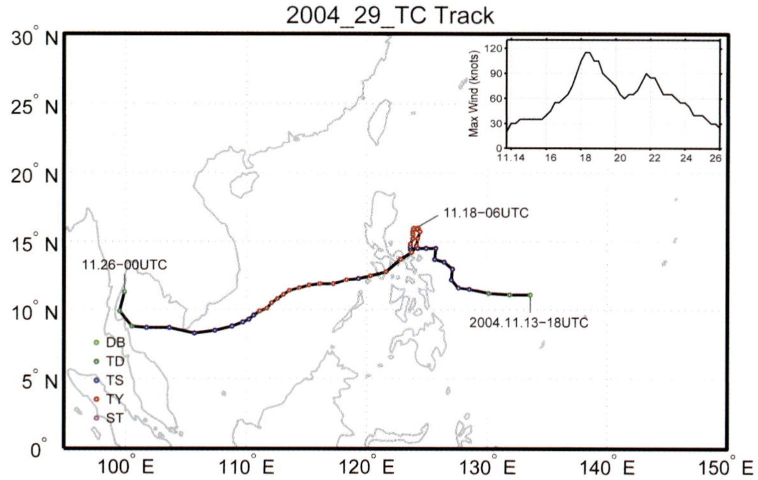

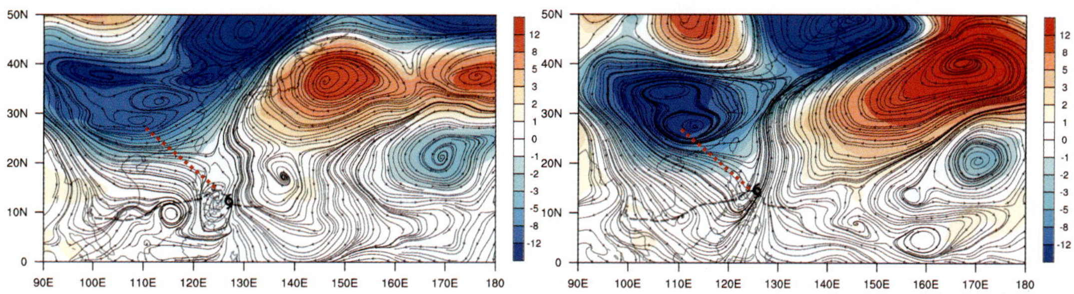

300 hPa flow/height anomalies at 0000 UTC 15 and 16 November 2004.

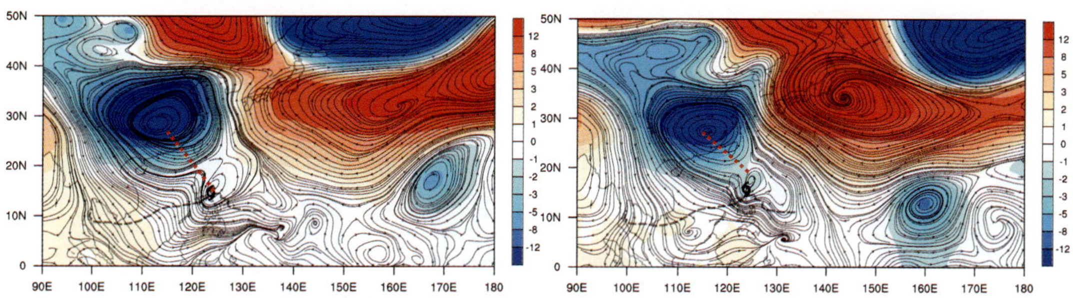

300 hPa flow/height anomalies at 0000 UTC 17 and 18 November 2004.

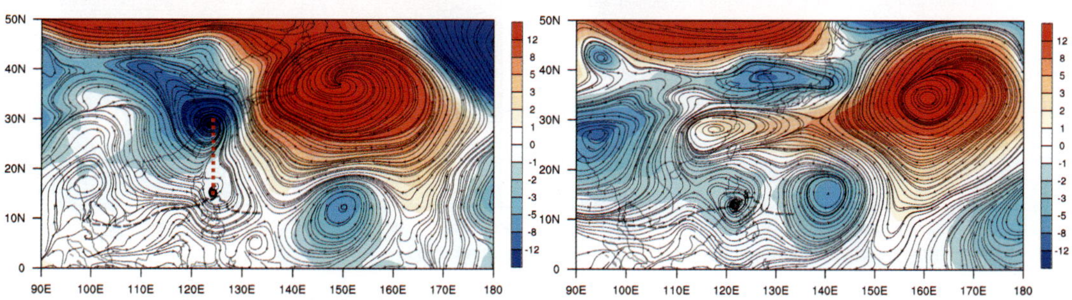

300 hPa flow/height anomalies at 0000 UTC 19 and 20 November 2004.

Figure 4.43. The typhoon (2004_No.29) generated in the tropical western Pacific on 13 November 2004 and moved northwestward along the climatic monsoon trough. Its intensity peaked on 18 November 2004 before landfall on the northern Philippines. Its specialty was a left turn with a stagnant state for about four days near the Philippines. This track was influenced by another anomalous vortex at 300 hPa.

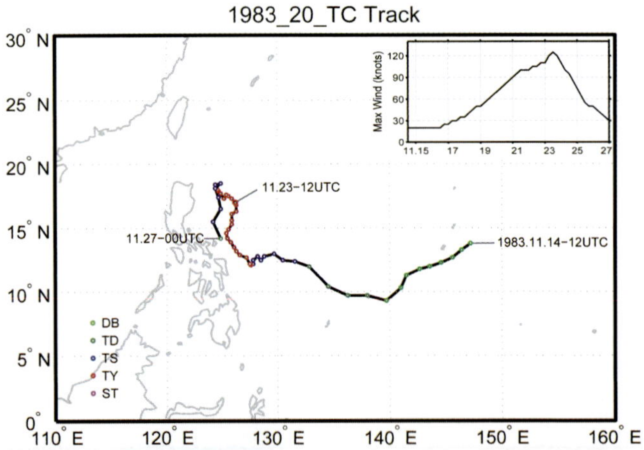

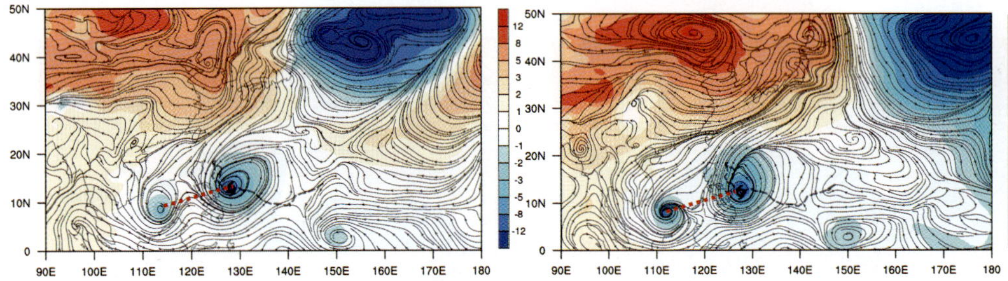

850 hPa flow/height anomalies at 0000 UTC 19 and 20 November 1983.

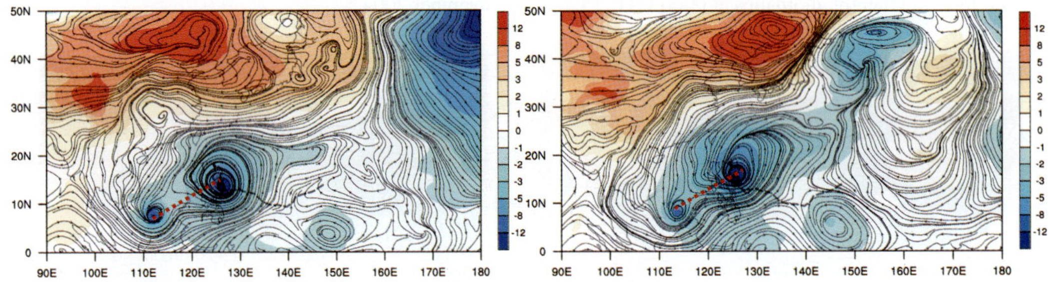

850 hPa flow/height anomalies at 0000 UTC 21and 22 November 1983.

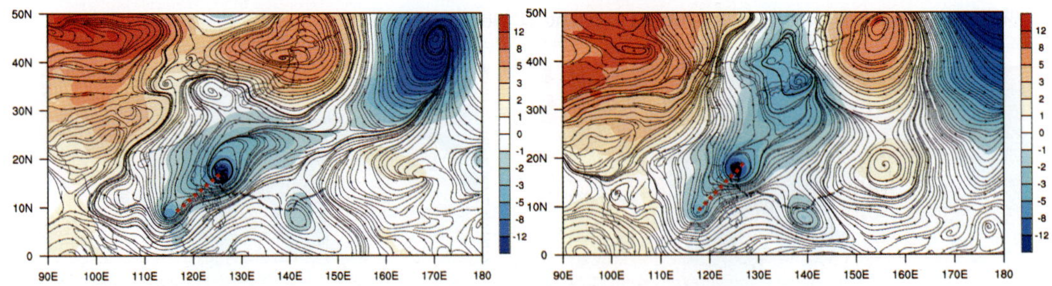

850 hPa flow/height anomalies at 0000 UTC 23 and 24 November 1983.

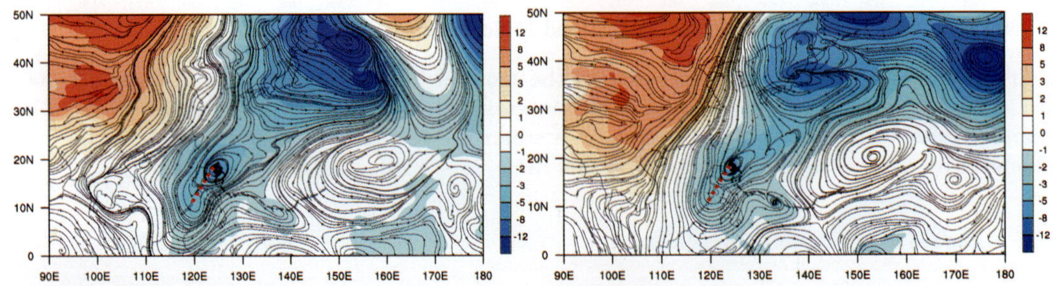

850 hPa flow/height anomalies at 0000 UTC 25 and 26 November 1983.

Figure 4.44. The typhoon (1983_No.20) generated in the tropical western Pacific on 14 November 1983 and moved westward. Its intensity peaked in the east sea of the Philippines on 23 November 1983 when it was arriving in the climatic monsoon trough. The right turn of its motion in the east sea of the Philippines was influenced by another anomalous vortex at 850 hPa.

Chapter 5

Intensity Peaked in the Taiwan Area

Many typhoons reaching their peak intensities near the Taiwan Island generated, respectively, in the South China Sea (SCS) (Sec. 5.1), in the Philippines (Sec. 5.2), in the equatorial western Pacific (Sec. 5.3), in the equatorial central Pacific (Sec. 5.4), in the tropical western Pacific (Sec. 5.5), and in the tropical central Pacific (Sec. 5.6).

5.1. Geneses in the South China Sea

Two typhoons generated in the SCS and their intensities peaked in the Taiwan area in June 2002 and September 1995, with the motions of turning left and turning right, respectively (Figs. 5.1 and 5.2).

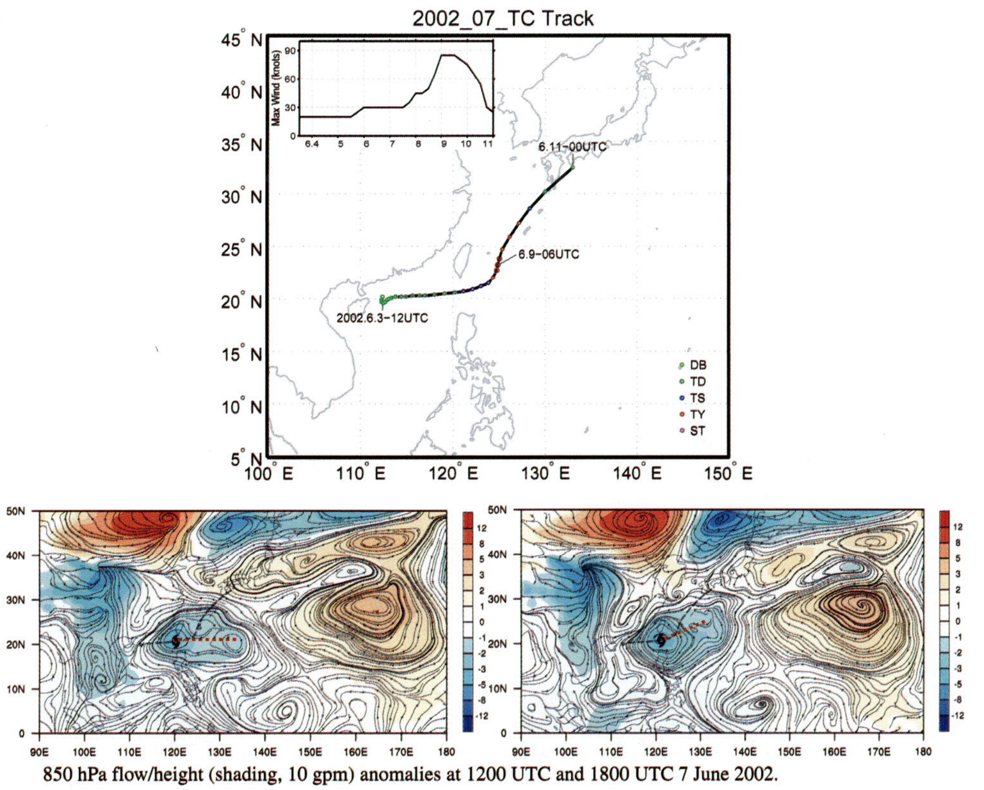

Figure 5.1. The typhoon (2002_No.07) generated in the northern SCS on 3 June 2002 and moved eastward. Its intensity peaked in the east sea of the Taiwan Island on 9 June 2002 when it was approaching the subtropical climatic monsoon trough. The left turn of its motion was influenced by anomalous vortices at 850 hPa. Symbols used are same as in Fig. 2.1.

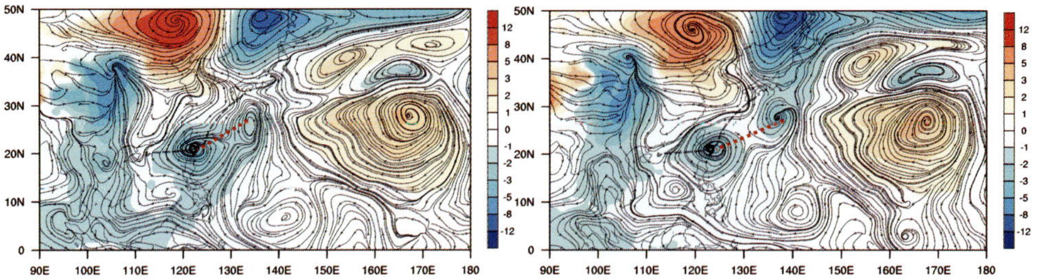

850 hPa flow/height anomalies at 0000 UTC and 0600 UTC 8 June 2002.

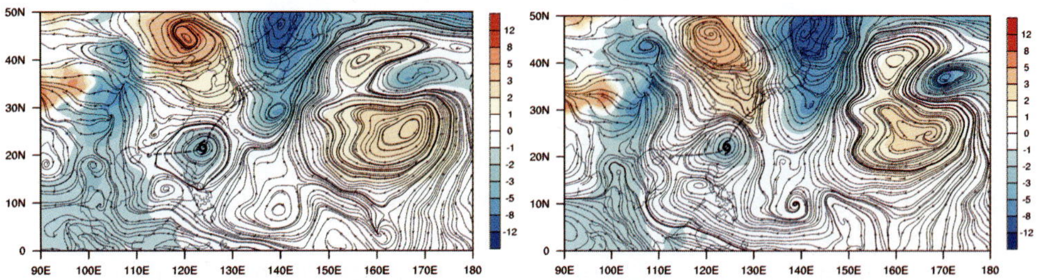

850 hPa flow/height anomalies at 1200 UTC and 1800 UTC 8 June 2002.

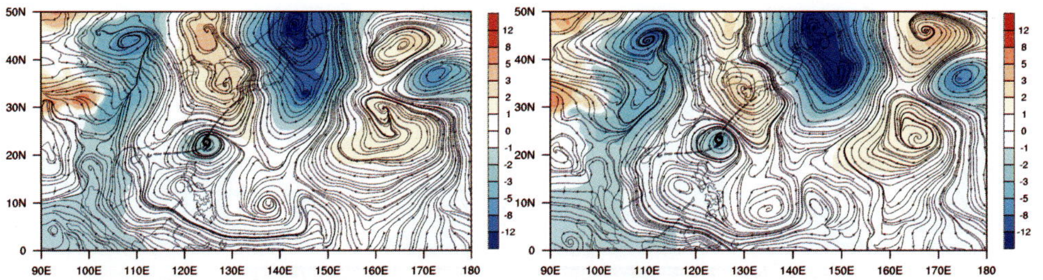

850 hPa flow/height anomalies at 0000 UTC and 0600 UTC 9 June 2002.

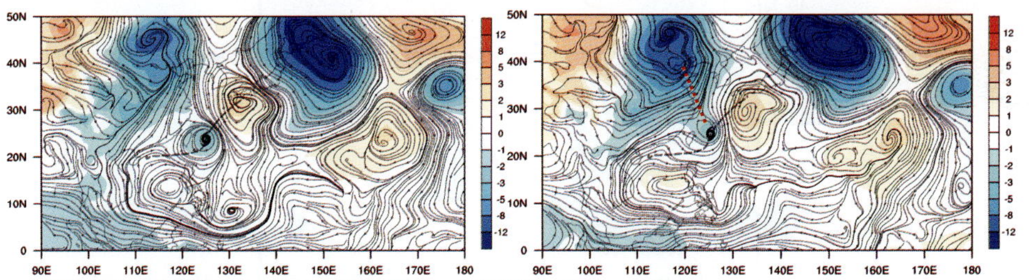

850 hPa flow/height anomalies at 1200 UTC and 1800 UTC 9 May 2002.

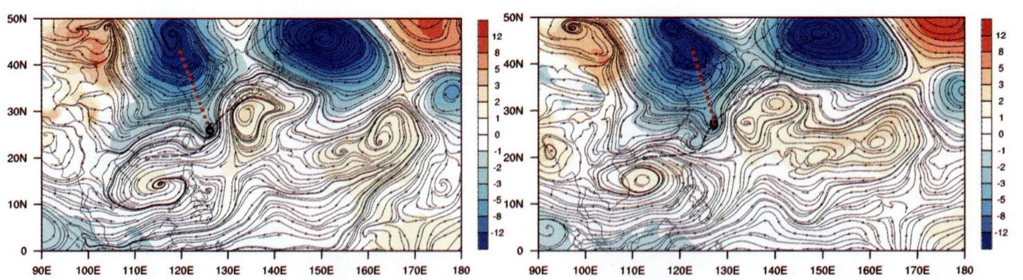

850 hPa flow/height anomalies at 0000 UTC and 0600 UTC 10 June 2002.

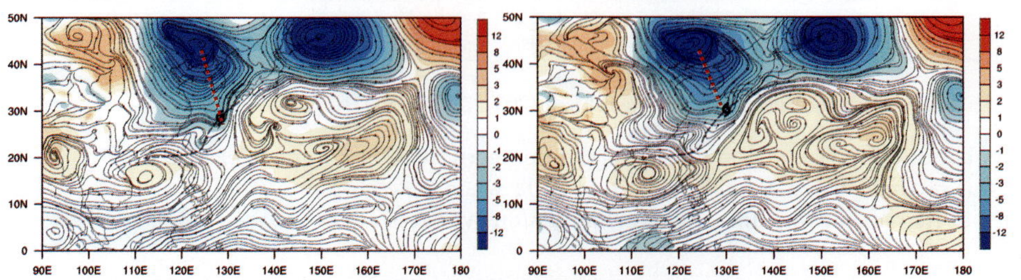

850 hPa flow/height anomalies at 1200 UTC and 1800 UTC 10 May 2002.

Figure 5.1. *(Continued)*

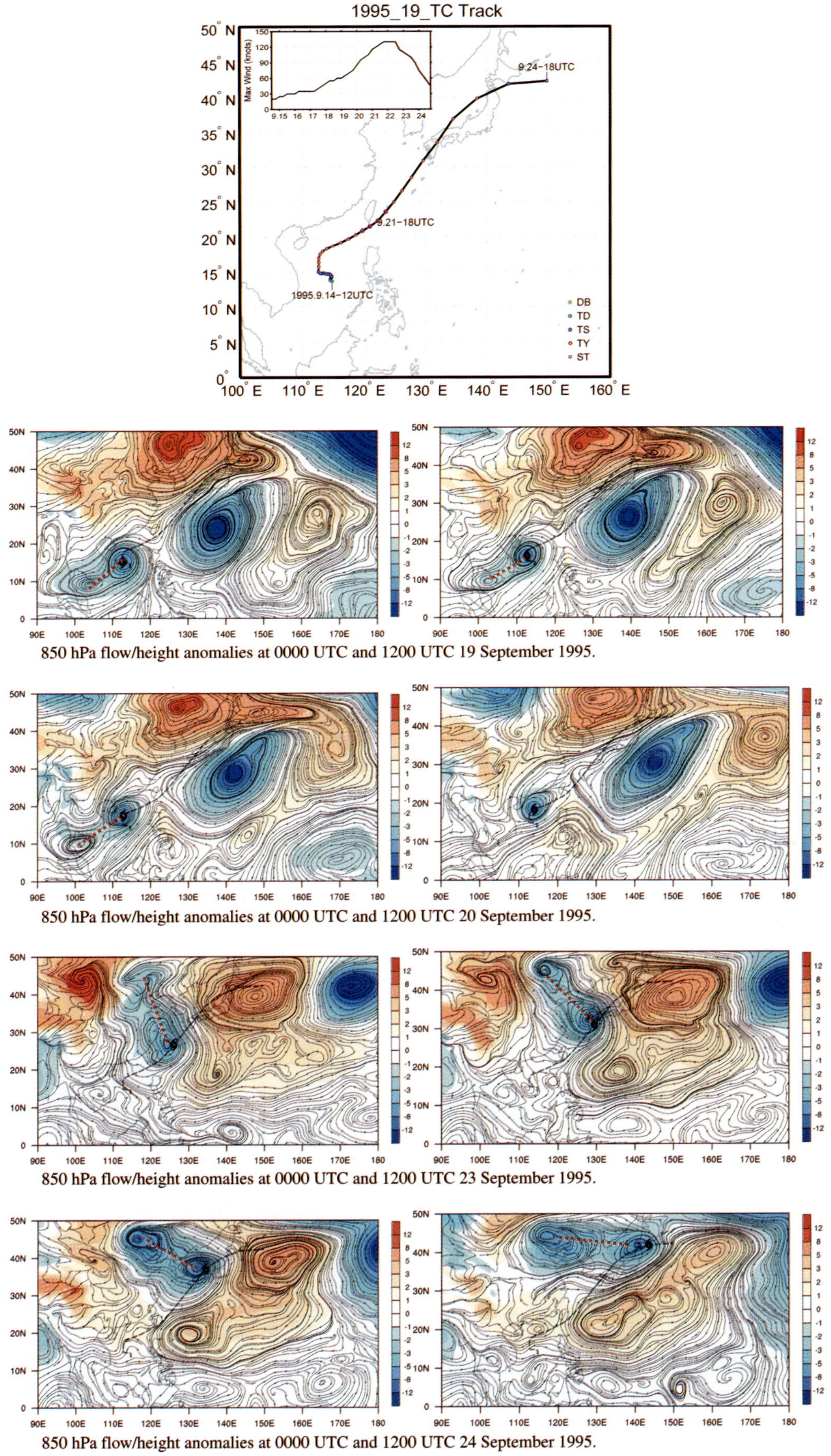

Figure 5.2. The typhoon (1995_No.19) generated in the central SCS on 14 September 1995. Its intensity peaked in the east sea of the Taiwan Island on 21 September 1995 when it was approaching the cross position of two climatic monsoon troughs. The right turn of its motion in the SCS was influenced by anomalous vortices at 850 hPa.

5.2. Geneses in the Philippines

Four typhoons generated in the Philippines and their intensities peaked in the Taiwan area. The right turns or left turns of their motions were influenced by anomalous vortices (Figs. 5.3–5.7).

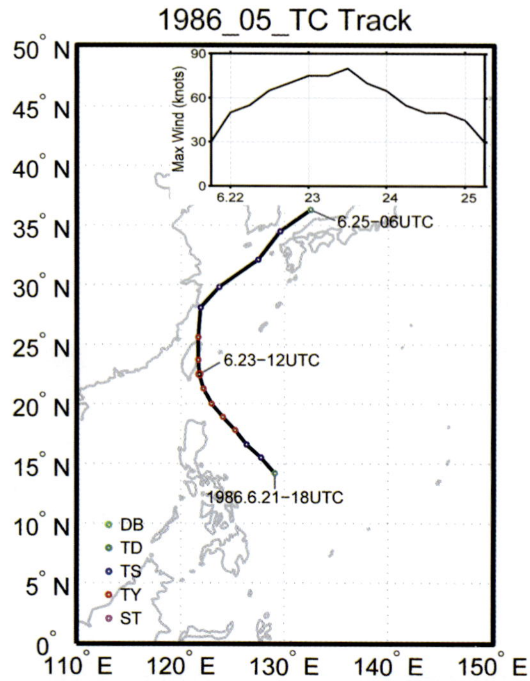

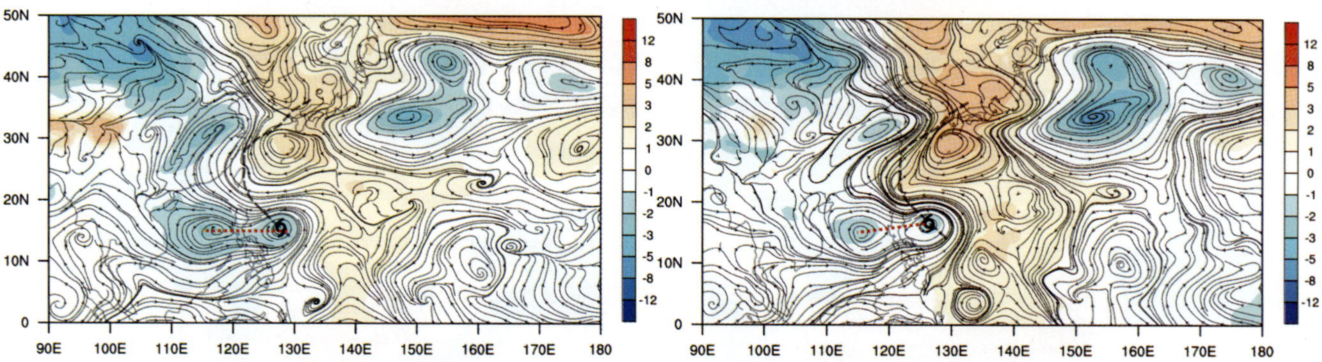

850 hPa flow/height (shading, 10 gpm) anomalies at 0000 UTC and 0600 UTC 22 June 1986.

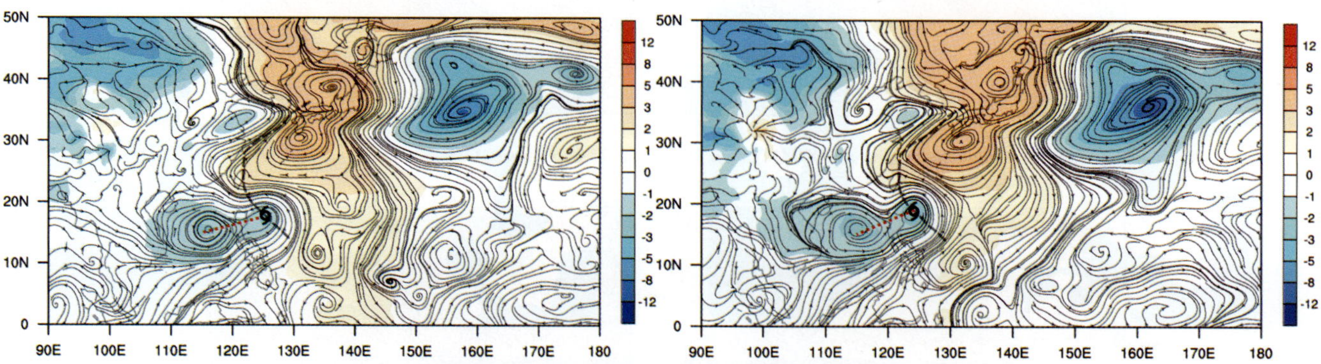

850 hPa flow/height anomalies at 1200 UTC and 1800 UTC 22 June 1986.

Figure 5.3. The typhoon (1986_No.05) generated in the east sea of Philippines on 21 June 1986 and moved northwestward along the climatic monsoon trough. Its intensity peaked near the Taiwan Island on 23 June 1986 before landfall on the island. The right turn of its motion in the east seas of China was influenced by another anomalous vortex at 850 hPa.

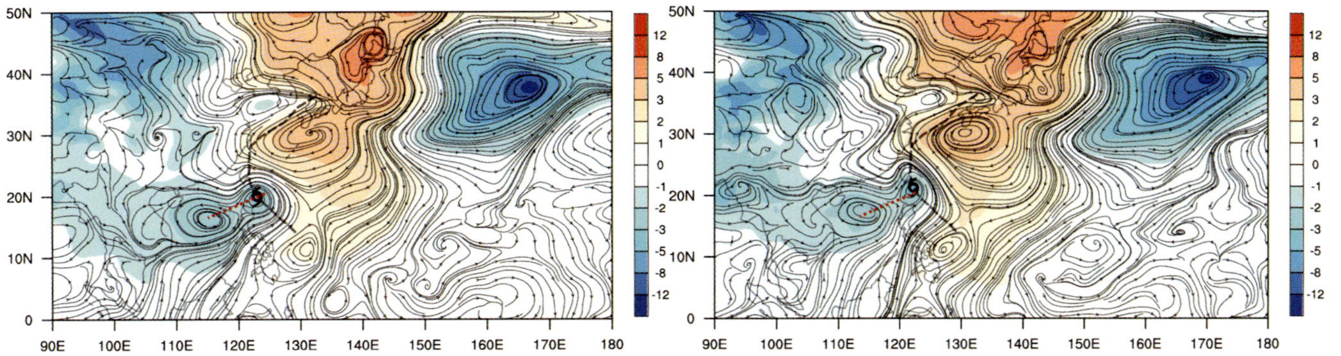

850 hPa flow/height anomalies at 0000 UTC and 0600 UTC 23 June 1986.

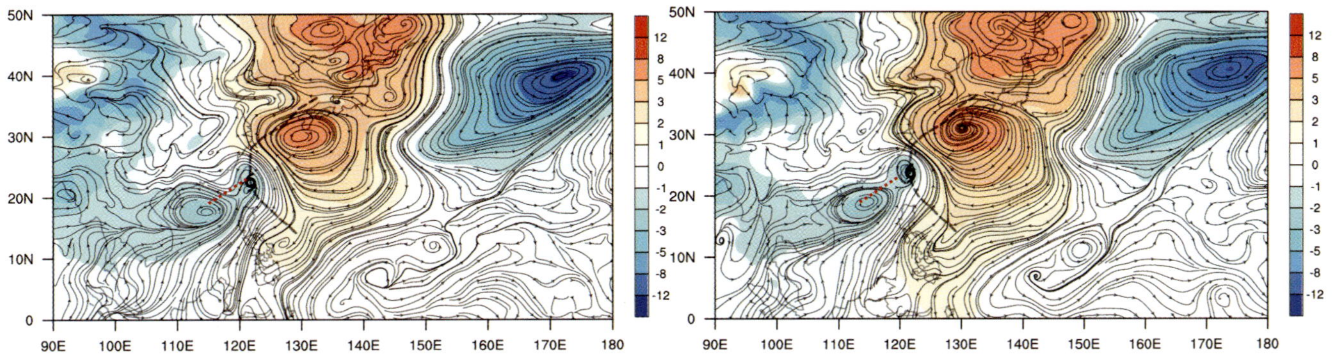

850 hPa flow/height anomalies at 1200 UTC and 1800 UTC 23 June 1986.

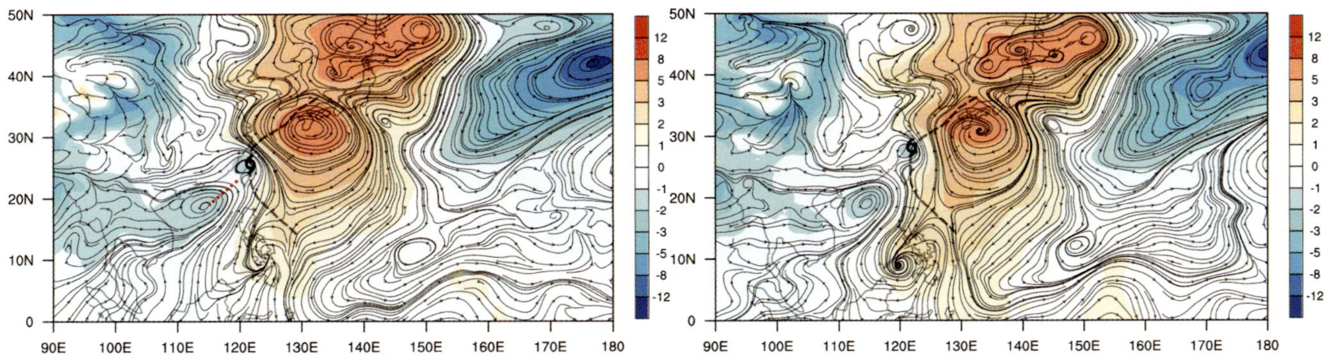

850 hPa flow/height anomalies at 0000 UTC and 0600 UTC 24 June 1986.

Figure 5.3. *(Continued)*

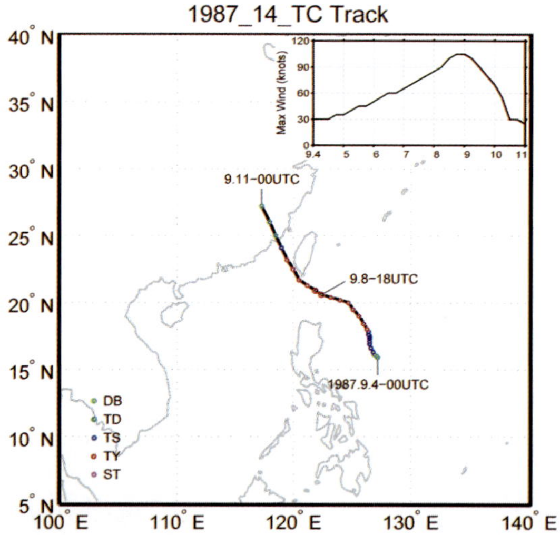

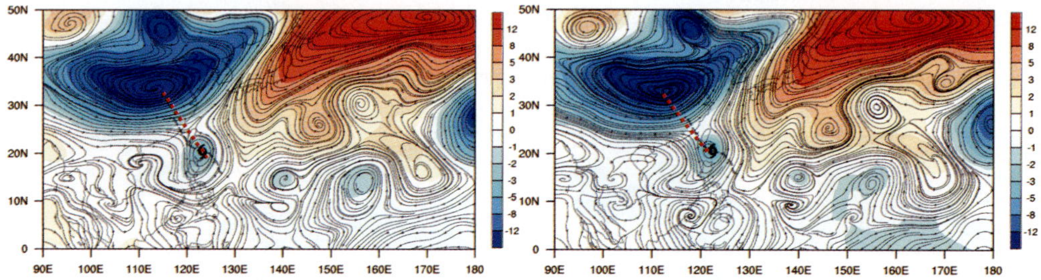

300 hPa flow/height anomalies at 1200 UTC and 1800 UTC 8 September 1987.

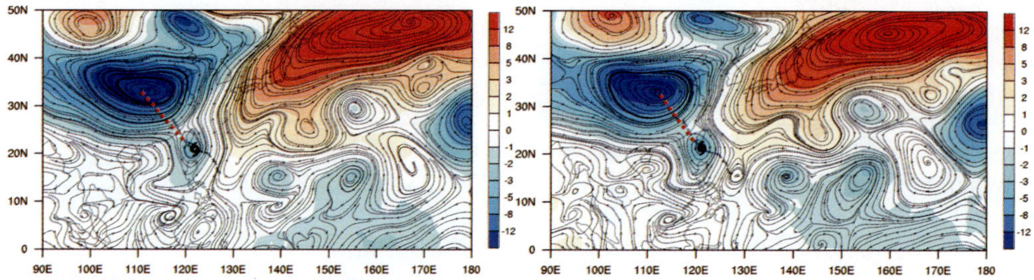

300 hPa flow/height anomalies at 0000 UTC and 0600 UTC 9 September 1987.

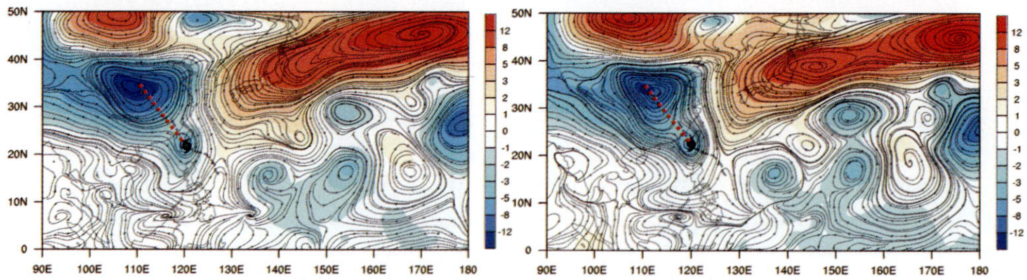

300 hPa flow/height anomalies at 1200 UTC and 1800 UTC 9 September 1987.

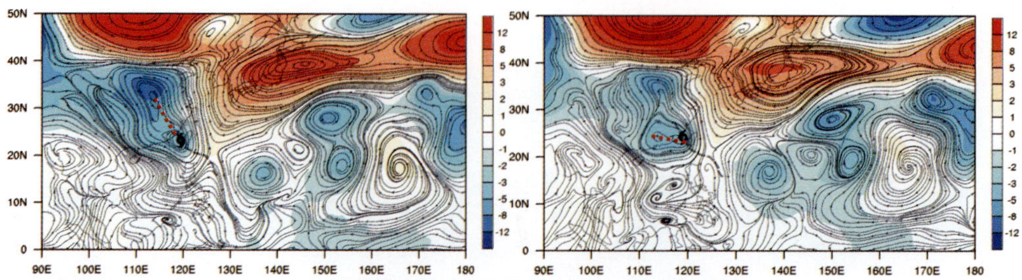

300 hPa flow/height anomalies at 0000 UTC and 0600 UTC 10 September 1987.

Figure 5.4. The typhoon (1987_No.14) generated in the east sea of the Philippines on 4 September 1987 and moved northwestward. Its intensity peaked in the southeast sea of the Taiwan Island on 8 September 1987 when it was approaching the climatic monsoon trough before landfall on the Southeast China coast. Its turning near the Taiwan Island was influenced by another anomalous vortex at 300 hPa.

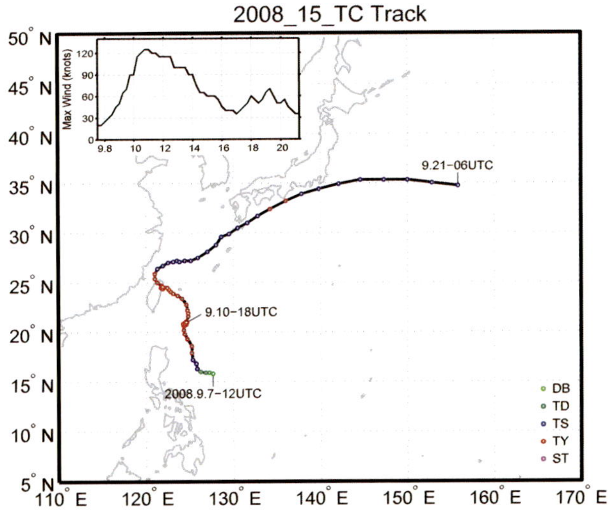

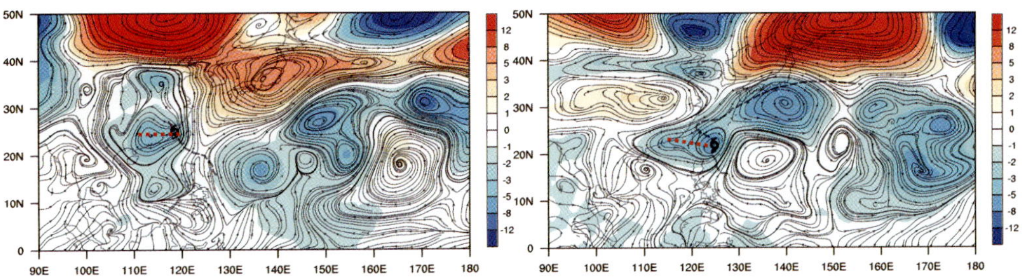

300 hPa flow/height anomalies at 1200 UTC 10 and 11 September 2008.

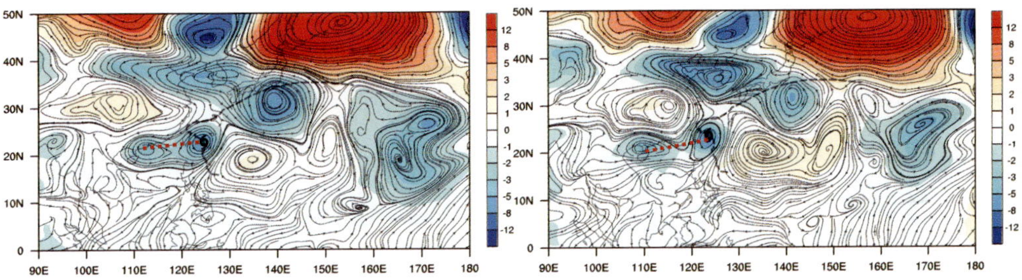

300 hPa flow/height anomalies at 0000 UTC and 1200 UTC 12 September 2008.

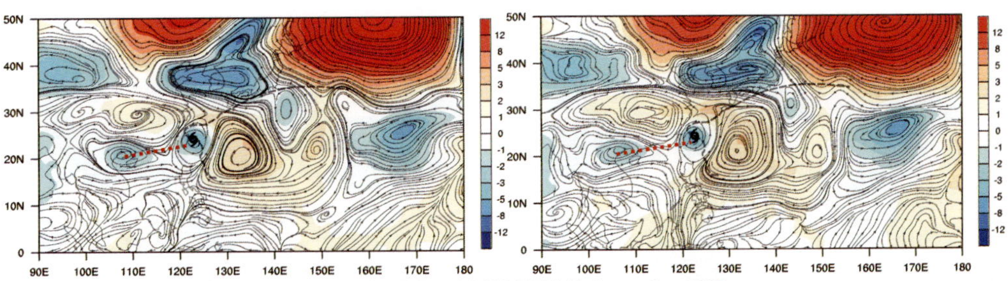

300 hPa flow/height anomalies at 0000 UTC and 0600 UTC 13 September 2008.

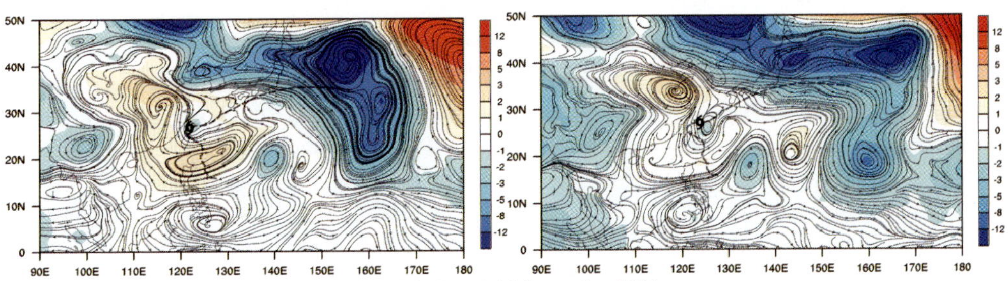

300 hPa flow/height anomalies at 0600 UTC 15 and 16 September 2008.

Figure 5.5. The typhoon (2008_No.15) generated in the east sea of the Philippines on 7 September 2008 and moved northward. Its intensity peaked in the southeast sea of the Taiwan Island on 10 September 2008 when it was approaching the climatic monsoon trough. The left turn of its motion in the southeast sea of the Taiwan Island was influenced by an anomalous vortex and the right turn in the East China Sea was influenced by anomalous anti-cyclonic circulations at 300 hPa.

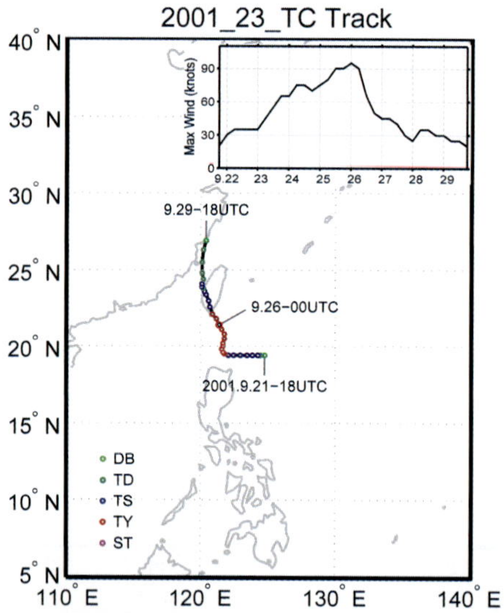

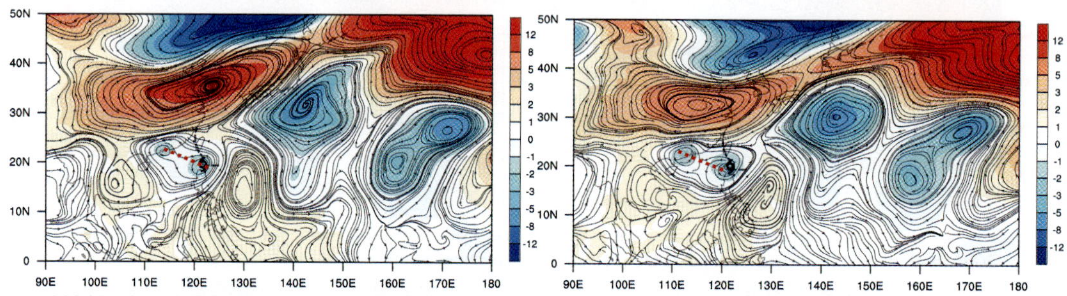

300 hPa flow/height anomalies at 0000 UTC and 1200 UTC 24 September 2001.

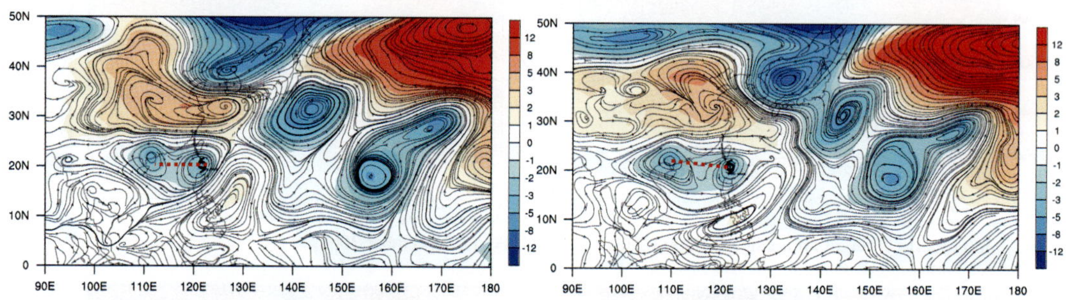

300 hPa flow/height anomalies at 0000 UTC and 1200 UTC 25 September 2001.

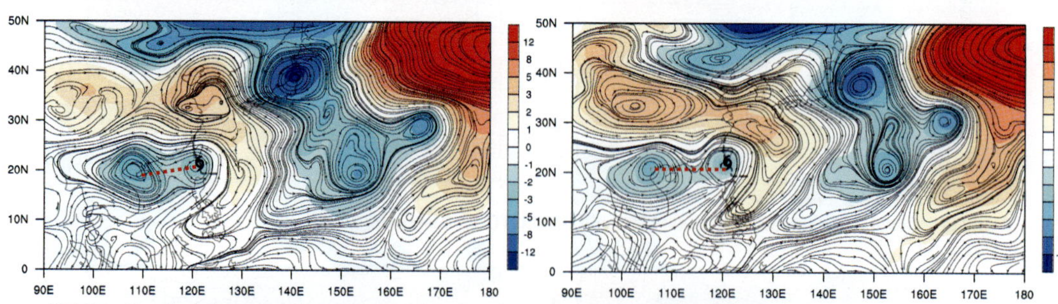

300 hPa flow/height anomalies at 0000 UTC and 1200 UTC 26 September 2001.

Figure 5.6. The typhoon (2001_No.23) generated in the northeast sea of the Philippines on 21 September 2001 and moved westward. Its intensity peaked on 26 September 2001 when it was approaching the climatic monsoon trough before landfall on the Taiwan Island. The right turn of its motion in the Balintang Channel was influenced by an anomalous vortex at 300 hPa.

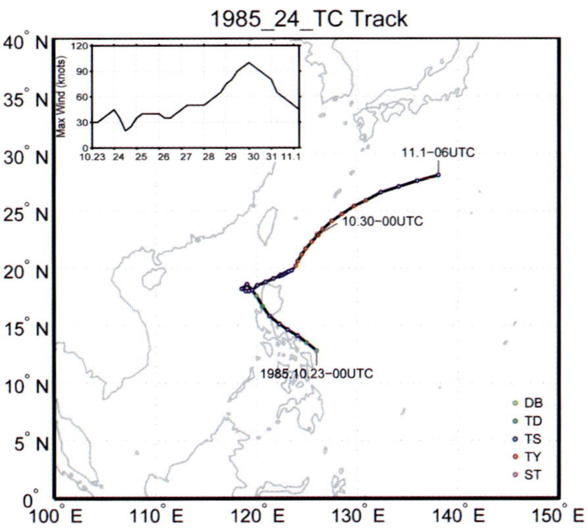

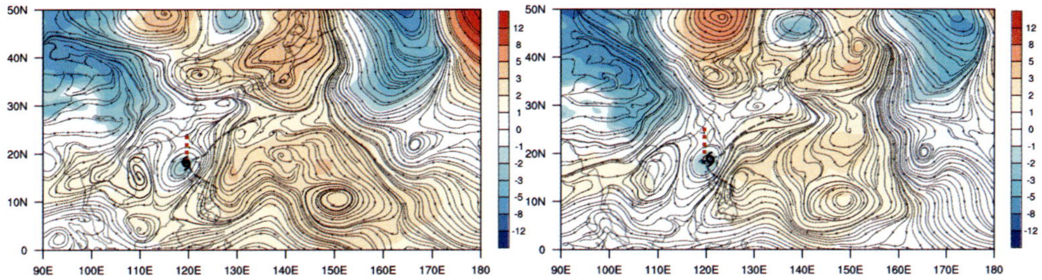

850 hPa flow/height anomalies at 0000 UTC and 1200 UTC 26 October 1985.

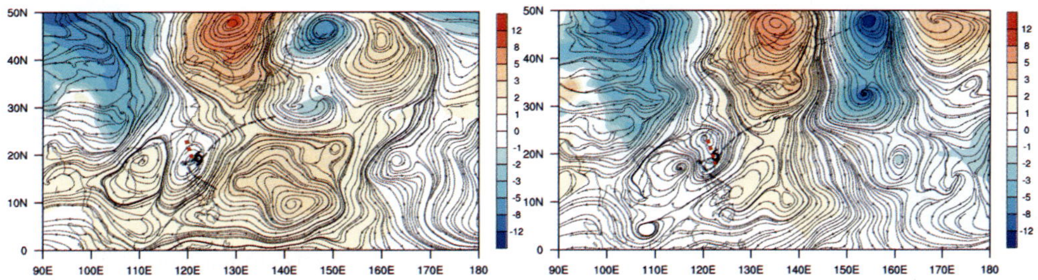

850 hPa flow/height anomalies at 0000 UTC and 1200 UTC 27 October 1985.

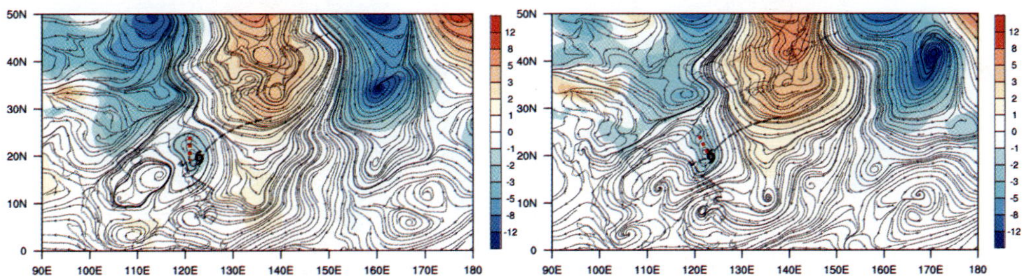

850 hPa flow/height anomalies at 0000 UTC and 1200 UTC 28 October 1985.

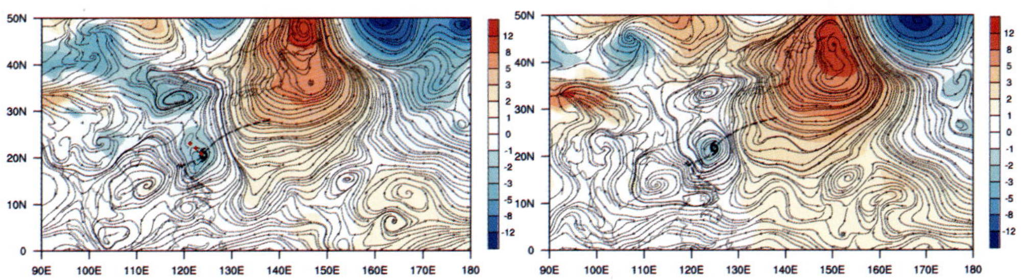

850 hPa flow/height anomalies at 0000 UTC and 1200 UTC 29 October 1985.

Figure 5.7. The typhoon (1985_No.24) generated in the east sea of the Philippines on 23 October 1985. Its intensity peaked in the east sea of the Taiwan Island on 30 October 1985 when it was approaching the climatic monsoon trough. The right turn of its motion in the northern Philippines was influenced by an anomalous vortex at 850 hPa.

5.3. Geneses in the Equatorial Western Pacific

Four typhoons generated in the equatorial western Pacific and their intensities peaked in the Taiwan area. The right turns or left turns of their motions were influenced by anomalous vortices (Figs. 5.8–5.11).

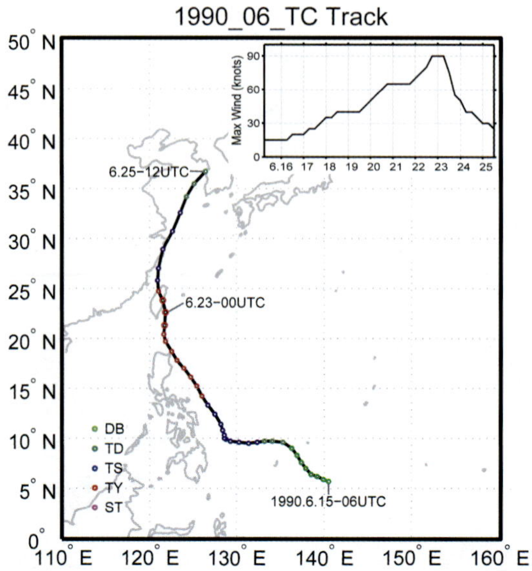

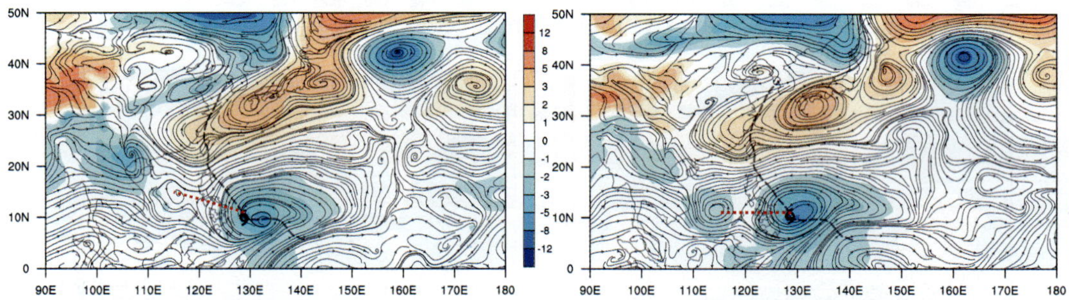

850 hPa flow/height (shading, 10 gpm) anomalies at 0000 UTC and 1200 UTC 19 June 1990.

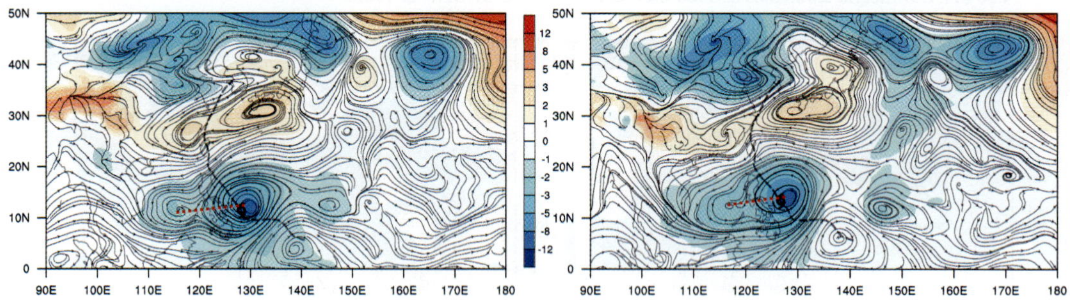

850 hPa flow/height anomalies at 0000 UTC and 1200 UTC 20 June 1990.

Figure 5.8. The typhoon (1990_No.06) generated in the equatorial western Pacific on 15 June 1990 and moved northwestward. Its intensity peaked near the Taiwan Island on 23 June 1990 when it was approaching the climatic monsoon trough before landfall on the island. The right turn of its motion was influenced by anomalous vortices at 850 hPa and 300 hPa.

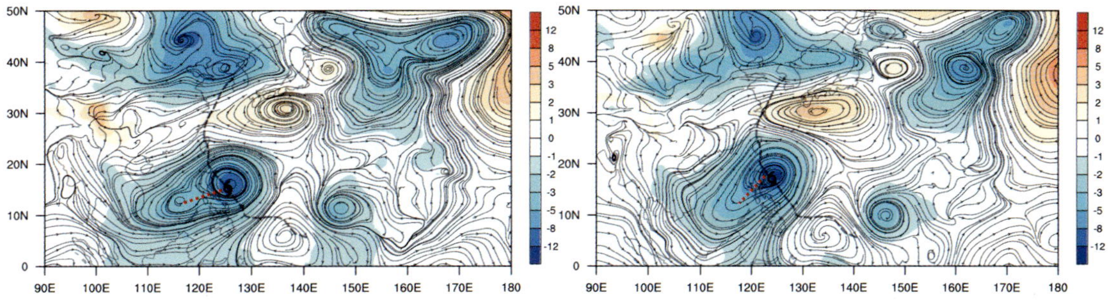

850 hPa flow/height anomalies at 0000 UTC and 1200 UTC 21 June 1990.

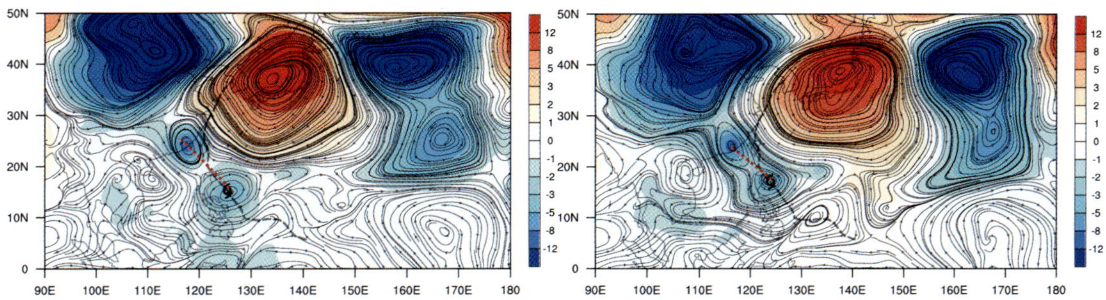

300 hPa flow/height anomalies at 0000 UTC and 1200 UTC 21 June 1990.

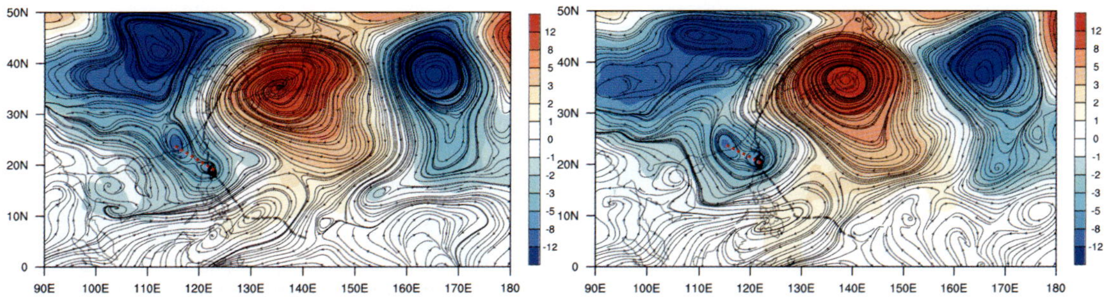

300 hPa flow/height anomalies at 0000 UTC and 1200 UTC 22 June 1990.

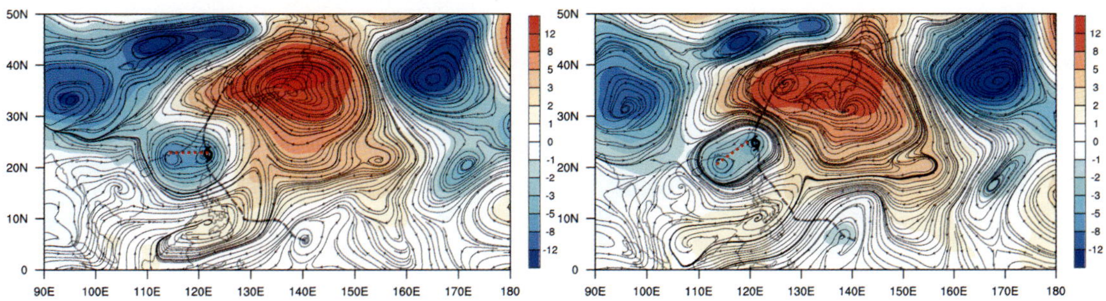

300 hPa flow/height anomalies at 0000 UTC and 1200 UTC 23 June 1990.

Figure 5.8. *(Continued)*

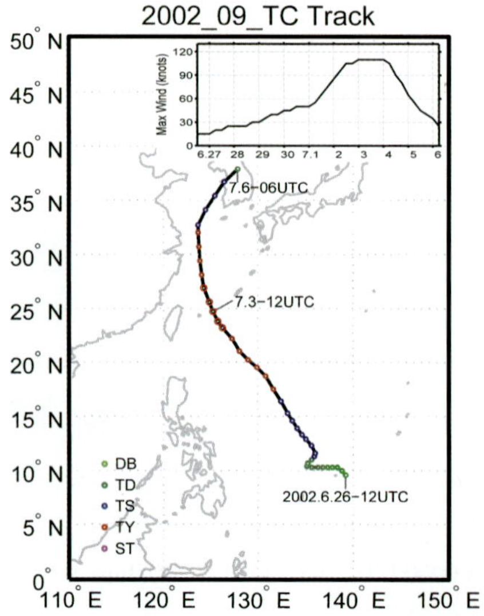

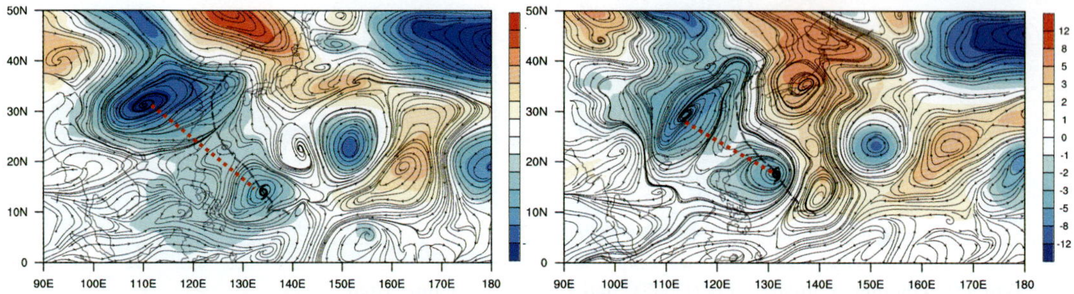

300 hPa flow/height anomalies at 1200 UTC 30 June and 1 July 2002.

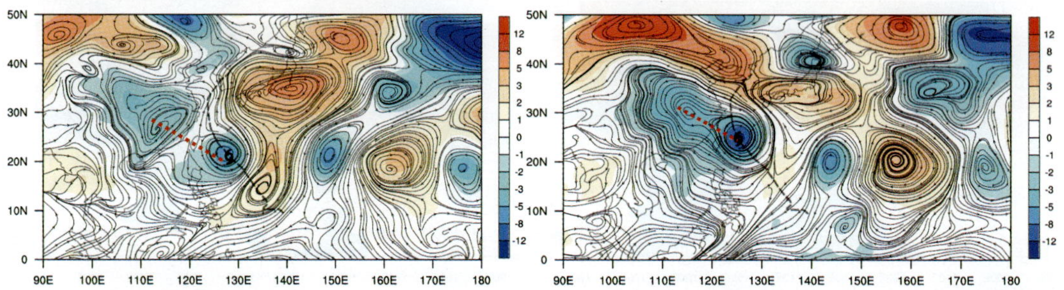

300 hPa flow/height anomalies at 1200 UTC 2 and 3 July 2002.

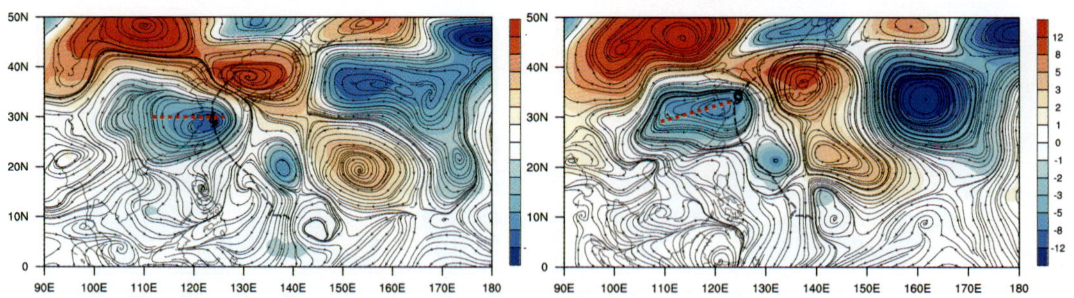

300 hPa flow/height anomalies at 1200 UTC 4 and 5 July 2002.

Figure 5.9. The typhoon (2002_No.09) generated in the equatorial western Pacific on 26 June 2002 and moved northwestward. Its maximum wind speed reached above 90 knots in the east sea of the Taiwan Island on 2 and 3 July 2002 when it was approaching the climatic monsoon trough. The right turn of its motion in the east seas of China was influenced by anomalous vortices at 300 hPa.

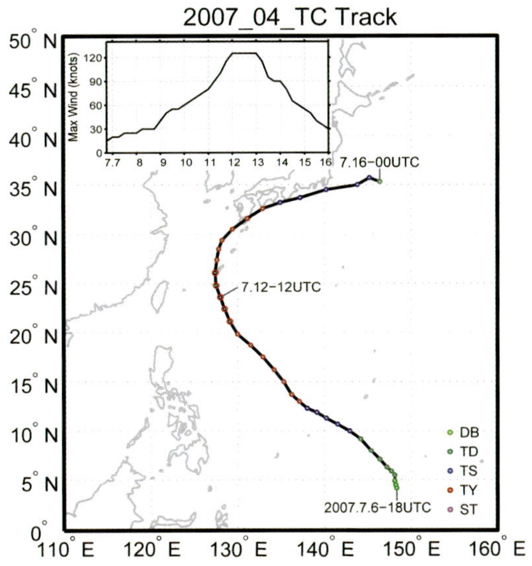

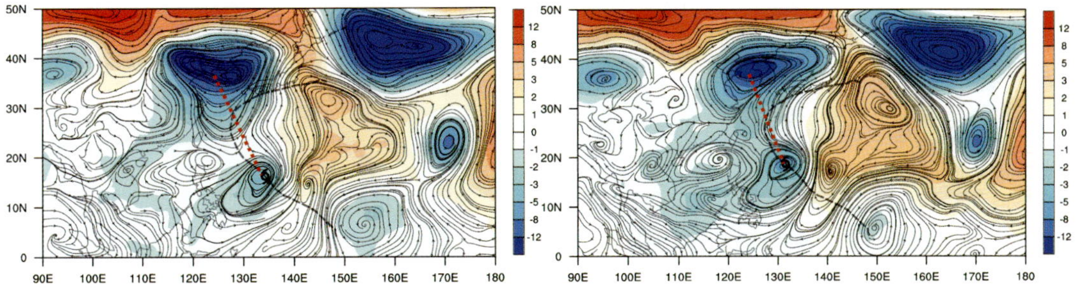

300 hPa flow/height anomalies at 0000 UTC and 1200 UTC 11 July 2007.

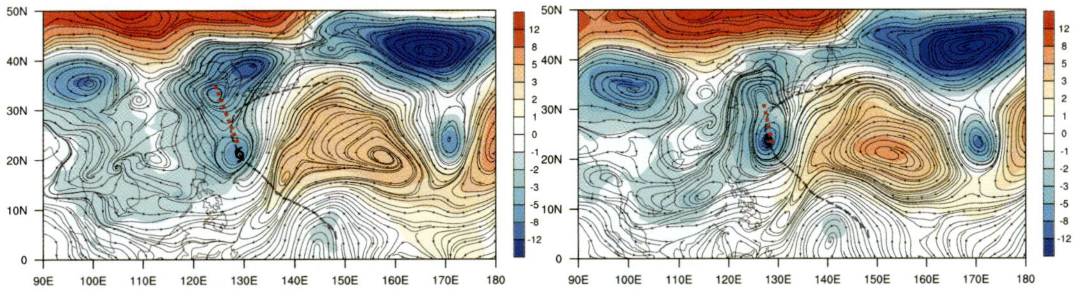

300 hPa flow/height anomalies at 0000 UTC and 1200 UTC 12 July 2007.

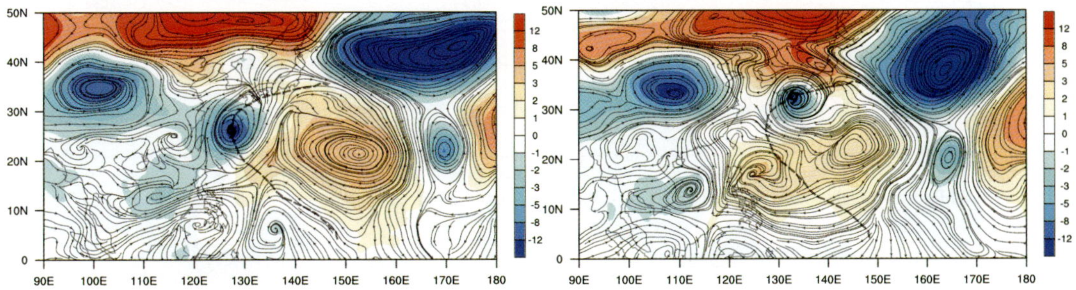

300 hPa flow/height anomalies at 1200 UTC 13 and 14 July 2007.

Figure 5.10. The typhoon (2007_No.04) generated in the equatorial western Pacific on 6 July 2007 and moved northwestward along the climatic monsoon trough. Its intensity peaked in the east sea of the Taiwan Island on 12 July 2007 before turning right. This right turn in the east seas of China was influenced by another anomalous vortex at 300 hPa. The anomalous vortex and the typhoon vortex also merged on 12 July 2007.

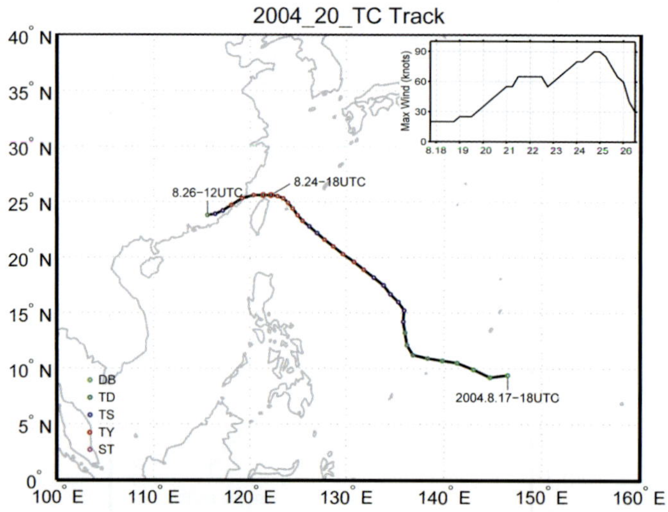

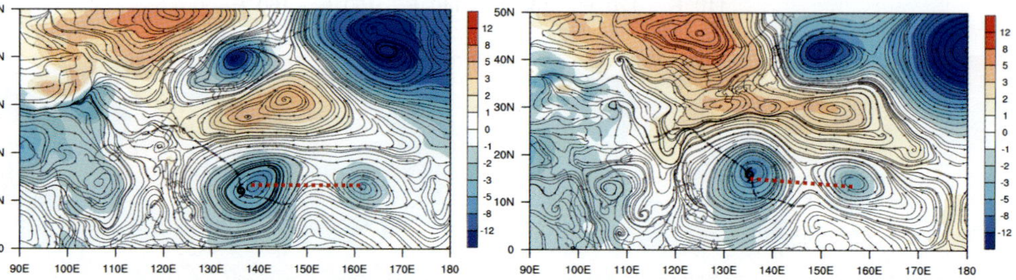

850 hPa flow/height anomalies at 1200 UTC 19 and 20 August 2004.

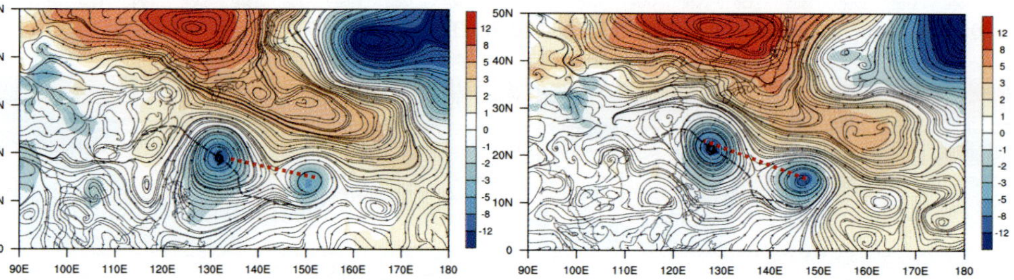

850 hPa flow/height anomalies at 1200 UTC 21 and 22 August 2004.

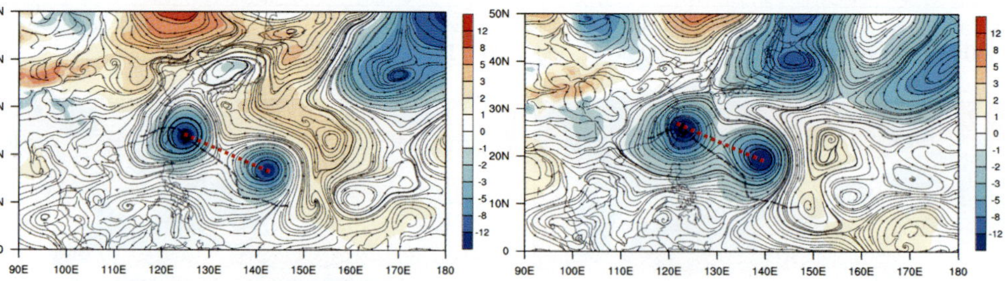

850 hPa flow/height anomalies at 1200 UTC 23 and 24 August 2004.

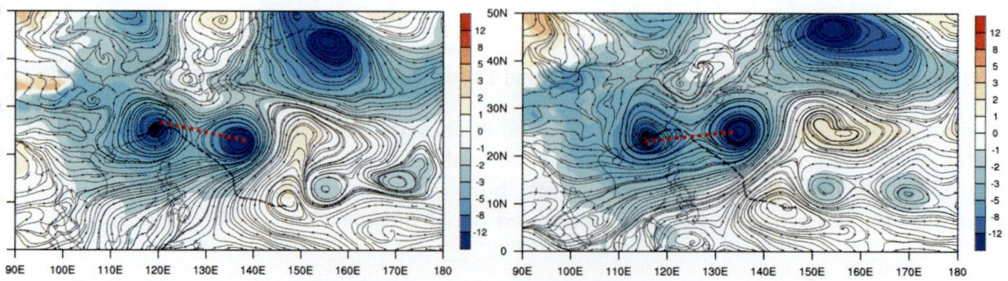

850 hPa flow/height anomalies at 1200 UTC 25 and 26 August 2004.

Figure 5.11. The typhoon (2004_No.20) generated in the equatorial western Pacific on 17 August 2004 and moved northwestward basically along the climatic monsoon trough. Its intensity peaked near the Taiwan Island on 24 August 2004 before landfall on the Southeast China coast. The right turn and left turn of its motion were influenced by another anomalous vortex and anomalous anti-cyclonic circulations at 850 hPa.

5.4. Geneses in the Equatorial Central Pacific

Four typhoons generated in the equatorial central Pacific and their intensities peaked in the Taiwan area. The right turns or left turns of their motions were influenced by anomalous vortices (Figs. 5.12–5.15).

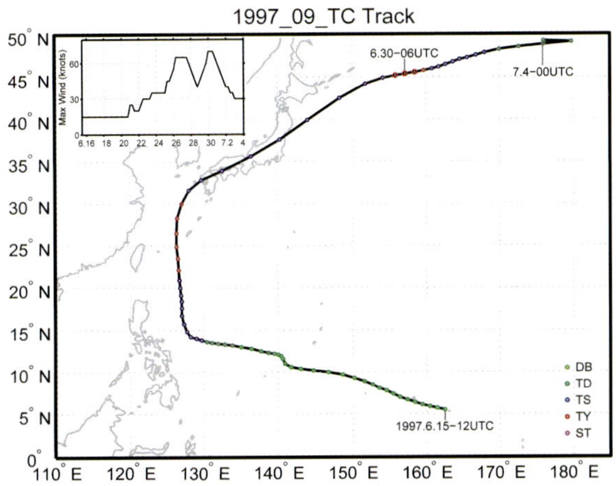

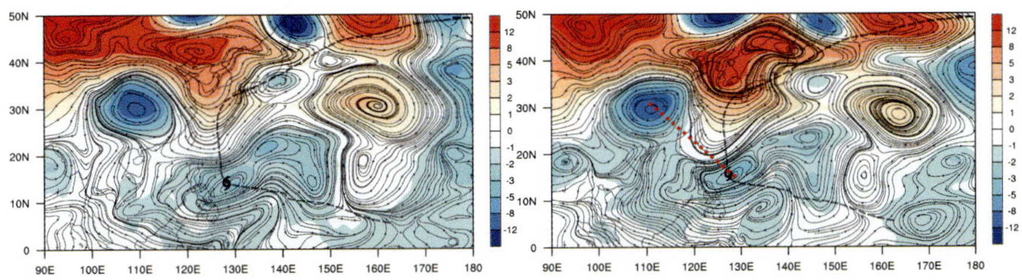

300 hPa flow/height (shading, 10 gpm) anomalies at 0000 UTC and 1200 UTC 24 June 1997.

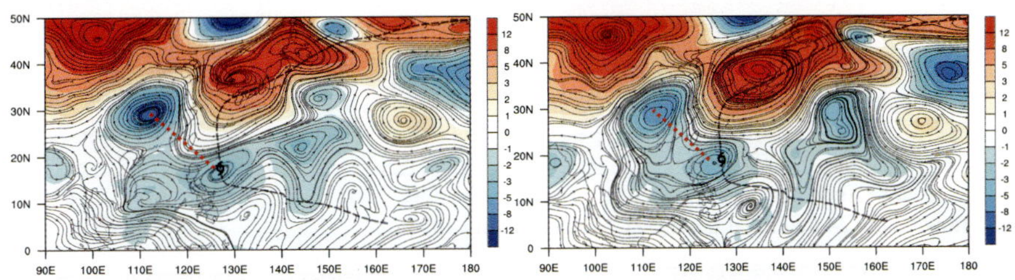

300 hPa flow/height anomalies at 0000 UTC and 1200 UTC 25 June 1997.

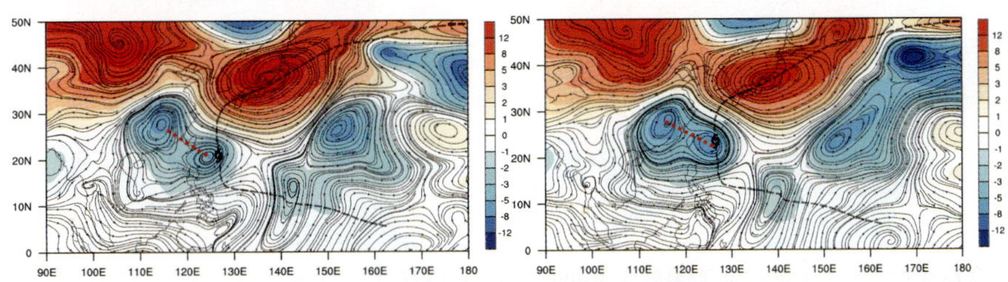

300 hPa flow/height anomalies at 0000 UTC and 1200 UTC 26 June 1997.

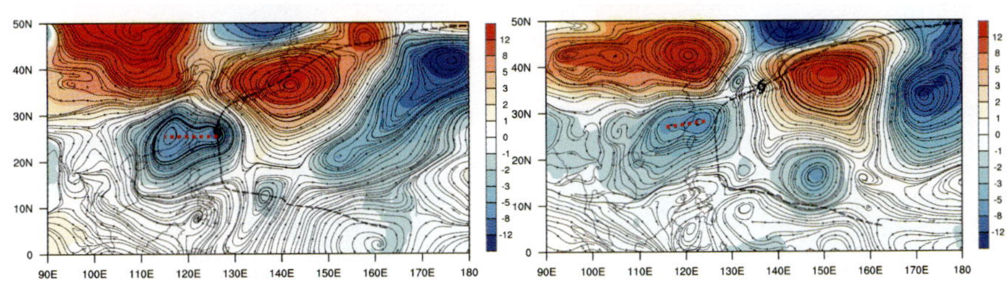

300 hPa flow/height anomalies at 0000 UTC and 1200 UTC 27 June 1997.

Figure 5.12. The typhoon (1997_No.09) generated in the equatorial central Pacific on 15 June 1997 and moved northwestward along the climatic monsoon trough. Its intensity first peaked in the east sea of the Taiwan Island on 26 June 1997 but then weakened and peaked again in the northeast sea of Japan when it was approaching the subtropical monsoon trough on 30 June 1997. The right turn of its motion was influenced by an anomalous vortex at 300 hPa.

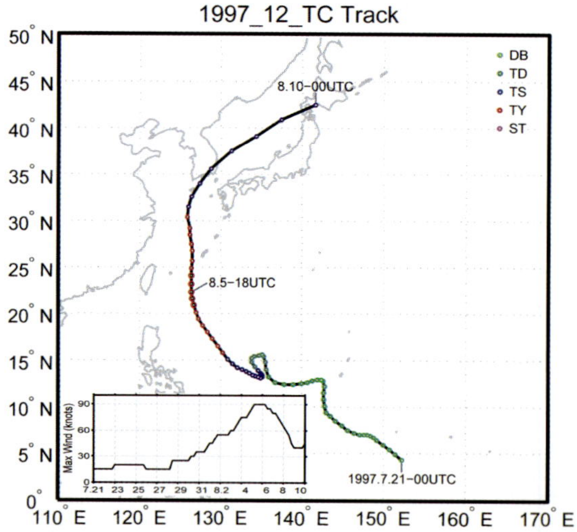

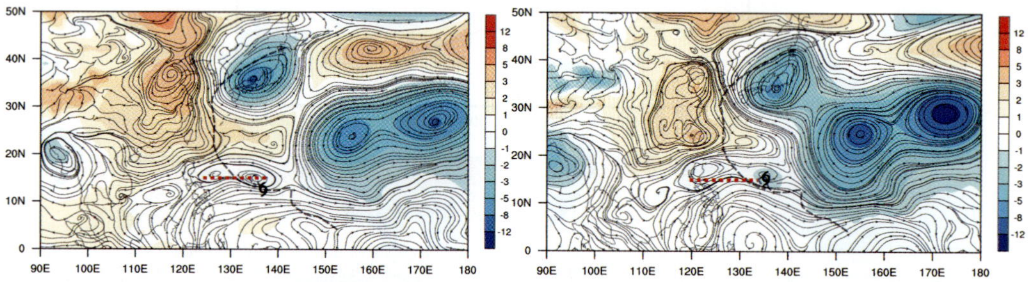

850 hPa flow/height anomalies at 0000 UTC 28 and 29 July 1997.

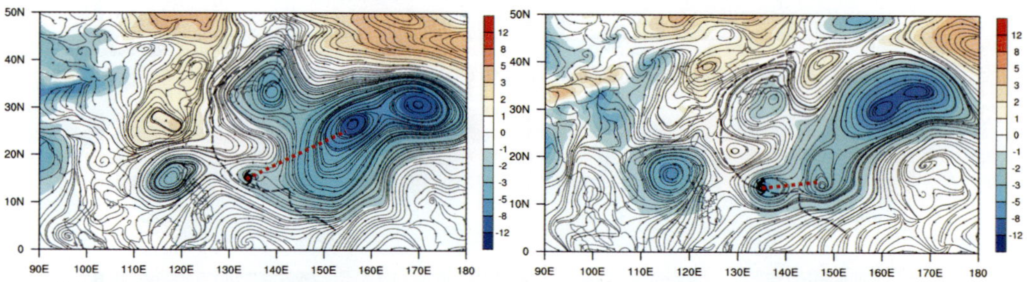

850 hPa flow/height anomalies at 0000 UTC 30 and 31 July 1997.

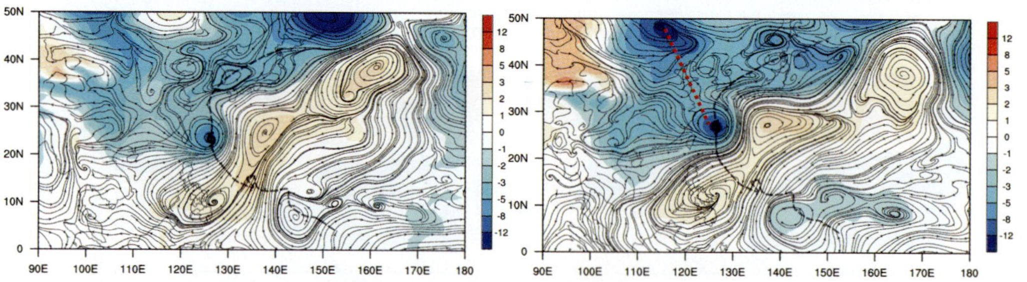

850 hPa flow/height anomalies at 0000 UTC 6 and 7 August 1997.

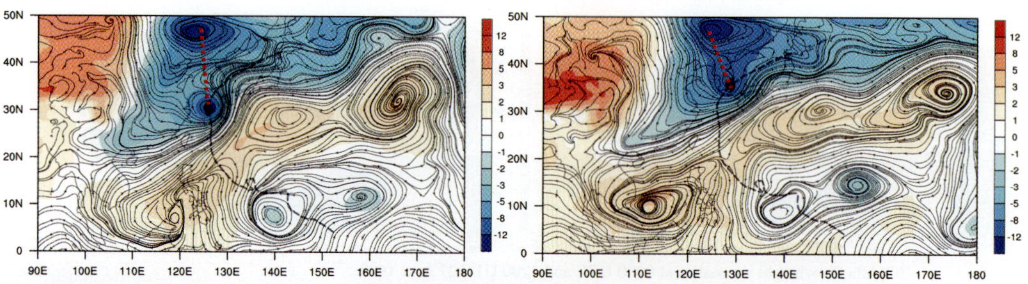

850 hPa flow/height anomalies at 0000 UTC 8 and 9 August 1997.

Figure 5.13. The typhoon (1997_No.12) generated in the equatorial central Pacific on 21 July 1997 and moved northwestward along the climatic monsoon trough. Its intensity peaked in the east sea of the Taiwan Island on 5 August 1997. The right turn of its motion with stagnant state for a few days was influenced by anomalous vortices at 850 hPa.

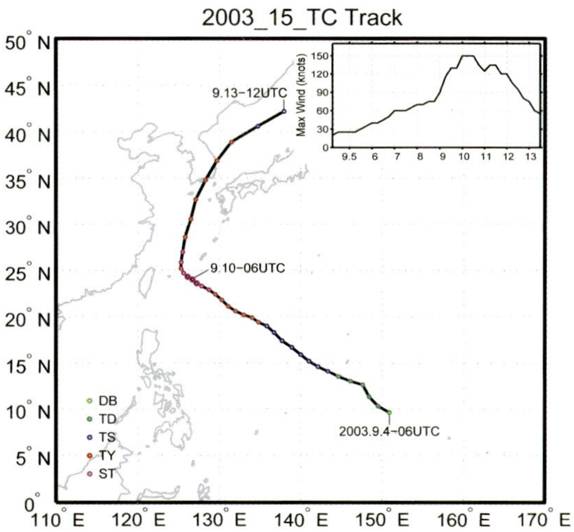

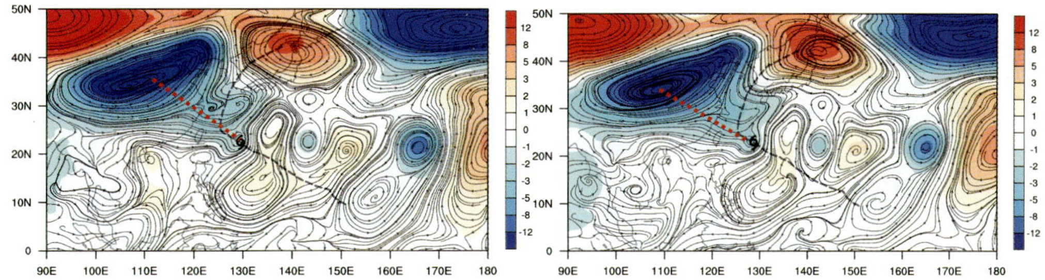

300 hPa flow/height anomalies at 0000 UTC and 1200 UTC 9 September 2003.

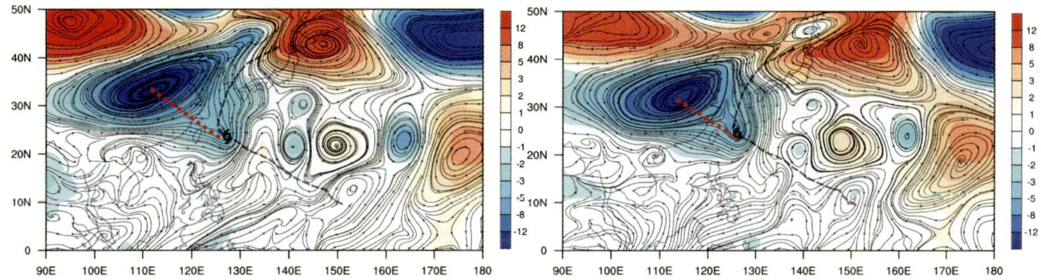

300 hPa flow/height anomalies at 0000 UTC and 1200 UTC 10 September 2003.

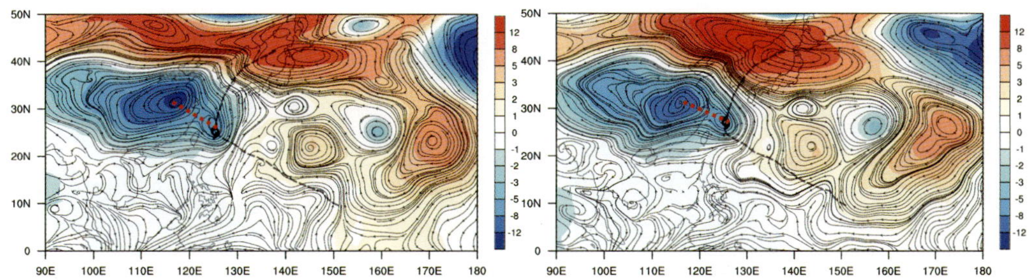

300 hPa flow/height anomalies at 0000 UTC and 1200 UTC 11 September 2003.

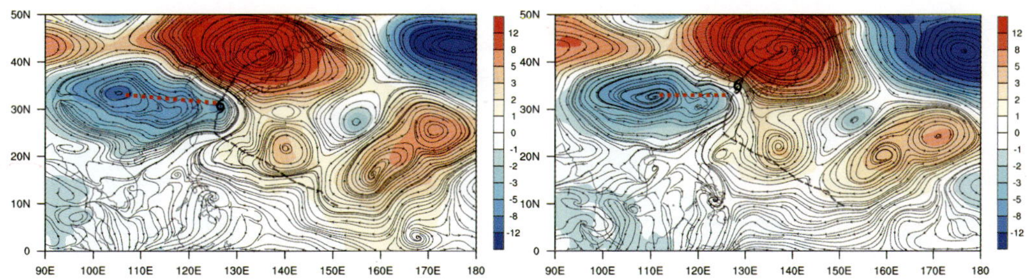

300 hPa flow/height anomalies at 0000 UTC and 1200 UTC 12 September 2003.

Figure 5.14. The typhoon (2003_No.15) generated in the equatorial central Pacific on 4 September 2003 and moved northwestward along the climatic monsoon trough. Its intensity peaked in the east sea of the Taiwan Island on 10 September 2003 before turning right. This right turn was influenced by an anomalous vortex at 300 hPa.

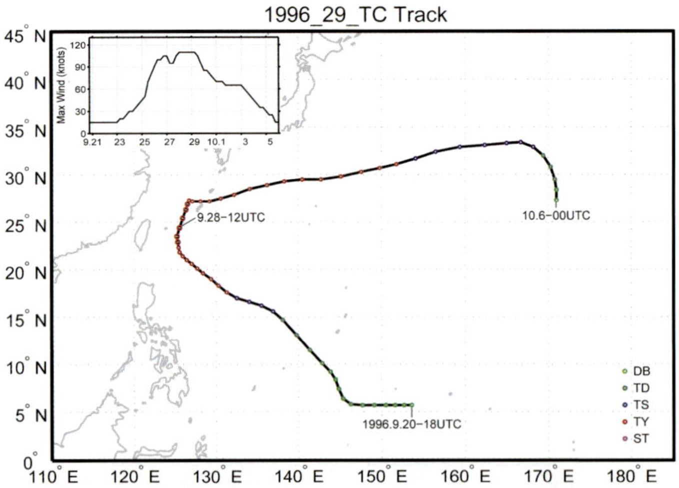

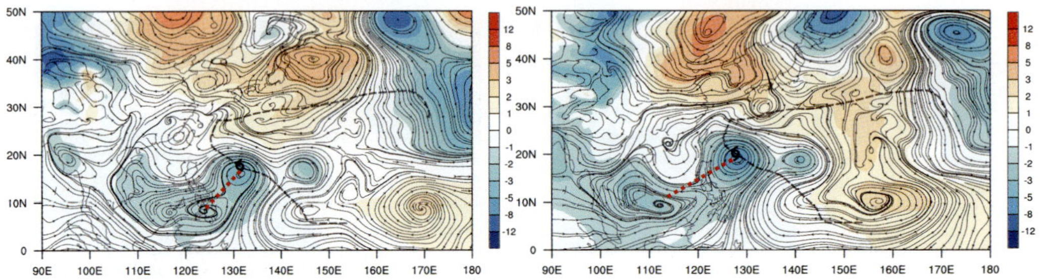

850 hPa flow/height anomalies at 1200 UTC 25 and 26 September 1996.

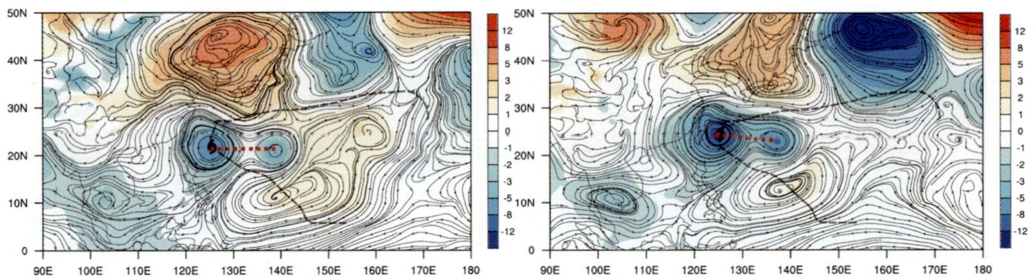

850 hPa flow/height anomalies at 1200 UTC 27 and 28 September 1996.

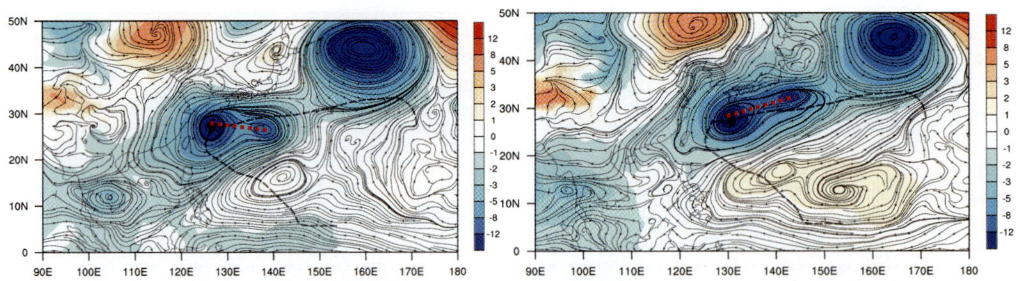

850 hPa flow/height anomalies at 1200 UTC 29 and 30 September 1996.

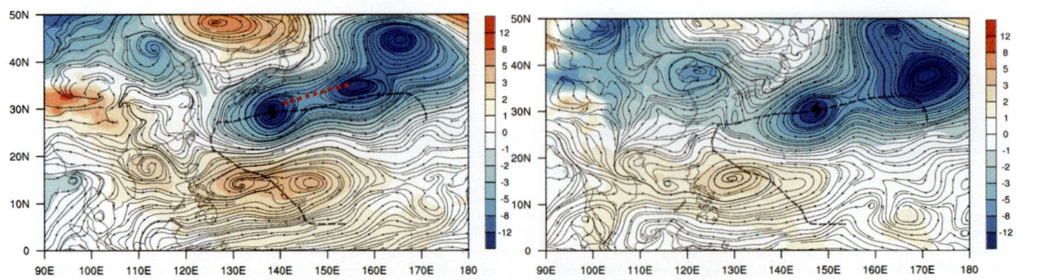

850 hPa flow/height anomalies at 1200 UTC 1 and 2 October 1996.

Figure 5.15. The typhoon (1996_No.29) generated in the equatorial central Pacific on 20 September 1996 and moved northwestward along the climatic monsoon trough. Its intensity peaked in the east sea of the Taiwan Island on 28 September 1996 when it turned right. This turning motion in the East China Sea was influenced by anomalous vortices at 850 hPa.

5.5. Geneses in the Tropical Western Pacific

There were 17 typhoons that generated in the tropical western Pacific and their intensities peaked in the Taiwan area. The right turns and left turns of their motions were influenced by anomalous vortices (Figs. 5.16–5.32).

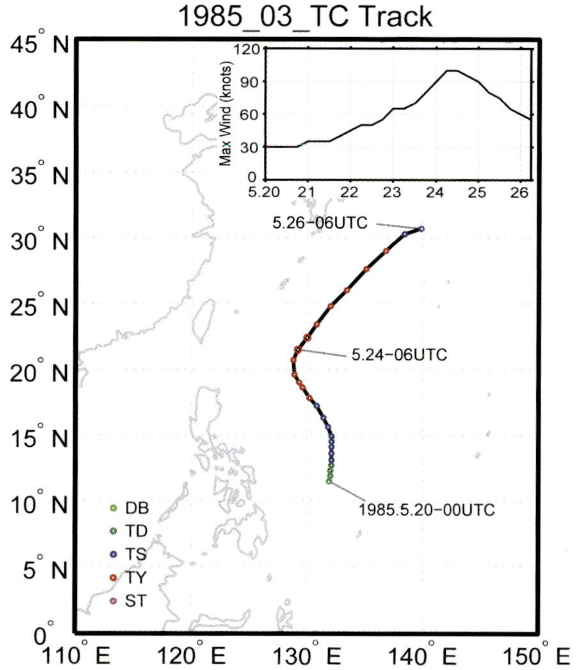

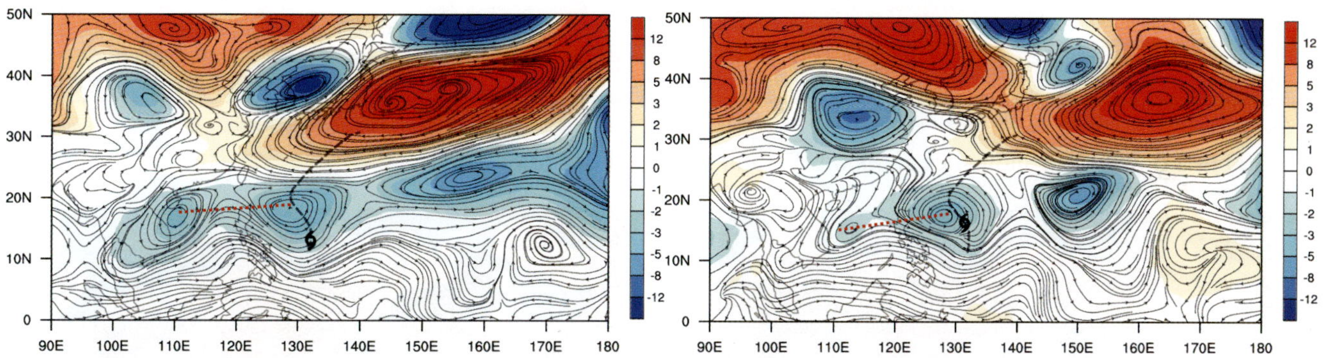

300 hPa flow/height (shading, 10 gpm) anomalies at 0000 UTC and 1200 UTC 21 May 1985.

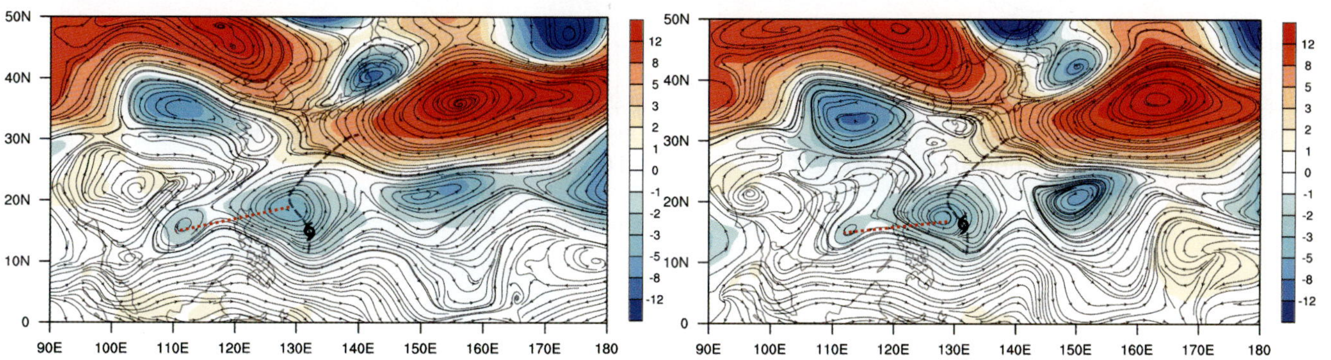

300 hPa flow/height anomalies at 0000 UTC and 1200 UTC 22 May 1985.

Figure 5.16. The typhoon (1985_No.03) generated in the tropical western Pacific on 20 May 1985 and moved northward. Its intensity peaked on 24 May 1985 when it was turning right in the subtropical western Pacific. This turning motion was influenced by anomalous vortices at 300 hPa.

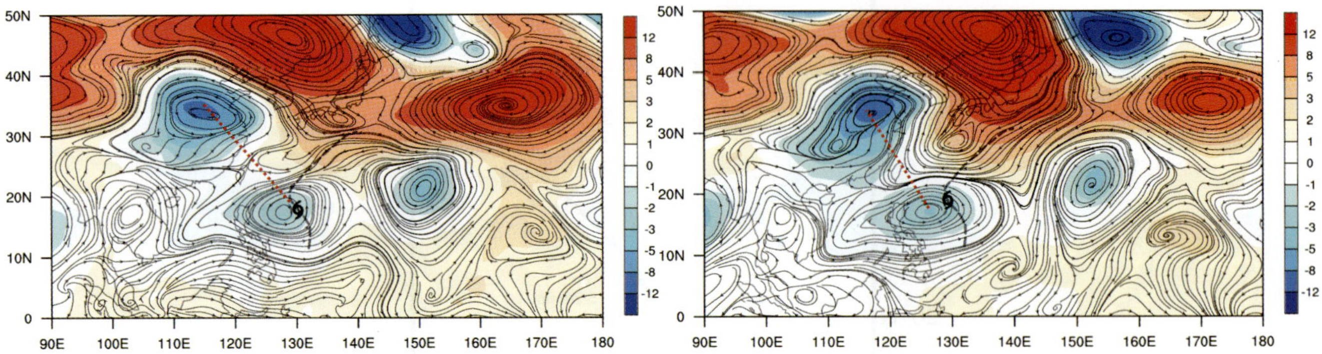

300 hPa flow/height anomalies at 0000 UTC and 1200 UTC 23 May 1985.

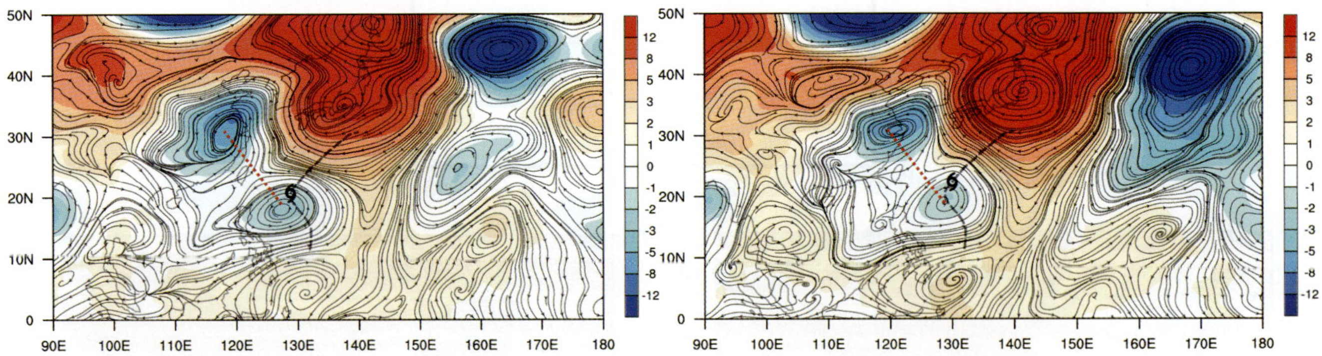

300 hPa flow/height anomalies at 0000 UTC and 1200 UTC 24 May 1985.

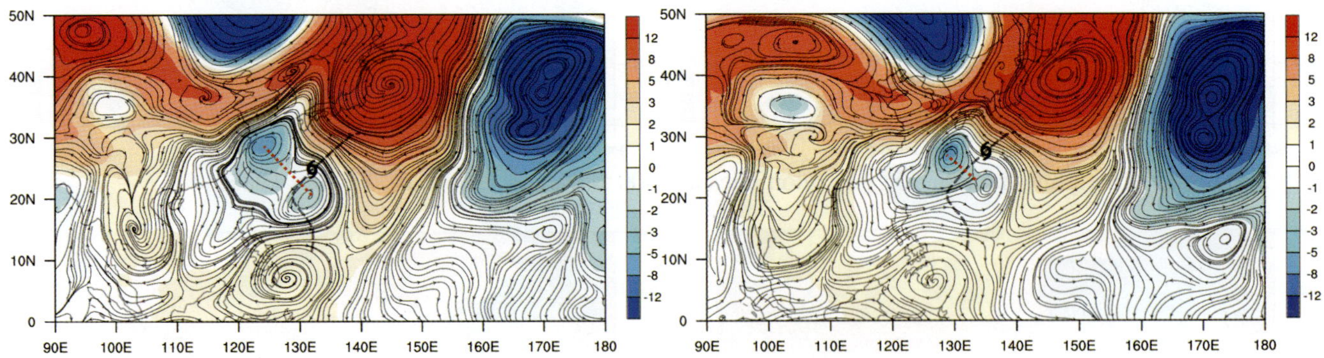

300 hPa flow/height anomalies at 0000 UTC and 1200 UTC 25 May 1985.

Figure 5.16. *(Continued)*

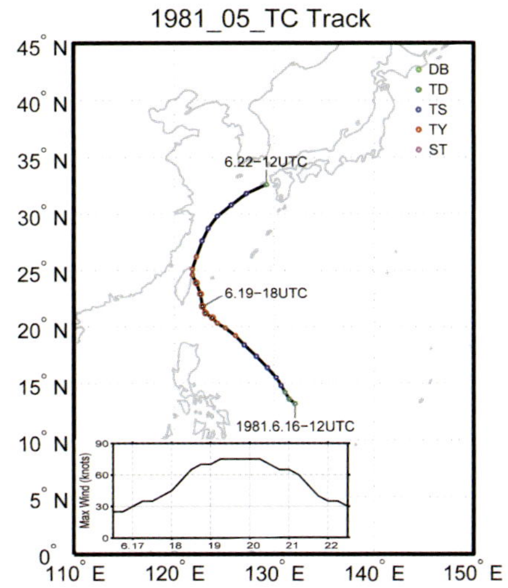

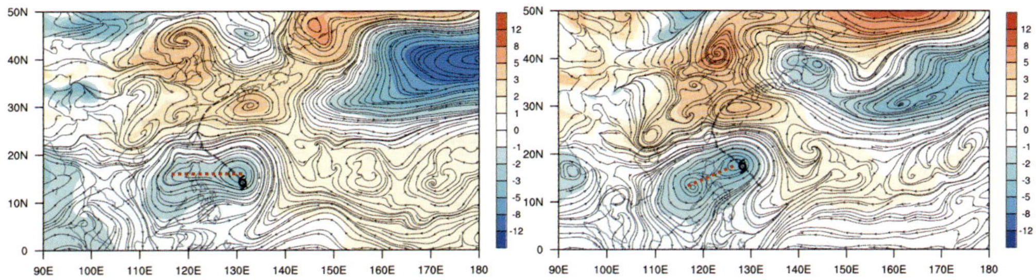

850 hPa flow/height anomalies at 0000 UTC 17 and 18 June 1981.

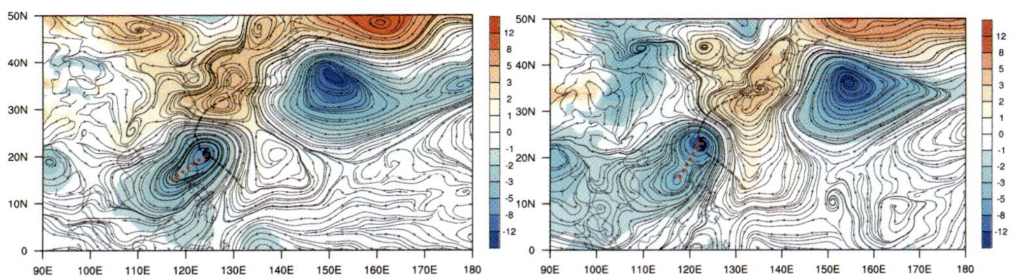

850 hPa flow/height anomalies at 0000 UTC 19 and 20 June 1981.

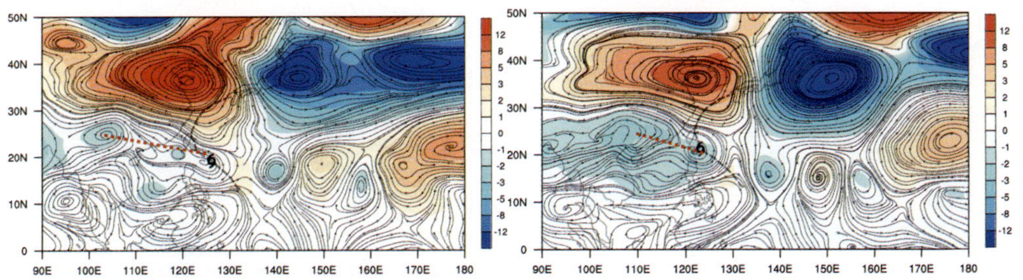

300 hPa flow/height anomalies at 1200 UTC 18 and 19 June 1981.

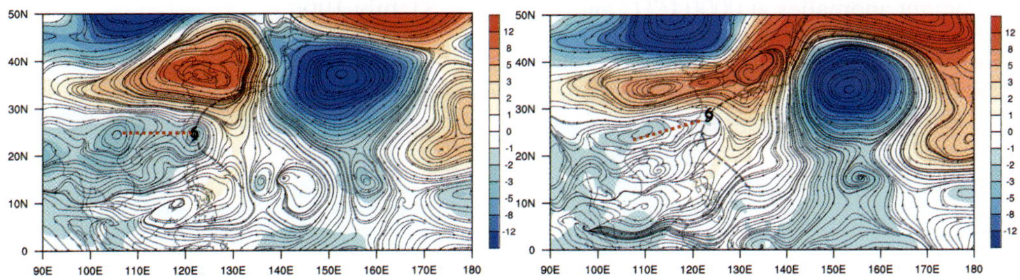

300 hPa flow/height anomalies at 1200 UTC 20 and 21 June 1981.

Figure 5.17. The typhoon (1981_No.05) generated in the tropical western Pacific on 16 June 1981 and moved northwestward along the climatic monsoon trough. Its intensity peaked in the southeast sea of the Taiwan Island on 19 June before landfall on the island. The right turn of its motion in the east seas of China was influenced by anomalous vortices at 850 hPa and 300 hPa.

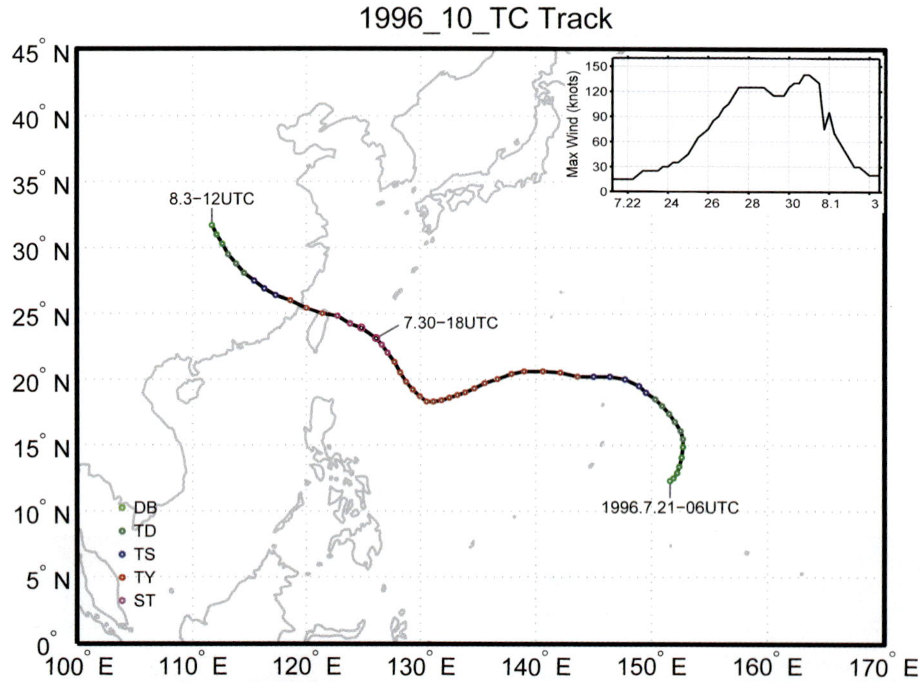

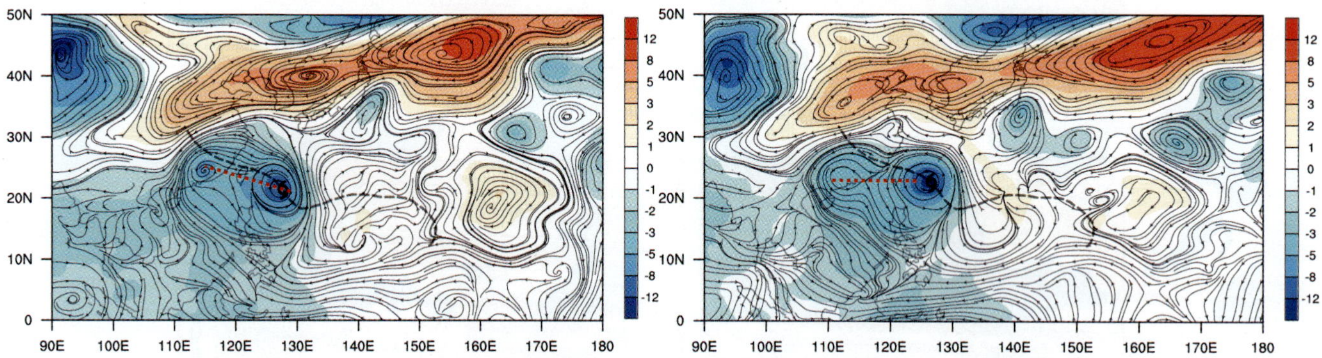

300 hPa flow/height anomalies at 0000 UTC and 1200 UTC 30 July 1996.

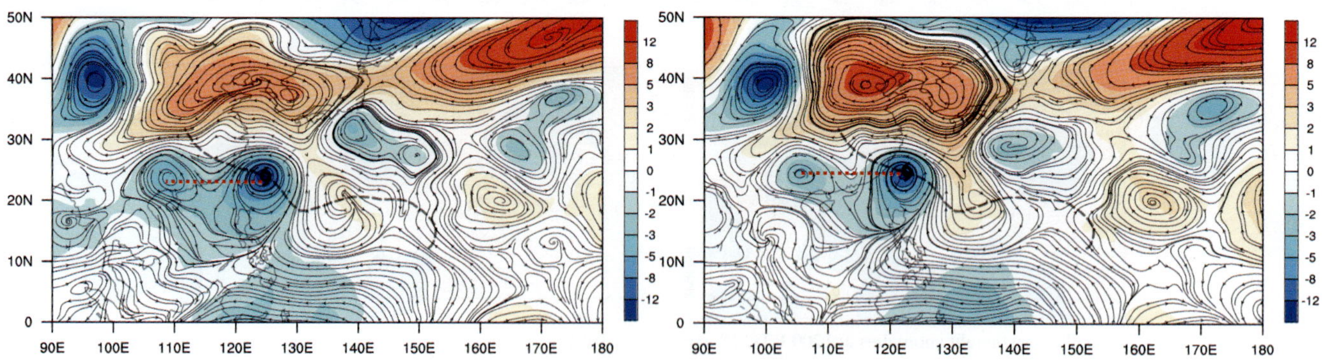

300 hPa flow/height anomalies at 0000 UTC and 1200 UTC 31 July 1996.

Figure 5.18. The typhoon (1996_No.10) generated in the tropical central Pacific on 21 July 1996. Its intensity peaked in the east sea of the Taiwan Island on 30 July 1996 when it was approaching the climatic monsoon trough before landfall on the island. The left turn of its motion near the Taiwan Island was influenced by an anomalous vortex at 300 hPa.

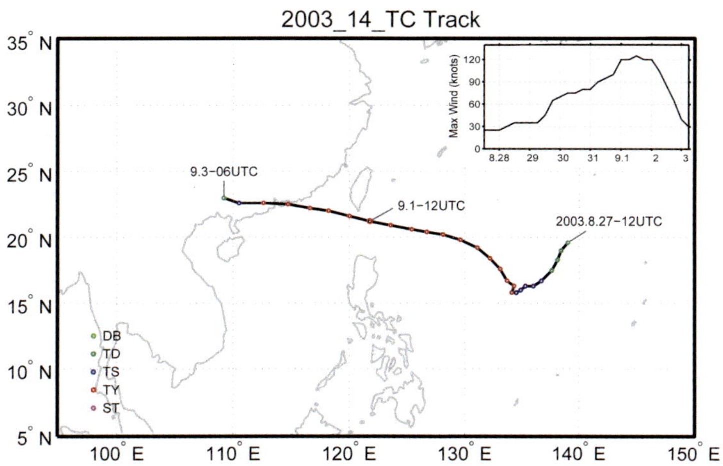

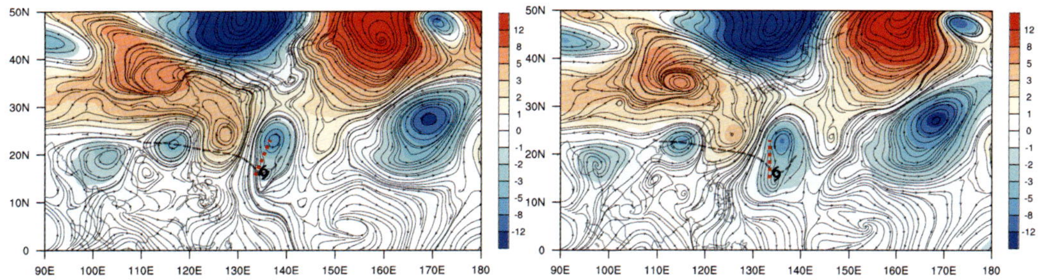

300 hPa flow/height anomalies at 0000 UTC and 0600 UTC 29 August 2003.

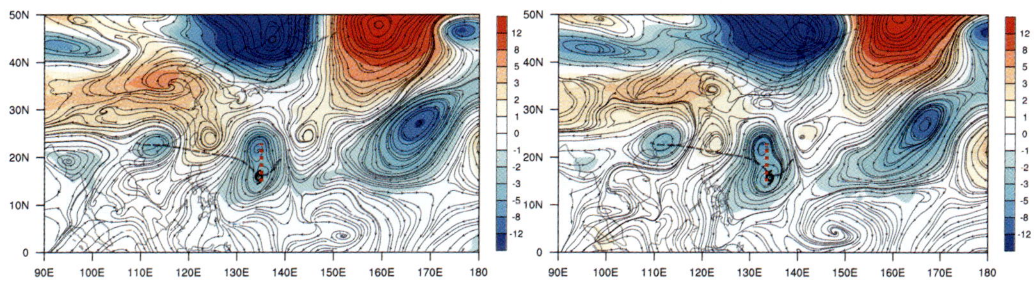

300 hPa flow/height anomalies at 1200 UTC and 1800 UTC 29 August 2003.

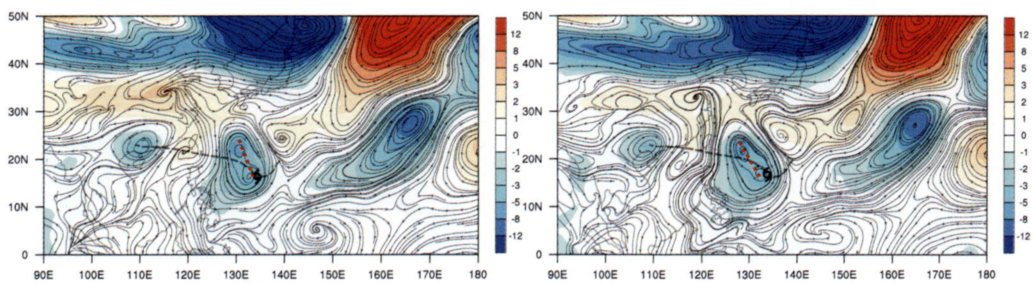

300 hPa flow/height anomalies at 0000 UTC and 0600 UTC 30 August 2003.

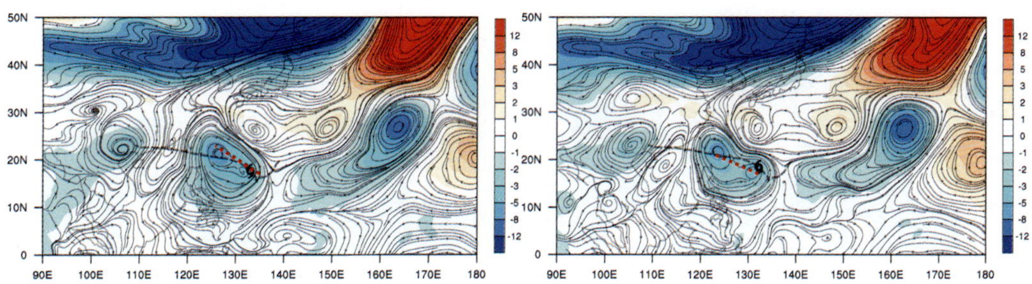

300 hPa flow/height anomalies at 1200 UTC and 1800 UTC 30 August 2003.

Figure 5.19. The typhoon (2003_No.14) generated in the tropical western Pacific on 27 August 2003 and moved southwestward at the beginning stage. Its intensity peaked on 1 September 2003 when it was approaching the climatic monsoon trough. The left turn of its motion near the Philippines was influenced by an anomalous vortex at 300 hPa.

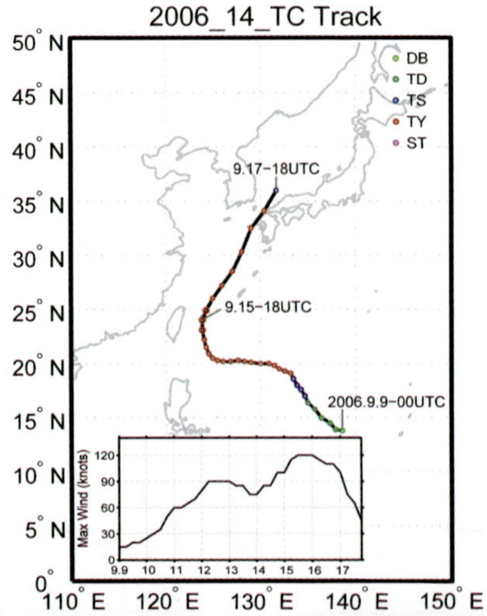

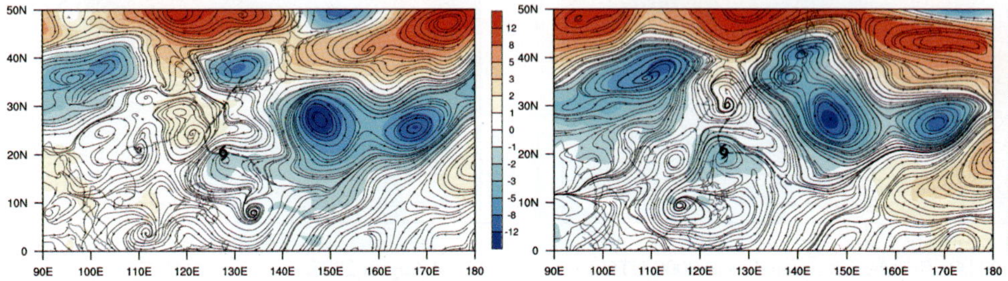

300 hPa flow/height anomalies at 1200 UTC 13 and 14 September 2006.

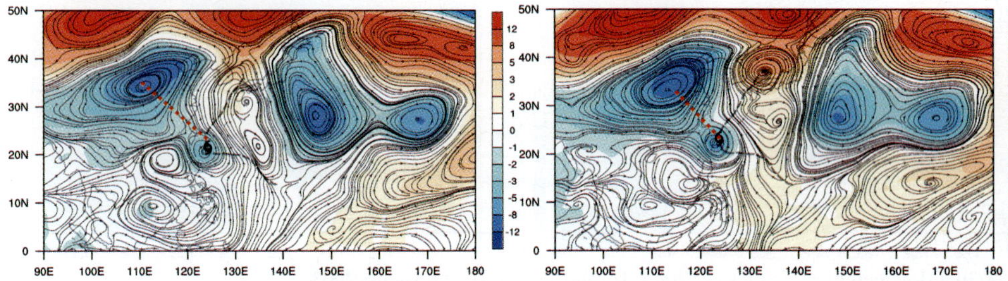

300 hPa flow/height anomalies at 0000 UTC and 1200 UTC 15 September 2006.

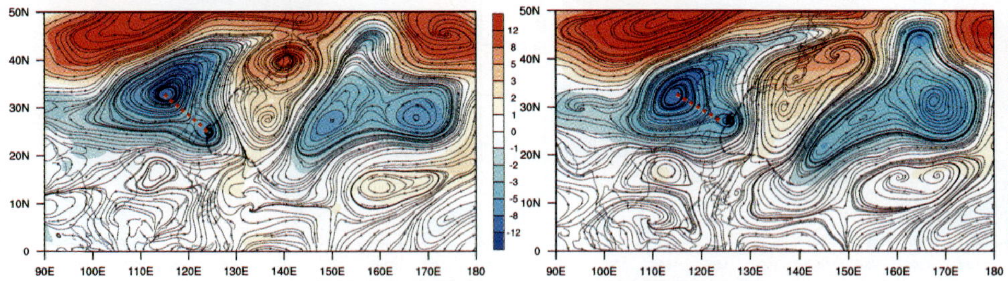

300 hPa flow/height anomalies at 0000 UTC and 1200 UTC 16 September 2006.

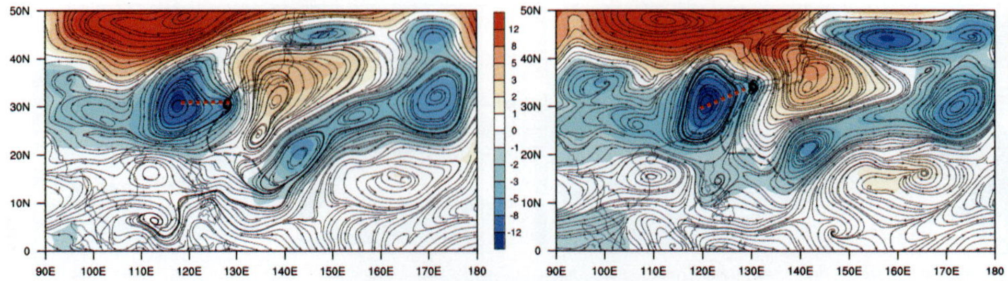

300 hPa flow/height anomalies at 0000 UTC and 1200 UTC 17 September 2006.

Figure 5.20. The typhoon (2006_No.14) generated in the tropical western Pacific on 9 September 2006 and moved northwestward along the climatic monsoon trough. Its intensity peaked in the east sea of the Taiwan Island on 15 September 2006, when it turned right. This turning motion near the Taiwan Island was influenced by an anomalous vortex in the China continent at 300 hPa.

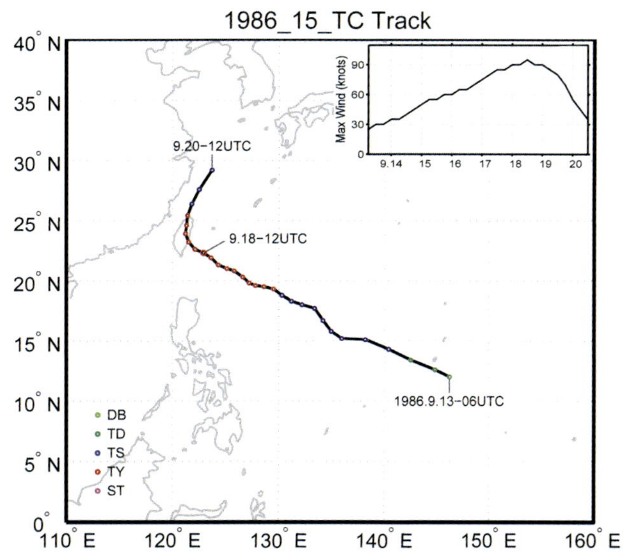

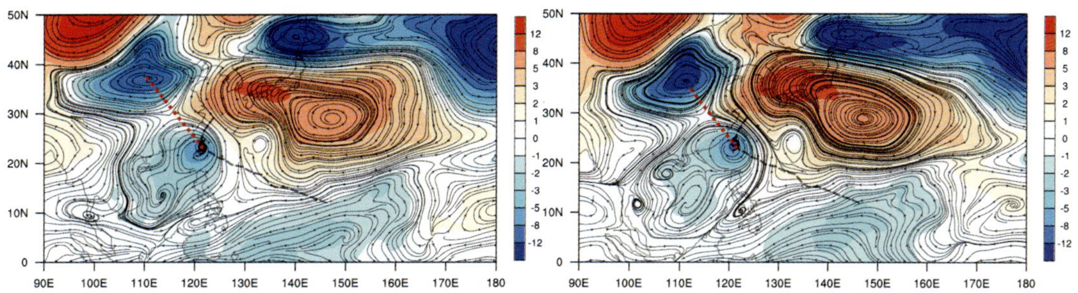

300 hPa flow/height anomalies at 0000 UTC and 0600 UTC 19 September 1986.

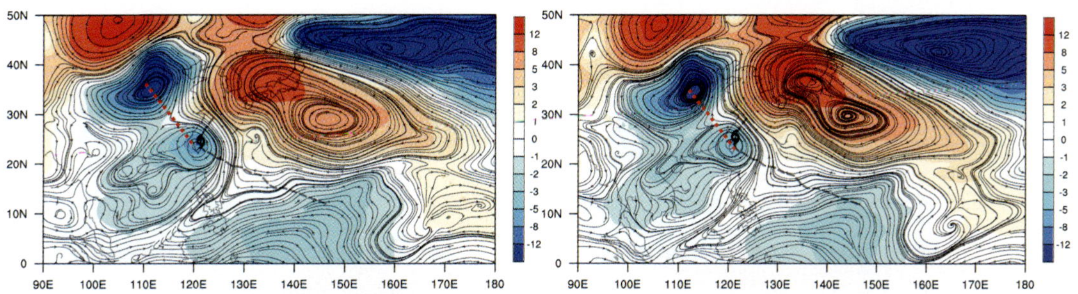

300 hPa flow/height anomalies at 1200 UTC and 1800 UTC 19 September 1986.

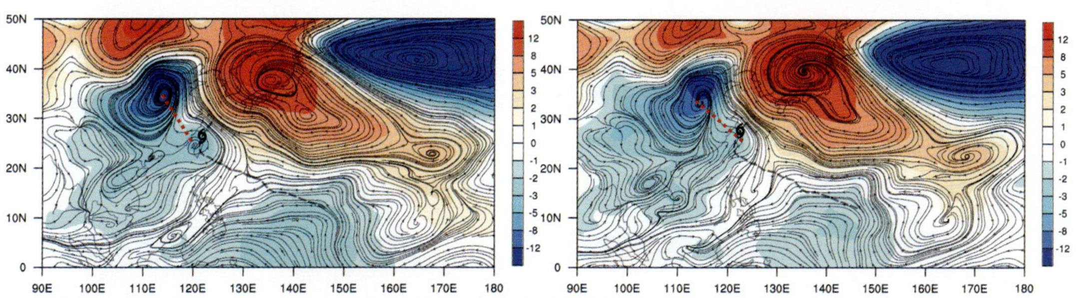

300 hPa flow/height anomalies at 0000 UTC and 0600 UTC 20 September 1986.

Figure 5.21. The typhoon (1986_No.15) generated in the tropical western Pacific on 13 September 1986 and moved northwestward along the climatic monsoon trough. Its intensity peaked in the east sea of the Taiwan Island on 18 September 1986 before landfall on the island. The right turn of its motion near the Taiwan Island was influenced by an anomalous vortex in the China continent at 300 hPa.

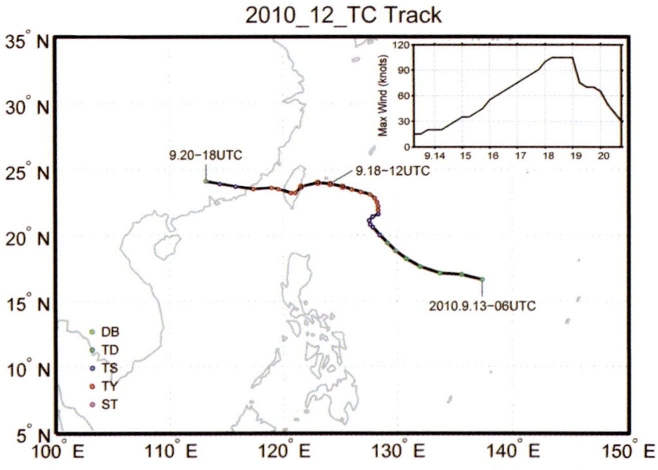

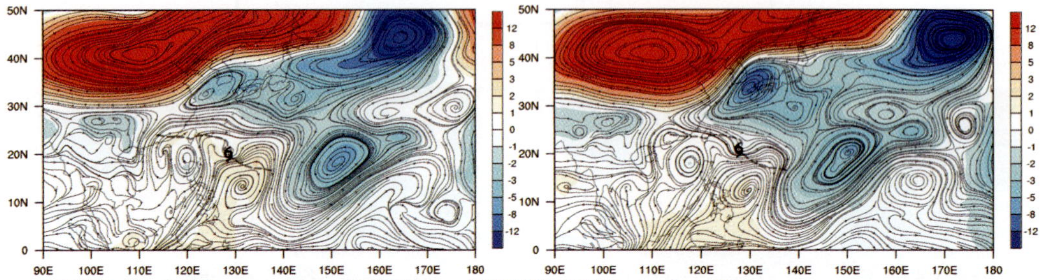

300 hPa flow/height anomalies at 0000 UTC and 1200 UTC 15 September 2010.

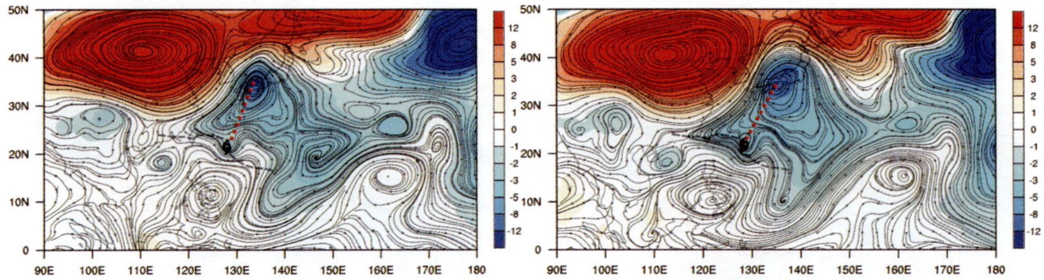

300 hPa flow/height anomalies at 0000 UTC and 1200 UTC 16 September 2010.

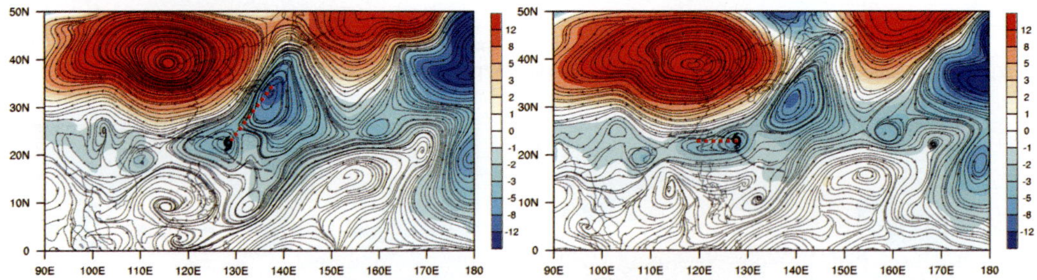

300 hPa flow/height anomalies at 0000 UTC and 1200 UTC 17 September 2010.

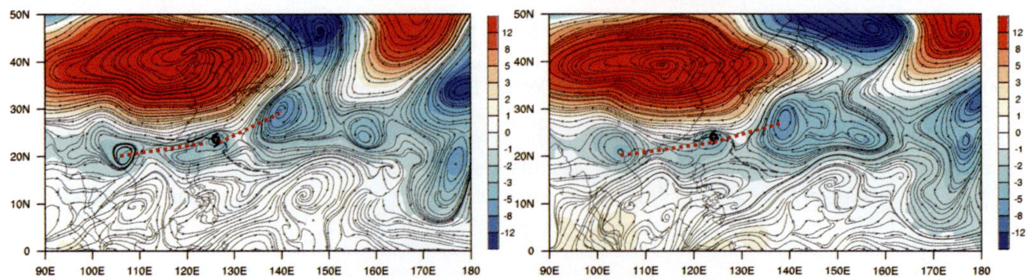

300 hPa flow/height anomalies at 0000 UTC and 1200 UTC 18 September 2010.

Figure 5.22. The typhoon (2010_No.12) generated in the tropical western Pacific on 13 September 2010 and moved northwestward along the climatic monsoon trough. Its intensity peaked in the east sea of the Taiwan Island on 18 September 2010 before landfall on the island and its complex track was influenced by anomalous vortices at 300 hPa.

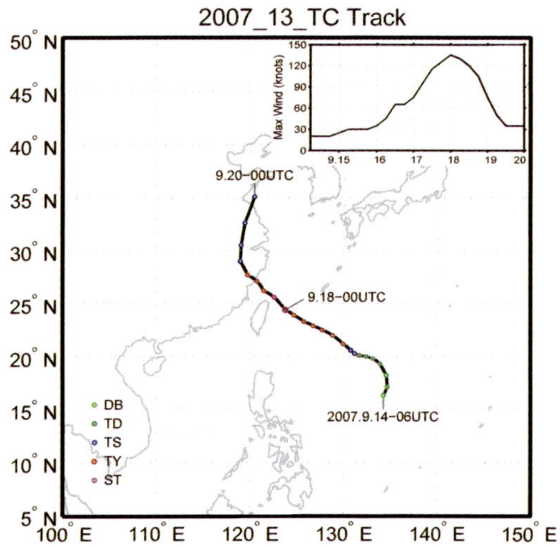

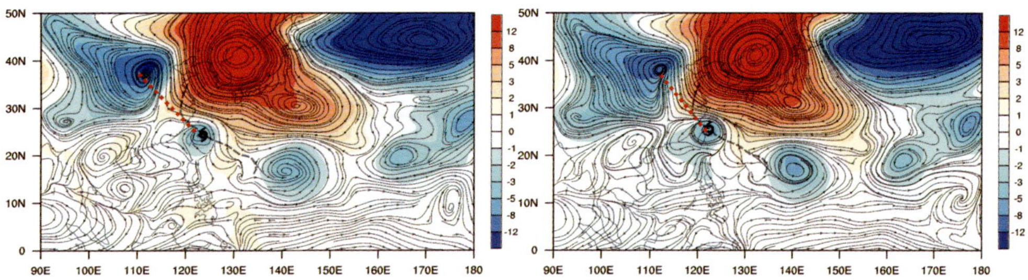

300 hPa flow/height anomalies at 0000 UTC and 0600 UTC 18 September 2007.

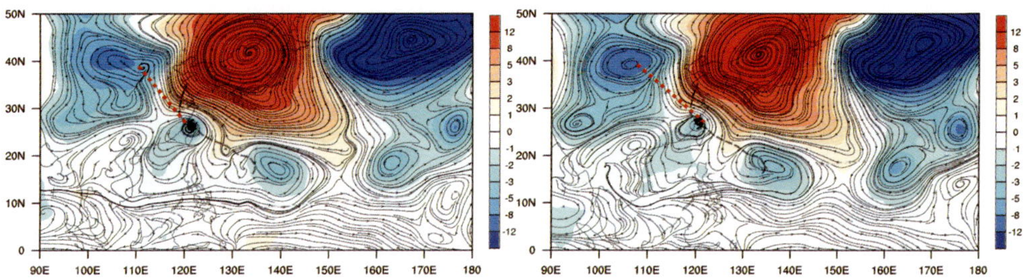

300 hPa flow/height anomalies at 1200 UTC and 1800 UTC 18 September 2007.

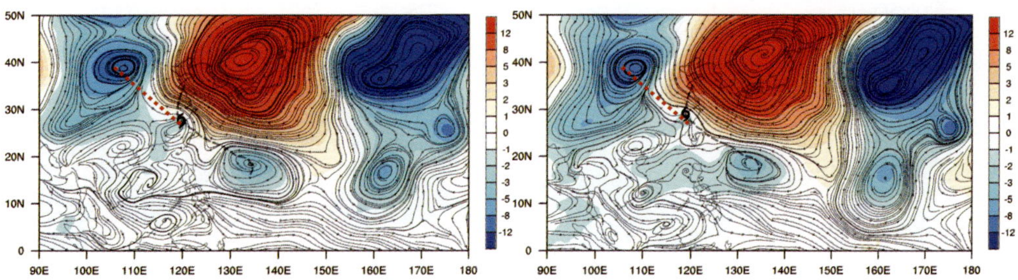

300 hPa flow/height anomalies at 0000 UTC and 0600 UTC 19 September 2007.

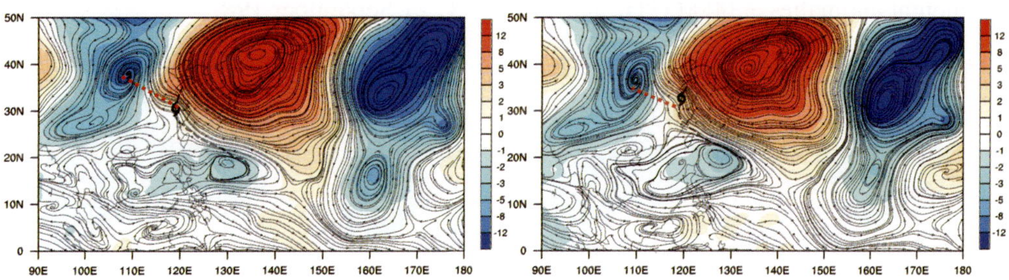

300 hPa flow/height anomalies at 1200 UTC and 1800 UTC 19 September 2007.

Figure 5.23. The typhoon (2007_No.13) generated in the tropical western Pacific on 14 September 2007 and moved northwestward along the climatic monsoon trough. Its intensity peaked in the east sea of the Taiwan Island on 18 September 2007 before landfall on the Southeast China coast. The right turn of its motion was influenced by an anomalous vortex from the China continent at 300 hPa.

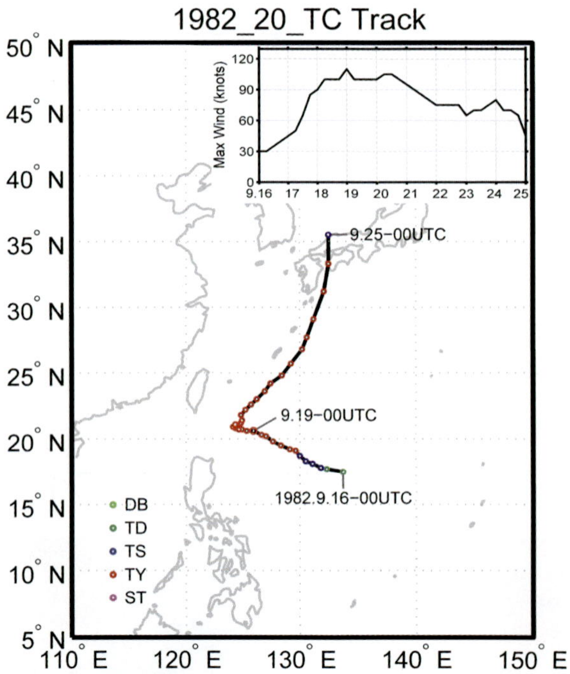

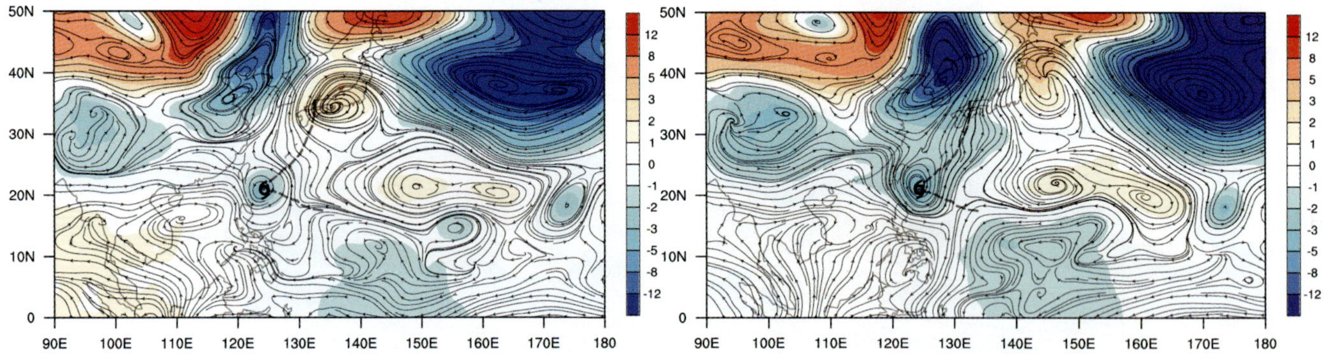

300 hPa flow/height anomalies at 0000 UTC and 1200 UTC 20 September 1982.

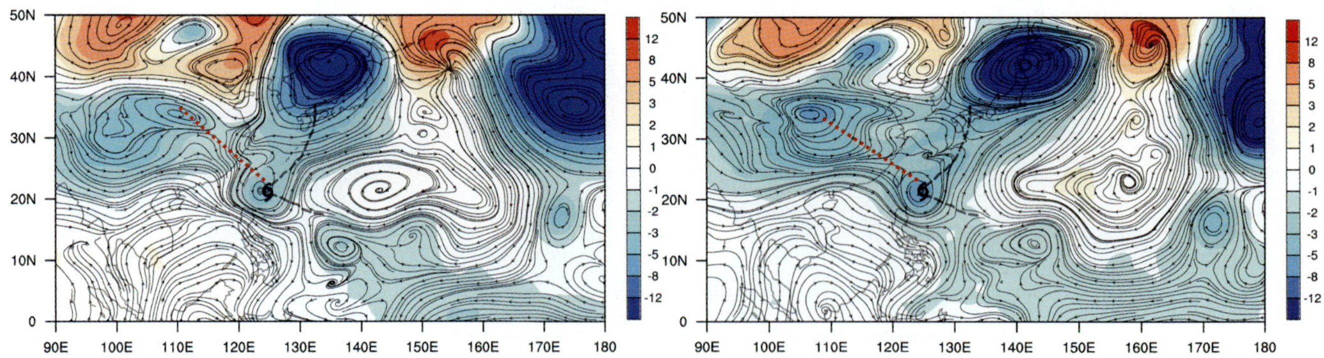

300 hPa flow/height anomalies at 0000 UTC and 1200 UTC 21 September 1982.

Figure 5.24. The typhoon (1982_No.20) generated in the tropical western Pacific on 16 September 1982 and moved northwestward along the climatic monsoon trough. Its intensity peaked in the southeast sea of the Taiwan Island on 19 September 1982 before turning right. This turning motion with a few-day stagnant state near the Philippines and the Taiwan Island was influenced by an anomalous vortex from the China continent at 300 hPa.

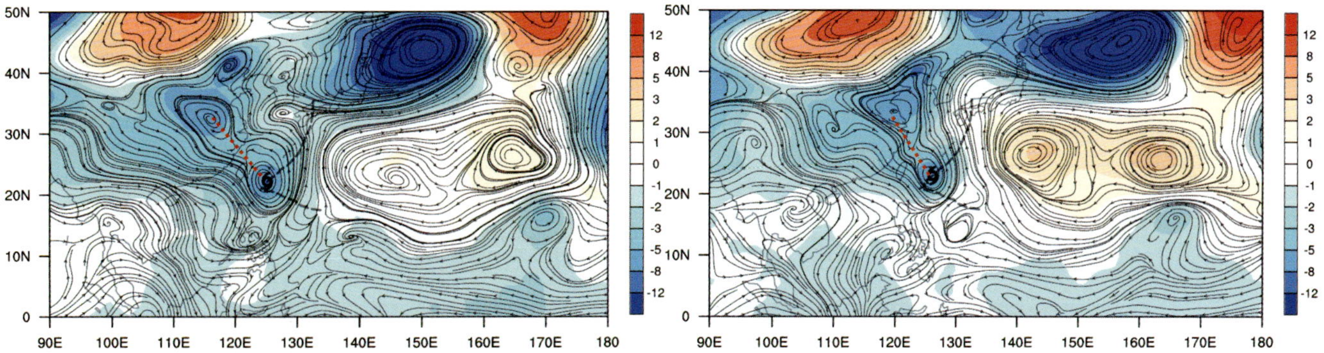

300 hPa flow/height anomalies at 0000 UTC and 1200 UTC 22 September 1982.

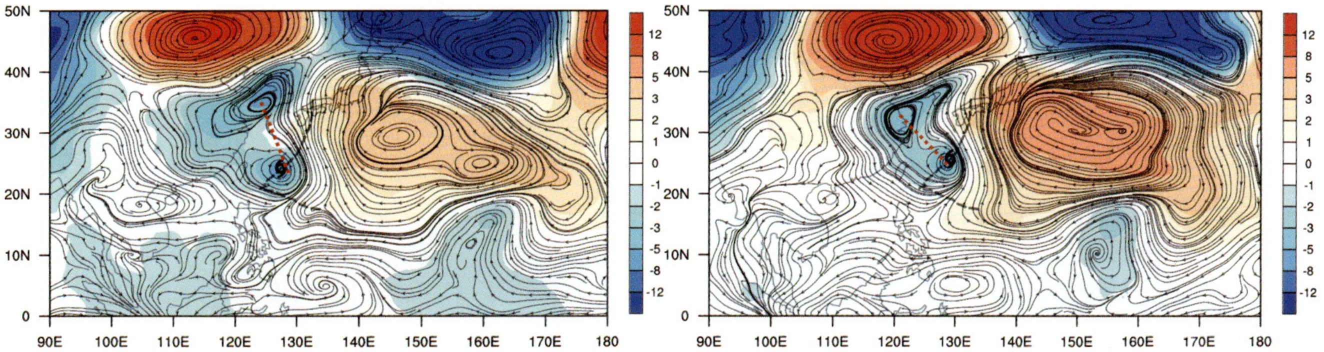

300 hPa flow/height anomalies at 0000 UTC and 1200 UTC 23 September 1982.

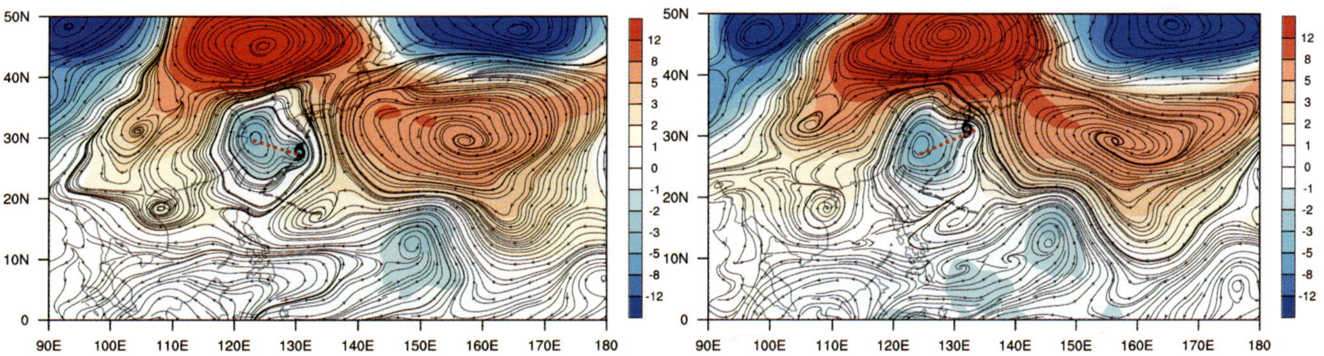

300 hPa flow/height anomalies at 0000 UTC and 1200 UTC 24 September 1982.

Figure 5.24. *(Continued)*

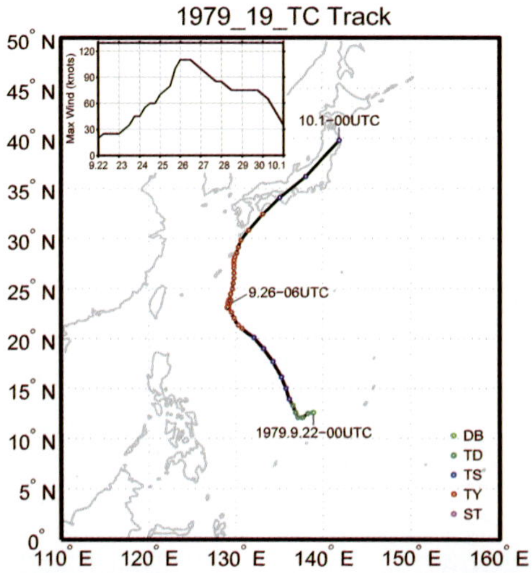

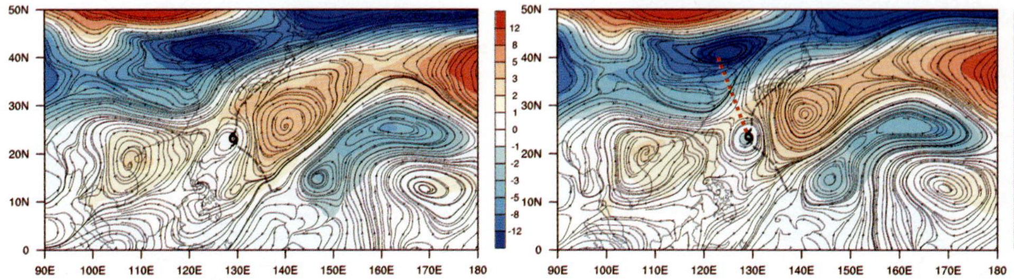

300 hPa flow/height anomalies at 0000 UTC and 0600 UTC 26 September 1979.

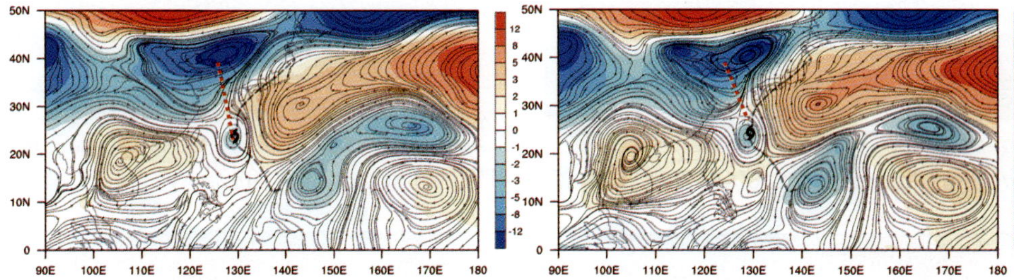

300 hPa flow/height anomalies at 1200 UTC and 1800 UTC 26 September 1979.

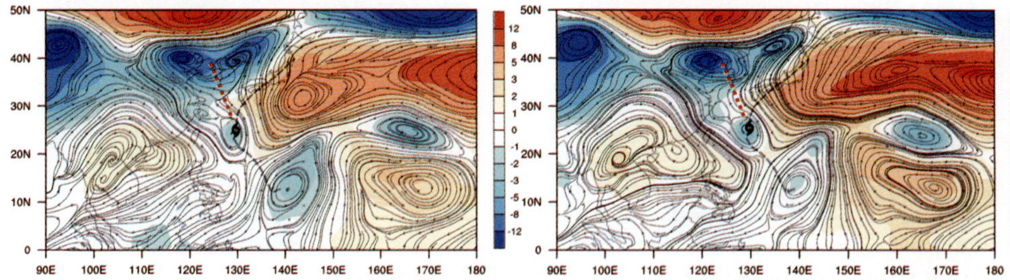

300 hPa flow/height anomalies at 0000 UTC and 0600 UTC 27 September 1979.

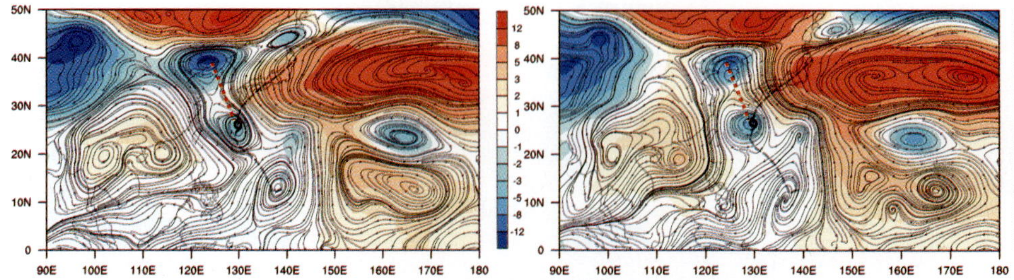

300 hPa flow/height anomalies at 1200 UTC and 1800 UTC 27 September 1979.

Figure 5.25. The typhoon (1979_No.19) generated in the tropical western Pacific on 22 September 1979 and moved westward then northwestward. Its intensity peaked in the east sea of the Taiwan Island on 26 September 1979 when it was approaching its turning point near the climatic monsoon trough. This turning motion was influenced by an anomalous vortex at 300 hPa.

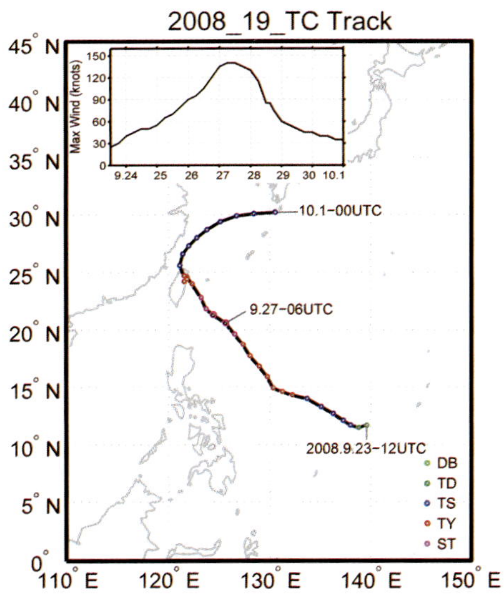

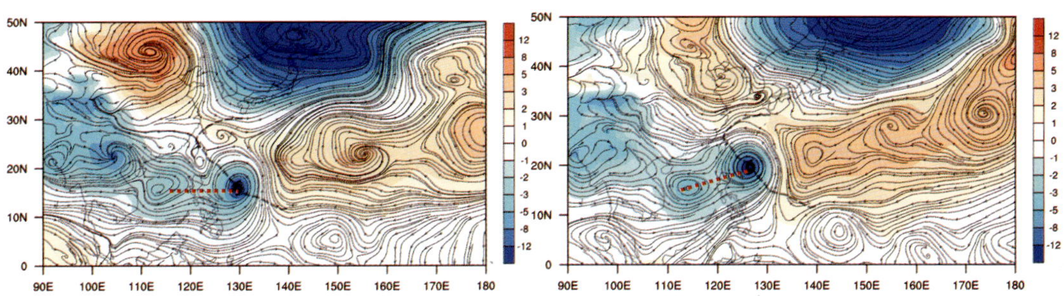

850 hPa flow/height anomalies at 0000 UTC 26 and 27 September 2008.

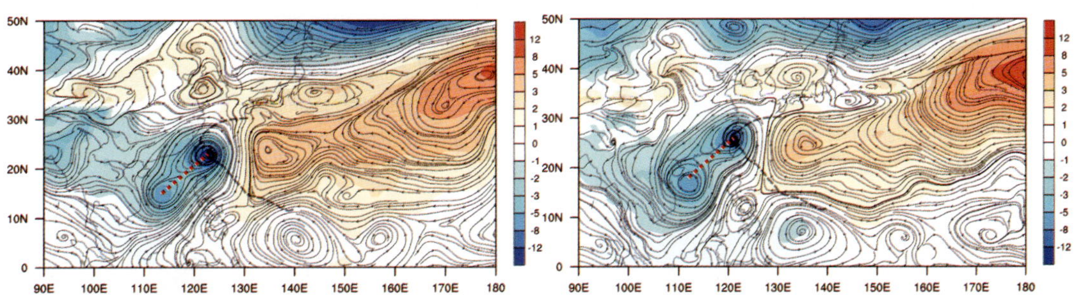

850 hPa flow/height anomalies at 0000 UTC 28 and 29 September 2008.

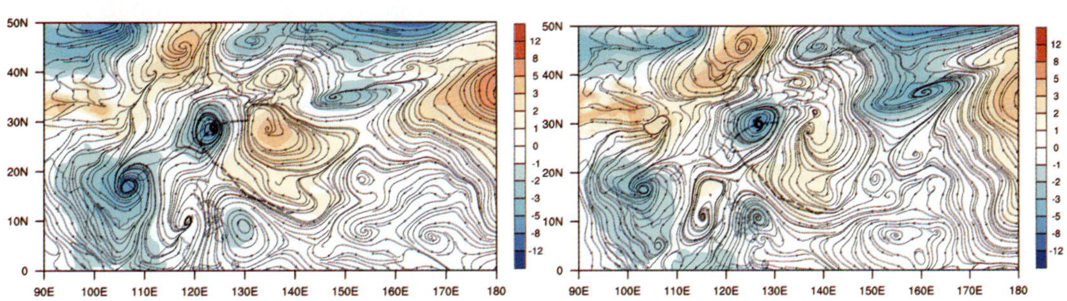

850 hPa flow/height anomalies at 0000 UTC and 1200 UTC 30 September 2008.

Figure 5.26. The typhoon (2008_No.19) generated in the tropical western Pacific on 23 September 2008 and moved northwestward along the climatic monsoon trough. Its intensity peaked in the southeast sea of the Taiwan Island on 27 September 2008 before landfall on the island. The right turn of its motion near the Taiwan Island was influenced by an anomalous vortex at 850 hPa.

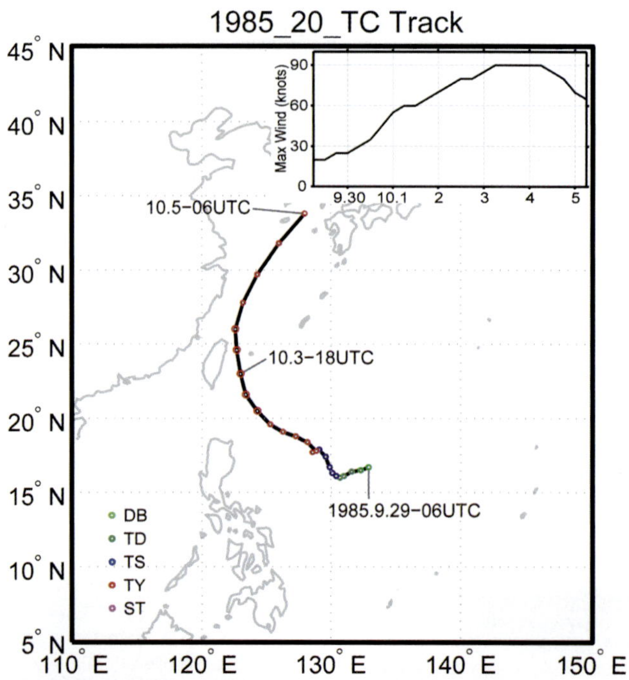

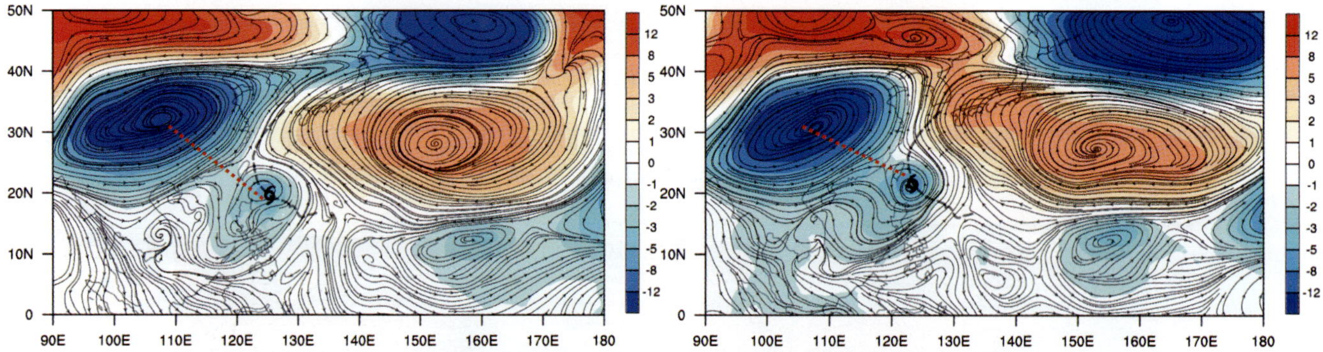

300 hPa flow/height anomalies at 0000 UTC and 1200 UTC 3 October 1985.

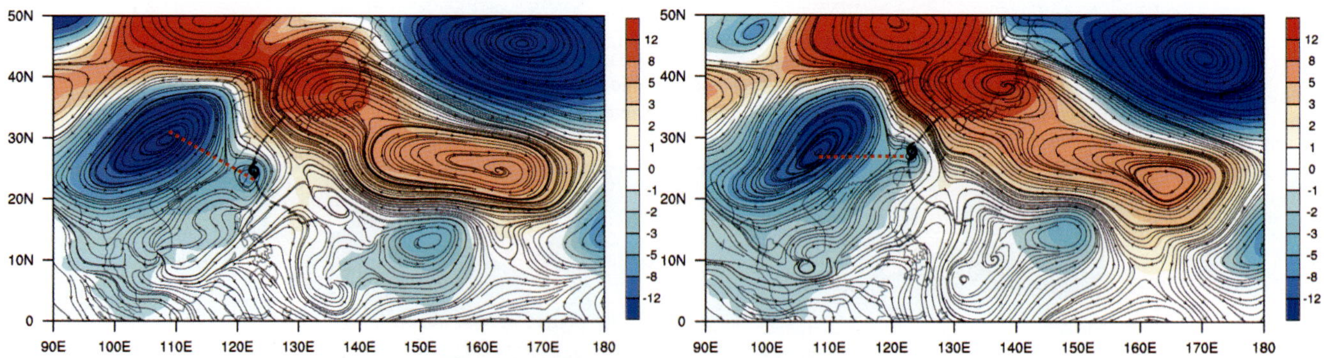

300 hPa flow/height anomalies at 0000 UTC and 1200 UTC 4 October 1985.

Figure 5.27. The typhoon (1985_No.20) generated in the tropical western Pacific on 29 September 1985 and moved northwestward along the climatic monsoon trough. Its intensity peaked in the east sea of the Taiwan Island on 3 October 1985, when it turned right. This turning motion was influenced by an anomalous vortex at 300 hPa.

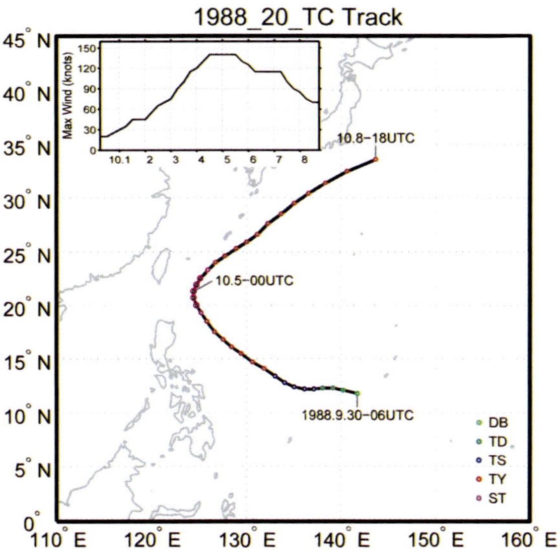

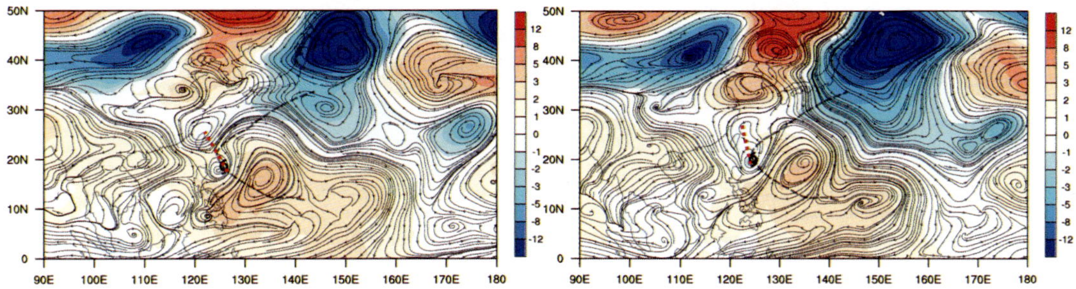

300 hPa flow/height anomalies at 0000 UTC and 1200 UTC 4 October 1988.

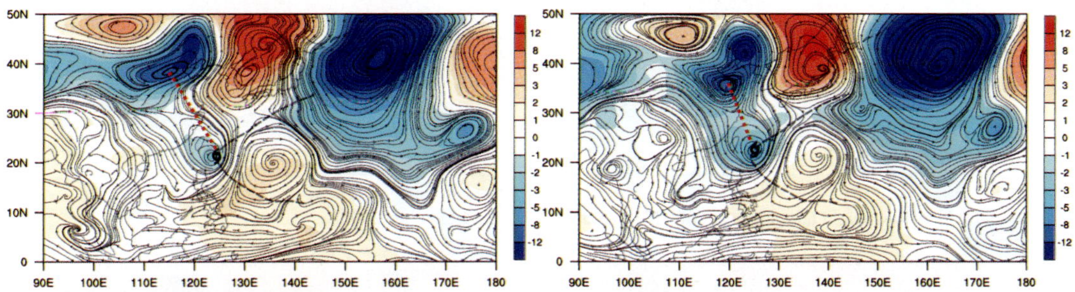

300 hPa flow/height anomalies at 0000 UTC and 1200 UTC 5 October 1988.

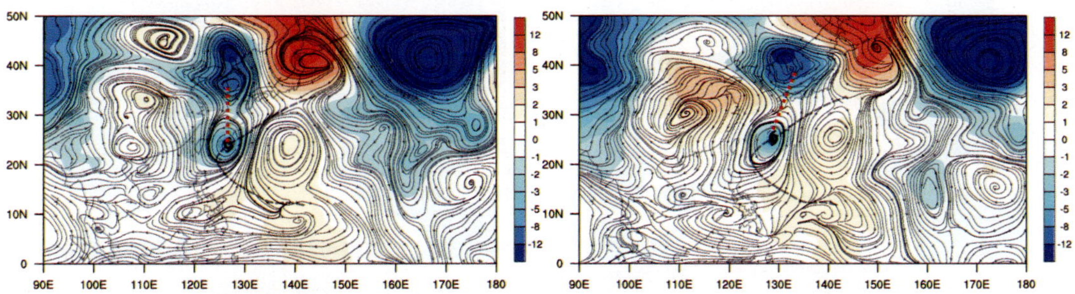

300 hPa flow/height anomalies at 0000 UTC and 1200 UTC 6 October 1988.

Figure 5.28. The typhoon (1988_No.20) generated in the tropical western Pacific on 30 September 1988 and moved northwestward along the climatic monsoon trough. Its intensity peaked in the southeast sea of the Taiwan Island on 5 October 1988, when it turned right. This turning motion was influenced by anomalous vortices at 300 hPa.

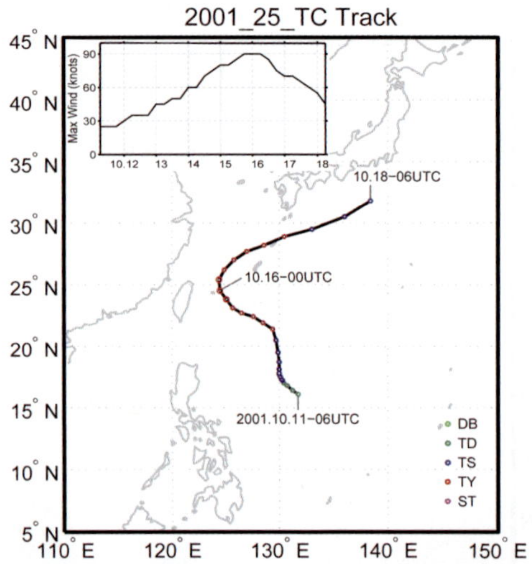

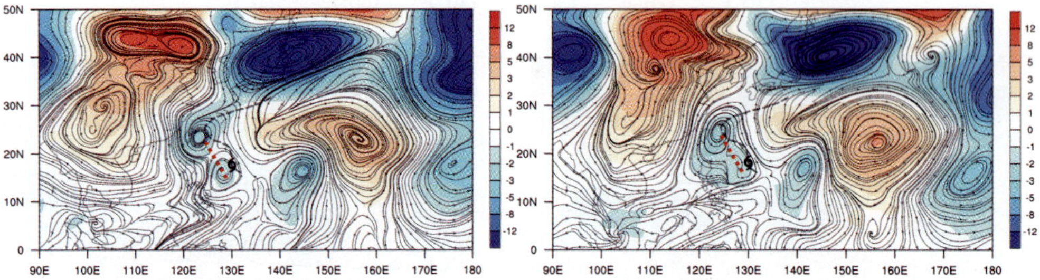

300 hPa flow/height anomalies at 0000 UTC and 1200 UTC 13 October 2001.

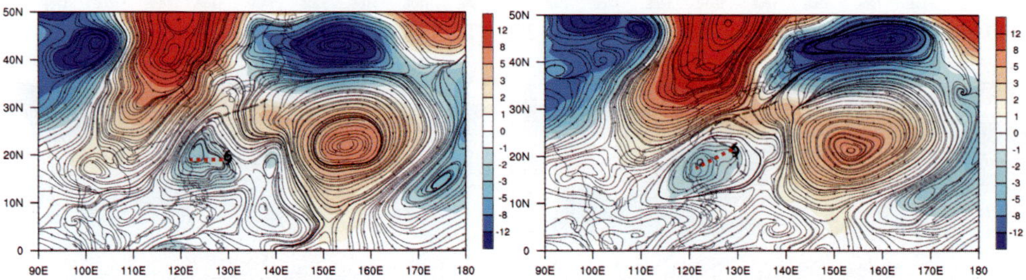

300 hPa flow/height anomalies at 0000 UTC and 1200 UTC 14 October 2001.

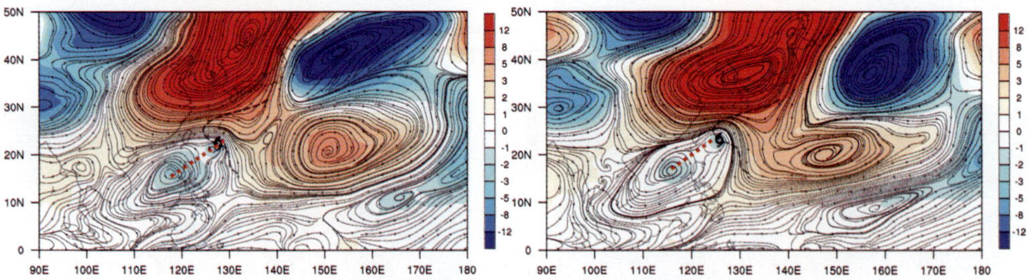

300 hPa flow/height anomalies at 0000 UTC and 1200 UTC 15 October 2001.

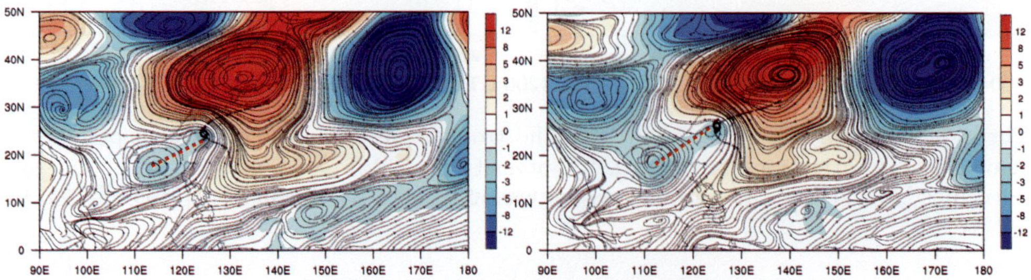

300 hPa flow/height anomalies at 0000 UTC and 1200 UTC 16 October 2001.

Figure 5.29. The typhoon (2001_No.25) generated in the tropical western Pacific on 11 October 2001 and moved northwestward when it was arriving in the climatic monsoon trough. Its intensity peaked in the east sea of the Taiwan Island on 16 October 2001, when it turned right. This turning motion was influenced by an anomalous vortex at 300 hPa.

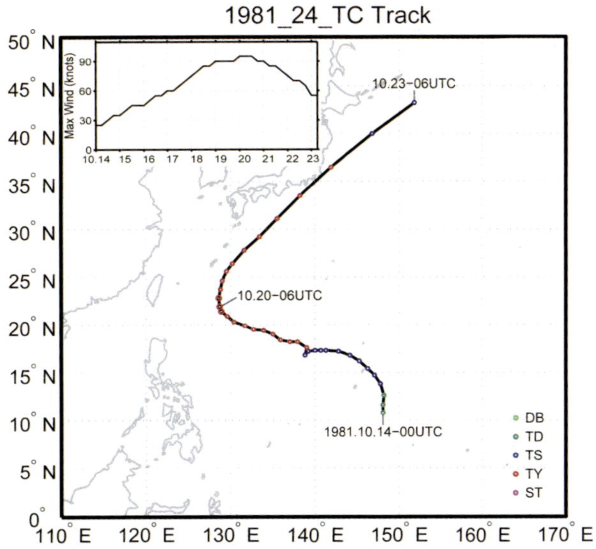

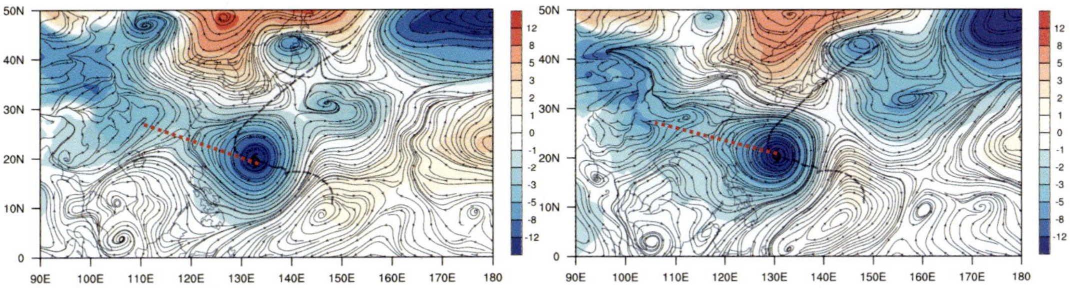
850 hPa flow/height anomalies at 0000 UTC and 1200 UTC 19 October 1981.

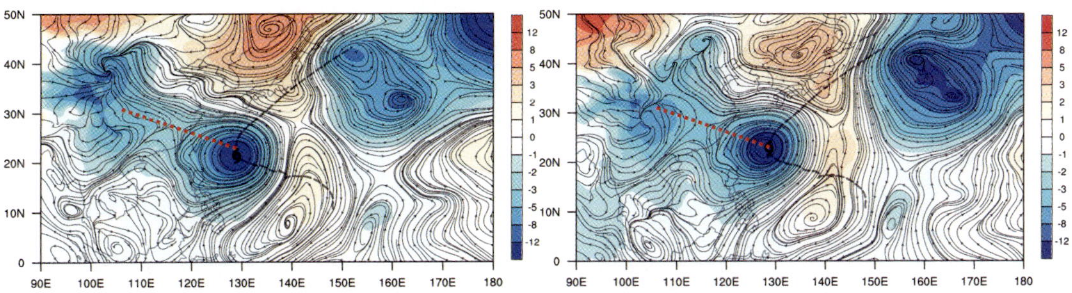
850 hPa flow/height anomalies at 0000 UTC and 1200 UTC 20 October 1981.

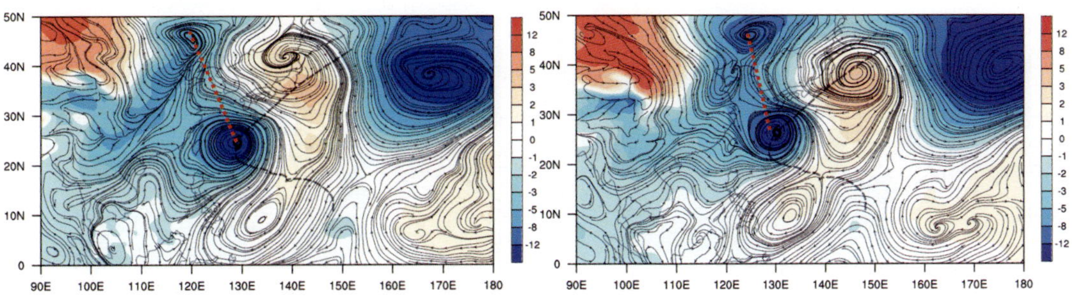
850 hPa flow/height anomalies at 0000 UTC and 1200 UTC 21 October 1981.

Figure 5.30. The typhoon (1981_No.24) generated in the tropical western Pacific on 14 October 1981 and moved northwestward along the climatic monsoon trough. Its intensity peaked in the east sea of the Taiwan Island on 20 October 1981, when it turned right. This turning motion was influenced by anomalous vortices at 850 hPa.

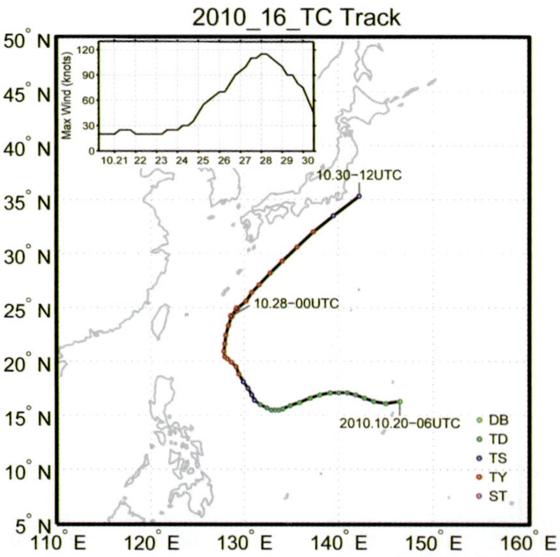

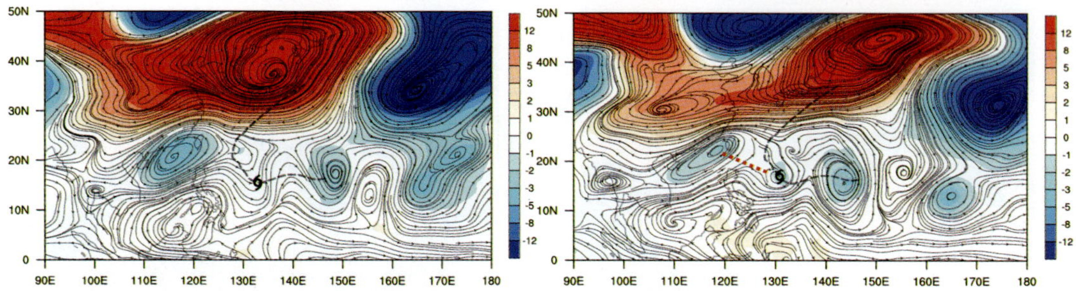

300 hPa flow/height anomalies at 0000 UTC 24 and 25 October 2010.

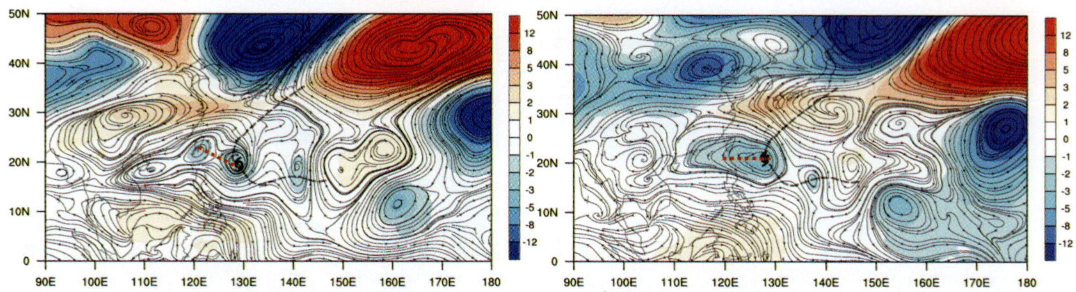

300 hPa flow/height anomalies at 0000 UTC 26 and 27 October 2010.

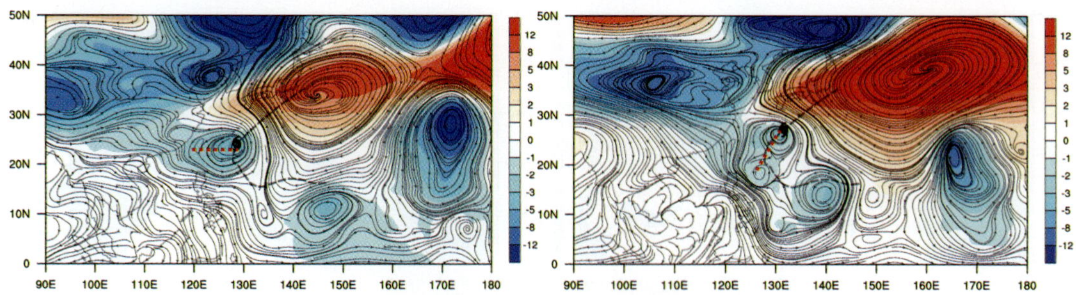

300 hPa flow/height anomalies at 0000 UTC 28 and 29 October 2010.

Figure 5.31. The typhoon (2010_No.16) generated in the tropical western Pacific on 20 October 2010 and moved westward near the climatic monsoon trough. Its intensity peaked in the east sea of the Taiwan Island on 28 October 2010. The right turn of its motion in the east seas of the Taiwan Island and the Philippines was influenced by an anomalous vortex at 300 hPa.

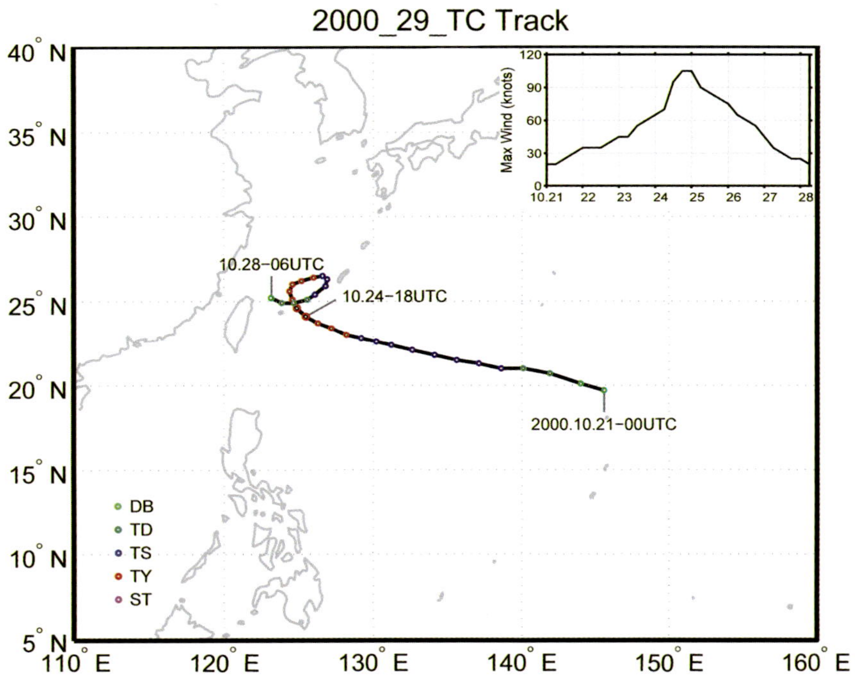

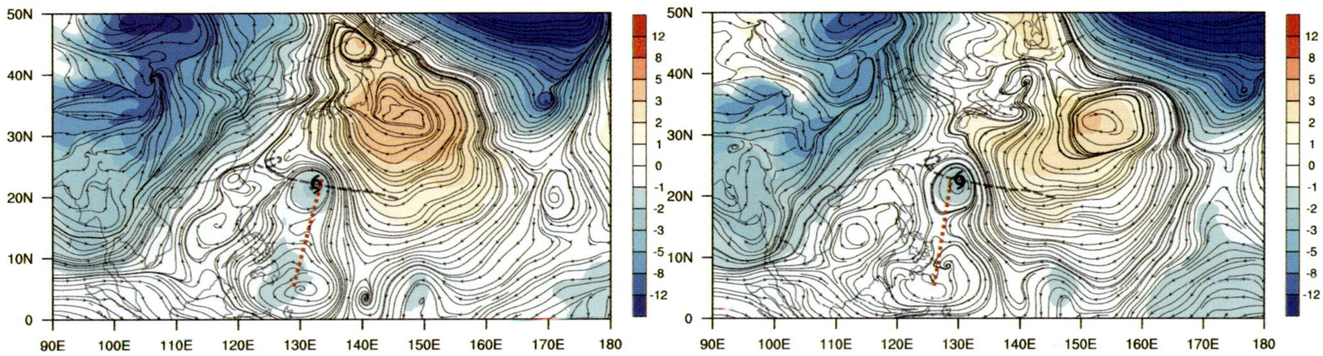

850 hPa flow/height anomalies at 0000 UTC and 1200 UTC 23 October 2000.

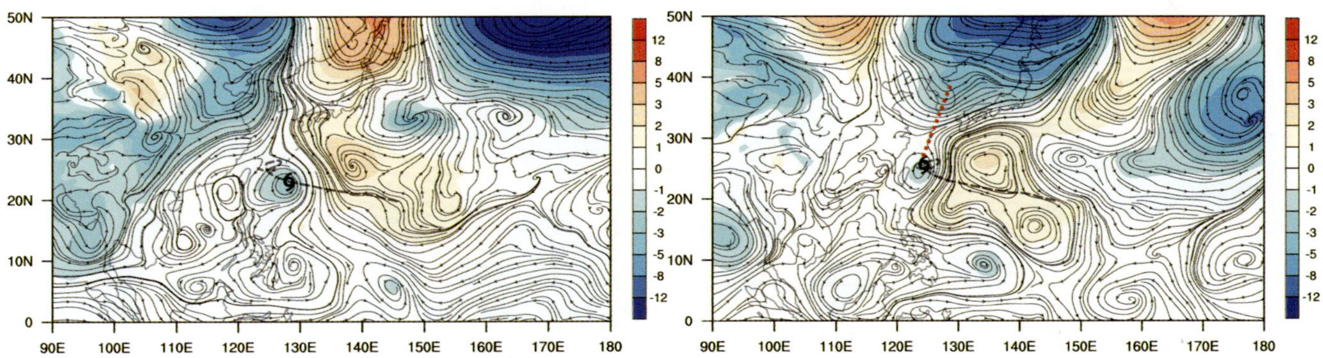

850 hPa flow/height anomalies at 0000 UTC and 1200 UTC 24 October 2000.

Figure 5.32. The typhoon (2000_No.29) generated in the tropical western Pacific on 21 October 2009 and moved westward. Its intensity peaked in the east sea of the Taiwan Island on 24 October 2000 near the climatic monsoon trough. Its circular motion near the Taiwan Island was influenced by anomalous vortices at 850 hPa.

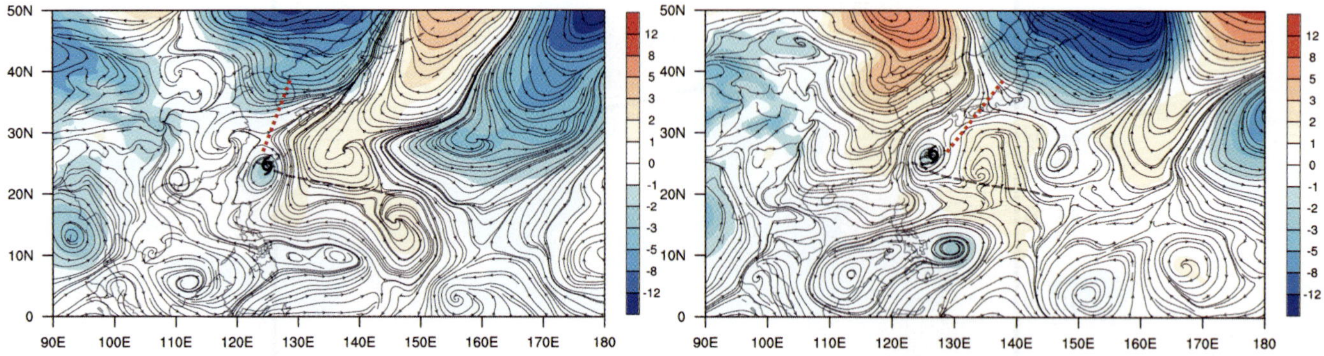

850 hPa flow/height anomalies at 0000 UTC and 1200 UTC 25 October 2000.

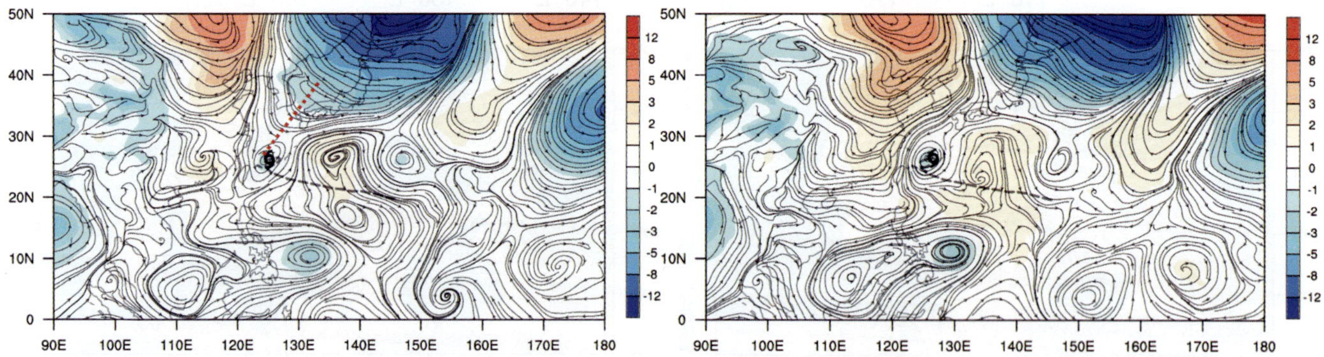

850 hPa flow/height anomalies at 0000 UTC and 1200 UTC 26 October 2000.

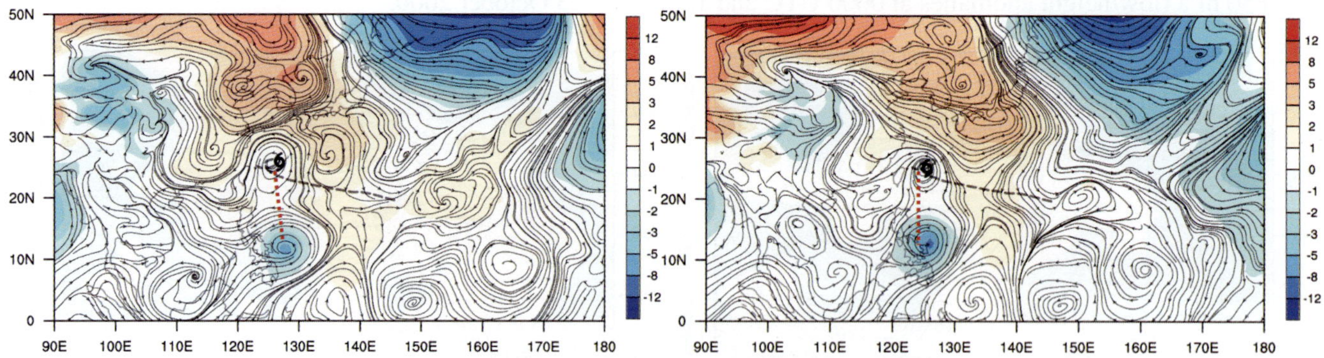

850 hPa flow/height anomalies at 0000 UTC and 1200 UTC 27 October 2000.

Figure 5.32. *(Continued)*

5.6. Geneses in the Tropical Central Pacific

Five typhoons generated in the tropical central Pacific and their intensities peaked in the Taiwan area. The right turns and left turns of their motions were influenced by anomalous vortices (Figs. 5.33–5.37).

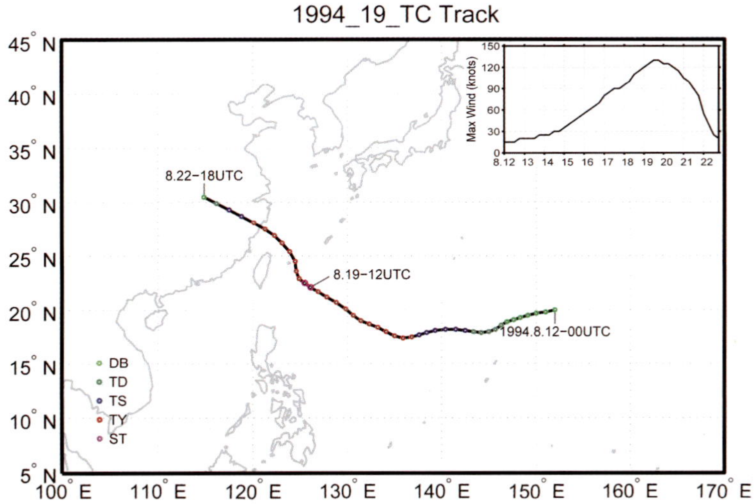

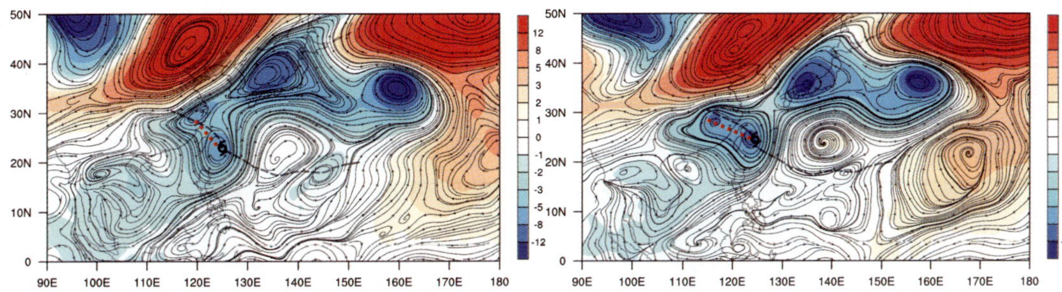

300 hPa flow/height (shading, 10 gpm) anomalies at 0000 UTC and 1200 UTC 20 August 1994.

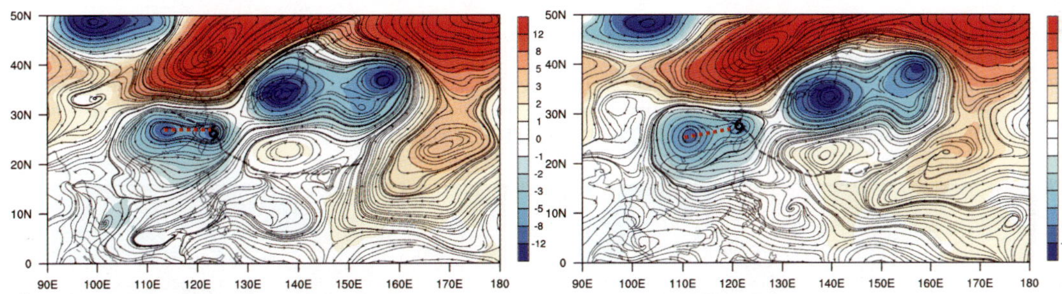

300 hPa flow/height anomalies at 0000 UTC and 1200 UTC 21 August 1994.

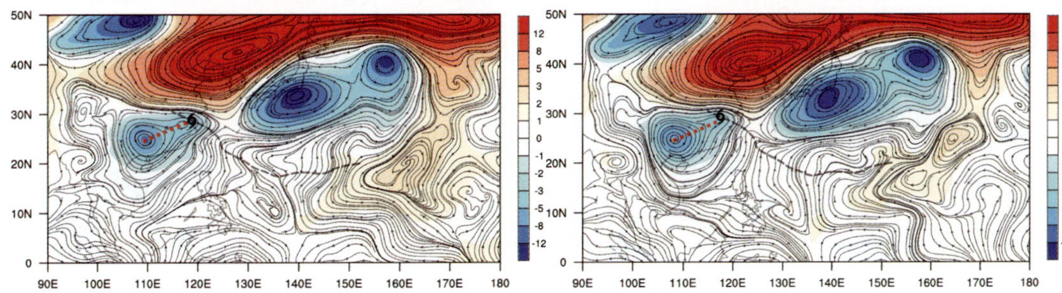

300 hPa flow/height anomalies at 0000 UTC and 0600 UTC 22 August 1994.

Figure 5.33. The typhoon (1994_No.19) generated in the tropical central Pacific on 12 August 1994 and moved westward. Its intensity peaked in the east sea of the Taiwan Island on 19 August 1994 when it was approaching the climatic monsoon trough. The right turn of its motion near the Taiwan Island was influenced by an anomalous vortex at 300 hPa.

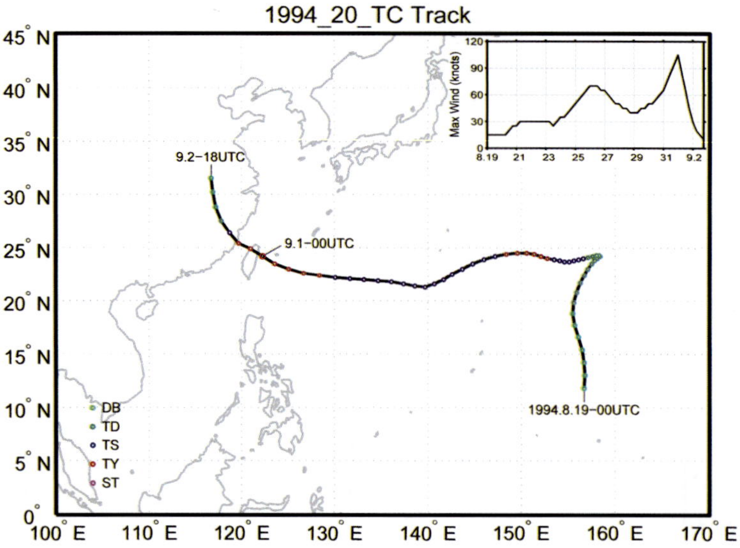

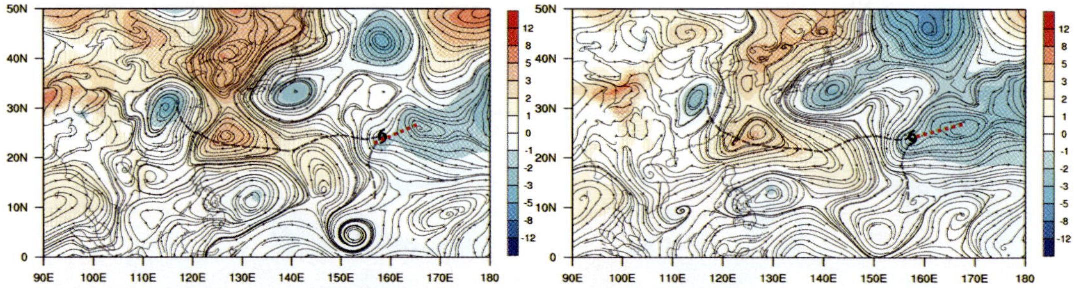

850 hPa flow/height anomalies at 0000 UTC and 1200 UTC 23 August 1994.

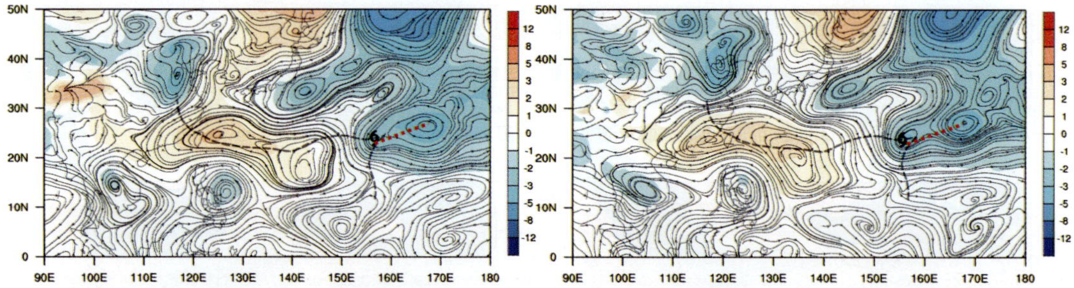

850 hPa flow/height anomalies at 0000 UTC and 1200 UTC 24 August 1994.

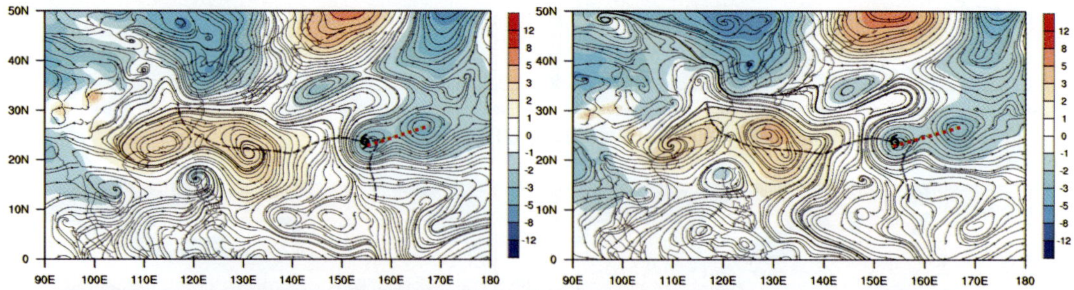

850 hPa flow/height anomalies at 0000 UTC and 0600 UTC 25 August 1994.

Figure 5.34. The typhoon (1994_No.20) generated in the tropical central Pacific on 19 August 1994. Its intensity peaked near the Taiwan Island on 1 September 1994 when it was approaching the subtropical monsoon trough before landfall on the Taiwan Island. The left turn of its motion with a stagnant state in the central Pacific was influenced by an anomalous vortex at 850 hPa.

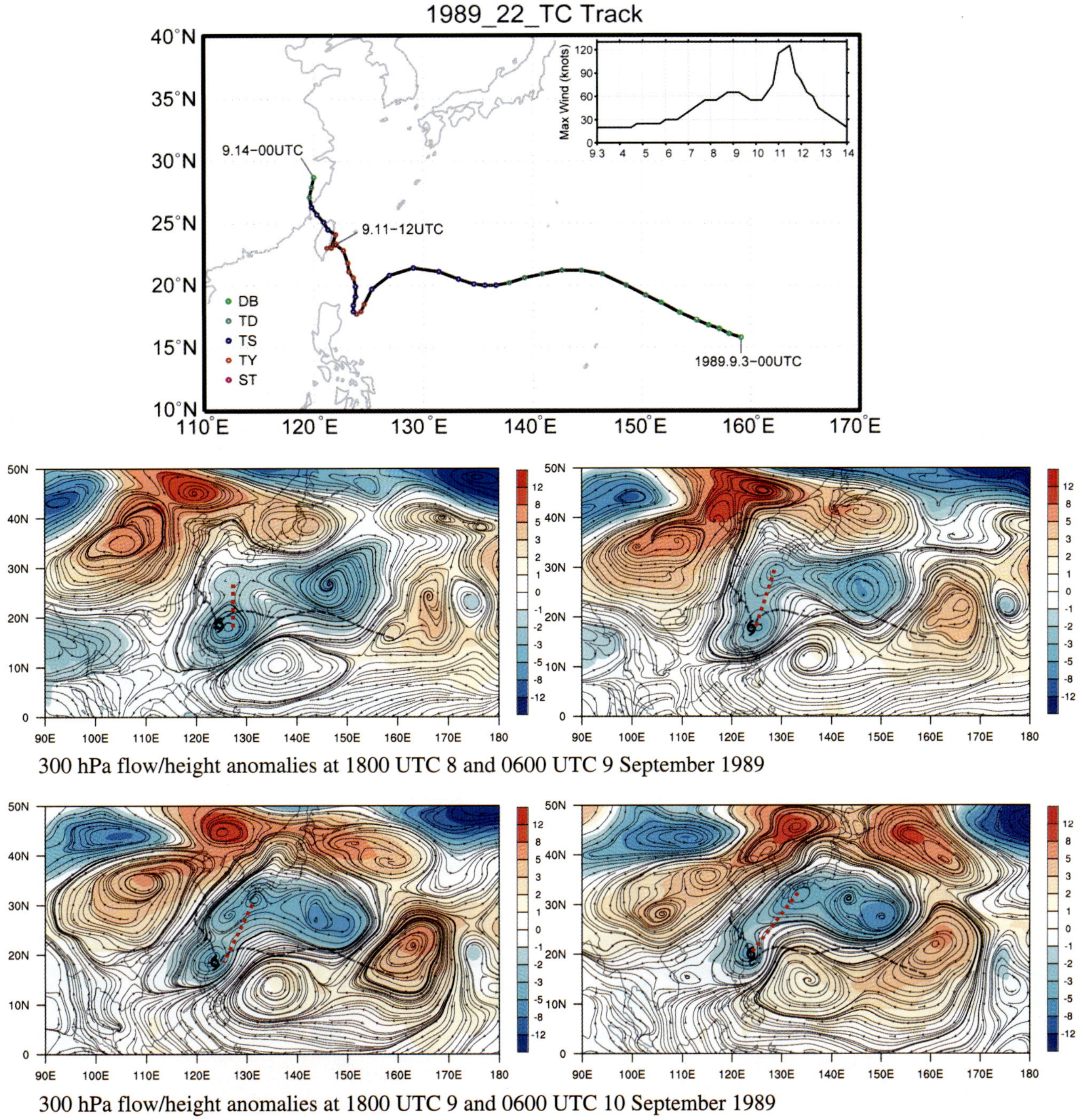

Figure 5.35. The typhoon (1989_No.22) generated in the tropical central Pacific on 3 September 1989. Its intensity peaked on 11 September 1989 before landfall on the Taiwan Island. Its right turn at the northeast sea of the Philippines was influenced by an anomalous vortex at 300 hPa.

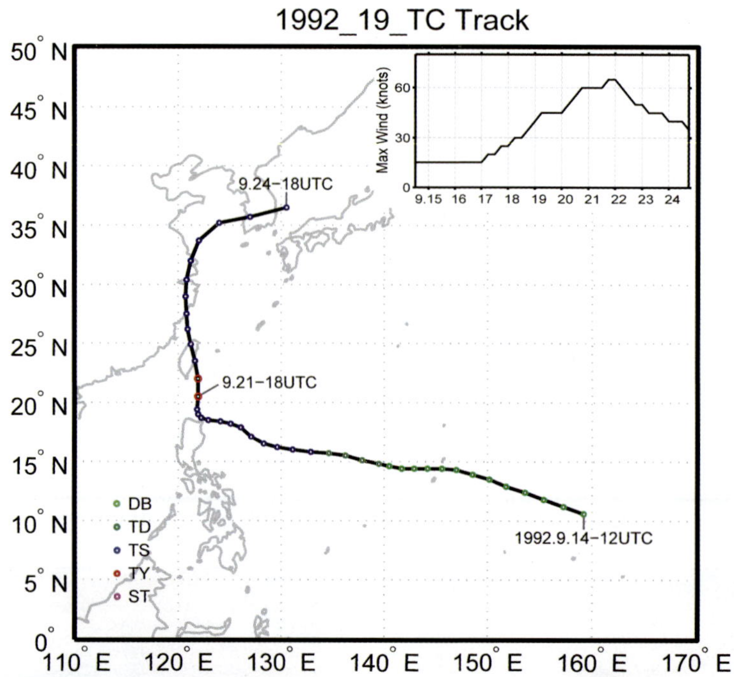

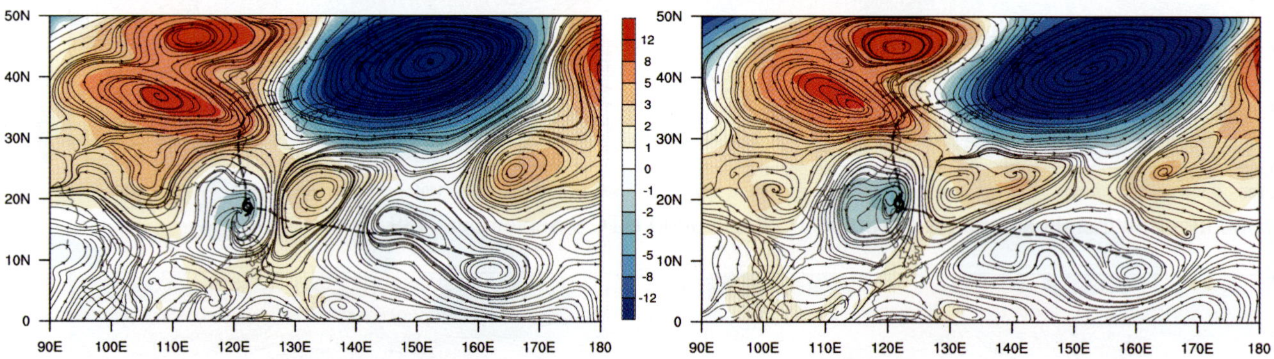

300 hPa flow/height anomalies at 0000 UTC and 1200 UTC 21 September 1992.

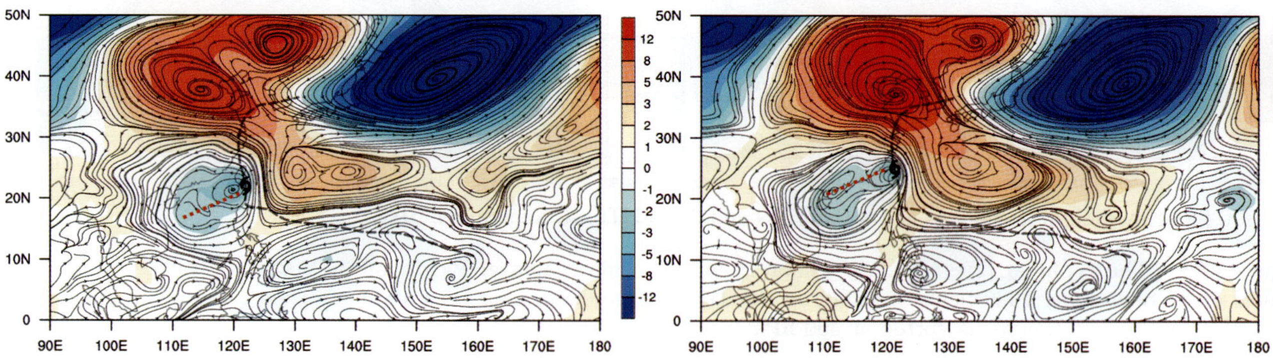

300 hPa flow/height anomalies at 0000 UTC and 1200 UTC 22 September 1992.

Figure 5.36. The typhoon (1992_No.19) generated in the tropical central Pacific on 14 September 1992 and moved northwestward. Its intensity peaked in the south sea of the Taiwan Island on 21 September 1992 when it was arriving in the cross position of two climatic monsoon troughs before landfall on the island. The right turn of its motion near the Philippines and the Taiwan Island was influenced by an anomalous vortex at 300 hPa.

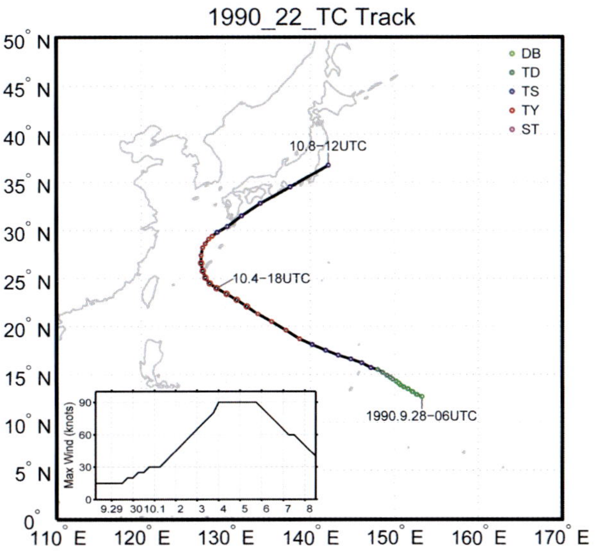

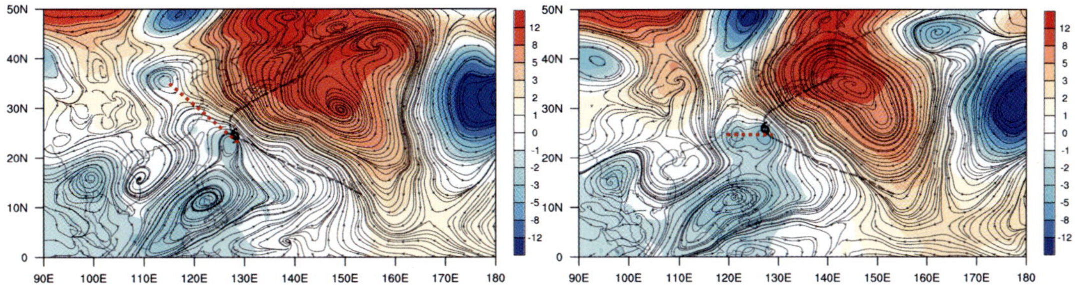

300 hPa flow/height anomalies at 0000 UTC and 1200 UTC 5 October 1990.

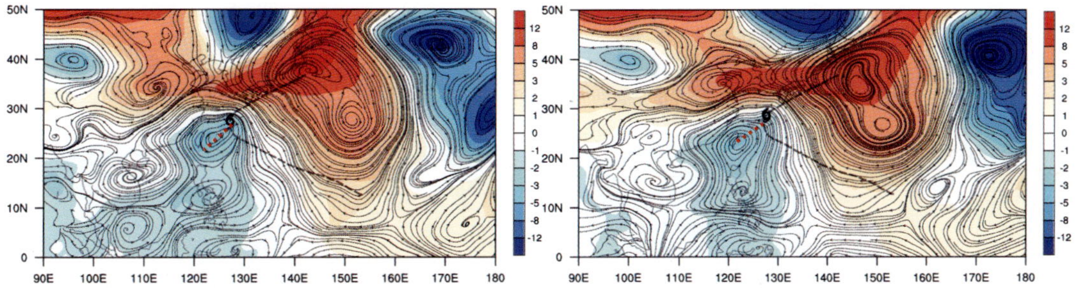

300 hPa flow/height anomalies at 0000 UTC and 1200 UTC 6 October 1990.

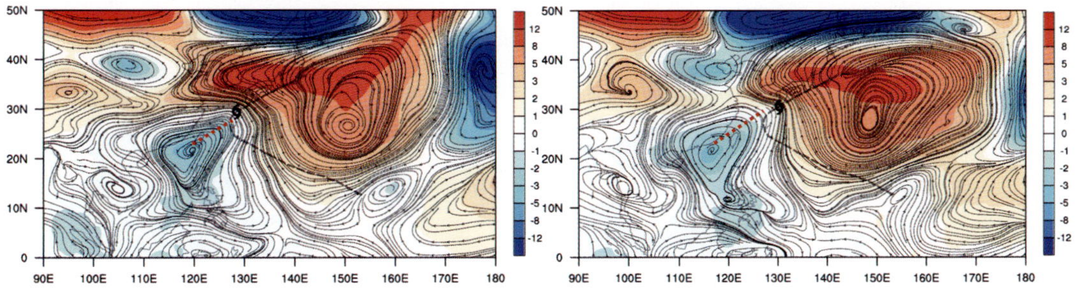

300 hPa flow/height anomalies at 0000 UTC and 1200 UTC 7 October 1990.

Figure 5.37. The typhoon (1990_No.22) generated in the tropical western Pacific on 28 September 1990 and moved northwestward along the climatic monsoon trough. Its intensity peaked in the east sea of the Taiwan Island on 4 October 1990 before turning right. This turning motion in the south of Japan was influenced by an anomalous vortex at 300 hPa.

Chapter 6

Intensity Peaked in the East Seas of China

Most typhoons reaching their peak intensities in the east seas of China generated from tropical disturbances in equatorial western Pacific (Sec. 6.1), in the equatorial central Pacific (Sec. 6.2), in the tropical western Pacific (Sec. 6.3), in the east seas of China (Sec. 6.4), and in the Northwest Pacific (Sec. 6.5), respectively.

6.1. Geneses in the Equatorial Western Pacific

Two typhoons generated in the equatorial western Pacific and their intensities peaked in the east seas of China. The right turns or left turns of their motions were influenced by anomalous vortices (Figs. 6.1 and 6.2).

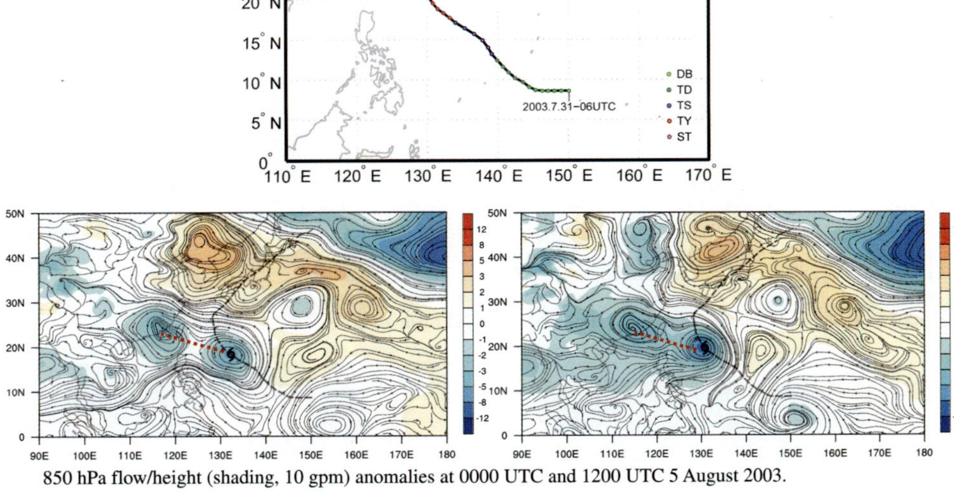

Figure 6.1. The typhoon (2003_No.11) generated in the equatorial western Pacific on 31 July 2003 and moved northwestward before its intensity peaked near the climatic monsoon trough in the East China Sea on 7 August 2003. The right turn of its motion in the east sea of the Taiwan Island was influenced by anomalous vortices at 850 hPa and 300 hPa. Symbols used are same as in Fig. 2.1.

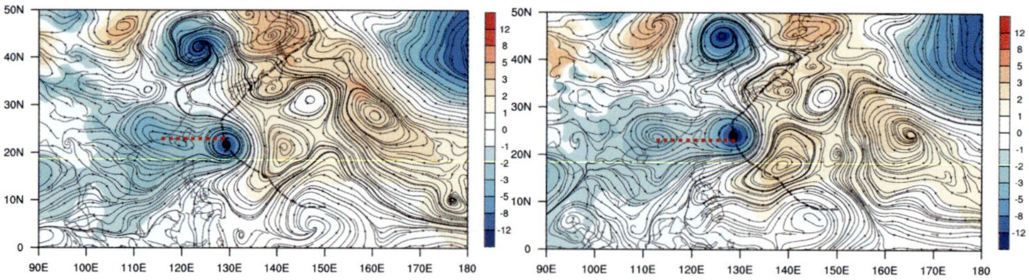

850 hPa flow/height anomalies at 0000 UTC and 1200 UTC 6 August 2003.

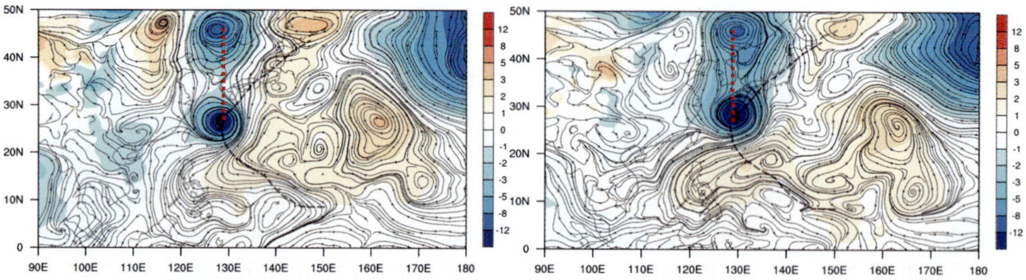

850 hPa flow/height anomalies at 0000 UTC and 1200 UTC 7 August 2003.

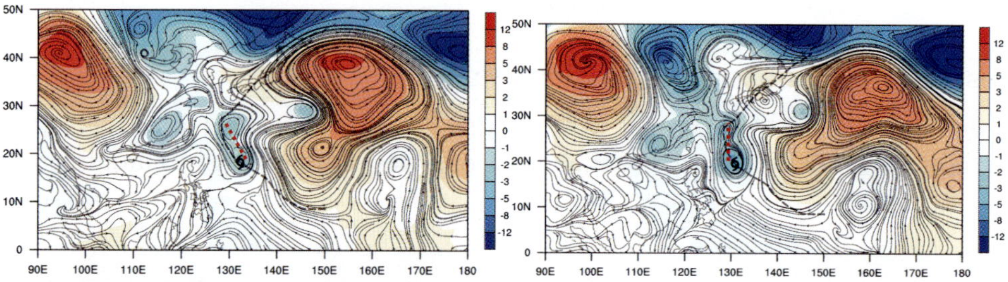

300 hPa flow/height anomalies at 0000 UTC and 1200 UTC 5 August 2003.

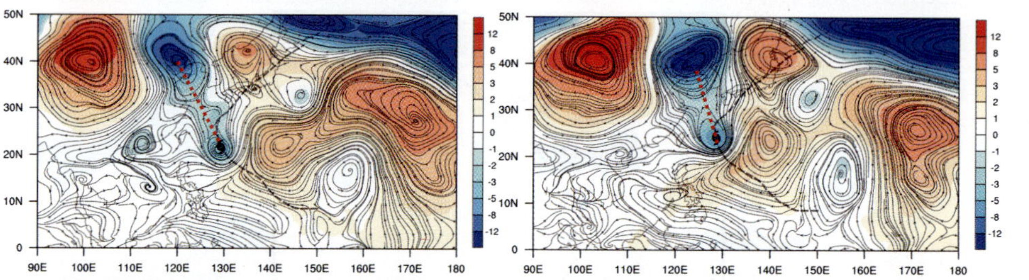

300 hPa flow/height anomalies at 0000 UTC and 1200 UTC 6 August 2003.

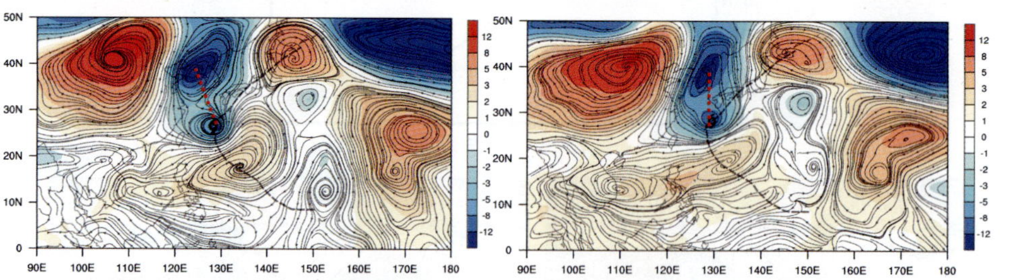

300 hPa flow/height anomalies at 0000 UTC and 1200 UTC 7 August 2003.

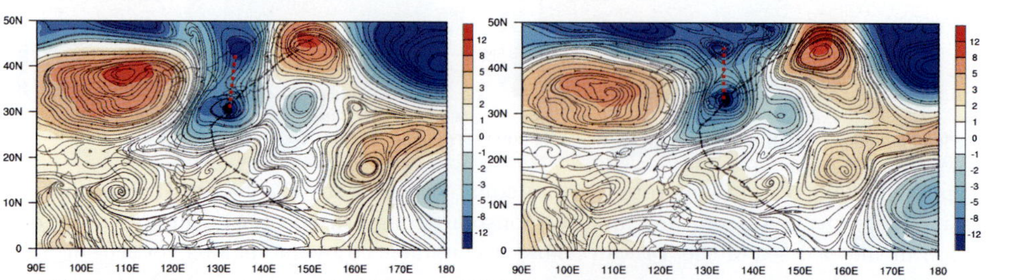

300 hPa flow/height anomalies at 0000 UTC and 1200 UTC 8 August 2003.

Figure 6.1. *(Continued)*

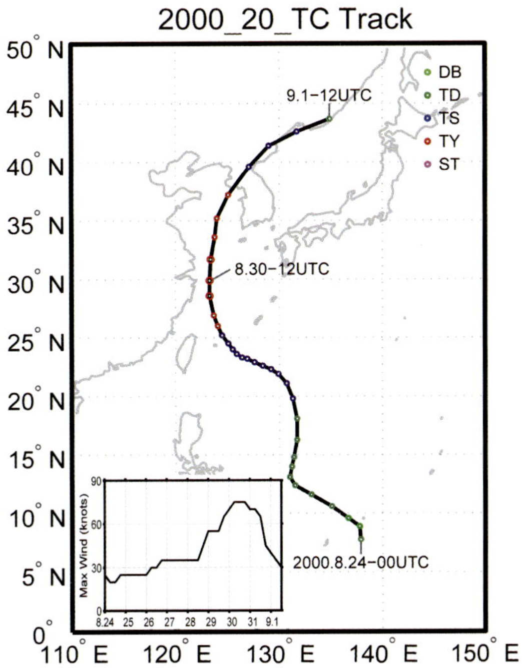

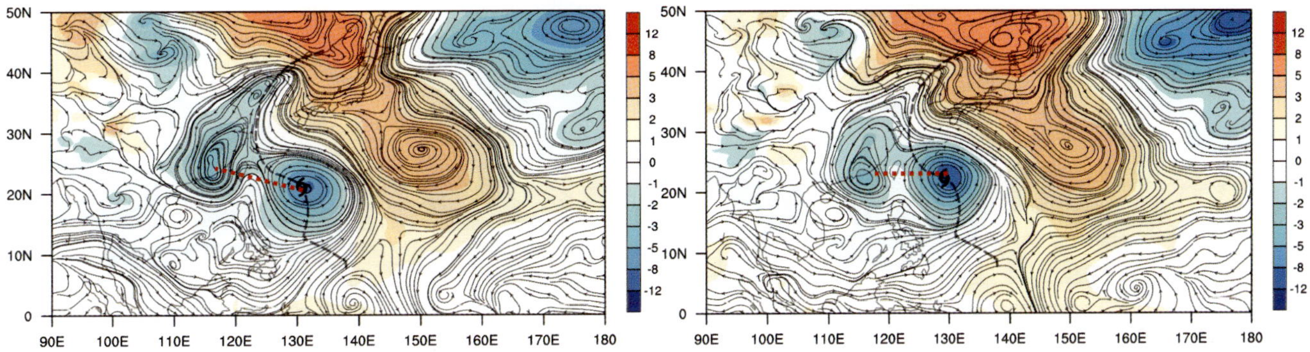

850 hPa flow/height anomalies at 0000 UTC and 1200 UTC 27 August 2000.

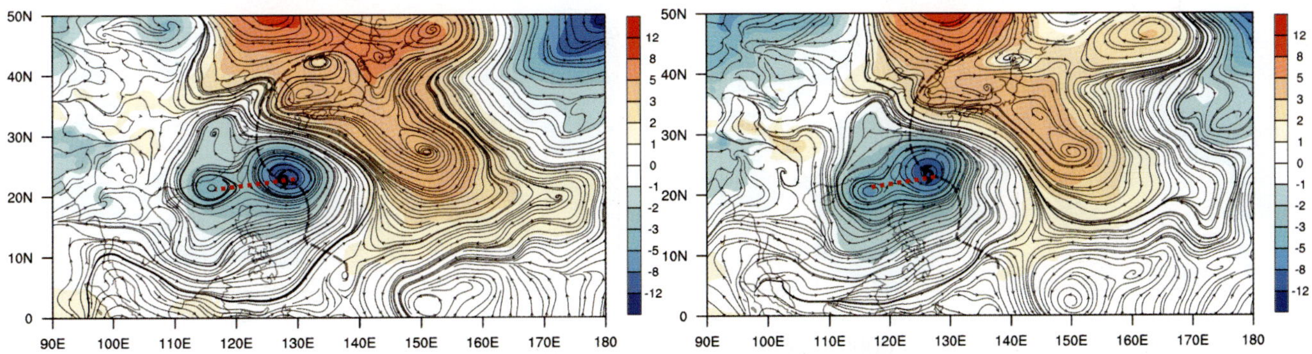

850 hPa flow/height anomalies at 0000 UTC and 1200 UTC 28 August 2000.

Figure 6.2. The typhoon (2000_No.20) generated in the equatorial western Pacific on 24 August 2000 and experienced a right turn and a left turn while moving northward. Its intensity peaked on 30 August 2000 when it was approaching the subtropical climatic monsoon trough and its motions were influenced by an anomalous vortex at 850 hPa.

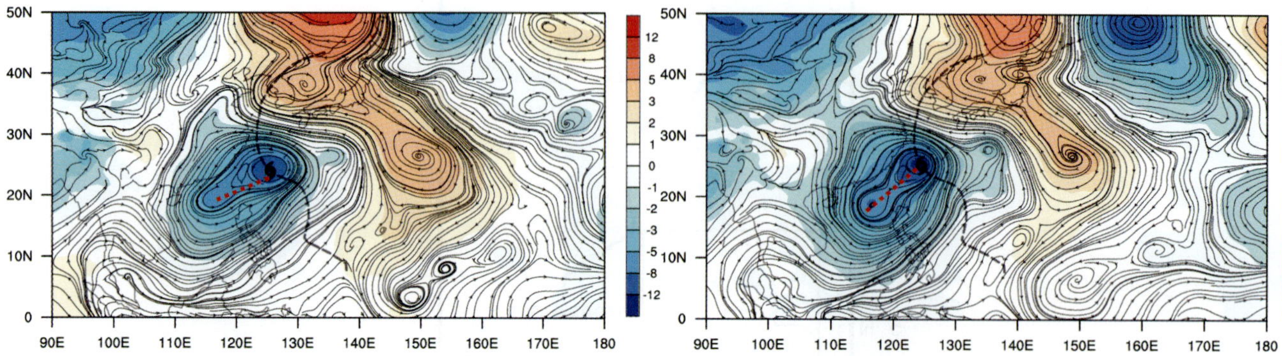

850 hPa flow/height anomalies at 0000 UTC and 1200 UTC 29 August 2000.

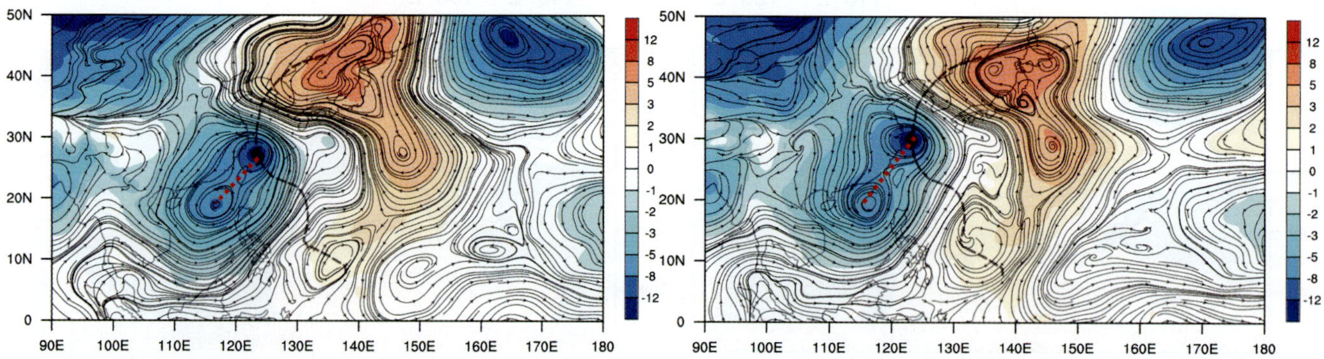

850 hPa flow/height anomalies at 0000 UTC and 1200 UTC 30 August 2000.

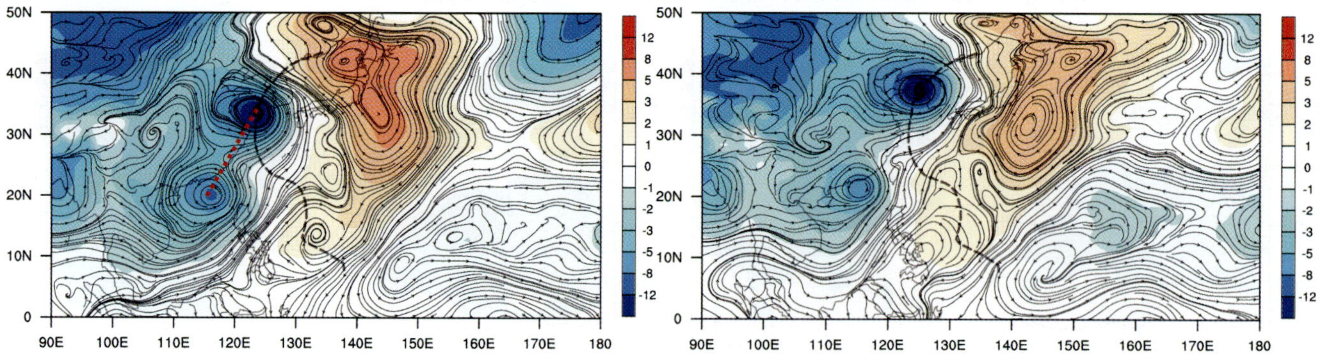

850 hPa flow/height anomalies at 0000 UTC and 1200 UTC 31 August 2000.

Figure 6.2. *(Continued)*

6.2. Geneses in the Equatorial Central Pacific

Five typhoons generated in the equatorial central Pacific and their intensities peaked in the east seas of China. The right turns and left turns of their motions were influenced by anomalous vortices (Figs. 6.3–6.7).

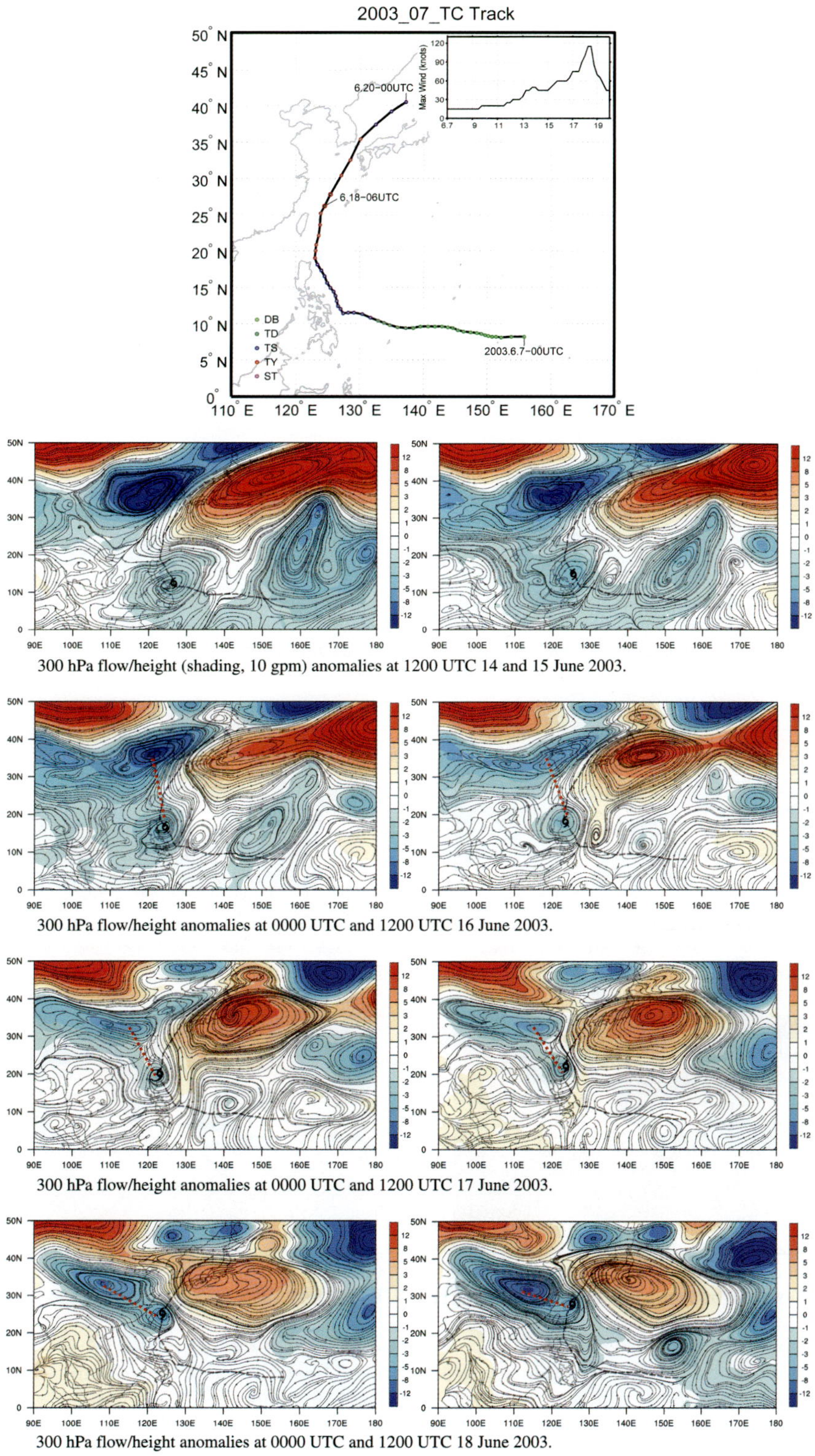

Figure 6.3. The typhoon (2003_No.07) generated in the equatorial central Pacific on 7 June 2003 and moved westward before turning right. Its intensity peaked in the East China Sea on 18 June 2003 when it was approaching the subtropical climatic monsoon trough. This right turn in the Philippines and the east seas of China was influenced by an anomalous vortex at 300 hPa.

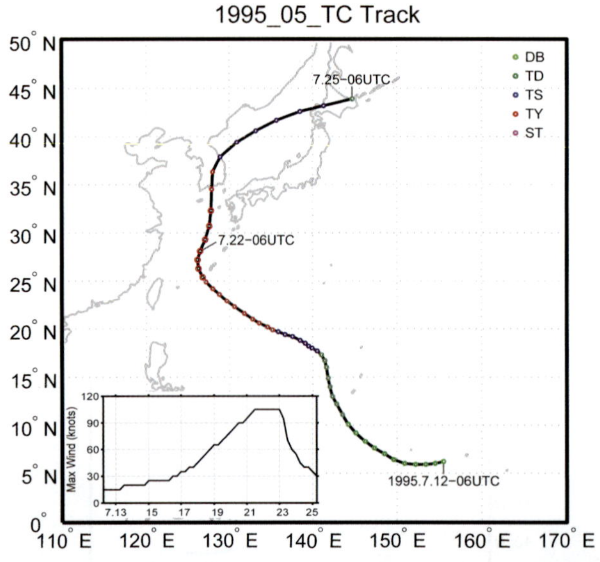

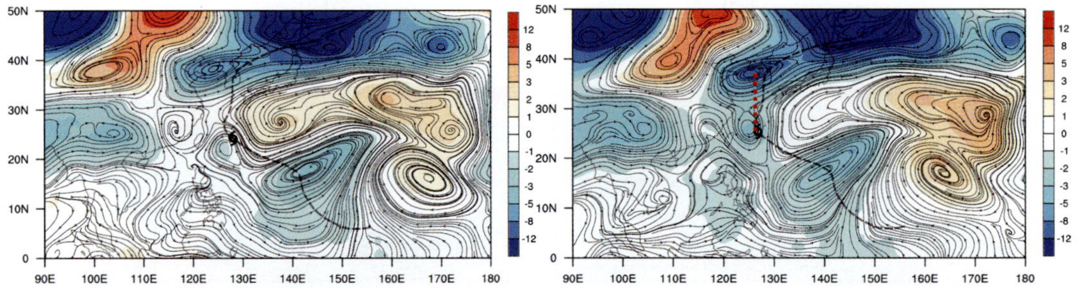

300 hPa flow/height anomalies at 0000 UTC and 1200 UTC 21 July 1995.

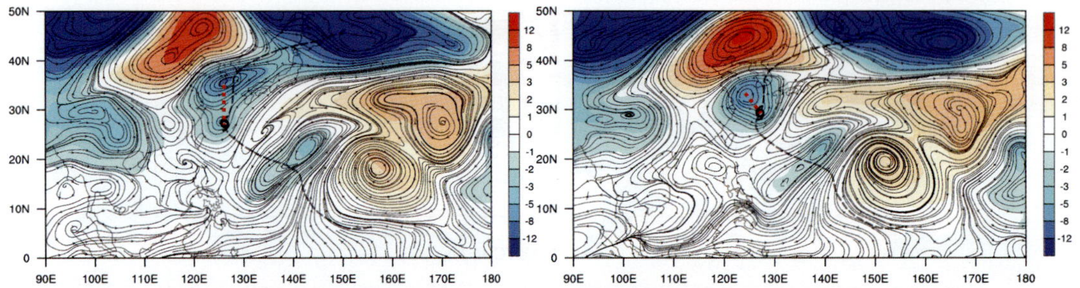

300 hPa flow/height anomalies at 0000 UTC and 1200 UTC 22 July 1995.

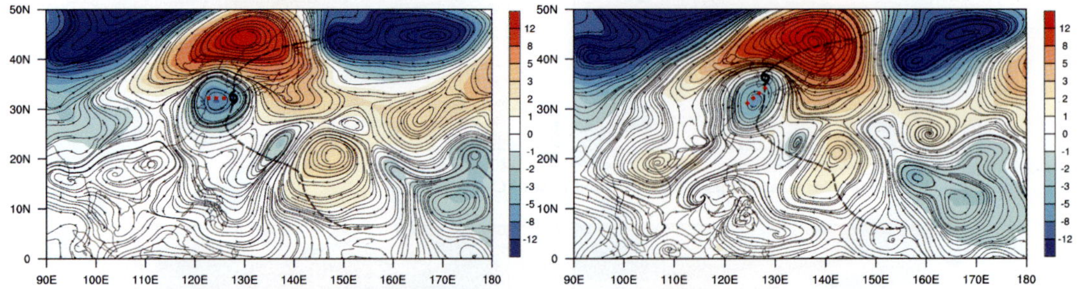

300 hPa flow/height anomalies at 0000 UTC and 1200 UTC 23 July 1995.

Figure 6.4. The typhoon (1995_No.05) generated in the equatorial central Pacific on 12 July 1995 and moved northwestward along the climatic monsoon trough. Its intensity peaked in the East China Sea on 22 July 1995 at the moment when it turned right. This turning motion in the east seas of China was influenced by an anomalous vortex at 300 hPa.

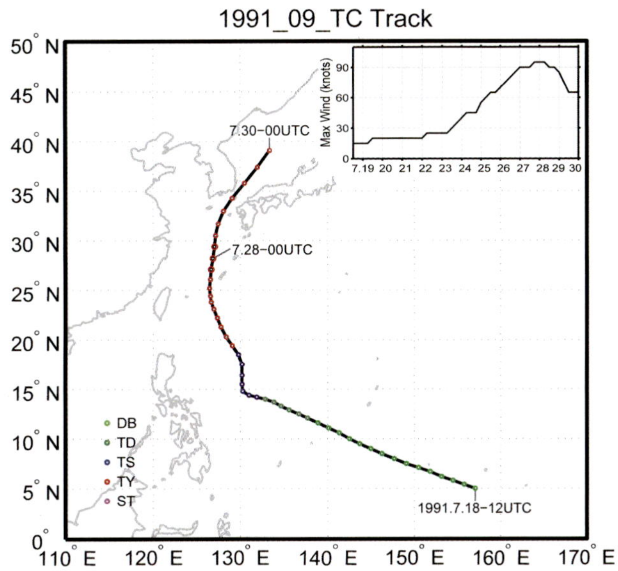

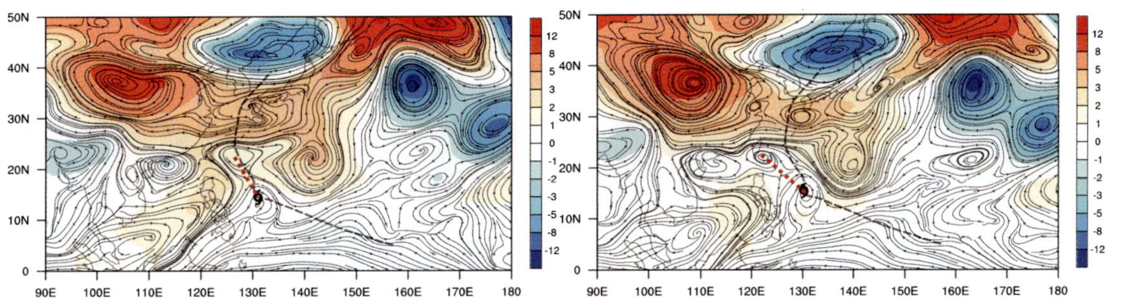

300 hPa flow/height anomalies at 0000 UTC and 1200 UTC 24 July 1991.

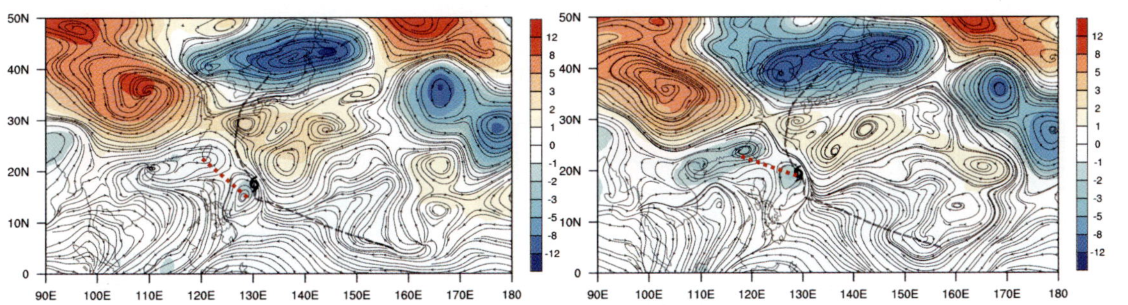

300 hPa flow/height anomalies at 0000 UTC and 1200 UTC 25 July 1991.

Figure 6.5. The typhoon (1991_No.09) generated in the equatorial central Pacific on 18 July 1991 and moved northwestward along the climatic monsoon trough. Its maximum wind speed reached above 90 knots in the east sea of the Taiwan Island on 27 and 28 July 1991. The right turn of its motion in the east seas of China was influenced by anomalous vortices at 300 hPa.

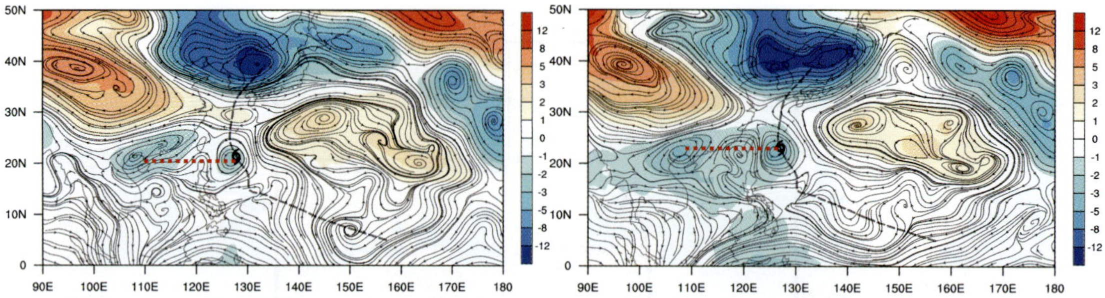

300 hPa flow/height anomalies at 0000 UTC and 1200 UTC 26 July 1991.

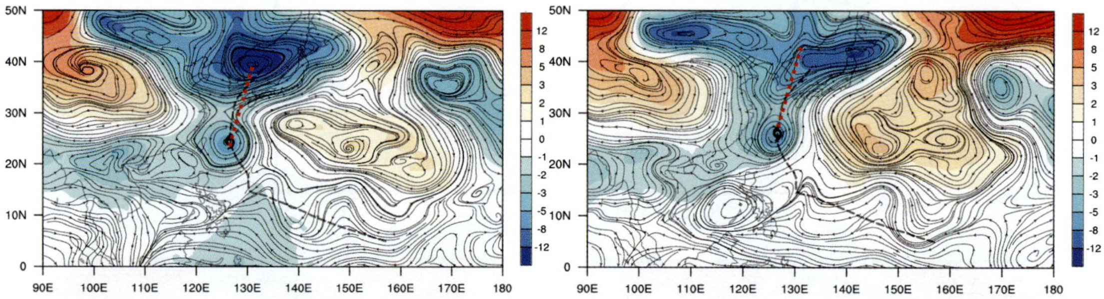

300 hPa flow/height anomalies at 0000 UTC and 1200 UTC 27 July 1991.

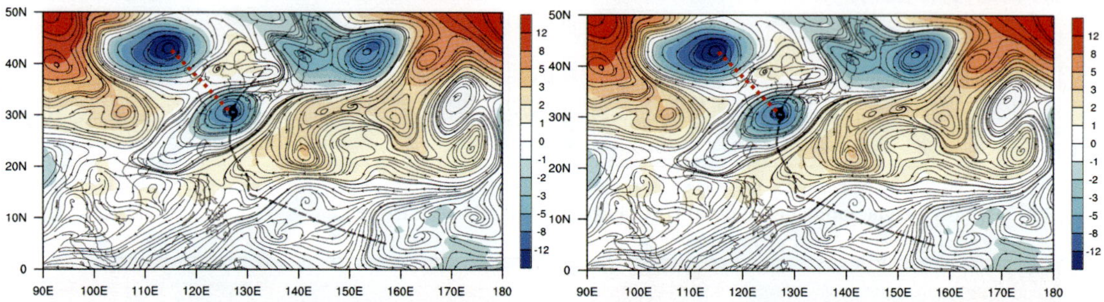

300 hPa flow/height anomalies at 0000 UTC and 1200 UTC 28 July 1991.

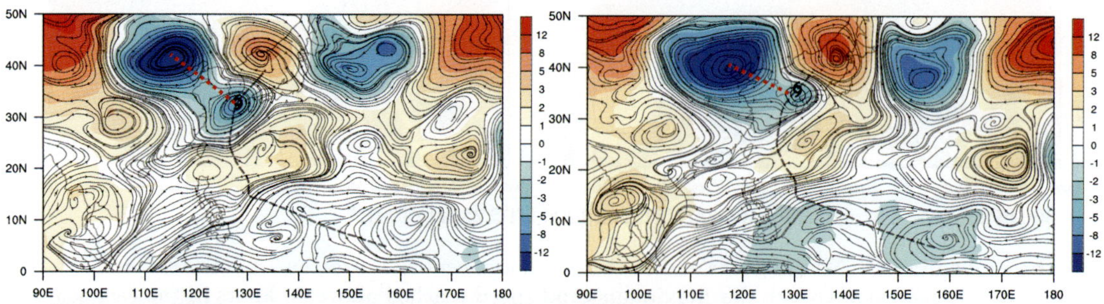

300 hPa flow/height anomalies at 0000 UTC and 1200 UTC 29 July 1991.

Figure 6.5. *(Continued)*

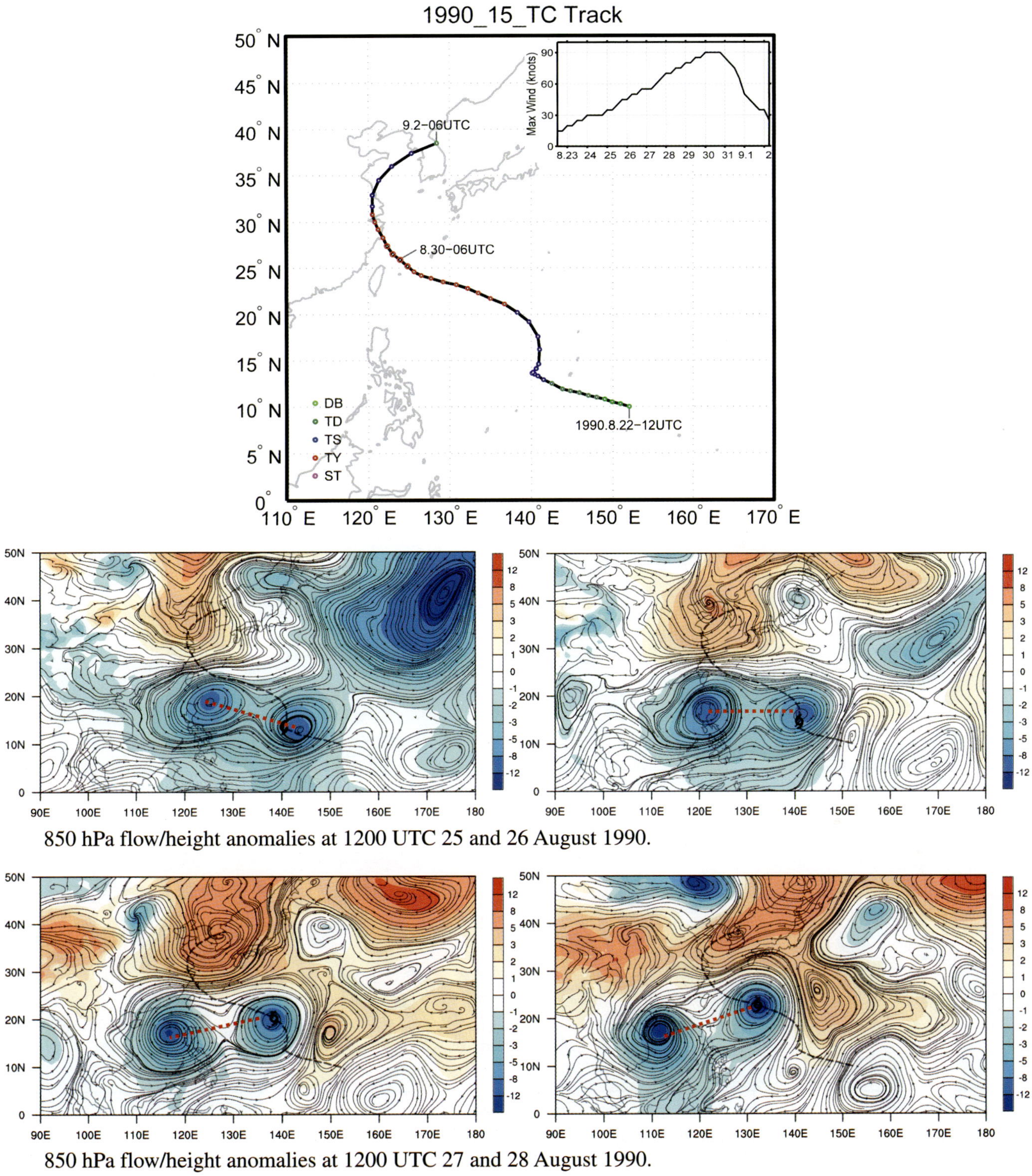

Figure 6.6. The typhoon (1990_No.15) generated in the equatorial central Pacific on 22 August 1990 and moved northwestward along the climatic monsoon trough. Its intensity peaked in the northeast sea of the Taiwan Island on 30 August 1990 before landfall on the East China coast. The left turn and right turn of its motion were influenced by anomalous vortices at 850 hPa and 300 hPa.

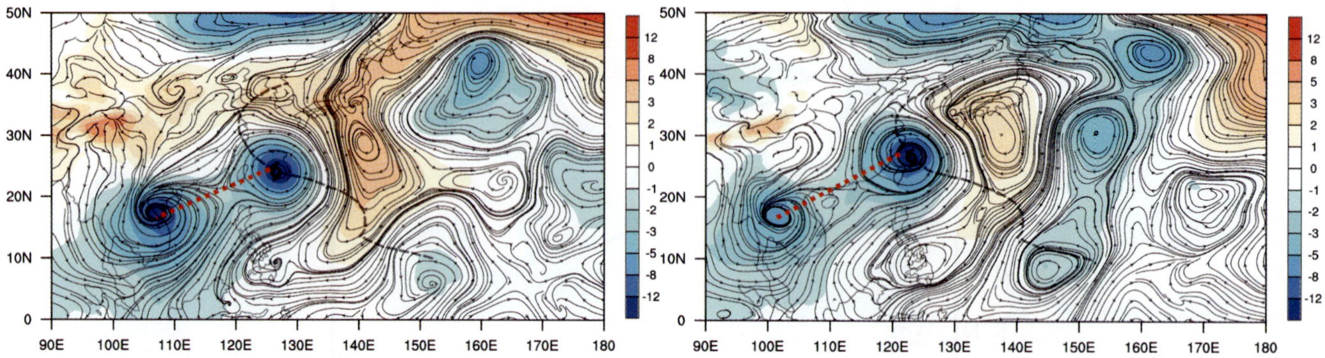

850 hPa flow/height anomalies at 1200 UTC 29 and 30 August 1990.

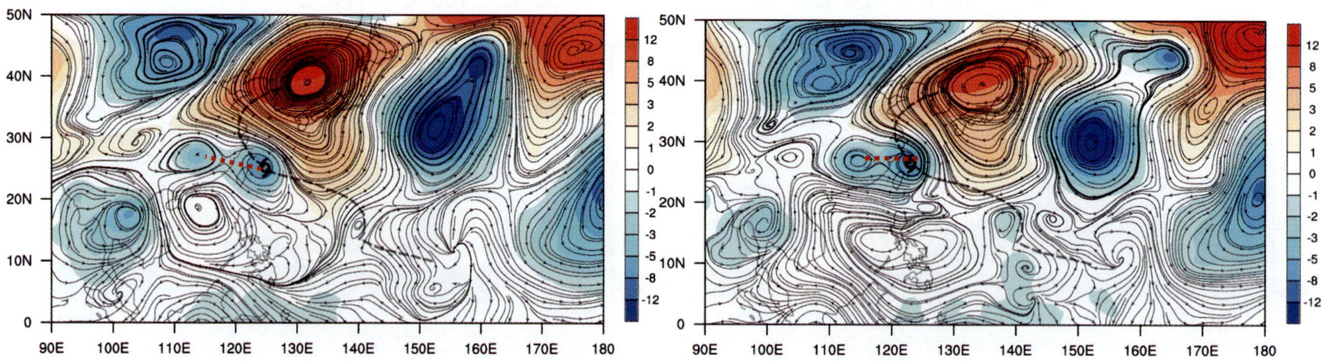

300 hPa flow/height anomalies at 0000 UTC and 1200 UTC 30 August 1990.

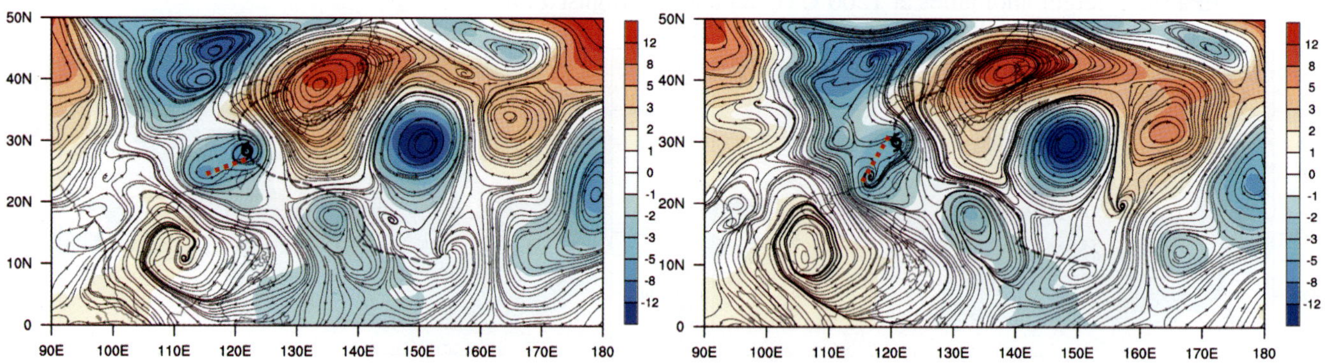

300 hPa flow/height anomalies at 0000 UTC and 1200 UTC 31 August 1990.

Figure 6.6. *(Continued)*

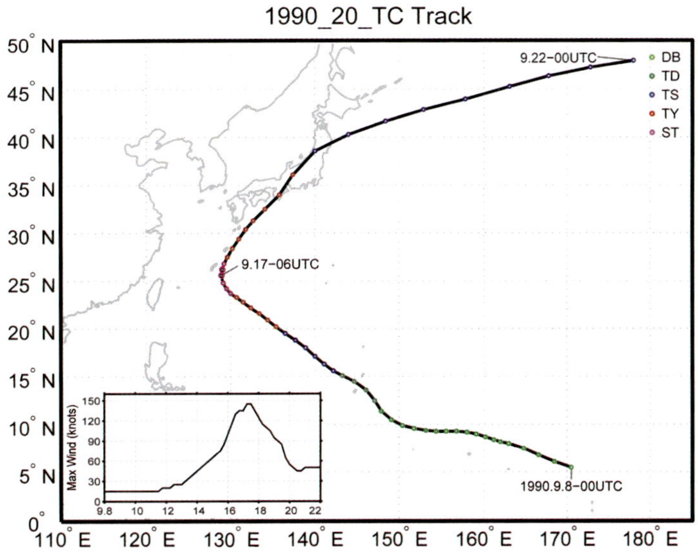

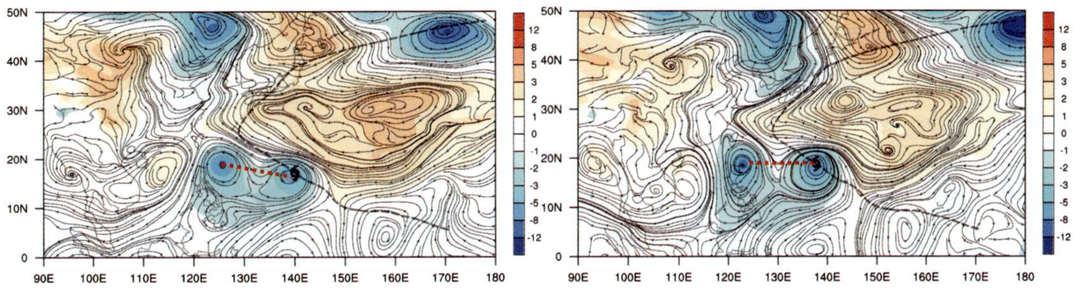

850 hPa flow/height anomalies at 0000 UTC and 1200 UTC 14 September 1990.

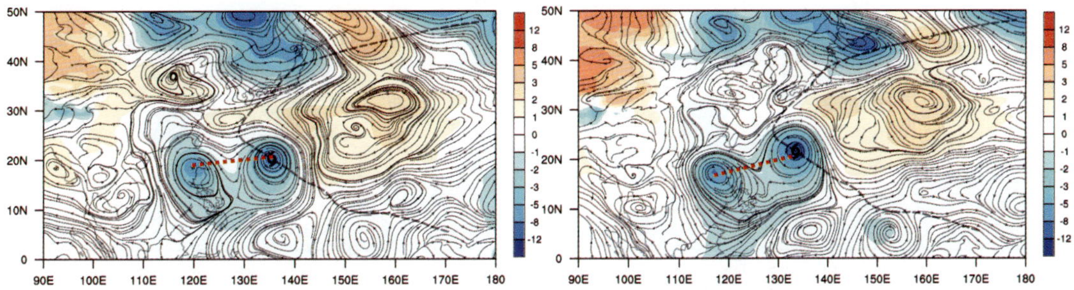

850 hPa flow/height anomalies at 0000 UTC and 1200 UTC 15 September 1990.

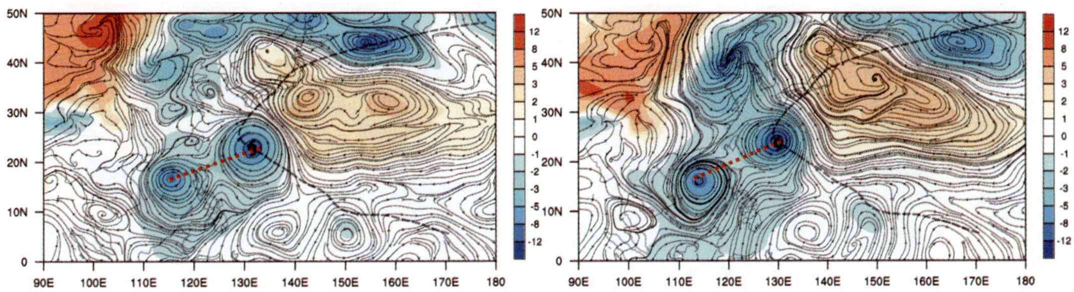

850 hPa flow/height anomalies at 0000 UTC and 1200 UTC 16 September 1990.

Figure 6.7. The typhoon (1990_No.20) generated in the equatorial central Pacific on 8 September 1990 and moved northwestward along the climatic monsoon trough. Its intensity peaked in the east sea of the Taiwan Island on 17 September 1990 at the moment when it turned right. This right turn in southern Japan was influenced by anomalous vortices at 850 hPa and 300 hPa.

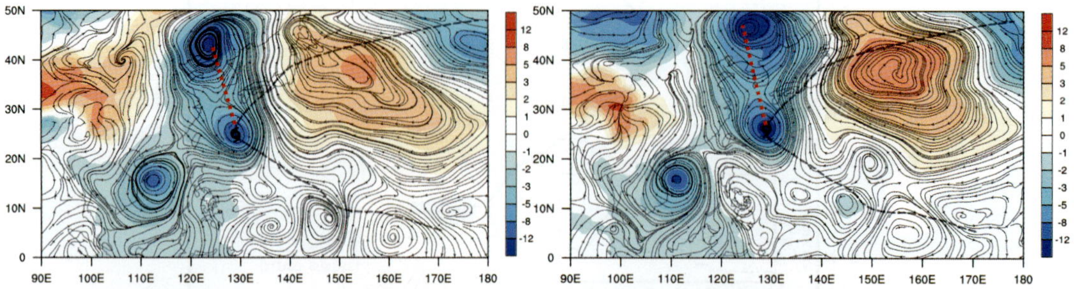

850 hPa flow/height anomalies at 0000 UTC and 1200 UTC 17 September 1990.

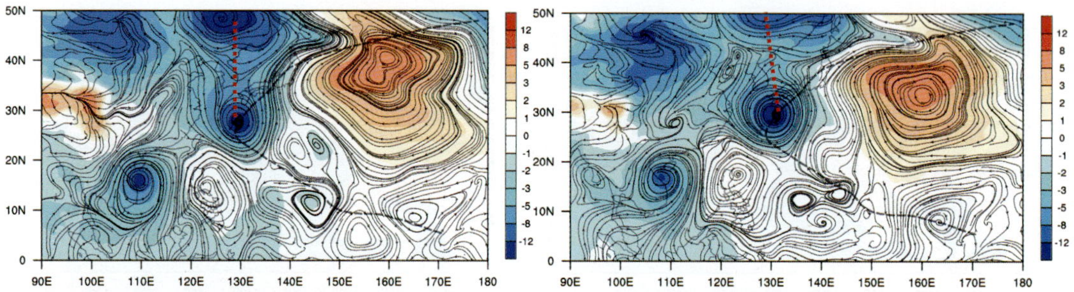

850 hPa flow/height anomalies at 0000 UTC and 1200 UTC 18 September 1990.

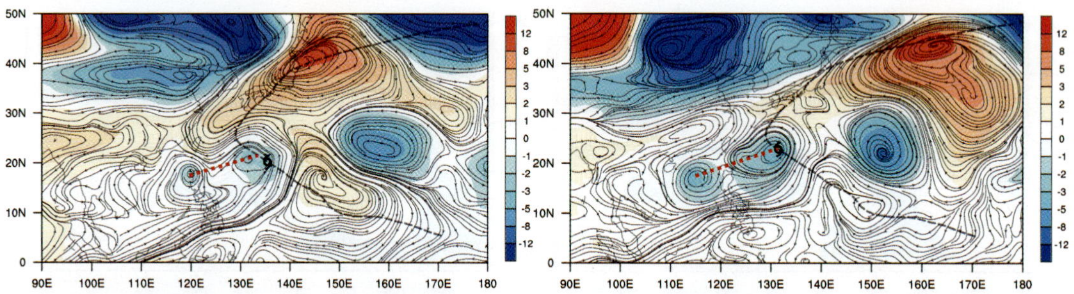

300 hPa flow/height anomalies at 0000 UTC 15 and 16 September 1990.

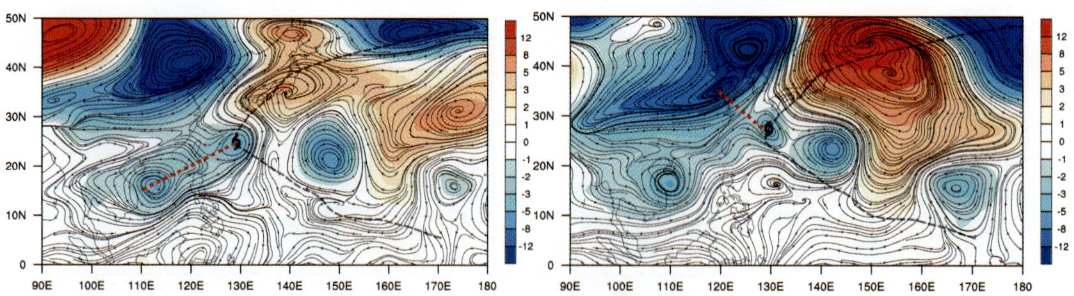

300 hPa flow/height anomalies at 0000 UTC 17 and 18 September 1990.

Figure 6.7. *(Continued)*

6.3. Geneses in the Tropical Western Pacific

There were 12 typhoons that generated in the tropical western Pacific and their intensities peaked in the east seas of China. The right turns or left turns of their motions were influenced by anomalous vortices (Figs. 6.8–6.19).

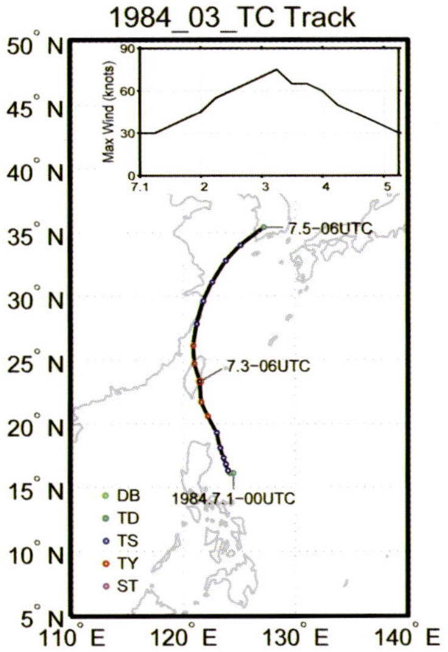

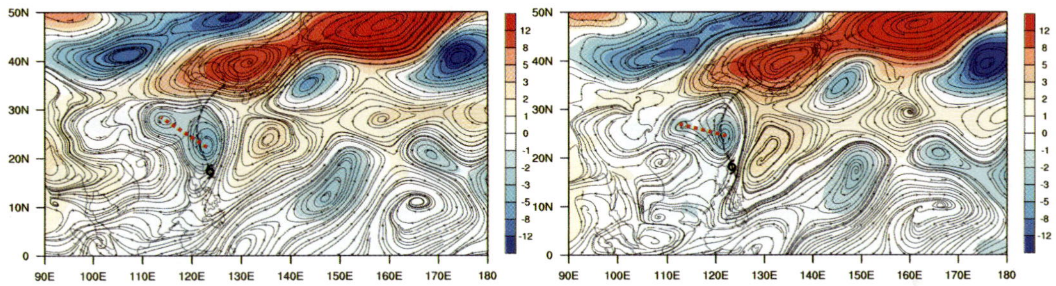

300 hPa flow/height (shading, 10 gpm) anomalies at 0000 UTC and 0600 UTC 2 July 1984.

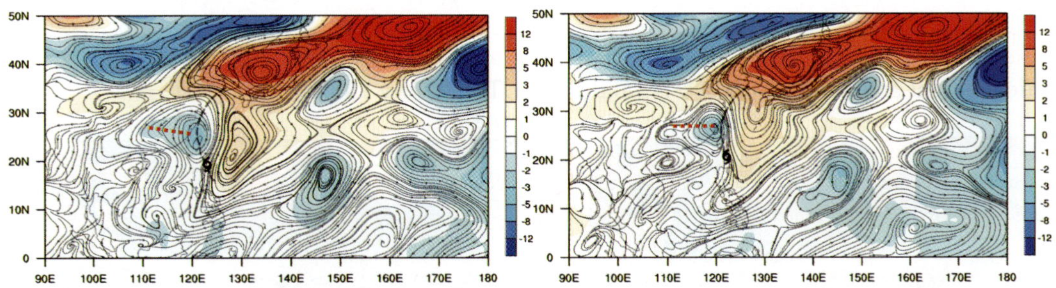

300 hPa flow/height anomalies at 1200 UTC and 1800 UTC 2 July 1984.

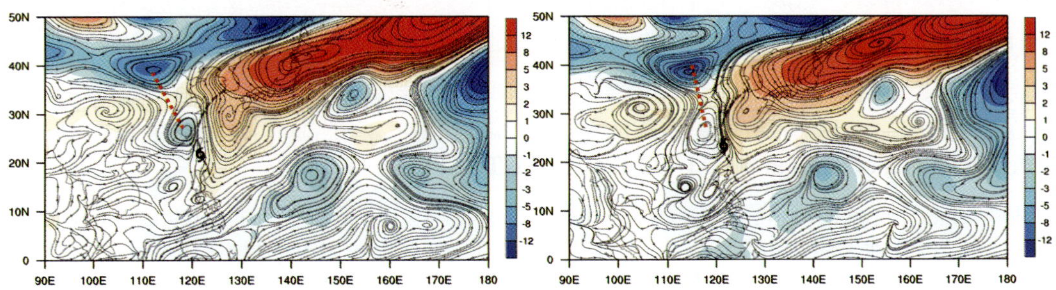

300 hPa flow/height anomalies at 0000 UTC and 0600 UTC 3 July 1984.

Figure 6.8. The typhoon (1984_No.03) generated in the east sea of the Philippines on 1 July 1984 and moved northward. Its intensity peaked on 3 July 1984 over the Taiwan Island when it was approaching the climatic monsoon trough. The right turn of its motion was influenced by anomalous vortices at 300 hPa in the east seas of China.

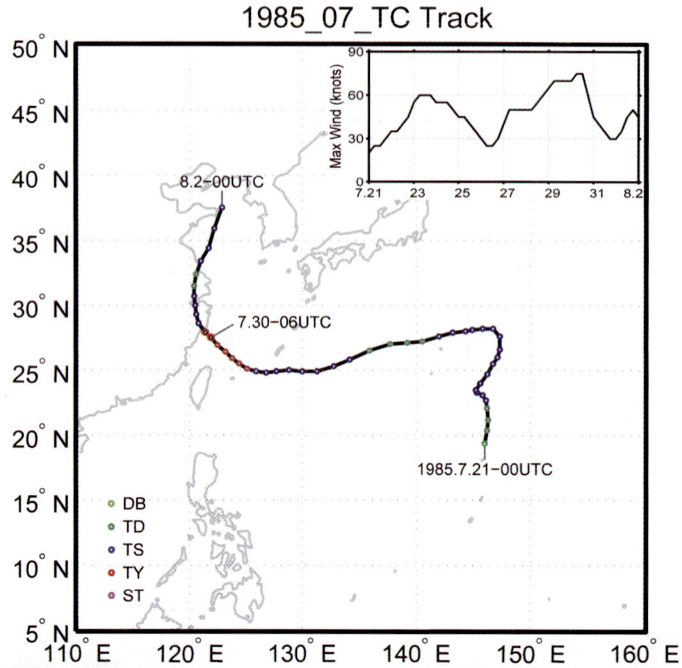

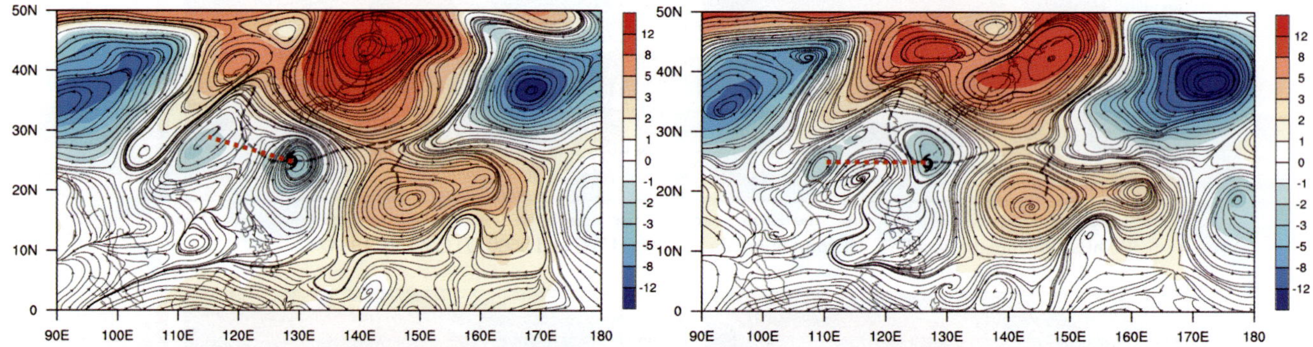

300 hPa flow/height anomalies at 0000 UTC and 1200 UTC 28 July 1985.

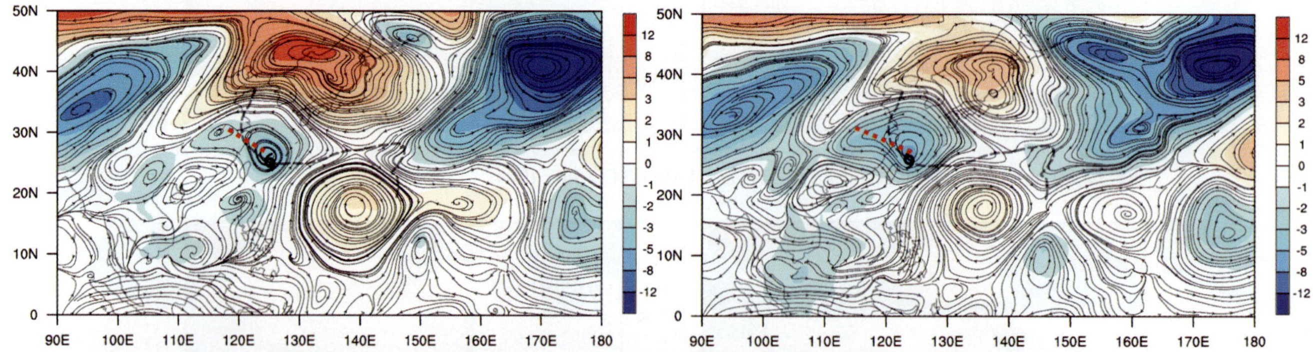

300 hPa flow/height anomalies at 0000 UTC and 1200 UTC 29 July 1985.

Figure 6.9. The typhoon (1985_No.07) generated in the tropical western Pacific on 21 July 1985 and moved northward at the beginning stage. Its intensity peaked in the East China Sea on 30 July 1985 when it was approaching the climatic monsoon trough before landfall on the East China coast. The right turn of its motion in the East China was influenced by anomalous vortices at 300 hPa.

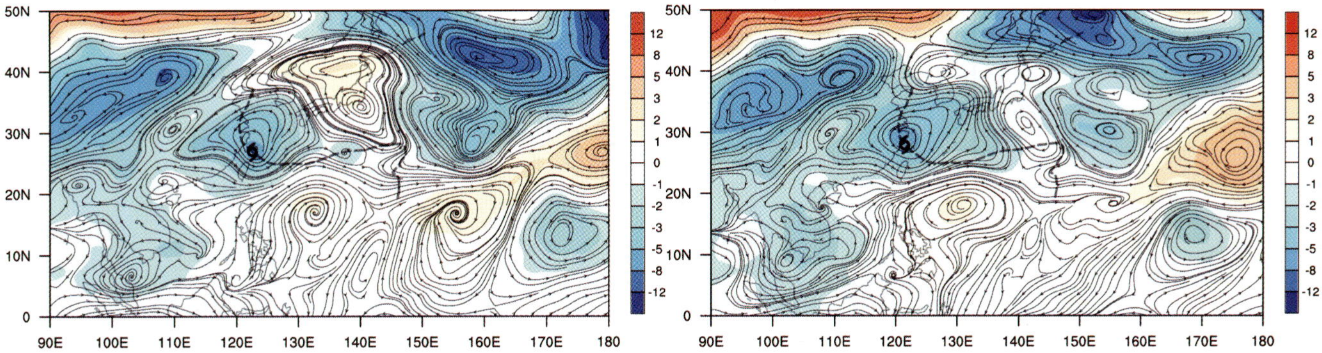

300 hPa flow/height anomalies at 0000 UTC and 1200 UTC 30 July 1985.

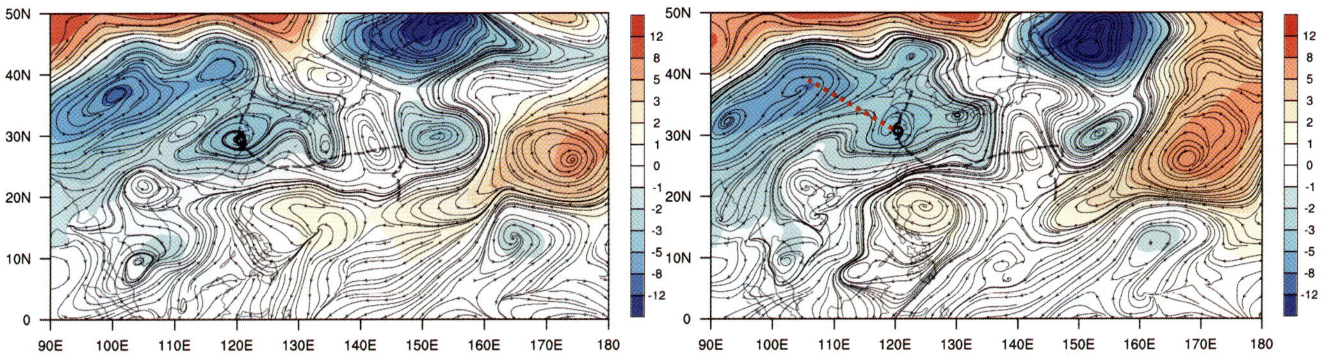

300 hPa flow/height anomalies at 0000 UTC and 1200 UTC 31 July 1985.

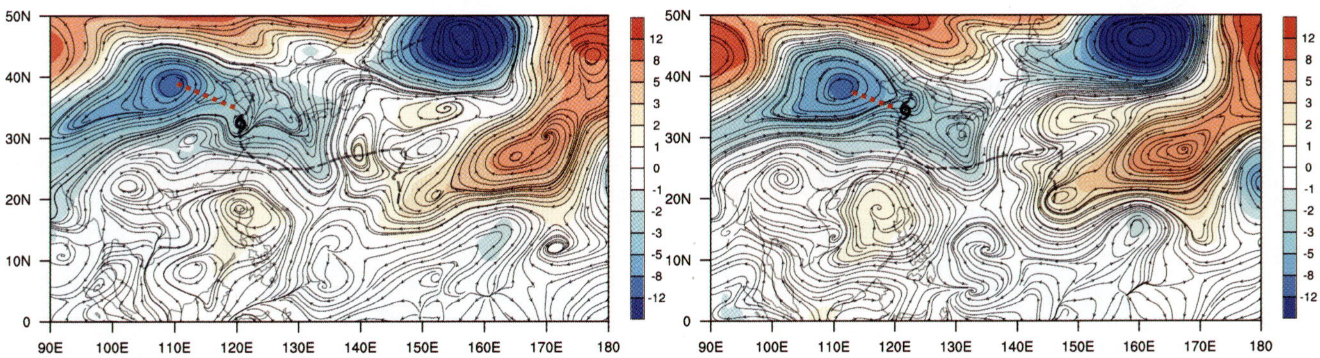

300 hPa flow/height anomalies at 0000 UTC and 1200 UTC 1 August 1985.

Figure 6.9. *(Continued)*

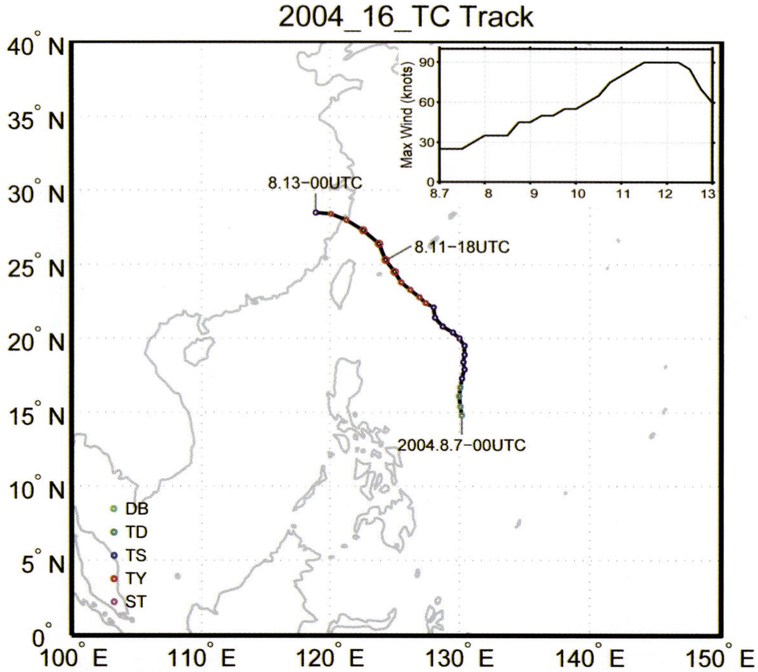

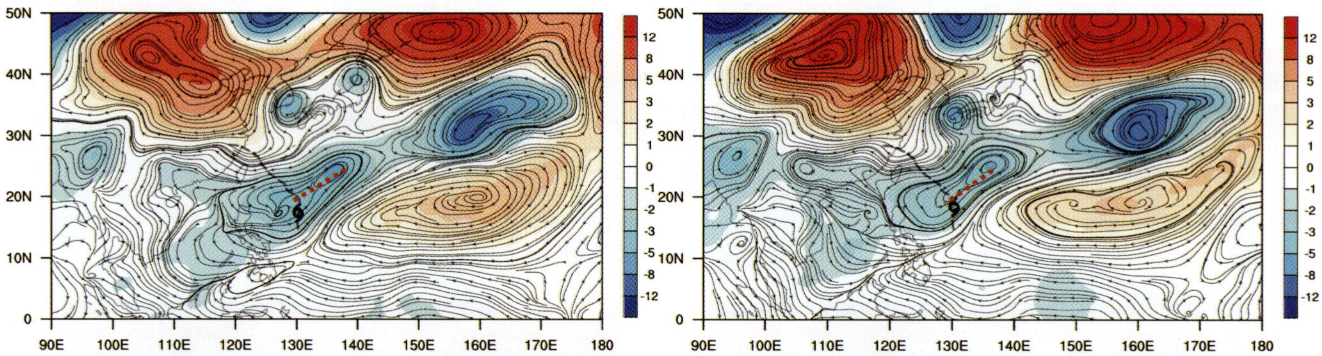

300 hPa flow/height anomalies at 0000 UTC and 1200 UTC 8 August 2004.

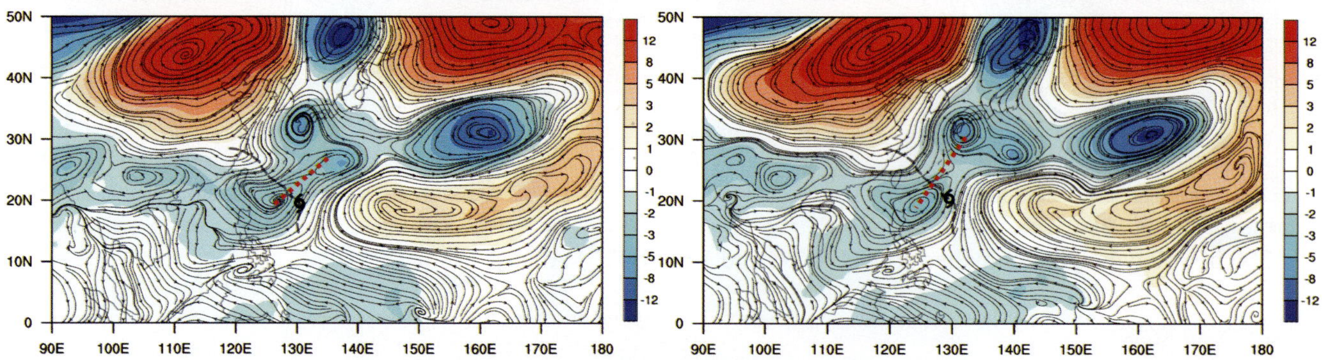

300 hPa flow/height anomalies at 0000 UTC and 1200 UTC 9 August 2004.

Figure 6.10. The typhoon (2004_No.16) generated in the east sea of the Philippines on 7 August 2004 and moved northward along the climatic monsoon trough. Its intensity peaked in the East China Sea on 11 August 2004 before landfall on the East China coast. The left turn of its motion was influenced by anomalous vortices at 300 hPa.

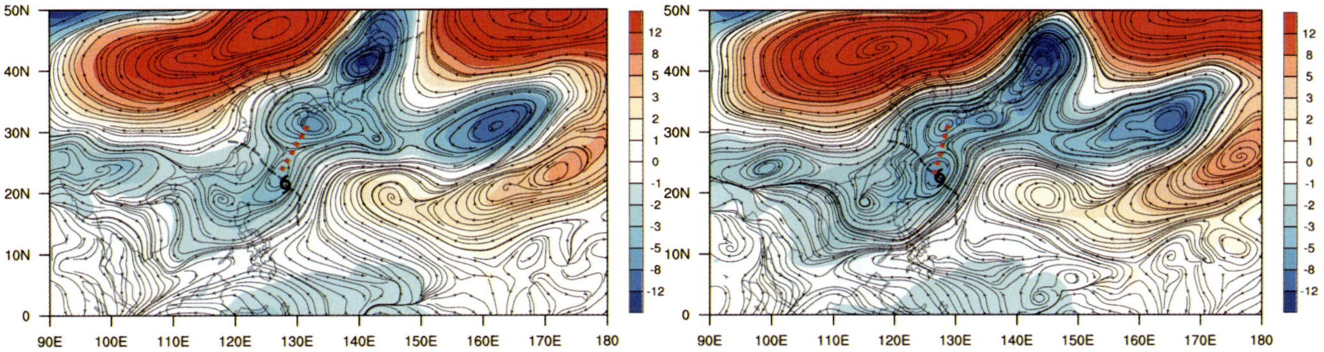

300 hPa flow/height anomalies at 0000 UTC and 1200 UTC 10 August 2004.

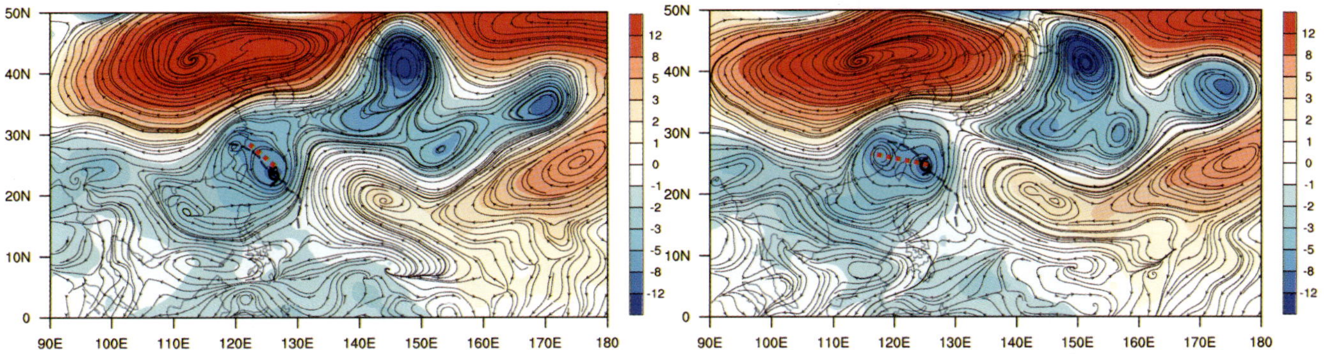

300 hPa flow/height anomalies at 0000 UTC and 1200 UTC 11 August 2004.

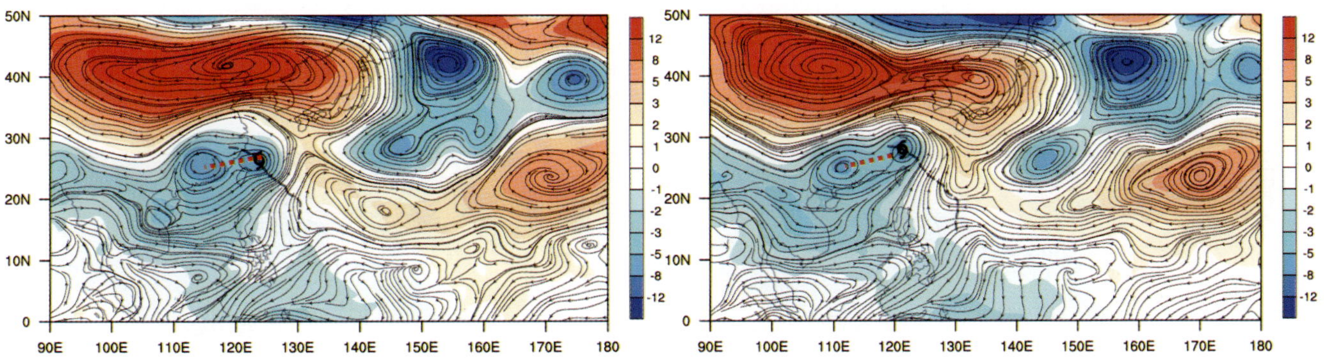

300 hPa flow/height anomalies at 0000 UTC and 1200 UTC 12 August 2004.

Figure 6.10. (*Continued*)

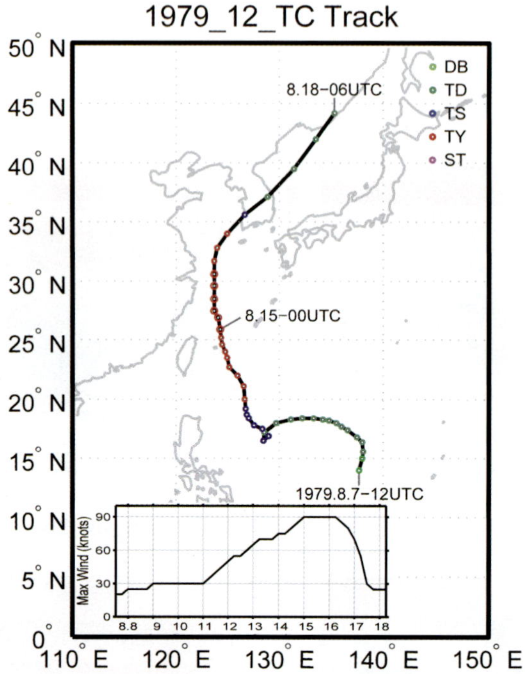

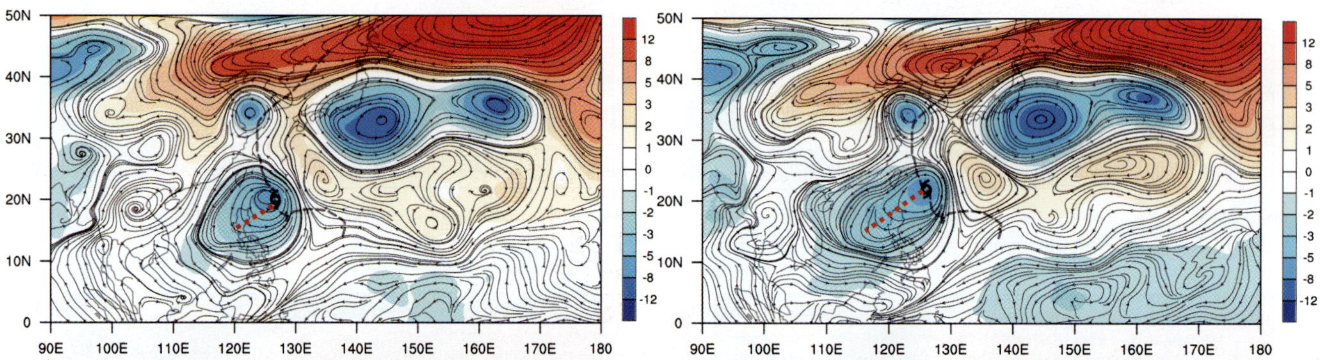

300 hPa flow/height anomalies at 0000 UTC and 1200 UTC 13 August 1979.

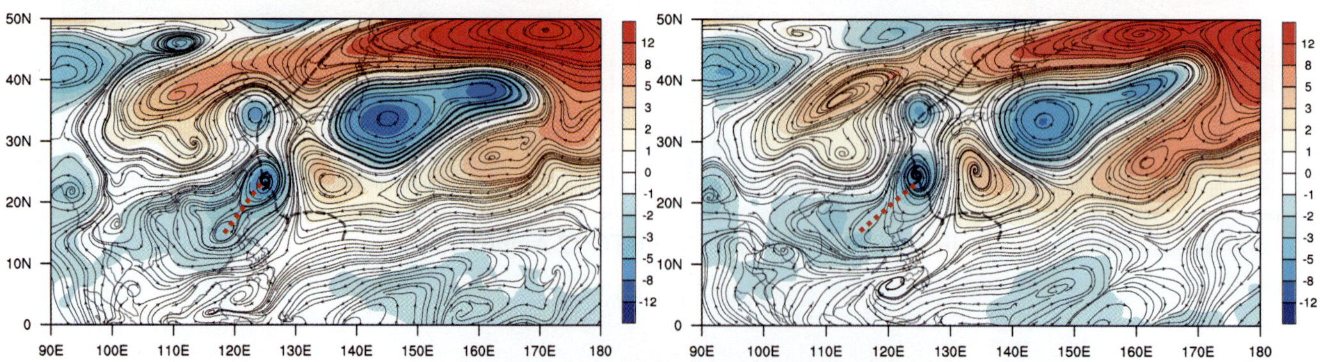

300 hPa flow/height anomalies at 0000 UTC and 1200 UTC 14 August 1979.

Figure 6.11. The typhoon (1979_No.12) generated in the tropical western Pacific on 7 August 1979 and moved northwestward near the climatic monsoon trough. Its intensity peaked in the northeast sea of the Taiwan Island on 15 August 1979. The right turn of its motion in the east seas of China was influenced by anomalous vortices at 300 hPa.

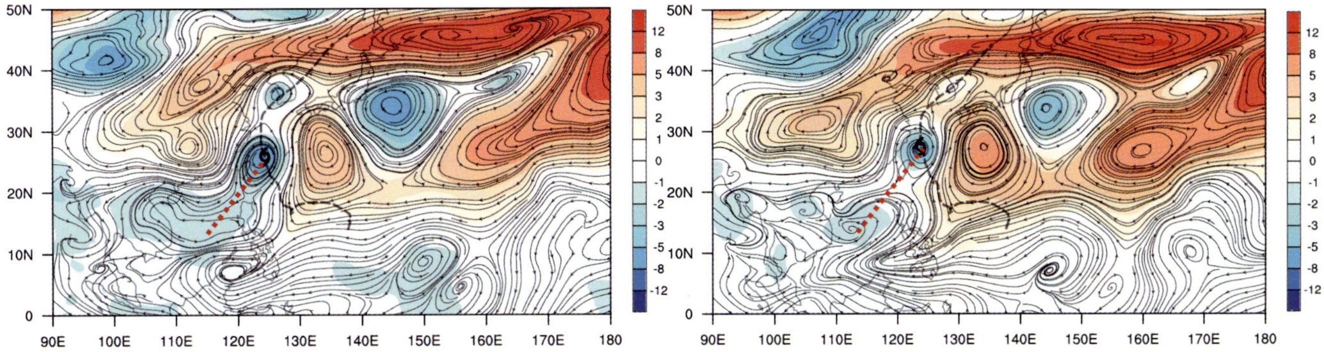

300 hPa flow/height anomalies at 0000 UTC and 1200 UTC 15 August 1979.

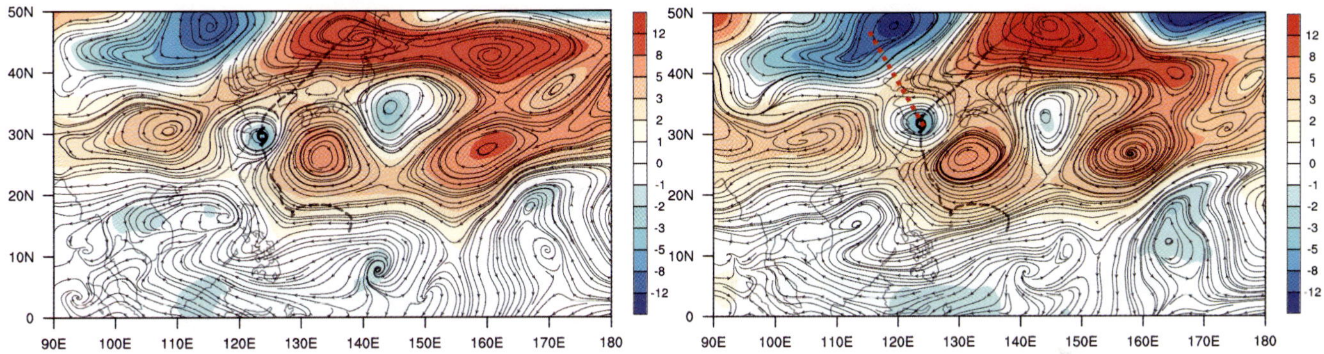

300 hPa flow/height anomalies at 0000 UTC and 1200 UTC 16 August 1979.

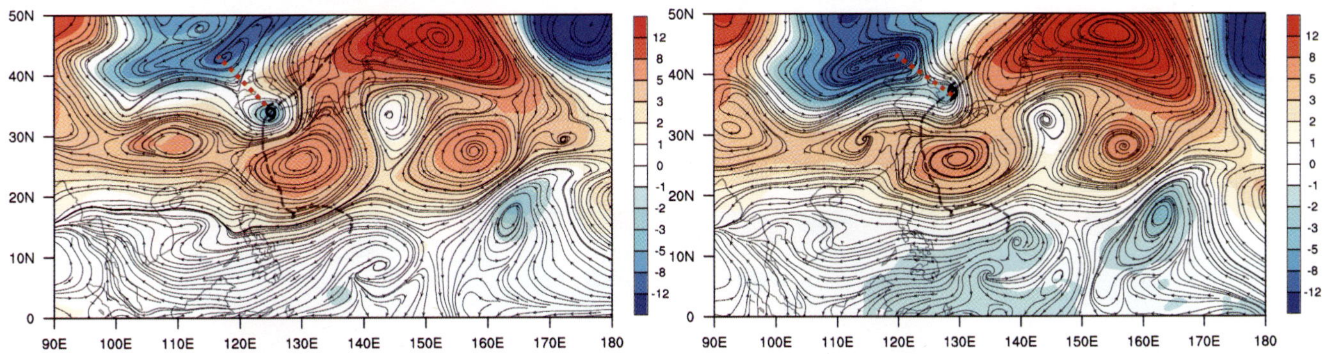

300 hPa flow/height anomalies at 0000 UTC and 1200 UTC 17 August 1979.

Figure 6.11. (*Continued*)

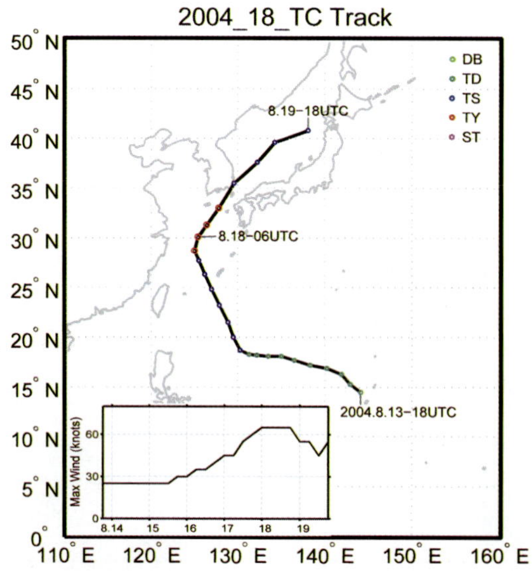

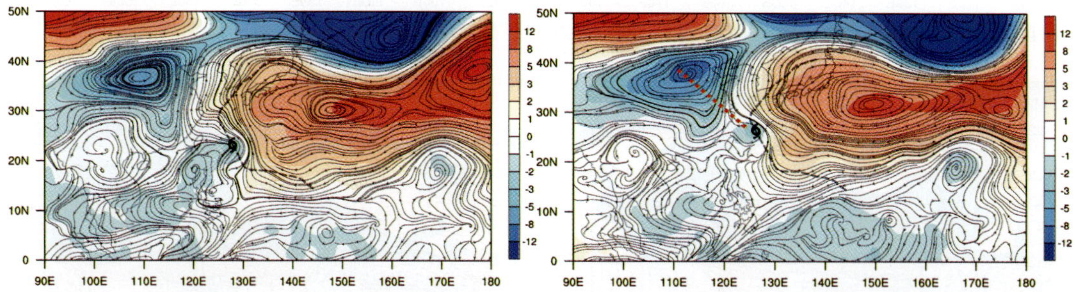

300 hPa flow/height anomalies at 0000 UTC and 1200 UTC 17 August 2004.

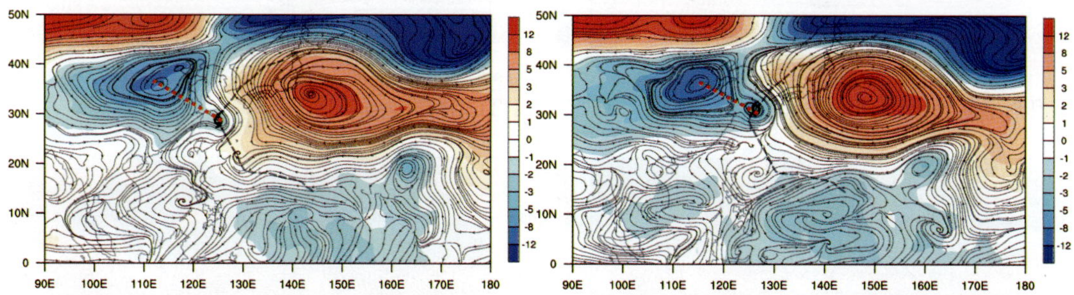

300 hPa flow/height anomalies at 0000 UTC and 1200 UTC 18 August 2004.

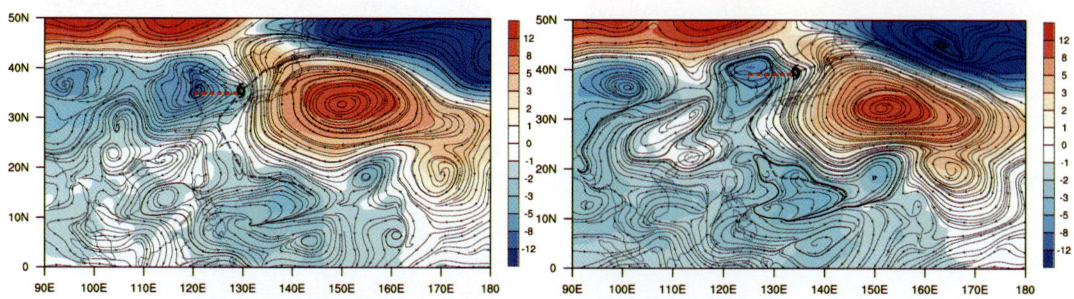

300 hPa flow/height anomalies at 0000 UTC and 1200 UTC 19 August 2004.

Figure 6.12. The typhoon (2004_No.18) generated in the tropical western Pacific on 13 August 2004 and moved northwestward along the climatic monsoon trough. Its intensity peaked in the East China Sea on 18 August 2004 when it turned right. The right turn in the east seas of China was influenced by an anomalous vortex from the China continent at 300 hPa.

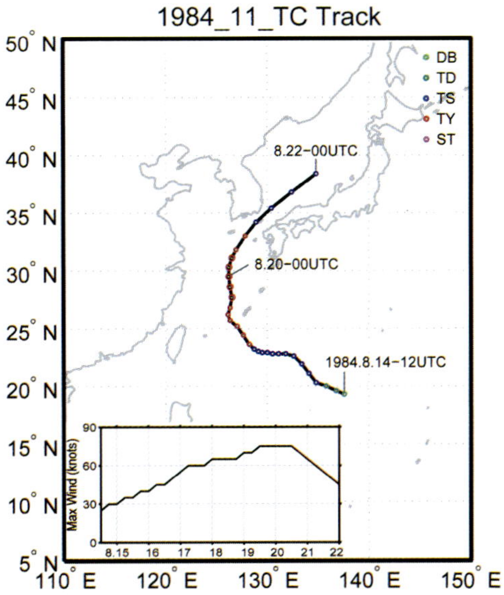

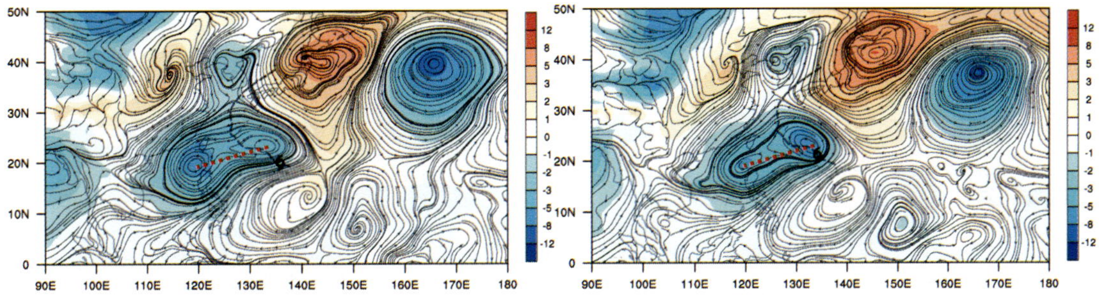

850 hPa flow/height anomalies at 0000 UTC and 1200 UTC 15 August 1984.

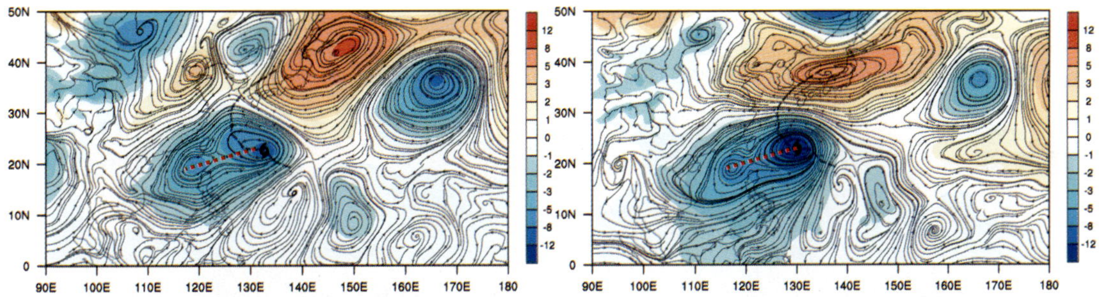

850 hPa flow/height anomalies at 0000 UTC 16 and 17 August 1984.

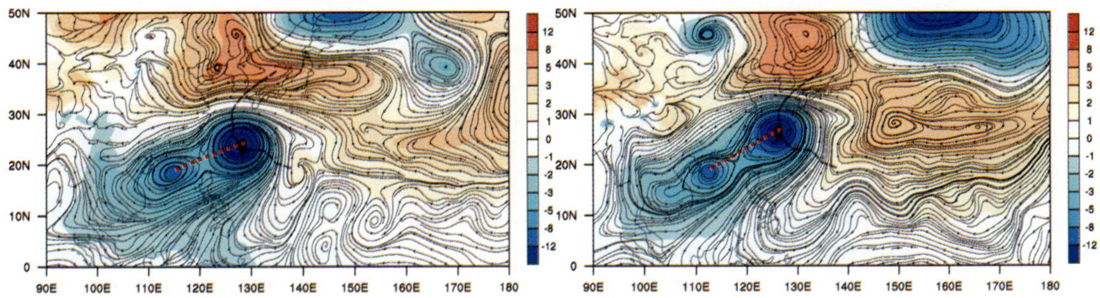

850 hPa flow/height anomalies at 0000 UTC 18 and 19 August 1984.

Figure 6.13. The typhoon (1984_No.11) generated in the tropical western Pacific on 14 August 1984 and moved northwestward along the climatic monsoon trough. Its intensity peaked in the East China Sea on 20 August 1984 when it turned right. The right turn in the east seas of China was influenced by anomalous vortices at 850 hPa and 300 hPa.

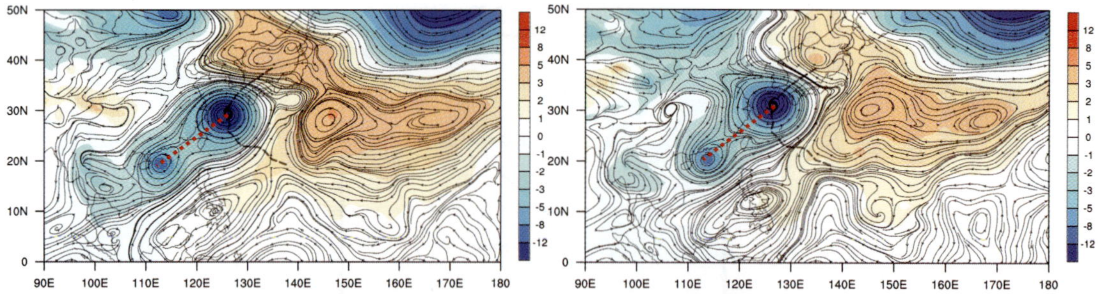

850 hPa flow/height anomalies at 0000 UTC and 1200 UTC 20 August 1984.

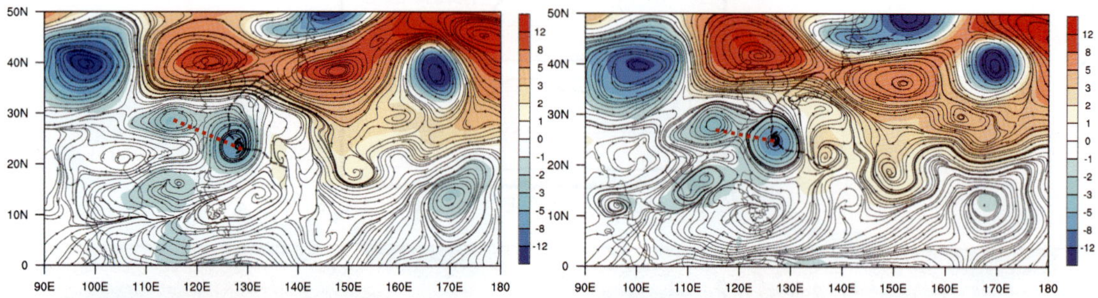

300 hPa flow/height anomalies at 0000 UTC and 1200 UTC 18 August 1984.

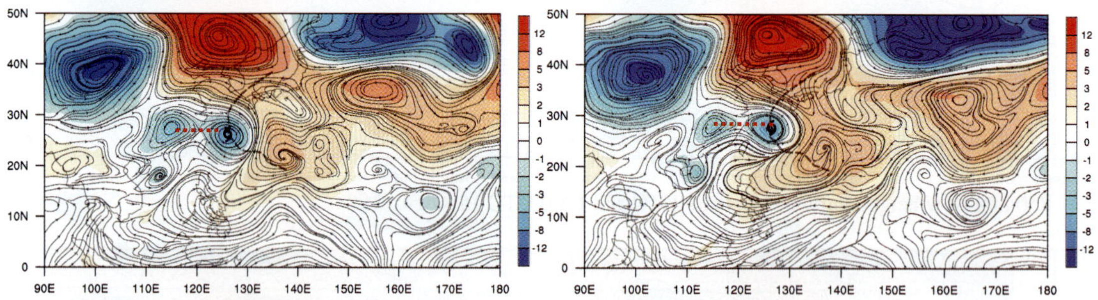

300 hPa flow/height anomalies at 0000 UTC and 1200 UTC 19 August 1984.

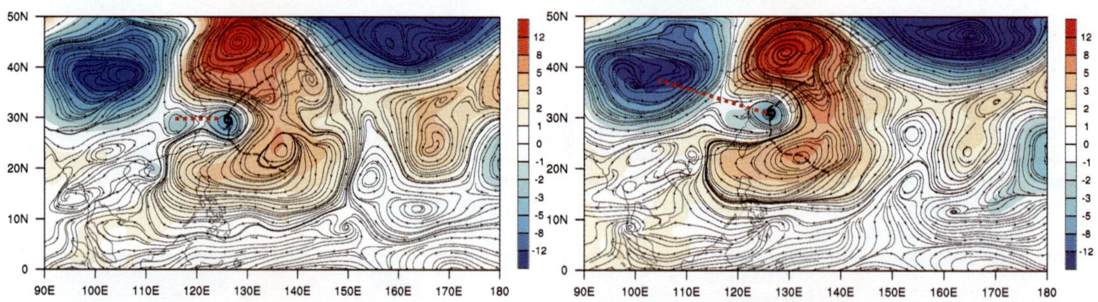

300 hPa flow/height anomalies at 0000 UTC and 1200 UTC 20 August 1984.

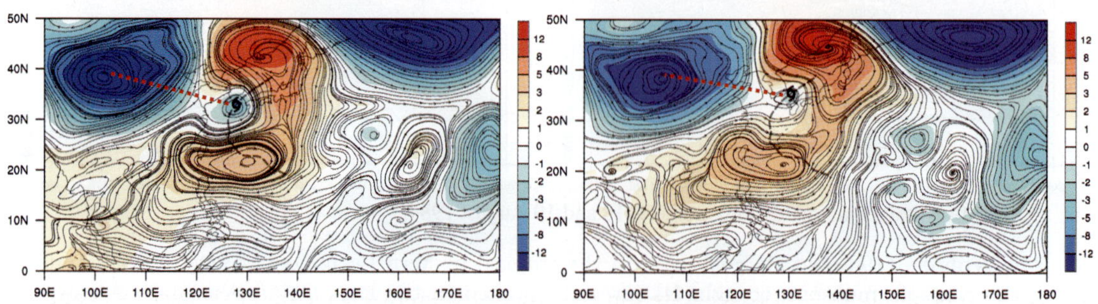

300 hPa flow/height anomalies at 0000 UTC and 1200 UTC 21 August 1984.

Figure 6.13. *(Continued)*

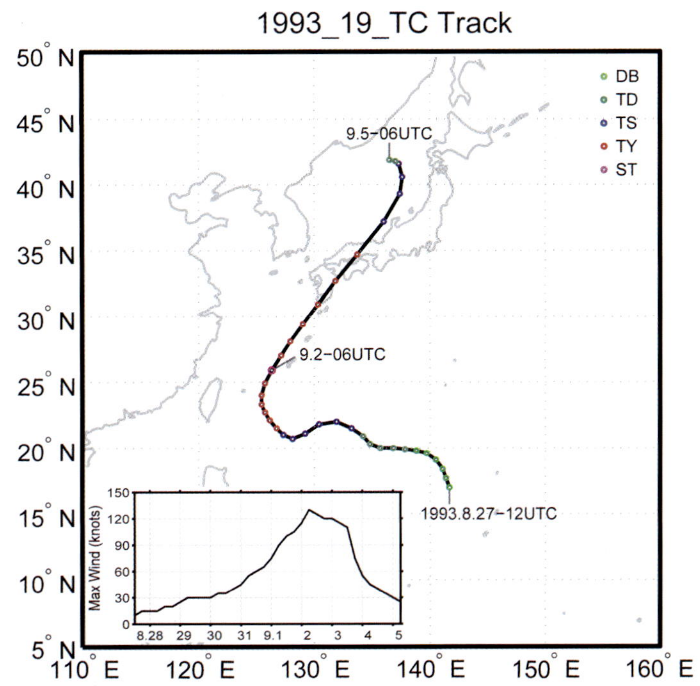

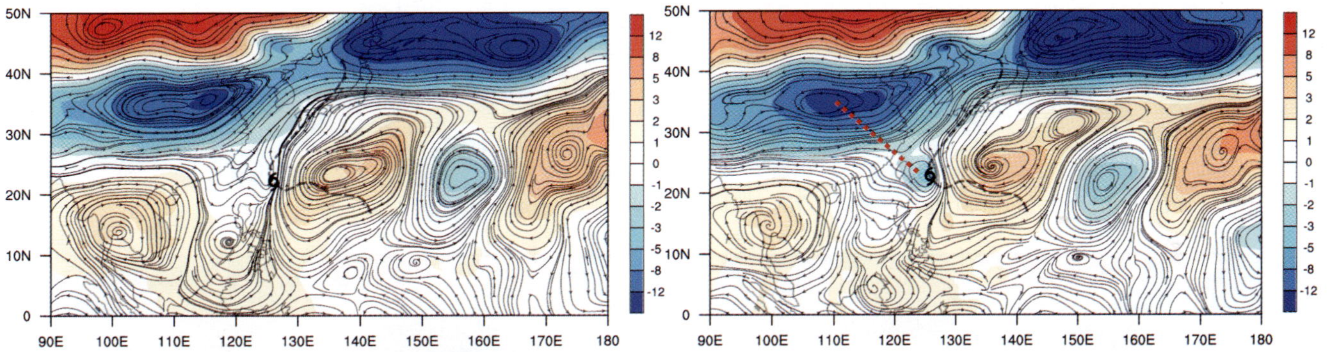

300 hPa flow/height anomalies at 0000 UTC and 0600 UTC 1 September 1993.

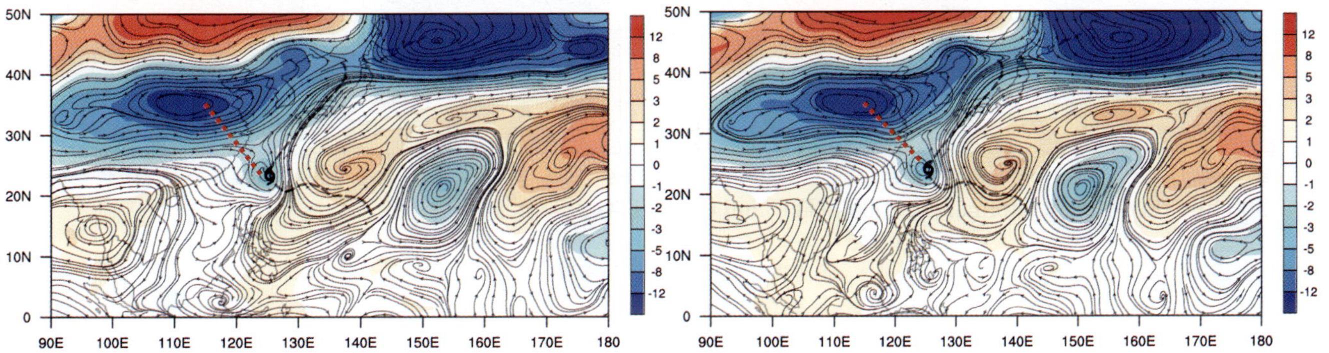

300 hPa flow/height anomalies at 1200 UTC and 1800 UTC 1 September 1993.

Figure 6.14. The typhoon (1993_No.19) generated in the tropical western Pacific on 27 August 1993 and moved northwestward near the climatic monsoon trough. Its intensity peaked in the East China Sea on 2 September 1993. The right turn of its motion in the south of Japan was influenced by an anomalous vortex at 300 hPa.

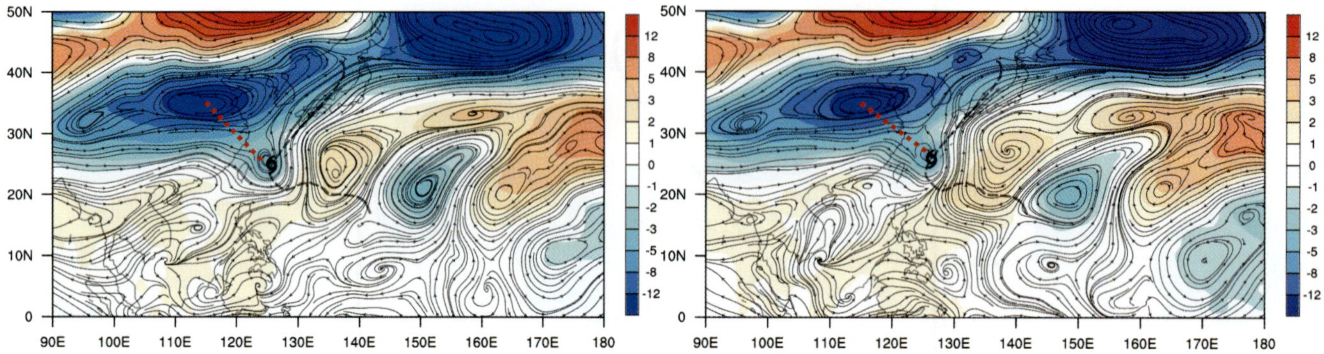

300 hPa flow/height anomalies at 0000 UTC and 0600 UTC 2 September 1993.

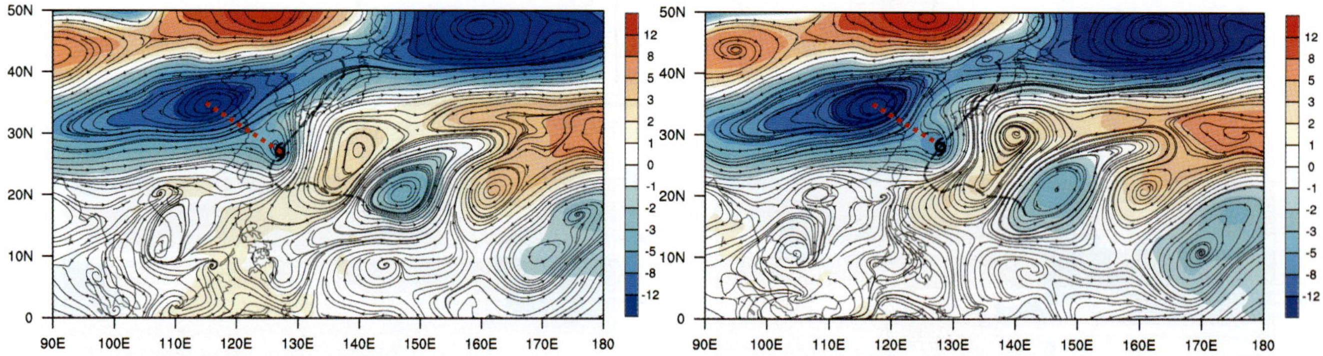

300 hPa flow/height anomalies at 1200 UTC and 1800 UTC 2 September 1993.

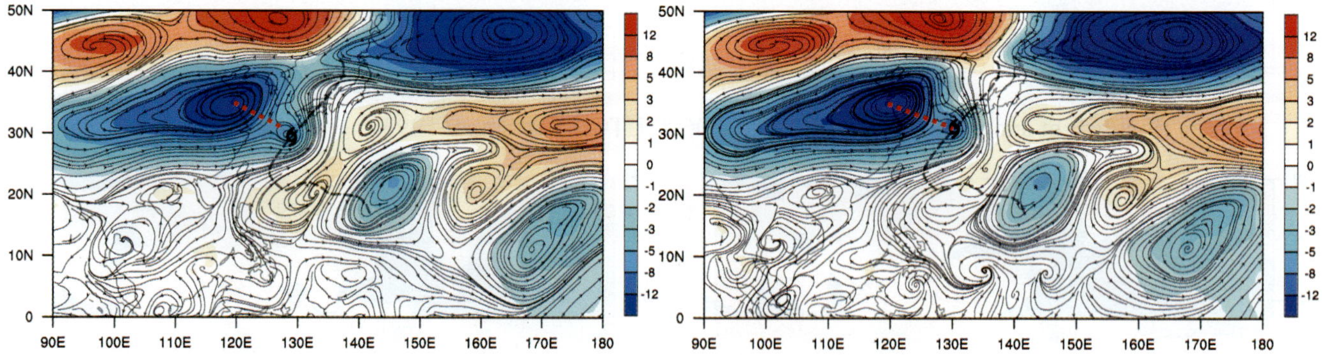

300 hPa flow/height anomalies at 0000 UTC and 0600 UTC 3 September 1993.

Figure 6.14. *(Continued)*

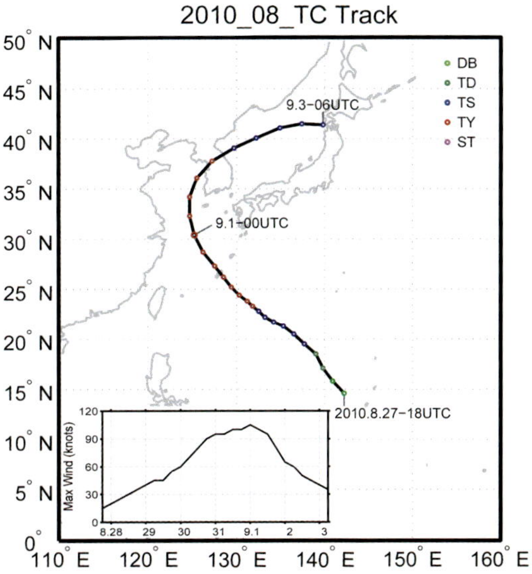

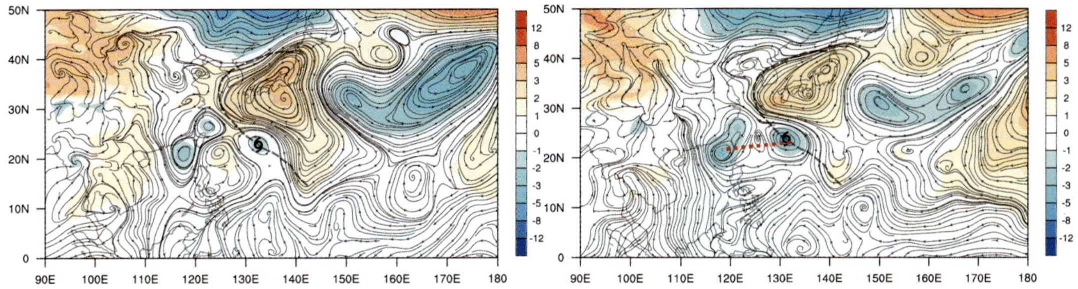

850 hPa flow/height anomalies at 0000 UTC and 1200 UTC 30 August 2010.

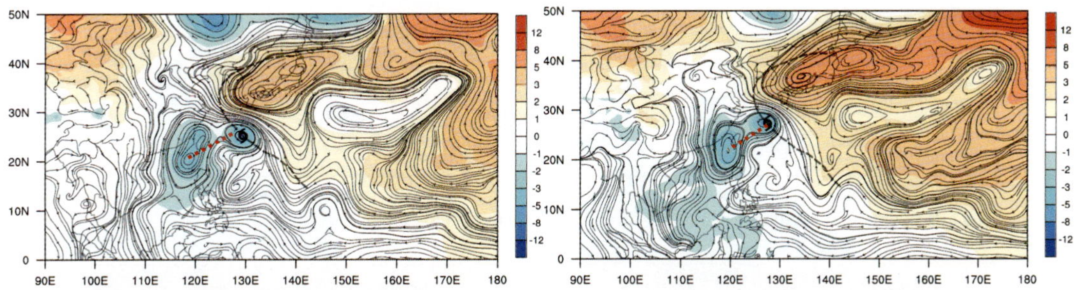

850 hPa flow/height anomalies at 0000 UTC and 1200 UTC 31 August 2010.

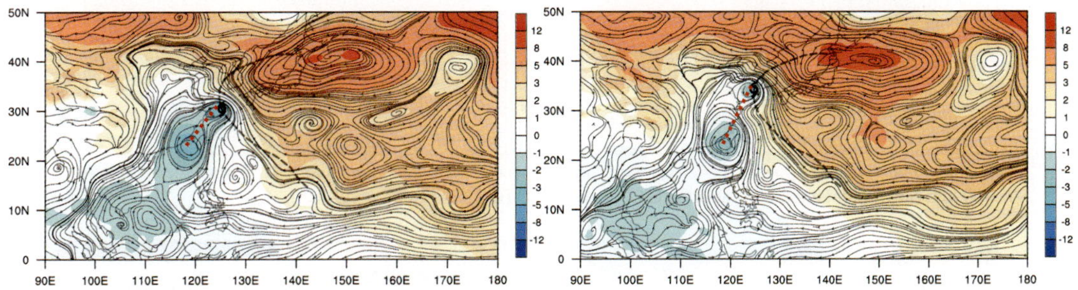

850 hPa flow/height anomalies at 0000 UTC and 1200 UTC 1 September 2010.

Figure 6.15. The typhoon (2010_No.08) generated in the tropical western Pacific on 27 August 2010 and moved northwestward along the climatic monsoon trough. Its intensity peaked in the East China Sea on 1 September 2010 when it turned right. The right turn in the east seas of China was influenced by an anomalous vortex at 850 hPa.

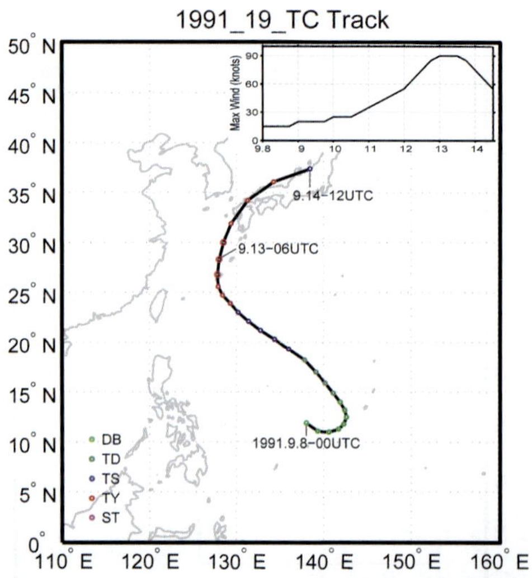

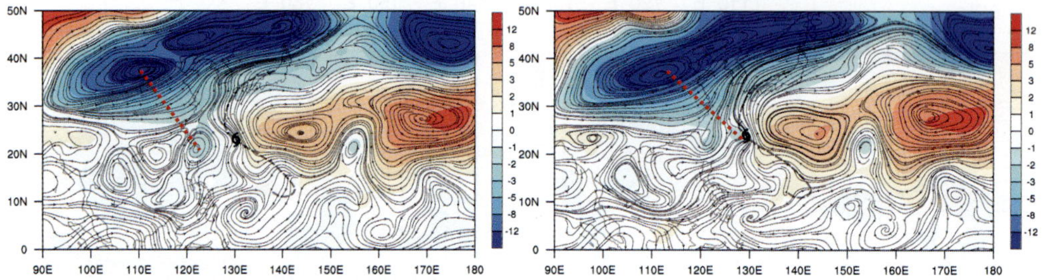

300 hPa flow/height anomalies at 0000 UTC and 0600 UTC 12 September 1991.

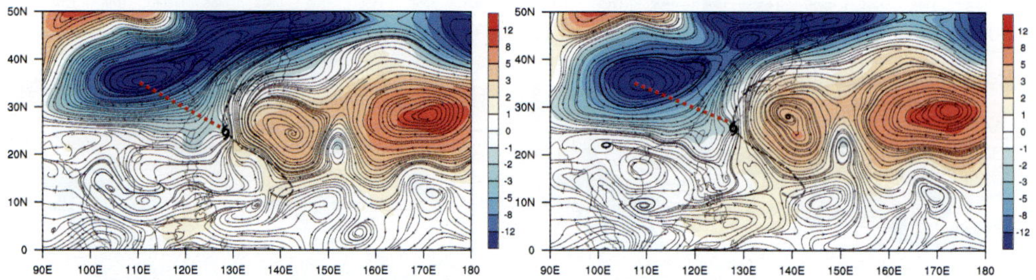

300 hPa flow/height anomalies at 1200 UTC and 1800 UTC 12 September 1991.

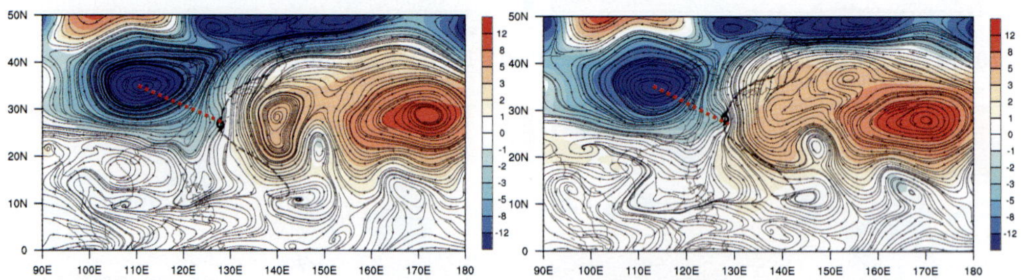

300 hPa flow/height anomalies at 0000 UTC and 0600 UTC 13 September 1991.

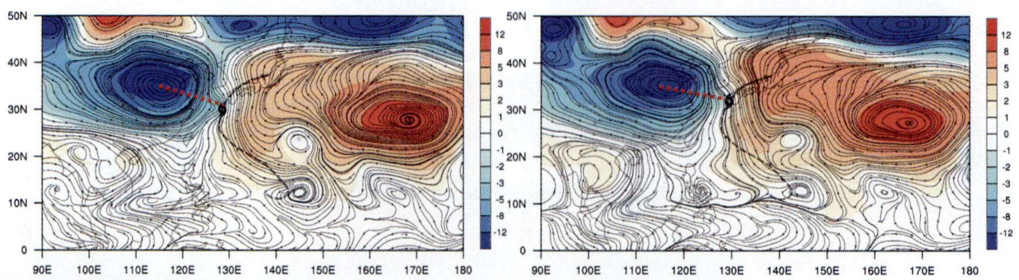

300 hPa flow/height anomalies at 1200 UTC and 1800 UTC 13 September 1991.

Figure 6.16. The typhoon (1991_No.19) generated in the tropical western Pacific on 8 September 1991. After a semicircular way, it moved northwestward along the climatic monsoon trough. Its intensity peaked in the East China Sea on 13 September 1991 when it turned right. The right turn in the east seas of China was influenced by an anomalous vortex that moved eastward in the China continent at 300 hPa.

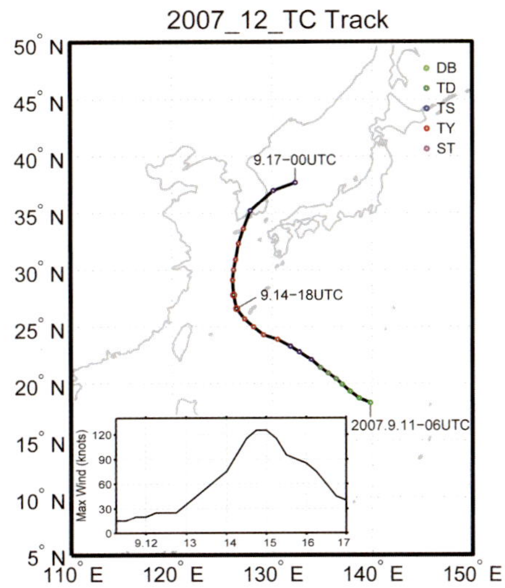

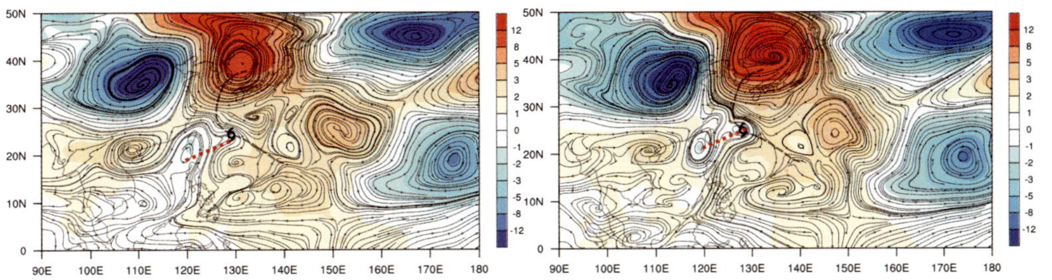

300 hPa flow/height anomalies at 0000 UTC and 0600 UTC 14 September 2007.

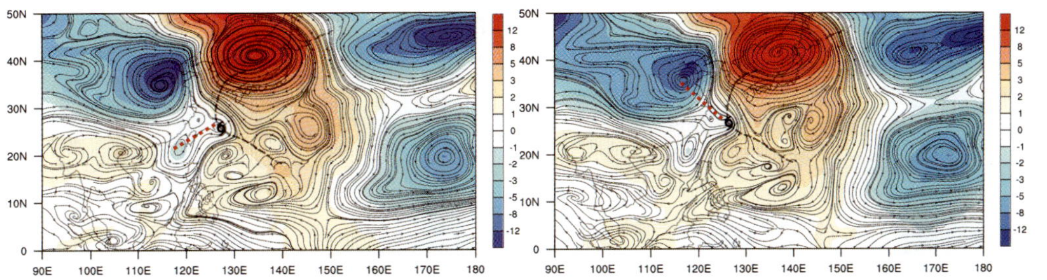

300 hPa flow/height anomalies at 1200 UTC and 1800 UTC 14 September 2007.

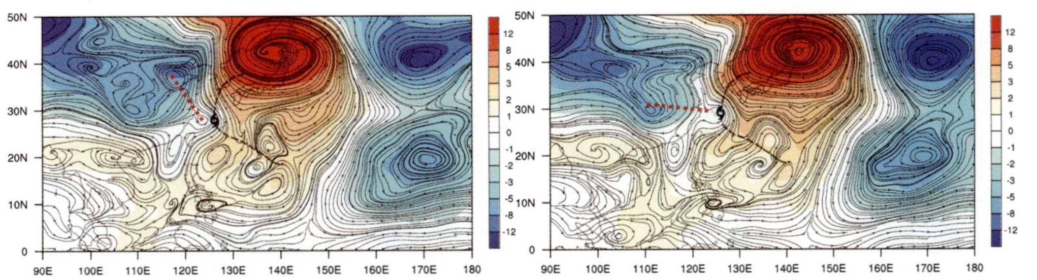

300 hPa flow/height anomalies at 0000 UTC and 0600 UTC 15 September 2007.

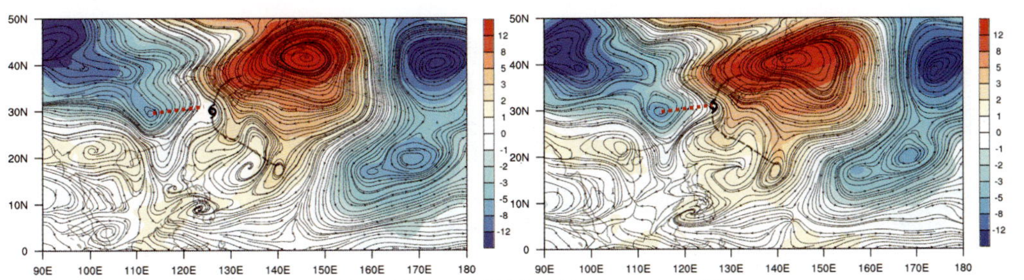

300 hPa flow/height anomalies at 1200 UTC and 1800 UTC 15 September 2007.

Figure 6.17. The typhoon (2007_No.12) generated in the tropical western Pacific on 11 September 2007 and moved northwestward along the climatic monsoon trough. Its intensity peaked in the East China Sea on 14 September 2007 when it turned right in the cross position of two climatic monsoon troughs. The right turn in the east seas of China was influenced by anomalous vortices at 300 hPa.

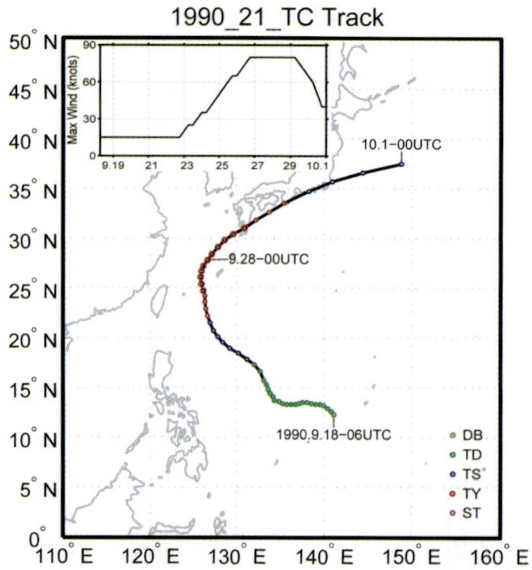

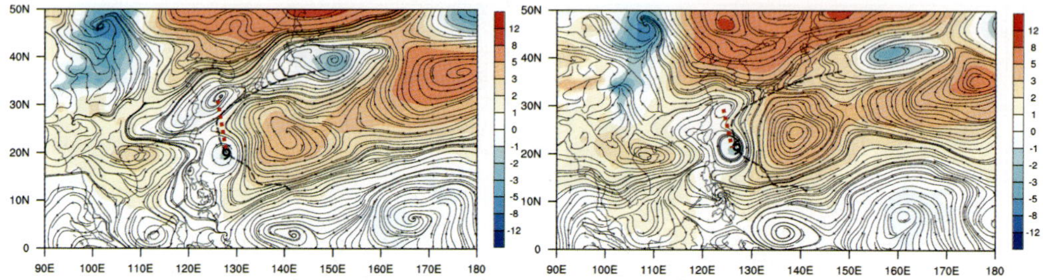
850 hPa flow/height anomalies at 0000 UTC and 1200 UTC 25 September 1990.

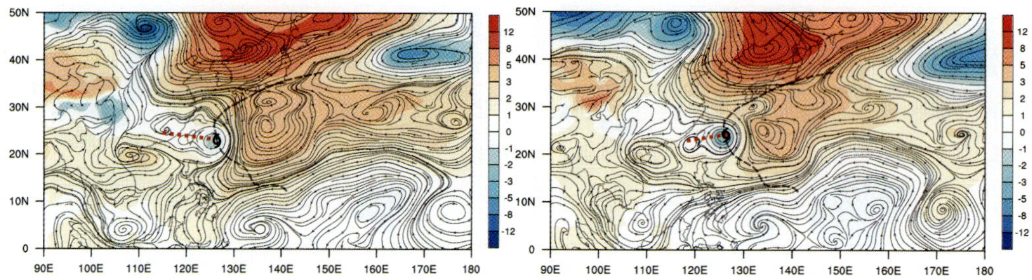
850 hPa flow/height anomalies at 0000 UTC and 1200 UTC 26 September 1990.

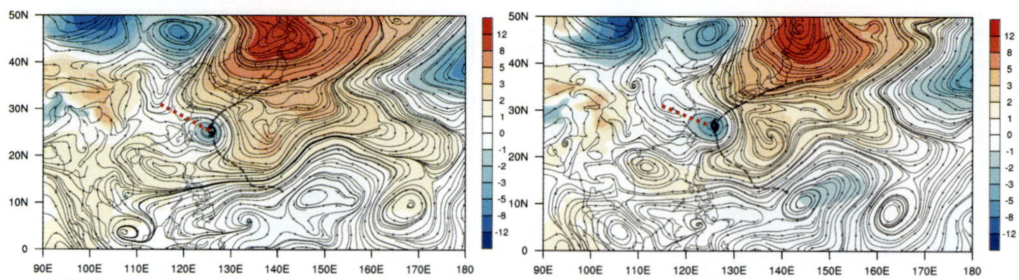
850 hPa flow/height anomalies at 0000 UTC and 1200 UTC 27 September 1990.

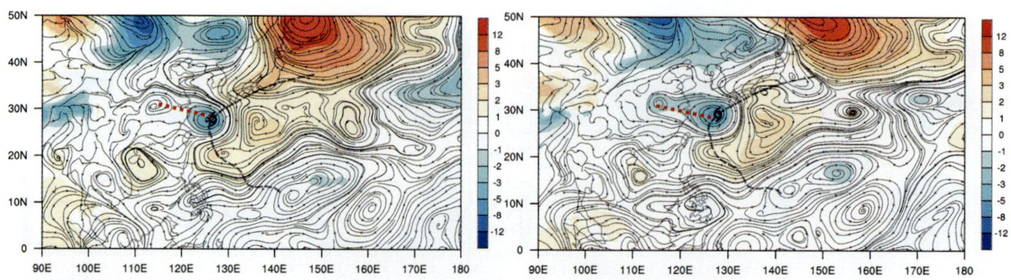
850 hPa flow/height anomalies at 0000 UTC and 1200 UTC 28 September 1990.

Figure 6.18. The typhoon (1990_No.21) generated in the tropical western Pacific on 18 September 1990 and moved northwestward along the climatic monsoon trough. Its intensity peaked in the East China Sea on 28 September 1990 when it turned right. The right turn in the east seas of China was influenced by anomalous vortices at 850 hPa.

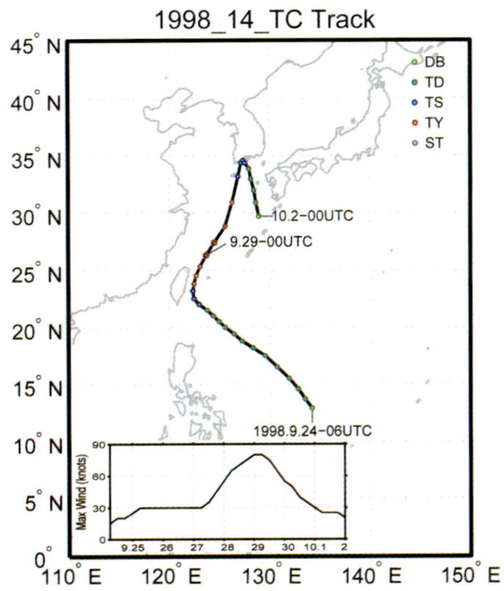

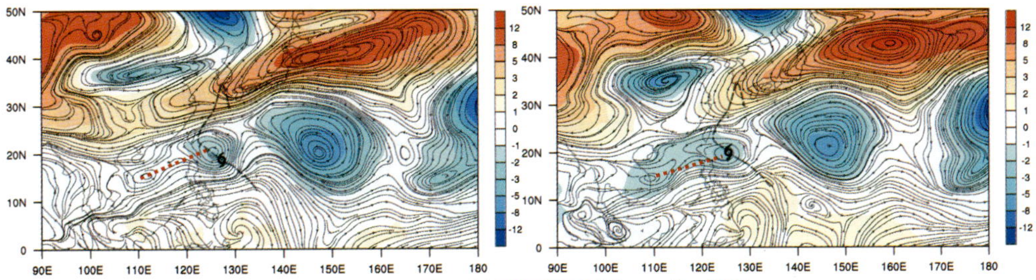

300 hPa flow/height anomalies at 0000 UTC and 1200 UTC 26 September 1998.

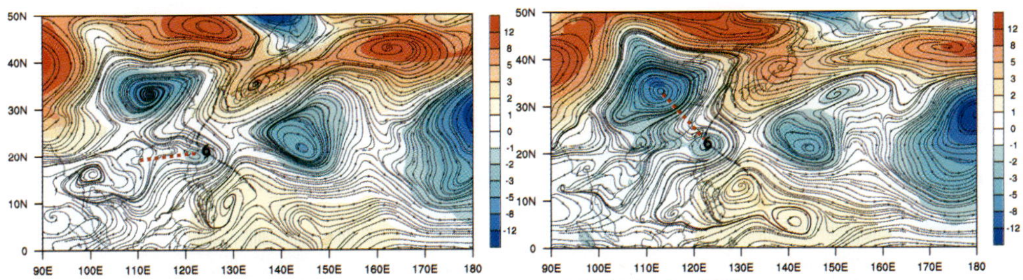

300 hPa flow/height anomalies at 0000 UTC and 1200 UTC 27 September 1998.

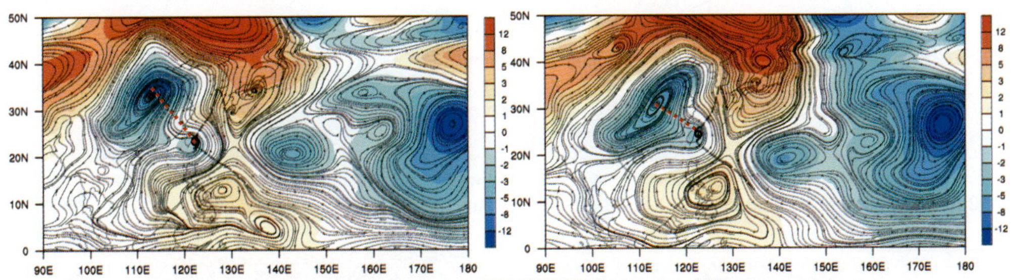

300 hPa flow/height anomalies at 0000 UTC and 1200 UTC 28 September 1998.

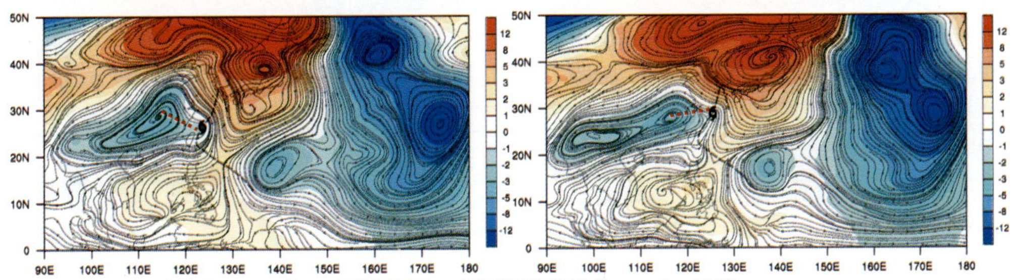

300 hPa flow/height anomalies at 0000 UTC and 1200 UTC 29 September 1998.

Figure 6.19. The typhoon (1998_No.14) generated in the tropical western Pacific on 24 September 1998 and moved northwestward. Its intensity peaked in the northeast sea of the Taiwan Island on 29 September 1998 when it was arriving in the cross position of two climatic monsoon troughs. The right turn of its motion in the east seas of China was influenced by anomalous vortices at 300 hPa.

6.4. Geneses in the East Seas of China

Six typhoons generated from tropical disturbances and their intensities peaked in the east seas of China. The right turn and left turn of their motions were influenced by anomalous vortices (Figs. 6.20–6.25).

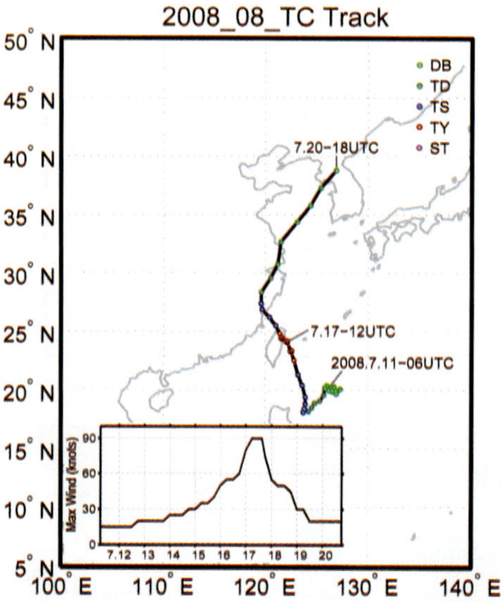

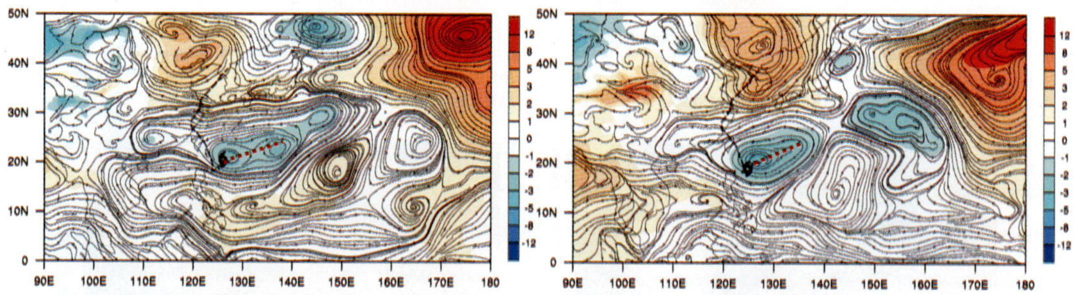

850 hPa flow/height (shading, 10 gpm) anomalies at 1200 UTC 13 and 14 July 2008.

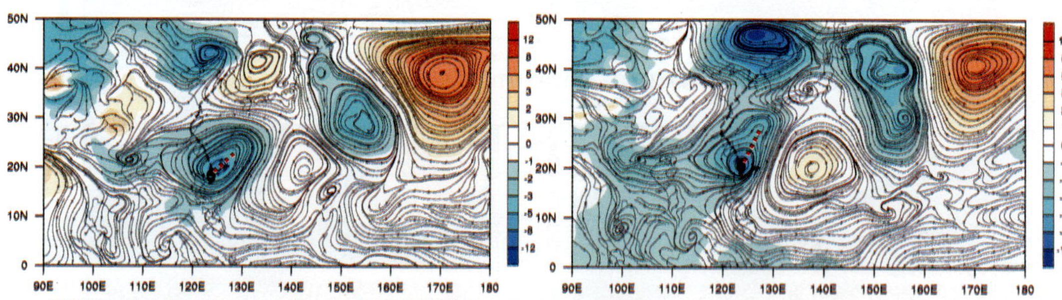

850 hPa flow/height anomalies at 1200 UTC 15 and 16 July 2008.

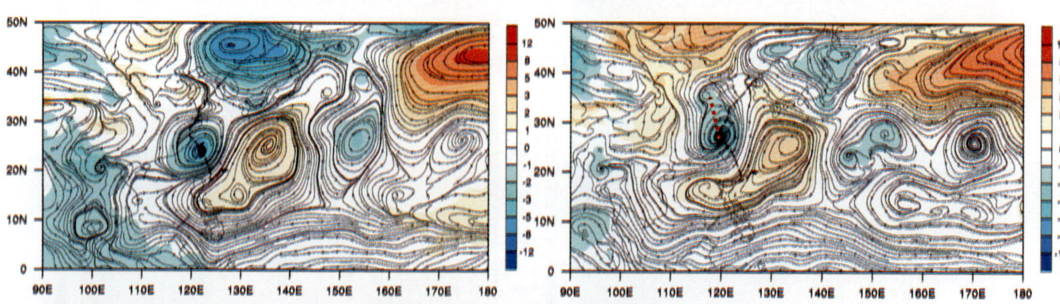

850 hPa flow/height anomalies at 1200 UTC 17 and 18 July 2008.

Figure 6.20. The typhoon (2008_No.08) generated in the northeast sea of the Philippines on 11 July 2008 and moved southwestward at the beginning stage. After a return, it reached its peak intensity in the east sea of the Taiwan Island on 17 July 2008 near the climatic monsoon trough. Its return and right turn of track were influenced by anomalous vortices at 300 hPa and 850 hPa.

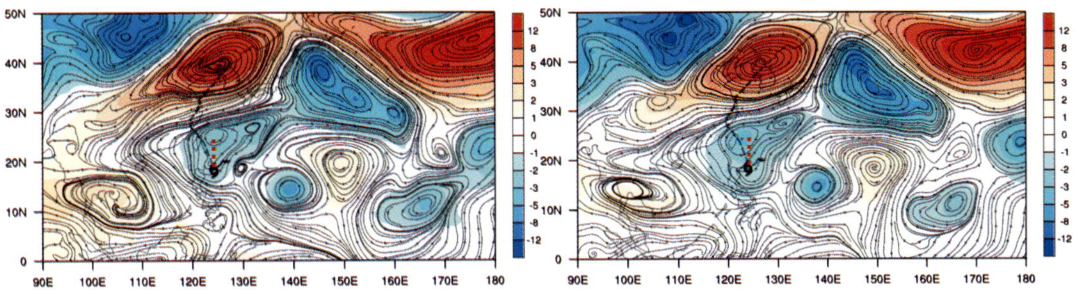

300 hPa flow/height anomalies at 0000 UTC and 0600 UTC 15 July 2008.

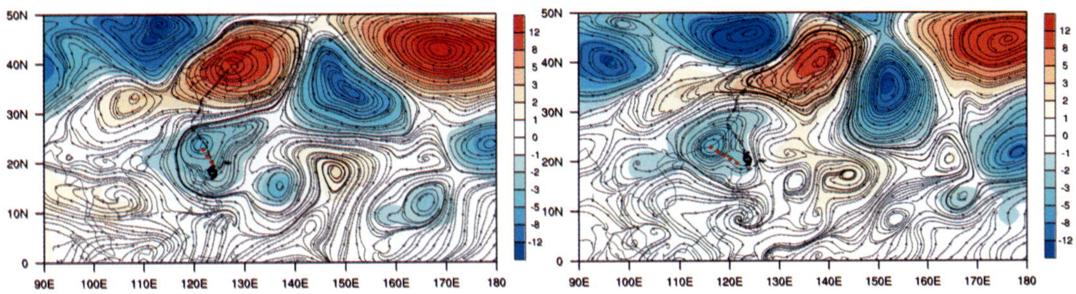

300 hPa flow/height anomalies at 1200 UTC 15 and 16 July 2008.

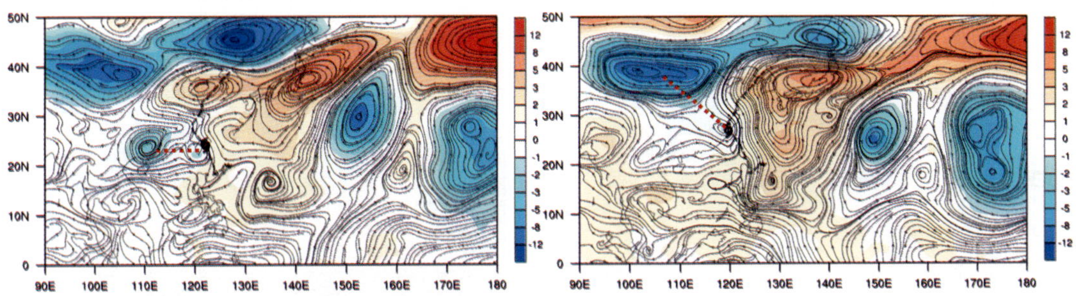

300 hPa flow/height anomalies at 1200 UTC 17 and 18 July 2008.

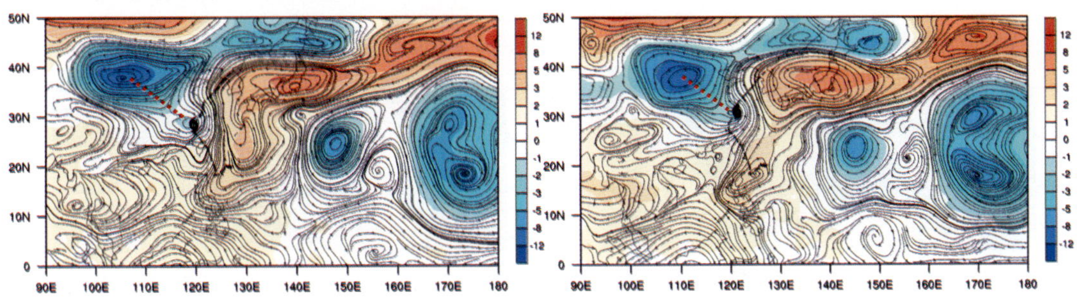

300 hPa flow/height anomalies at 0000 UTC and 1200 UTC 19 July 2008.

Figure 6.20. *(Continued)*

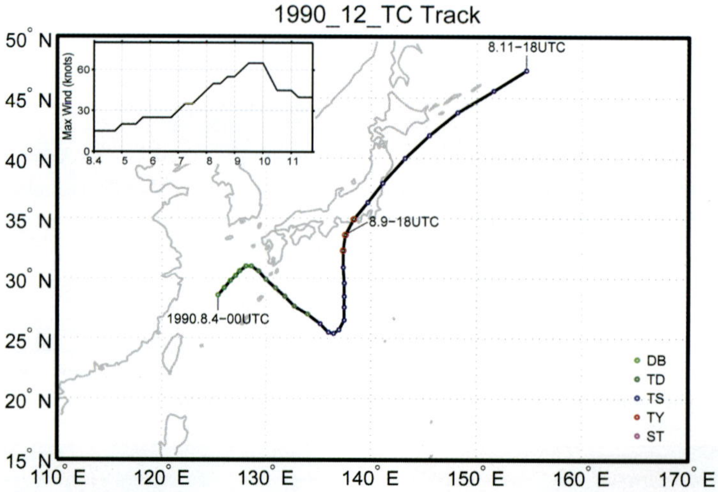

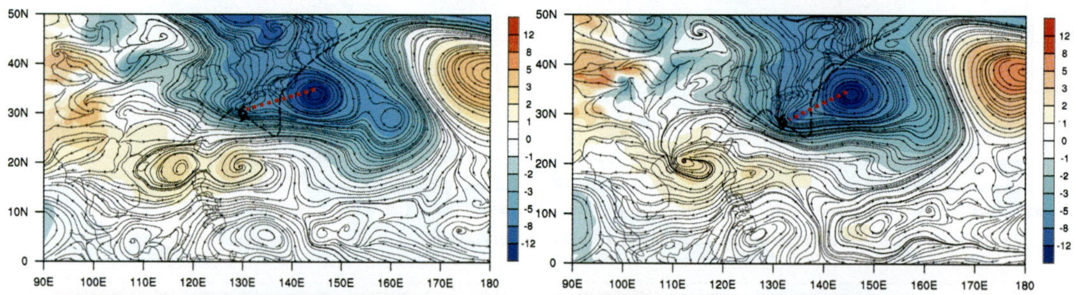

850 hPa flow/height anomalies at 0000 UTC and 1200 UTC 6 August 1990.

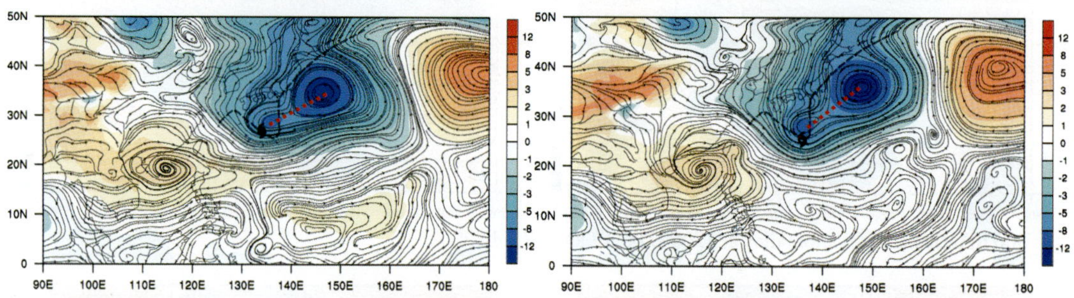

850 hPa flow/height anomalies at 0000 UTC and 1200 UTC 7 August 1990.

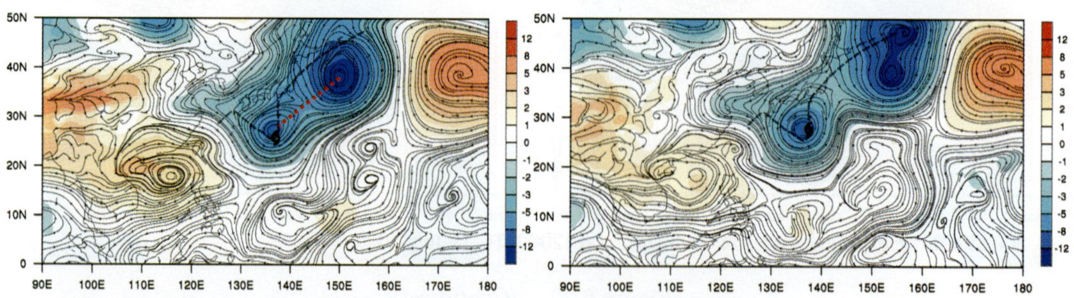

850 hPa flow/height anomalies at 0000 UTC and 1200 UTC 8 August 1990.

Figure 6.21. The typhoon (1990_No.12) generated in the East China Sea on 4 August 1990 and moved northeastward then southeastward. Its intensity peaked on 9 August 1990 before landfall on the central Japan Island and its turning motion near Japan was influenced by anomalous vortices at 850 hPa and 300 hPa.

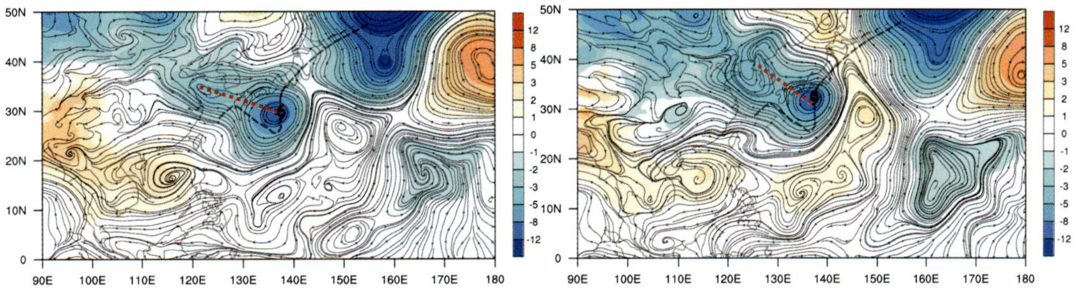

850 hPa flow/height anomalies at 0000 UTC and 1200 UTC 9 August 1990.

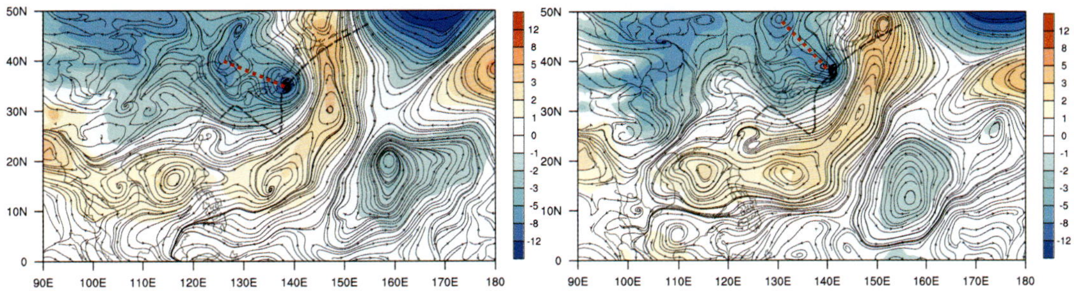

850 hPa flow/height anomalies at 0000 UTC and 1200 UTC 10 August 1990.

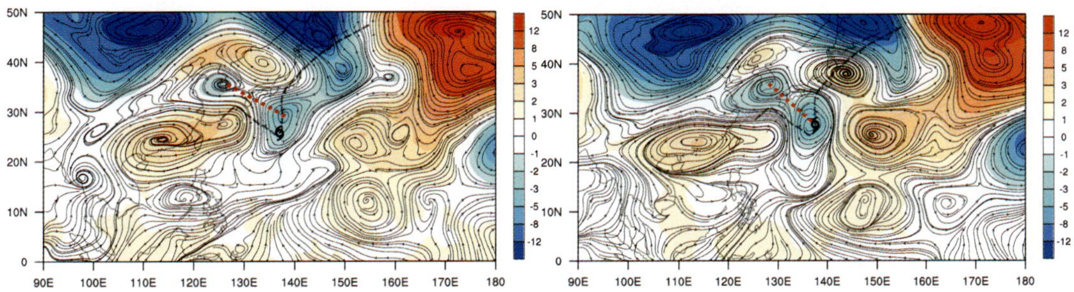

300 hPa flow/height anomalies at 0000 UTC and 1200 UTC 8 August 1990.

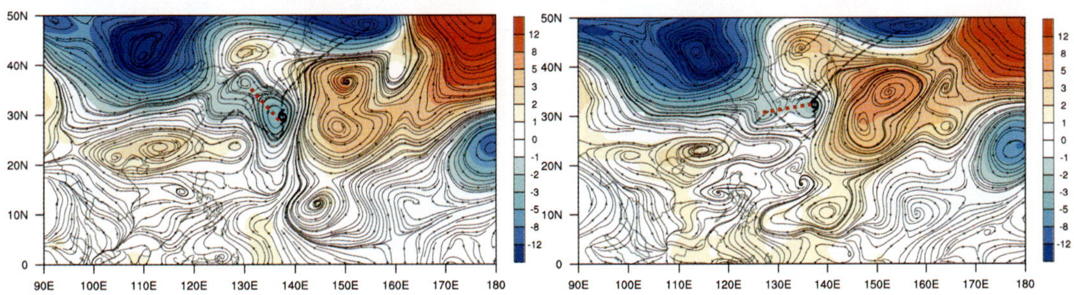

300 hPa flow/height anomalies at 0000 UTC and 1200 UTC 9 August 1990.

Figure 6.21. *(Continued)*

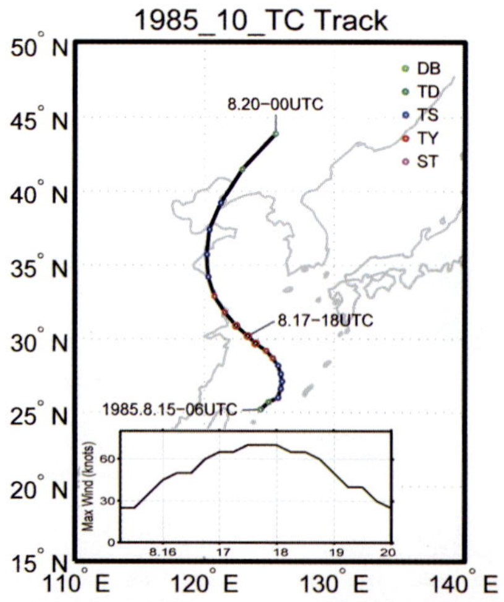

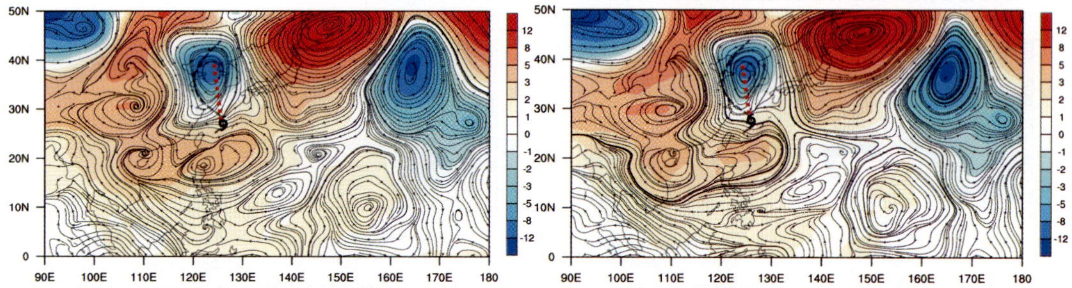

300 hPa flow/height anomalies at 0000 UTC and 1200 UTC 16 August 1985.

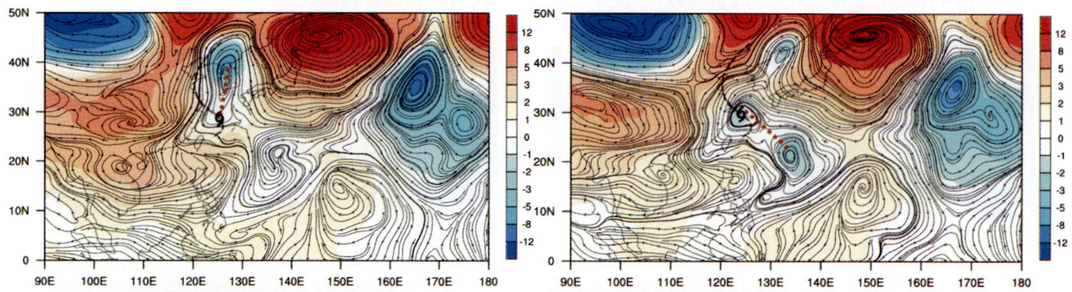

300 hPa flow/height anomalies at 0000 UTC and 1200 UTC 17 August 1985.

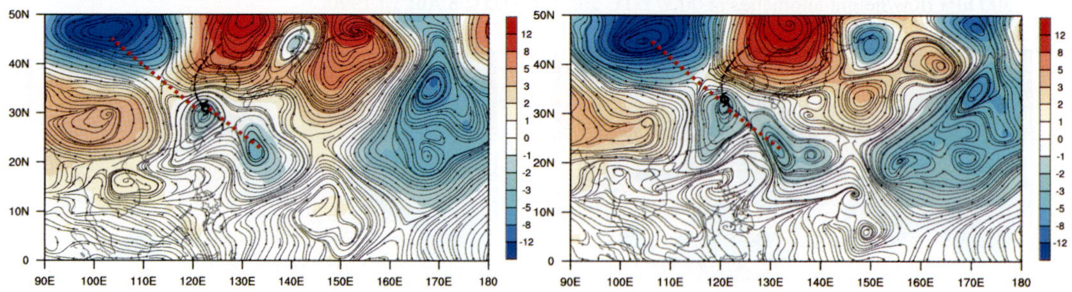

300 hPa flow/height anomalies at 0000 UTC and 1200 UTC 18 August 1985.

Figure 6.22. The typhoon (1985_No.10) generated in the east sea of the Taiwan Island on 15 August 1985 and its track showed a left turn and a right turn. Its intensity peaked in the East China Sea on 17 August 1985 when it was approaching the climatic monsoon trough before landfall on the East China coast. The right turn in the east seas of China was influenced by anomalous vortices at 300 hPa.

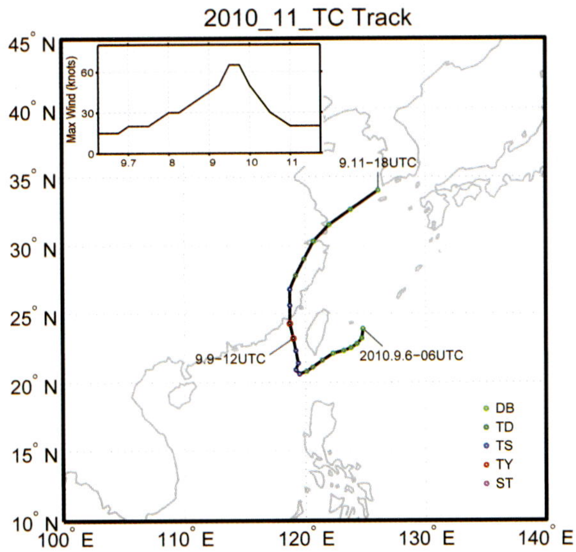

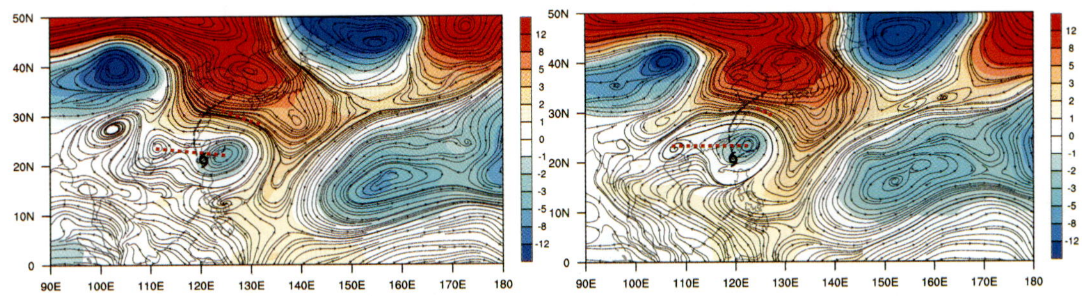

300 hPa flow/height anomalies at 0000 UTC and 1200 UTC 8 September 2010.

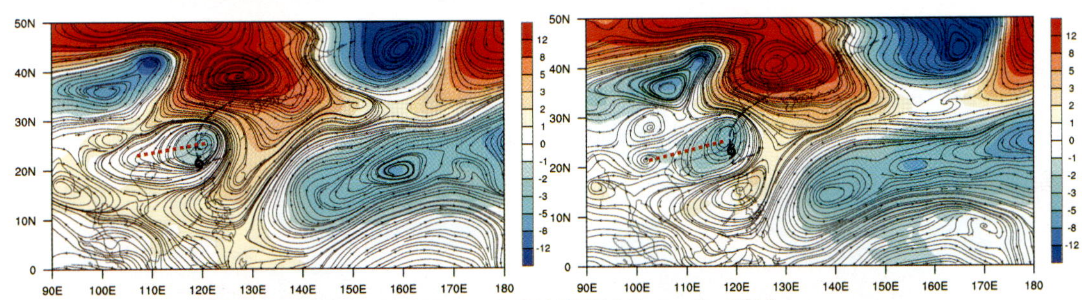

300 hPa flow/height anomalies at 0000 UTC and 1200 UTC 9 September 2010.

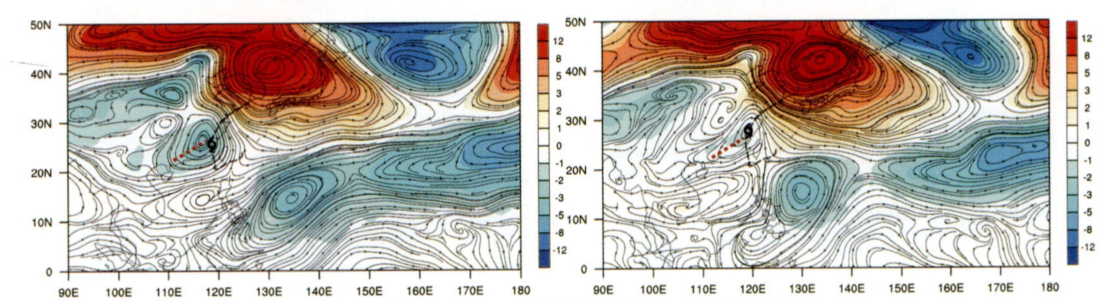

300 hPa flow/height anomalies at 0000 UTC and 1200 UTC 10 September 2010.

Figure 6.23. The typhoon (2010_No.11) generated in the east sea of the Taiwan Island on 6 September 2010 and moved southwestward at the beginning stage. It turned back to northward in the southwest sea of the Taiwan Island and its intensity peaked in the Taiwan Strait on 9 September 2010 when it was approaching the climatic monsoon trough before landfall on the Southeast China coast. This turning motion was influenced by an anomalous vortex at 300 hPa.

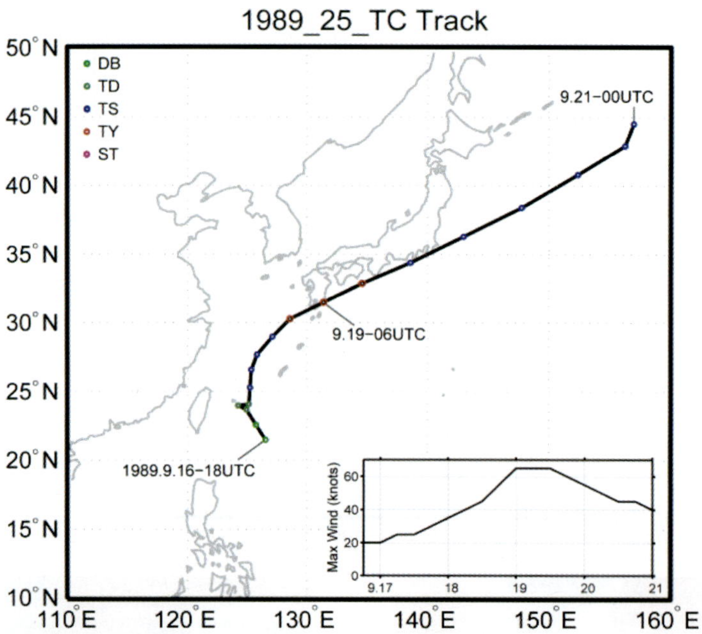

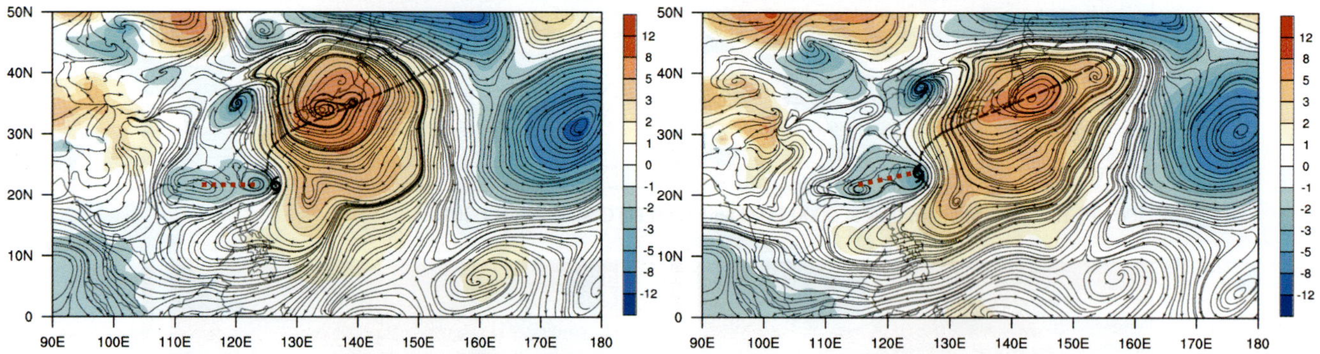

850 hPa flow/height anomalies at 1800 UTC 16 and 0600 UTC 17 September 1989

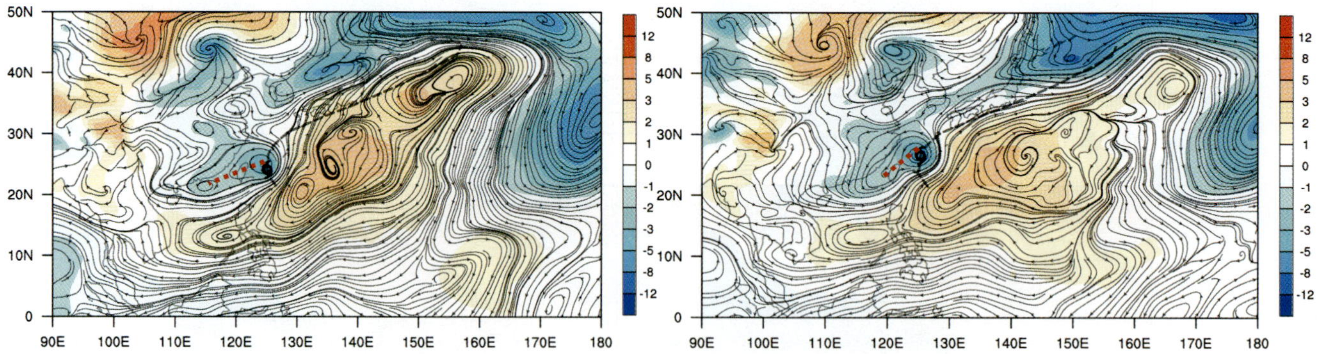

850 hPa flow/height anomalies at 1800 UTC 17 and 0600 UTC 18 September 1989

Figure 6.24. The typhoon (1989_No.25) generated in the southeast sea of Taiwan on 16 September 1989 and its motion was basically along the climatic monsoon trough. Its intensity peaked on 19 September 1989 when landfall on the Japan Island. Its right turn in the east sea of Taiwan was influenced by an anomalous vortex at 850 hPa.

6.5. Geneses in the Northwest Pacific

Eight typhoons generated in the Northwest Pacific and their intensities peaked in the east seas of China. The right turns or left turns of their motions were influenced by anomalous vortices (Figs. 6.26–6.33).

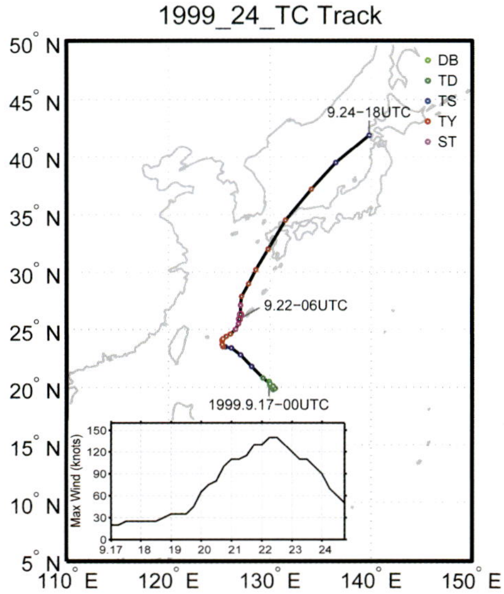

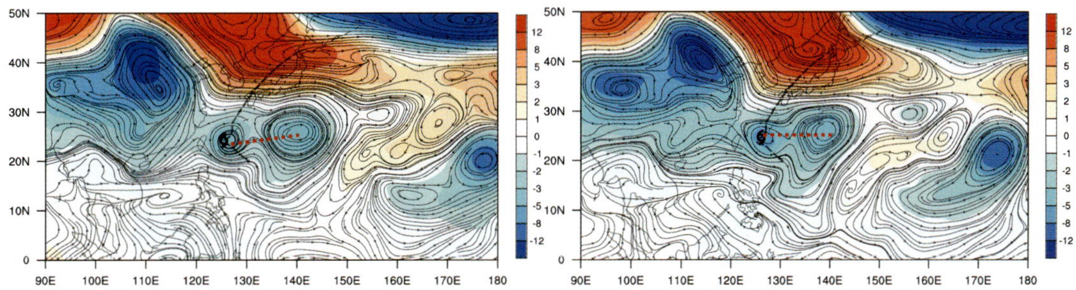

300 hPa flow/height anomalies at 0000 UTC and 1200 UTC 21 September 1999.

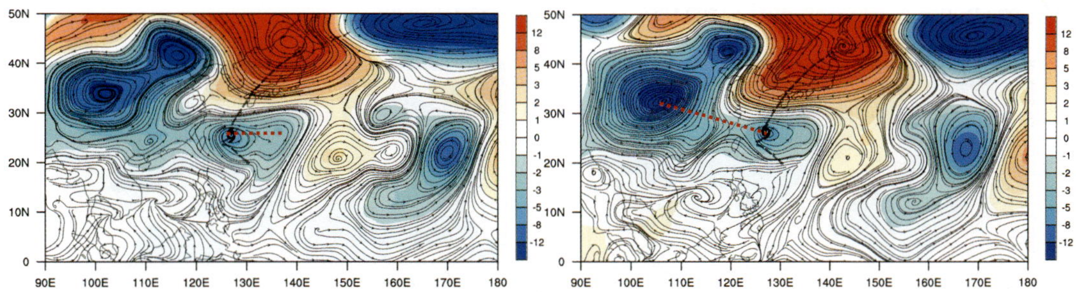

300 hPa flow/height anomaliesat 0000 UTC and 1200 UTC 22 September 1999.

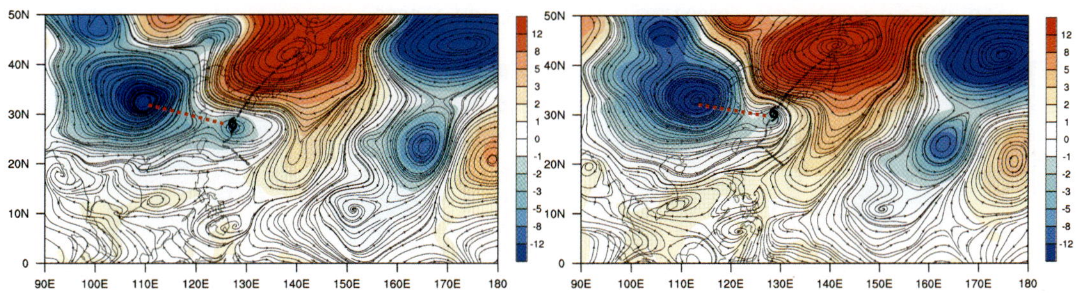

300 hPa flow/height anomalies at 0000 UTC and 1200 UTC 23 September 1999.

Figure 6.25. The typhoon (1999_No.24) generated in the southeast sea of the Taiwan Island on 17 September 1999. After a stagnant state, it moved northwestward along the tropical climatic monsoon trough, and then northeastward along the subtropical climatic monsoon trough. Its intensity peaked in the northeast sea of the Taiwan Island on 22 September 1999. The right turn in the east sea of the Taiwan Island was influenced by anomalous vortices at 300 hPa.

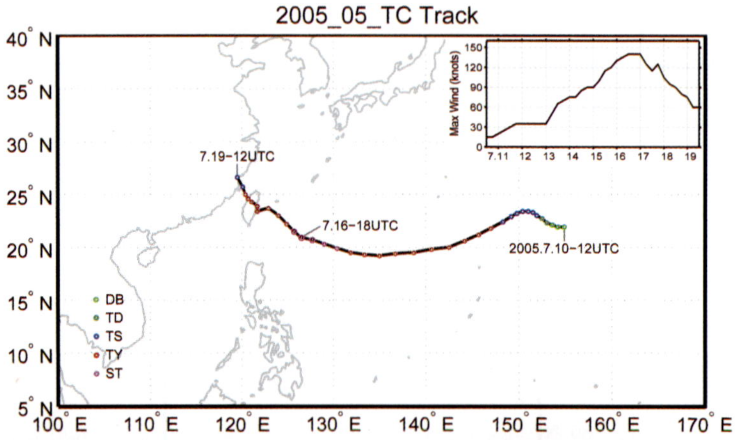

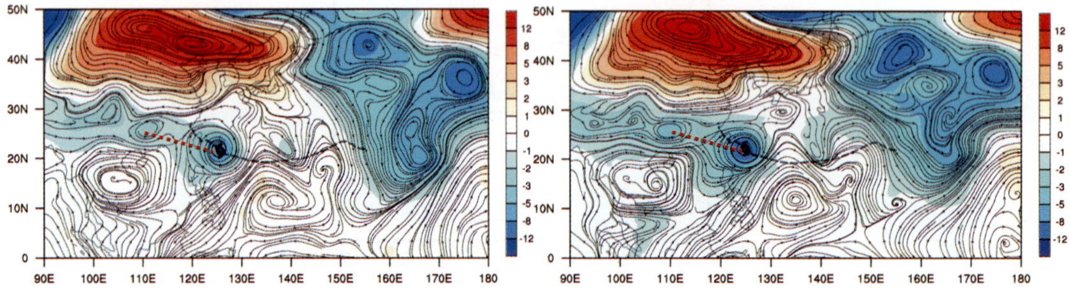

300 hPa flow/height (shading, 10 gpm) anomalies at 0000 UTC and 0600 UTC 17 July 2005.

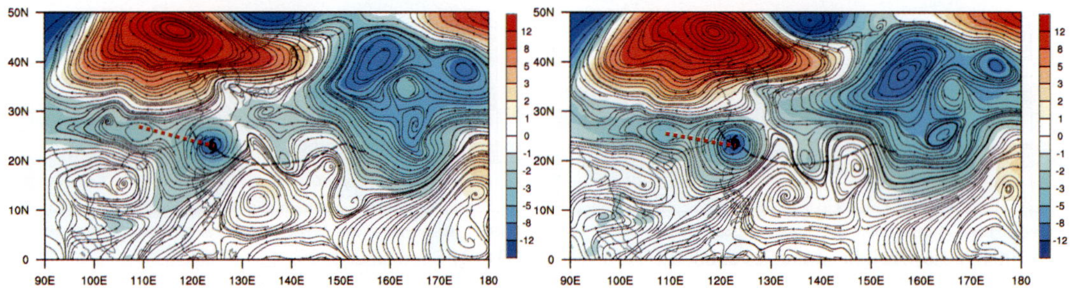

300 hPa flow/height anomalies at 1200 UTC and 1800 UTC 17 July 2005.

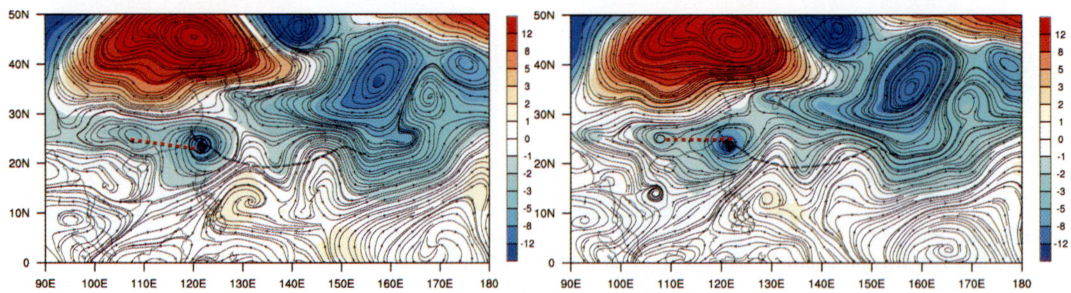

300 hPa flow/height anomalies at 0000 UTC and 0600 UTC 18 July 2005.

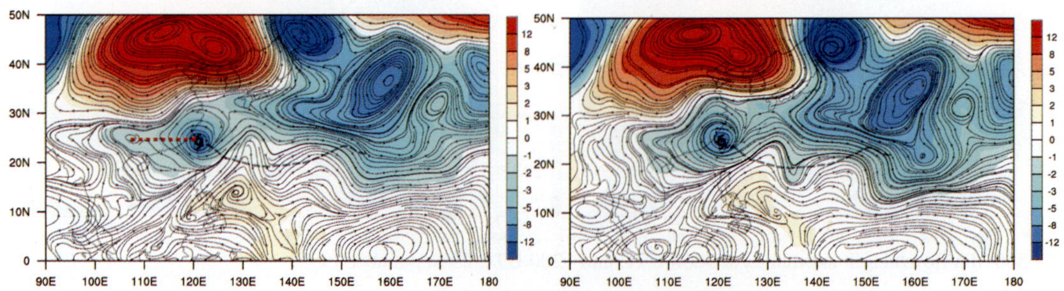

300 hPa flow/height anomalies at 1200 UTC and 1800 UTC 18 July 2005.

Figure 6.26. The typhoon (2005_No.05) generated in the subtropical Northwest Pacific on 10 July 2005 and moved westward. Its intensity peaked in the southeast sea of the Taiwan Island on 16 July 2005 when it was approaching the climatic monsoon trough before landfall on the island. The right turn of its motion in the southeast of the Taiwan Island was influenced by an anomalous vortex at 300 hPa.

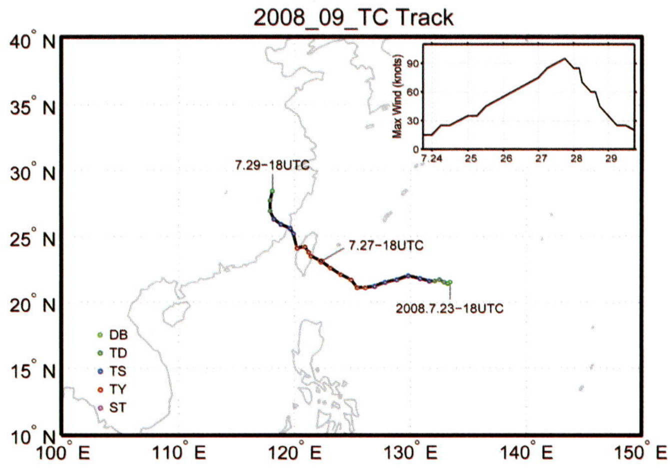

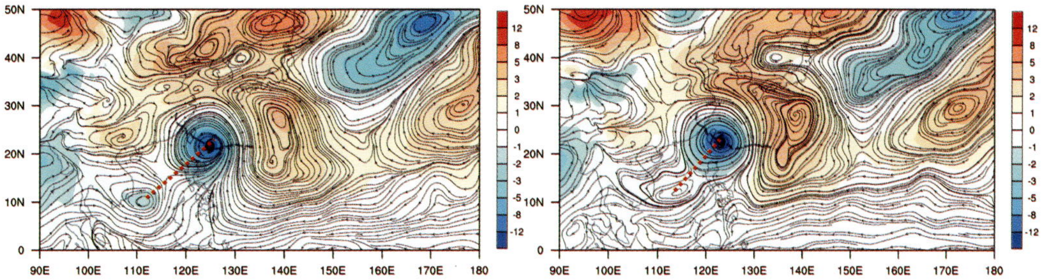

850 hPa flow/height anomalies at 0000 UTC and 1200 UTC 27 July 2008.

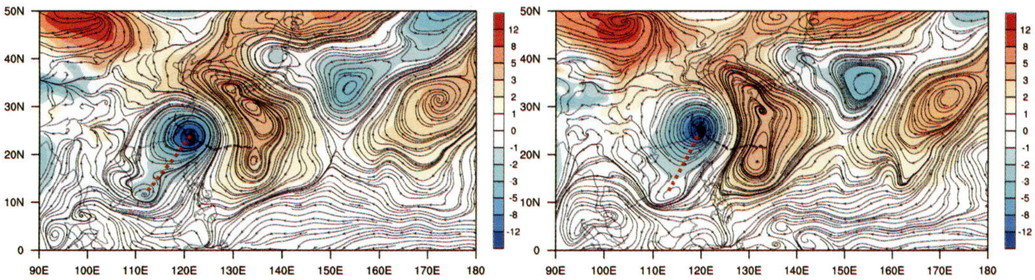

850 hPa flow/height anomalies at 0000 UTC and 1200 UTC 28 July 2008.

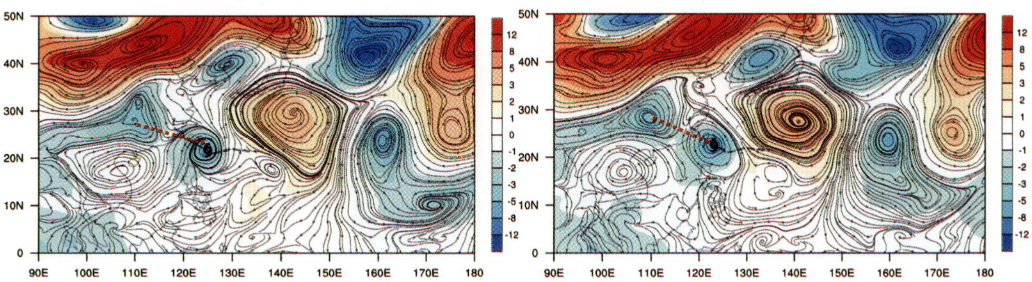

300 hPa flow/height anomalies at 0000 UTC and 1200 UTC 27 July 2008.

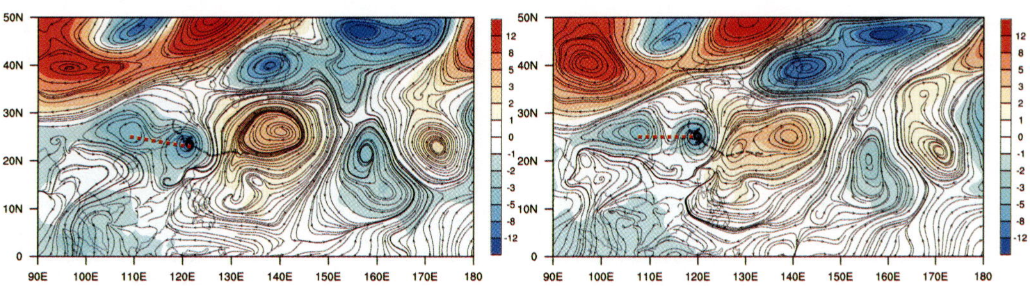

300 hPa flow/height anomalies at 0000 UTC and 1200 UTC 28 July 2008.

Figure 6.27. The typhoon (2008_No.09) generated in the subtropical Northwest Pacific on 23 July 2008 and moved northwestward along the climatic monsoon trough. Its intensity peaked on 27 July 2008 before landfall on the Taiwan Island. The right turn of its motion in the east sea of the Taiwan Island was influenced by anomalous vortices at 850 hPa and 300 hPa.

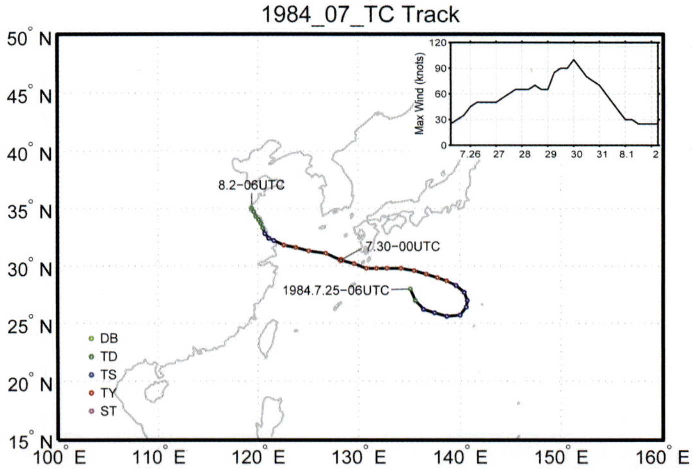

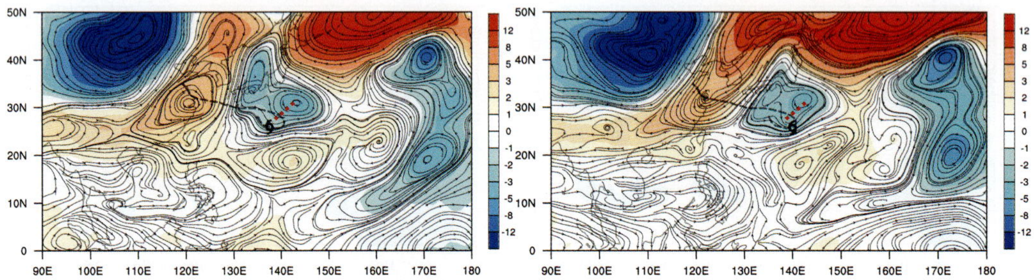

300 hPa flow/height anomalies at 0000 UTC and 1200 UTC 26 July 1984.

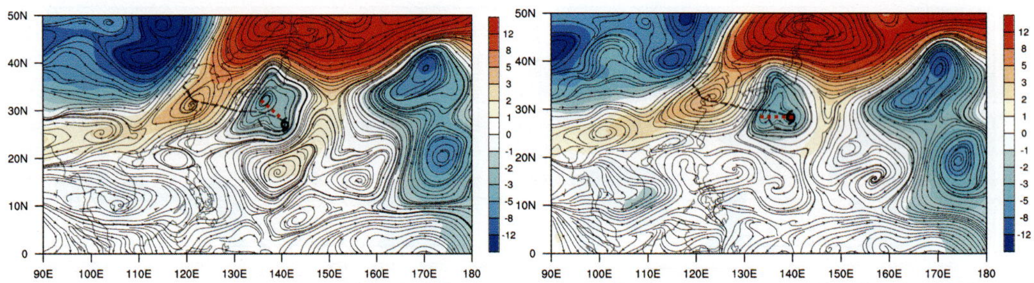

300 hPa flow/height anomalies at 0000 UTC and 1200 UTC 27 July 1984.

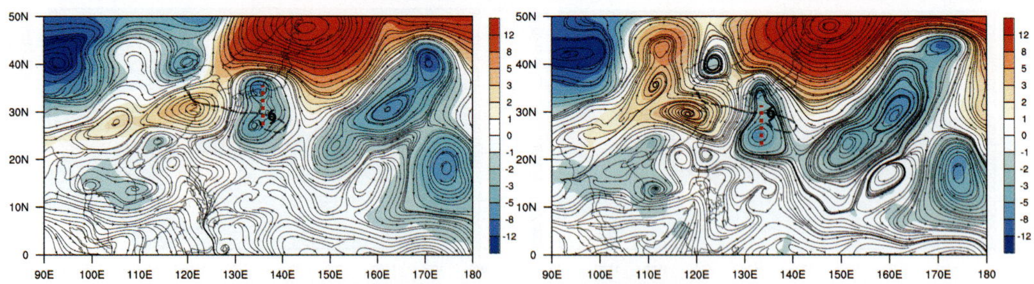

300 hPa flow/height anomalies at 0000 UTC and 1200 UTC 28 July 1984.

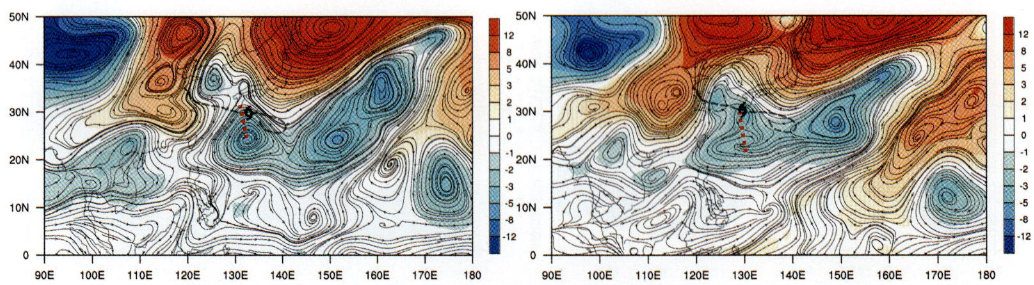

300 hPa flow/height anomalies at 0000 UTC and 1200 UTC 29 July 1984.

Figure 6.28. The typhoon (1984_No.07) generated in the subtropical Northwest Pacific on 25 July 1984. It moved in a circular way at the beginning stage. Its intensity peaked on 30 July 1984 in the south sea of Japan when it moved near the subtropical climatic monsoon trough. Its motion was influenced by anomalous vortices at 300 hPa.

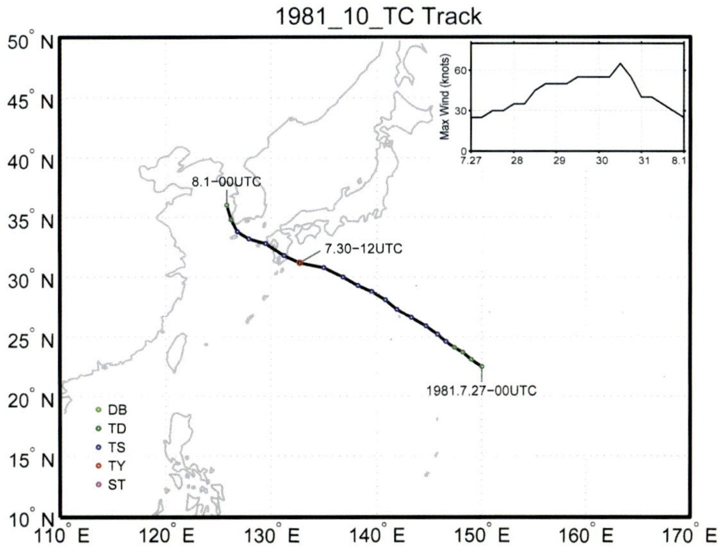

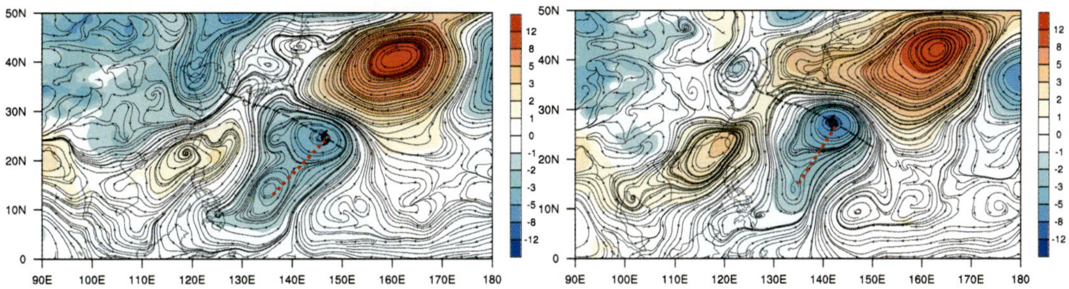

850 hPa flow/height anomalies at 0000 UTC 28 and 29 July 1981.

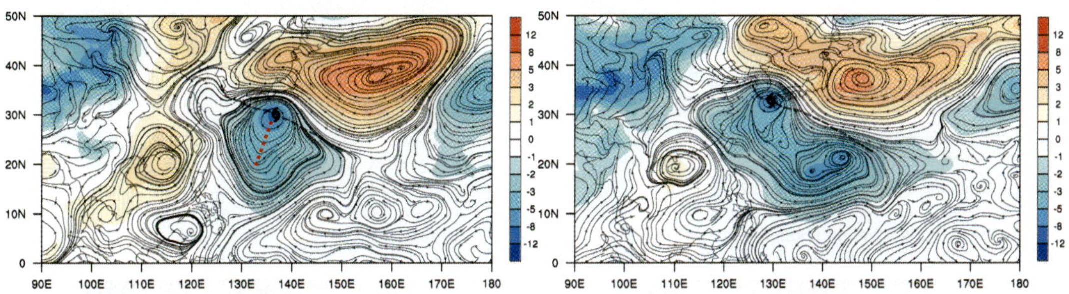

850 hPa flow/height anomalies at 0000 UTC 30 and 31 July 1981.

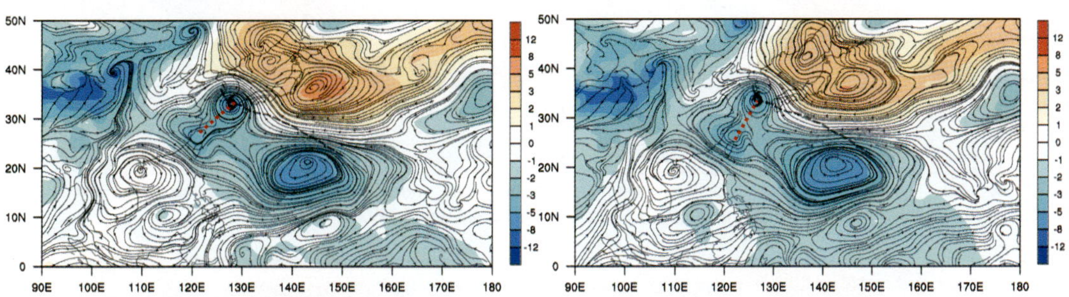

850 hPa flow/height anomalies at 0600 UTC and 1200 UTC 31 July 1981.

Figure 6.29. The typhoon (1981_No.10) generated in the subtropical Northwest Pacific on 27 July 1981 and moved northwestward. Its intensity peaked on 30 July 1981 before landfall on the southern Japan. A small right turn of its motion in the south of Korea was influenced by anomalous vortices at 850 hPa.

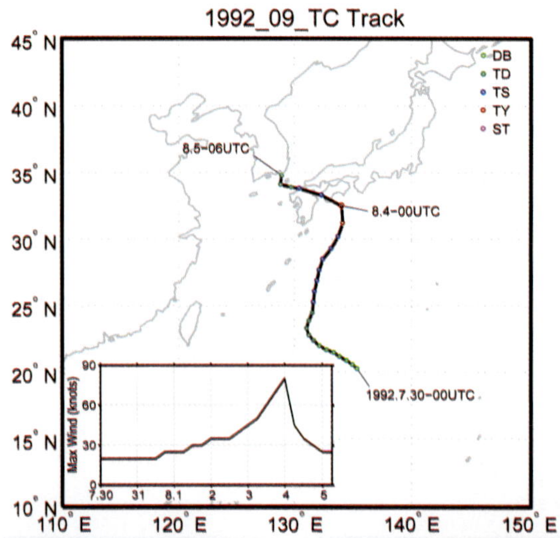

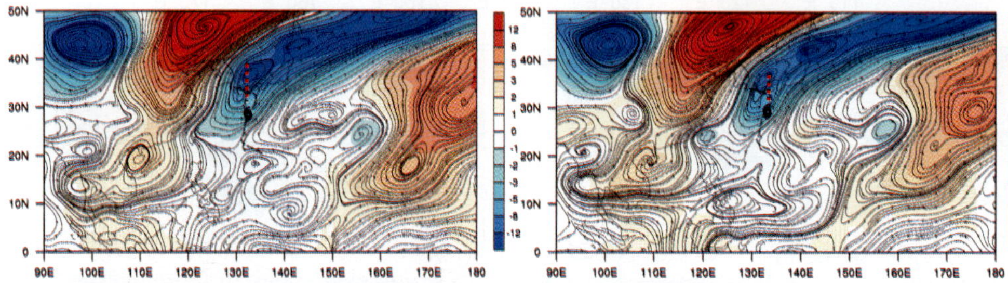

300 hPa flow/height anomalies at 0000 UTC and 0600 UTC 3 August 1992.

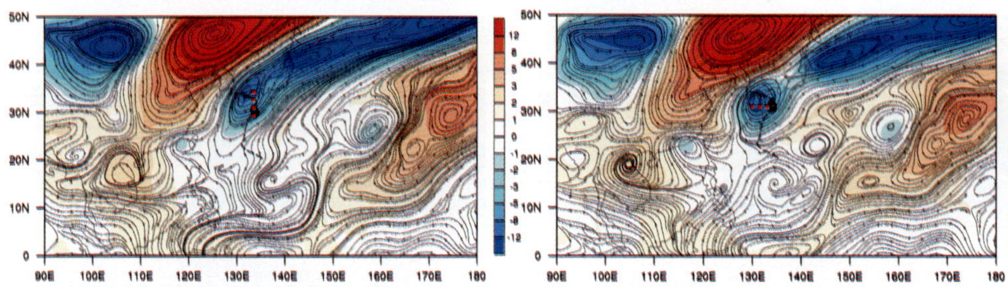

300 hPa flow/height anomalies at 1200 UTC and 1800 UTC 3 August 1992.

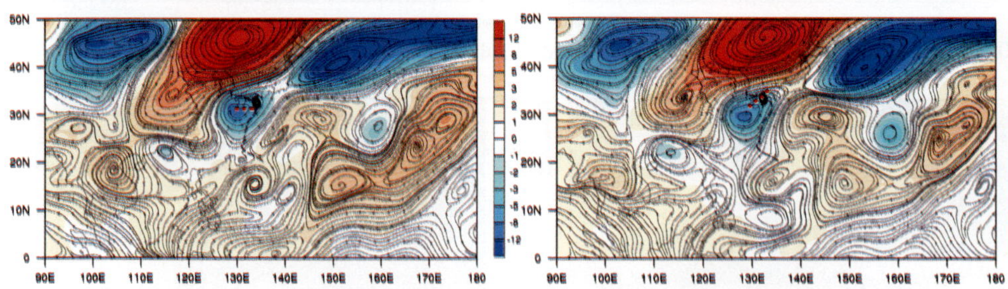

300 hPa flow/height anomalies at 0000 UTC and 0600 UTC 4 August 1992.

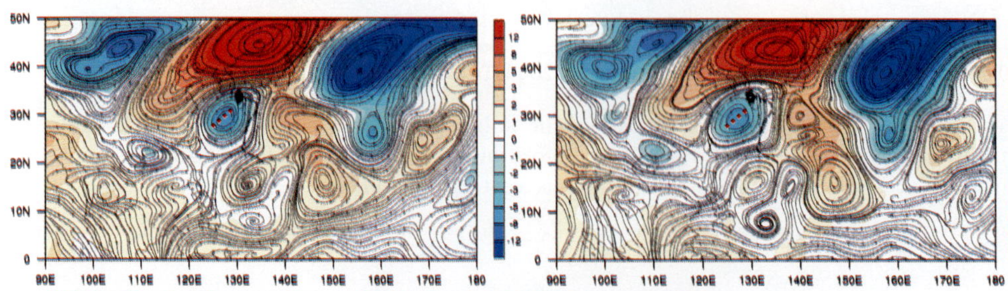

300 hPa flow/height anomalies at 1200 UTC and 1800 UTC 4 August 1992.

Figure 6.30. The typhoon (1992_No.09) generated in the subtropical Northwest Pacific on 30 July 1992 and moved northward with a right turn and a left turn. Its intensity peaked on 4 August 1992 before landfall on the Japan Island. The left turn of its motion near the south of Japan was influenced by an anomalous vortex at 300 hPa.

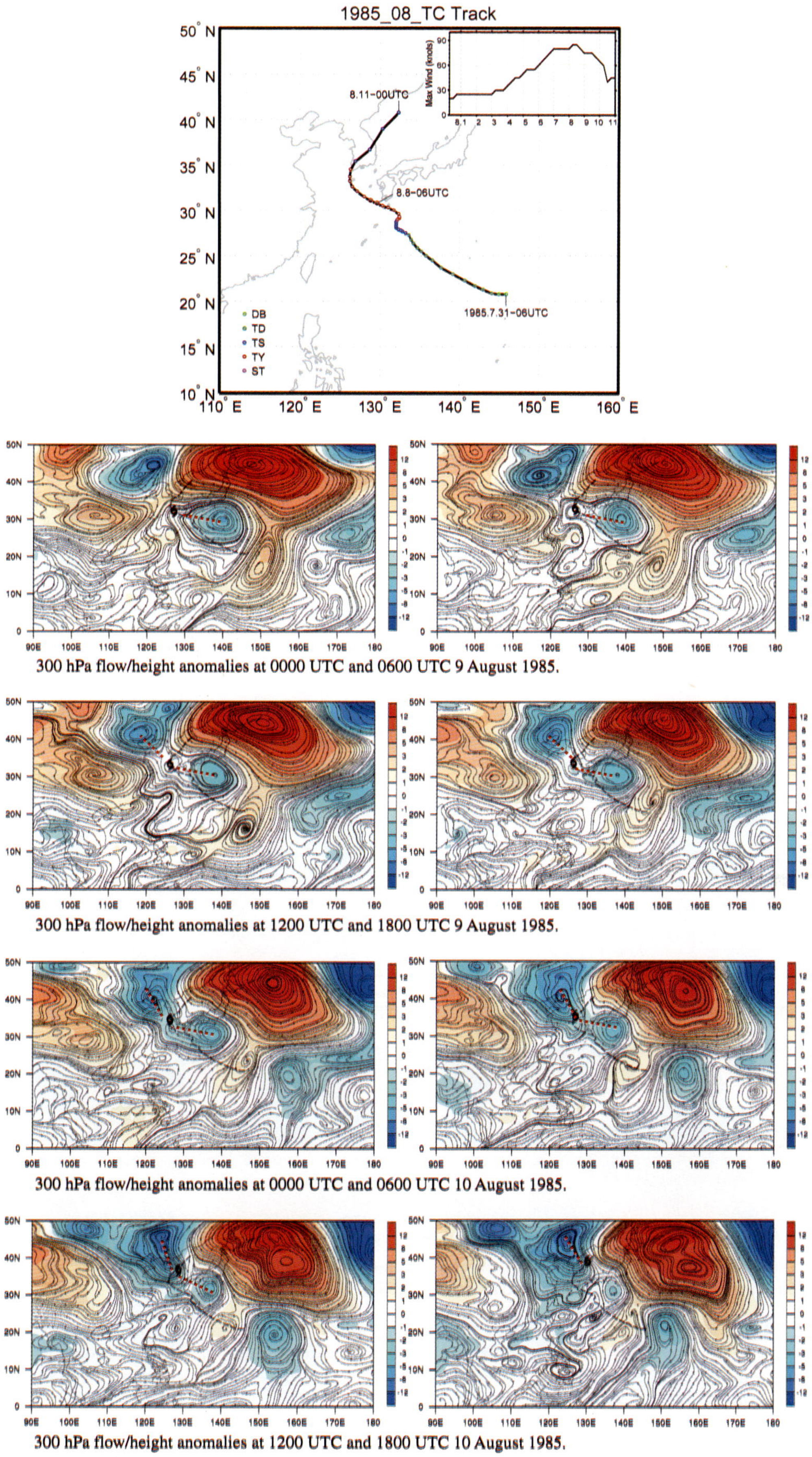

Figure 6.31. The typhoon (1985_No.08) generated in the subtropical Northwest Pacific on 31 July 1985 and moved northwestward. Its intensity peaked near southern Japan on 8 August 1985 before landfall on the South Korea. The right turn of its motion in the seas between China and Japan was influenced by anomalous vortices at 300 hPa.

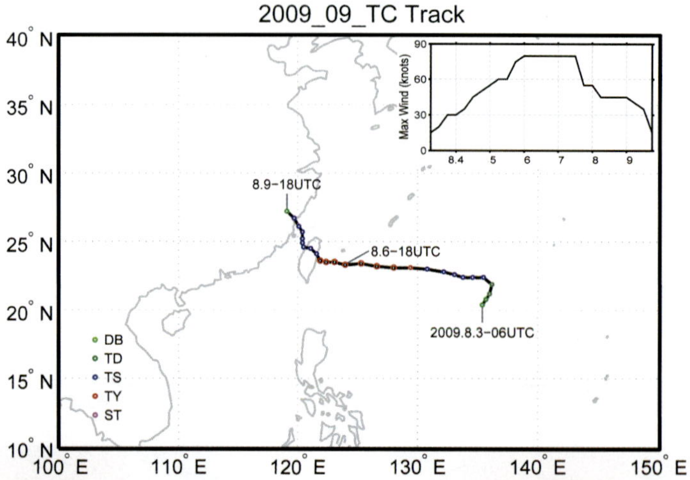

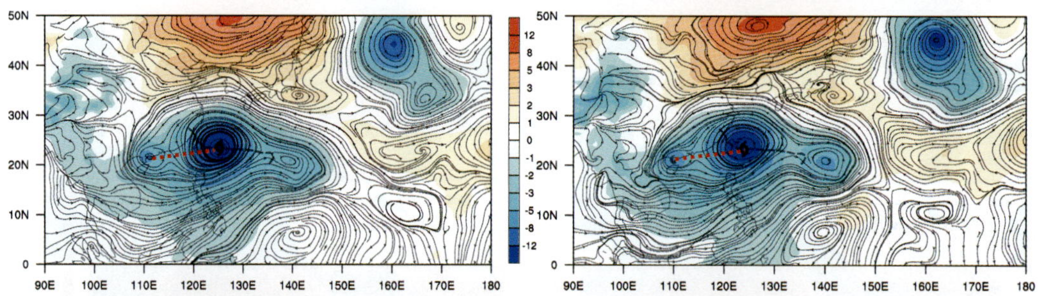

850 hPa flow/height anomalies at 1200 UTC and 1800 UTC 6 August 2009.

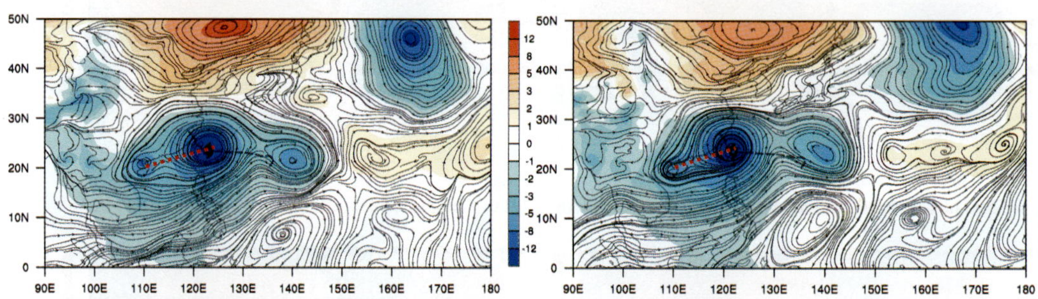

850 hPa flow/height anomalies at 0000 UTC and 1200 UTC 7 August 2009.

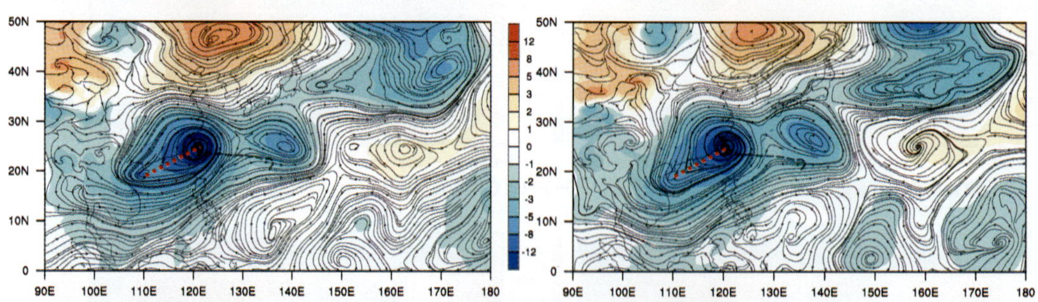

850 hPa flow/height anomalies at 0000 UTC and 0600 UTC 8 August 2009.

Figure 6.32. The typhoon (2009_No.09) generated in the subtropical Northwest Pacific on 3 August 2009. Its intensity peaked on 6 August 2009 when it was approaching the climatic monsoon trough before landfall on the Taiwan Island. Its track had a left turn and a right turn. The right turn in the east sea of the Taiwan Island was influenced by anomalous vortices at 850 hPa and 300 hPa.

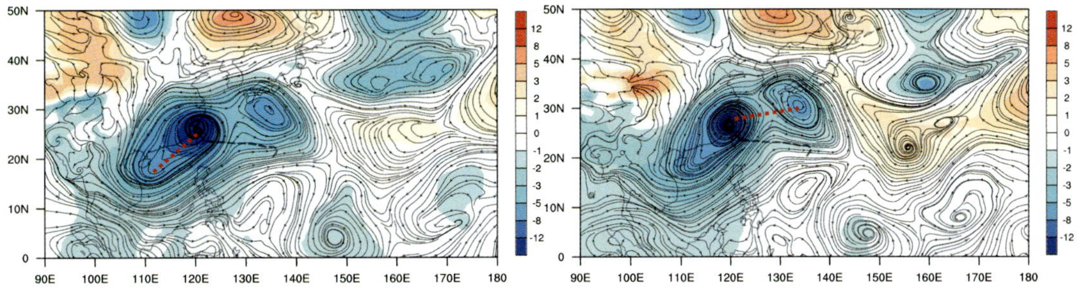

850 hPa flow/height anomalies at 0000 UTC and 1200 UTC 9 August 2009.

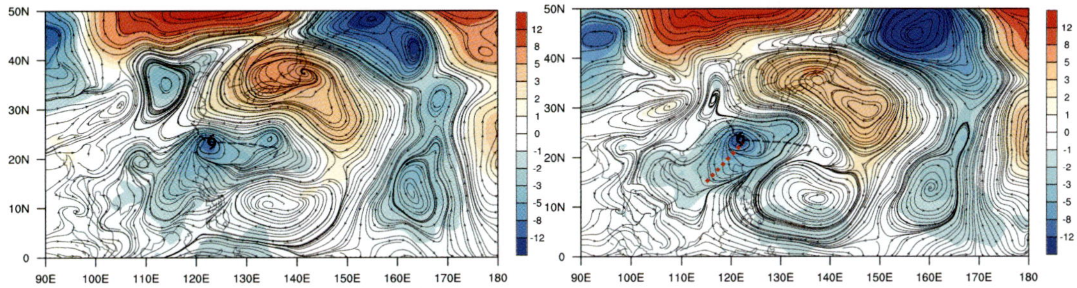

300 hPa flow/height anomalies at 0000 UTC and 1200 UTC 7 August 2009.

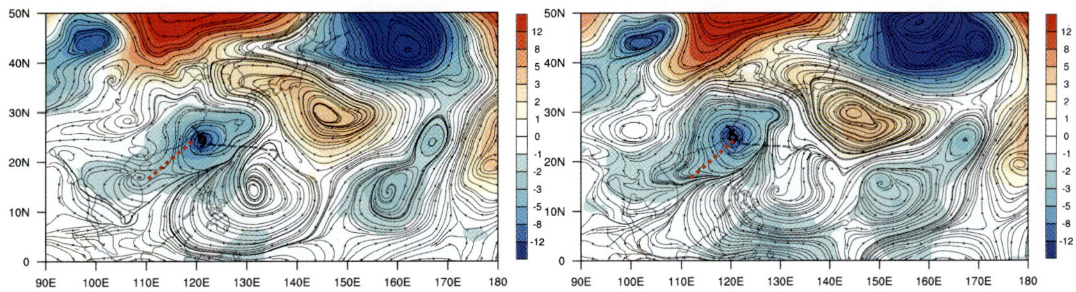

300 hPa flow/height anomalies at 0000 UTC and 1200 UTC 8 August 2009.

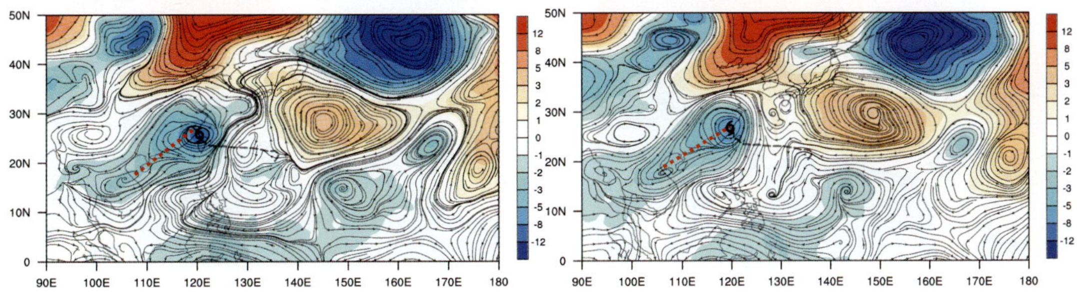

300 hPa flow/height anomalies at 0000 UTC and 1200 UTC 9 August 2009.

Figure 6.32. *(Continued)*

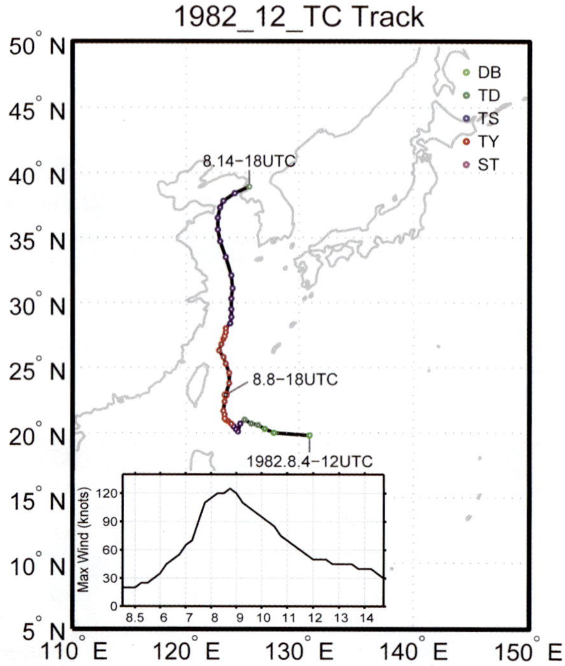

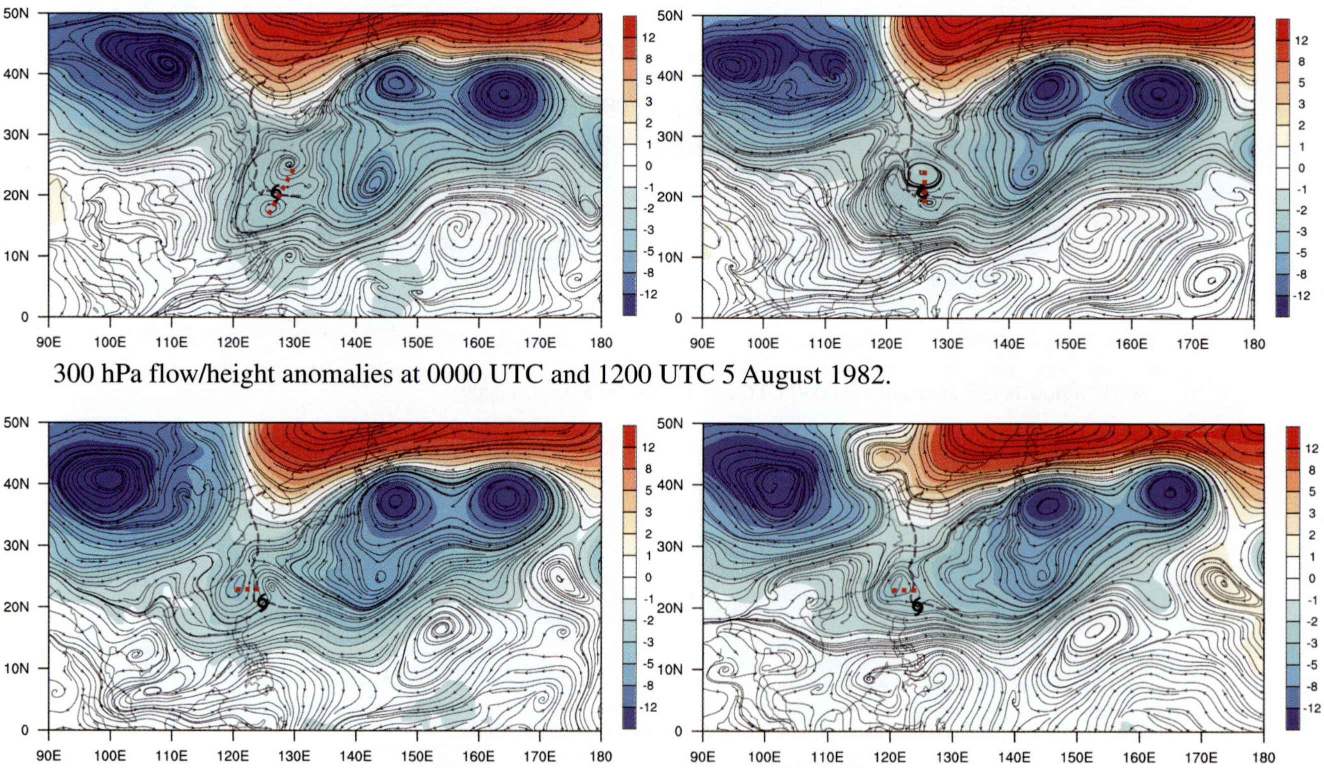

300 hPa flow/height anomalies at 0000 UTC and 1200 UTC 5 August 1982.

300 hPa flow/height anomalies at 0000 UTC and 1200 UTC 6 August 1982.

Figure 6.33. The typhoon (1982_No.12) generated in the tropical western Pacific on 4 August 1982 and moved westward at the beginning stage. Its intensity peaked in the east sea of the Taiwan Island on 8 August 1982 when it was approaching the climatic monsoon trough. The right turn in the east sea of the Taiwan Island was influenced by anomalous vortices at 300 hPa.

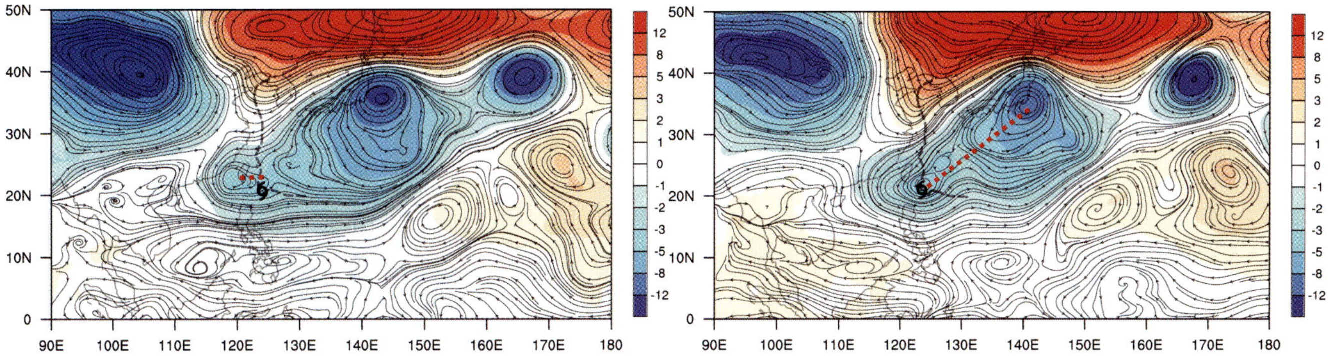

300 hPa flow/height anomalies at 0000 UTC and 1200 UTC 7 August 1982.

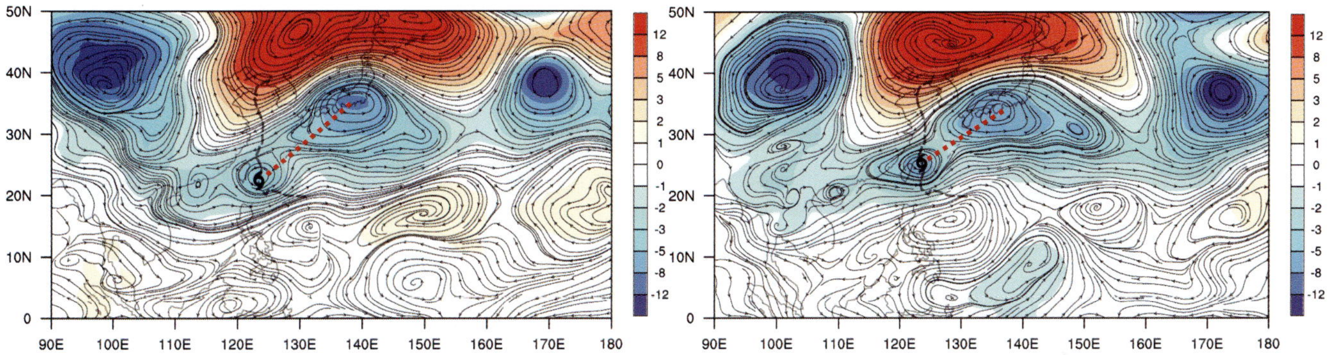

300 hPa flow/height anomalies at 1200 UTC 8 and 9 August 1982.

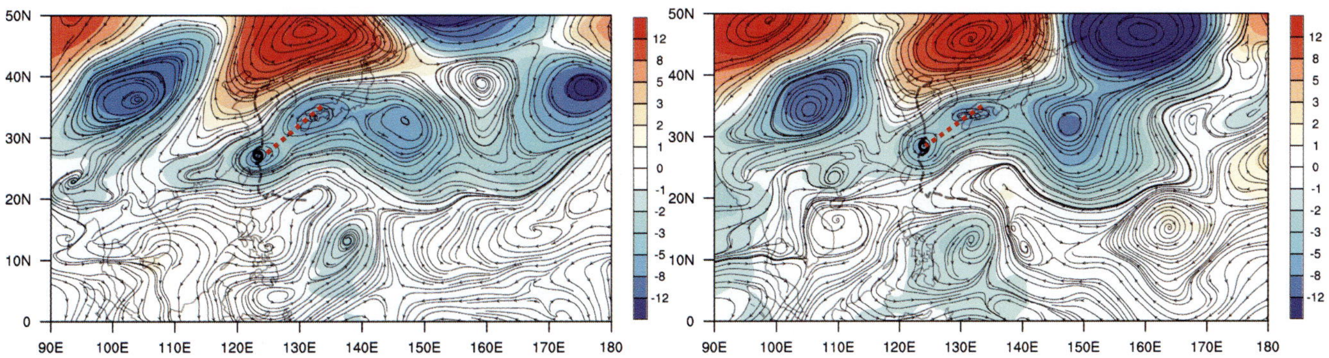

300 hPa flow/height anomalies at 1200 UTC 10 and 11 August 1982.

Figure 6.33. *(Continued)*

Chapter 7

Intensity Peaked in the Pacific

Some typhoons moved for long distances during their lifetime. These typhoons' intensities peaked in the equatorial Pacific (Sec. 7.1) and in the Northwest Pacific (Sec. 7.2), as well as generated in the western Pacific (Sec. 7.3) and central Pacific (Sec. 7.4), and finally entered the east seas of China. Some cases showed twin typhoons in the ocean (Sec. 7.5) and some typhoons with complex motions (Sec. 7.6).

7.1. Intensity Peaked in the Equatorial Pacific

There were 8 typhoons that traveled for long distances and their intensities peaked in the equatorial Pacific. Their tracks were influenced by anomalous vortices (Figs. 7.1–7.8).

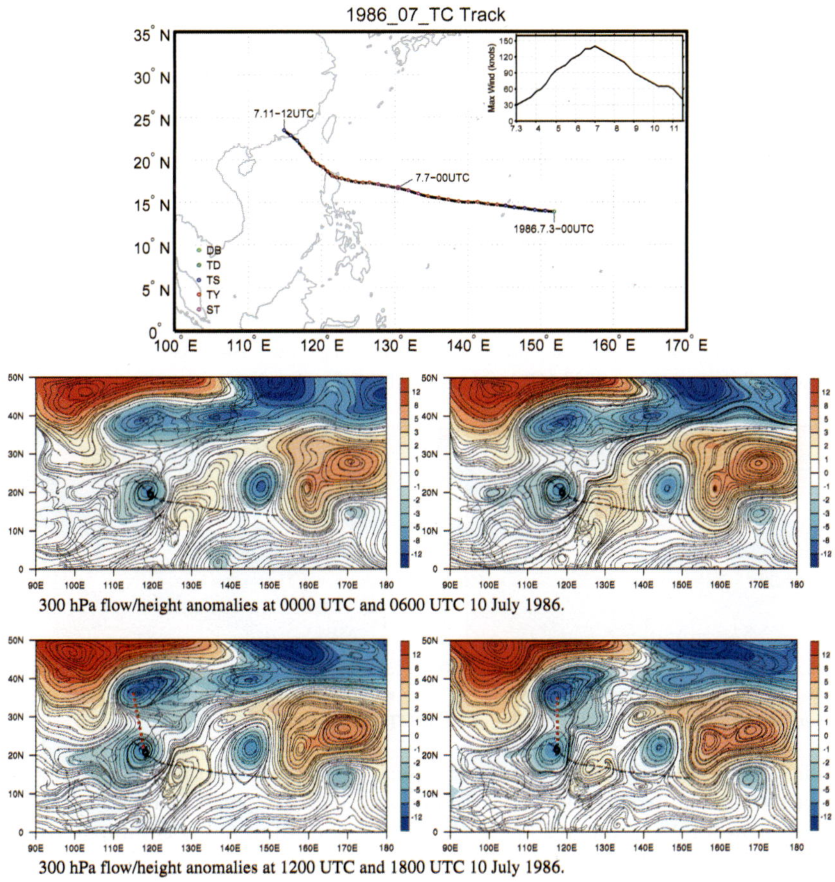

Figure 7.1. The typhoon (1986_No.07) generated in the tropical central Pacific on 3 July 1986 and moved westward. Its intensity peaked in the Northwest Pacific on 7 July 1986 when it was approaching the climatic monsoon trough. The right turn in the northern Philippines to northeastern South China Sea (SCS) was influenced by an anomalous vortex and anomalous anti-cyclonic circulations at 300 hPa. Symbols used are same as in Fig. 2.1.

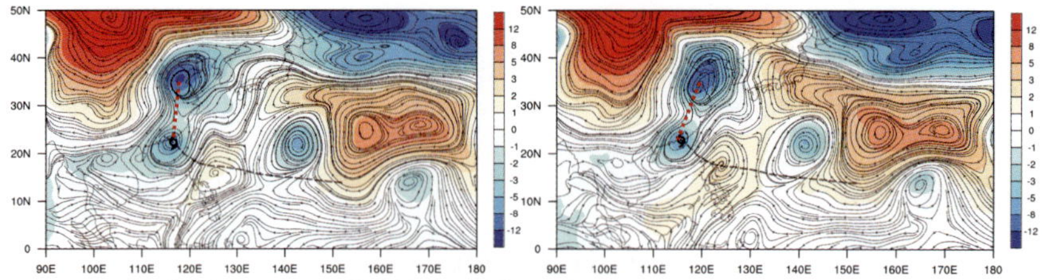

300 hPa flow/height anomalies at 0000 UTC and 0600 UTC 11 July 1986.

Figure 7.1. (*Continued*)

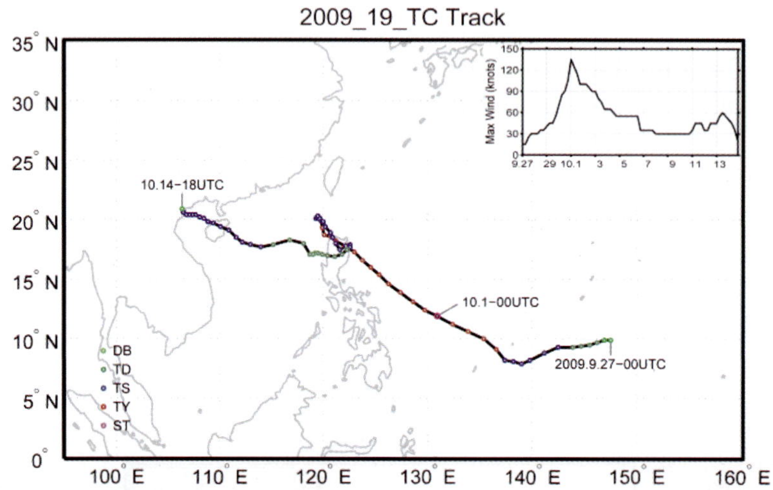

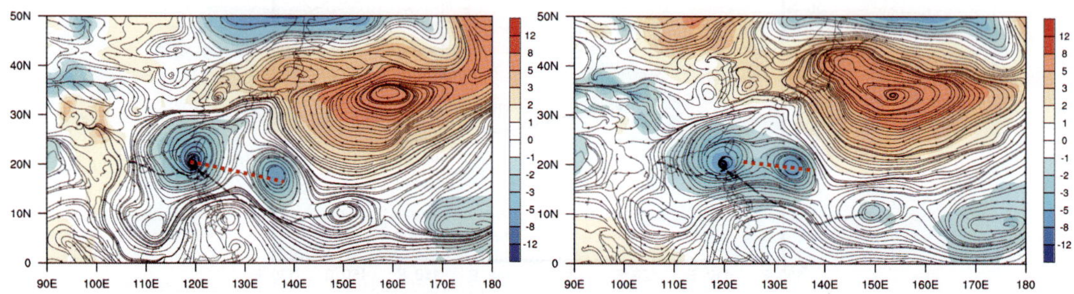

850hPa flow/height anomalies at 0000 UTC and 1200 UTC 5 October 2009.

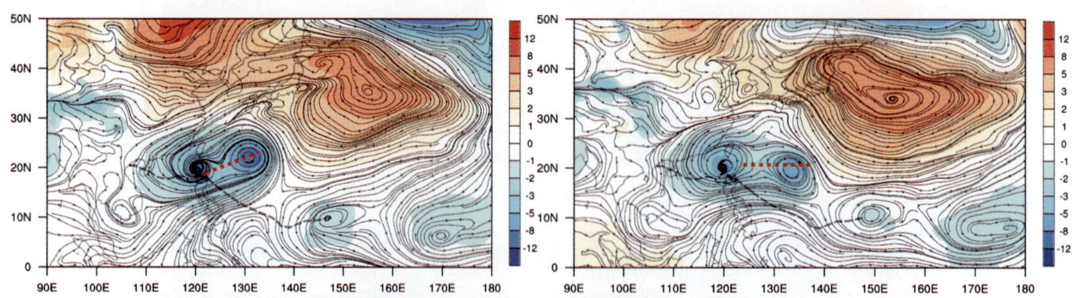

850hPa flow/height anomalies at 0000 UTC and 1200 UTC 6 October 2009.

Figure 7.2. The typhoon (2009_No.19) generated in the tropical central Pacific on 27 September 2009 and moved northwestward. Its intensity peaked in the tropical western Pacific on 1 October 2009 when it was near the climatic monsoon trough. The return motion from 3 to 8 October 2009 in the northern Philippines was influenced by anomalous vortices at 850 hPa and 300 hPa.

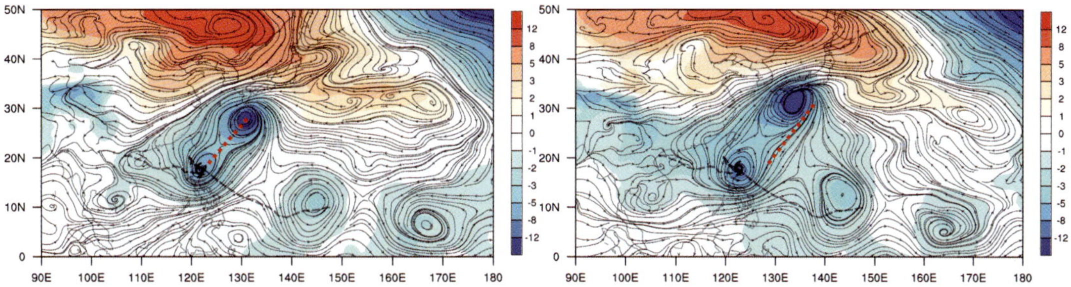

850hPa flow/height anomalies at 0000 UTC and 1200 UTC 7 October 2009.

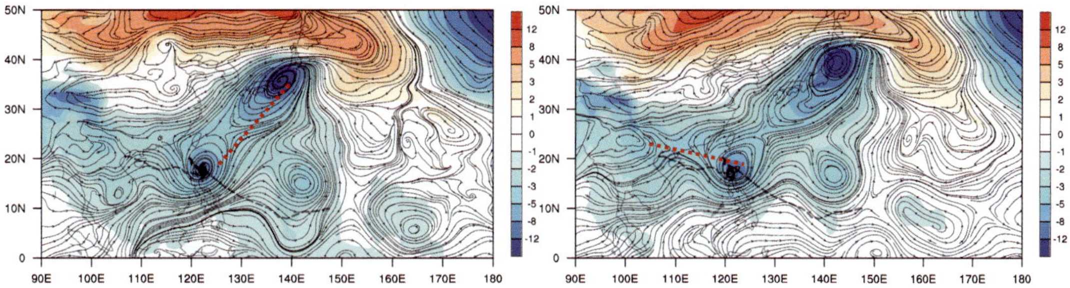

850hPa flow/height anomalies at 0000 UTC and 1200 UTC 8 October 2009.

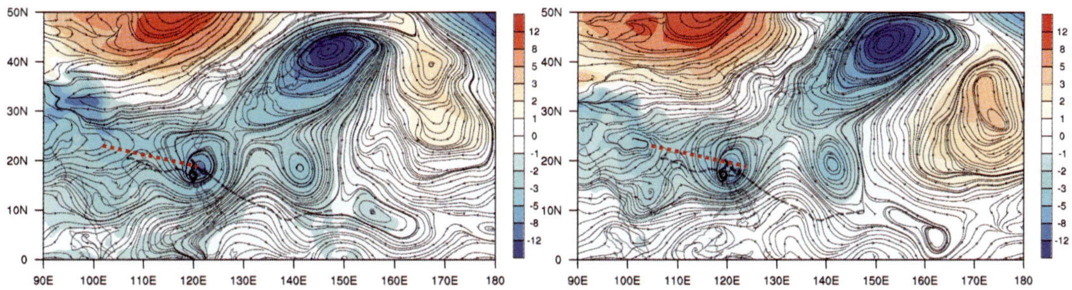

850hPa flow/height anomalies at 0000 UTC and 1200 UTC 9 October 2009.

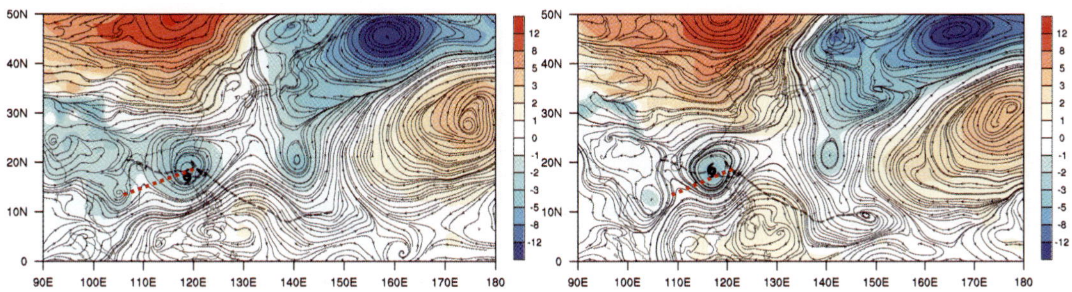

850hPa flow/height anomalies at 0000 UTC and 1200 UTC 10 October 2009.

Figure 7.2. *(Continued)*

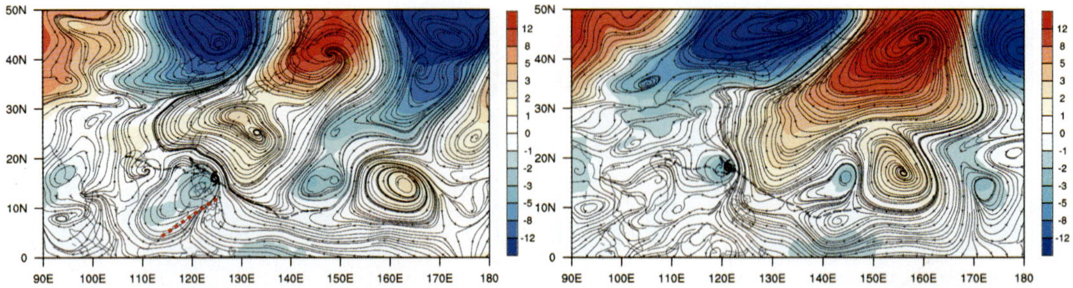

300hPa flow/height anomalies at 1200 UTC 2 and 3 October 2009.

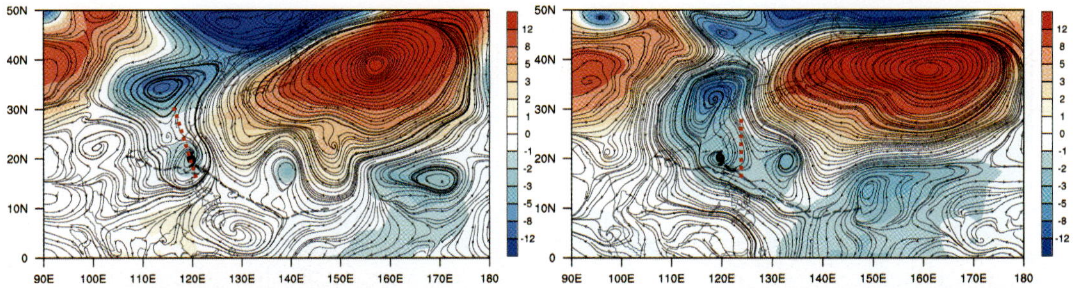

300hPa flow/height anomalies at 1200 UTC 4 and 5 October 2009.

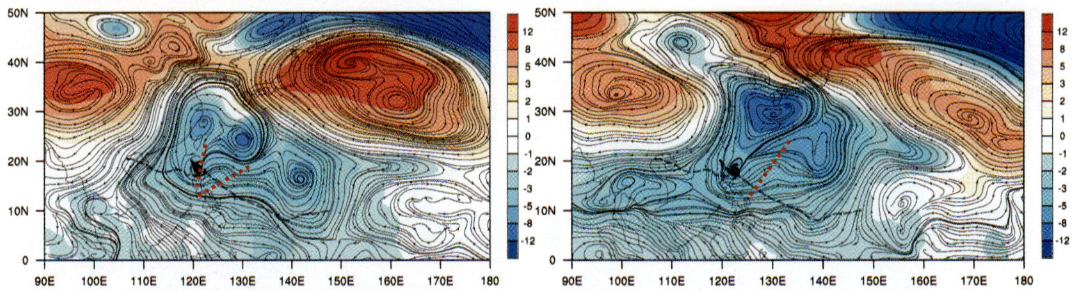

300hPa flow/height anomalies at 1200 UTC 6 and 7 October 2009.

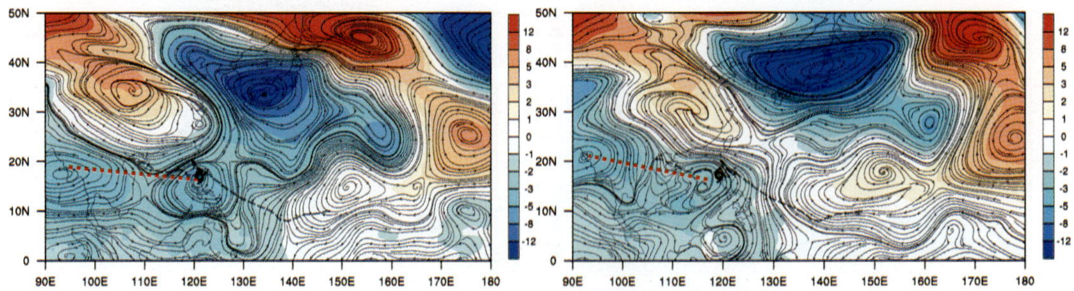

300hPa flow/height anomalies at 1200 UTC 8 and 9 October 2009.

Figure 7.2. (*Continued*)

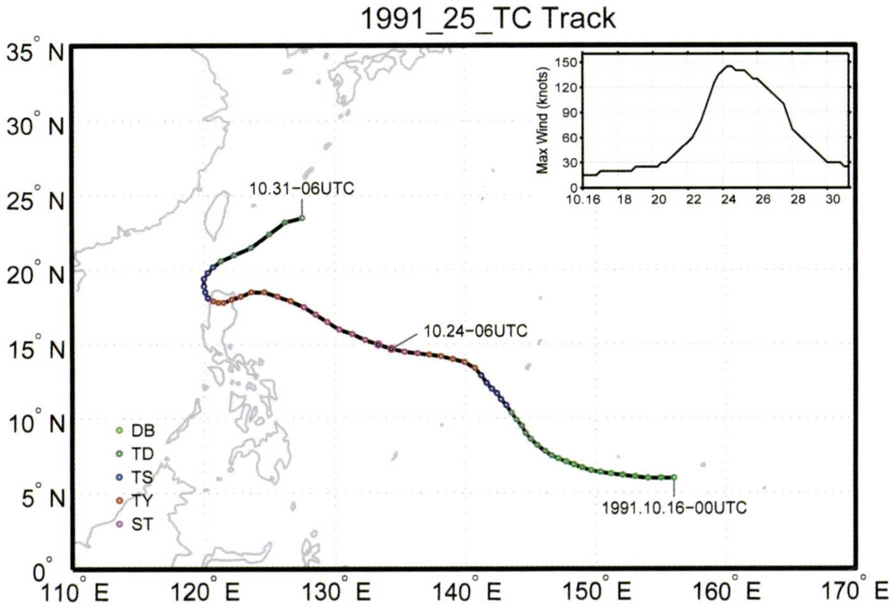

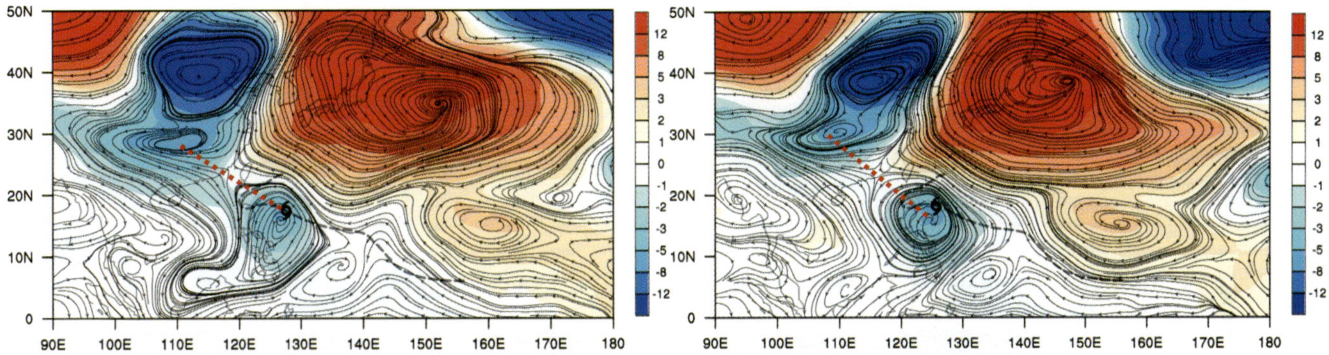

300 hPa flow/height anomalies at 0000 UTC and 1200 UTC 26 October 1991.

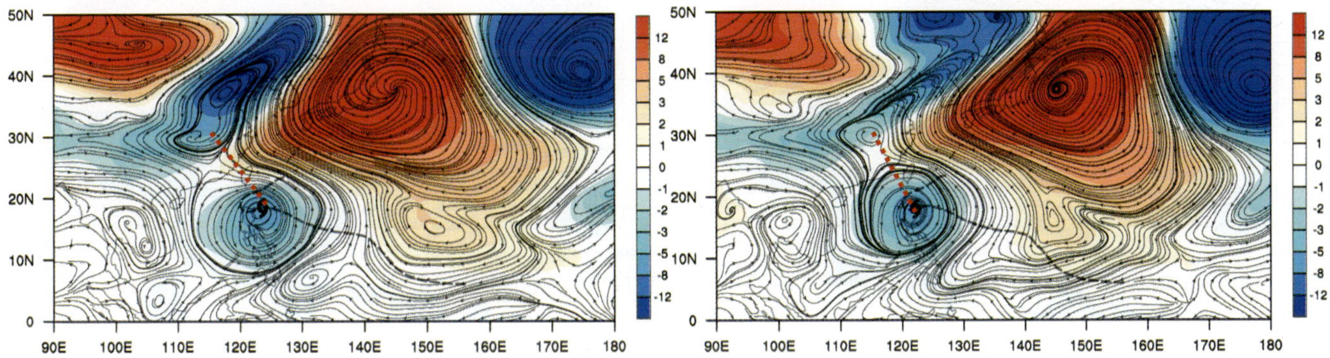

300 hPa flow/height anomalies at 0000 UTC and 1200 UTC 27 October 1991.

Figure 7.3. The typhoon (1991_No.25) generated in the equatorial central Pacific on 16 October 1991 and moved northwestward near the climatic monsoon trough. Its intensity peaked in the tropical western Pacific on 24 October 1991. The right turn near the Philippines was influenced by an anomalous vortex at 300 hPa.

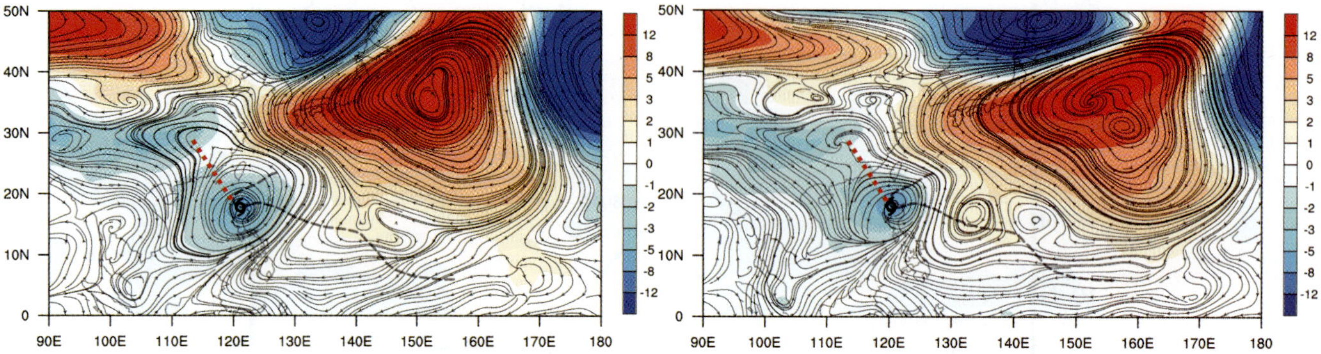

300 hPa flow/height anomalies at 0000 UTC and 1200 UTC 28 October 1991.

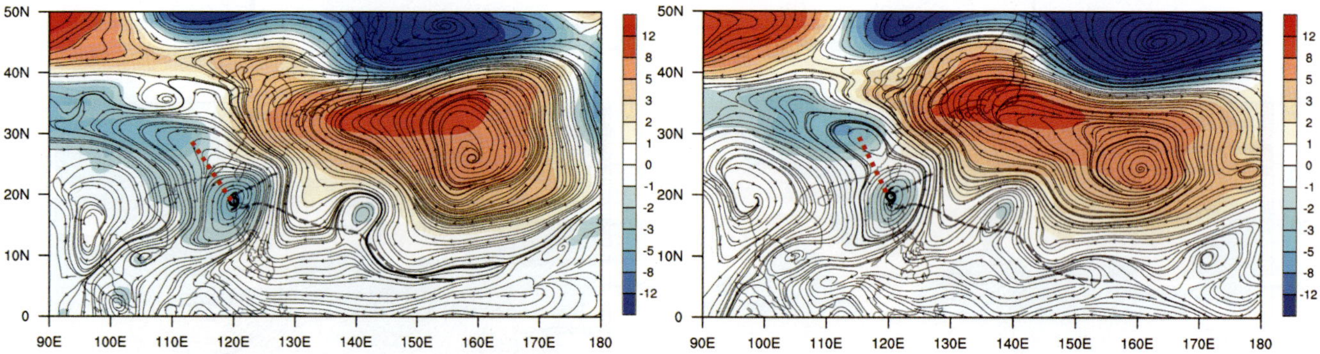

300 hPa flow/height anomalies at 0000 UTC and 1200 UTC 29 October 1991.

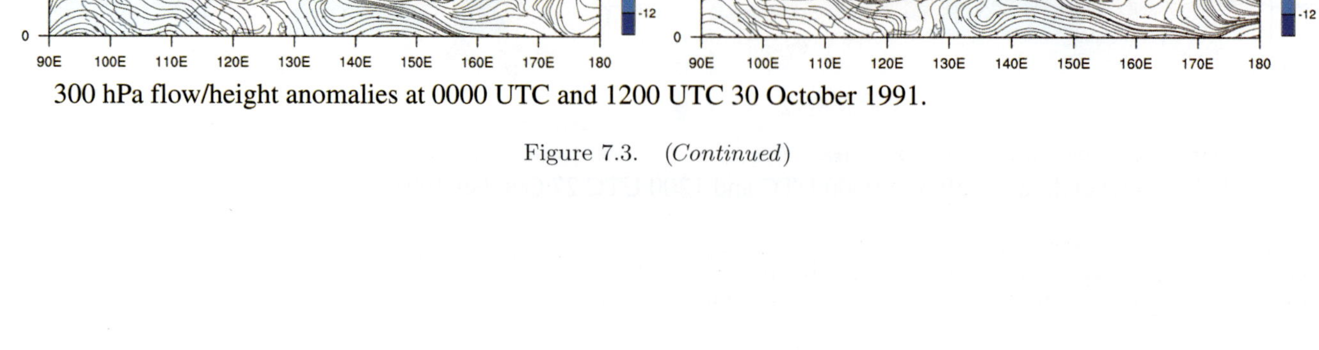

300 hPa flow/height anomalies at 0000 UTC and 1200 UTC 30 October 1991.

Figure 7.3. (*Continued*)

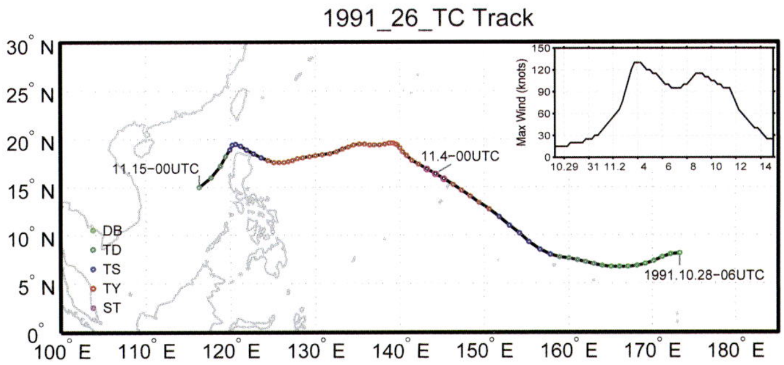

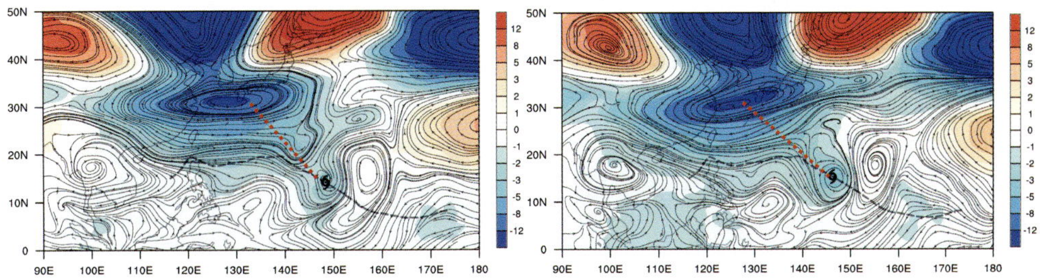

300 hPa flow/height anomalies at 0000 UTC and 1200 UTC 3 November 1991.

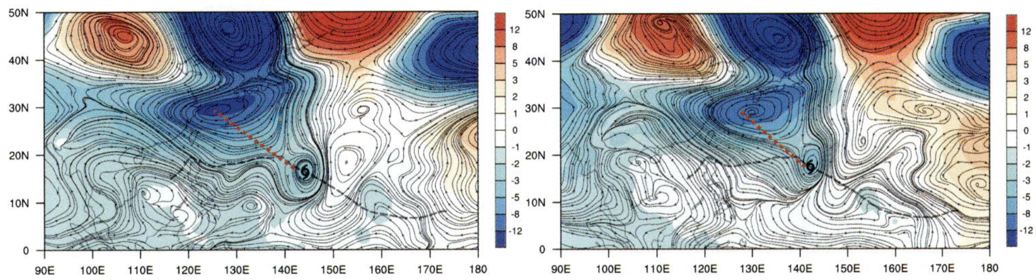

300 hPa flow/height anomalies at 0000 UTC and 1200 UTC 4 November 1991.

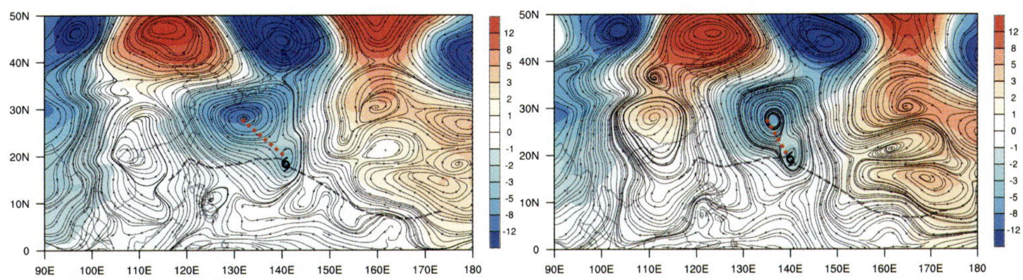

300 hPa flow/height anomalies at 0000 UTC and 1200 UTC 5 November 1991.

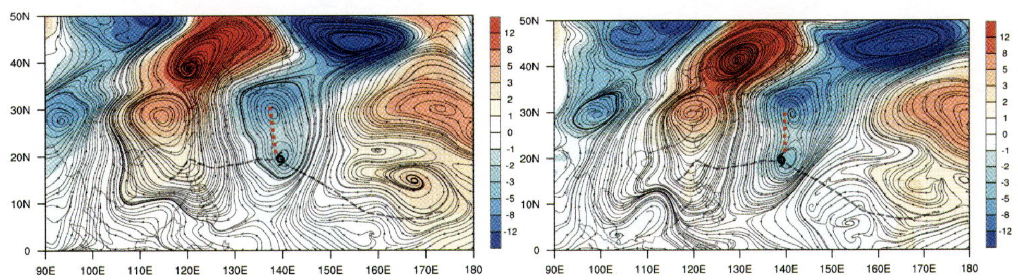

300 hPa flow/height anomalies at 0000 UTC and 1200 UTC 6 November 1991.

Figure 7.4. The typhoon (1991_No.26) generated in the equatorial central Pacific on 28 October 1991 and moved westward near the climatic monsoon trough. Its intensity peaked in the tropical western Pacific on 4 November 1991. The left turn of its motion in the Northwest Pacific was influenced by anomalous vortices and anomalous anti-cyclonic circulations at 300 hPa and 850 hPa.

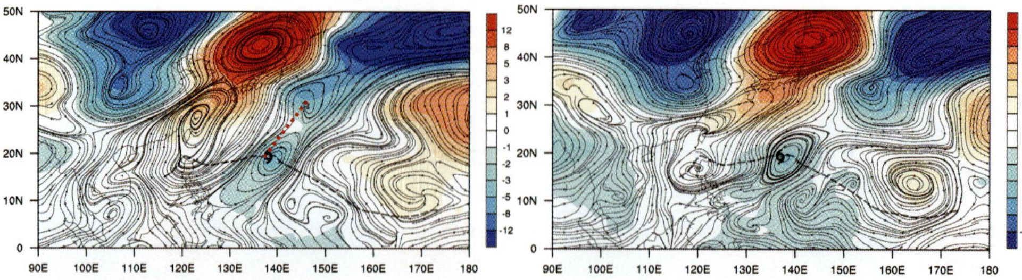

300 hPa flow/height anomalies at 0000 UTC and 1200 UTC 7 November 1991.

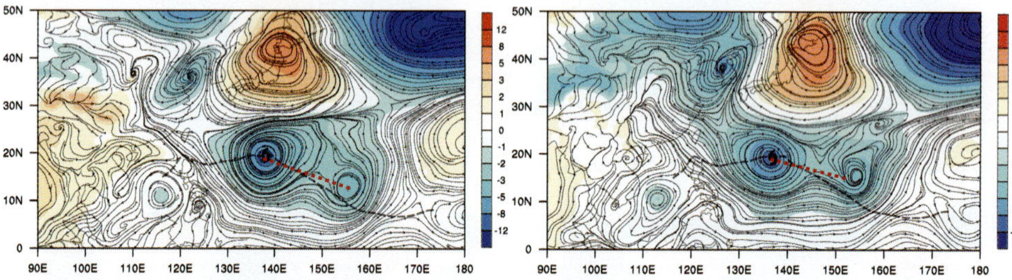

850 hPa flow/height anomalies at 0000 UTC and 1200 UTC 7 November 1991.

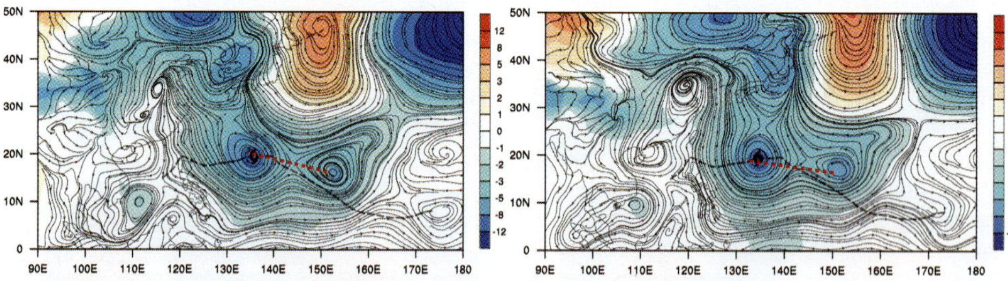

850 hPa flow/height anomalies at 0000 UTC and 1200 UTC 8 November 1991.

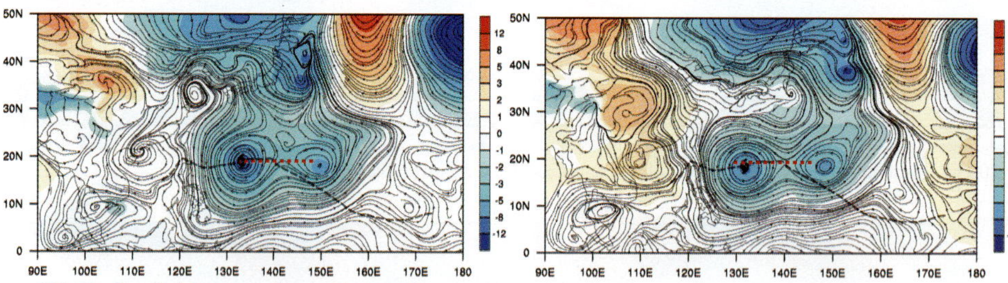

850 hPa flow/height anomalies at 0000 UTC and 1200 UTC 9 November 1991.

Figure 7.4. (*Continued*)

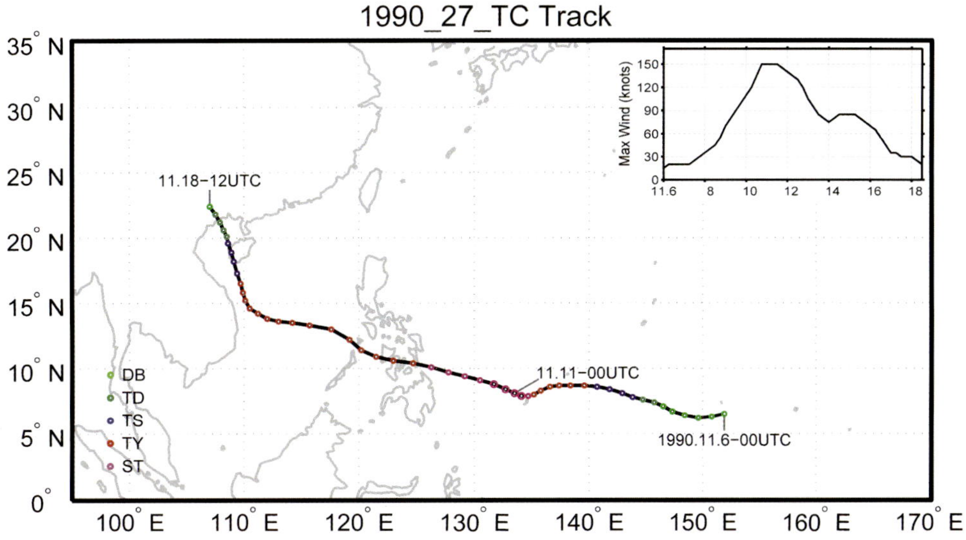

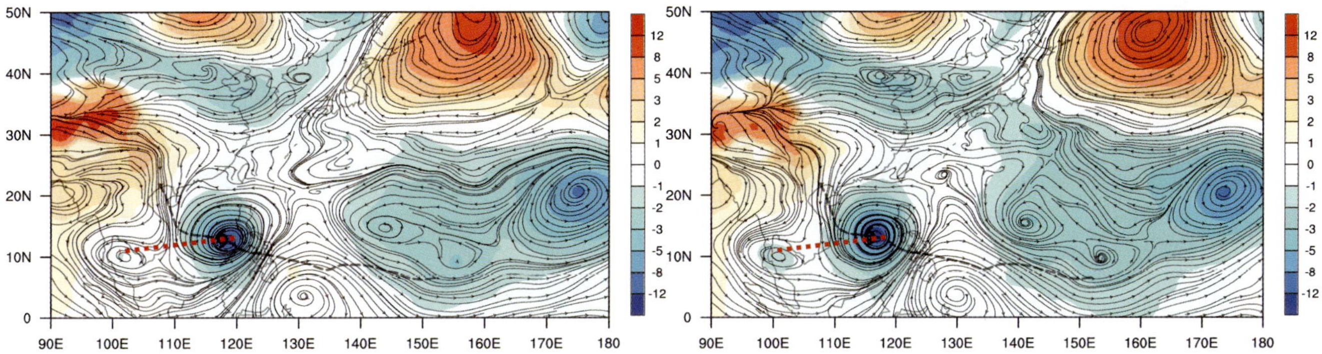

850 hPa flow/height anomalies at 0000 UTC and 0600 UTC 14 November 1990.

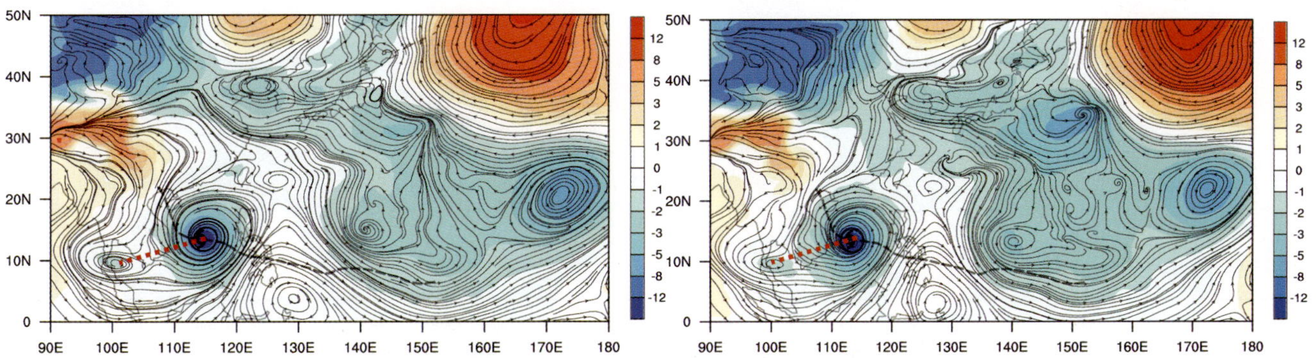

850 hPa flow/height anomalies at 1200 UTC and 1800 UTC 14 November 1990.

Figure 7.5. The typhoon (1990_No.27) generated in the equatorial central Pacific on 6 November 1990 and moved west-northwestward near the climatic monsoon trough. Its intensity peaked in the equatorial western Pacific on 11 November 1990. The right turn of its motion in the northwestern SCS was influenced and attracted by anomalous vortices at 850 hPa and 300 hPa.

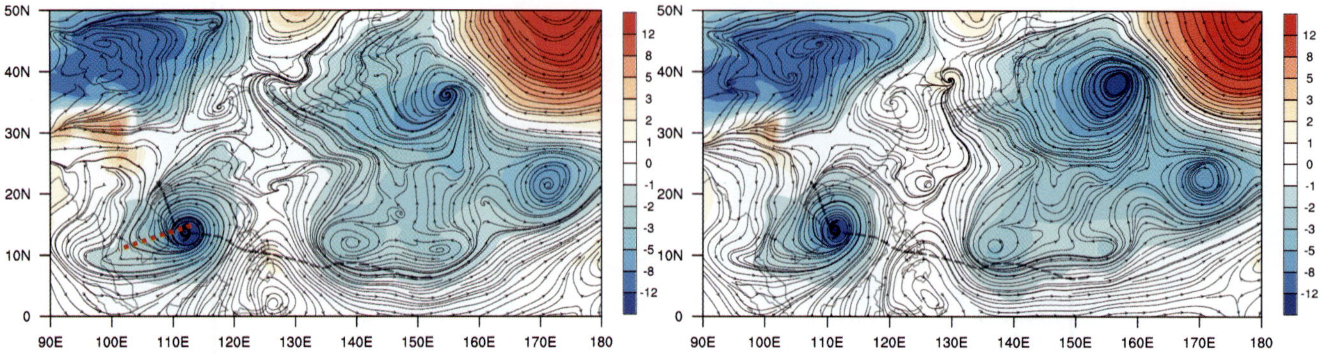

850 hPa flow/height anomalies at 0000 UTC and 0600 UTC 15 November 1990.

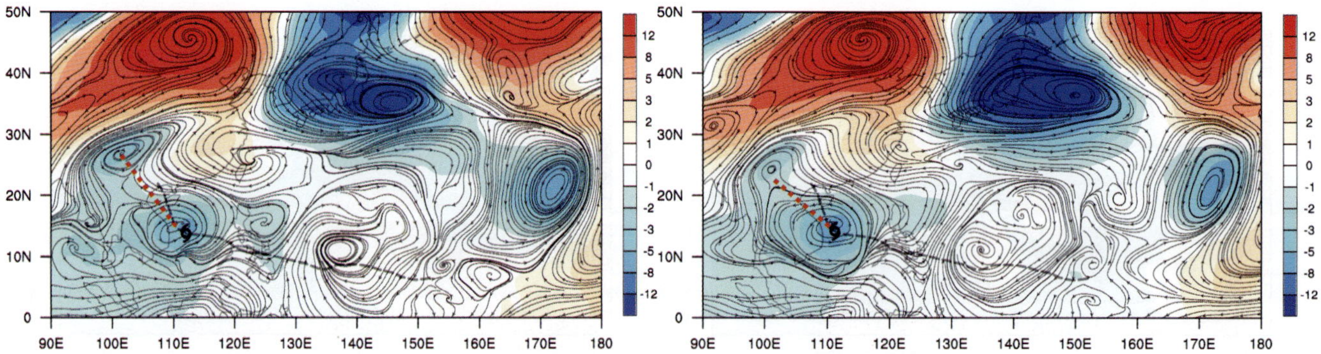

300 hPa flow/height anomalies at 0000 UTC and 0600 UTC 15 November 1990.

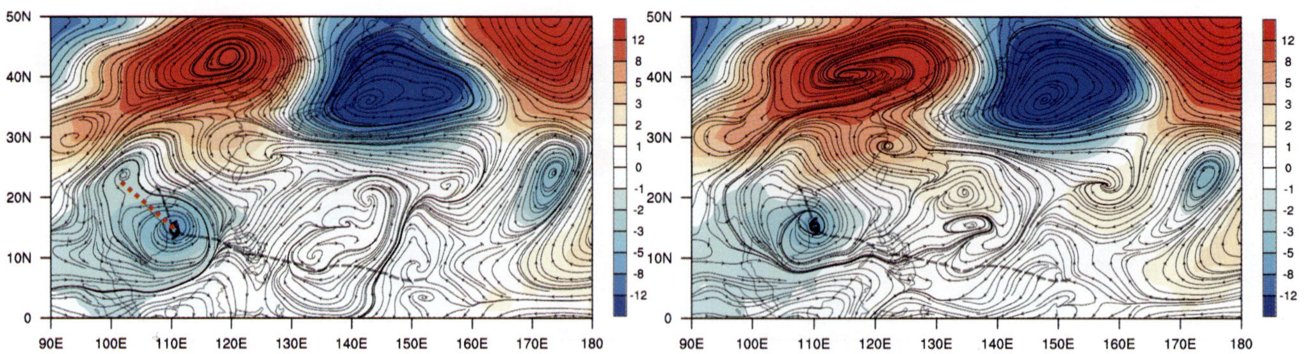

300 hPa flow/height anomalies at 1200 UTC and 1800 UTC 15 November 1990.

Figure 7.5. *(Continued)*

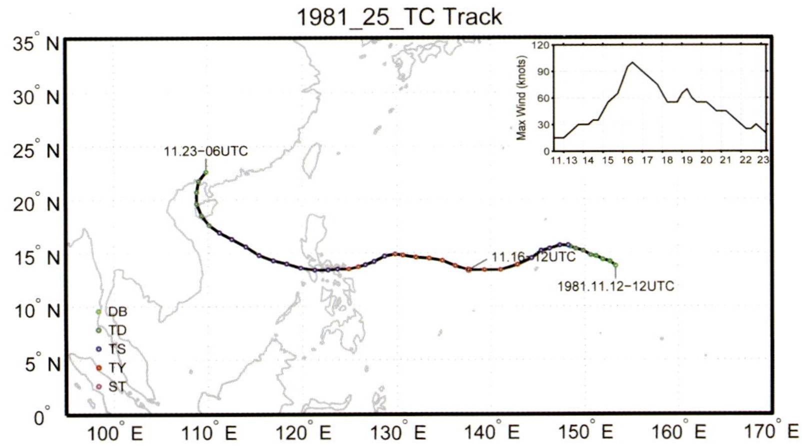

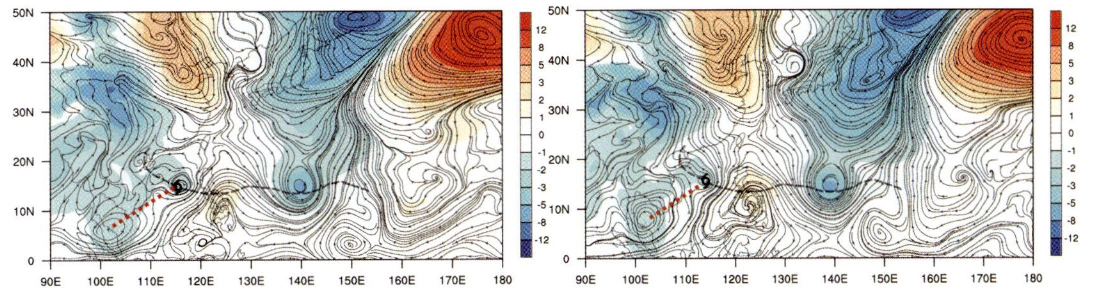

850 hPa flow/height anomalies at 0000 UTC and 0600 UTC 21 November 1981.

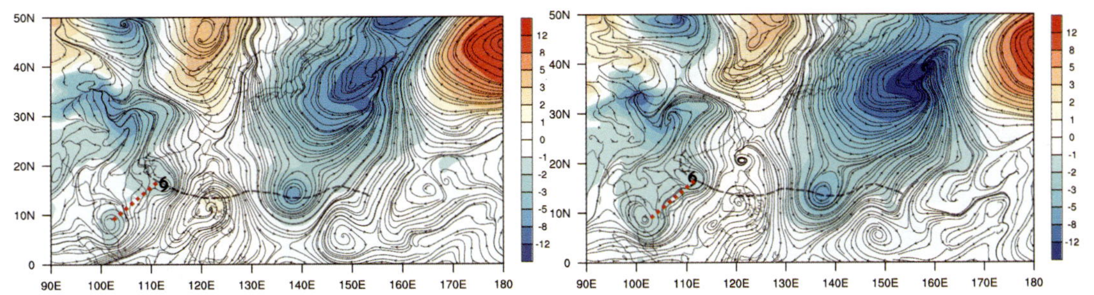

850 hPa flow/height anomalies at 1200 UTC and 1800 UTC 21 November 1981.

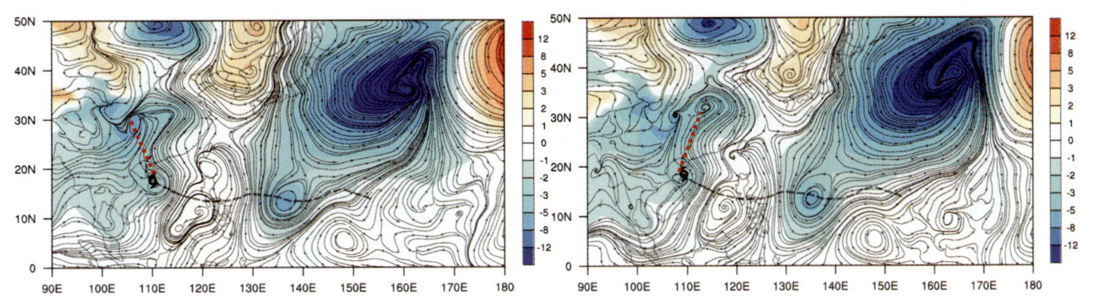

850 hPa flow/height anomalies at 0000 UTC and 0600 UTC 22 November 1981.

Figure 7.6. The typhoon (1981_No.25) generated in the tropical central Pacific on 12 November 1981 and moved westward near the climatic monsoon trough. Its intensity peaked in the tropical western Pacific on 16 November 1981. The right turn of its motion in the SCS was influenced by anomalous vortices at 850 hPa.

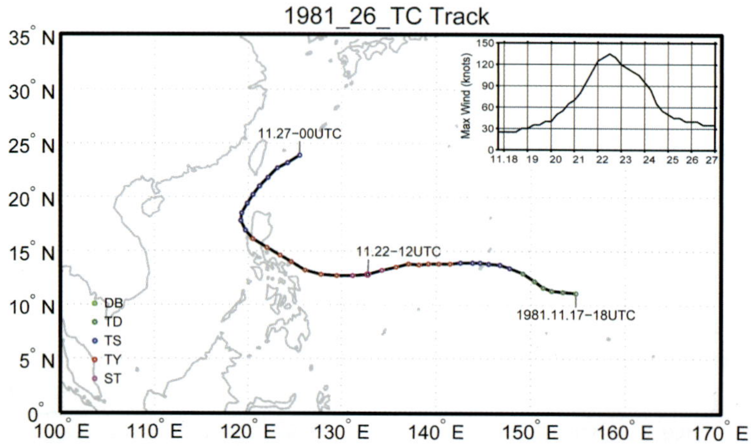

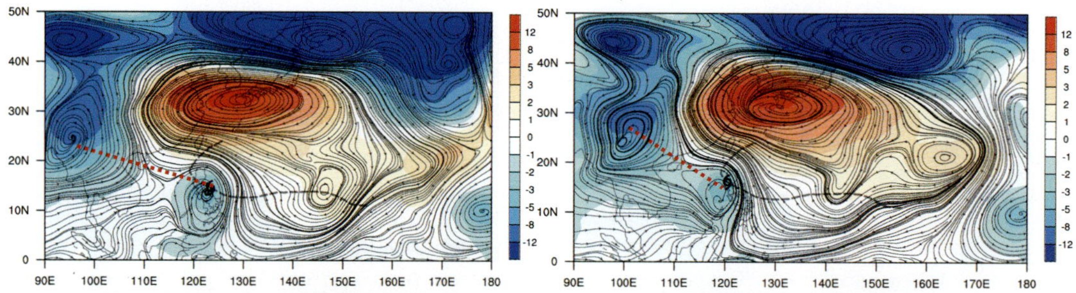

300 hPa flow/height anomalies at 0000 UTC and 1200 UTC 24 November 1981.

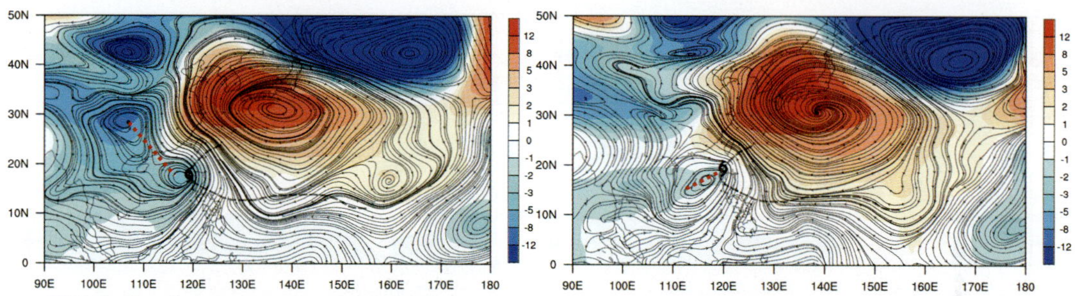

300 hPa flow/height anomalies at 0000 UTC and 1200 UTC 25 November 1981.

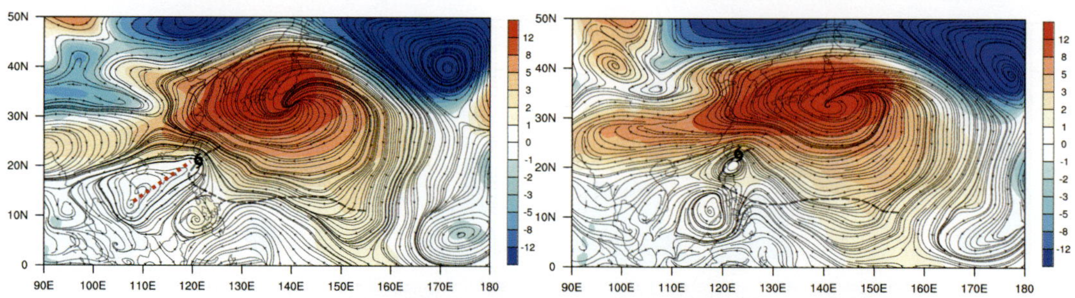

300 hPa flow/height anomalies at 0000 UTC and 1200 UTC 26 November 1981.

Figure 7.7. The typhoon (1981_No.26) generated in the tropical central Pacific on 17 November 1981 and moved westward near the climatic monsoon trough. Its intensity peaked in the tropical western Pacific on 22 November 1981. The right turn near the Philippines was influenced by an anomalous vortex at 300 hPa.

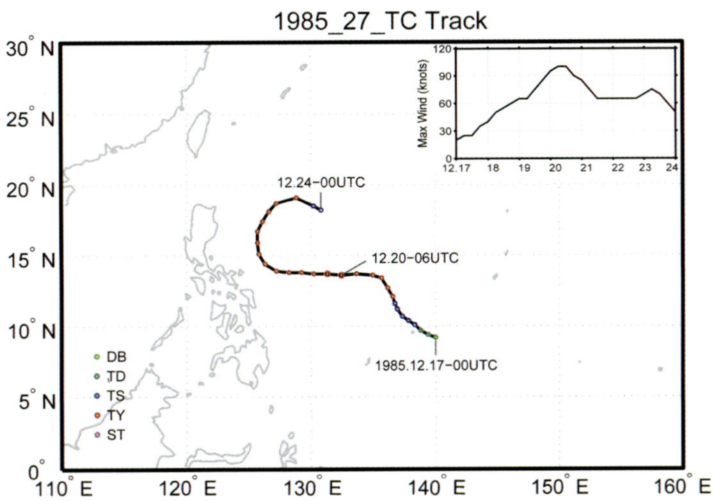

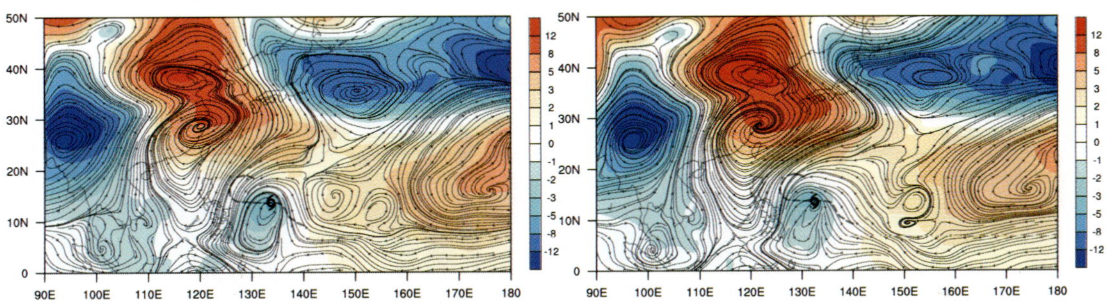

300 hPa flow/height anomalies at 0000 UTC and 0600 UTC 20 December 1985.

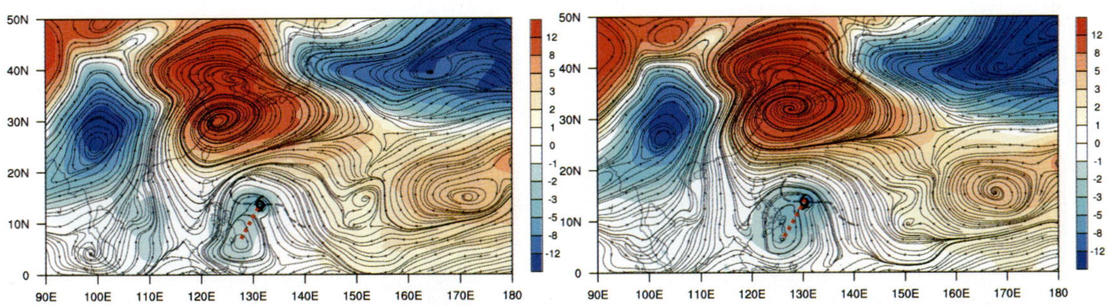

300 hPa flow/height anomalies at 1200 UTC and 1800 UTC 20 December 1985.

Figure 7.8. The typhoon (1985_No.27) generated in the equatorial western Pacific on 17 December 1985 and had a complex motion. Its intensity peaked in the tropical western Pacific on 17 December 1985 when it was near the climatic monsoon trough. The right turn of its motion in the east sea of the Philippines was influenced by anomalous vortices at 300 hPa.

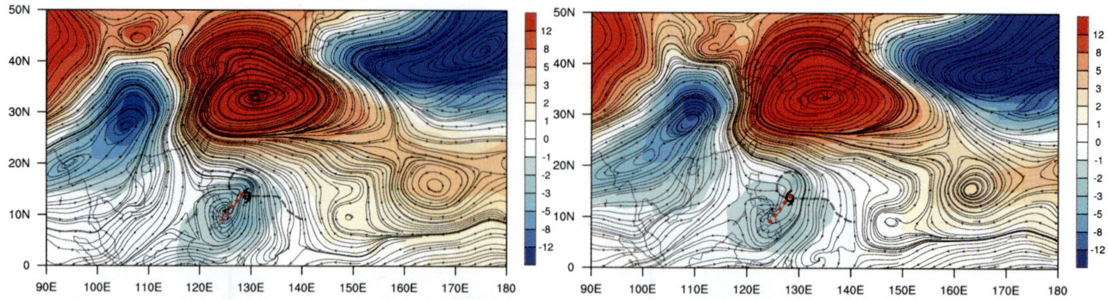

300 hPa flow/height anomalies at 0000 UTC and 0600 UTC 21 December 1985.

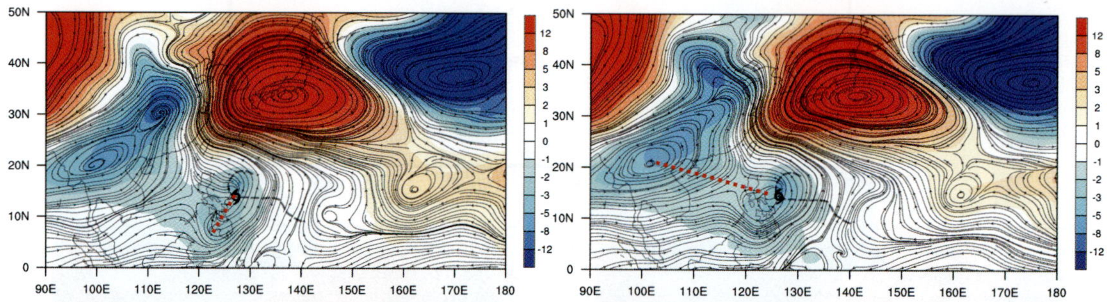

300 hPa flow/height anomalies at 1200 UTC and 1800 UTC 21 December 1985.

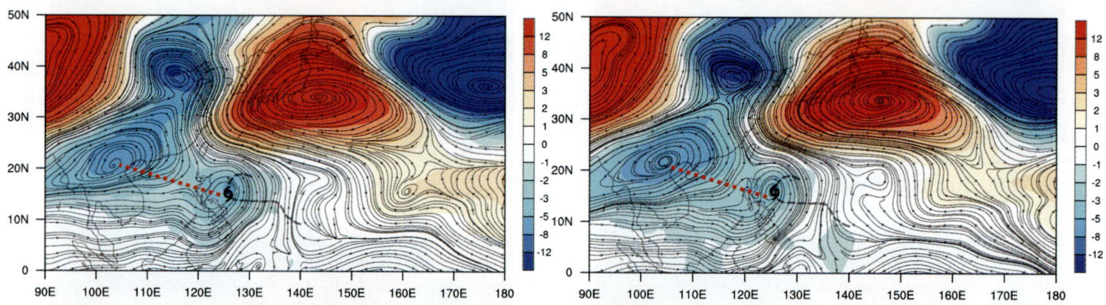

300 hPa flow/height anomalies at 0000 UTC and 0600 UTC 22 December 1985.

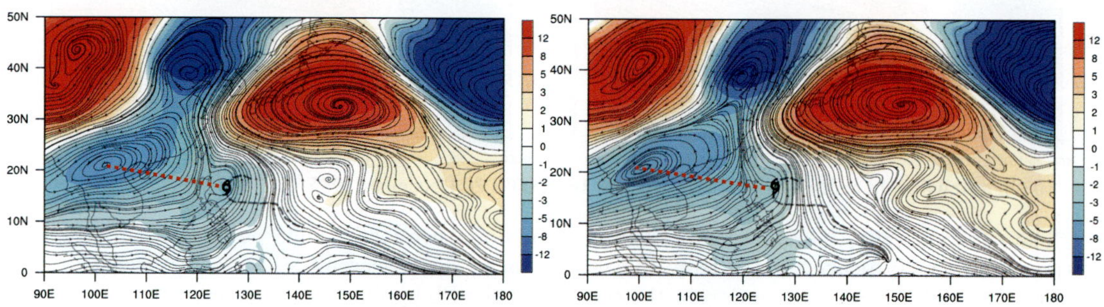

300 hPa flow/height anomalies at 1200 UTC and 1800 UTC 22 December1985.

Figure 7.8. *(Continued)*

7.2. Intensity Peaked in the Northwest Pacific

There were 19 typhoons that peaked their intensities in the Northwest Pacific. The right turns and left turns of their motions were influenced by anomalous vortices (Figs. 7.9–7.27).

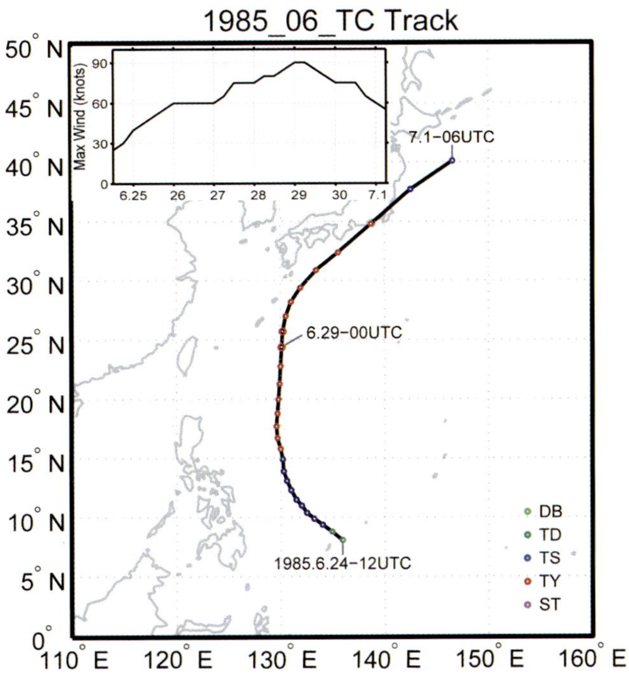

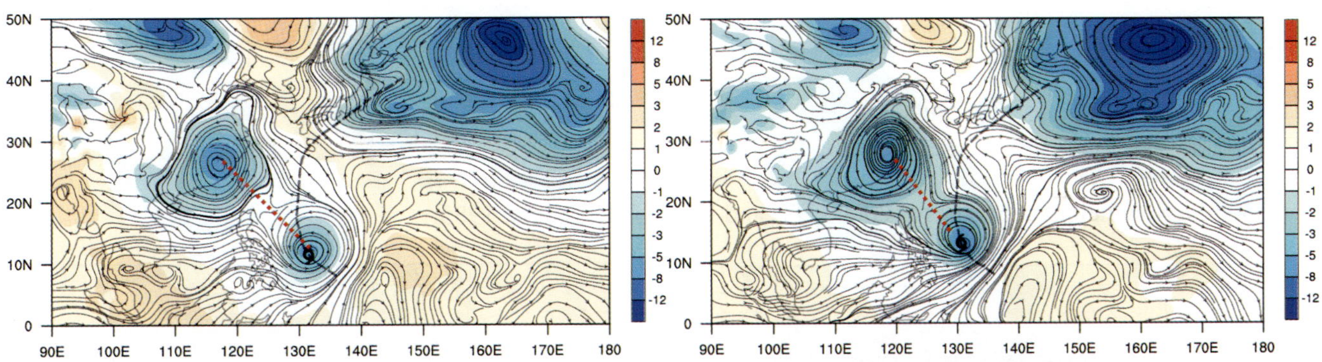

850 hPa flow/height (shading, 10 gpm) anomalies at 0000 UTC and 1200 UTC 26 June 1985.

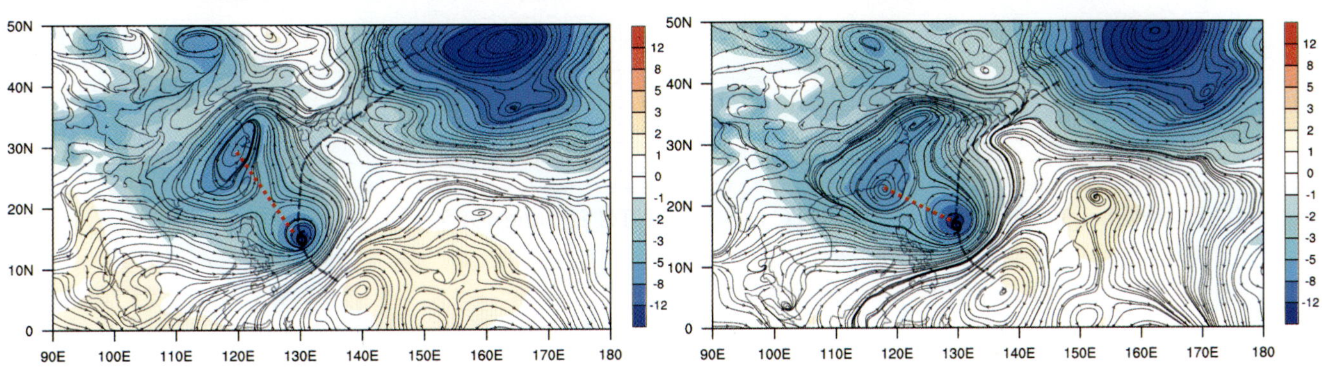

850 hPa flow/height anomalies at 0000 UTC and 1200 UTC 27 June 1985.

Figure 7.9. The typhoon (1985_No.06) generated in the equatorial western Pacific on 24 June 1985 and moved northward. Its intensity peaked in southern Japan on 29 June 1985 before turning northeastward. The right turn of its motion in southern Japan was influenced by an anomalous vortex at 850 hPa.

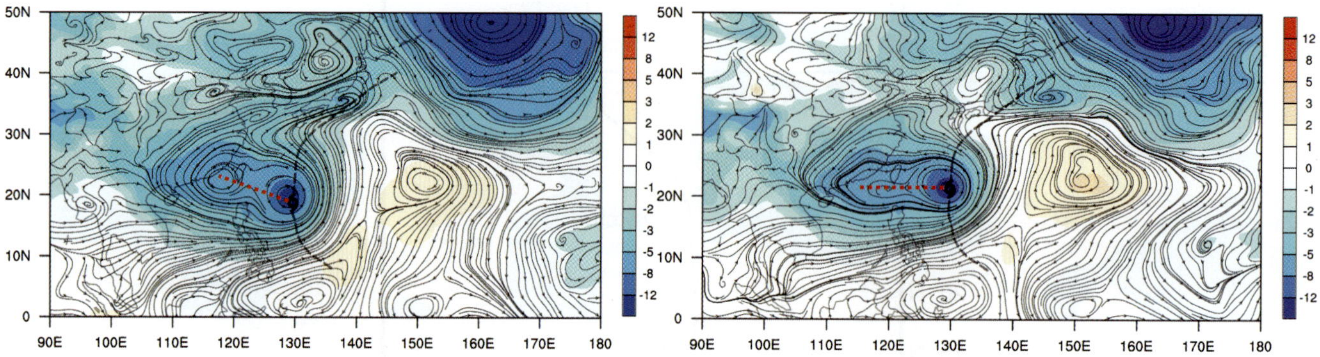

850 hPa flow/height anomalies at 0000 UTC and 1200 UTC 28 June 1985.

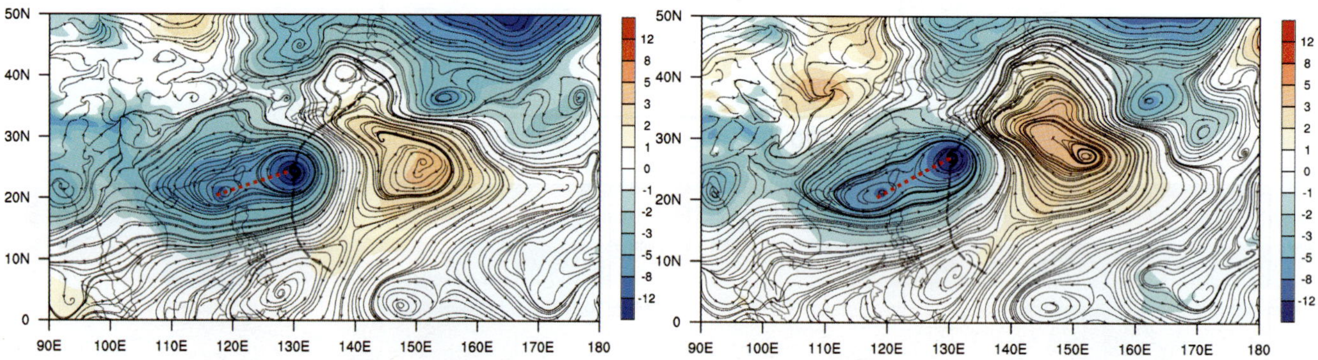

850 hPa flow/height anomalies at 0000 UTC and 1200 UTC 29 June 1985.

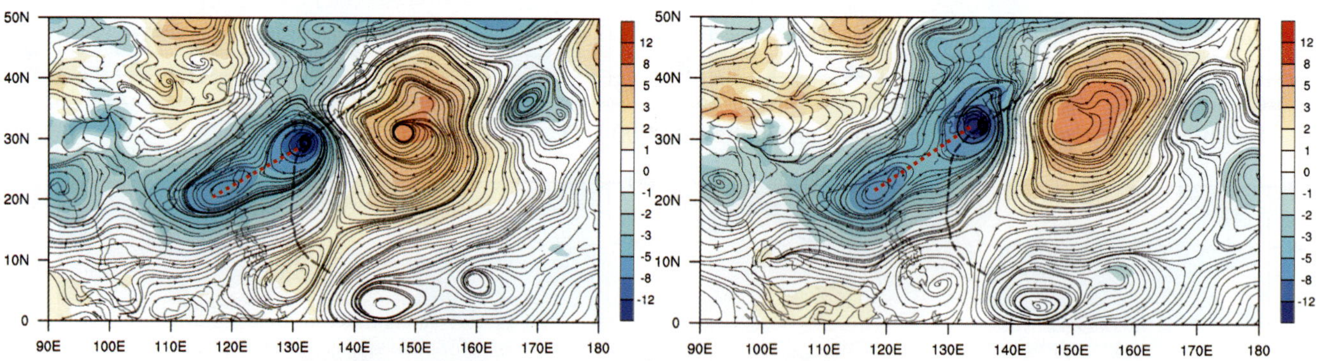

850 hPa flow/height anomalies at 0000 UTC and 1200 UTC 30 June 1985.

Figure 7.9. *(Continued)*

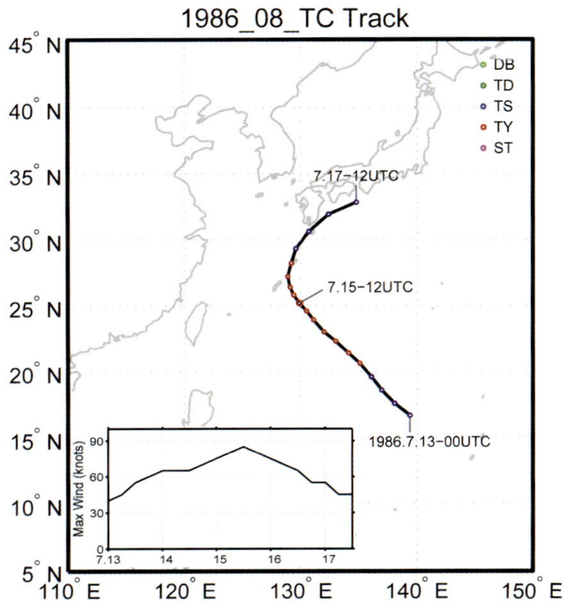

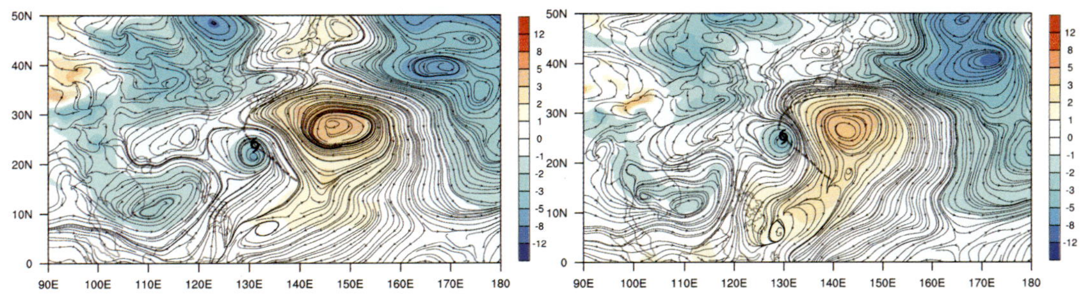

850 hPa flow/height anomalies at 0000 UTC and 1200 UTC 15 July 1986.

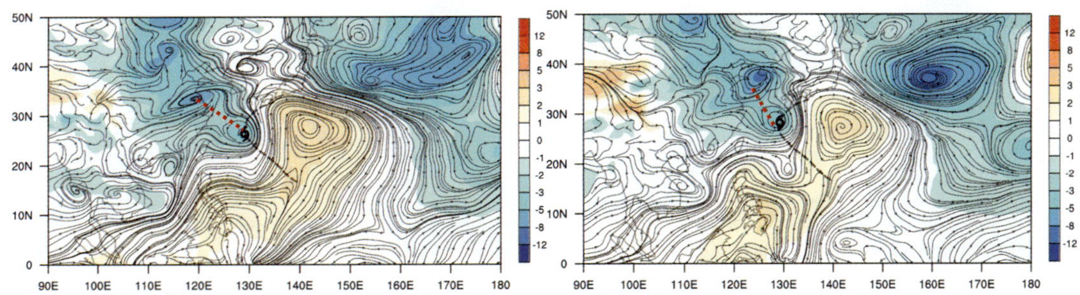

850 hPa flow/height anomalies at 0000 UTC and 1200 UTC 16 July 1986.

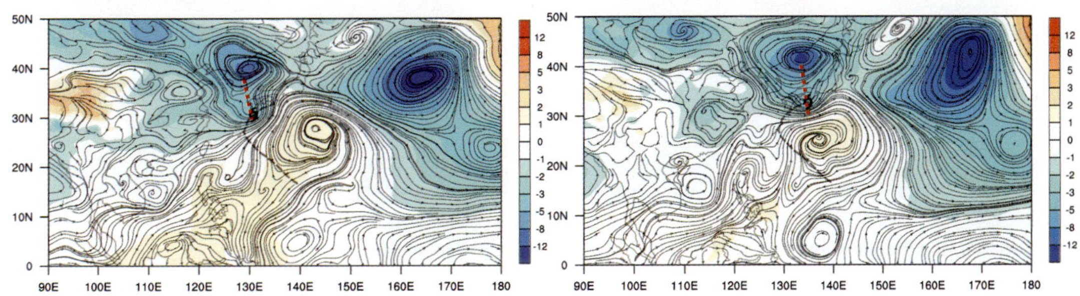

850 hPa flow/height anomalies at 0000 UTC and 1200 UTC 17 July 1986.

Figure 7.10. The typhoon (1986_No.08) generated in the tropical western Pacific on 13 July 1986 and moved northwestward. Its intensity peaked in the east sea of the Taiwan Island on 15 July 1986 before turning northeastward. The right turn of its motion in the south seas of Japan was influenced by an anomalous vortex at 850 hPa.

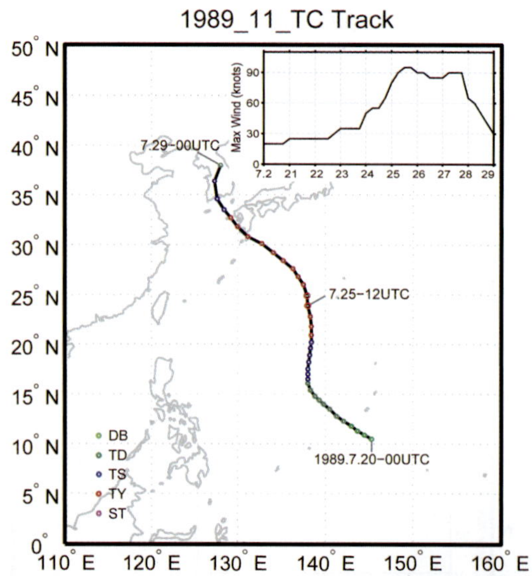

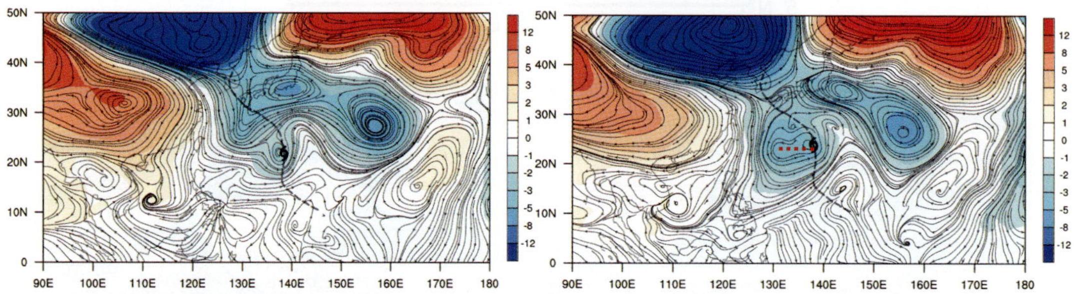

300 hPa flow/height anomalies at 0000 UTC and 1200 UTC 25 July 1989.

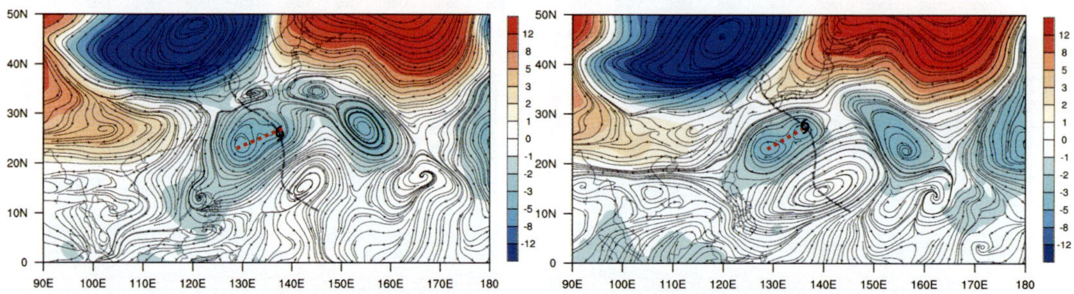

300 hPa flow/height anomalies at 0000 UTC and 1200 UTC 26 July 1989.

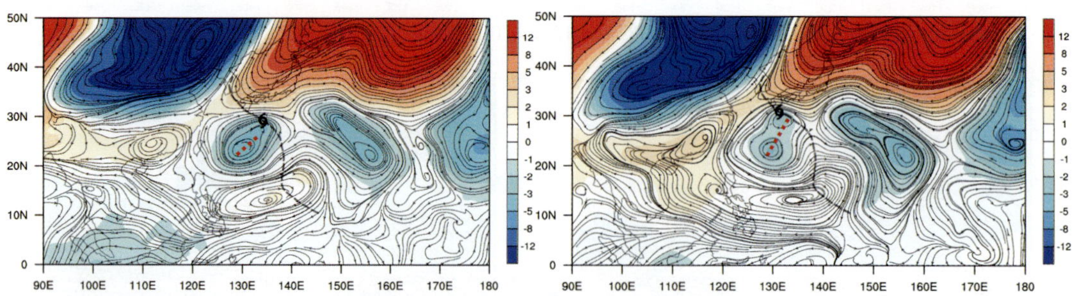

300 hPa flow/height anomalies at 0000 UTC and 1200 UTC 27 July 1989.

Figure 7.11. The typhoon (1989_No.11) generated in the tropical western Pacific on 20 July 1989 and moved northwestward. Its intensity peaked on 25 July 1989 near the climatic monsoon trough. The left turn of its motion in the south of Japan was influenced by an anomalous vortex at 300 hPa.

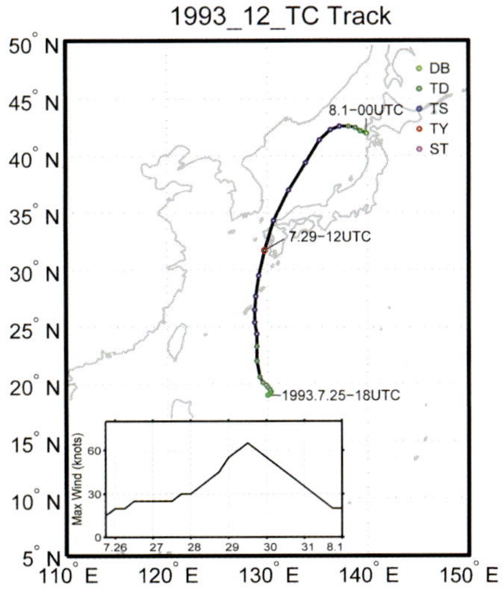

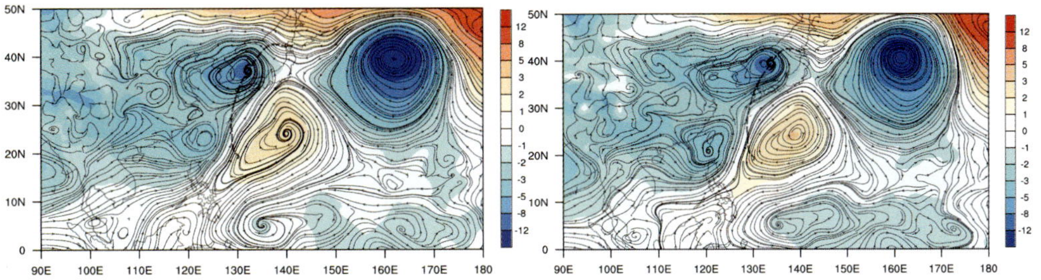

850 hPa flow/height anomalies at 0000 UTC and 0600 UTC 30 July 1983.

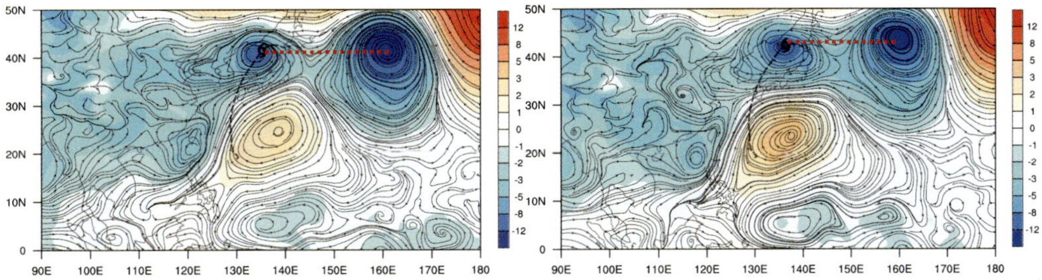

850 hPa flow/height anomalies at 1200 UTC and 1800 UTC 30 July 1983.

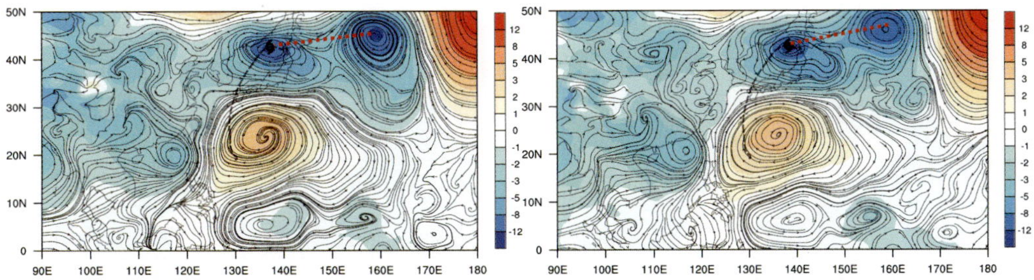

850 hPa flow/height anomalies at 0000 UTC and 0600 UTC 31 July 1983.

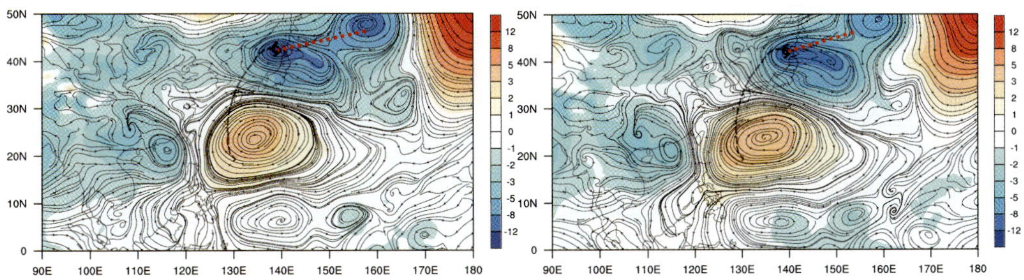

850 hPa flow/height anomalies at 1200 UTC and 1800 UTC 31 July 1983.

Figure 7.12. The typhoon (1993_No.12) generated in the northeast sea of the Philippines on 25 July 1993 and moved northward. Its intensity peaked on 29 July 1993 before landfall on the islands of Japan. The right turn in the Japan Sea was due to the attraction of an anomalous vortex at 850 hPa.

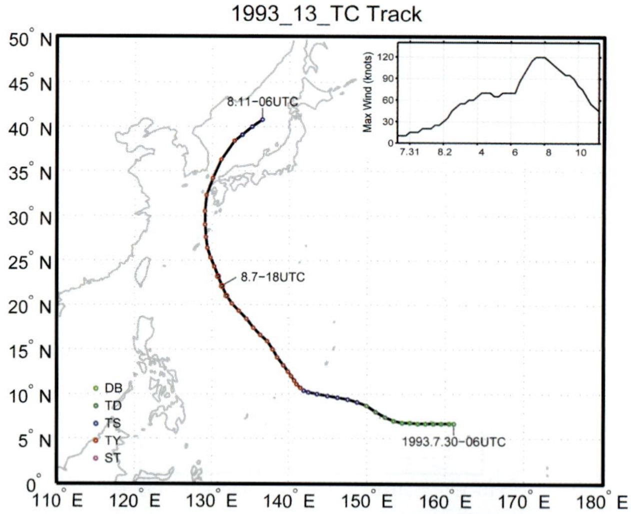

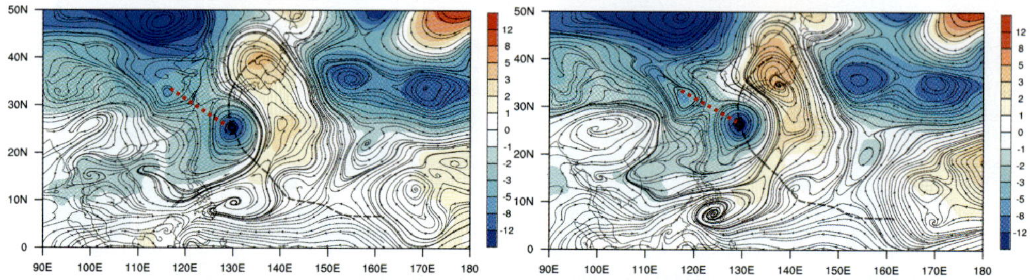

300 hPa flow/height anomalies at 1200 UTC and 1800 UTC 8 August 1993.

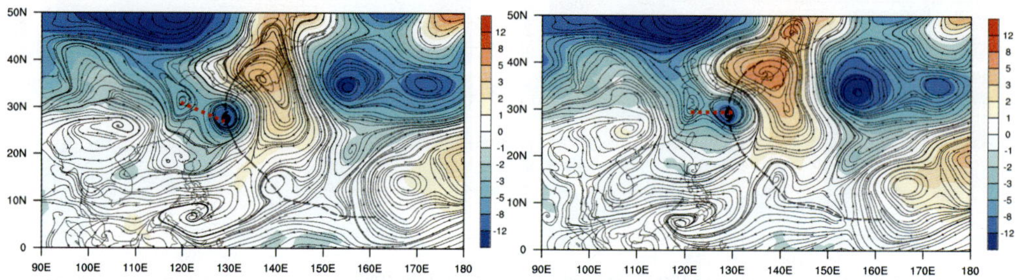

300 hPa flow/height anomalies at 0000 UTC and 0600 UTC 9 August 1993.

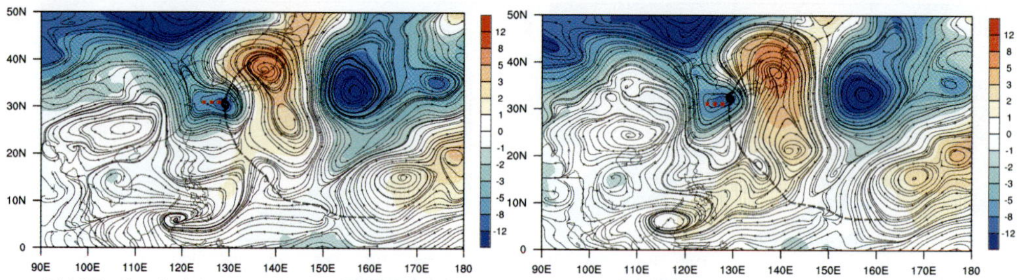

300 hPa flow/height anomalies at 1200 UTC and 1800 UTC 9 August 1993.

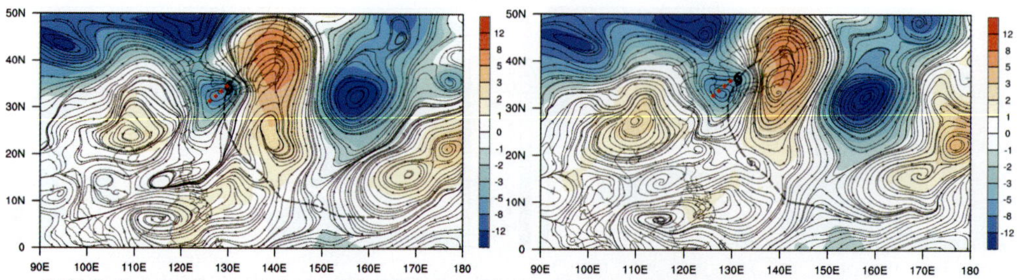

300 hPa flow/height anomalies at 0000 UTC and 0600 UTC 10 August 1993.

Figure 7.13. The typhoon (1993_No.13) generated in the equatorial central Pacific on 30 July 1993 and moved along the climatic monsoon trough. Its intensity peaked in the south of Japan on 7 August 1993 before turning northeastward. The right turn in the east seas of China was influenced by an anomalous vortex at 300 hPa.

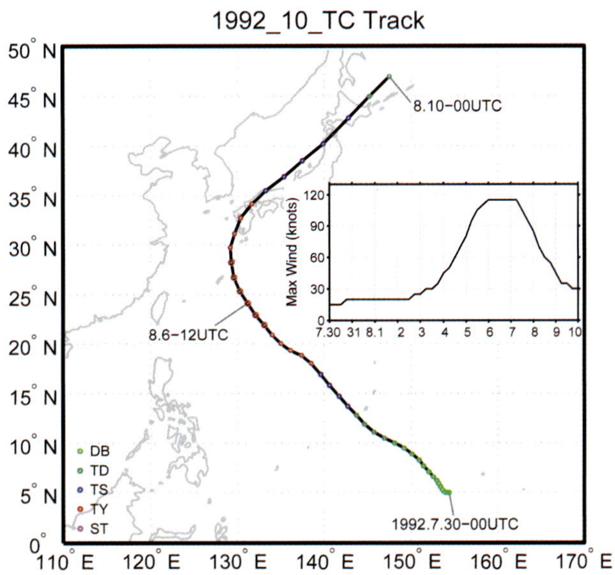

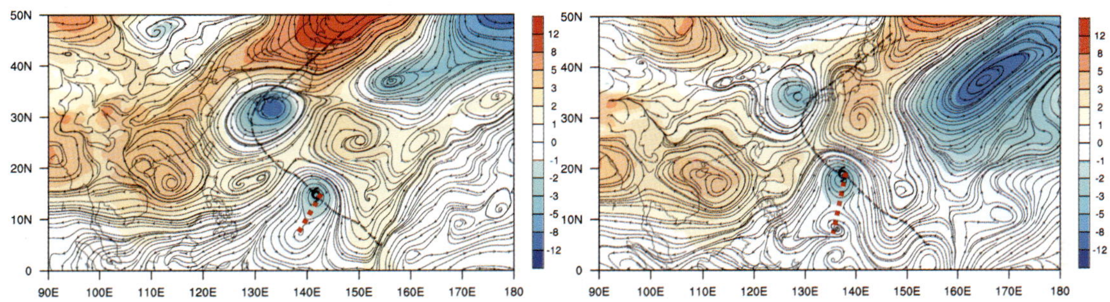

850 hPa flow/height anomalies at 0000 UTC 4 and 5 August 1992.

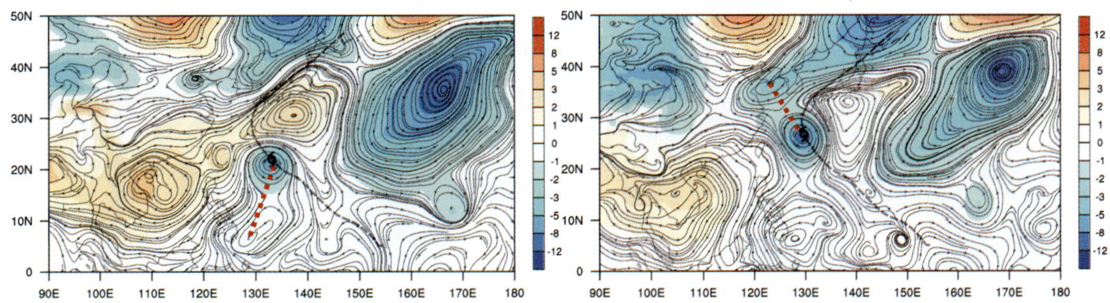

850 hPa flow/height anomalies at 0000 UTC 6 and 7 August 1992.

Figure 7.14. The typhoon (1992_No.10) generated in the equatorial central Pacific on 30 July 1992 and moved northwestward along the climatic monsoon trough. Its intensity peaked in the south of Japan on 6 August 1992 before turning northeastward. This right turn was influenced by anomalous vortices at 300 hPa.

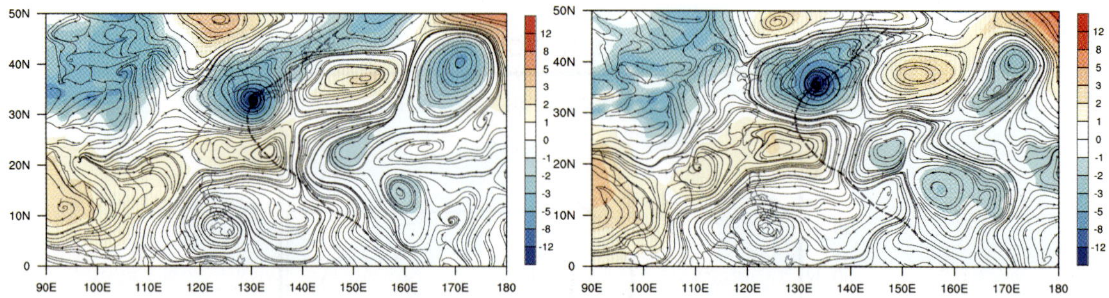

850 hPa flow/height anomalies at 0000 UTC and 1200 UTC 8 August 1992.

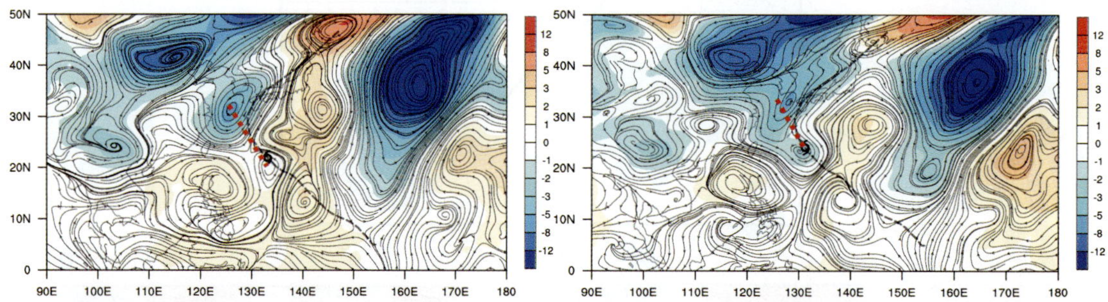

300 hPa flow/height anomalies at 0000 UTC and 1200 UTC 6 August 1992.

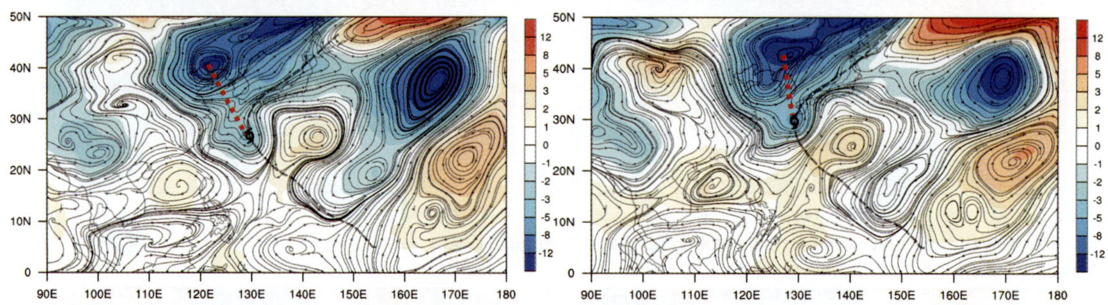

300 hPa flow/height anomalies at 0000 UTC and 1200 UTC 7 August 1992.

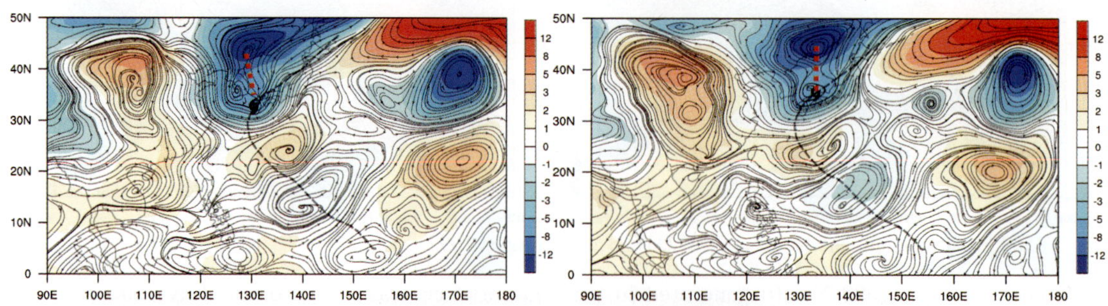

300 hPa flow/height anomalies at 0000 UTC and 1200 UTC 8 August 1992.

Figure 7.14. *(Continued)*

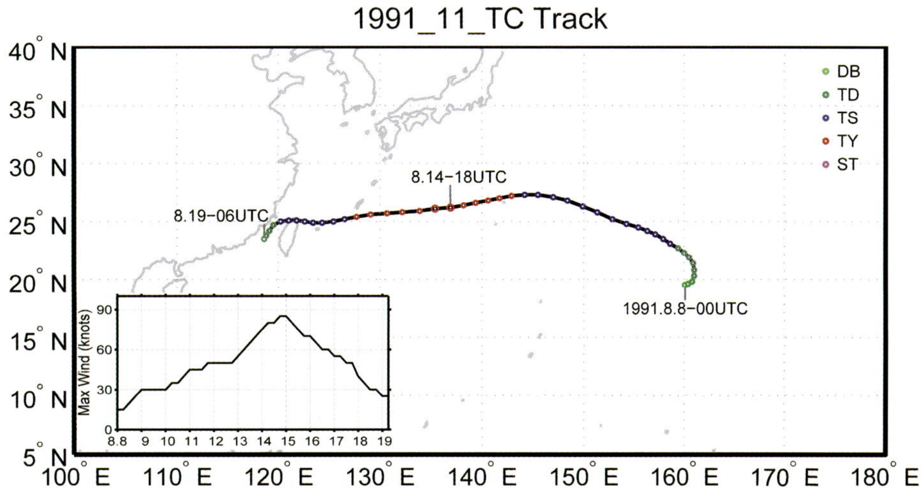

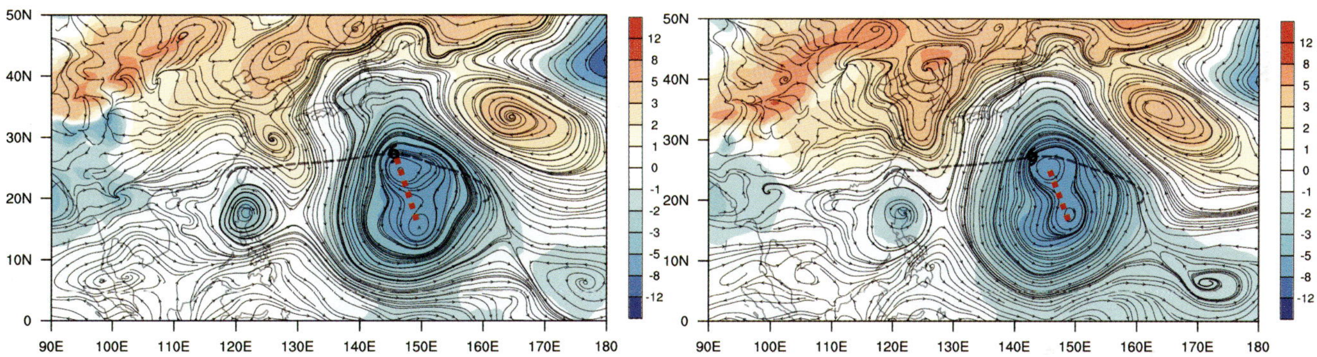

850 hPa flow/height anomalies at 0000 UTC and 1200 UTC 13 August 1991.

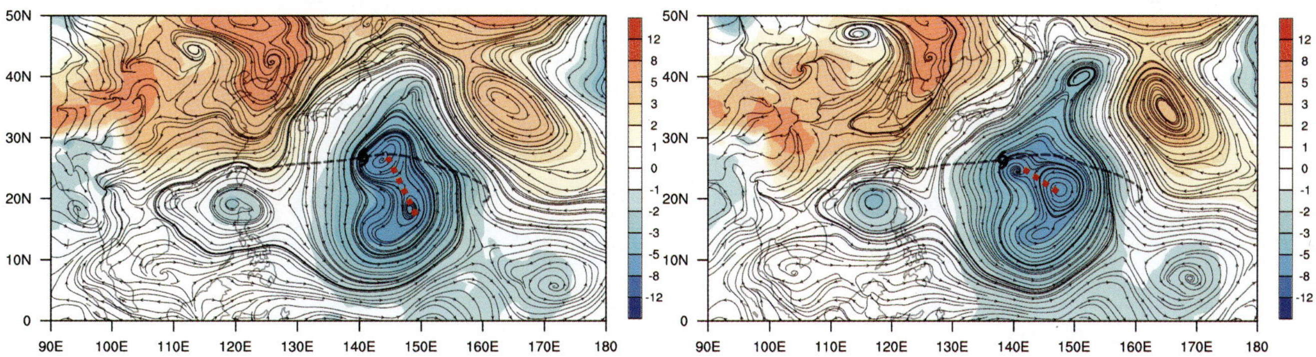

850 hPa flow/height anomalies at 0000 UTC and 1200 UTC 14 August 1991.

Figure 7.15. The typhoon (1991_No.11) generated in the tropical central Pacific on 8 August 1991. After a local curvature motion at the beginning stage, it moved westward. Its intensity peaked in the south of Japan on 14 August 1991 when it was approaching the climatic monsoon trough. The left turn in the south of Japan was influenced by an anomalous vortex at 850 hPa.

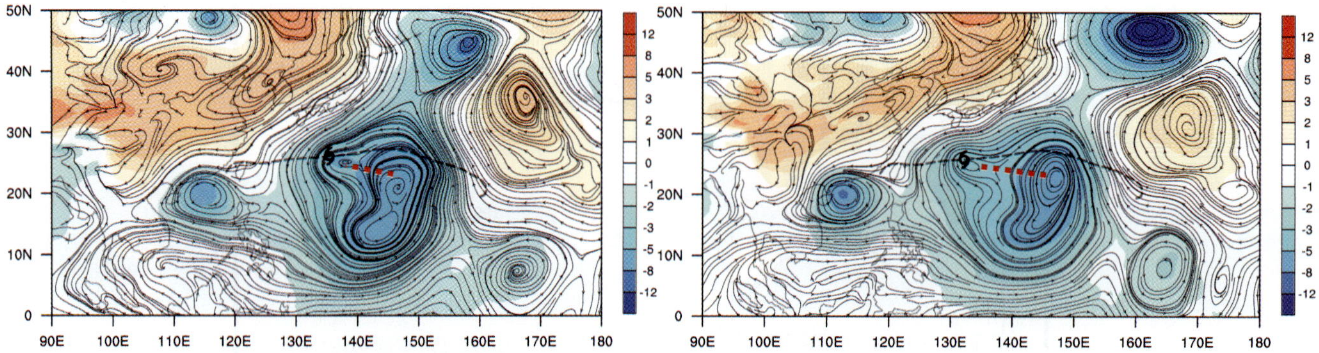

850 hPa flow/height anomalies at 0000 UTC and 1200 UTC 15 August 1991.

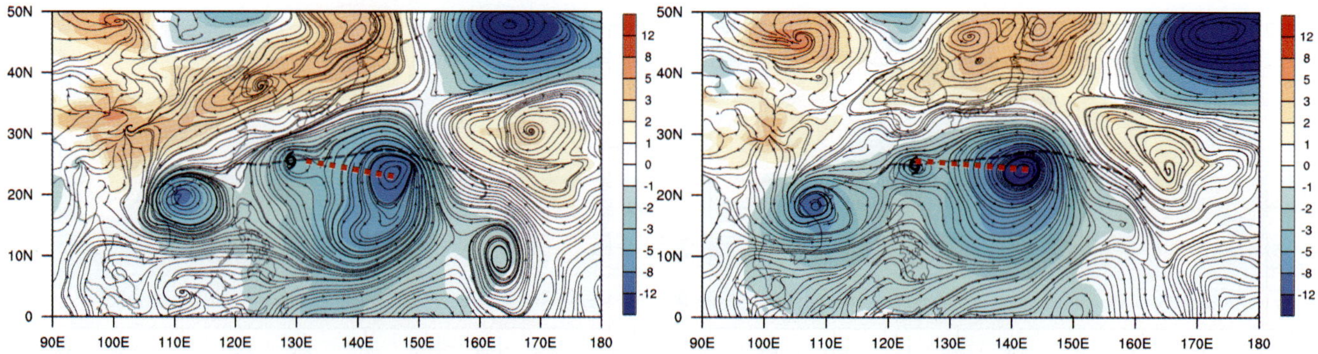

850 hPa flow/height anomalies at 0000 UTC 16 and 17 August 1991.

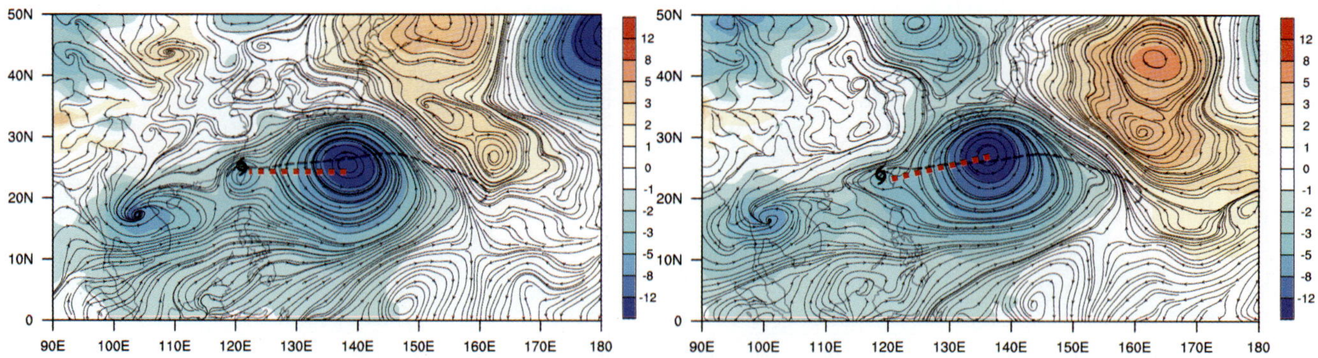

850 hPa flow/height anomalies at 0000 UTC 18 and 19 August 1991.

Figure 7.15. (*Continued*)

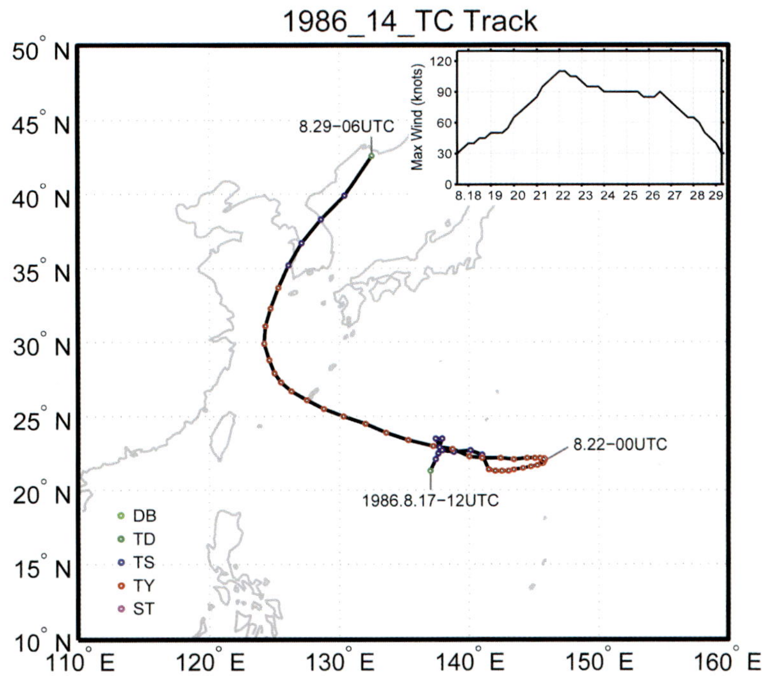

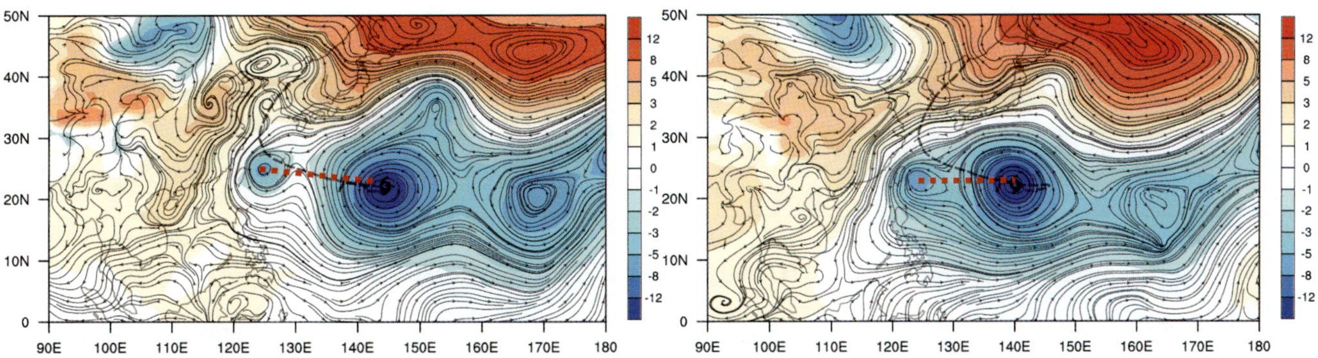

850 hPa flow/height anomalies at 0000 UTC 23 and 24 August 1986.

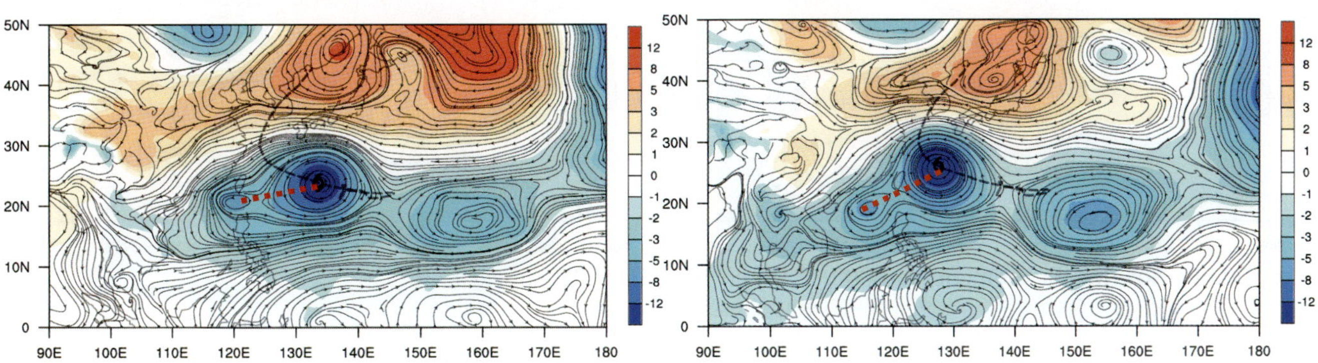

850 hPa flow/height anomalies at 0000 UTC 25 and 26 August 1986.

Figure 7.16. The typhoon (1986_No.14) generated in the subtropical Northwest Pacific on 17 August 1986. After a complex local return, it moved northwestward along the climatic monsoon trough and its maximum wind speed maintained at above 90 knots from 21 to 26 August 1986. The right turn in the east seas of China was influenced by anomalous vortices at 850 hPa.

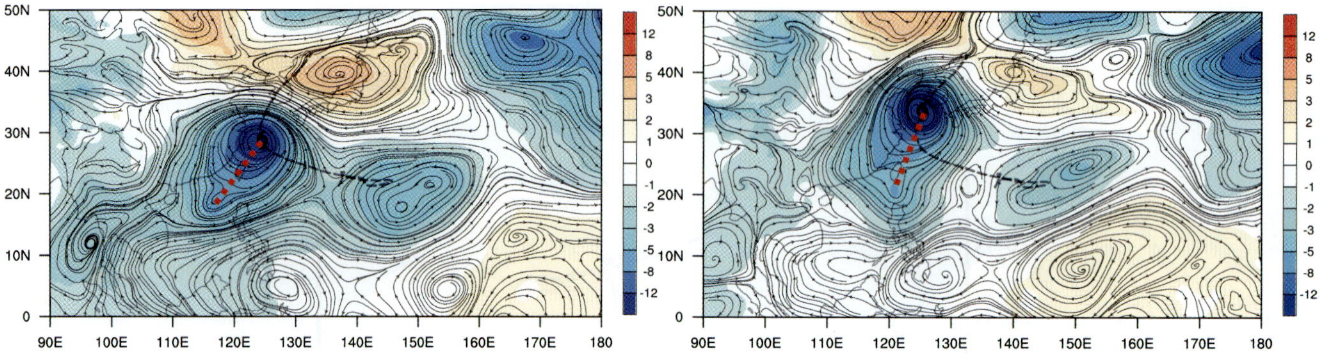

850 hPa flow/height anomalies at 0000 UTC 27 and 28 August 1986.

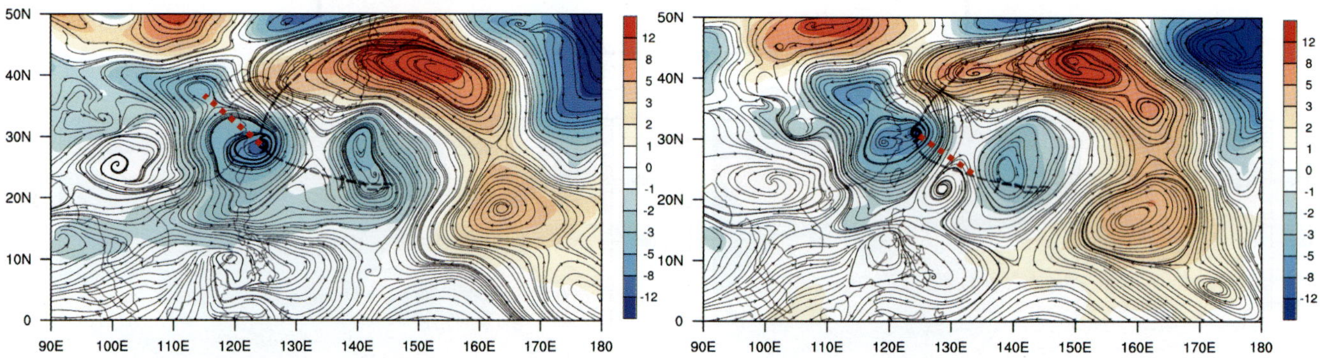

300 hPa flow/height anomalies at 0000 UTC and 1200 UTC 27 August 1986.

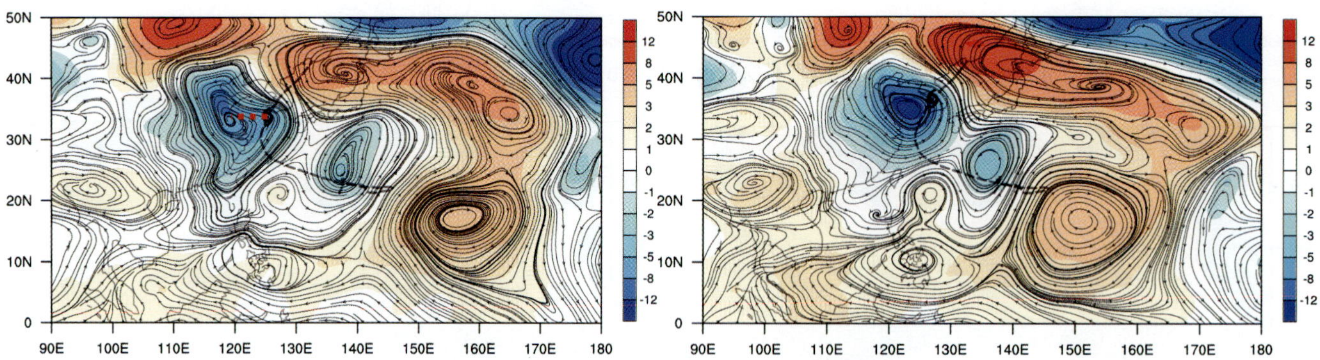

300 hPa flow/height anomalies at 0000 UTC and 1200 UTC 28 August 1986.

Figure 7.16. *(Continued)*

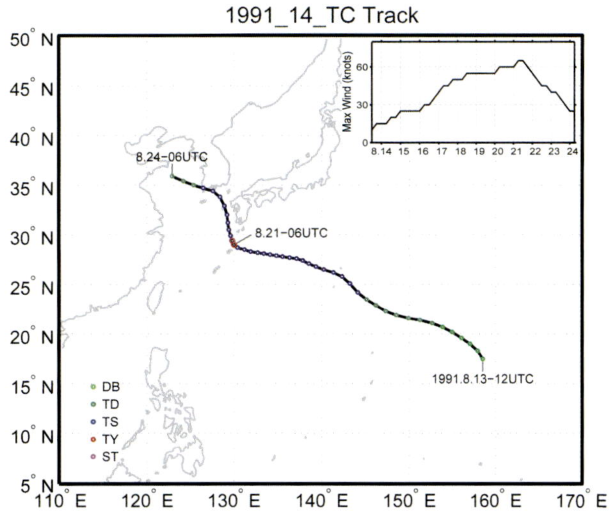

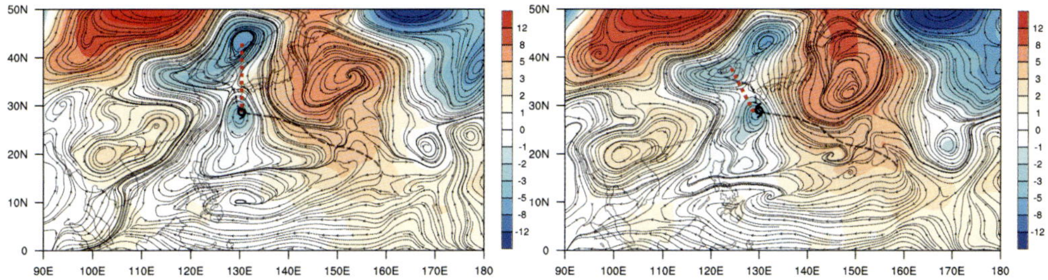

300 hPa flow/height anomalies at 0000 UTC and 0600 UTC 21 August 1991.

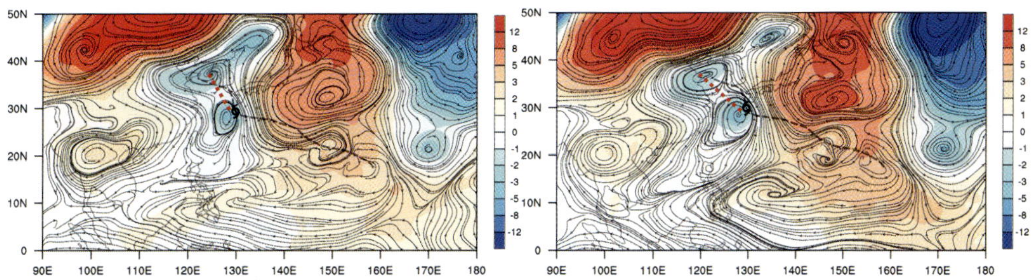

300 hPa flow/height anomalies at 1200 UTC and 1800 UTC 21 August 1991.

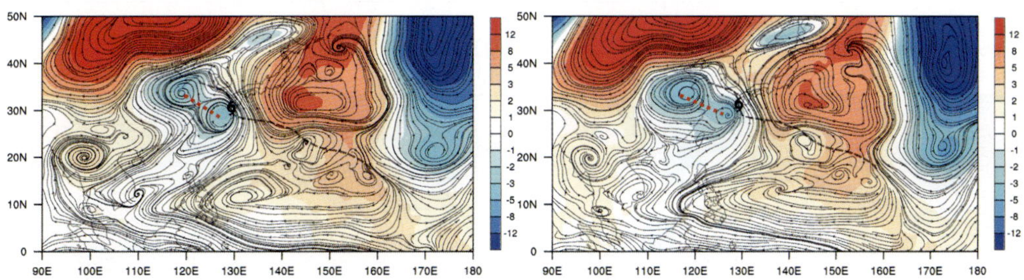

300 hPa flow/height anomalies at 0000 UTC and 0600 UTC 22 August 1991.

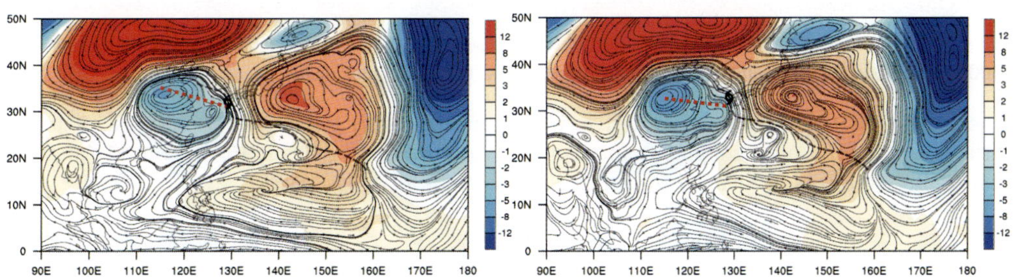

300 hPa flow/height anomalies at 1200 UTC and 1800 UTC 22 August 1991.

Figure 7.17. The typhoon (1991_No.14) generated in the tropical central Pacific on 13 August 1991 and moved northwestward. Its intensity peaked in the southern Japan on 21 August 1991 when it was approaching the climatic monsoon trough. The left turn in the Yellow Sea was influenced by an anomalous vortex at 300 hPa.

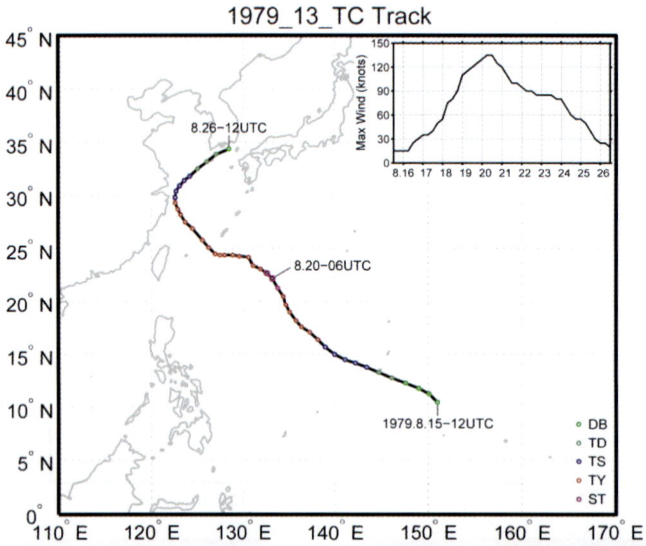

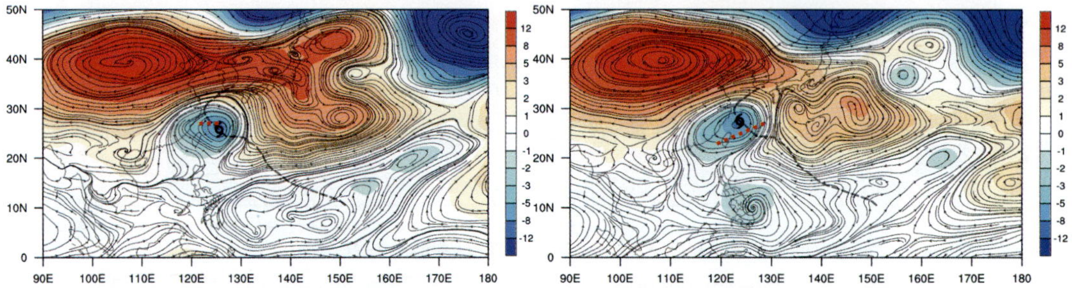

300hPa flow/height anomalies at 0000 UTC and 1200 UTC 23 August 1979.

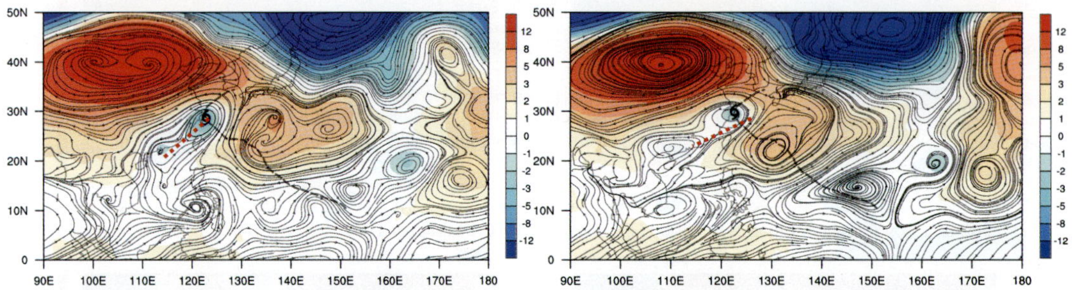

300hPa flow/height anomalies at 0000 UTC and 1200 UTC 24 August 1979.

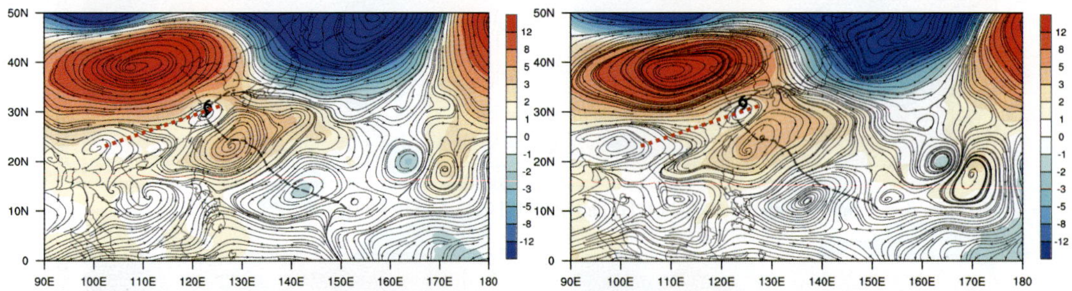

300hPa flow/height anomalies at 0000 UTC and 1200 UTC 25 August 1979.

Figure 7.18. The typhoon (1979_No.13) generated in the tropical central Pacific on 15 August 1979 and moved northwestward along the climatic monsoon trough. Its intensity peaked in the Northwest Pacific on 20 August 1979. The right turn in the east seas of China was influenced by anomalous vortices at 300 hPa.

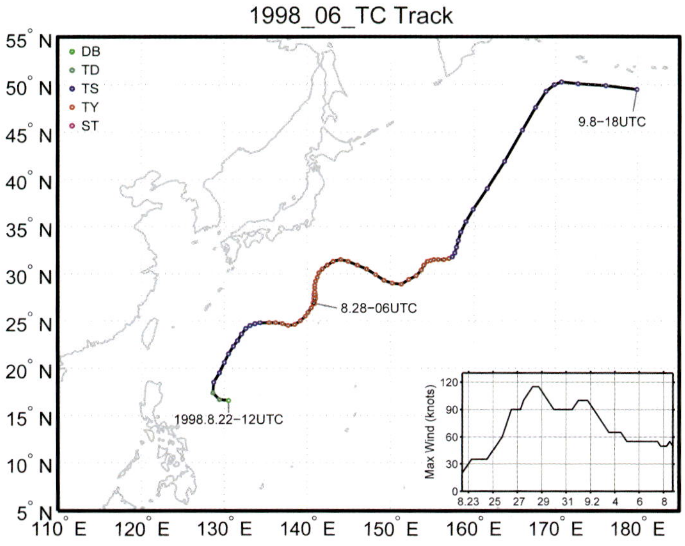

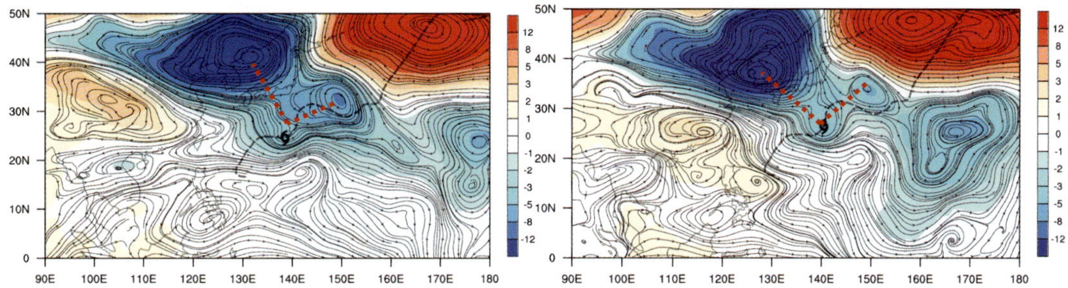

300 hPa flow/height anomalies at 0000 UTC 27 and 28 August 1998.

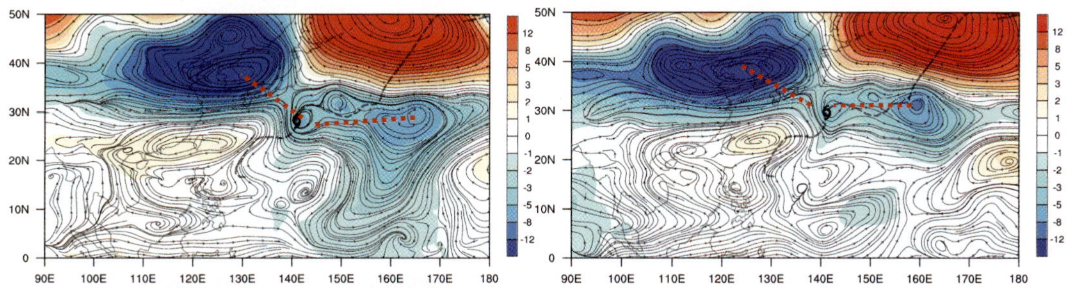

300 hPa flow/height anomalies at 0000 UTC 29 and 30 August 1998.

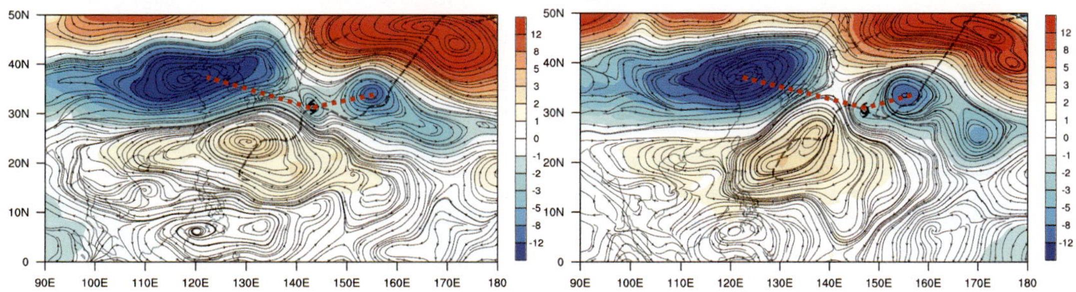

300 hPa flow/height anomalies at 0000 UTC 31 August and 1 September 1998.

Figure 7.19. The typhoon (1998_No.06) generated in the east sea of the Philippines on 22 August 1998 and had a complex motion. Its intensity peaked in the Northwest Pacific on 28 August 1998 when it was approaching the climatic monsoon trough. This complex motion can be explained by the interaction with anomalous vortices at 300 hPa.

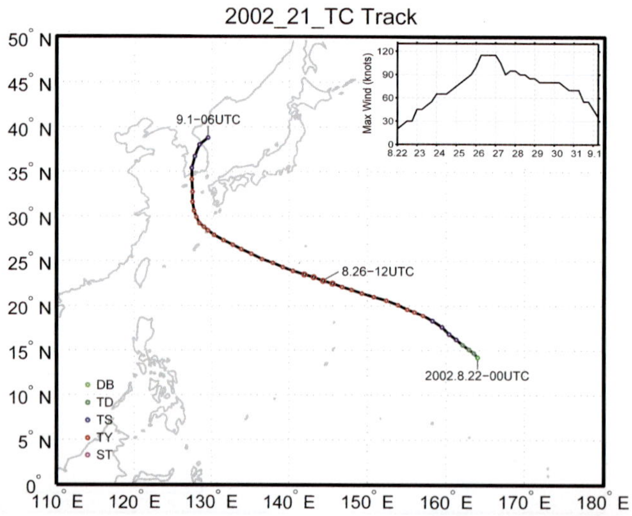

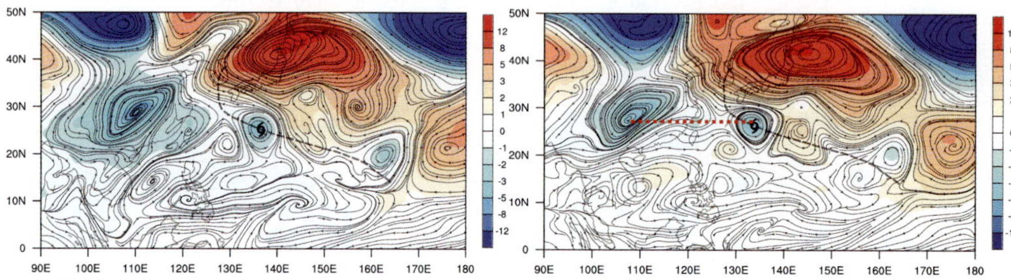

300 hPa flow/height anomalies at 0000 UTC and 1200 UTC 28 August 2002.

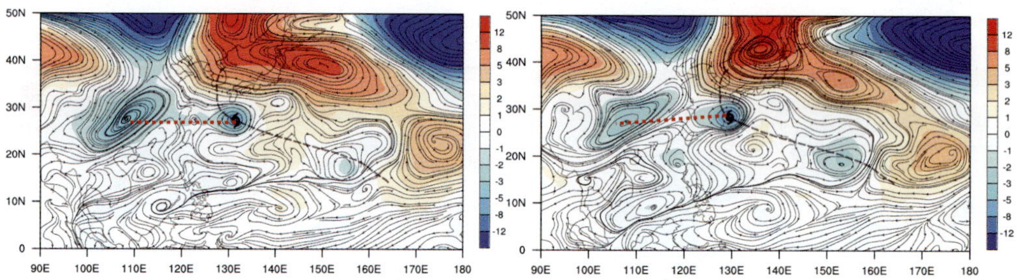

300 hPa flow/height anomalies at 0000 UTC and 1200 UTC 29 August 2002.

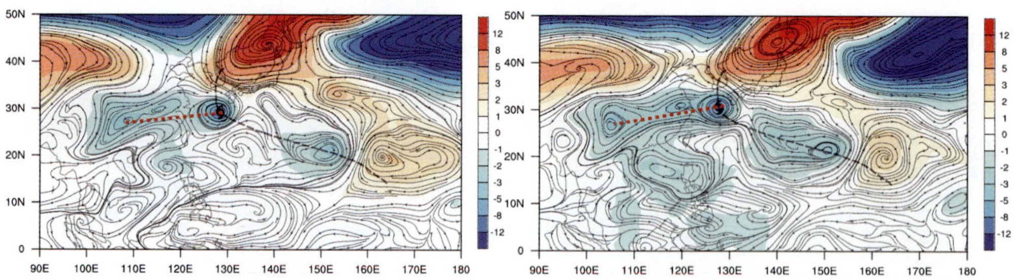

300 hPa flow/height anomalies at 0000 UTC and 1200 UTC 30 August 2002.

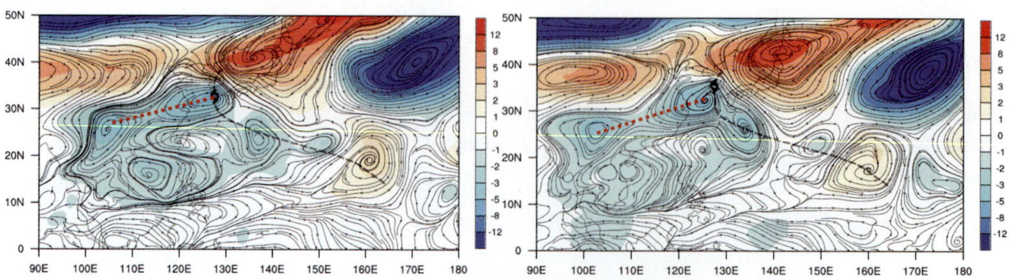

300 hPa flow/height anomalies at 0000 UTC and 1200 UTC 31 August 2002.

Figure 7.20. The typhoon (2002_No.21) generated in the tropical central Pacific on 22 August 2008 and moved northwestward near the climatic monsoon trough. Its intensity peaked on 26 August 2002. The right turn in the east seas of China was influenced by an anomalous vortex at 300 hPa.

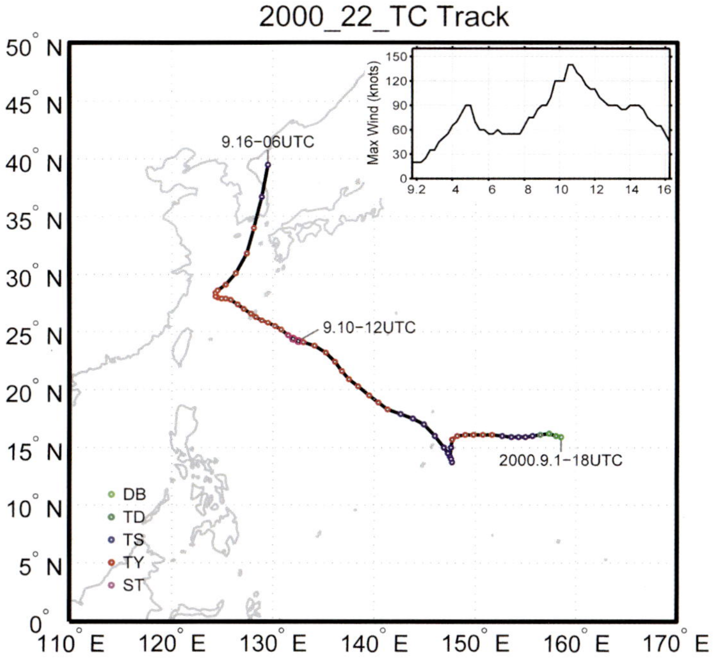

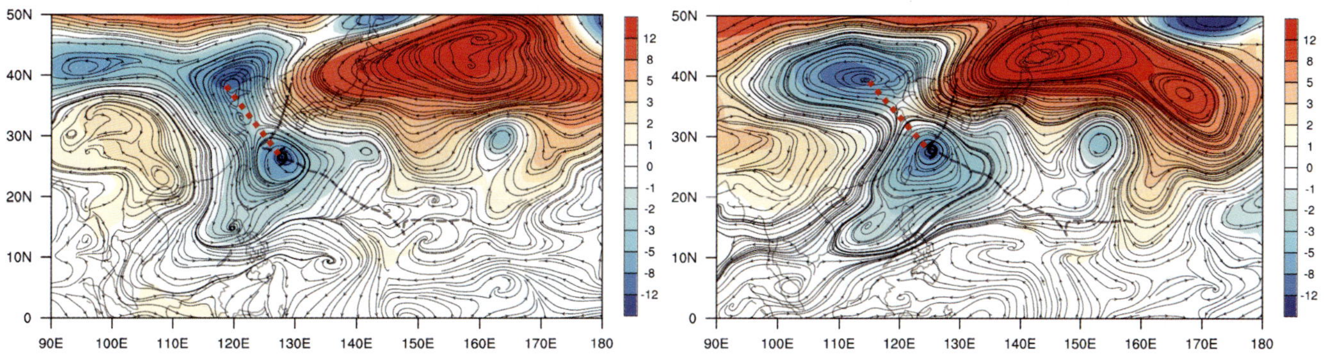

300 hPa flow/height anomalies at 1200 UTC 12 and 13 September 2000.

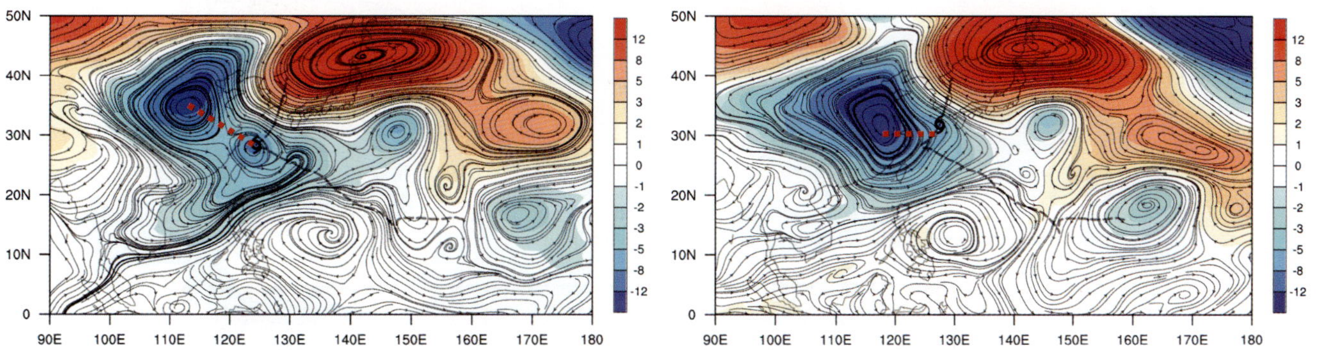

300 hPa flow/height anomalies at 1200 UTC 14 and 15 September 2000.

Figure 7.21. The typhoon (2000_No.22) generated in the tropical central Pacific on 1 September 2000 and moved northwestward along the climatic monsoon trough. Its maximum wind speed reached higher than 120 knots in the east seas of China from 10 to 11 October 2000. The turning right in the east seas of China was influenced by an anomalous vortex at 300 hPa.

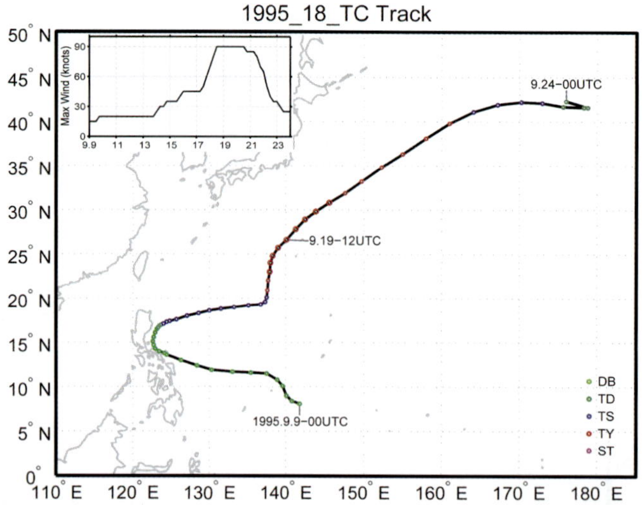

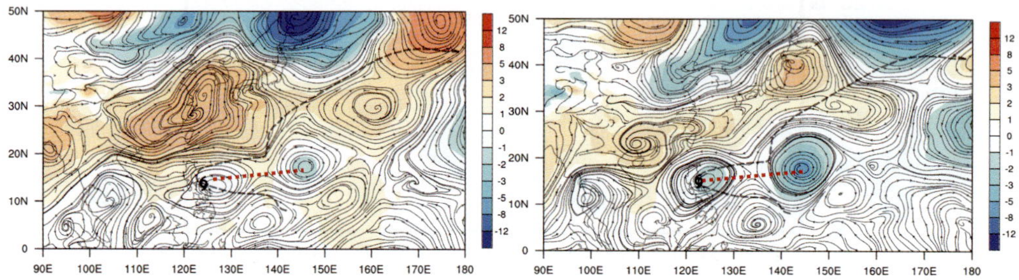

850 hPa flow/height anomalies at 0000 UTC 12 and 13 September 1995.

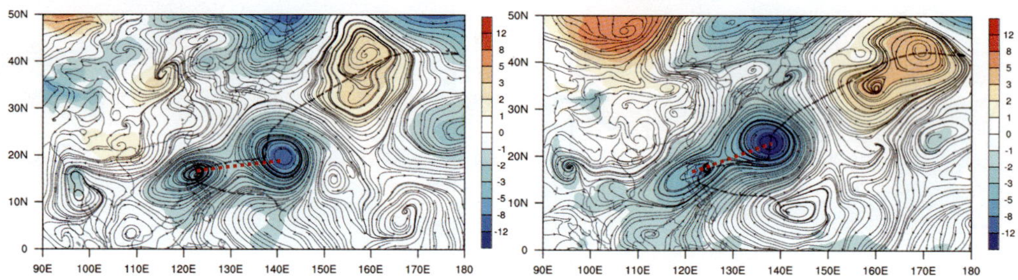

850 hPa flow/height anomalies at 0000 UTC 14 and 15 September 1995.

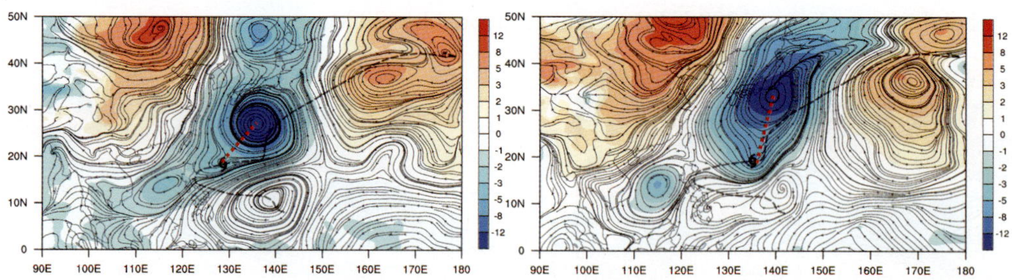

850 hPa flow/height anomalies at 0000 UTC 16 and 17 September 1995.

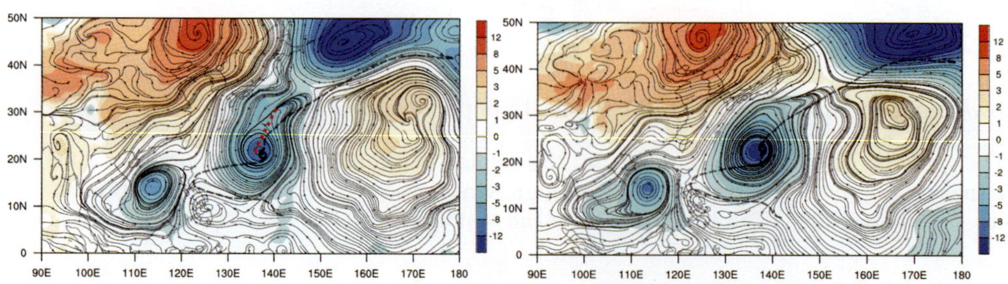

850 hPa flow/height anomalies at 0000 UTC and 1200 UTC 18 September 1995.

Figure 7.22. The typhoon (1995_No.18) generated in the equatorial western Pacific on 9 September 1995 and had a complex motion. Its intensity peaked in the south of Japan and was influenced by the climatic monsoon trough on 19 September 1995. The right turn in the east sea of the Philippines was influenced by an anomalous vortex at 850 hPa.

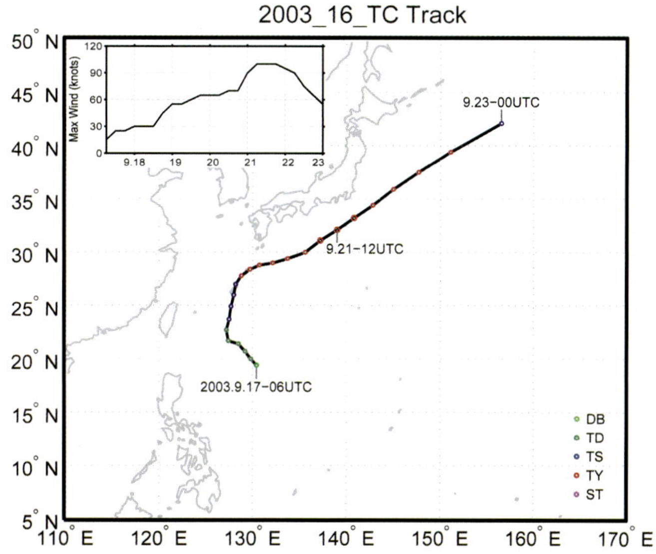

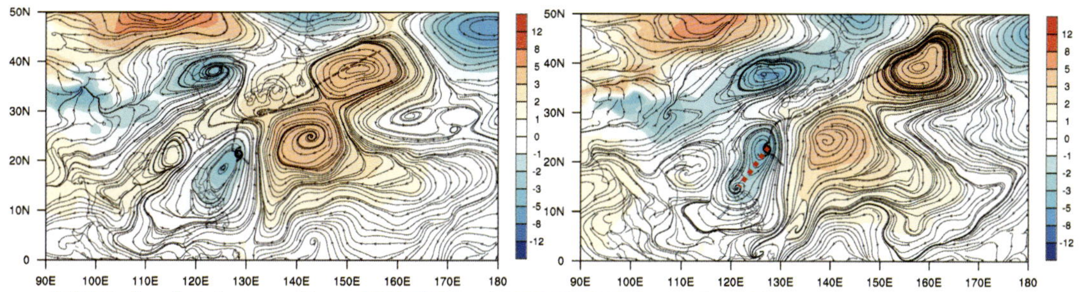

850 hPa flow/height anomalies at 0000 UTC and 1200 UTC 18 September 2003.

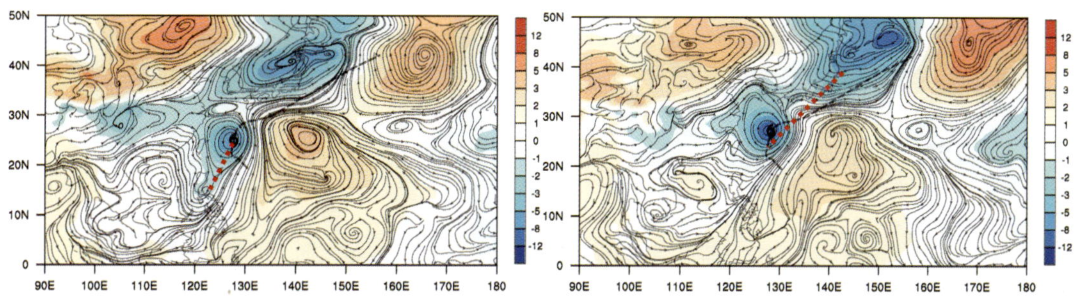

850 hPa flow/height anomalies at 0000 UTC and 1200 UTC 19 September 2003.

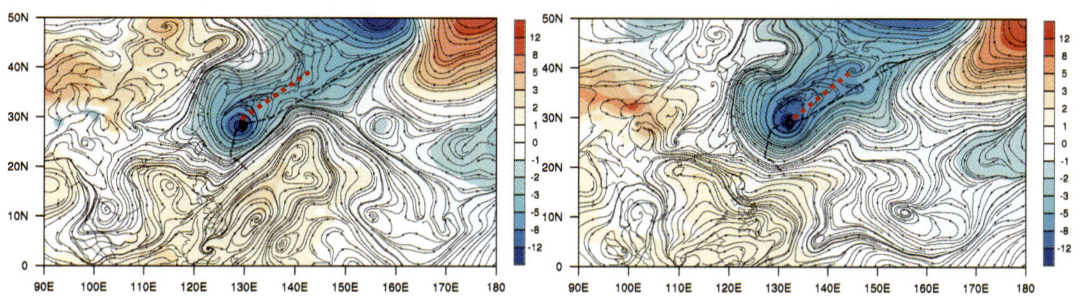

850 hPa flow/height anomaliesat 0000 UTC and 1200 UTC 20 September 2003.

Figure 7.23. The typhoon (2003_No.16) generated in the tropical western Pacific on 17 September 2003 and moved northward at the beginning stage. At the late stage, it moved northeastward near the subtropical climatic monsoon trough and its intensity peaked in the southern seas of Japan on 21 September 2003. The right turn in the east sea of the Taiwan Island was influenced by anomalous vortices and anomalous anti-cyclonic circulations at 850 hPa.

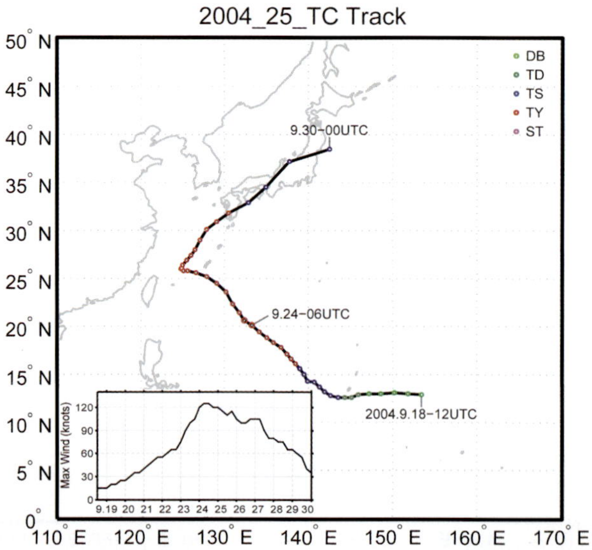

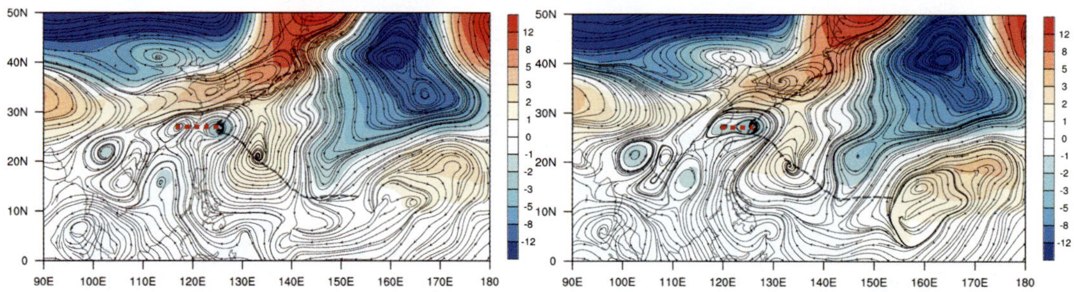

300 hPa flow/height anomalies at 1200 UTC and 1800 UTC 27 September 2004.

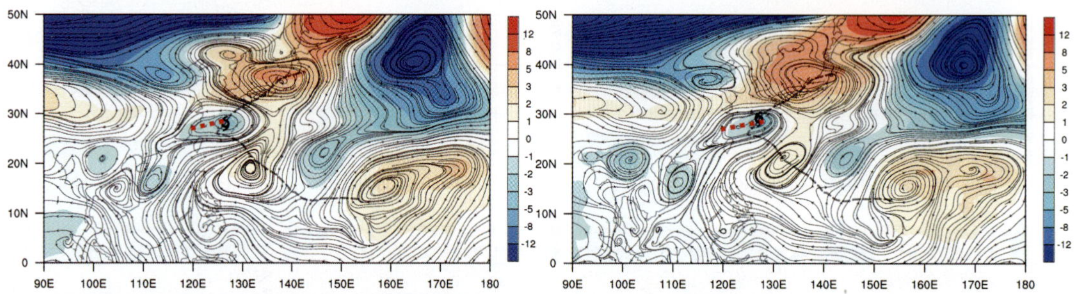

300 hPa flow/height anomalies at 0000 UTC and 0600 UTC 28 September 2004.

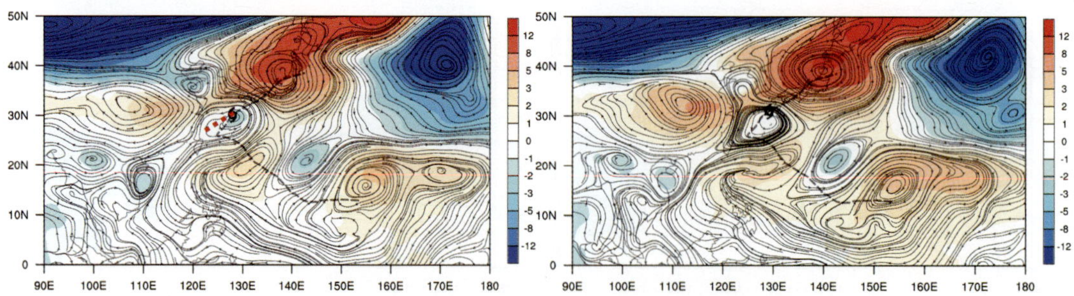

300 hPa flow/height anomalies at 1200 UTC and 1800 UTC 28 September 2004.

Figure 7.24. The typhoon (2004_No.25) generated in the tropical central Pacific on 18 September 2004 and moved west-northwestward along the climatic monsoon trough. Its maximum wind speed reached higher than 100 knots in the east seas of China from 24 to 26 September 2004. The right turn in the east seas of China was influenced by anomalous vortices at 300 hPa.

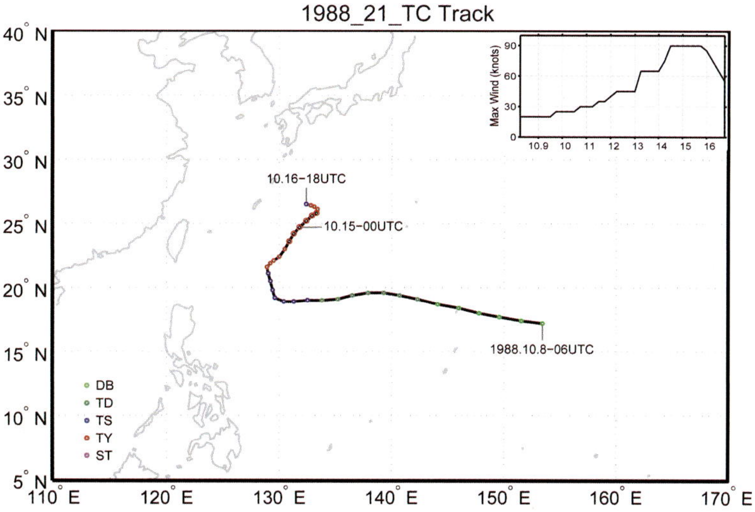

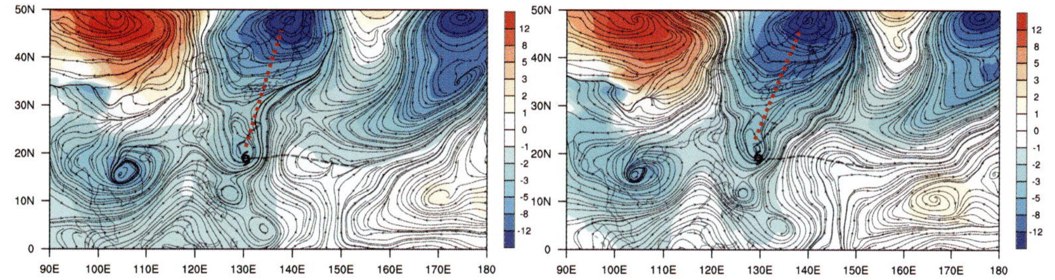

850 hPa flow/height anomalies at 0000 UTC and 0600 UTC 12 October 1988.

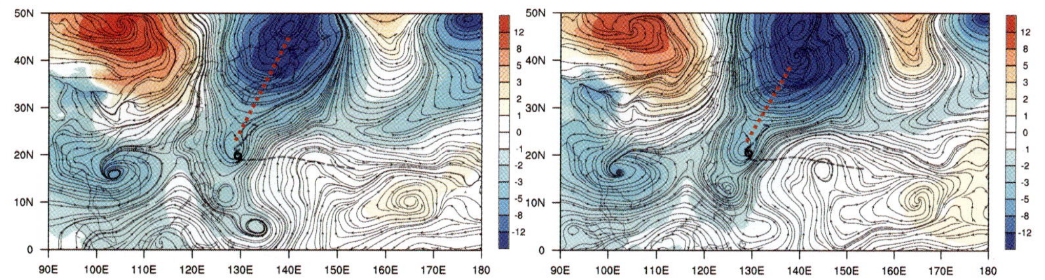

850 hPa flow/height anomalies at 1200 UTC and 1800 UTC 12 October 1988.

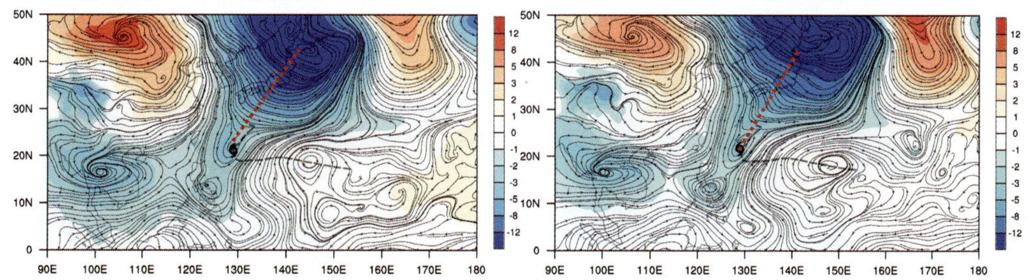

850 hPa flow/height anomalies at 0000 UTC and 0600 UTC 13 October 1988.

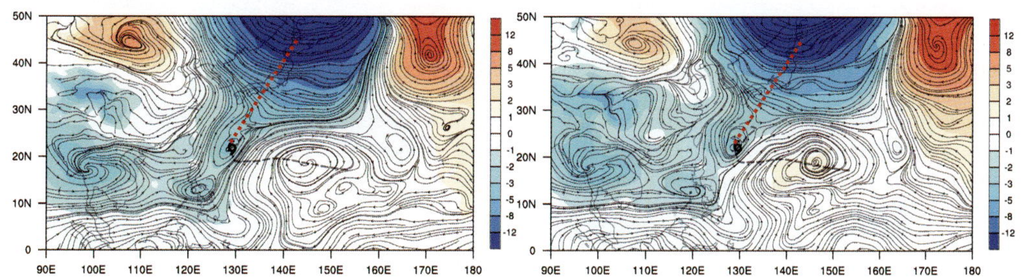

850 hPa flow/height anomalies at 1200 UTC and 1800 UTC 13 October 1988.

Figure 7.25. The typhoon (1988_No.21) generated in the tropical central Pacific on 8 October 1988 and moved westward. After turning right, it entered the climatic monsoon trough and its intensity peaked in the southern seas of Japan on 15 October 1988. This right turn was influenced by a large anomalous vortex at 850 hPa.

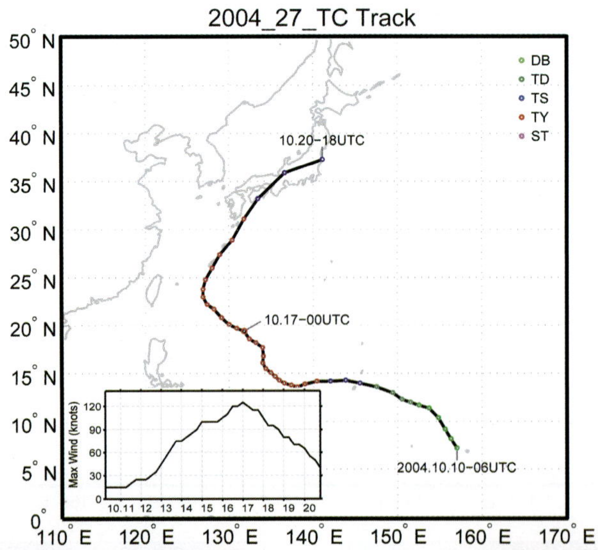

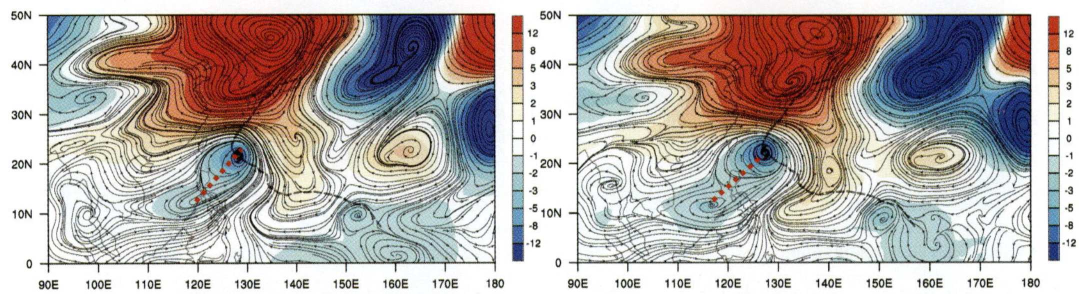

300 hPa flow/height anomalies at 0000 UTC and 0600 UTC 18 October 2004.

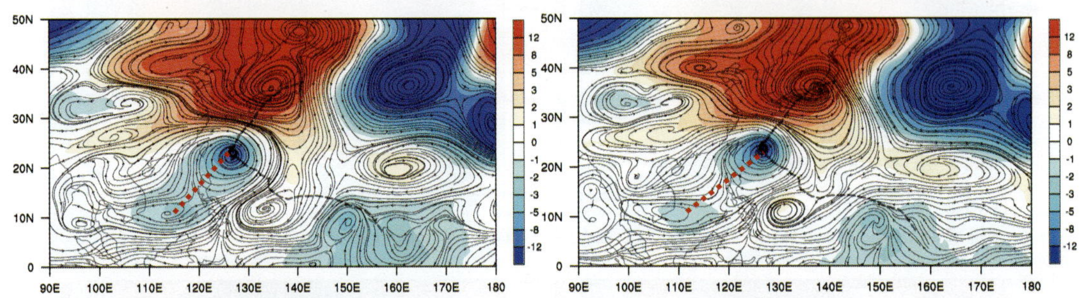

300 hPa flow/height anomalies at 1200 UTC and 1800 UTC 18 October 2004.

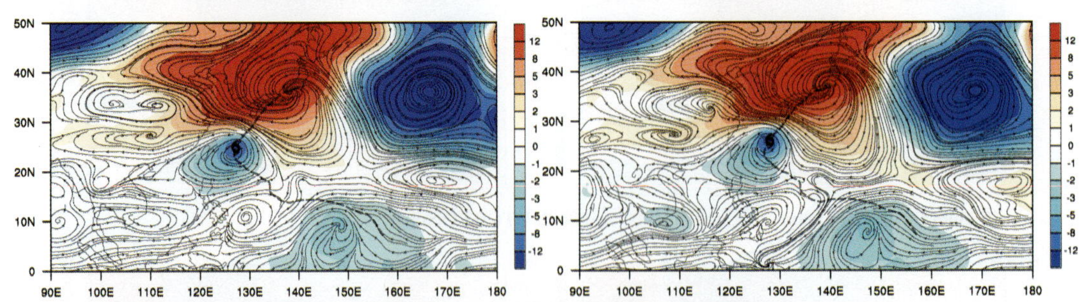

300 hPa flow/height anomalies at 0000 UTC and 0600 UTC 19 October 2004.

Figure 7.26. The typhoon (2004_No.27) generated in the equatorial central Pacific on 10 October 2004 and moved northwestward along the climatic monsoon trough. Its intensity peaked in the southeast sea of Taiwan Island on 17 October 2004 before turning northeastward. The right turn in the east sea of the Taiwan Island was influenced by an anomalous vortex at 300 hPa.

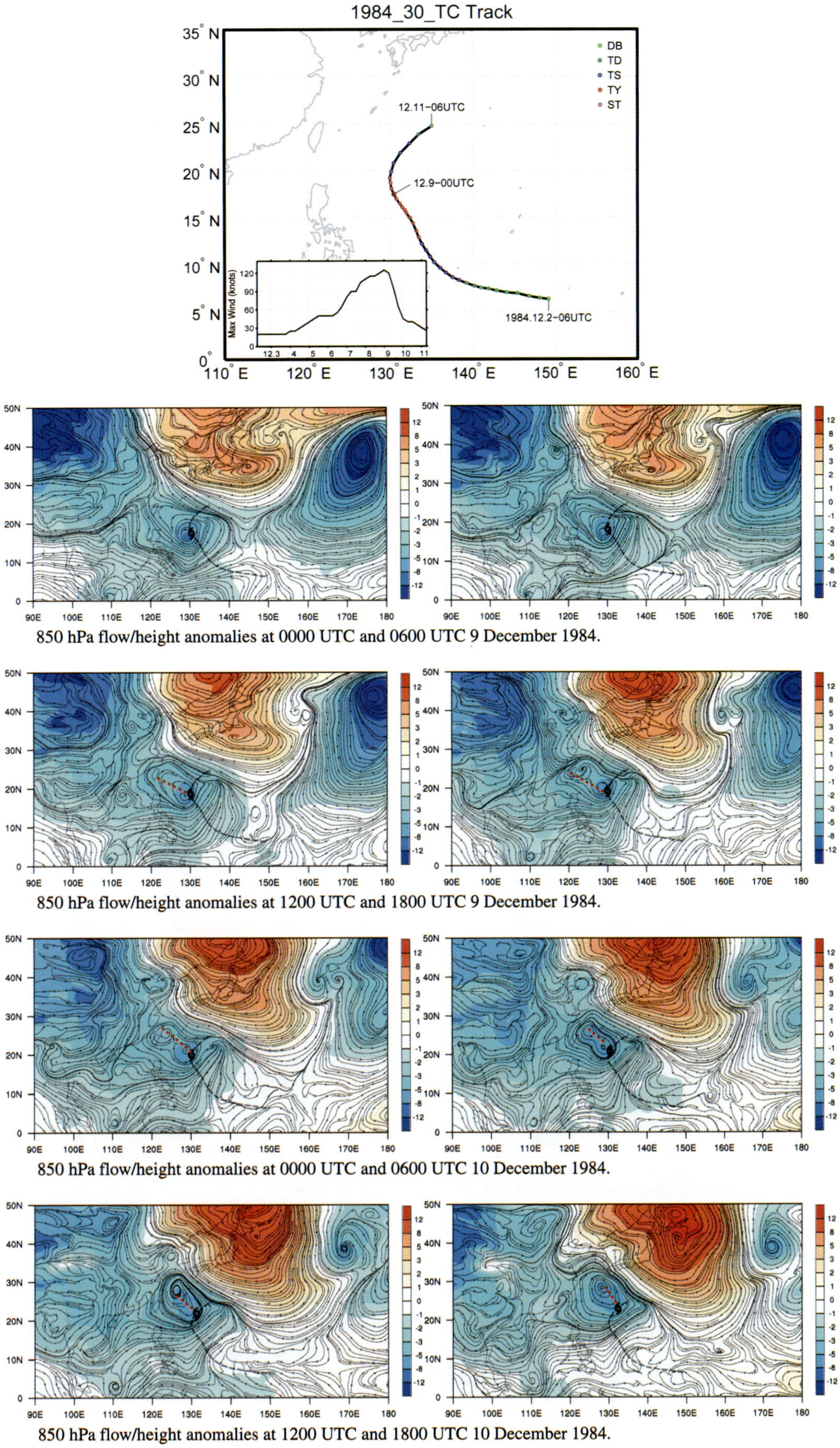

Figure 7.27. The typhoon (1984_No.30) generated in the equatorial western Pacific on 2 December 1984 and moved northwestward. Its intensity peaked on 9 December 1984 before turning northeastward. The right turn in the subtropical Northwest Pacific was influenced by an anomalous vortex at 300 hPa.

7.3. Geneses in the Western Pacific and Entering the East Seas of China

Three typhoons generated in the western Pacific and entered the east seas of China. Their motions were influenced by anomalous vortices (Figs. 7.28–7.30).

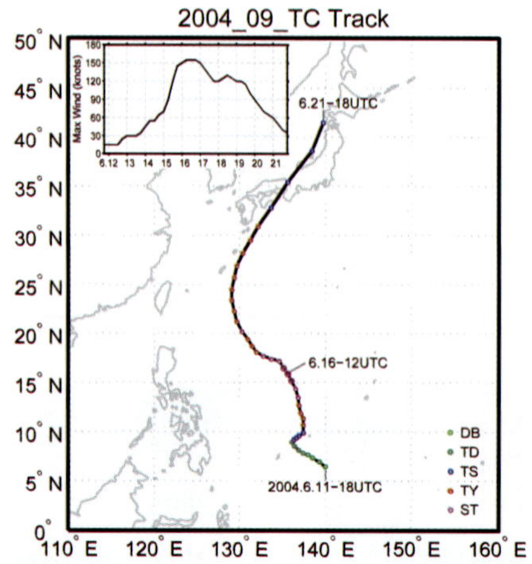

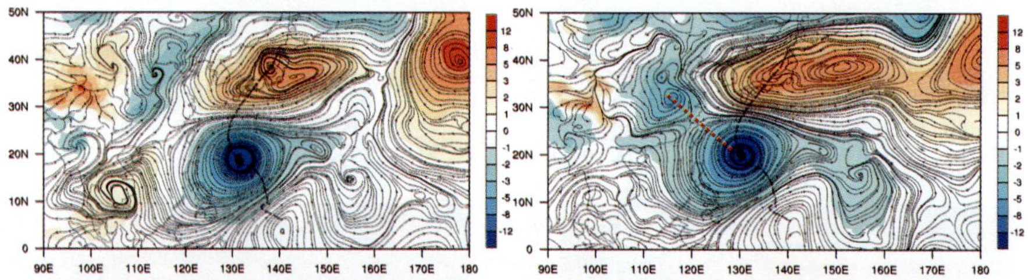

850 hPa flow/height (shading, 10 gpm) anomalies at 0000 UTC and 1200 UTC 18 June 2004.

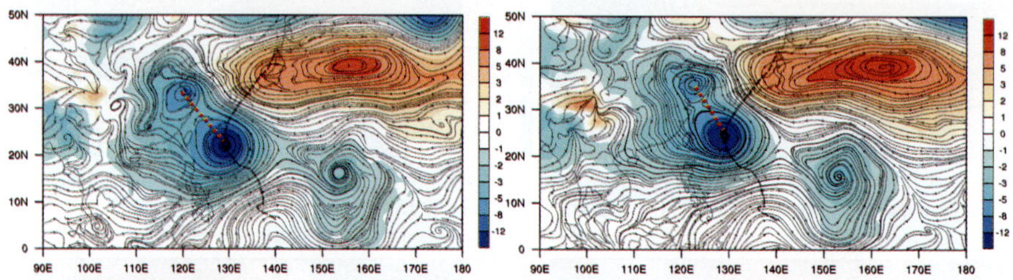

850 hPa flow/height anomalies at 0000 UTC and 1200 UTC 19 June 2004.

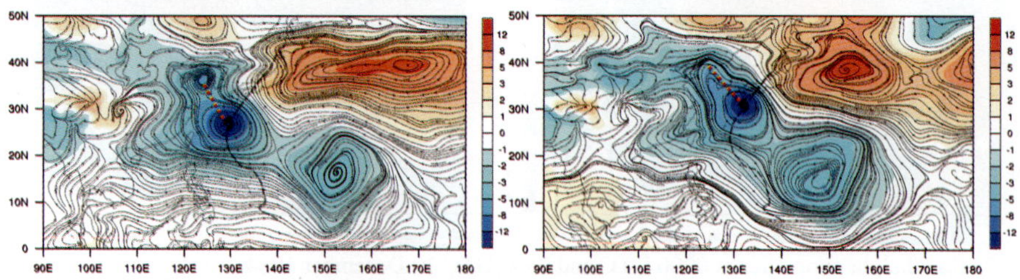

850 hPa flow/height anomalies at 0000 UTC and 1200 UTC 20 June 2004.

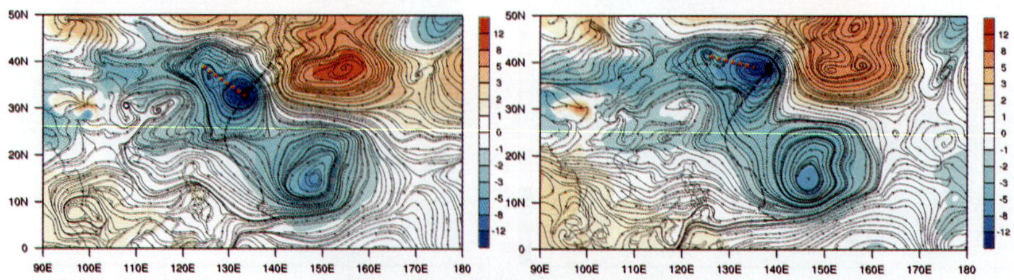

850 hPa flow/height anomalies at 0000 UTC and 1200 UTC 21 June 2004.

Figure 7.28. The typhoon (2004_No.09) generated in the equatorial western Pacific on 11 June 2004 and moved northward. Its intensity peaked in the Northwest Pacific on 16 June 2004 when it was approaching the climatic monsoon trough. The right turn in the Northwest Pacific was influenced by an anomalous vortex at 850 hPa.

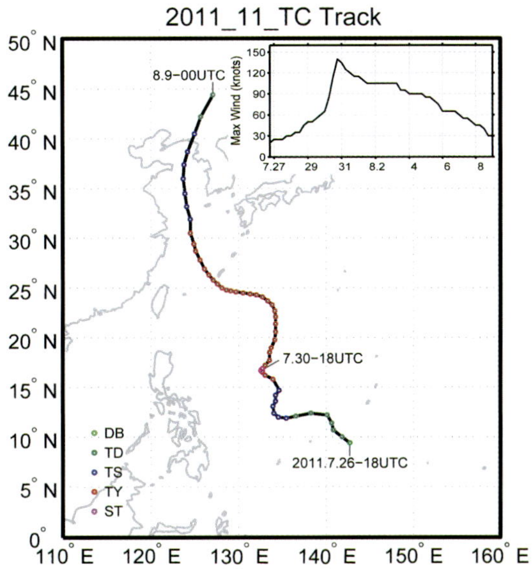

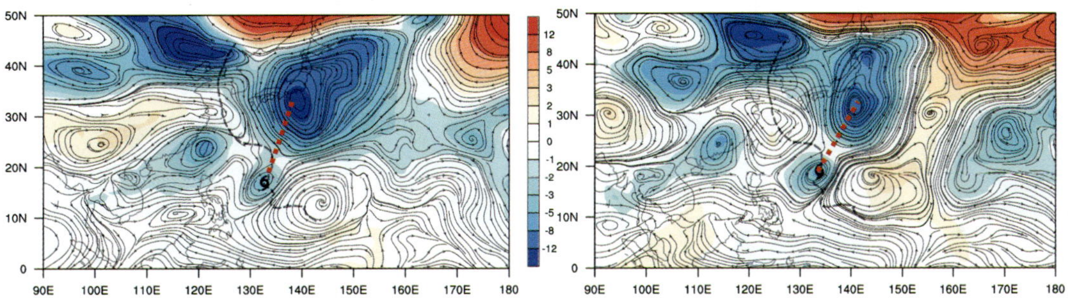

300 hPa flow/height anomalies at 0000 UTC 31 July and 1 August 2011.

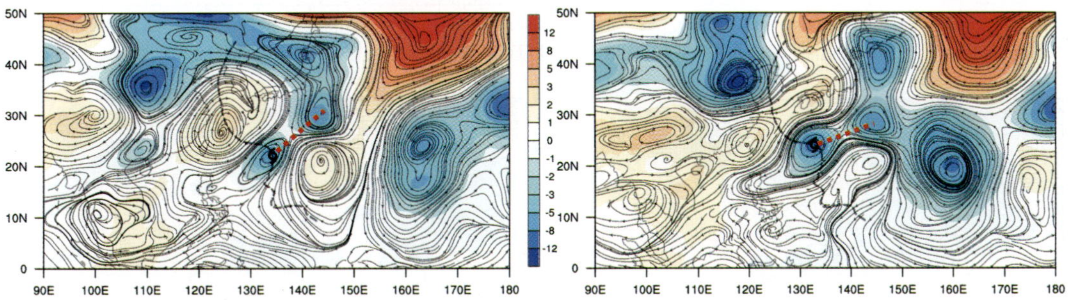

300 hPa flow/height anomalies at 0000 UTC 2 and 3 August 2011.

Figure 7.29. The typhoon (2011_No.11) generated in the equatorial western Pacific on 26 July 2011 and moved with a complex track. Its intensity peaked in the Northwest Pacific on 30 July 2011 when it was approaching the climatic monsoon trough. This complex motion was influenced by anomalous vortices at 300 hPa.

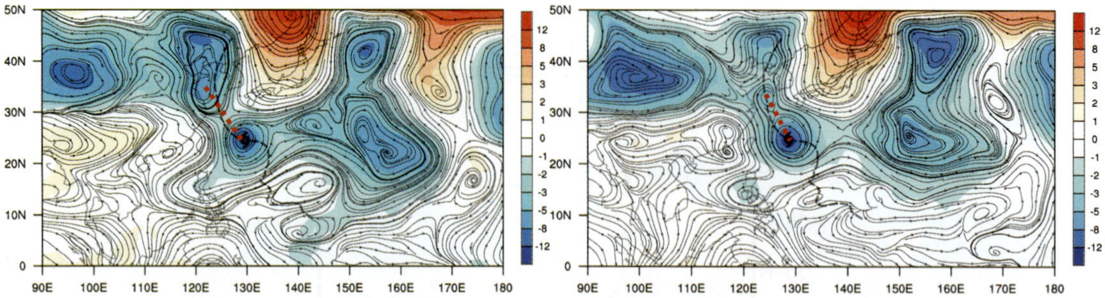

300 hPa flow/height anomaliesat 0000 UTC and 1200 UTC 4 August 2011.

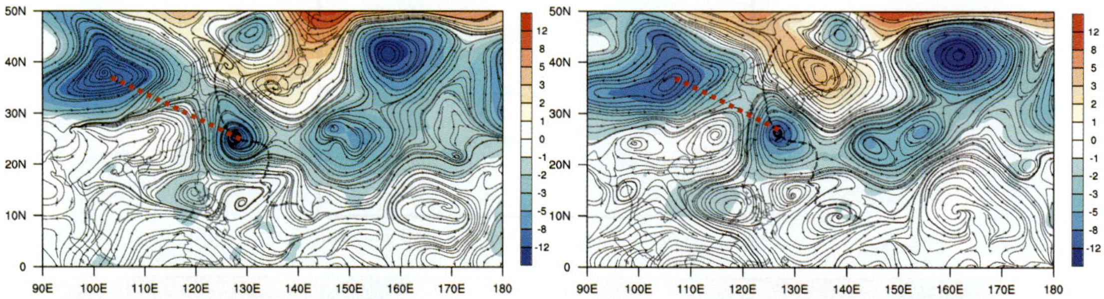

300 hPa flow/height anomalies at 0000 UTC and 1200 UTC 5 August 2011.

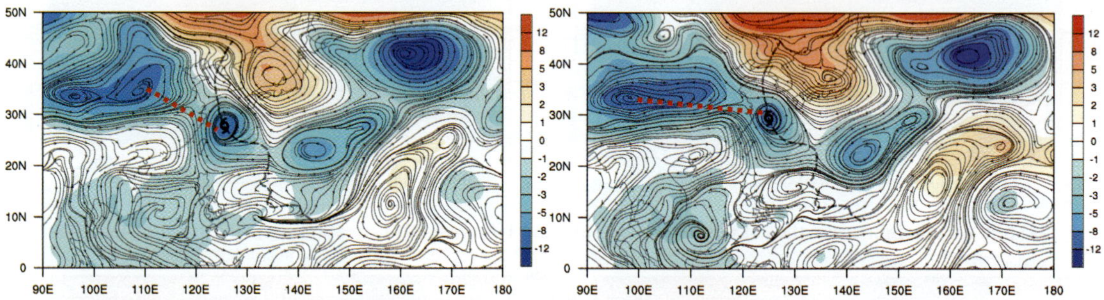

300 hPa flow/height anomalies at 0000 UTC and 1200 UTC 6 August 2011.

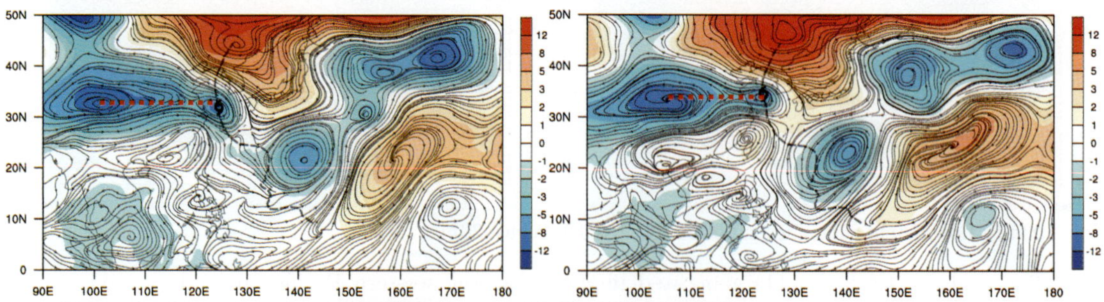

300 hPa flow/height anomalies at 0000 UTC and 1200 UTC 7 August 2011.

Figure 7.29. *(Continued)*

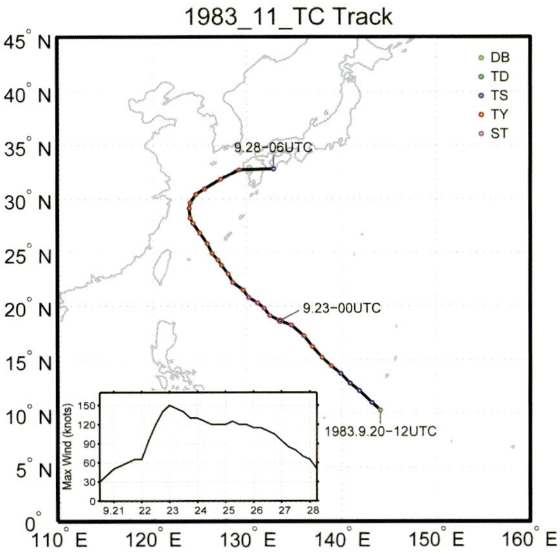

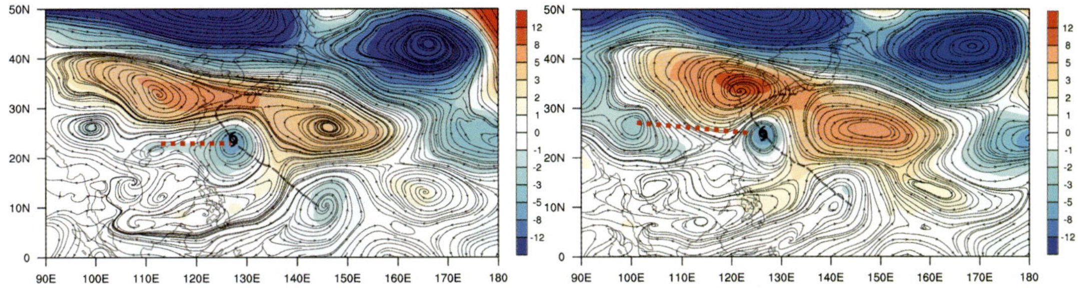

300 hPa flow/height anomalies at 0000 UTC and 1200 UTC 25 September 1983.

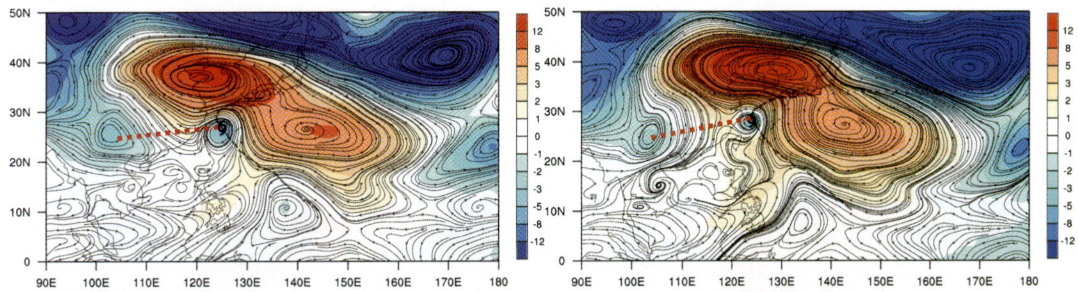

300 hPa flow/height anomalies at 0000 UTC and 1200 UTC 26 September 1983.

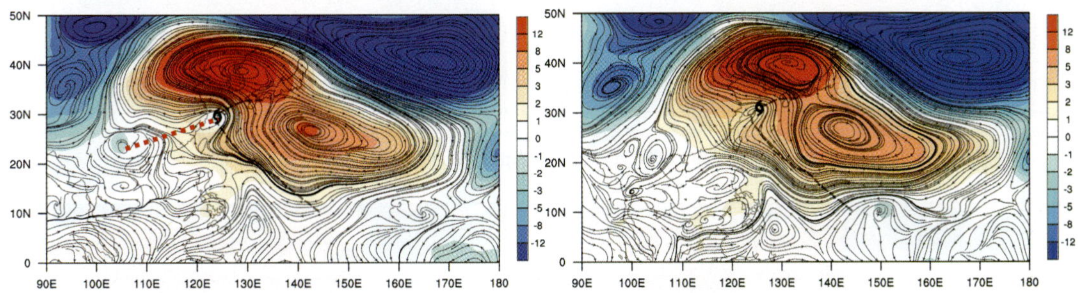

300 hPa flow/height anomalies at 0000 UTC and 1200 UTC 27 September 1983.

Figure 7.30. The typhoon (1983_No.11) generated in the tropical western Pacific on 20 September 1983 and moved northwestward along the climatic monsoon trough. Its intensity peaked in the Northwest Pacific on 23 September 1983. The right turn in the east seas of China was influenced by an anomalous vortex at 300 hPa.

7.4. Geneses in the Central Pacific and Entering the East Seas of China

There were 13 typhoons that generated in the central Pacific and entered the east seas of China. Their motions were influenced by anomalous vortices (Figs. 7.31–7.43).

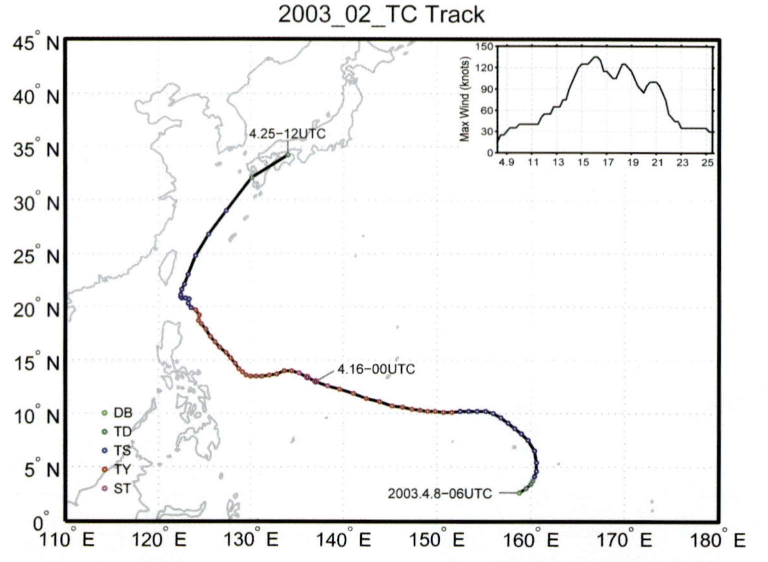

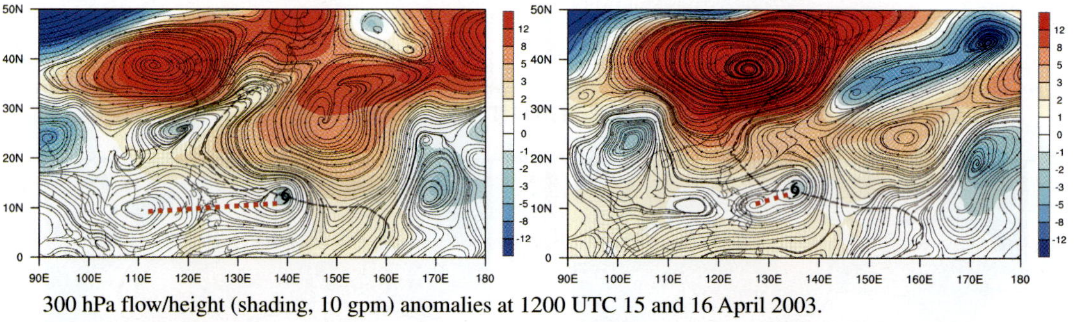

300 hPa flow/height (shading, 10 gpm) anomalies at 1200 UTC 15 and 16 April 2003.

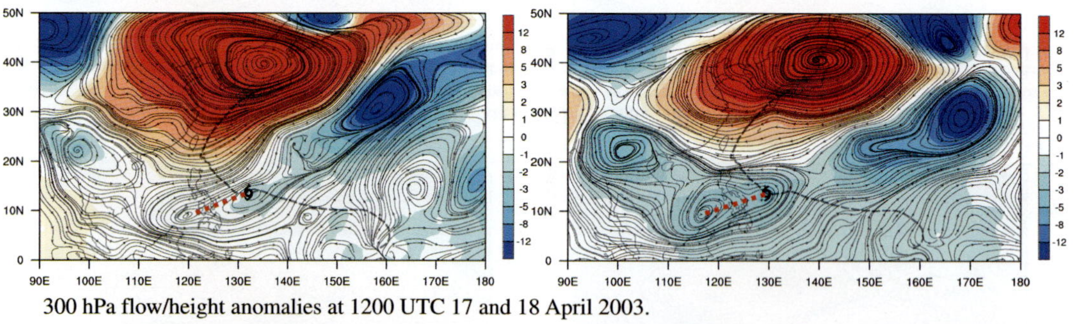

300 hPa flow/height anomalies at 1200 UTC 17 and 18 April 2003.

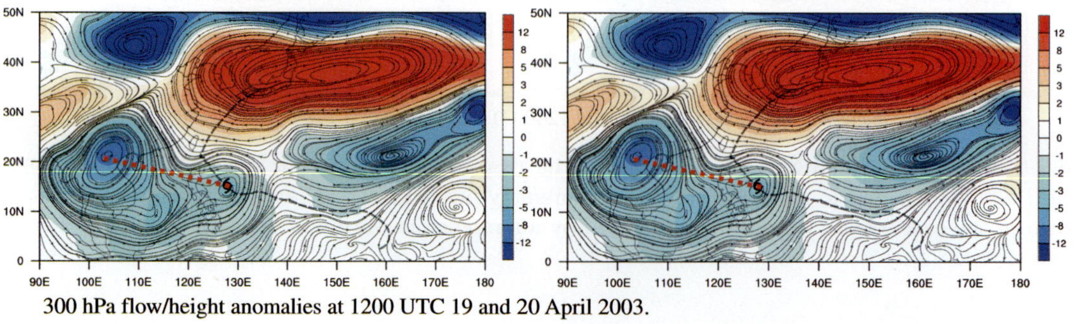

300 hPa flow/height anomalies at 1200 UTC 19 and 20 April 2003.

Figure 7.31. The typhoon (2003_No.02) generated in the equatorial central Pacific on 8 April 2003 and had a left turn and a right turn in its track. Its intensity peaked on 16 April 2003 when it was approaching the climatic monsoon trough. The right turn in the east sea of the Taiwan Island was influenced by several anomalous vortices at 850 hPa and 300 hPa.

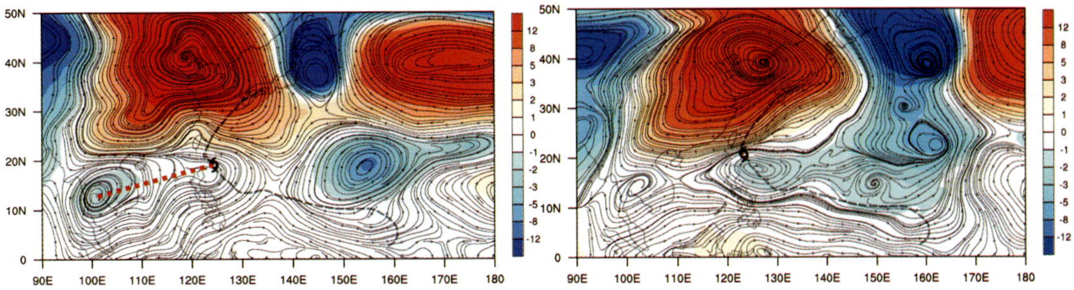

300 hPa flow/height anomalies at 1200 UTC 21 and 22 April 2003.

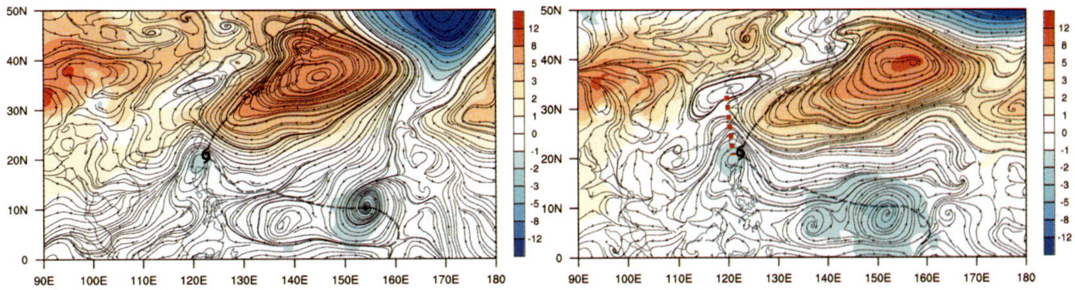

850 hPa flow/height anomalies at 0000 UTC and 1200 UTC 23 April 2003.

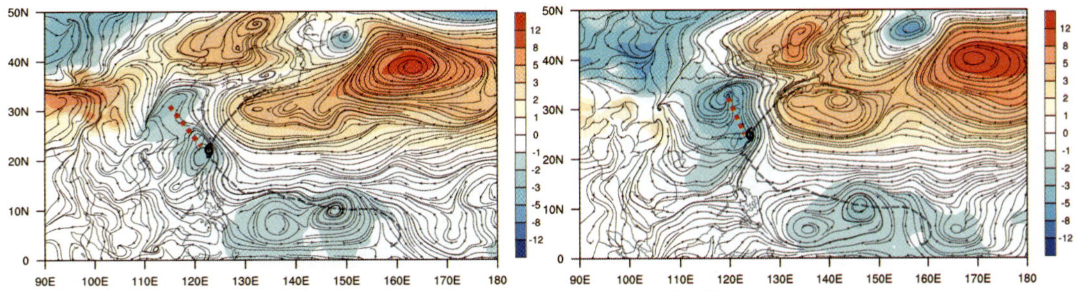

850 hPa flow/height anomalies at 0000 UTC and 1200 UTC 24 April 2003.

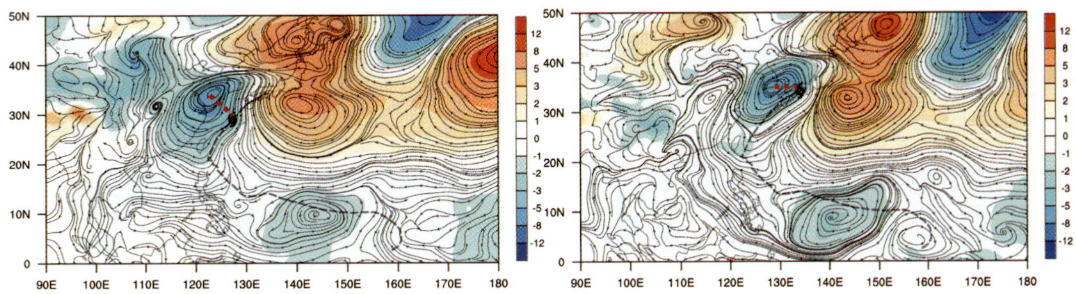

850 hPa flow/height anomalies at 0000 UTC and 1200 UTC 25 April 2003.

Figure 7.31. *(Continued)*

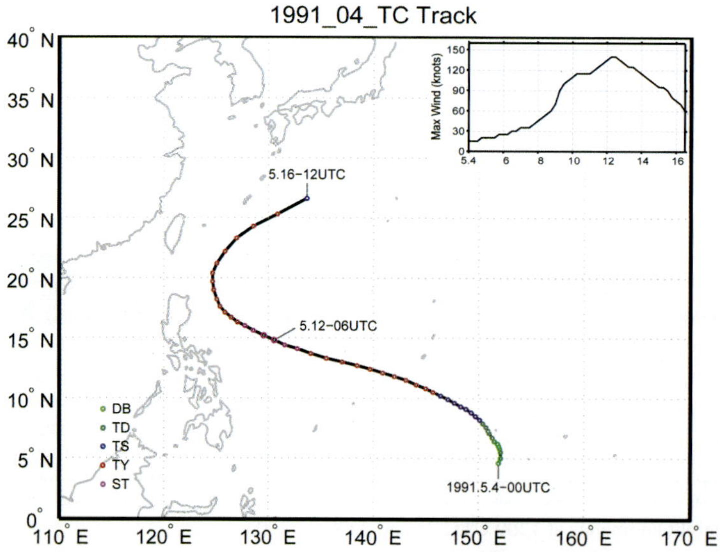

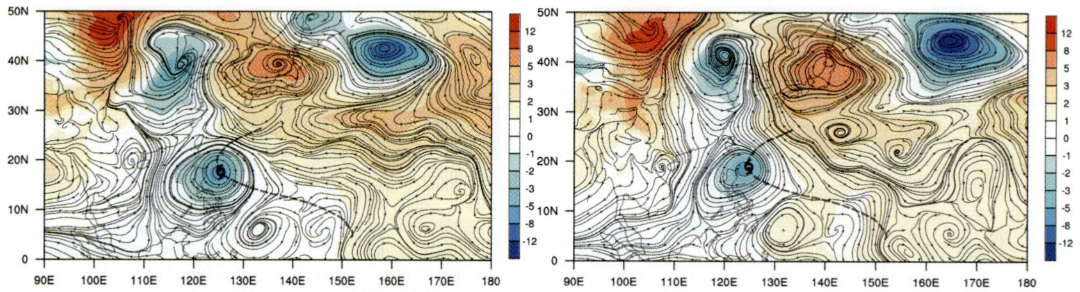

850 hPa flow/height anomalies at 0000 UTC and 1200 UTC 14 May 1991.

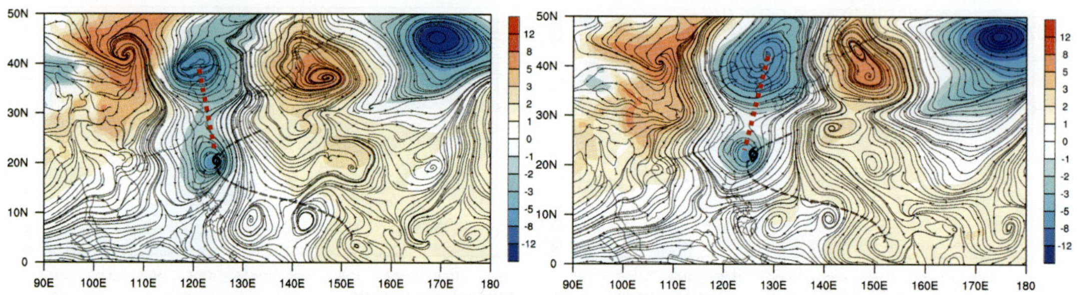

850 hPa flow/height anomalies at 0000 UTC and 1200 UTC 15 May 1991.

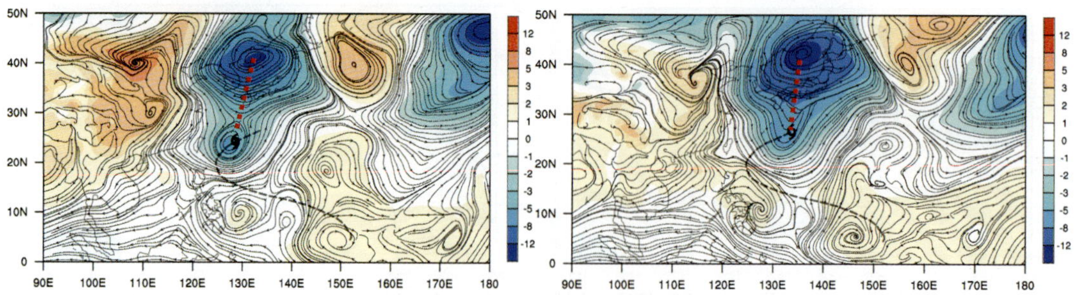

850 hPa flow/height anomalies at 0000 UTC and 1200 UTC 16 May 1991.

Figure 7.32. The typhoon (1991_No.04) generated in the equatorial central Pacific on 4 May 1991 and moved northwestward along the climatic monsoon trough. Its intensity peaked in the Northwest Pacific on 12 May 1991. The right turn in the east sea of the Philippines was influenced by an anomalous vortex at 850 hPa.

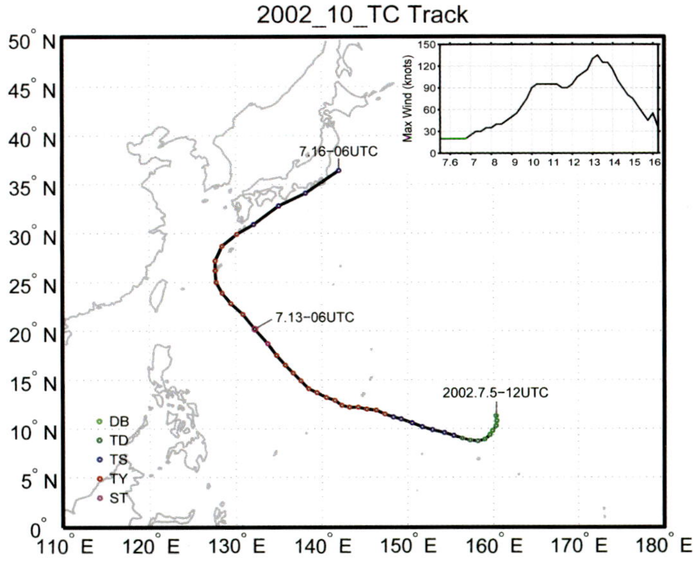

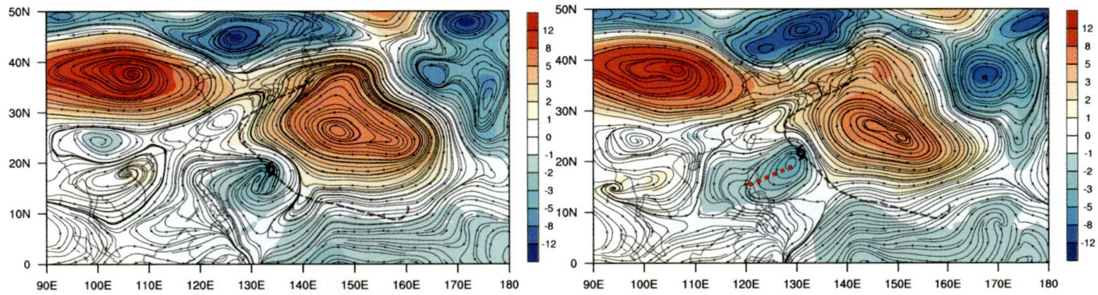

300 hPa flow/height anomalies at 0000 UTC and 1200 UTC 13 July 2002.

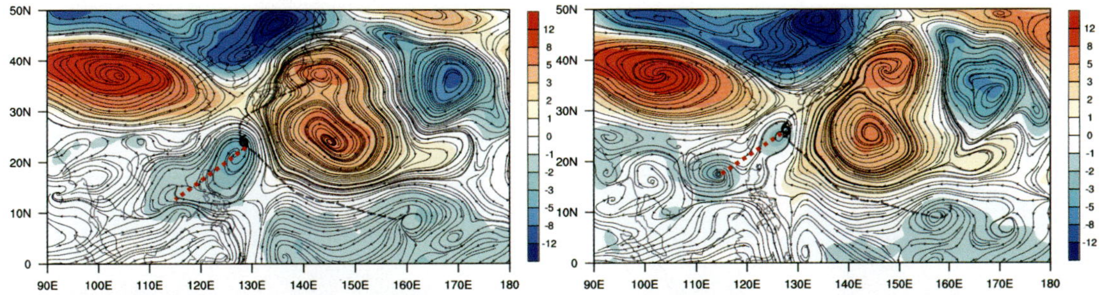

300 hPa flow/height anomalies at 0000 UTC and 1200 UTC 14 July 2002.

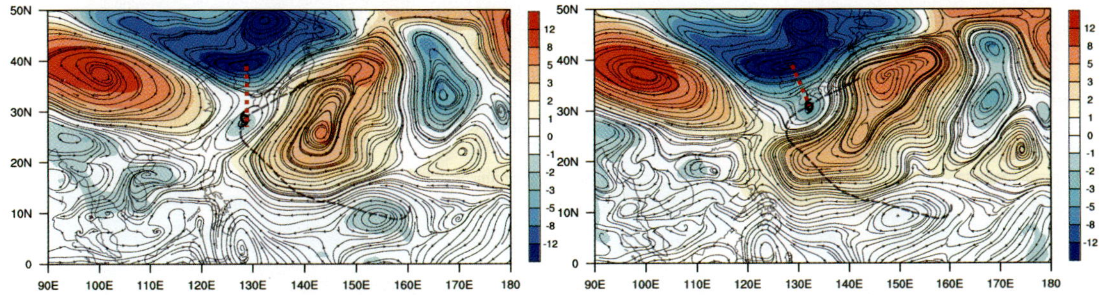

300 hPa flow/height anomalies at 0000 UTC and 1200 UTC 15 July 2002.

Figure 7.33. The typhoon (2002_No.10) generated in the tropical central Pacific on 5 July 2002. After a local curvature motion, it moved northwestward along the climatic monsoon trough and its intensity peaked in the Northwest Pacific on 13 July 2002. The right turn in the south seas of Japan was influenced by anomalous vortices at 300 hPa and 850 hPa.

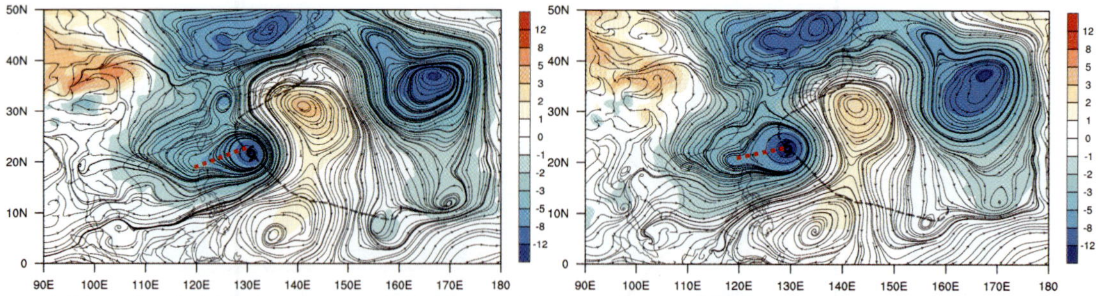

850 hPa flow/height anomalies at 1200 UTC and 1800 UTC 13 July 2002.

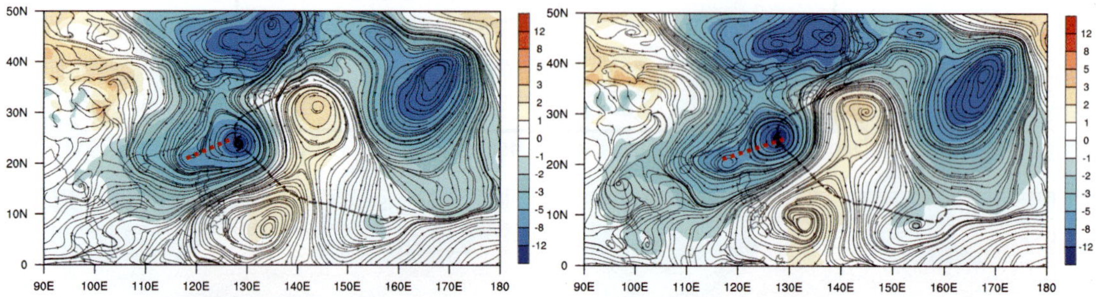

850 hPa flow/height anomalies at 0000 UTC and 0600 UTC 14 July 2002.

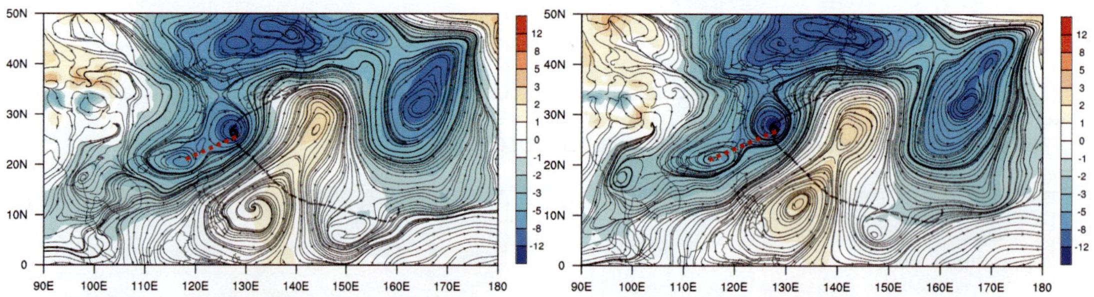

850 hPa flow/height anomalies at 1200 UTC and 1800 UTC 14 July 2002.

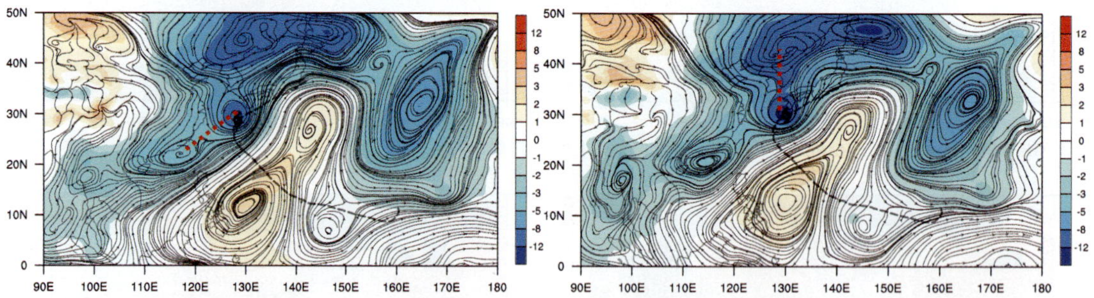

850 hPa flow/height anomalies at 0000 UTC and 0600 UTC 15 July 2002.

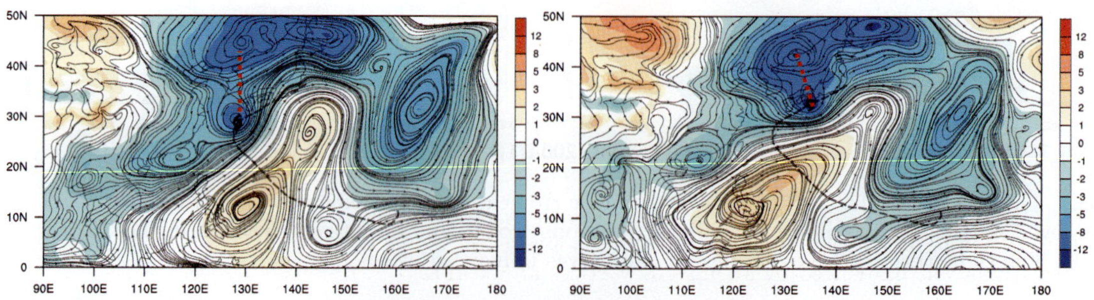

850 hPa flow/height anomalies at 1200 UTC and 1800 UTC 15 July 2002.

Figure 7.33. *(Continued)*

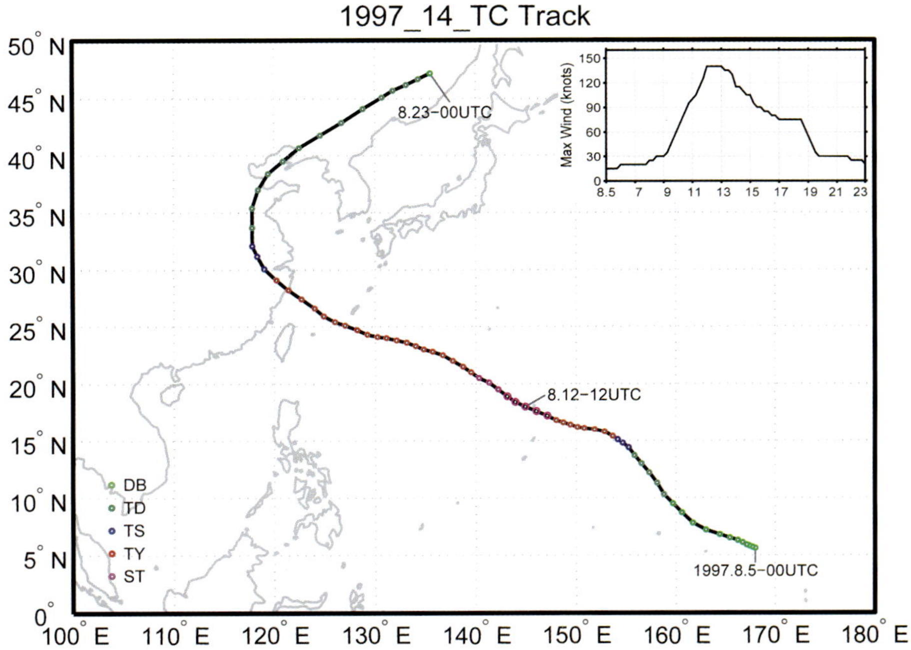

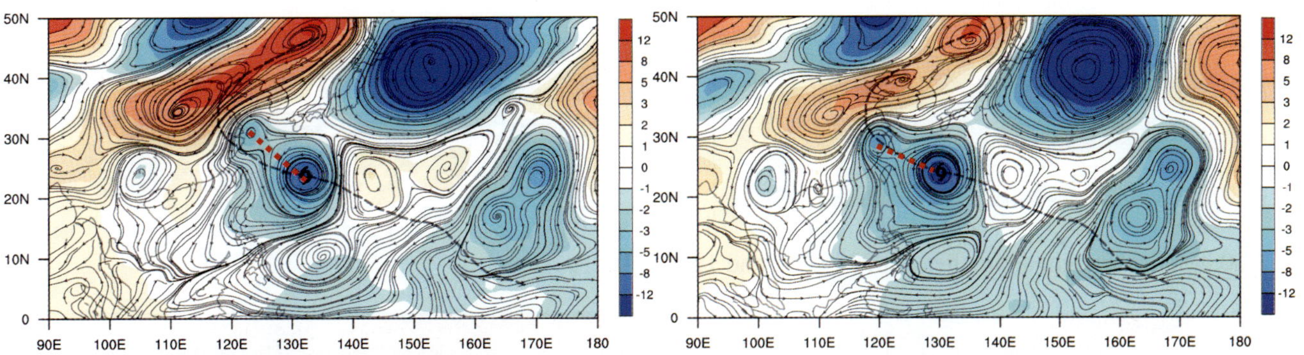

300 hPa flow/height anomalies at 0000 UTC and 1200 UTC 16 August 1997.

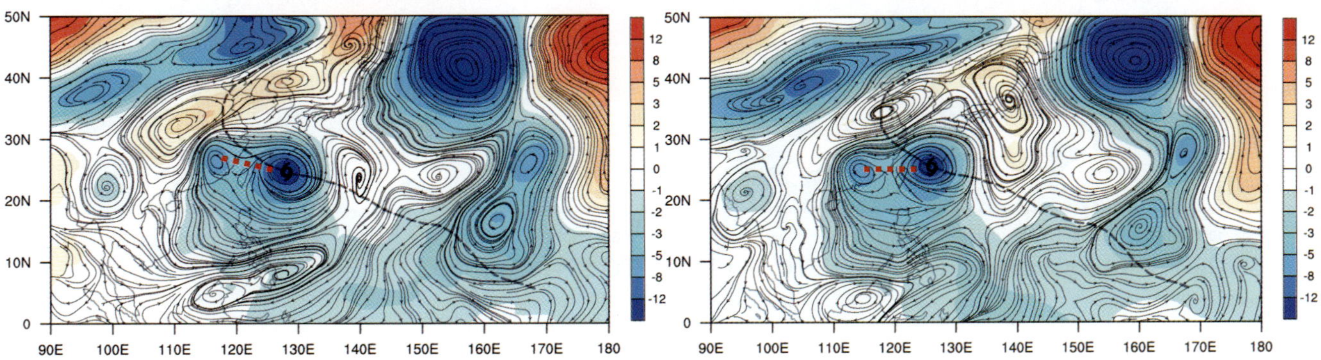

300 hPa flow/height anomalies at 0000 UTC and 1200 UTC 17 August 1997.

Figure 7.34. The typhoon (1997_No.14) generated in the equatorial central Pacific on 5 August 1997 and moved northwestward along the climatic monsoon trough. Its intensity peaked in the Northwest Pacific on 12 August 1997. The right turn in eastern China was influenced by anomalous vortices at 300 hPa.

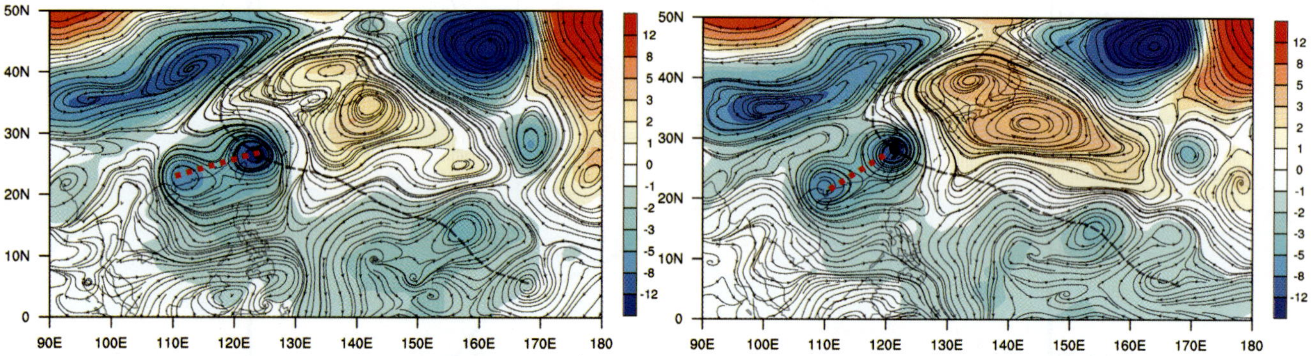

300 hPa flow/height anomalies at 0000 UTC and 1200 UTC 18 August 1997.

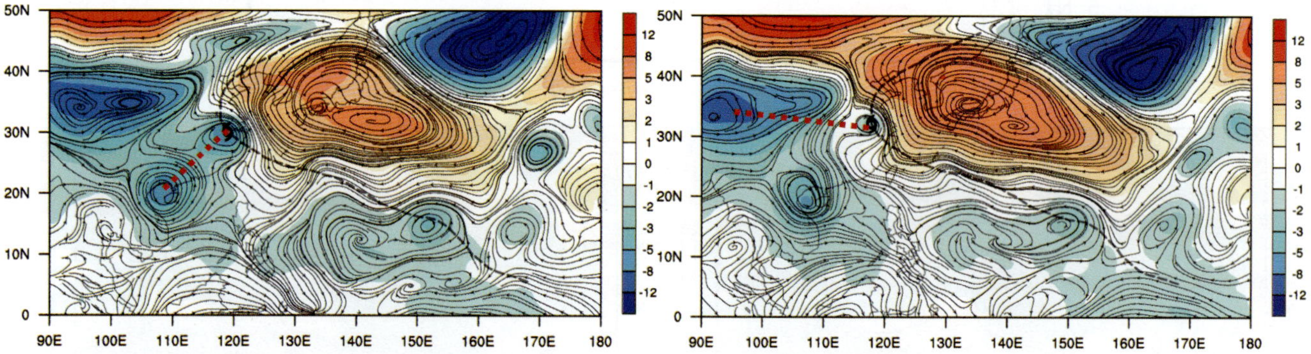

300 hPa flow/height anomalies at 0000 UTC and 1200 UTC 19 August 1997.

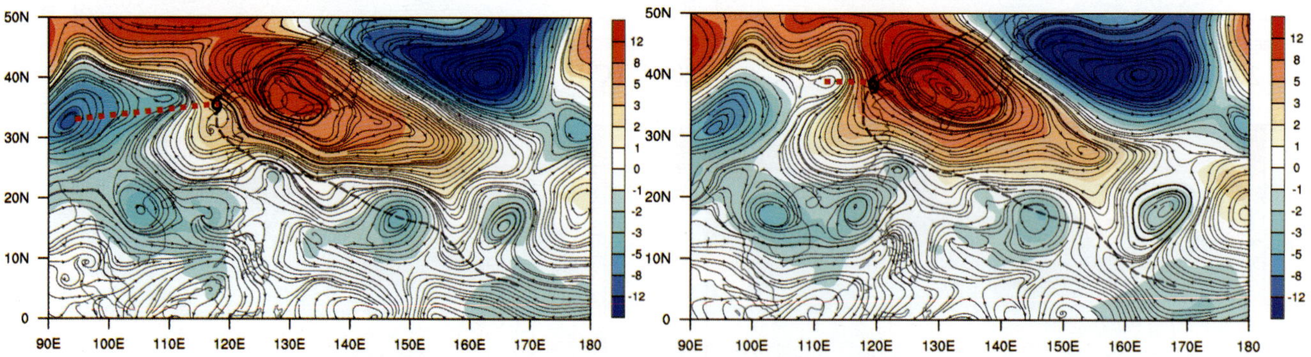

300 hPa flow/height anomalies at 0000 UTC and 1200 UTC 20 August 1997.

Figure 7.34. *(Continued)*

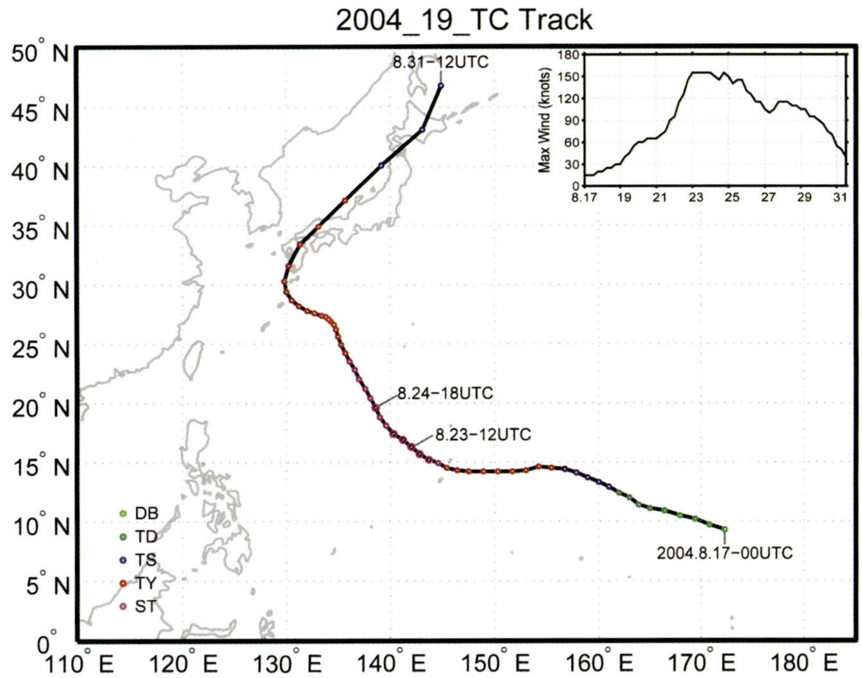

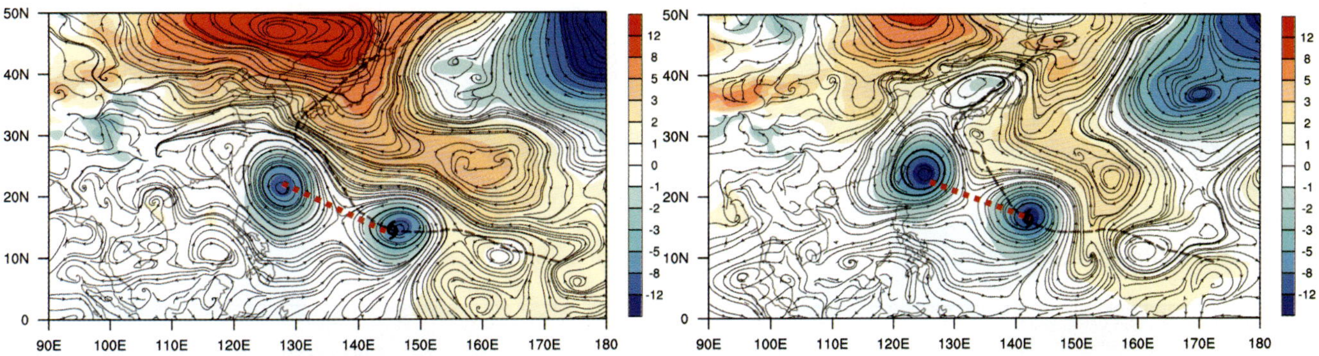

850 hPa flow/height anomalies at 1200 UTC 22 and 23 August 2004.

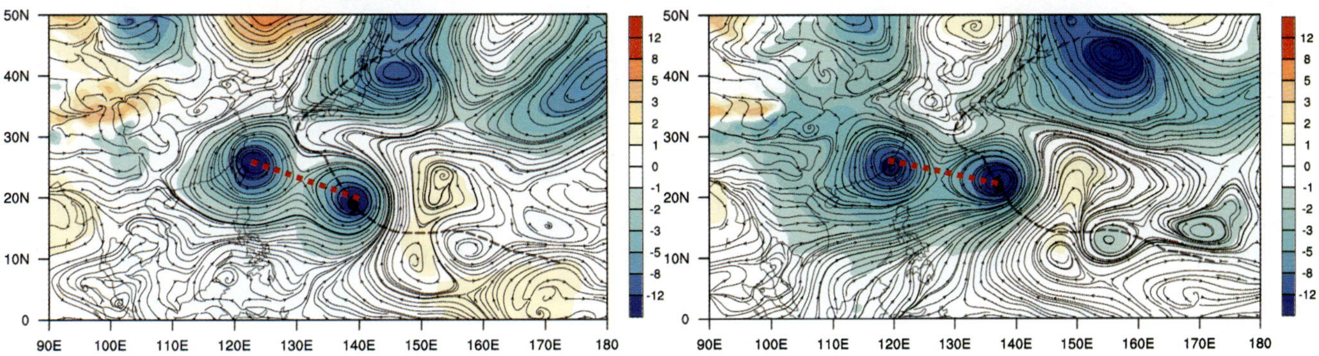

850 hPa flow/height anomalies at 1200 UTC 24 and 25 August 2004.

Figure 7.35. The super typhoon (2004_No.19) generated in the equatorial central Pacific on 17 August 2004 and moved northwestward along the climatic monsoon trough. Its intensity peaked in the Northwest Pacific on 23–24 August 2004. The right turn in southern Japan was influenced by anomalous vortices at 850 hPa.

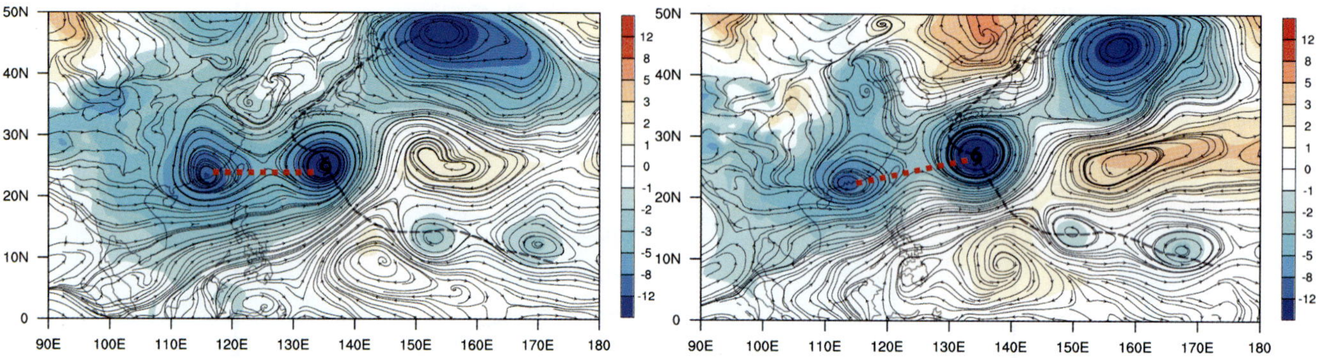

850 hPa flow/height anomalies at 1200 UTC 26 and 27 August 2004.

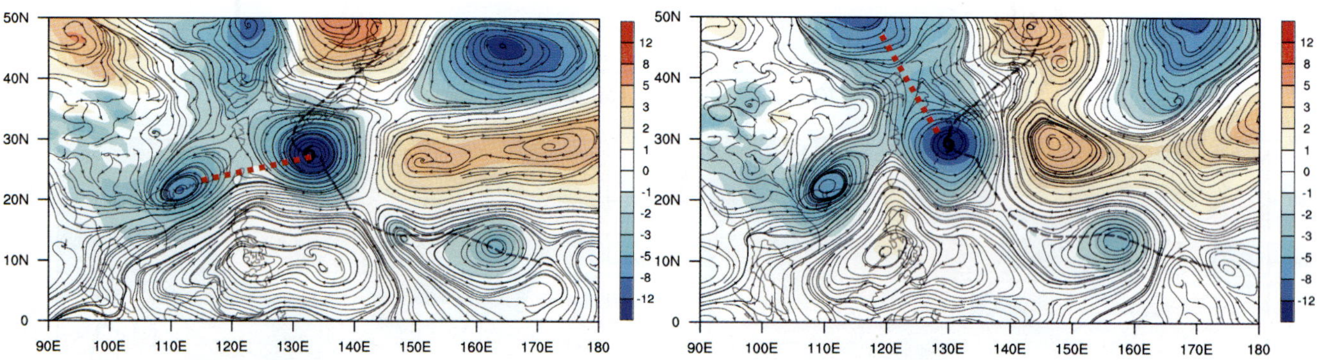

850 hPa flow/height anomalies at1200 UTC 28 and 29 August 2004.

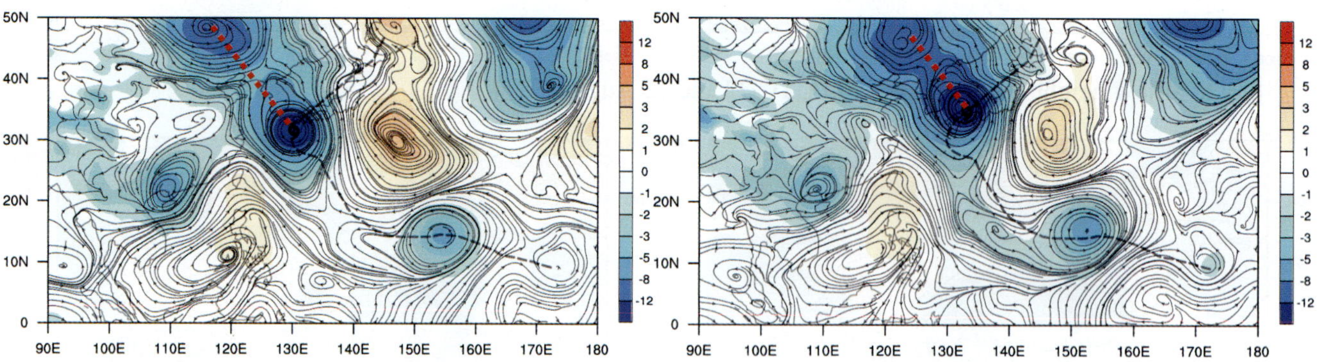

850 hPa flow/height anomalies at 0000 UTC and 1200 UTC 30 August 2004.

Figure 7.35. *(Continued)*

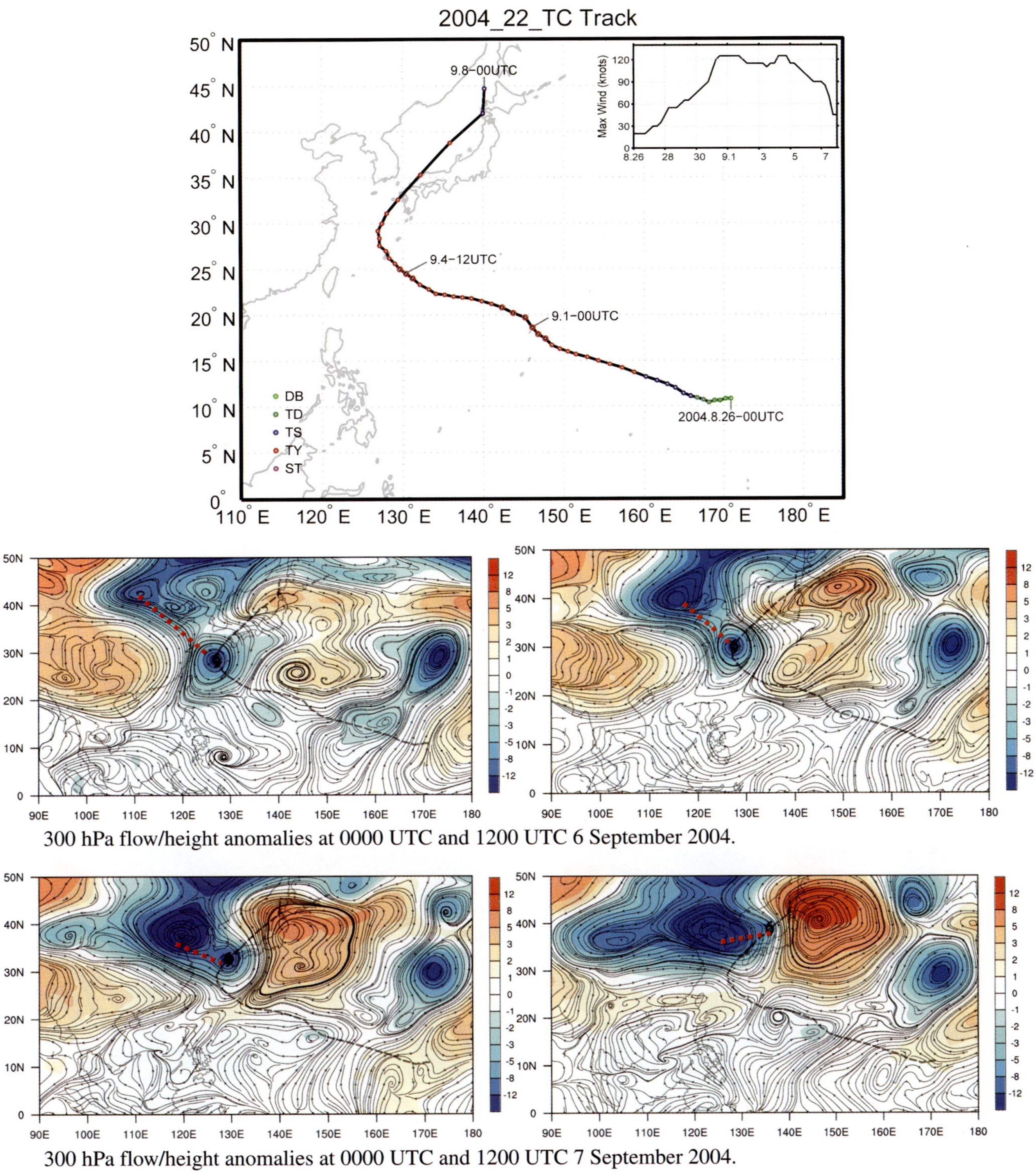

Figure 7.36. The typhoon (2004_No.22) generated in the tropical central Pacific on 26 August 2004 and moved northwestward along the climatic monsoon trough. Its intensity peaked in the Northwest Pacific on 1–4 September 2004. The right turn in the east seas of China was influenced by an anomalous vortex from the China continent at 300 hPa.

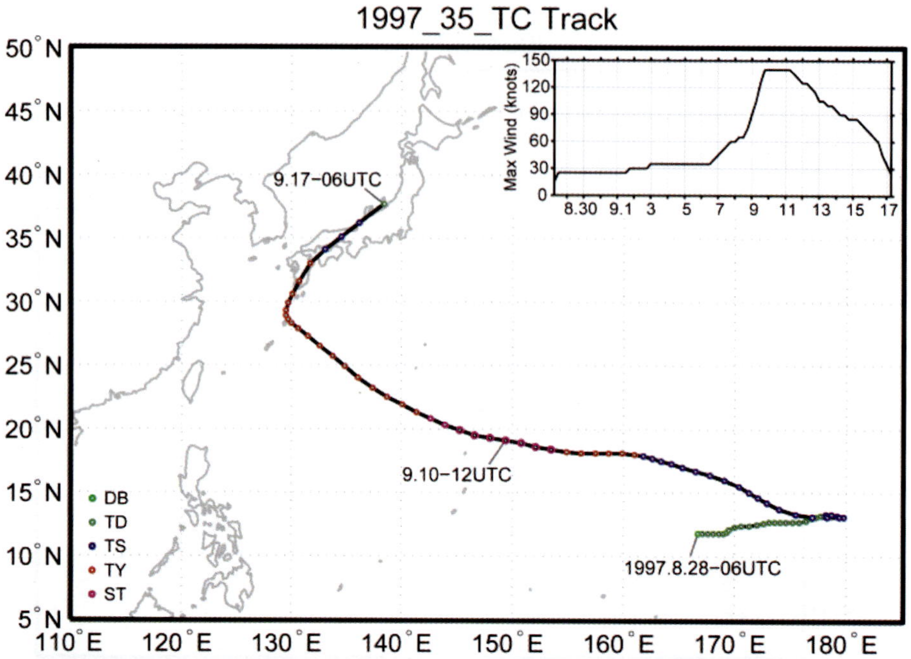

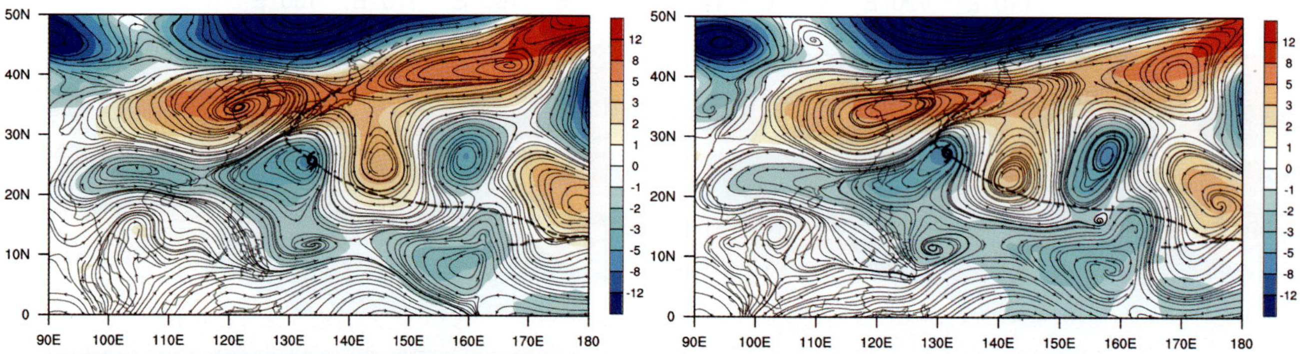

300 hPa flow/height anomalies at 0000 UTC and 1200 UTC 13 September 1997

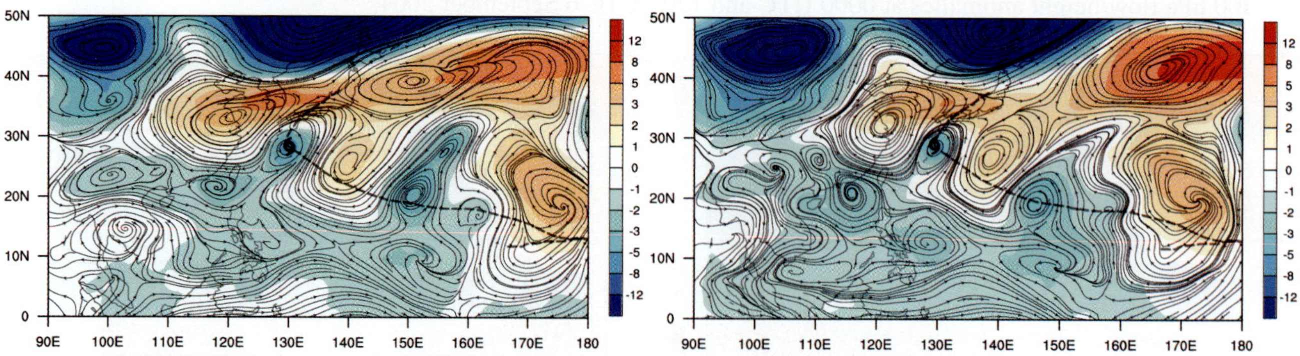

300 hPa flow/height anomalies at 0000 UTC and 1200 UTC 14 September 1997

Figure 7.37. The typhoon (1997_No.35) generated in the tropical central Pacific on 28 August 1997 and had a track basically along the climatic monsoon trough except a return motion at beginning. Its intensity peaked on 10 September 1997. Its right turn in the south of Japan was influenced by another vortex at 300 hPa.

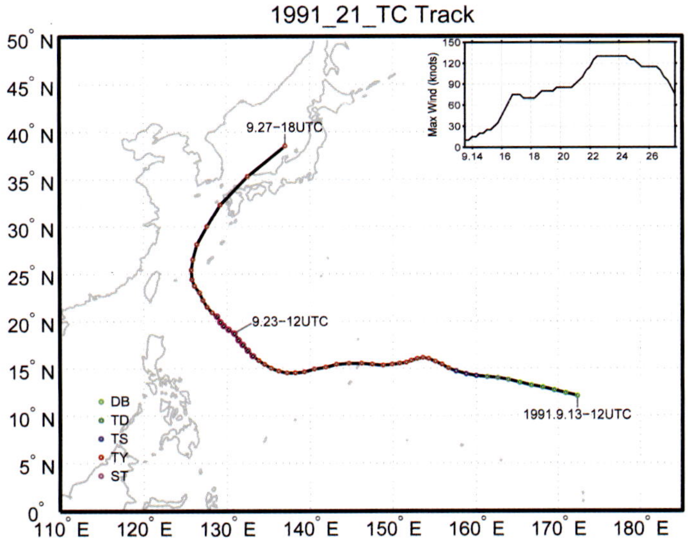

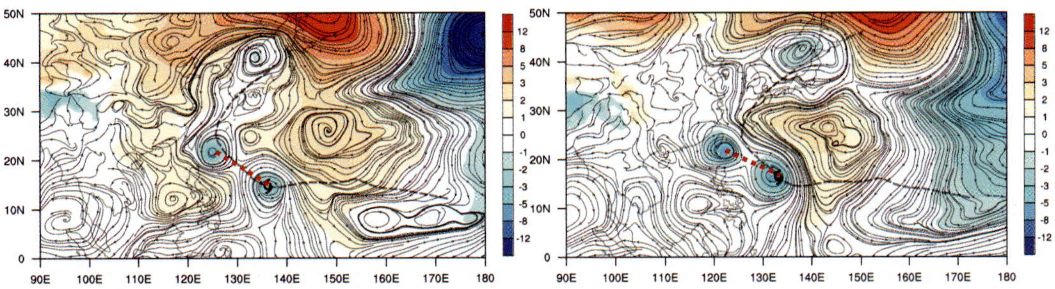

850 hPa flow/height anomalies at 1200 UTC 21 and 22 September 1991.

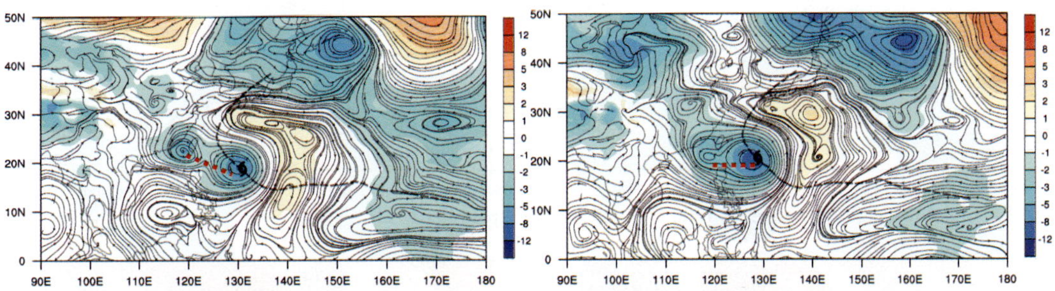

850 hPa flow/height anomalies at 1200 UTC 23 and 24 September 1991.

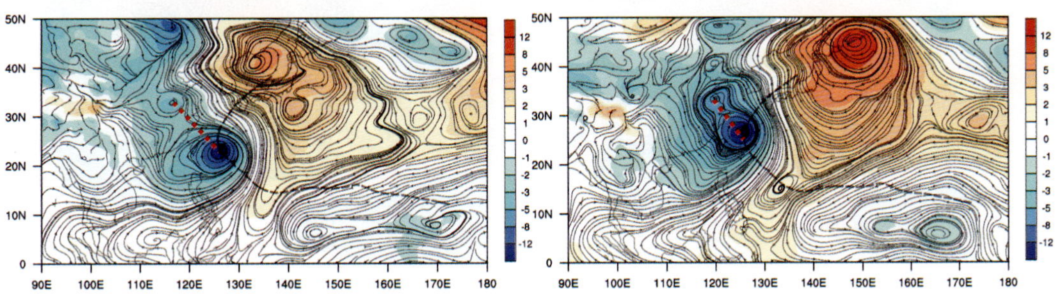

850 hPa flow/height anomalies at 1200 UTC 25 and 26 September 1991.

Figure 7.38. The typhoon (1991_No.21) generated in the tropical central western Pacific on 13 September 1991 and moved westward before turning right. In the western Pacific, its track was basically along the climatic monsoon trough and its intensity peaked in the northeast sea of the Philippines on 23 September 1991. The right turn in the east sea of the Taiwan Island was influenced by anomalous vortices at 850 hPa and 300 hPa.

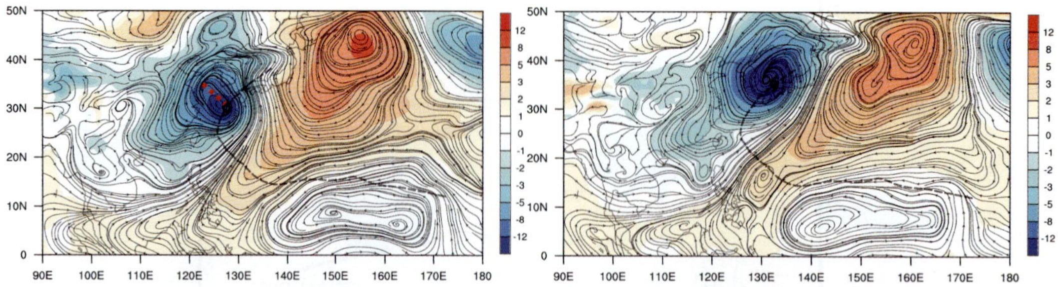

850 hPa flow/height anomalies at 0000 UTC and 1200 UTC 27 September 1991.

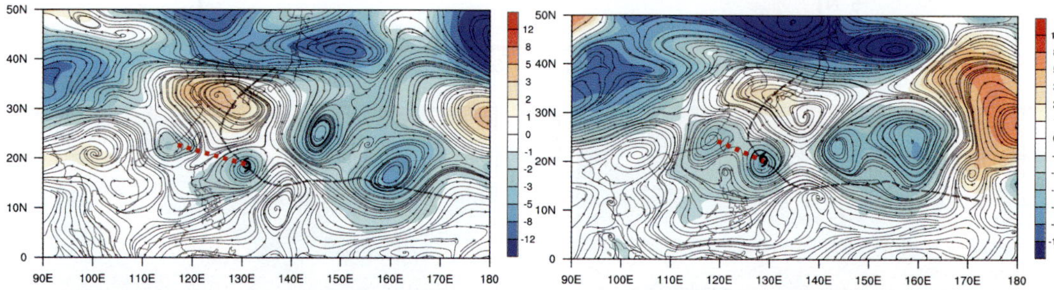

300 hPa flow/height anomalies at 1200 UTC 23 and 24 September 1991.

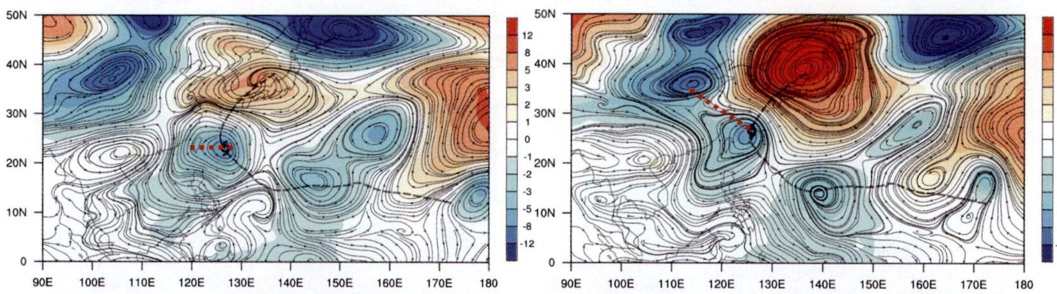

300 hPa flow/height anomalies at 1200 UTC 25 and 26 September 1991.

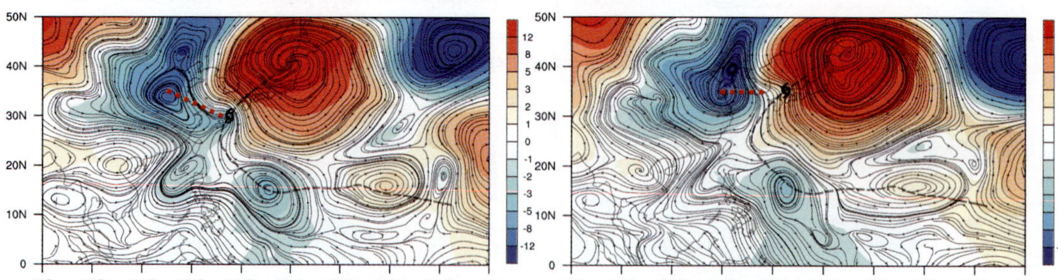

300 hPa flow/height anomalies at 0000 UTC and 1200 UTC 27 September 1991.

Figure 7.38. *(Continued)*

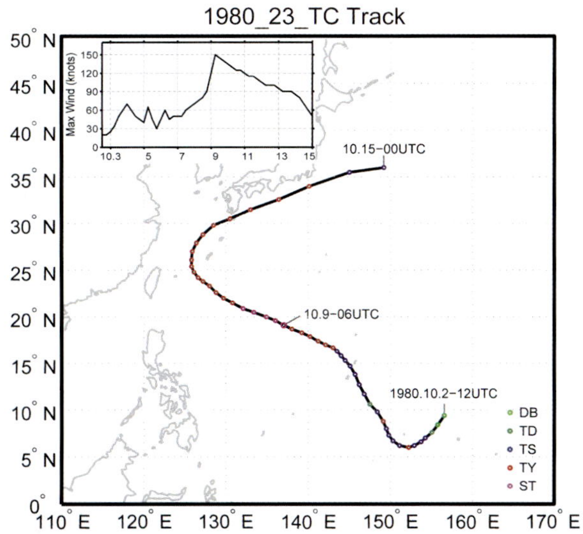

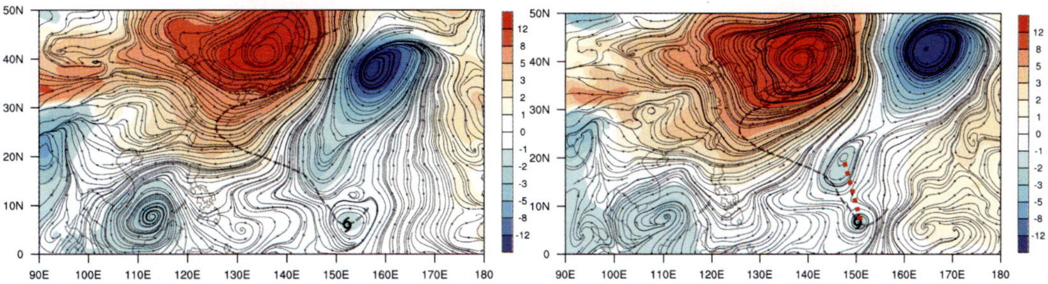

850 hPa flow/height anomalies at 0000 UTC and 1200 UTC 4 October 1980.

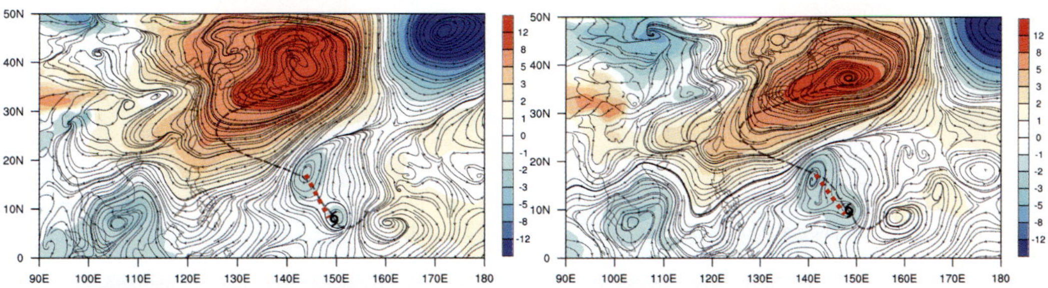

850 hPa flow/height anomalies at 0000 UTC and 1200 UTC 5 October 1980.

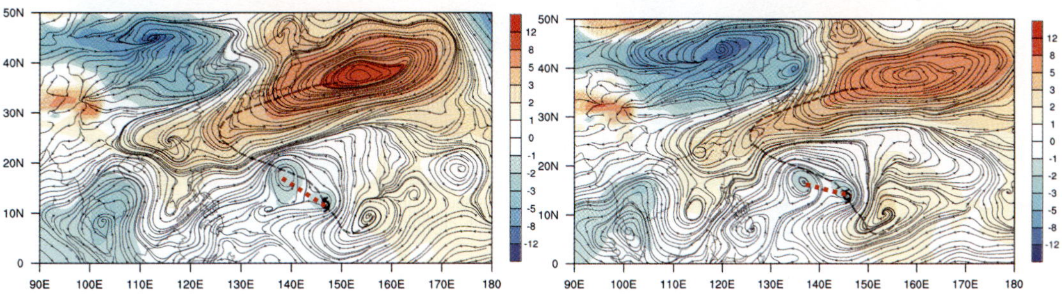

850 hPa flow/height anomalies at 0000 UTC and 1200 UTC 6 October 1980.

Figure 7.39. The typhoon (1980_No.23) generated in the equatorial central Pacific on 2 October 1980. After a right turn at the beginning stage, the typhoon moved along the climatic monsoon trough and its intensity peaked in the Northwest Pacific on 9 October 1980. The right turn in the east sea of China was influenced by anomalous vortices at 850 hPa and 300 hPa.

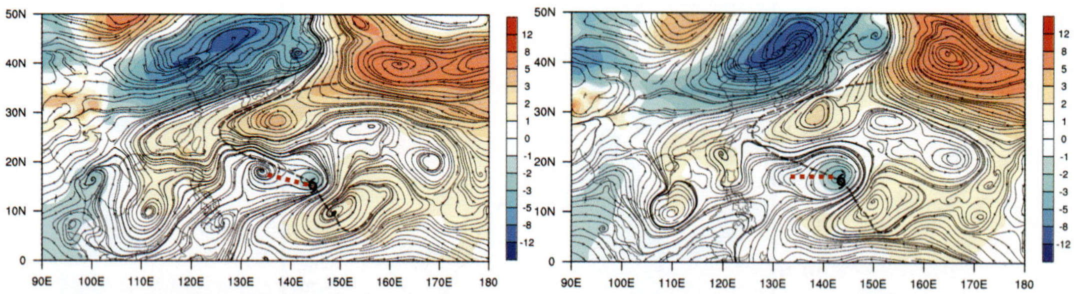

850 hPa flow/height anomalies at 0000 UTC and 1200 UTC 7 October 1980.

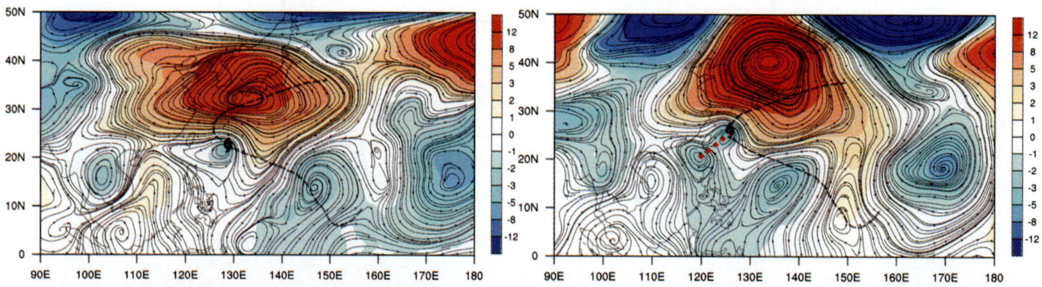

300 hPa flow/height anomalies at 0000 UTC and 1200 UTC 11 October 1980.

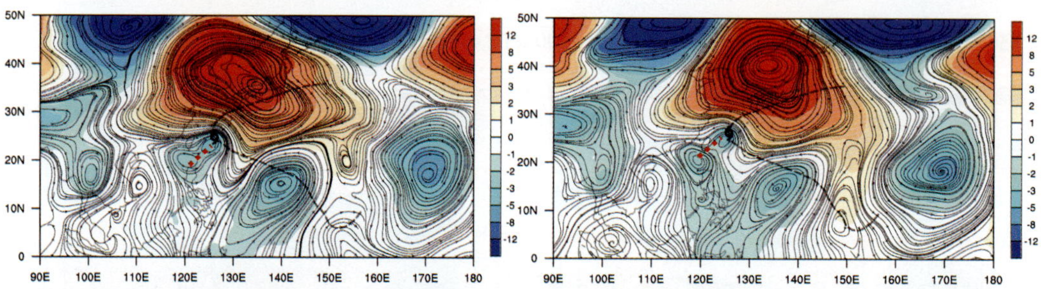

300 hPa flow/height anomalies at 0000 UTC and 1200 UTC 12 October 1980.

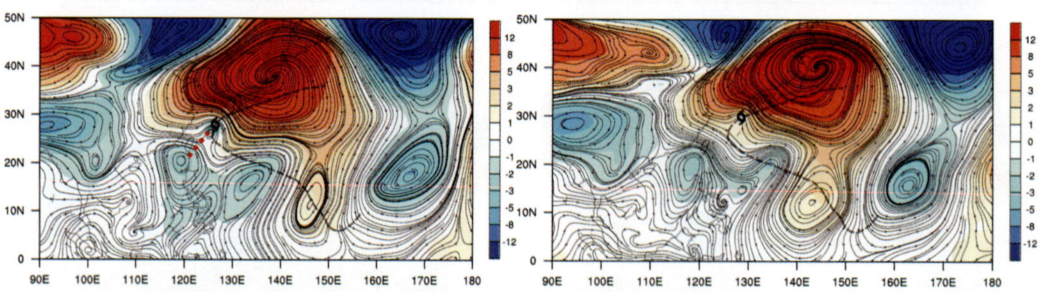

300 hPa flow/height anomalies at 0000 UTC and 1200 UTC 13 October 1980.

Figure 7.39. (*Continued*)

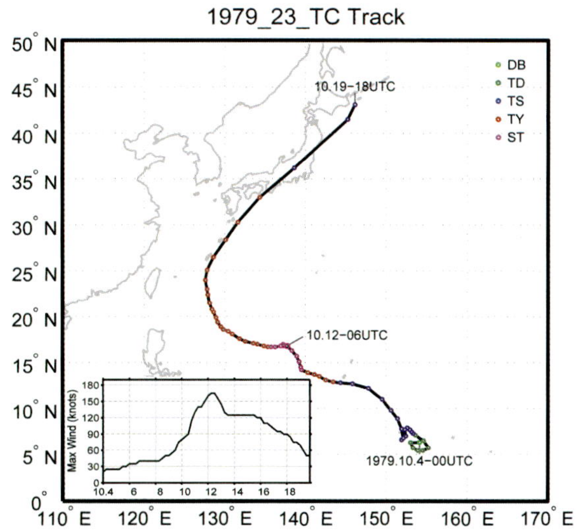

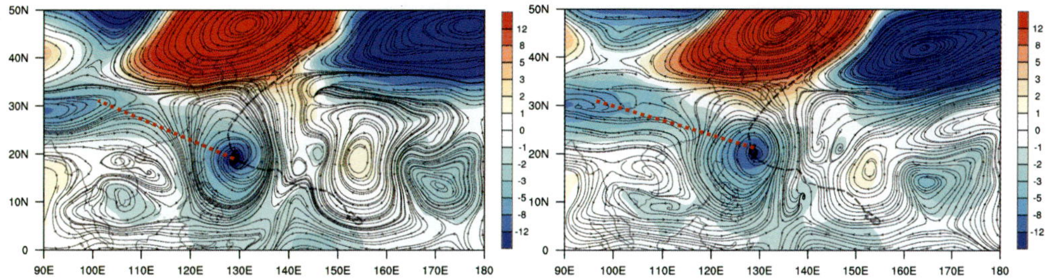

300 hPa flow/height anomalies at 0000 UTC and 1200 UTC 16 October 1979.

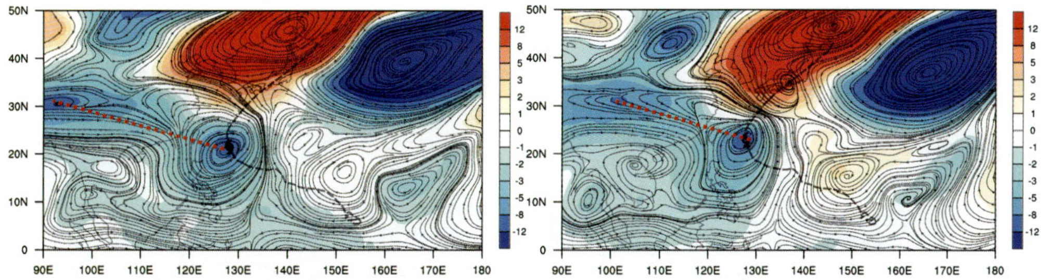

300 hPa flow/height anomalies at 0000 UTC and 1200 UTC 17 October 1979.

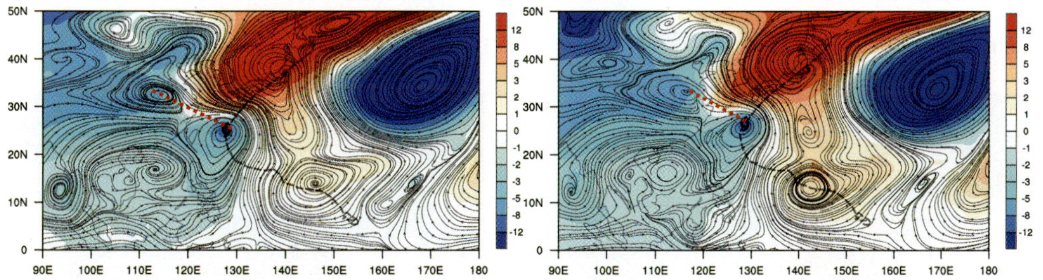

300 hPa flow/height anomalies at 0000 UTC and 0600 UTC 18 October 1979.

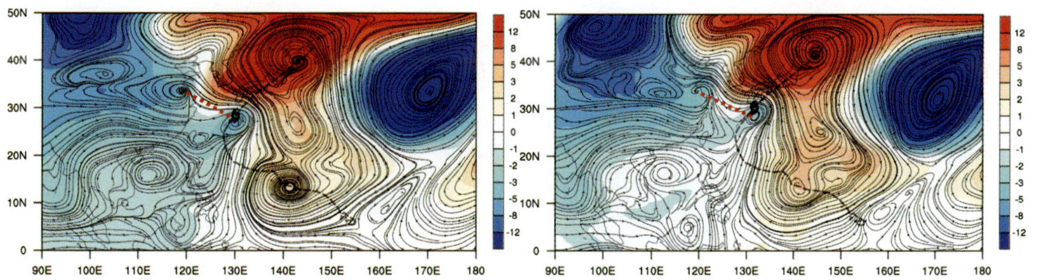

300 hPa flow/height anomalies at 1200 UTC and 1800 UTC 18 October 1979.

Figure 7.40. The typhoon (1979_No.23) generated in the equatorial central Pacific on 4 October 1979 and experienced a stagnant state for a few days then moved northwestward along the climatic monsoon trough. Its intensity peaked in the Northwest Pacific on 12 October 1979. It turned right in the Northwest Pacific due to the influence of anomalous vortices at 300 hPa.

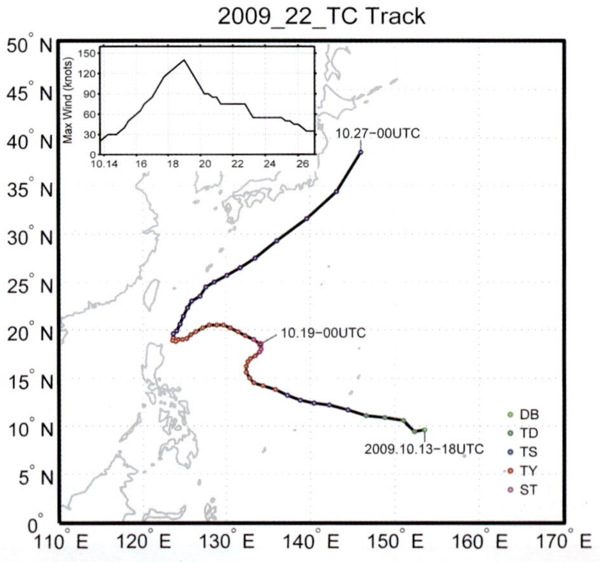

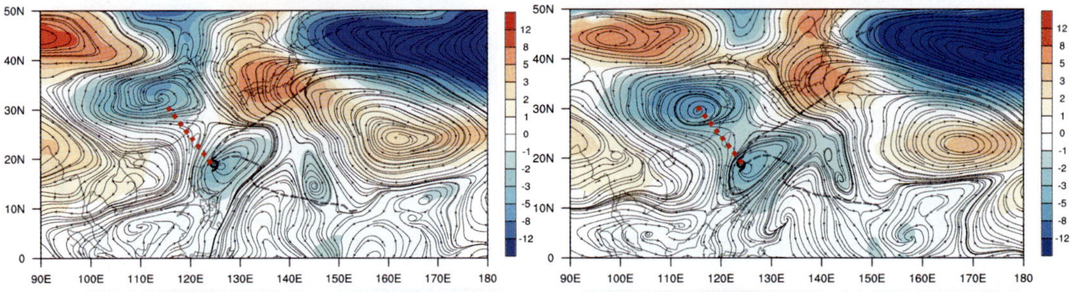

300 hPa flow/height anomalies at 0000 UTC and 1200 UTC 22 October 2009.

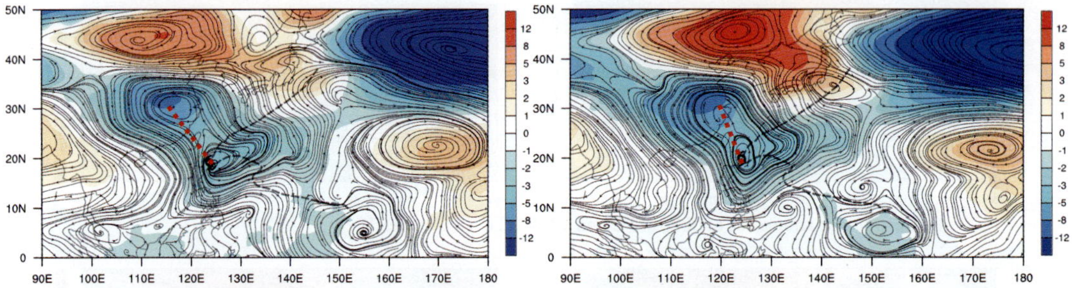

300 hPa flow/height anomalies at 0000 UTC and 1200 UTC 23 October 2009.

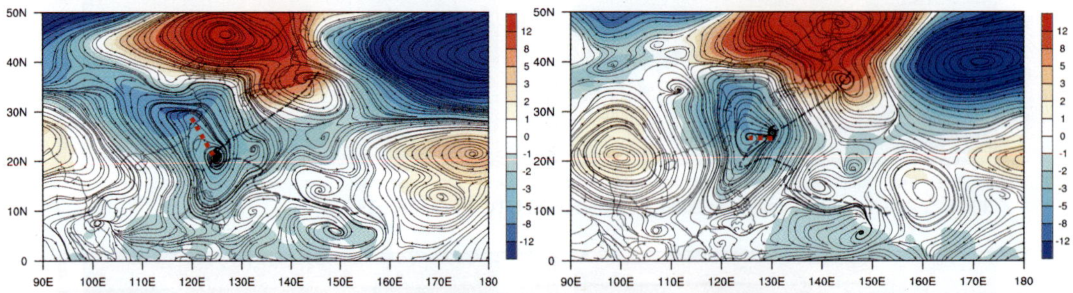

300 hPa flow/height anomalies at 0000 UTC and 1200 UTC 24 October 2009.

Figure 7.41. The typhoon (2009_No.22) generated in the equatorial central Pacific on 13 October 2009 and had a complex motion. Its intensity peaked in the Northwest Pacific on 19 October 2009 when it was approaching the climatic monsoon trough. This right turn near the Philippines was influenced by an anomalous vortex at 300 hPa.

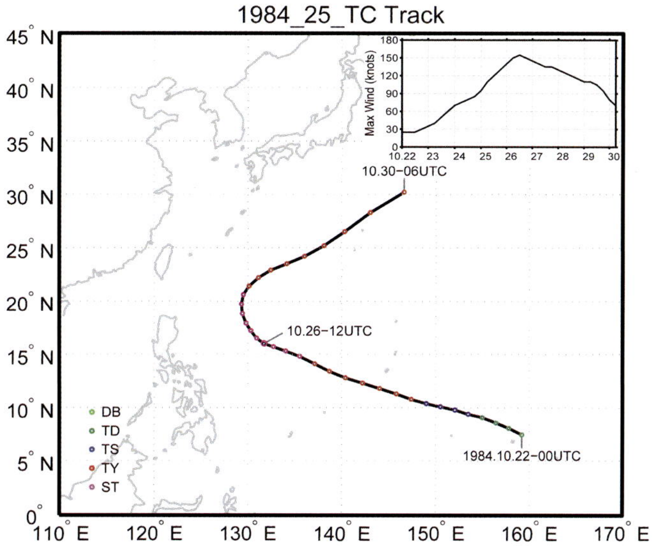

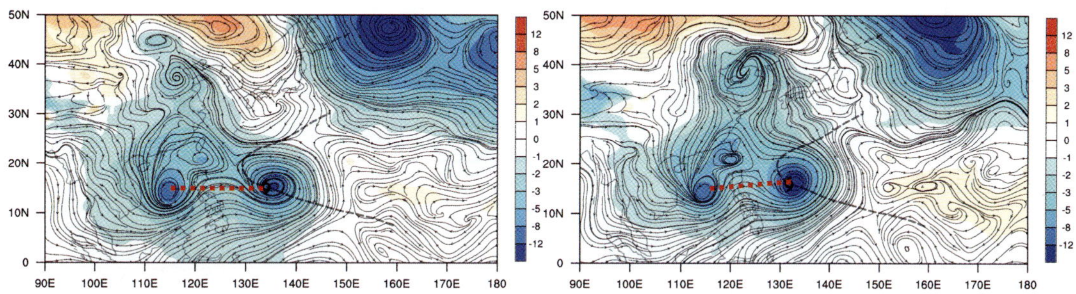

850 hPa flow/height anomalies at 0000 UTC and 1200 UTC 26 October 1984.

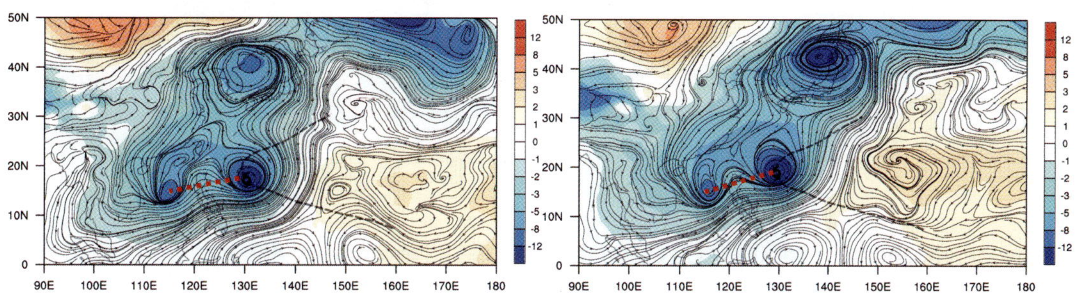

850 hPa flow/height anomalies at 0000 UTC and 1200 UTC 27 October 1984.

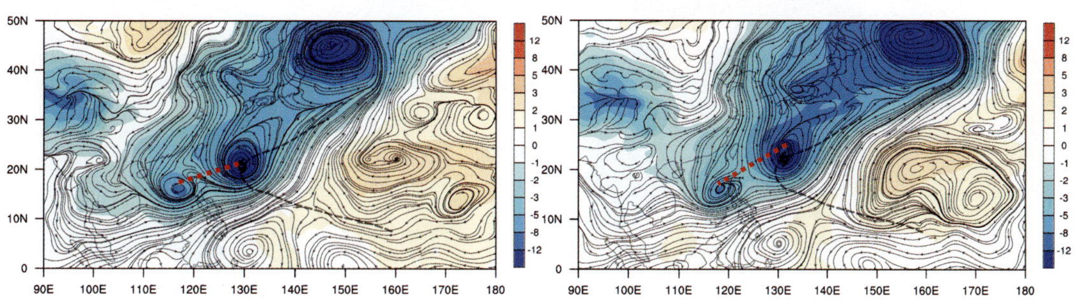

850 hPa flow/height anomalies at 0000 UTC and 1200 UTC 28 October 1984.

Figure 7.42. The typhoon (1984_No.25) generated in the equatorial central Pacific on 22 October 1984 and moved west-northwestward along the climatic monsoon trough. Its intensity peaked in the Northwest Pacific on 26 October 1984 before it turned right. This right turn was influenced by an anomalous vortex at 850 hPa.

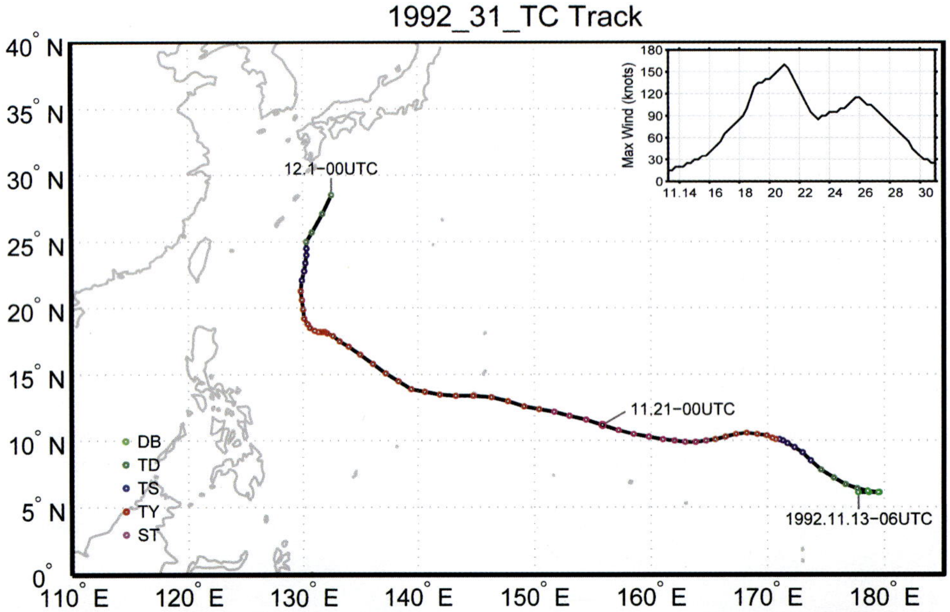

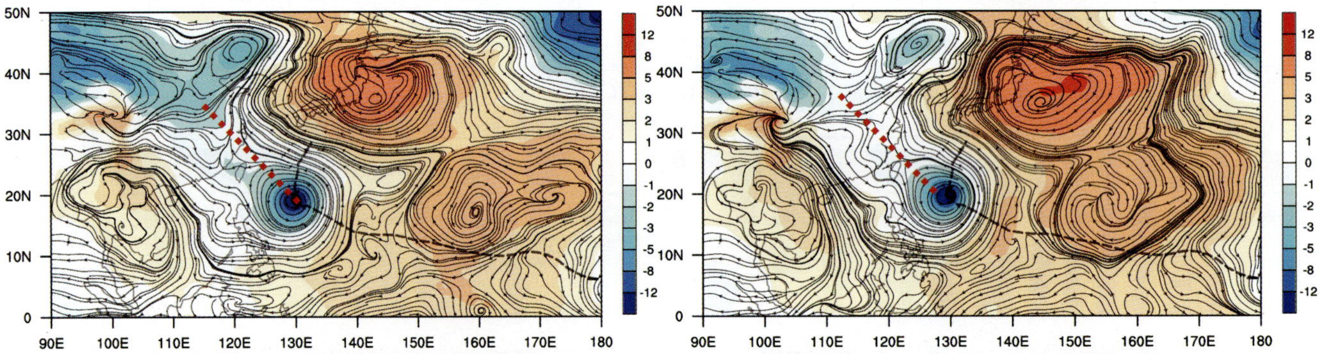

850 hPa flow/height anomalies at 0000 UTC and 0600 UTC 28 November 1992.

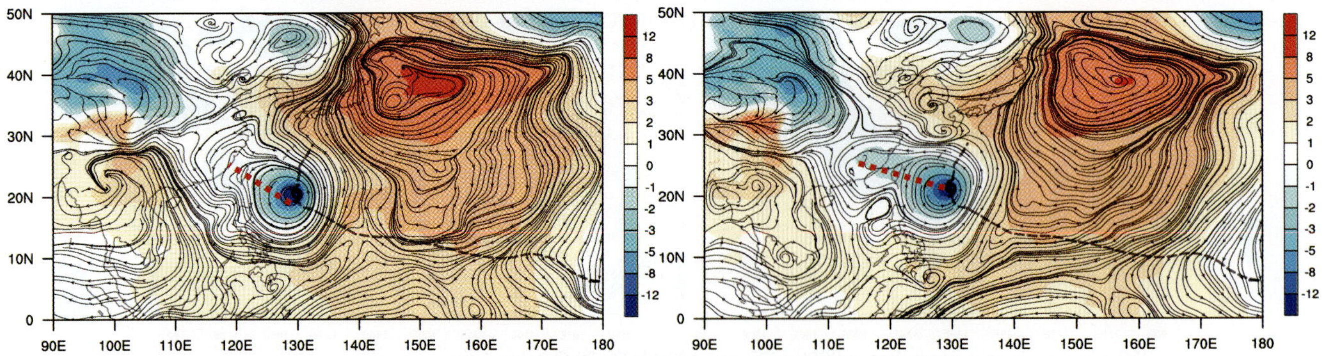

850 hPa flow/height anomalies at 1200 UTC and 1800 UTC 28 November 1992.

Figure 7.43. The typhoon (1992_No.31) generated in the equatorial central Pacific on 13 November 1992 and moved west-northwestward along the climatic monsoon trough. Its intensity peaked in the tropical central Pacific on 21 November 1992. The right turn in the south seas of Japan was influenced by an anomalous vortex at 850 hPa.

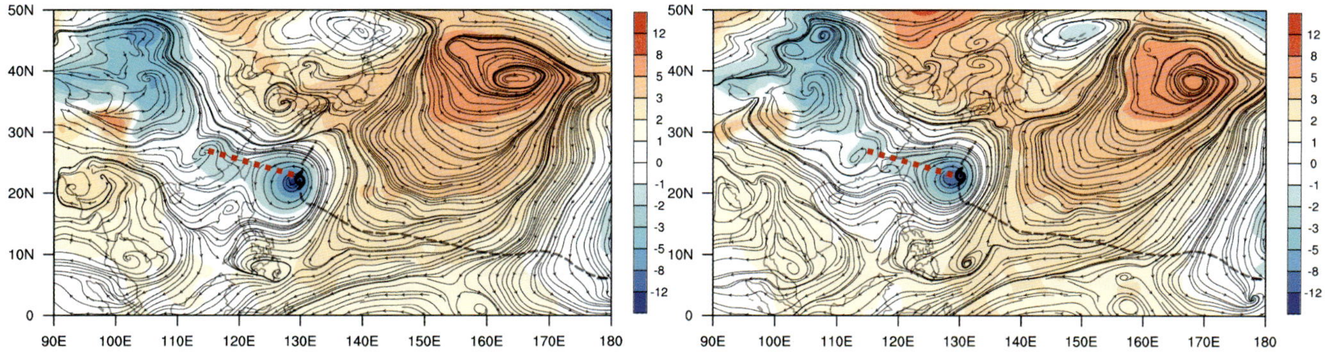

850 hPa flow/height anomalies at 0000 UTC and 0600 UTC 29 November 1992.

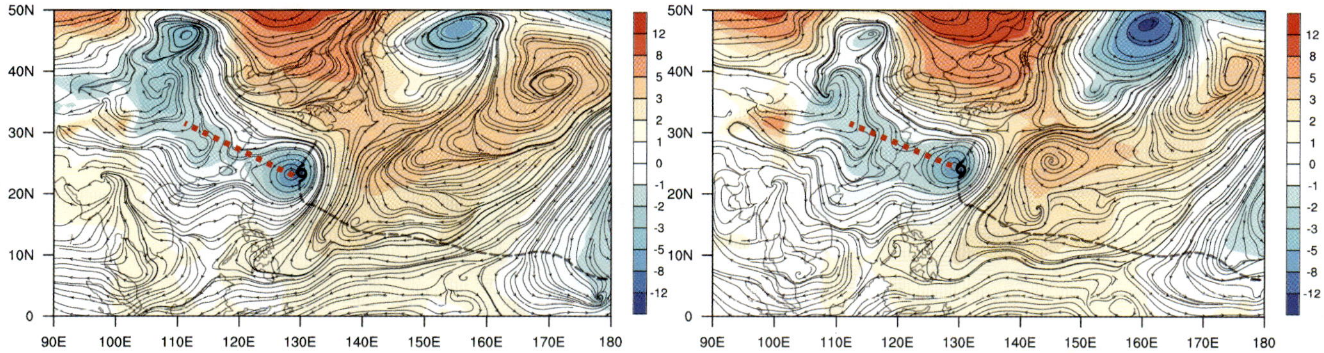

850 hPa flow/height anomalies at 1200 UTC and 1800 UTC 29 November 1992.

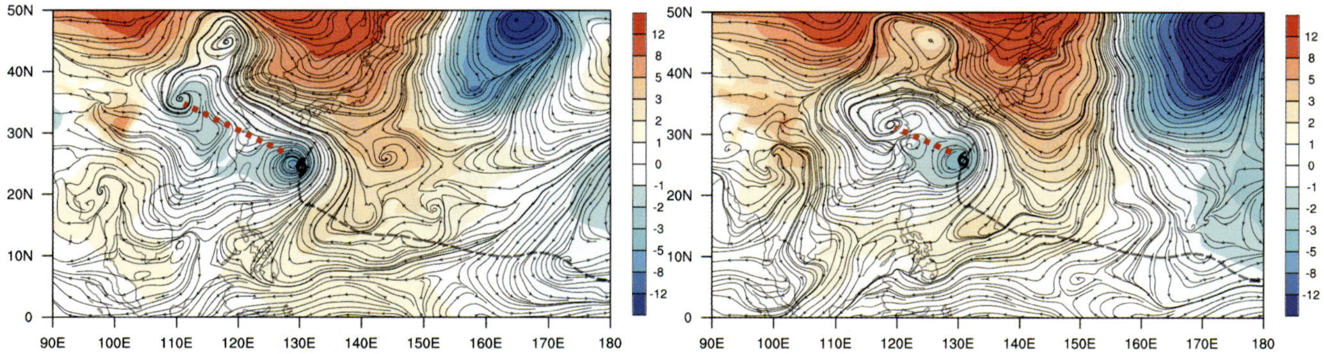

850 hPa flow/height anomalies at 0000 UTC and 1200 UTC 30 November 1992.

Figure 7.43. *(Continued)*

7.5. Twin Typhoons

There were 13 twin typhoons with complex motions due to the influence of anomalous vortices (Figs. 7.44–7.56).

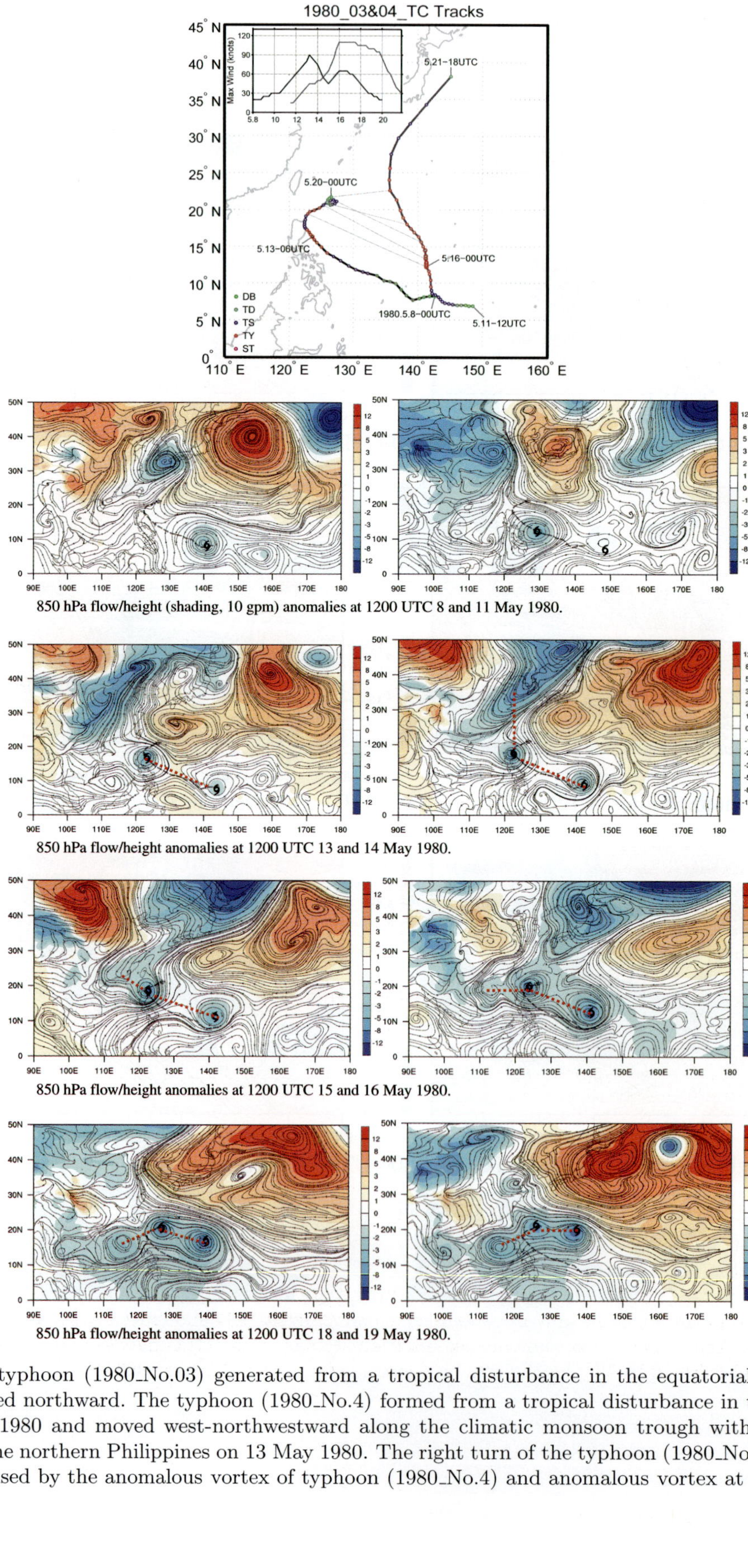

Figure 7.44. The typhoon (1980_No.03) generated from a tropical disturbance in the equatorial western Pacific on 8 May 1980 and moved northward. The typhoon (1980_No.4) formed from a tropical disturbance in the equatorial western Pacific on 11 May 1980 and moved west-northwestward along the climatic monsoon trough with its intensity peaking before landfall on the northern Philippines on 13 May 1980. The right turn of the typhoon (1980_No.03) near the northern Philippines was caused by the anomalous vortex of typhoon (1980_No.4) and anomalous vortex at 850 hPa.

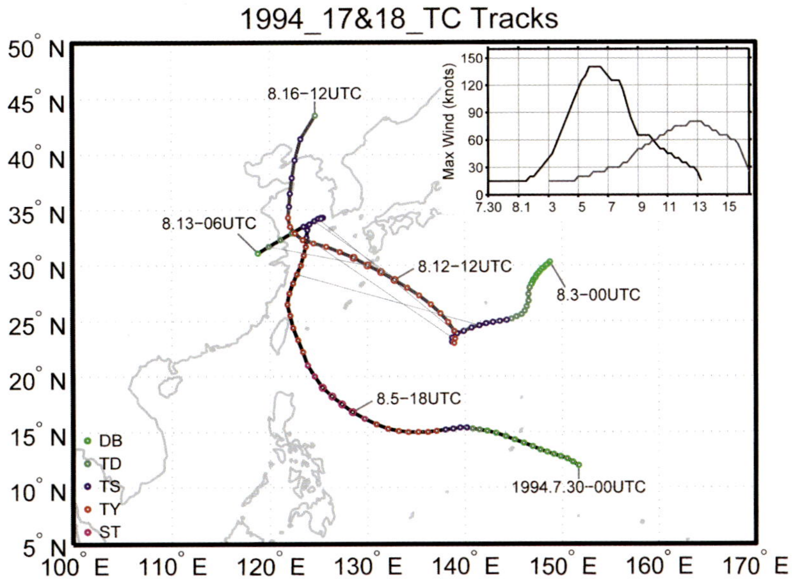

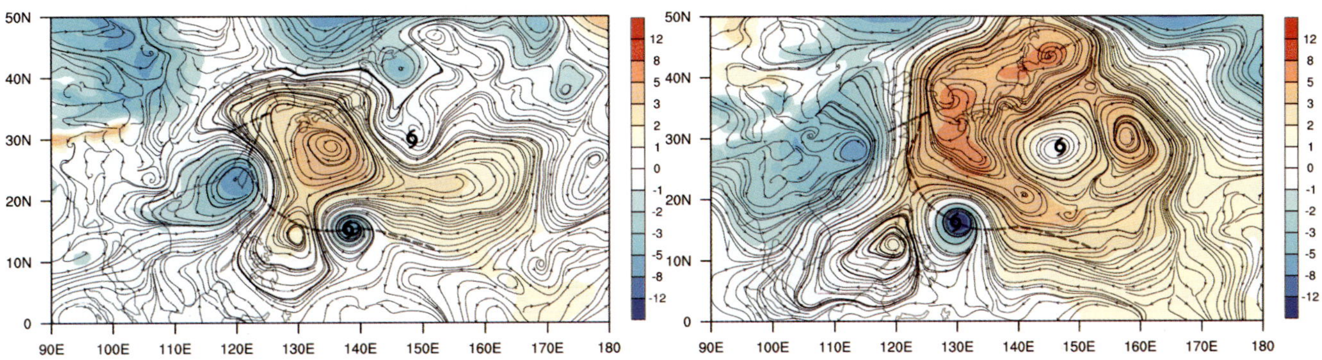

850 hPa flow/height anomalies at 1200 UTC 3 and 5 August 1994.

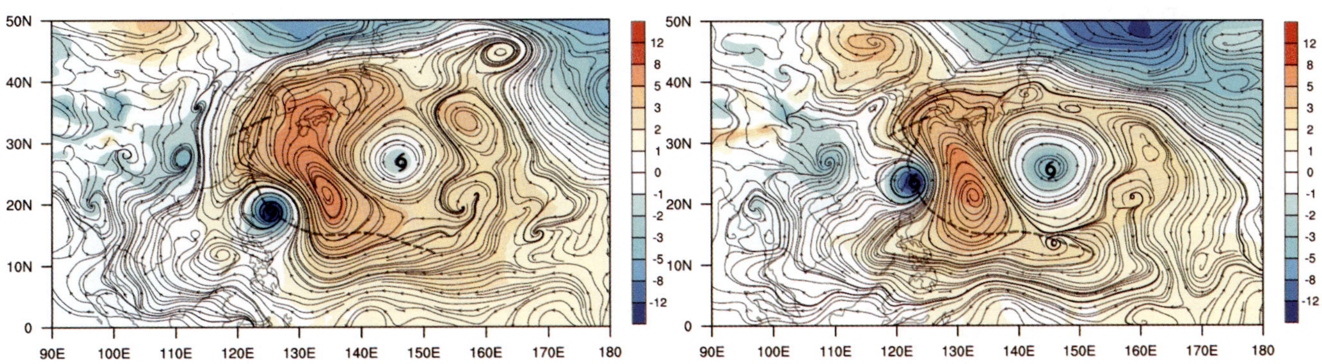

850 hPa flow/height anomalies at 1200 UTC 6 and 7 August 1994.

Figure 7.45. The typhoon (1994_No.17) generated from a tropical disturbance in the tropical central Pacific on 30 July 1994 and moved west-northwestward along the climatic monsoon trough with its intensity peaking in the northeast sea of the Philippines on 5 August 1994. Its right turn from the northeast sea of the Philippines to the East China Sea and a "T"-type track in the Yellow Sea were influenced by anomalous vortices at 850 hPa and 300 hPa. The typhoon (1994_No.18) generated in the subtropical Northwest Pacific on 3 August 1994 and moved westward with a stagnant state and a right turn in the Yellow Sea. Two typhoons experienced a binary interaction between two anomalous vortices that can be clearly observed from 10 to 13 August 1994 by anomalous synoptic charts at 850 hPa.

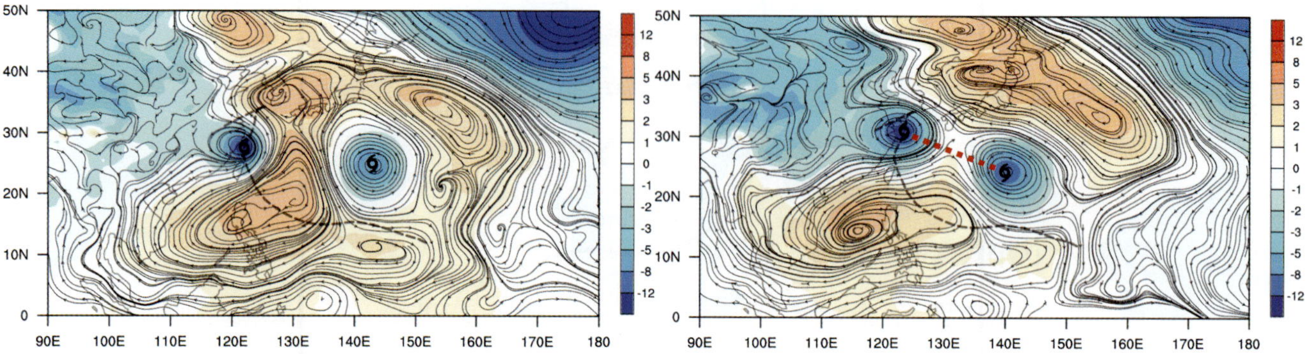

850 hPa flow/height anomalies at 1200 UTC 8 and 9 August 1994.

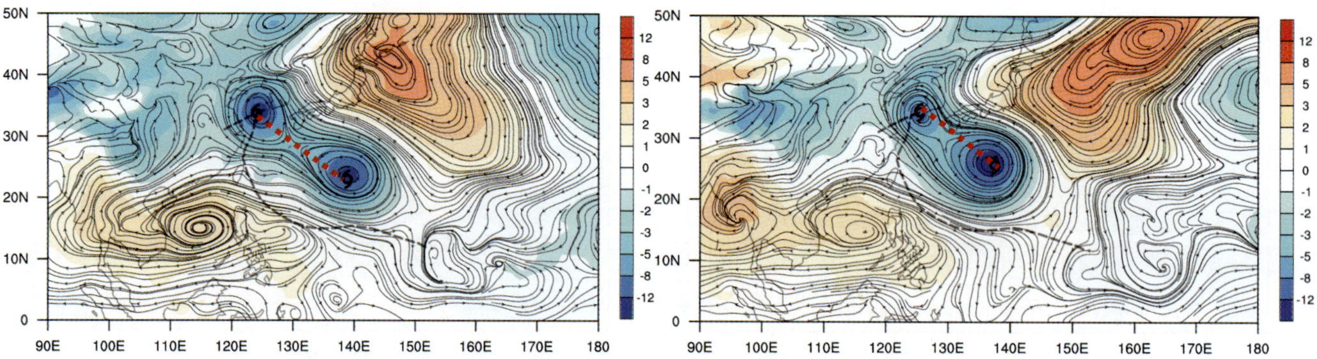

850 hPa flow/height anomalies at 1200 UTC 10 and 11 August 1994.

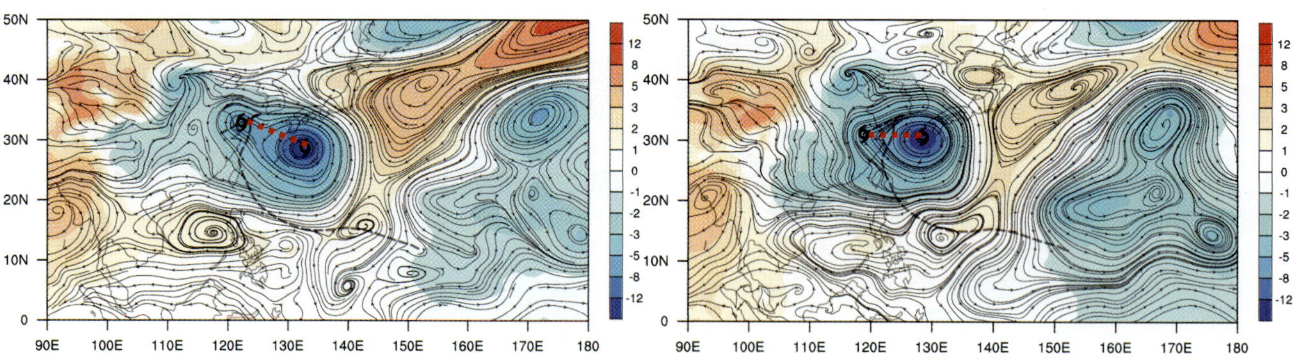

850 hPa flow/height anomalies at 1200 UTC 12 August 1994 and 0600 UTC 13 August 1994.

Figure 7.45. (*Continued*)

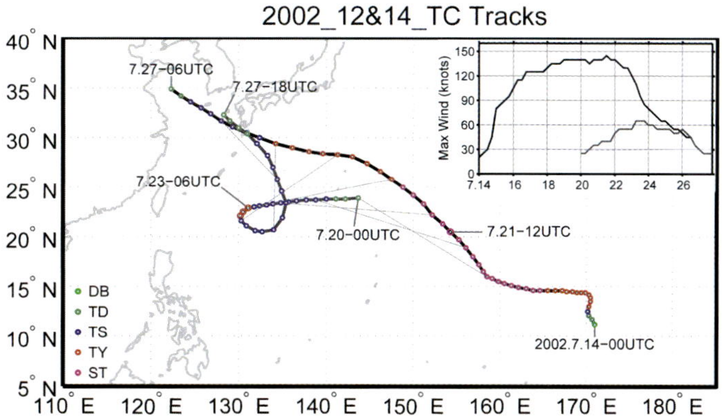

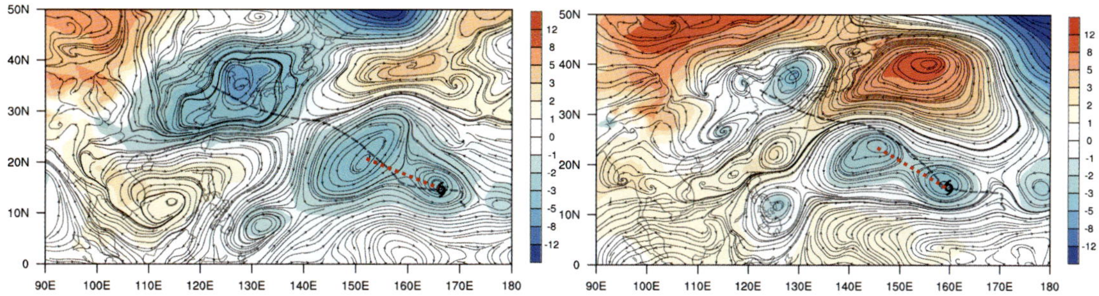

850 hPa flow/height anomalies at 1200 UTC 17 and 19 July 2002.

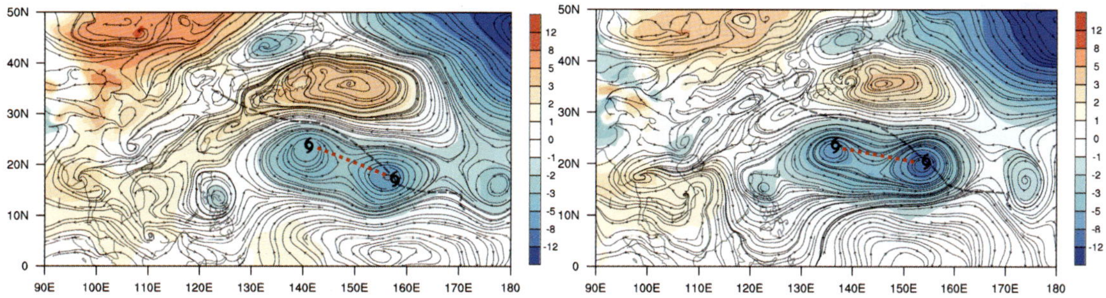

850 hPa flow/height anomalies at 1200 UTC 20 and 21 July 2002.

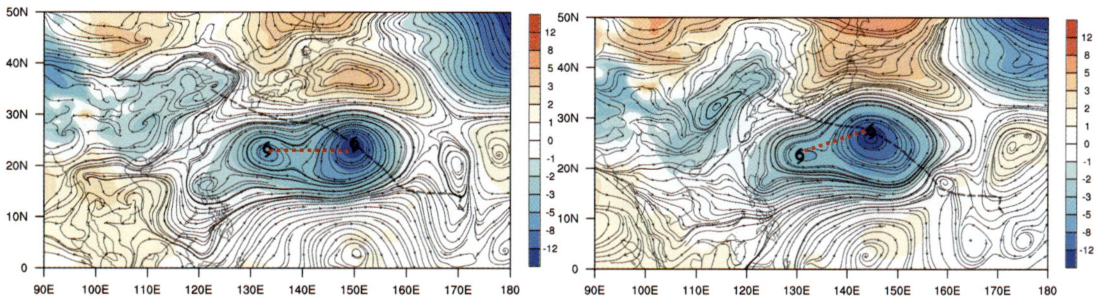

850 hPa flow/height anomalies at 1200 UTC 22 and 23 July 2002.

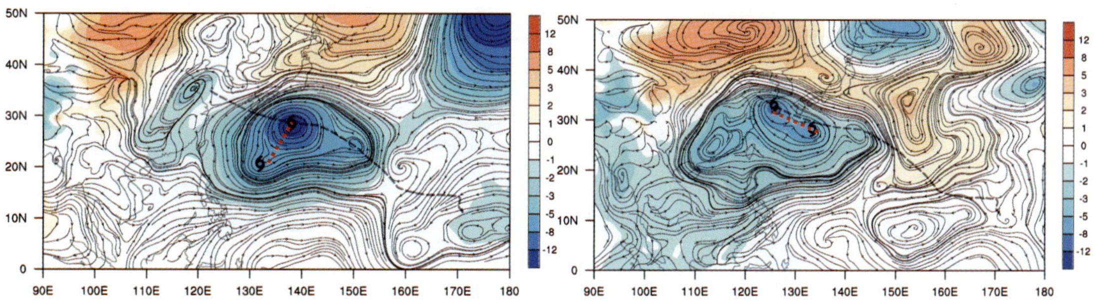

850 hPa flow/height anomalies at 1200 UTC 24 and 26 July 2002.

Figure 7.46. The typhoon (2002_No.12) generated in the tropical central Pacific on 14 July 2002 and moved northwestward. The typhoon (2002_No.14) generated in the subtropical Northwest Pacific on 20 July 2002 and had a local circular motion with its peak intensity near the climatic monsoon trough in the subtropical Northwest Pacific on 23 July 2002. The two typhoons experienced a binary interaction between two anomalous vortices that can be clearly observed from 20 to 26 July 2002 by anomalous synoptic charts at 850 hPa.

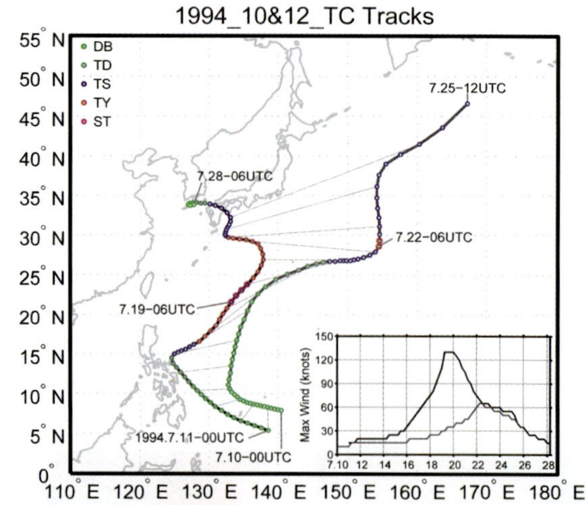

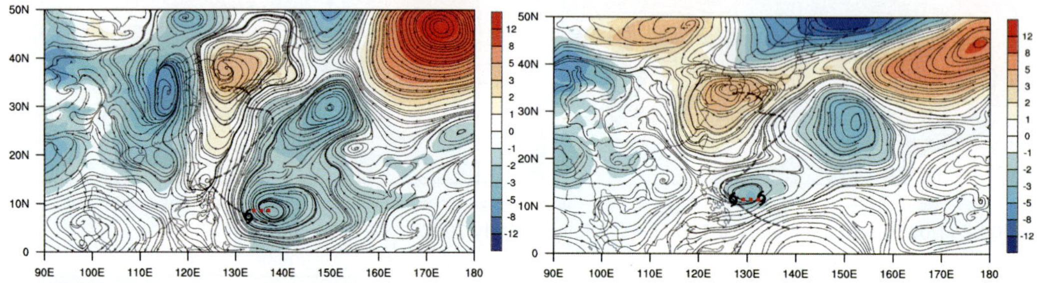

850 hPa flow/height anomalies at 1200 UTC 12 and 14 July 1994.

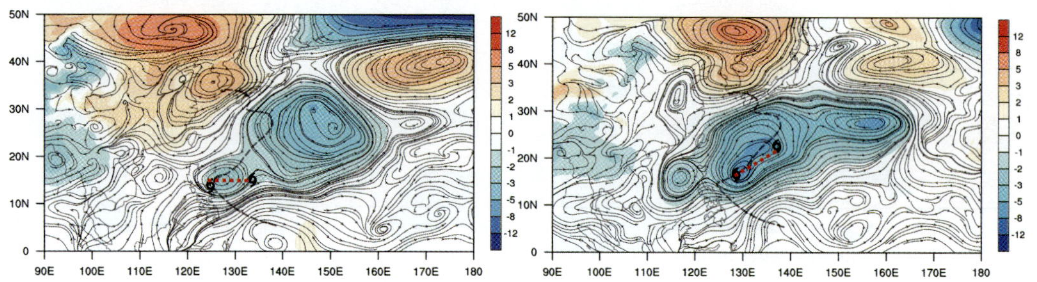

850 hPa flow/height anomalies at 1200 UTC 15 and 17 July 1994.

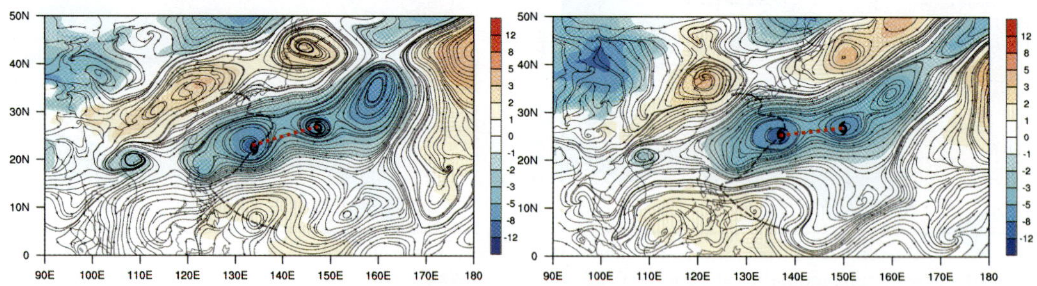

850 hPa flow/height anomalies at 1200 UTC 19 and 20 July 1994.

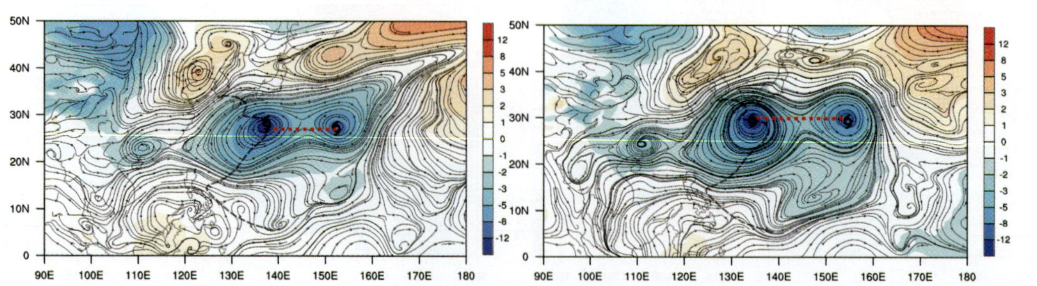

850 hPa flow/height anomalies at 1200 UTC 21 and 22 July 1994.

Figure 7.47. Two typhoons (1994_No.10 and 1994_No.12) generated in the equatorial western Pacific on 10 and 11 July 1994, respectively, and moved northward closely. The two typhoons experienced a binary interaction between two anomalous vortices that can be clearly observed from 12 to 22 July 1994 at 850 hPa.

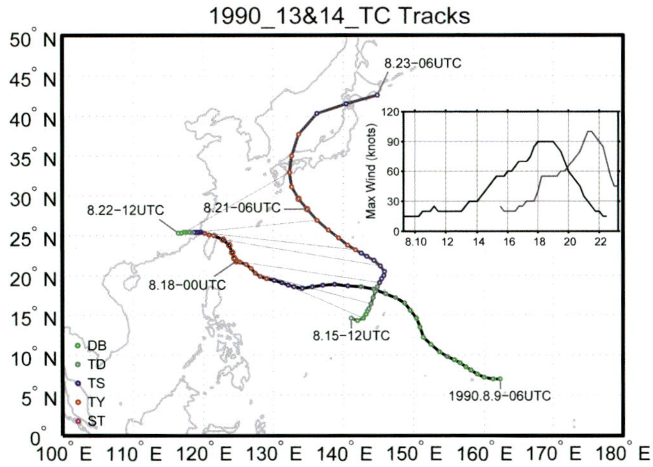

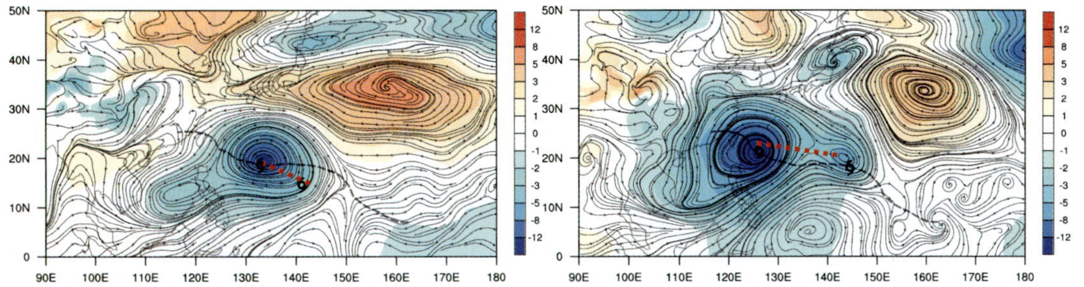

850 hPa flow/height anomalies at 1200 UTC 15 and 17 August 1990.

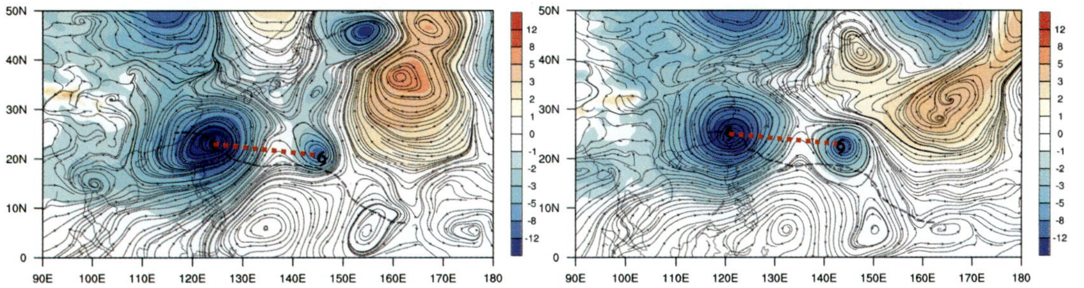

850 hPa flow/height anomalies at 1200 UTC 18 and 19 August 1990.

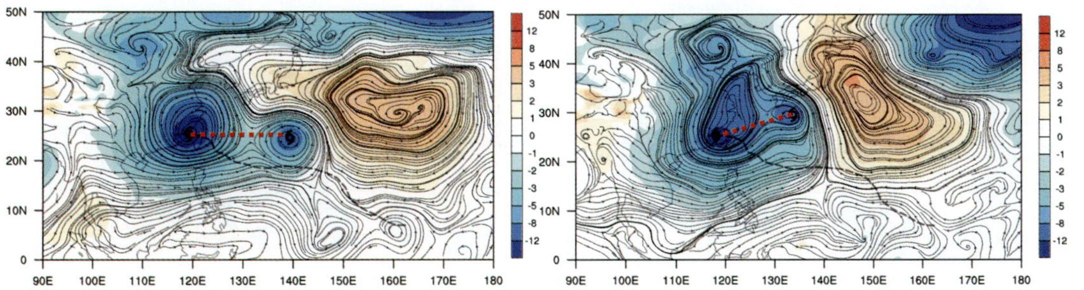

850 hPa flow/height anomalies at 1200 UTC 20 and 21 August 1990.

Figure 7.48. Two typhoons (1990_No. 13 and 14) formed from tropical disturbances on 9 and 15 August 1990 in the equatorial central Pacific and tropical western Pacific, respectively. The typhoon (1990_No.13) moved northwestward, without changing its direction, along the climatic monsoon trough and its intensity peaked in the east sea of the Taiwan Island on 18 August 1990 before landfall on the island. Two typhoons experienced a binary interaction between two anomalous vortices that can be clearly observed on 15–21 August 1990 at 850 hPa.

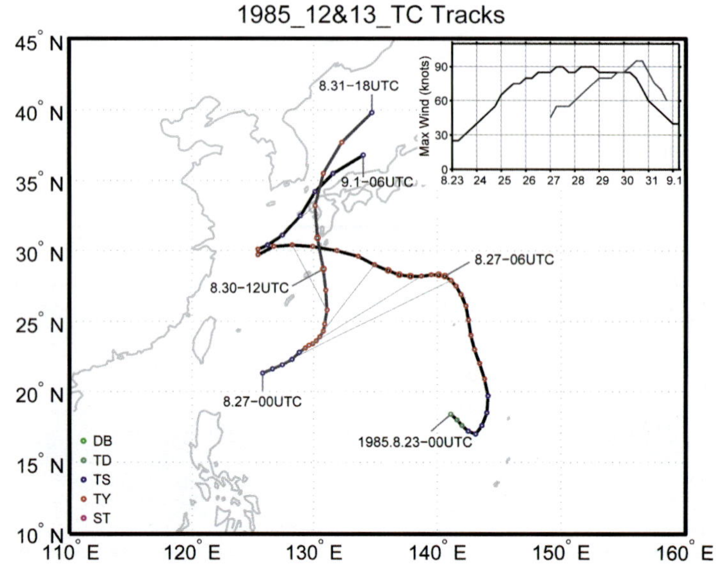

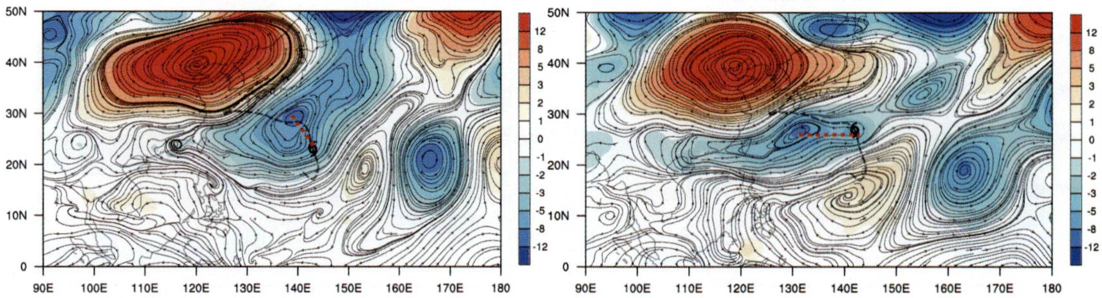

300 hPa flow/height anomalies at 1200 UTC 25 and 26 August 1985.

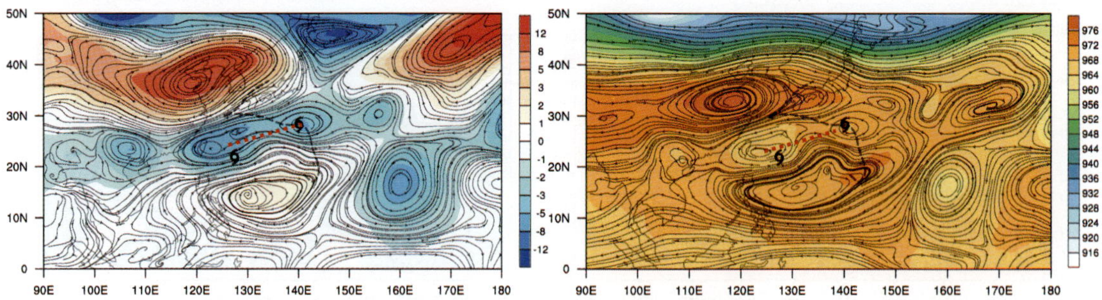

300 hPa flow/height anomalies and total flow/height at 1200 UTC 27 August 1985.

Figure 7.49. The typhoon (1985_No.12) generated in the tropical western Pacific on 23 August 1985 and the typhoon (1985_No.13) generated in the southeast sea of the Taiwan Island on 27 August 1985. The intensity of the typhoon (1985_No.13) peaked in southern Japan when it was approaching the climatic monsoon trough on 30 August 1985. Two typhoons experienced a binary interaction and their motions were also influenced by other anomalous vortices at 300 hPa and 850 hPa.

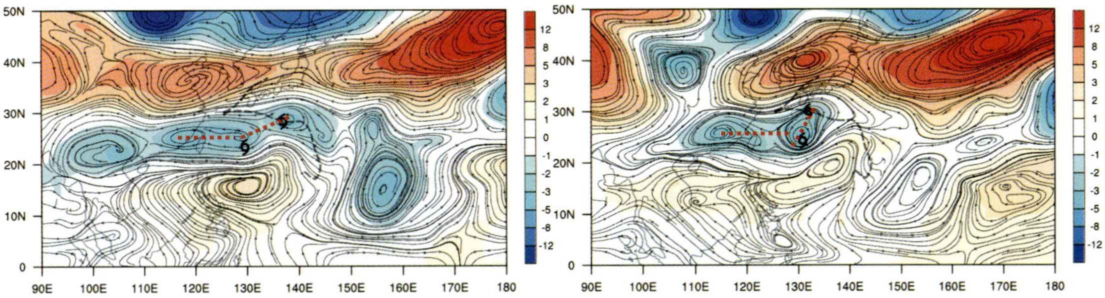

300 hPa flow/height anomalies at 1200 UTC 28 and 29 August 1985.

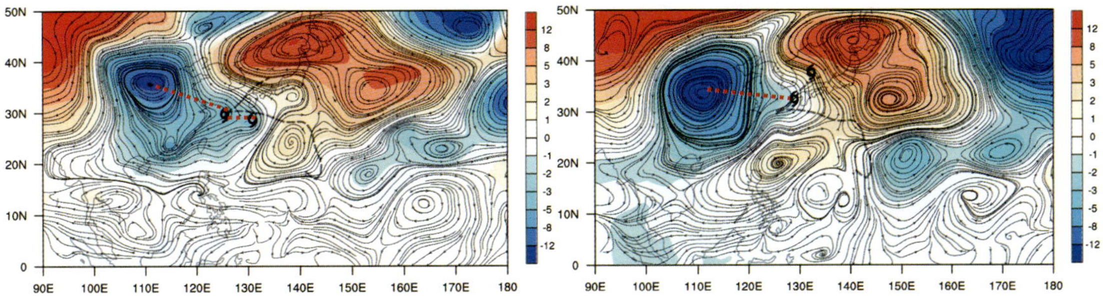

300 hPa flow/height anomalies at 1200 UTC 30 and 31 August 1985.

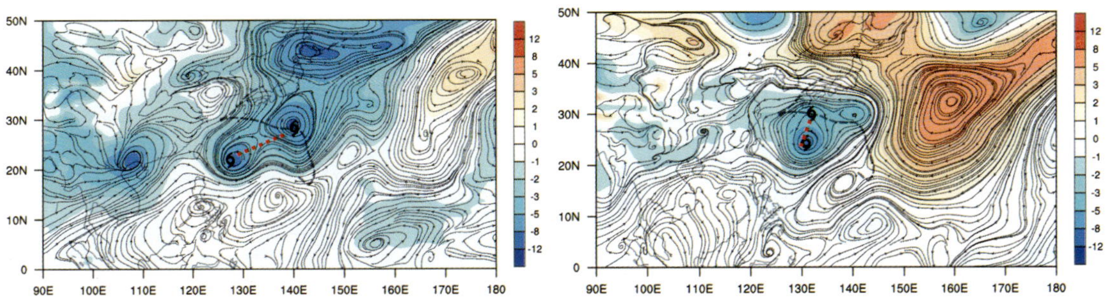

850 hPa flow/height anomalies at 1200 UTC 27 and 29 August 1985.

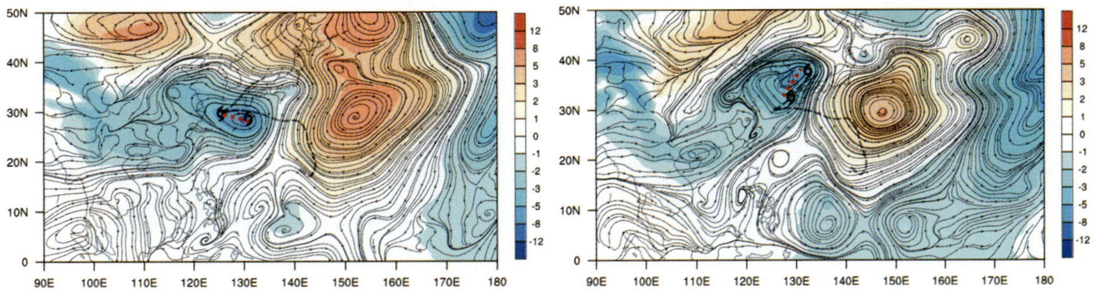

850 hPa flow/height anomalies at 1200 UTC 30 and 31 August 1985.

Figure 7.49. *(Continued)*

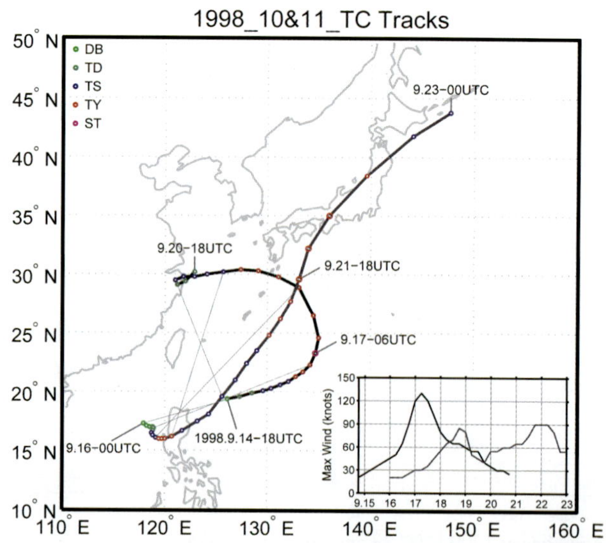

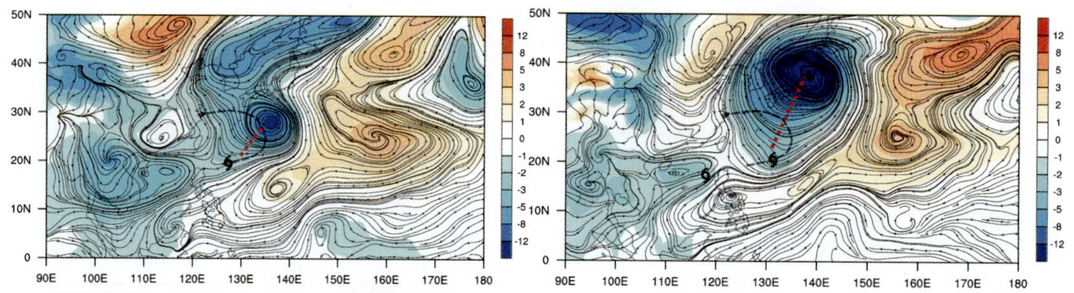

850 hPa flow/height anomalies at 0000 UTC 15 and 16 September 1998.

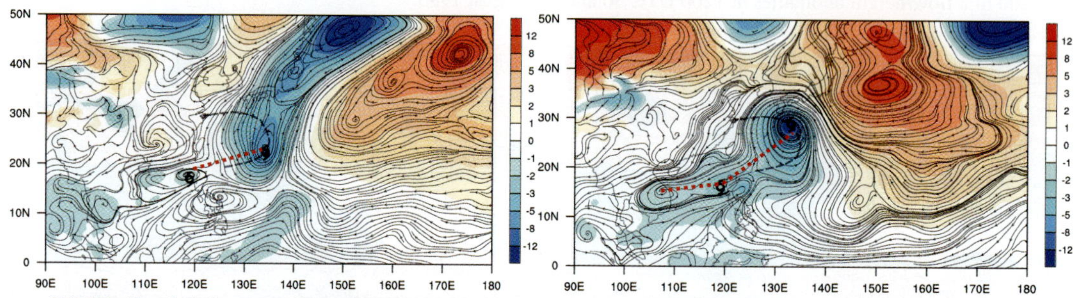

850 hPa flow/height anomalies at 0000 UTC 17 and 18 September 1998.

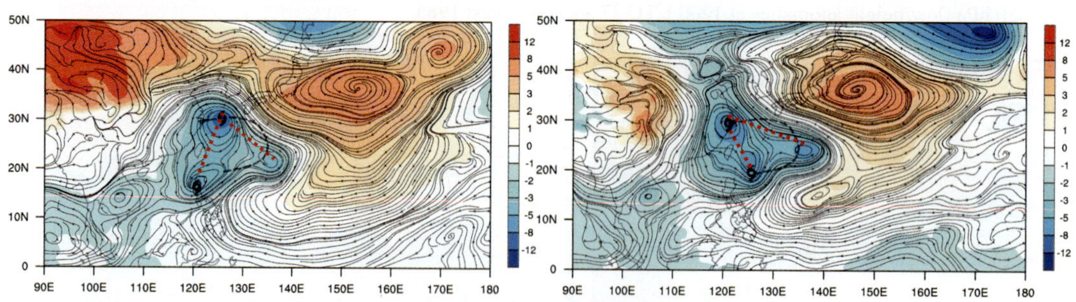

850 hPa flow/height anomalies at 0000 UTC 19 and 20 September 1998.

Figure 7.50. Two typhoons (1998_No. 10 and 11) generated in the northeast sea of the Philippines on 14 September 1998 and the eastern SCS on 16 September 1998, respectively. Their anomalous vortices interacted from 17 to 20 September 1998 and were influenced by other anomalous vortices during their lifetimes at 850 hPa. After landfall in Zhejiang Province in the East China coast on 19 September 1998, the typhoon (1998_No.10) returned to the East China Sea again on 20 September 1998 due to the influence of other anomalous vortices at 850 hPa.

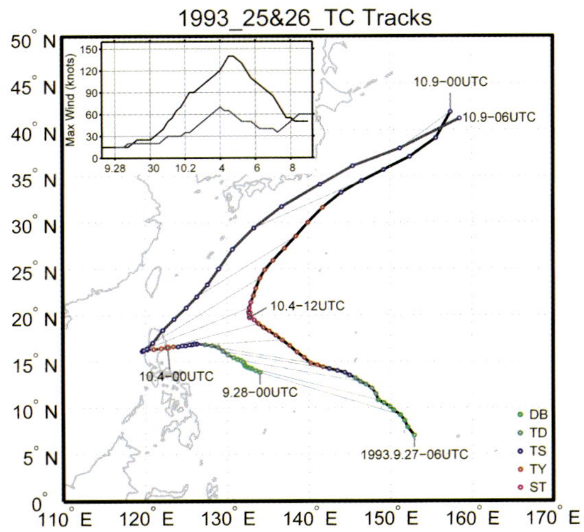

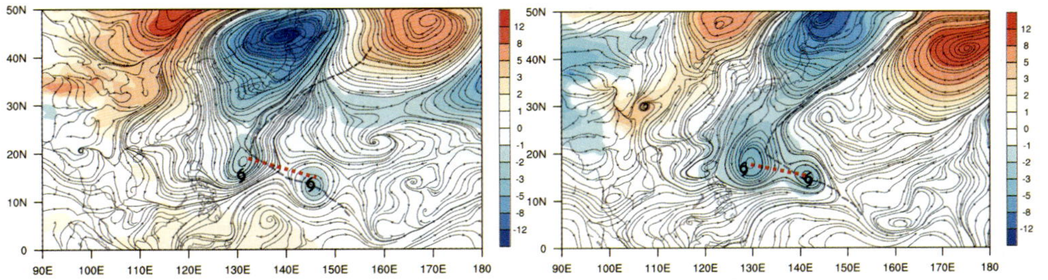

850 hPa flow/height anomalies at 1200 UTC 30 September and 1 October 1993.

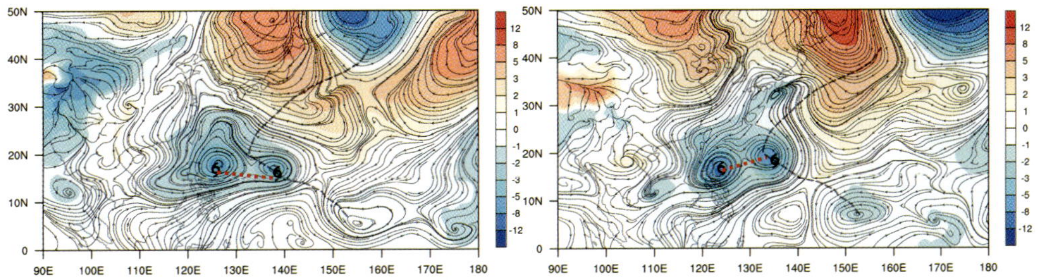

850 hPa flow/height anomalies at 1200 UTC 2 and 3 October 1993.

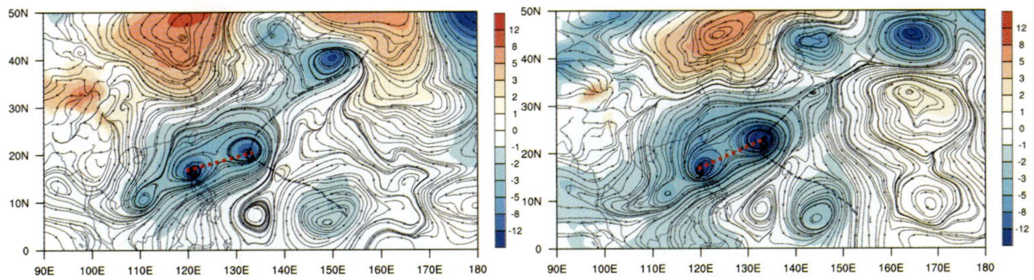

850 hPa flow/height anomalies at 1200 UTC 4 and 5 October 1993.

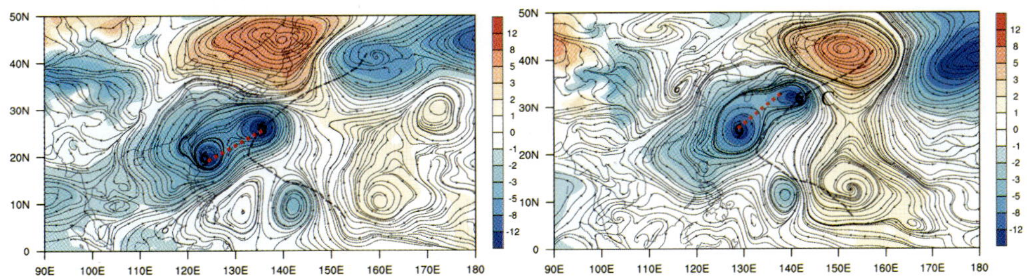

850 hPa flow/height anomalies at 1200 UTC 6 and 7 October 1993.

Figure 7.51. Two typhoons (1993_No.25 and 1993_No.26) generated in the equatorial central Pacific and tropical western Pacific on 27 and 28 September 1993, respectively. They moved northwestward near the climatic monsoon trough and their intensities both peaked on 4 October 1993. Two typhoons experienced a binary interaction by two anomalous vortices at 850 hPa.

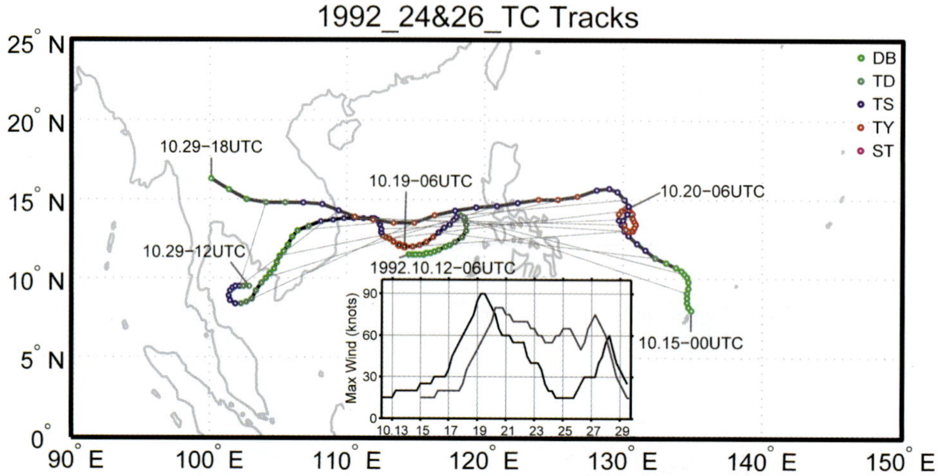

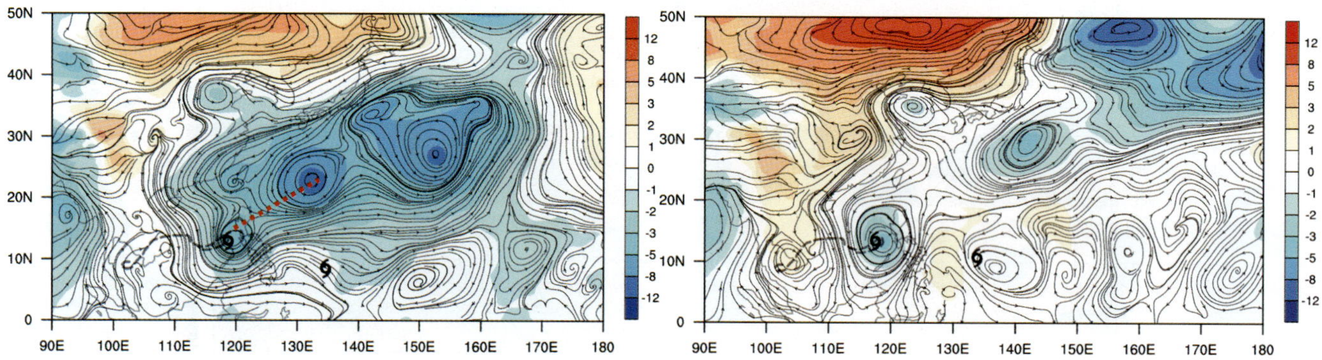

850 hPa flow/height anomalies at 1200 UTC 15 and 17 October 1992.

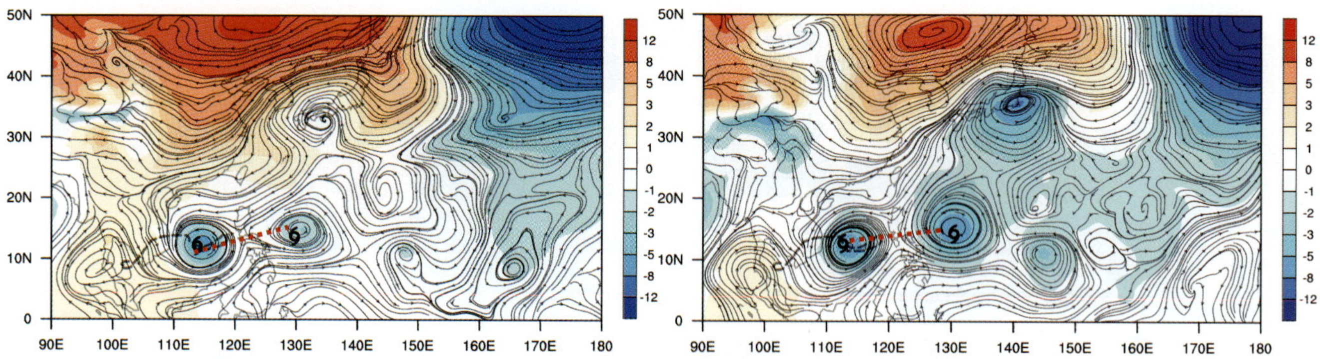

850 hPa flow/height anomalies at 1200 UTC 19 and 20 October 1992.

Figure 7.52. Two typhoons (1992_No. 24 and 26) generated in the southern SCS and the equatorial western Pacific on 12 and 15 October 1992, respectively. After formation, they experienced complex circular motions and their intensities peaked on 19 and 20 October 1992 in the climatic monsoon trough. The binary interaction of the two typhoons and the influence of other anomalous vortices can be observed at 850 hPa.

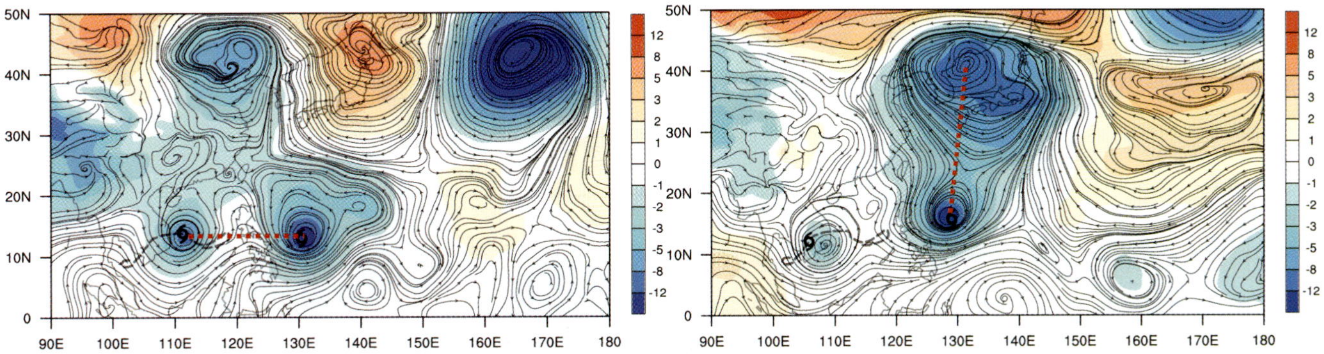

850 hPa flow/height anomalies at 1200 UTC 22 and 24 October 1992.

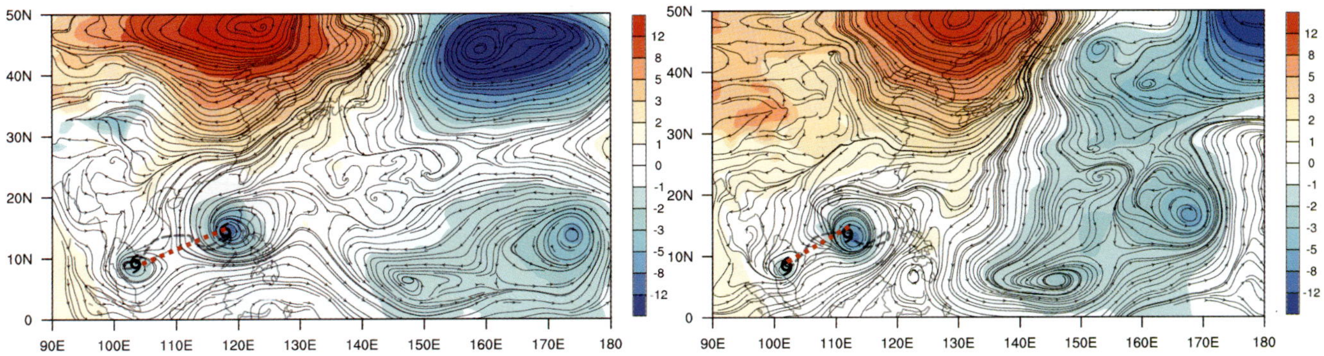

850 hPa flow/height anomalies at 1200 UTC 26 and 27 October 1992.

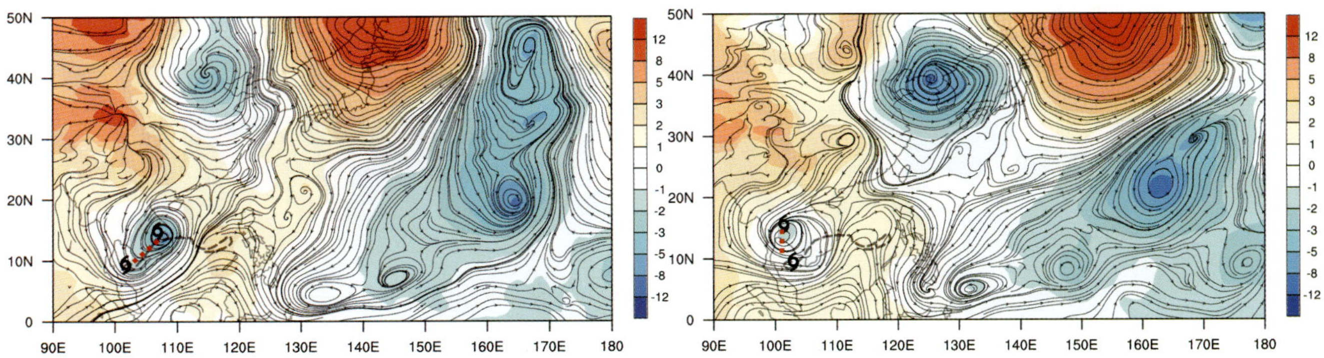

850 hPa flow/height anomalies at 1200 UTC 28 and 29 October 1992.

Figure 7.52. *(Continued)*

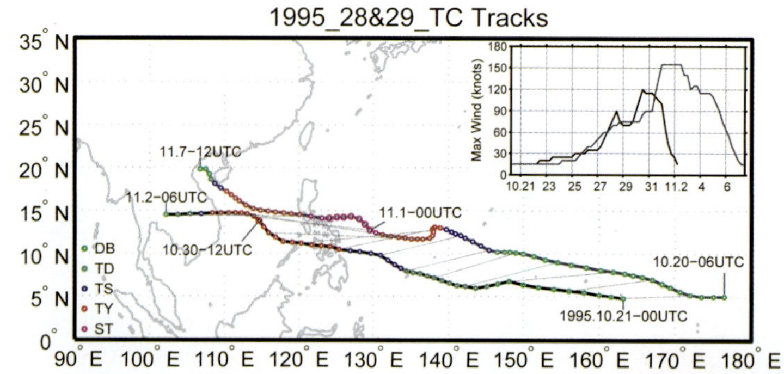

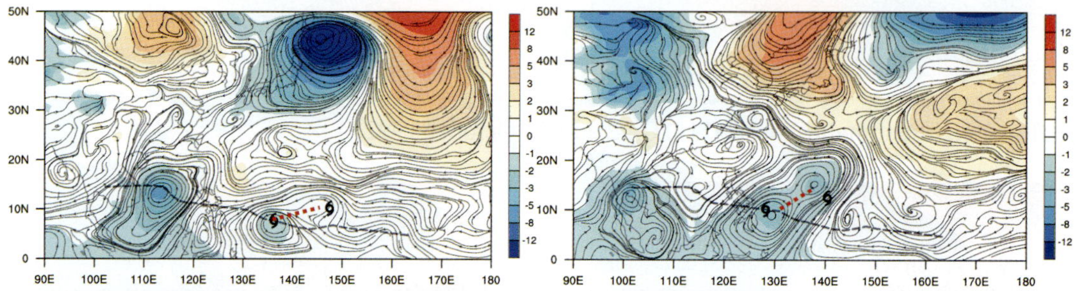

850 hPa flow/height anomalies at 1200 UTC 25 and 27 October 1995.

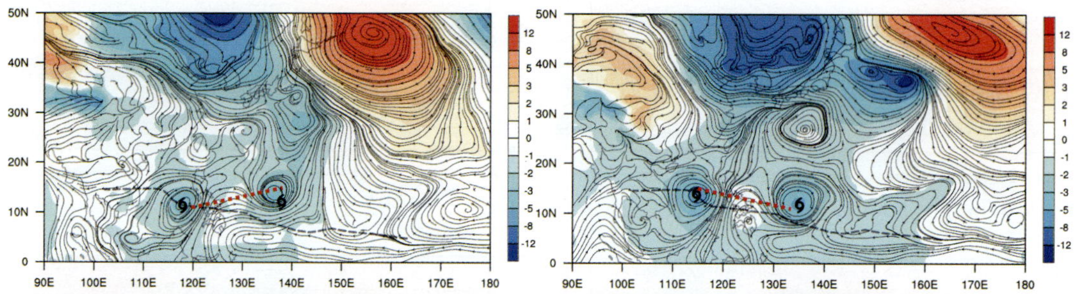

850 hPa flow/height anomalies at 1200 UTC 29 and 30 October 1995.

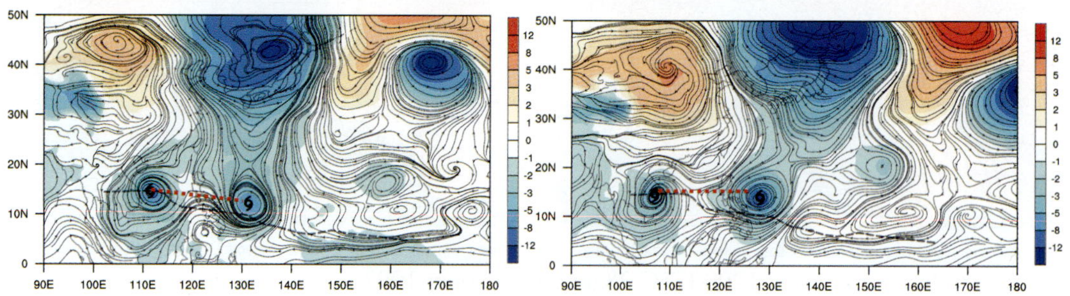

850 hPa flow/height anomalies at 1200 UTC 31 October and 1 November 1995.

Figure 7.53. Two typhoons (1995_No.28 and 1995_No.29) generated in equatorial central Pacific on 20 and 21 October and moved westward along the climatic monsoon trough. Their intensities peaked in the east sea of the Philippines on 1 November and the SCS on 30 October 1995, respectively. Two typhoons moved westward in parallel on October 21 to November 2 in 1995. The binary interaction between the two typhoons can be observed on 25–27 October 1995 at 850 hPa.

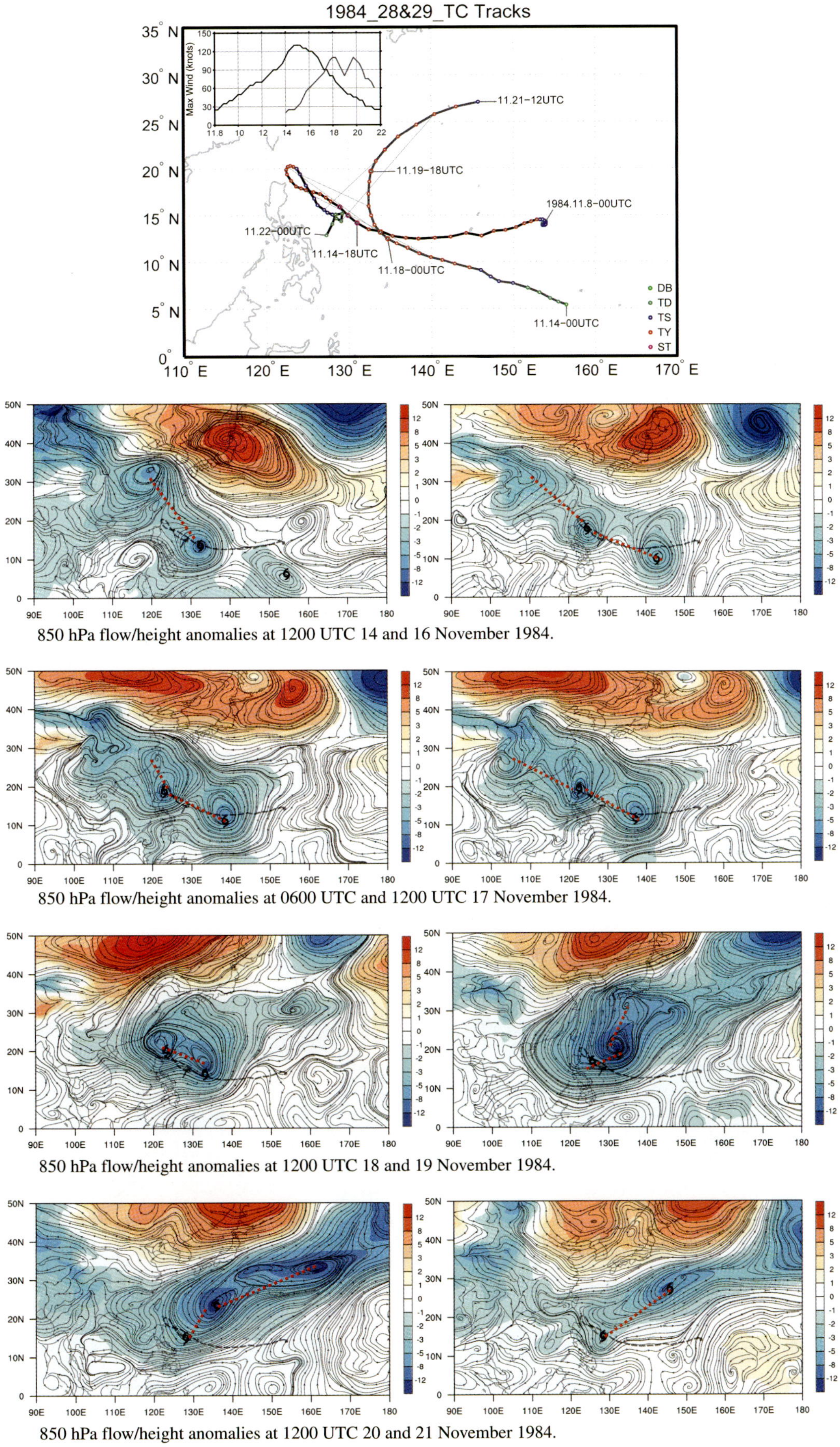

Figure 7.54. Two typhoons (1984_No.28 and 1984_No.29) generated in the tropical and equatorial central Pacific on 8 and 14 November 1984. The typhoon (1984_No.28) experienced a stagnant state near the Philippines for a few days due to the binary interaction between the two typhoons and the influence of anomalous vortices at 850 hPa.

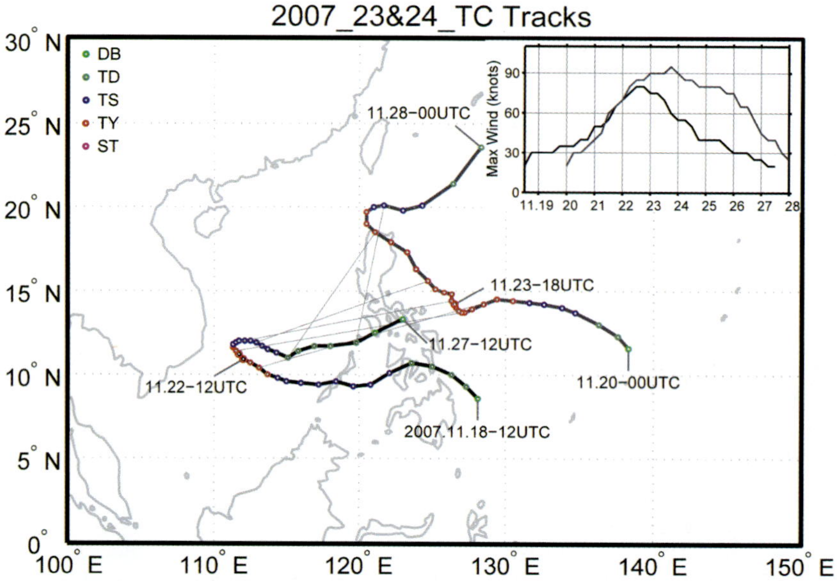

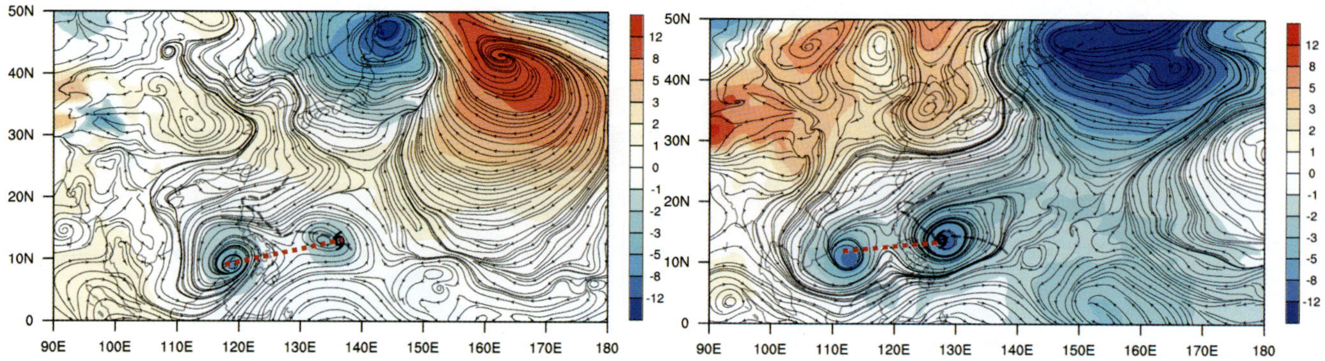

850 hPa flow/height anomalies at 1200 UTC 20 and 21 November 2007.

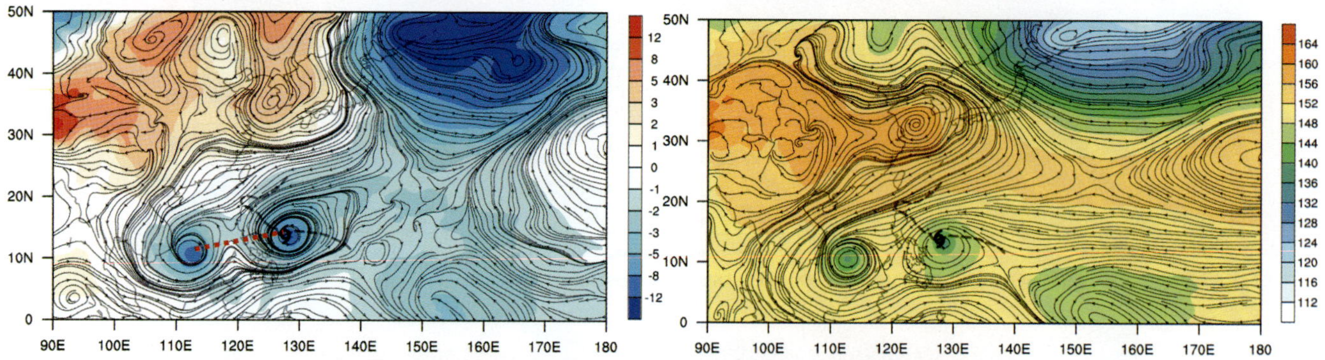

850 hPa flow/height anomalies and total flow/height at 1200 UTC 22 November 2007.

Figure 7.55. The typhoon (2007_No.23) generated in the southeast sea of the Philippines on 18 November 2007 and moved westward along the climatic monsoon trough in the SCS with its intensity peaking in the southwestern SCS on 22 November 2007. The typhoon (2007_No.24) generated in the tropical western Pacific on 20 December 2007 and entered the climatic monsoon trough with its intensity peaking in the east sea of the Philippines on 23 November 2007. The binary interaction between the two typhoons can be clearly observed through two anomalous vortices at 850 hPa.

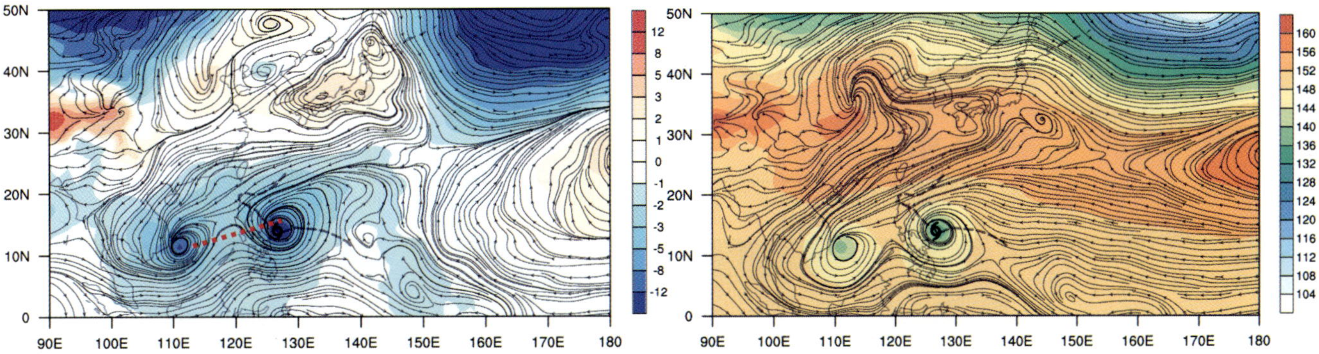

850 hPa flow/height anomalies and total flow/height at 1200 UTC 23 November 2007.

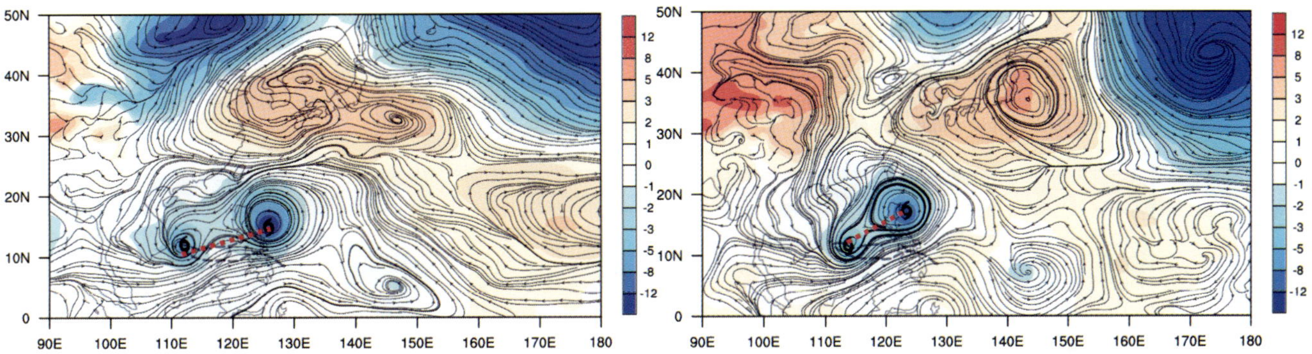

850 hPa flow/height anomalies at 1200 UTC 24 and 25 November 2007.

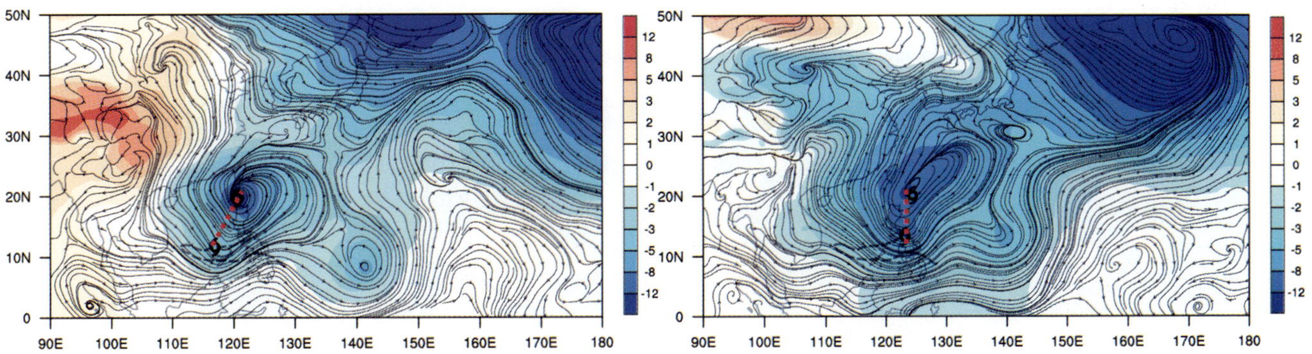

850 hPa flow/height anomalies at 1200 UTC 26 and 27 November 2007.

Figure 7.55. *(Continued)*

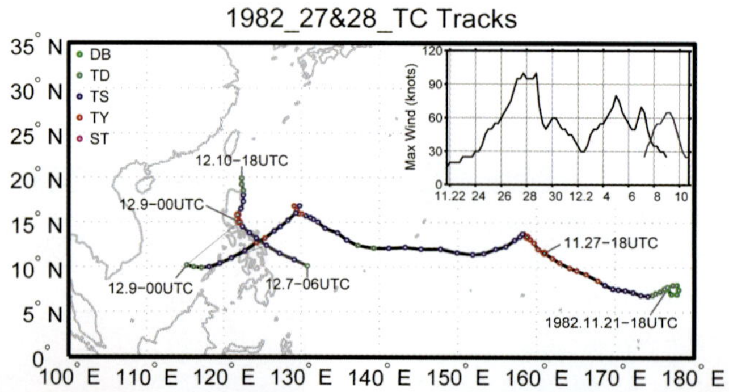

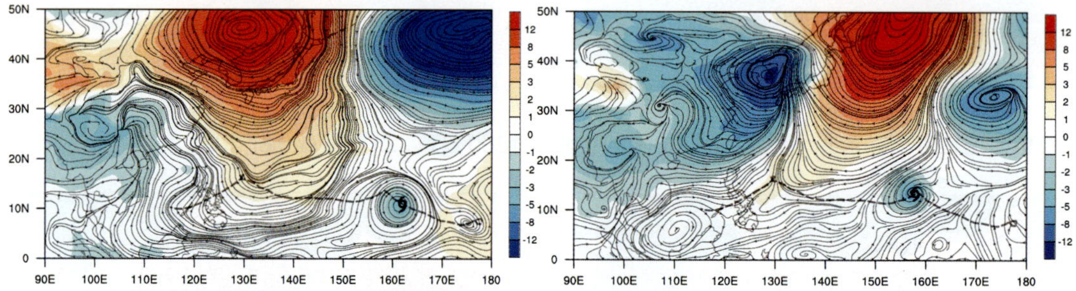

850 hPa flow/height anomalies at 1200 UTC 27 and 29 November 1982.

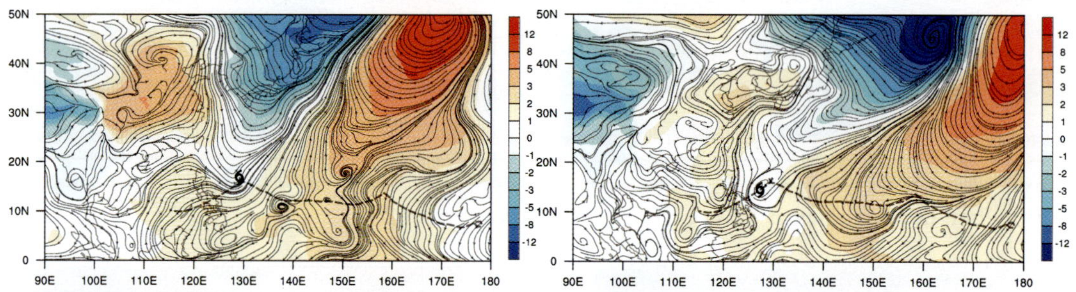

850 hPa flow/height anomalies at 1200 UTC 5 and 6 December 1982.

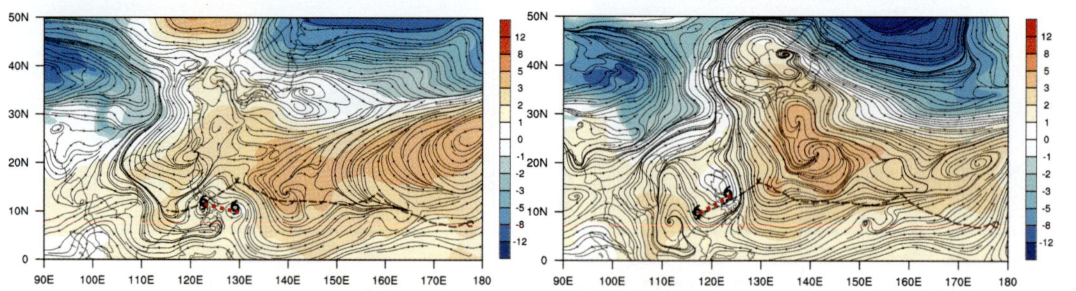

850 hPa flow/height anomalies at 1200 UTC 7 and 8 December 1982.

Figure 7.56. The typhoon (1982_No.27) generated in the equatorial central Pacific on 21 November 1982 and moved westward along the equatorial western Pacific monsoon trough with its intensity peaking in the tropical central Pacific on 27 November 1982. The typhoon (1982_No.28) generated in the tropical western Pacific on 7 December 1982 and moved northwestward near the climatic monsoon trough with its intensity peaking in the northern Philippines on 9 December 1982. Two typhoons were close together only on 7 and 8 December 1982. Two anomalous vortices of the two typhoons were weak at 850 hPa.

7.6. Complex Tracks

There were 12 typhoons moving with complex motions after they generated. Some of them experienced several stagnant states that were influenced by anomalous vortices (Figs. 7.57–7.68).

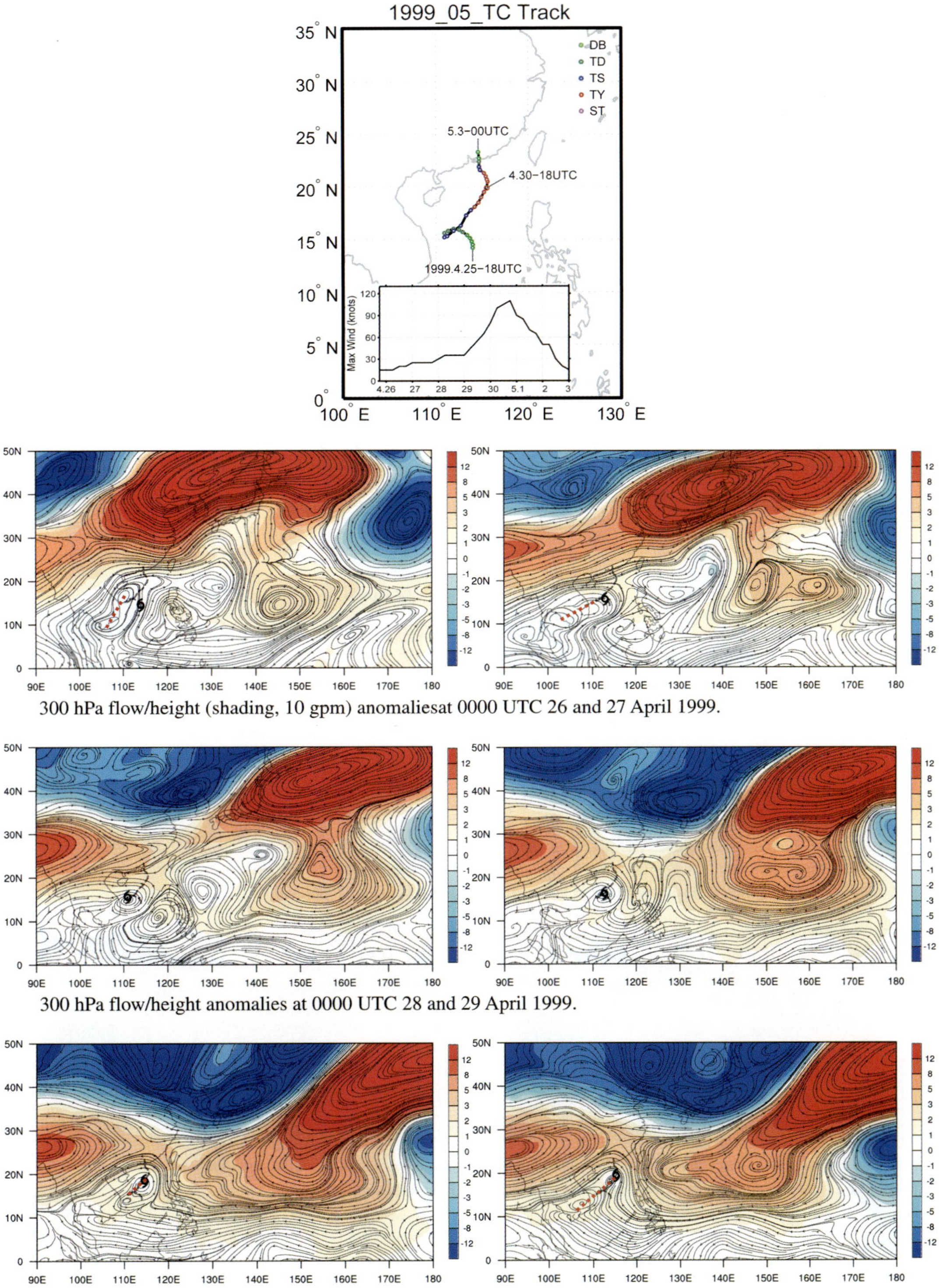

Figure 7.57. The typhoon (1999_No.05) generated in the central SCS on 25 April 1999 and experienced a complex motion. Its intensity peaked on 30 April 1999 before landfall on the South China coast. Its track was influenced by anomalous vortices at 300 hPa and 850 hPa.

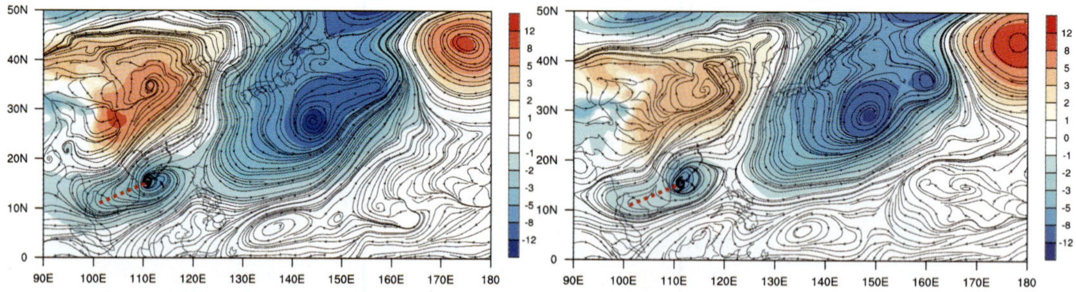

850 hPa flow/height anomalies at 0000 UTC and 1200 UTC 28 April 1999.

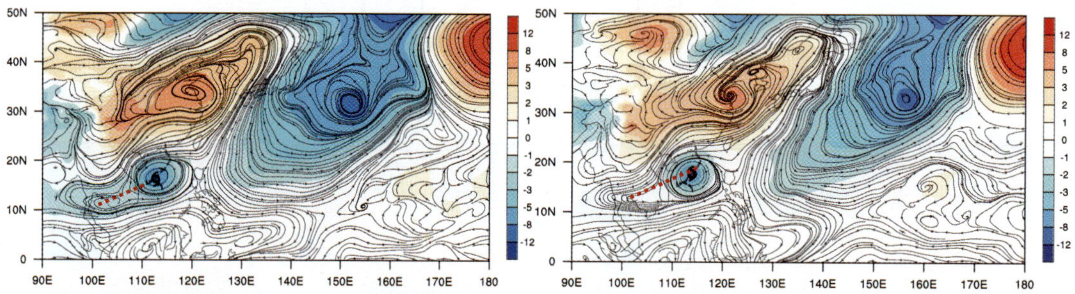

850 hPa flow/height anomalies at 0000 UTC and 1200 UTC 29 April 1999.

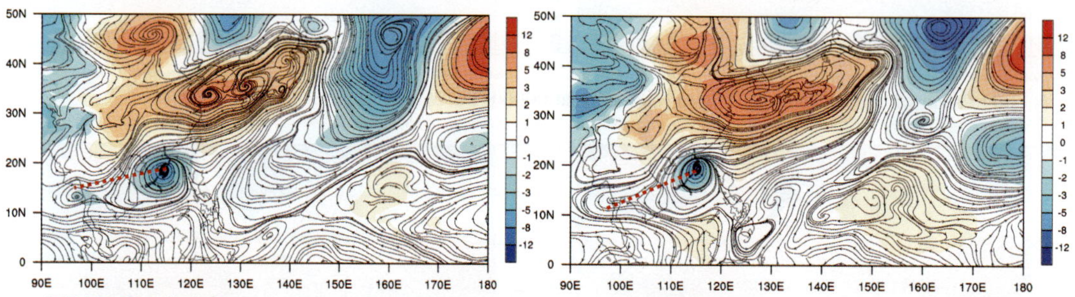

850 hPa flow/height anomalies at 0000 UTC and 1200 UTC 30 April 1999.

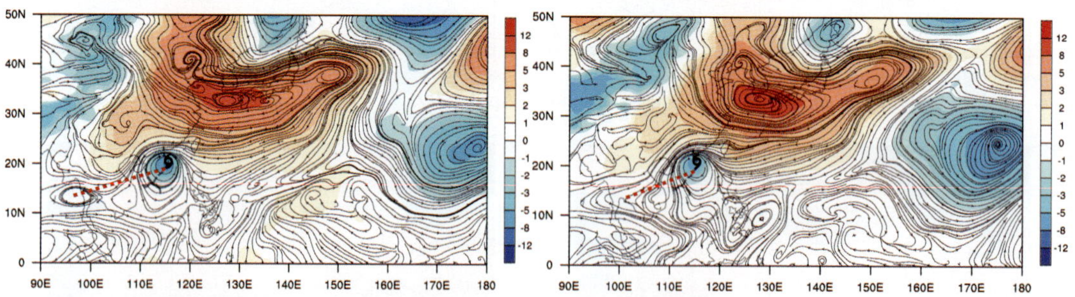

850 hPa flow/height anomalies at 0000 UTC and 1200 UTC 1 May 1999.

Figure 7.57. *(Continued)*

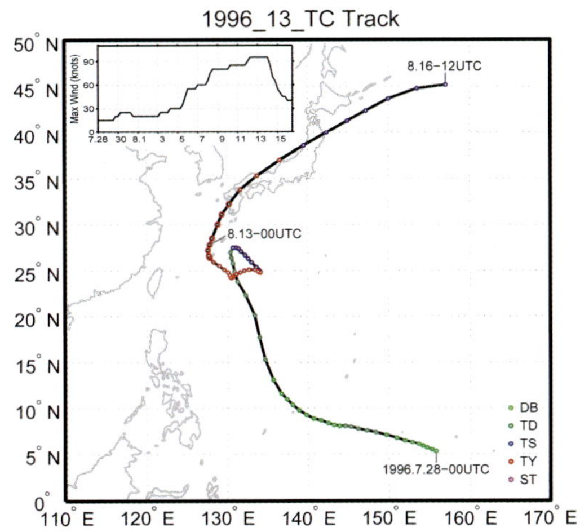

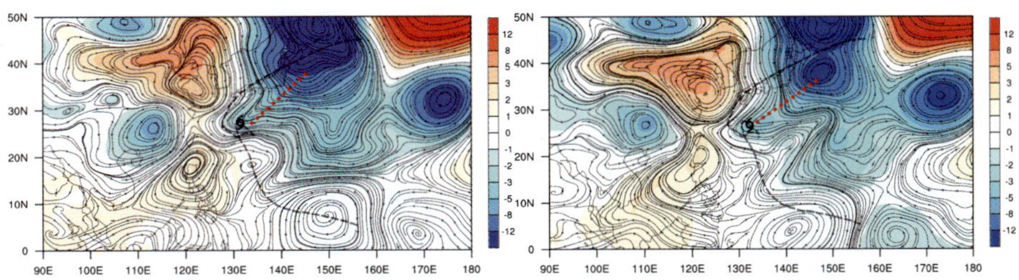

300 hPa flow/height anomalies at 0000 UTC and 1200 UTC 6 August 1996.

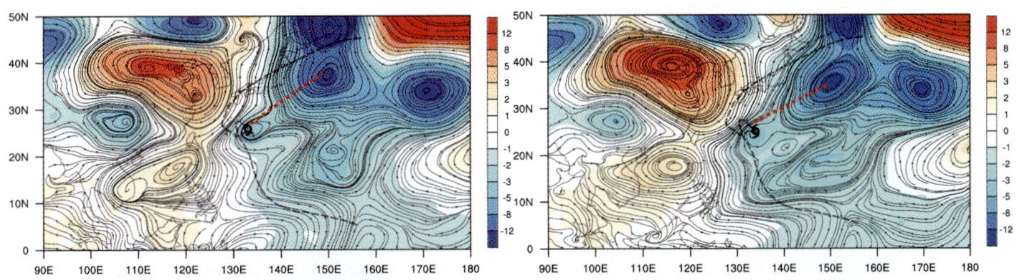

300 hPa flow/height anomalies at 0000 UTC and 1200 UTC 7 August 1996.

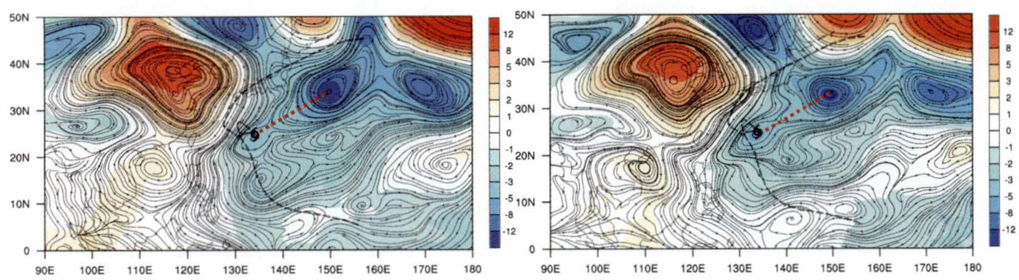

300 hPa flow/height anomalies at 0000 UTC and 1200 UTC 8 August 1996.

Figure 7.58. The typhoon (1996_No.13) generated in the equatorial central Pacific on 28 July 1996 and had a local circular motion. Its intensity peaked in the East China Sea on 13 August 1996 when it was approaching the climatic monsoon trough. This stagnant state and right turn in the East China Sea were influenced by anomalous vortices at 300 hPa.

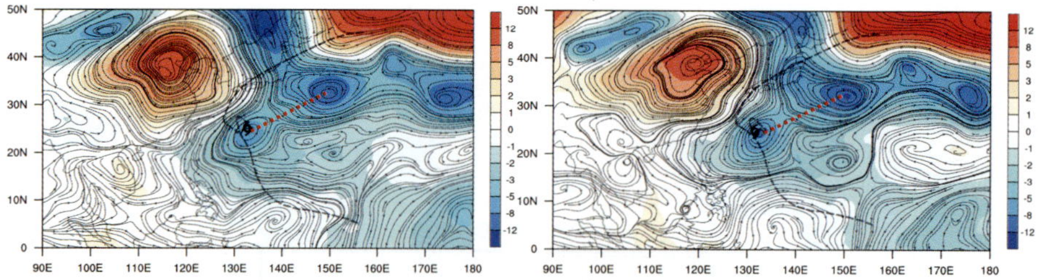

300 hPa flow/height anomalies at 0000 UTC and 1200 UTC 9 August 1996.

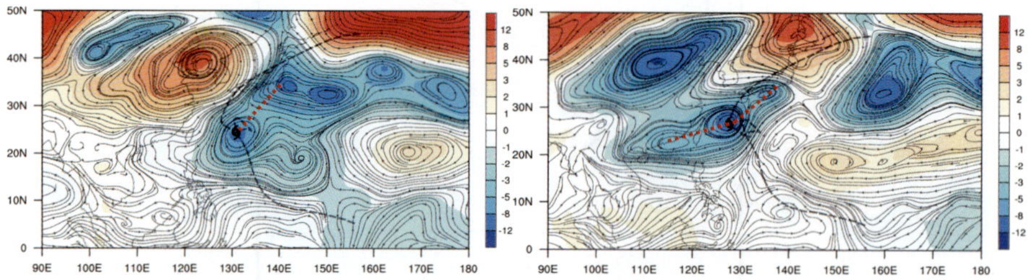

300 hPa flow/height anomalies at 0000 UTC and 1200 UTC 10 August 1996.

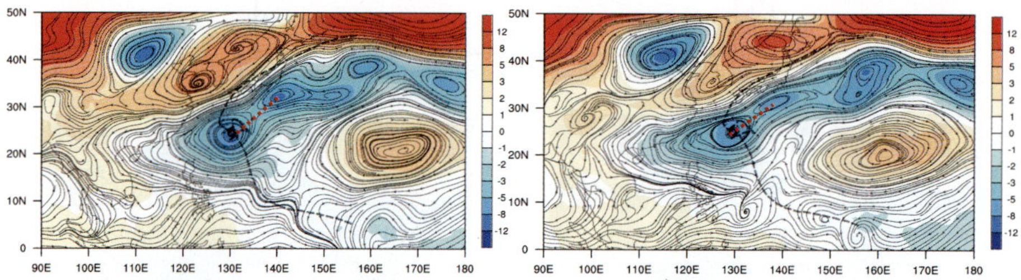

300 hPa flow/height anomalies at 0000 UTC and 1200 UTC 11 August 1996.

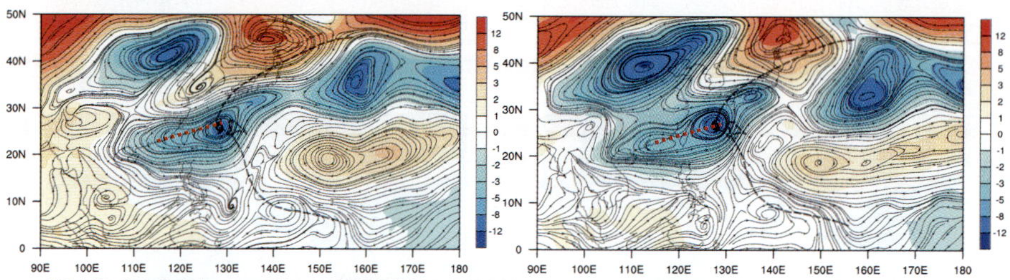

300 hPa flow/height anomalies at 0000 UTC and 1200 UTC 12 August 1996.

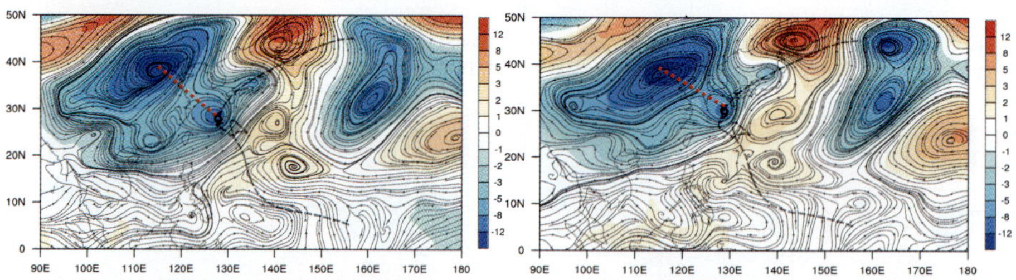

300 hPa flow/height anomalies at 0000 UTC and 1200 UTC 13 August 1996.

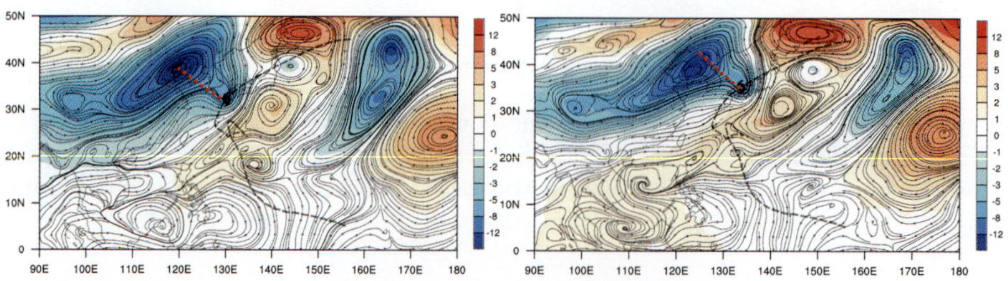

300 hPa flow/height anomalies at 0000 UTC and 1200 UTC 14 August 1996.

Figure 7.58. *(Continued)*

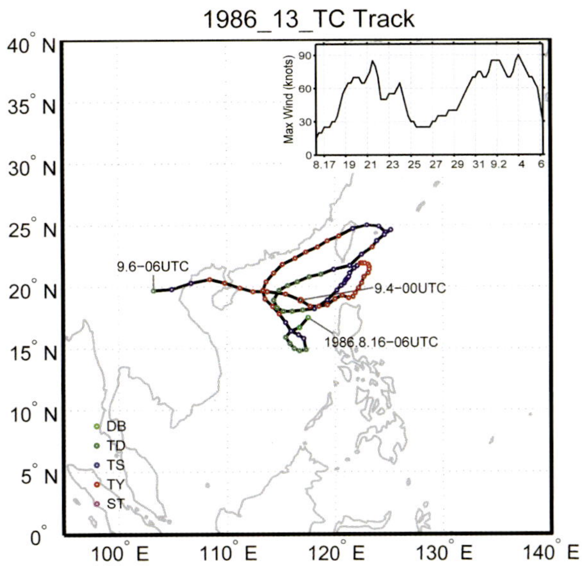

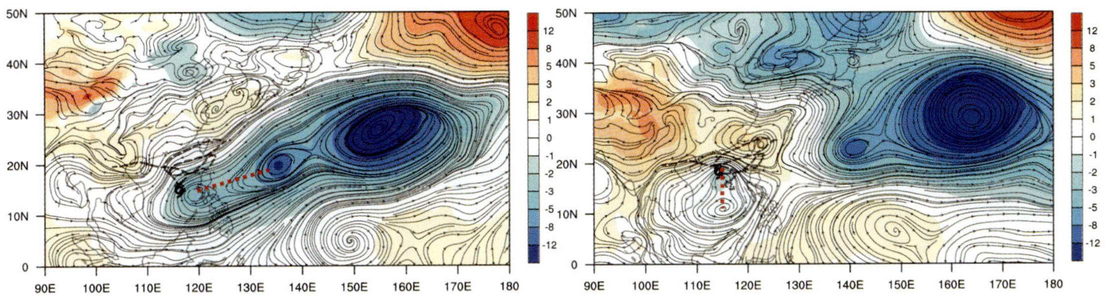

850 hPa flow/height anomalies at 1200 UTC 17 and 19 August 1986.

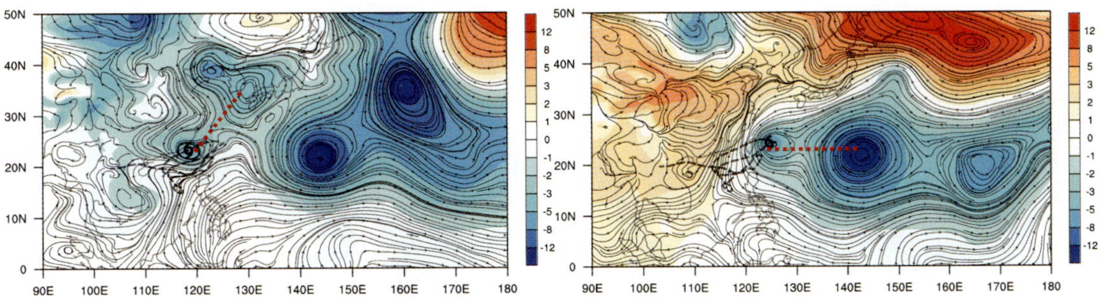

850 hPa flow/height anomalies at 1200 UTC 21 and 23 August 1986.

Figure 7.59. The typhoon (1986_No.13) generated in the northern SCS on 16 August 1986. Its motion experienced three circles and two strengthening processes during its lifetime of 22 days. Its intensity peaked in the northeast SCS from 2 to 4 September 1986 when it was located in the climatic monsoon trough. These circular motions were influenced complexly by anomalous vortices at 850 hPa.

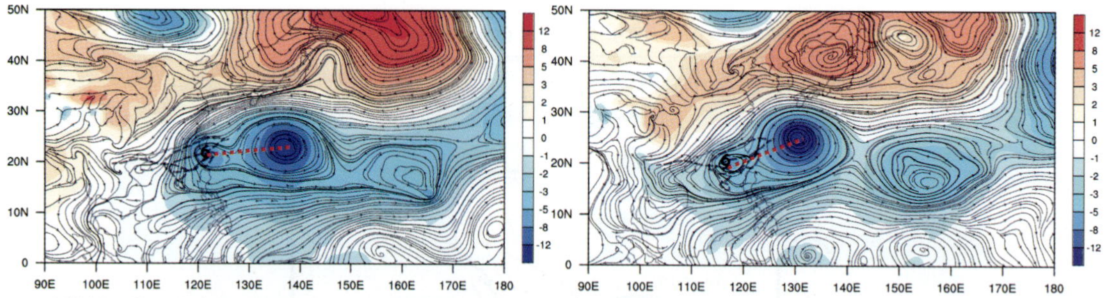

850 hPa flow/height anomalies at 1200 UTC 24 and 25 August 1986.

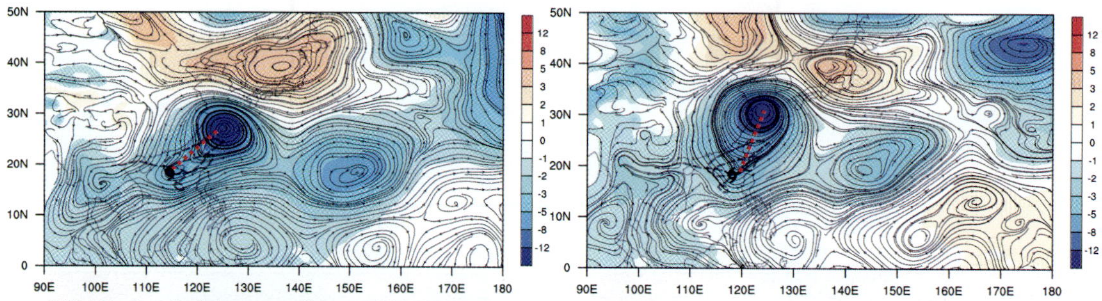

850 hPa flow/height anomalies at 1200 UTC 26 and 27 August 1986.

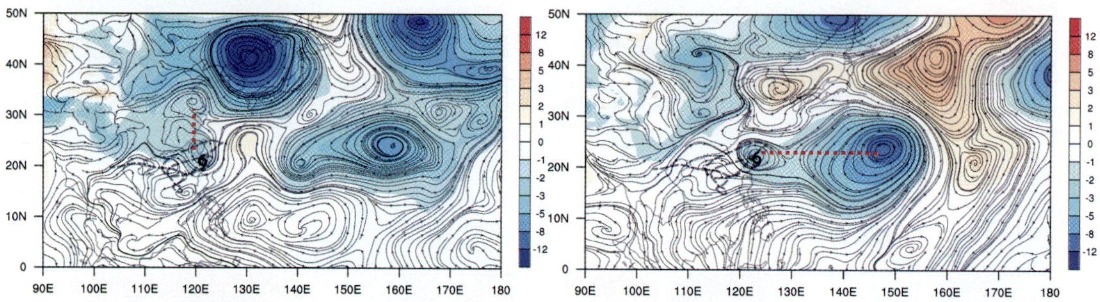

850 hPa flow/height anomalies at 1200 UTC 29 and 31 August 1986.

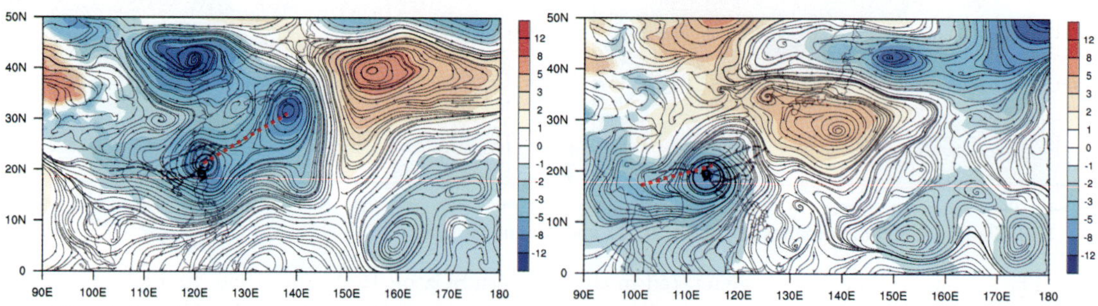

850 hPa flow/height anomalies at 1200 UTC 2 and 4 September 1986.

Figure 7.59. *(Continued)*

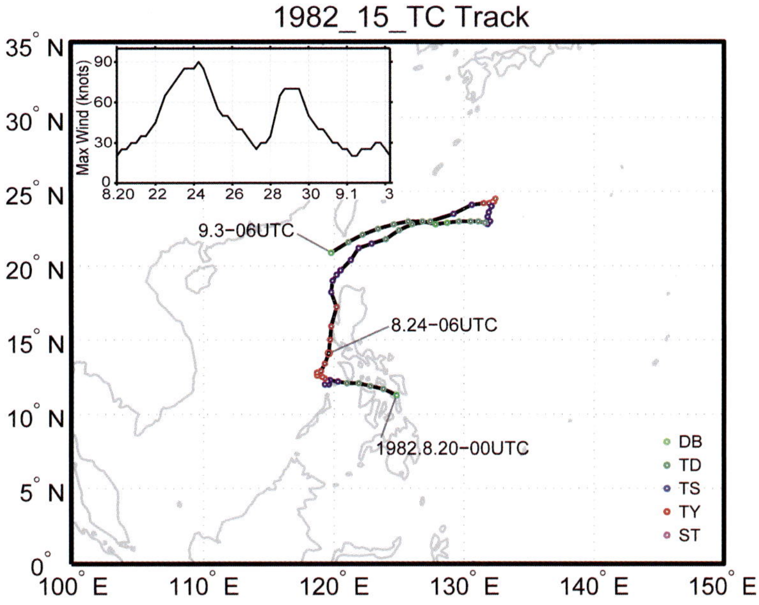

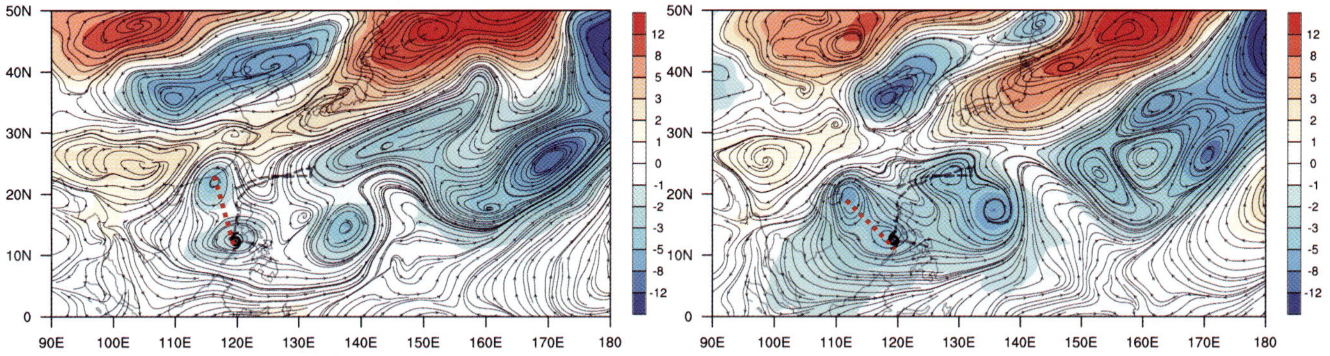

300 hPa flow/height anomalies at 1200 UTC 21 and 22 August 1982.

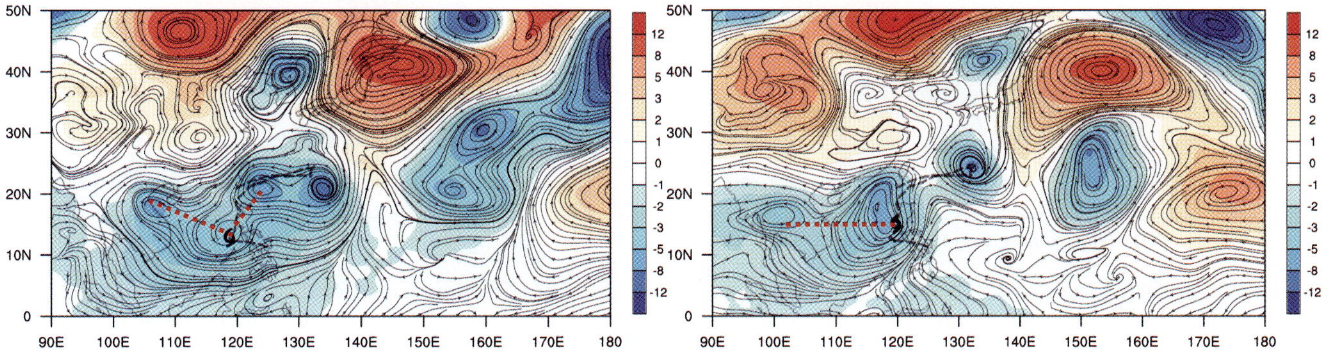

300 hPa flow/height anomalies at 1200 UTC 23 and 24 August 1982.

Figure 7.60. The typhoon (1982_No.15) generated on 20 August 1982 in the central Philippines. Its intensity reached two peaks in the central eastern SCS on 24 August 1982 and in the subtropical Northwest Pacific on 29 August 1982 when it was approaching the climatic monsoon troughs. The right turn in the west sea of the Philippines was influenced by an anomalous vortex at 300 hPa.

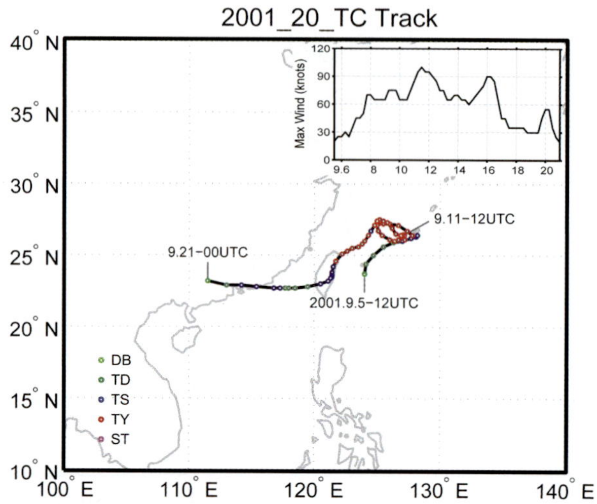

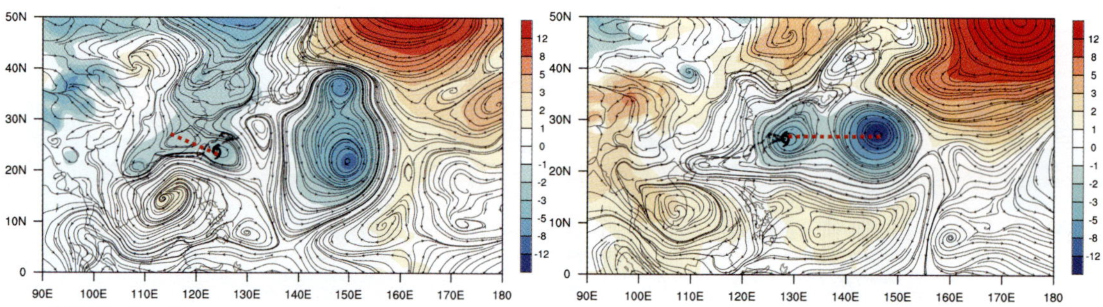

850 hPa flow/height anomalies at 1200 UTC 5 and 7 September 2001.

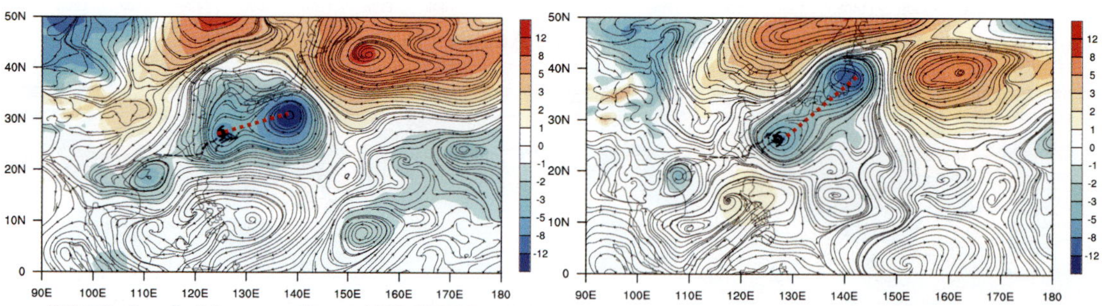

850 hPa flow/height anomalies at 1200 UTC 9 and 11 September 2001.

Figure 7.61. The typhoon (2001_No.20) generated in the east sea of the Taiwan Island on 5 September 2001. Its local circular motion and the peak intensity position on 11 September 2001 were confined within the climatic monsoon trough. This circular motion was influenced by anomalous circulations and interacted with other anomalous vortices at 850 hPa and 300 hPa.

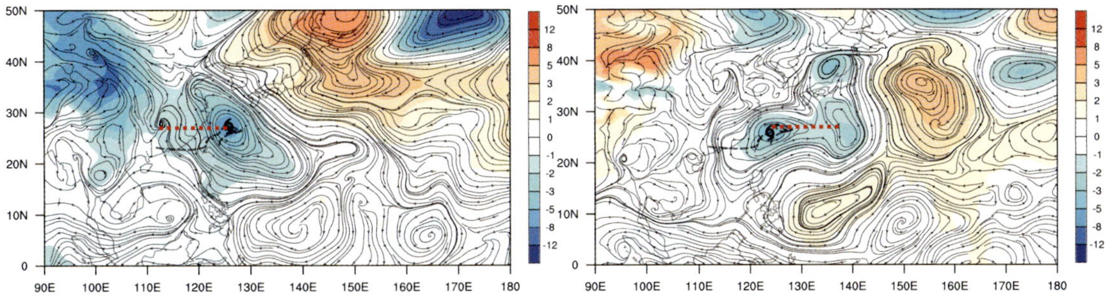

850 hPa flow/height anomalies at 1200 UTC 13 and 15 September 2001.

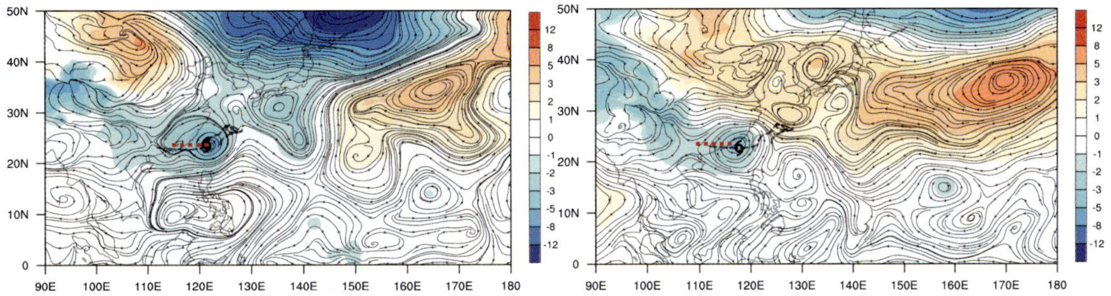

850 hPa flow/height anomalies at 1200 UTC 17 and 19 September 2001.

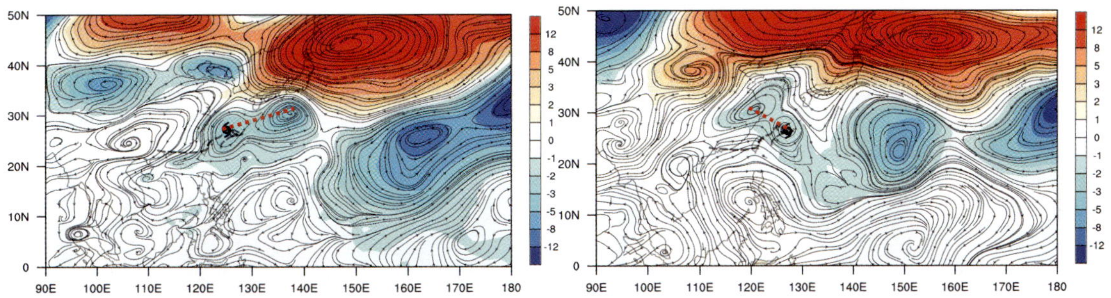

300 hPa flow/height anomalies at 1200 UTC 9 and 11 September 2001.

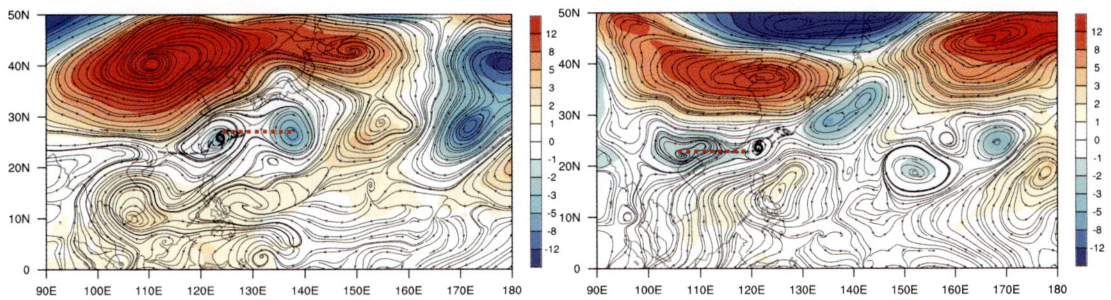

300 hPa flow/height anomalies at 1200 UTC 15 and 17 September 2001.

Figure 7.61. *(Continued)*

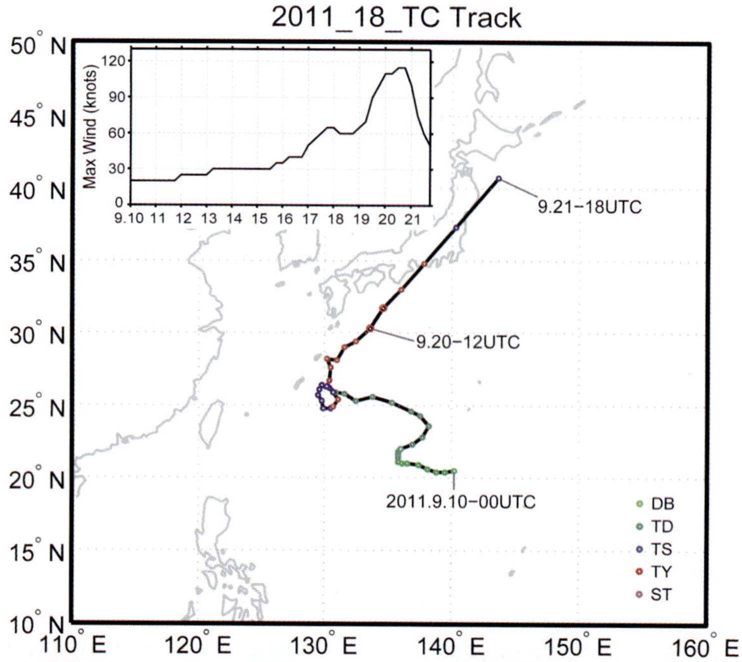

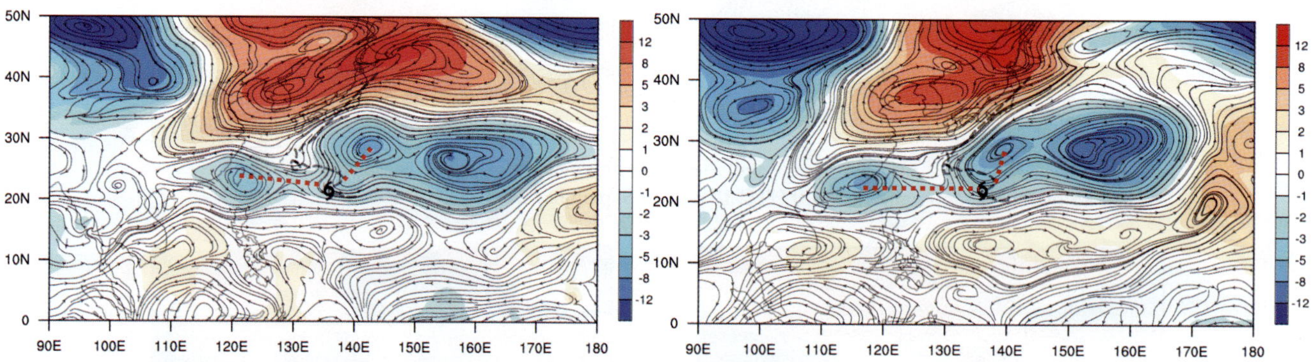

300 hPa flow/height anomalies at 0000 UTC 12 and 13 September 2011.

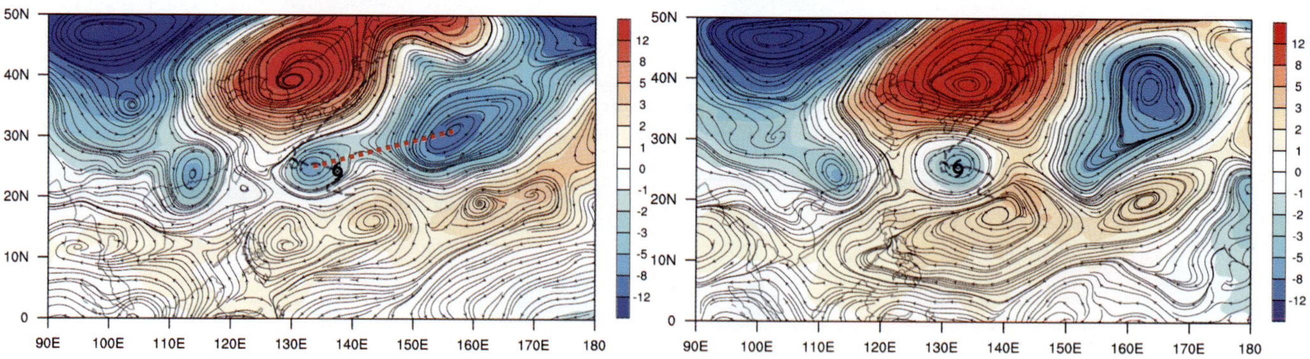

300 hPa flow/height anomalies at 0000 UTC 14 and 15 September 2011.

Figure 7.62. The typhoon (2011_No.18) generated in the subtropical Northwest Pacific on 10 September 2011 and had a complex motion during its lifetime period of 12 days. Its intensity peaked in the south seas of Japan on 20 September 2011 when it was approaching the climatic monsoon trough. The right turn and circular motion were influenced by anomalous vortices at 300 hPa.

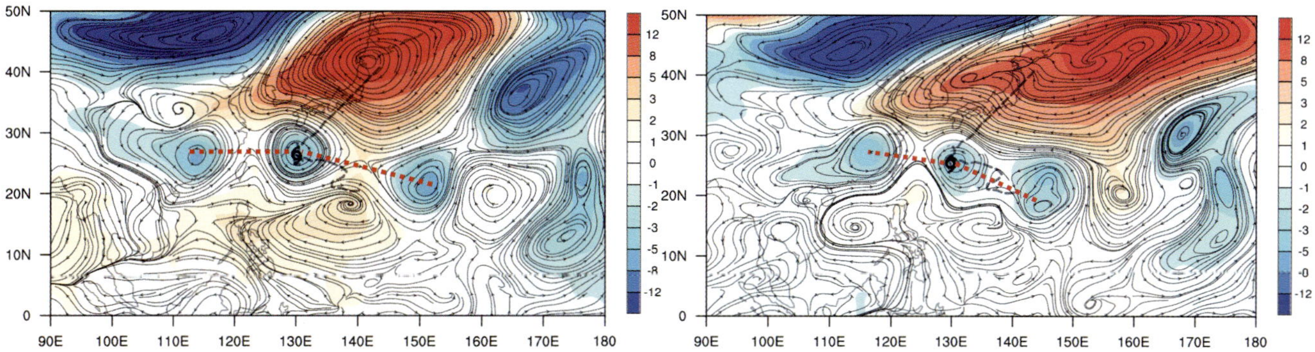

300 hPa flow/height anomalies at 0000 UTC 16 and 17 September 2011.

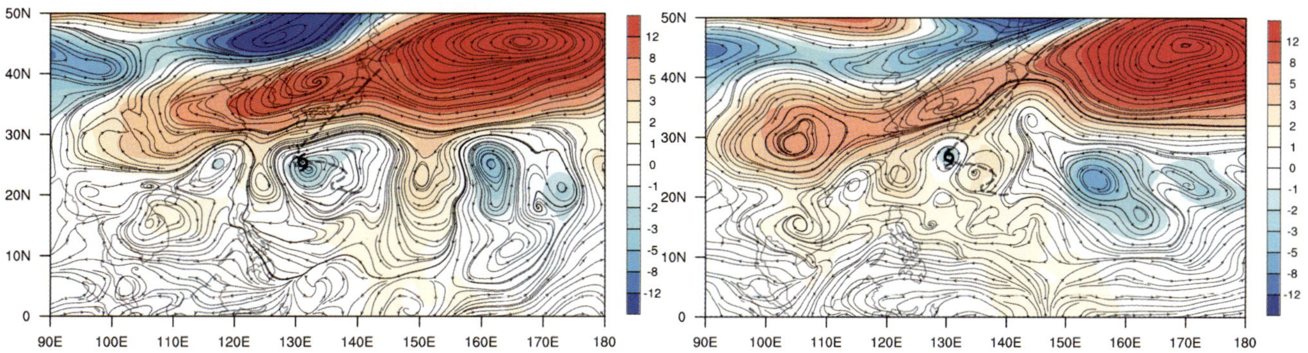

300 hPa flow/height anomalies at 0000 UTC 18 and 19 September 2011.

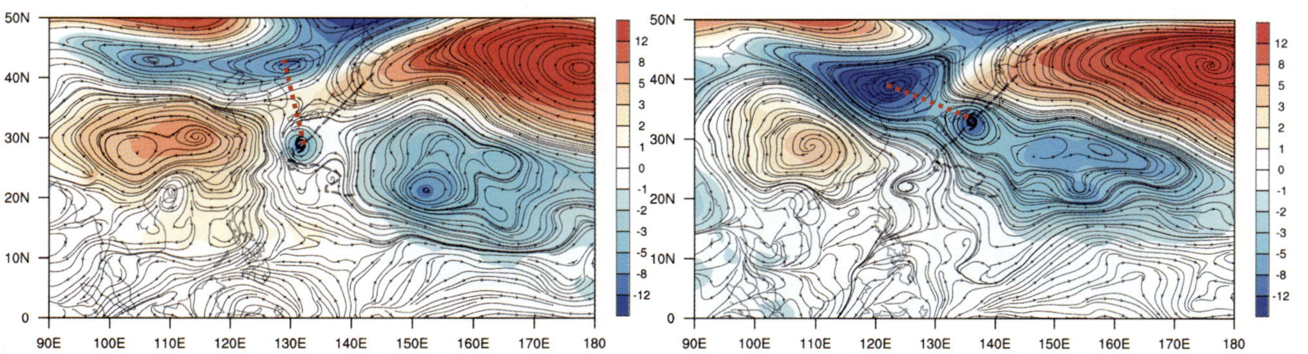

300 hPa flow/height anomalies at 0000 UTC 20 and 21 September 2011.

Figure 7.62. *(Continued)*

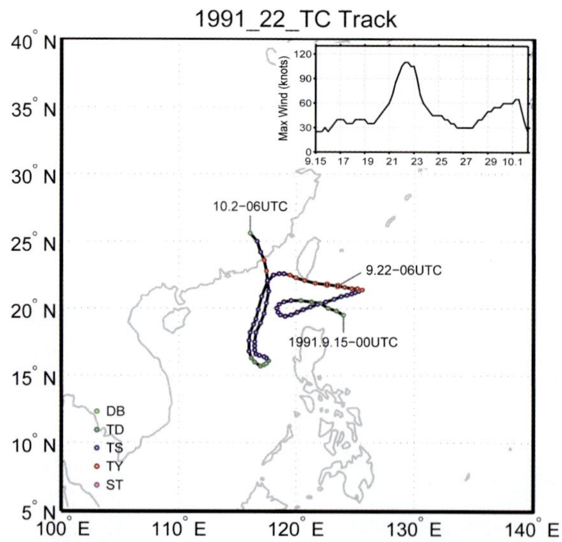

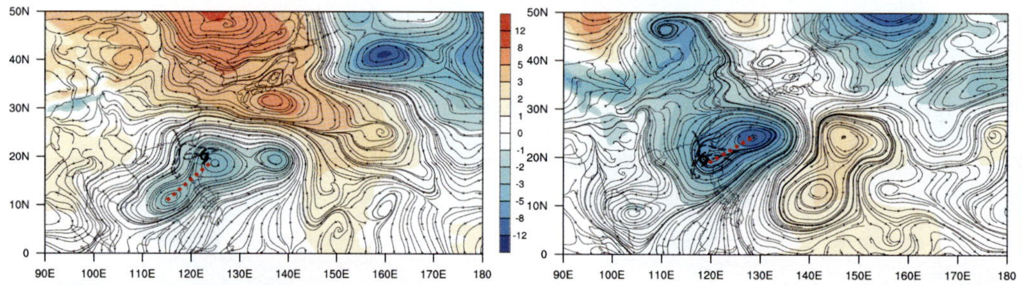

850 hPa flow/height anomalies at 1200 UTC 15 and 17 September 1991.

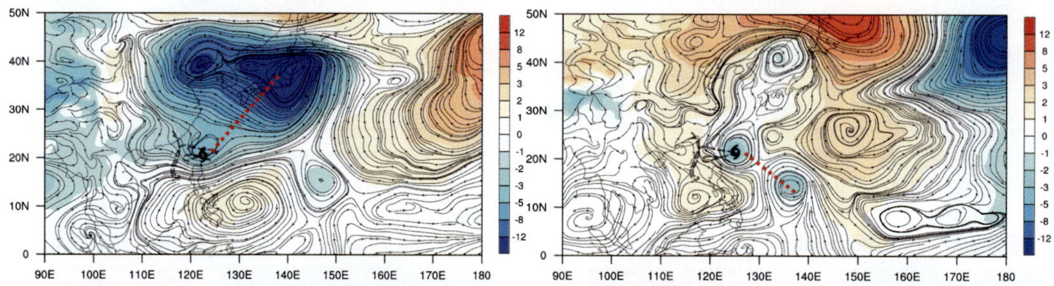

850 hPa flow/height anomalies at 1200 UTC 19 and 21 September 1991.

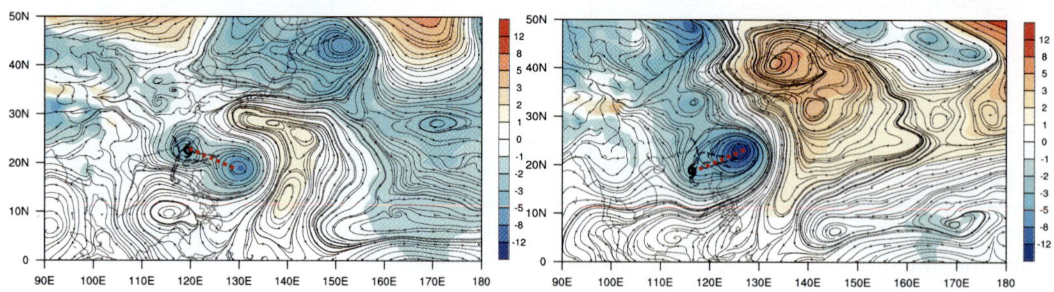

850 hPa flow/height anomalies at 1200 UTC 23 and 25 September 1991.

Figure 7.63. The typhoon (1991_No.22) generated in the northeast sea of the Philippines on 15 September 1991 and had two circle-liked motions and two strengthening processes during its lifetime of 18 days. Its local circular motions and the position of peak intensity on 22 September 1991 were confined within the climatic monsoon trough. This complex motion was influenced by anomalous circulations and interacted with other anomalous vortices at 850 hPa and 300 hPa.

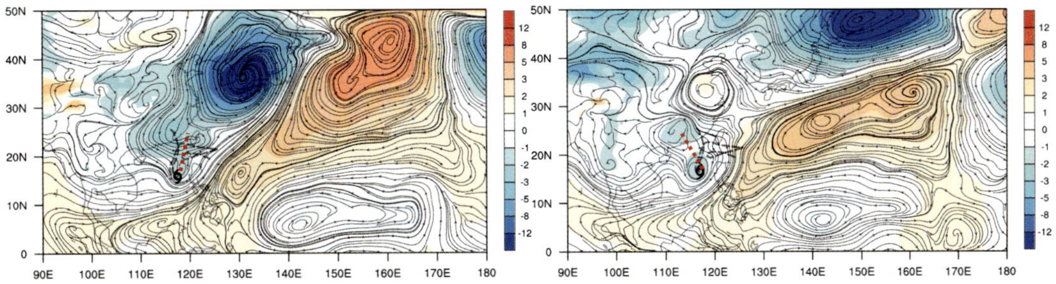

850 hPa flow/height anomalies at 1200 UTC 27 and 28 September 1991.

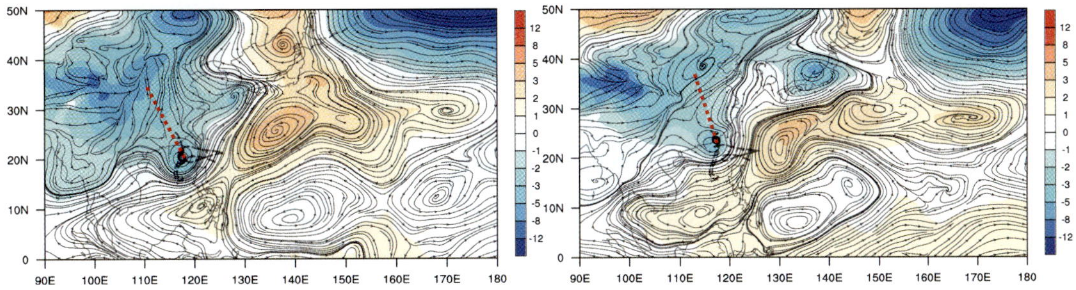

850 hPa flow/height anomalies at 1200 UTC 30 September and 1 October 1991.

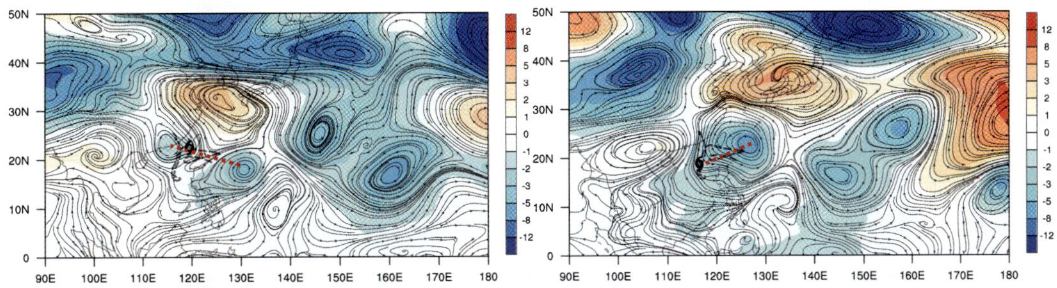

300 hPa flow/height anomalies at 1200 UTC 23 and 25 September 1991.

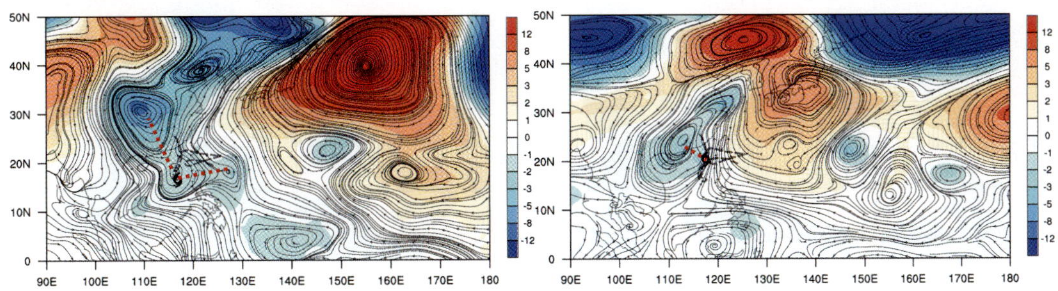

300 hPa flow/height anomalies at 1200 UTC 28 and 30 September 1991.

Figure 7.63. *(Continued)*

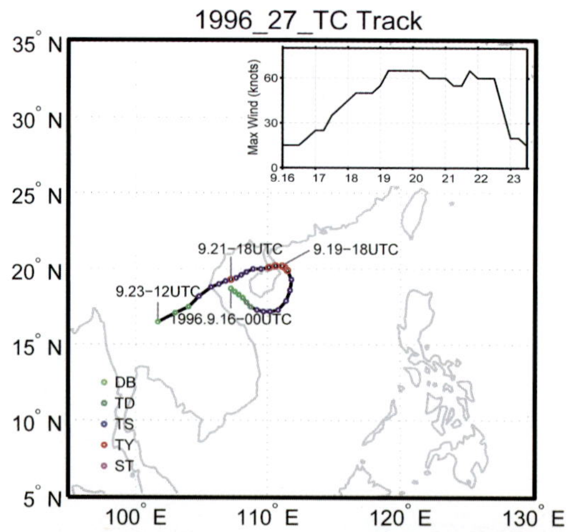

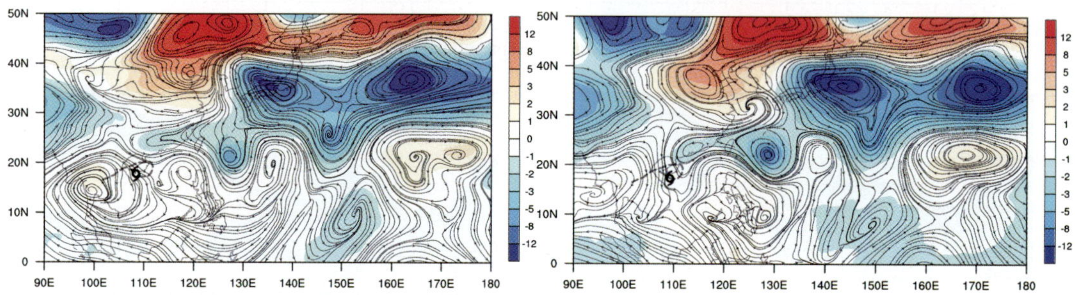

300 hPa flow/height anomalies at 0000 UTC and 1200 UTC 17 September 1996.

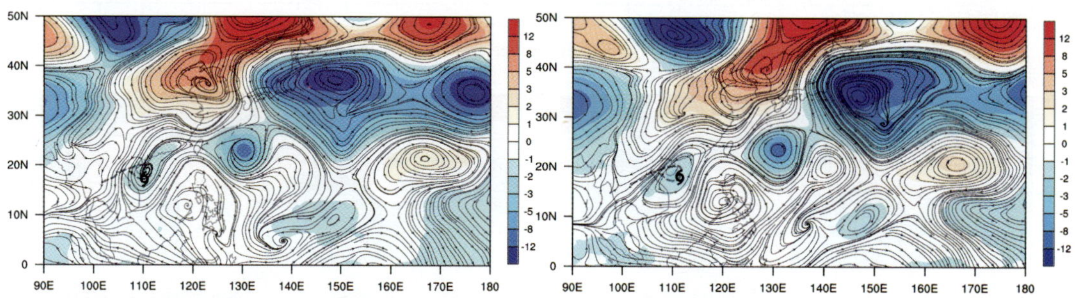

300 hPa flow/height anomalies at 0000 UTC and 1200 UTC 18 September 1996.

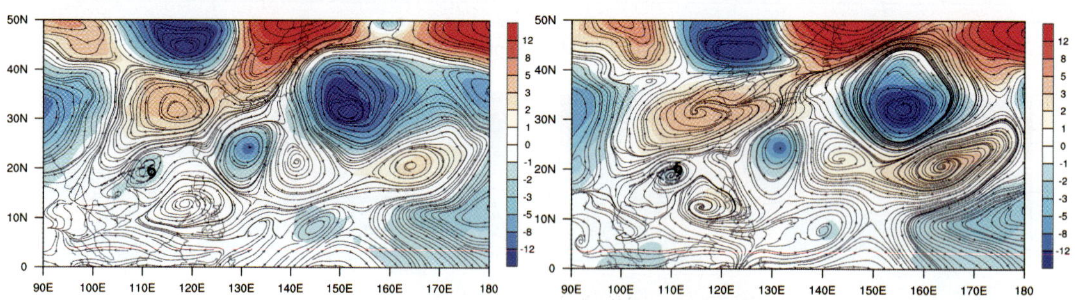

300 hPa flow/height anomalies at 0000 UTC and 1200 UTC 19 September 1996.

Figure 7.64. The typhoon (1996_No.27) generated in the west sea of the Hainan Island on 16 September 1996 and had a circle-liked motion. Its intensity peaked in the northernmost SCS on 19 September 1996 when it was approaching the climatic monsoon trough. This circular motion was influenced by an anomalous vortex that stood near the Hainan Island due to stable other anomalous circulations at 300 hPa.

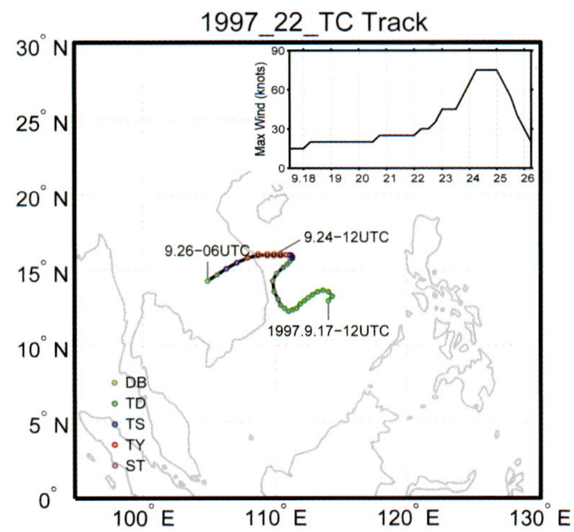

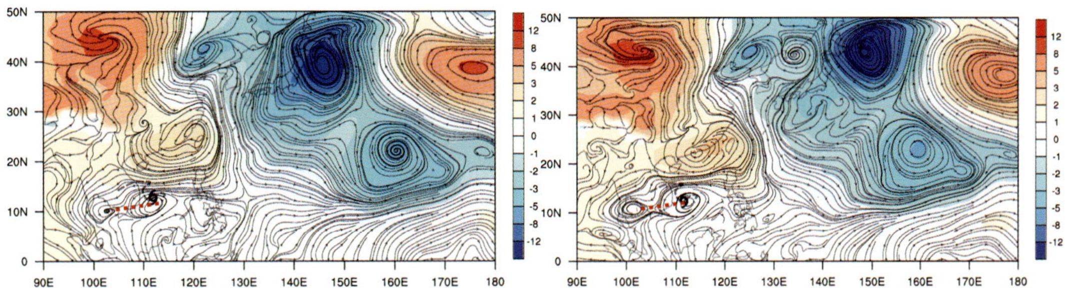

850 hPa flow/height anomalies at 1200 UTC and 1800 UTC 19 September 1997.

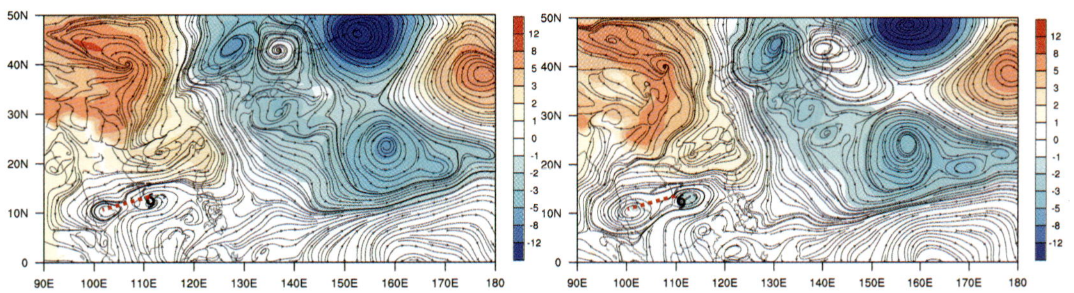

850 hPa flow/height anomalies at 0000 UTC and 0600 UTC 20 September 1997.

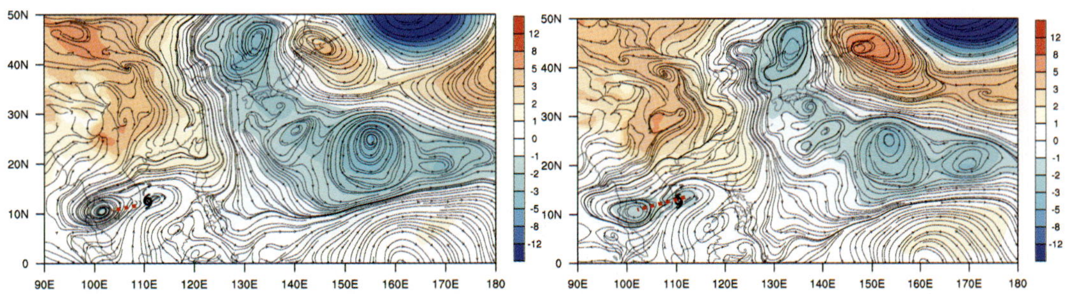

850 hPa flow/height anomalies at 1200 UTC and 1800 UTC 20 September 1997.

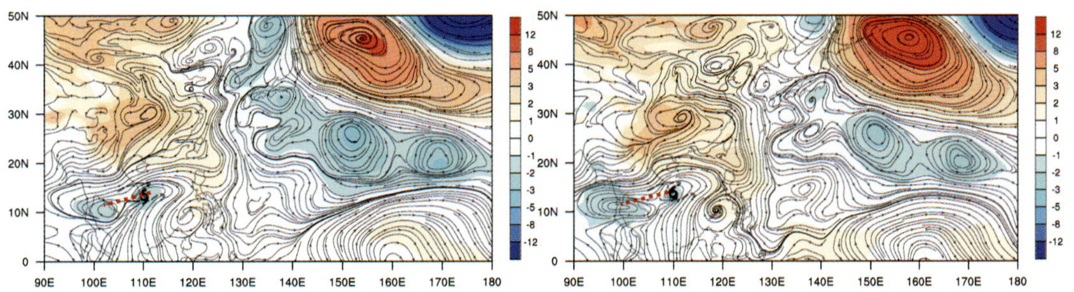

850 hPa flow/height anomalies at 0000 UTC and 0600 UTC 21 September 1997.

Figure 7.65. The typhoon (1997_No.22) generated in the central SCS on 17 September 1997 and experienced a complex motion in the SCS. Its intensity peaked in the central western SCS on 24 September 1997 when it was approaching the climatic monsoon trough. This complex motion was influenced by anomalous vortices at 850 hPa.

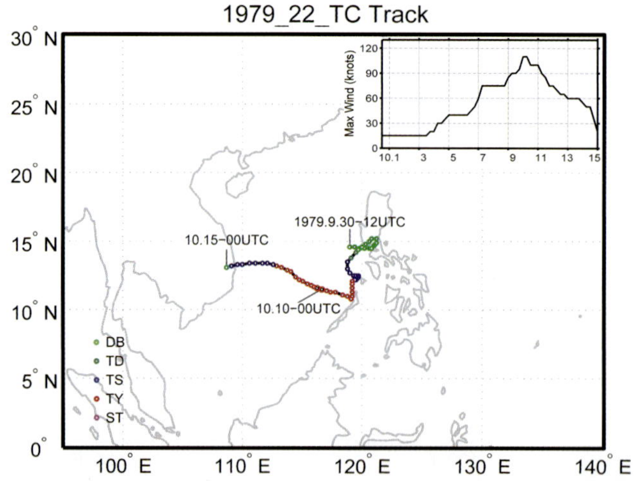

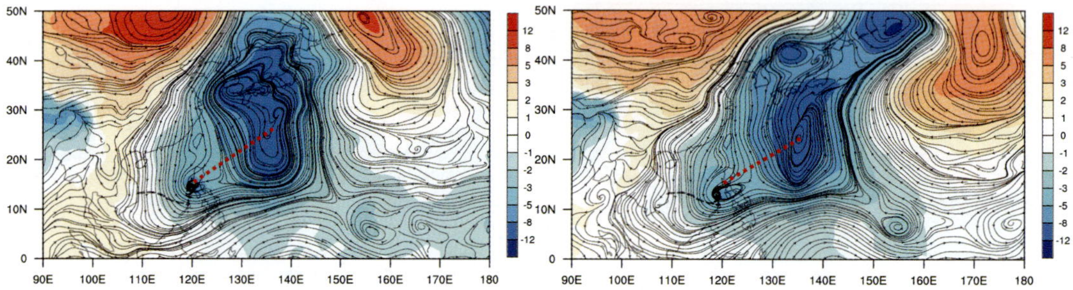

850 hPa flow/height anomalies at 0000 UTC 4 and 5 October 1979.

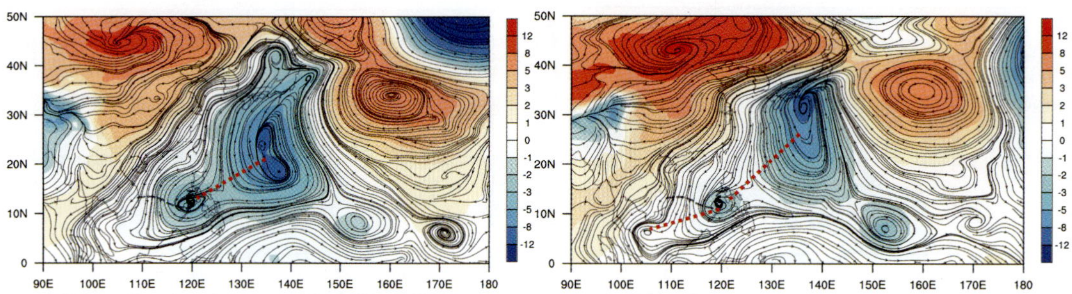

850 hPa flow/height anomalies at 0000 UTC 6 and 7 October 1979.

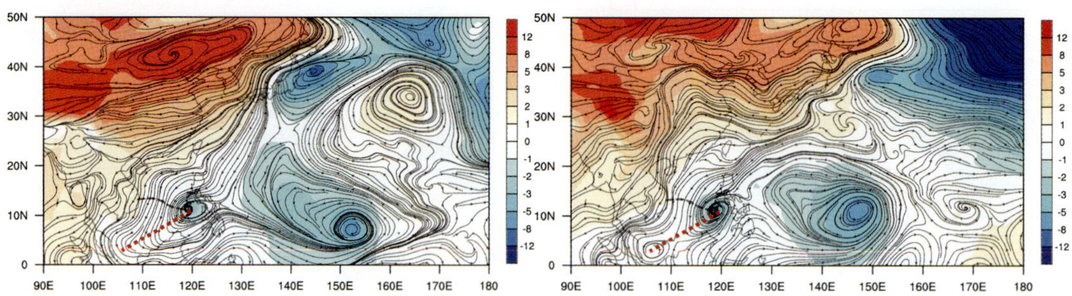

850 hPa flow/height anomalies at 0000 UTC 8 and 9 October 1979.

Figure 7.66. The typhoon (1979_No.22) generated in the central eastern SCS on 30 September 1979 and experienced two circle-liked motions and stagnant states in the central eastern SCS. Its intensity peaked in the central eastern SCS on 10 October 1979 when it was approaching the climatic monsoon trough. This complex motion was influenced by anomalous vortices at 850 hPa.

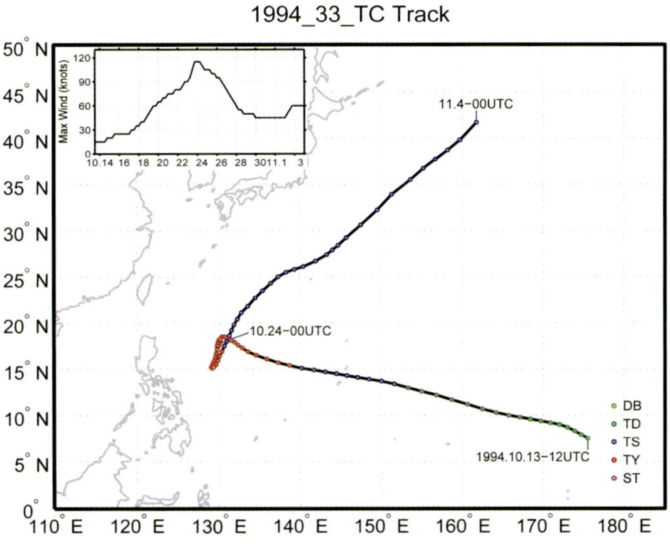

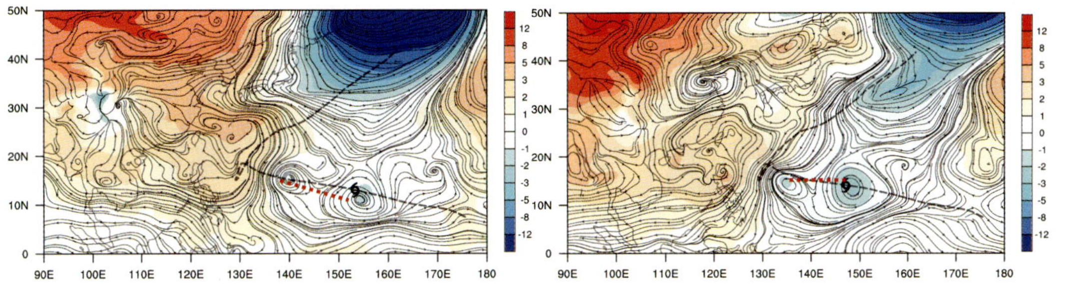

850 hPa flow/height anomalies at 1200 UTC 17 and 18 October 1994.

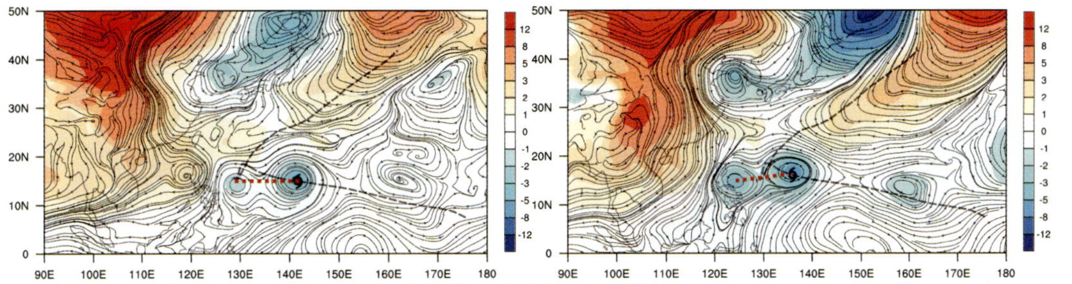

850 hPa flow/height anomalies at 1200 UTC 19 and 20 October 1994.

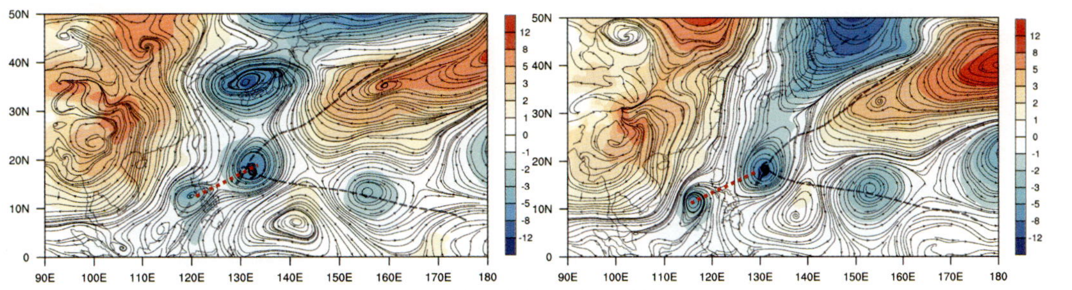

850 hPa flow/height anomalies at 1200 UTC 21 and 22 October 1994.

Figure 7.67. The typhoon (1994_No.33) generated in the equatorial central Pacific on 13 October 1994 and moved west-northwestward along the climatic monsoon trough. Its intensity peaked in the east sea of the Philippines on 24 October 1994 before a circle-liked motion (or stagnant state) and turned northeastward. This unusual right turn and stagnant state were influenced by anomalous vortices at 850 hPa.

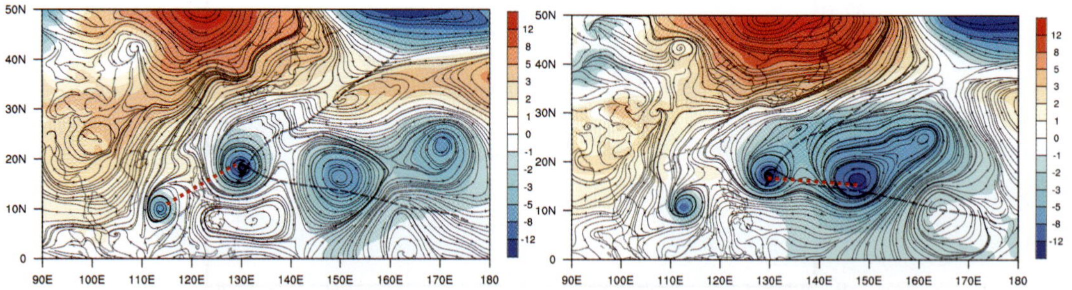

850 hPa flow/height anomalies at 1200 UTC 23 and 24 October 1994.

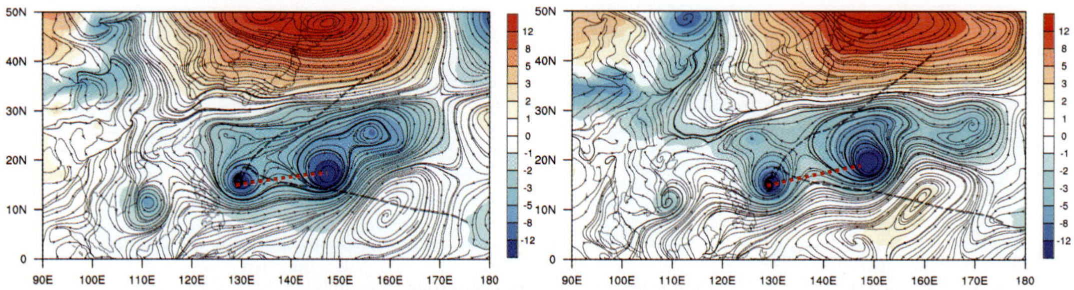

850 hPa flow/height anomalies at 1200 UTC 25 and 26 October 1994.

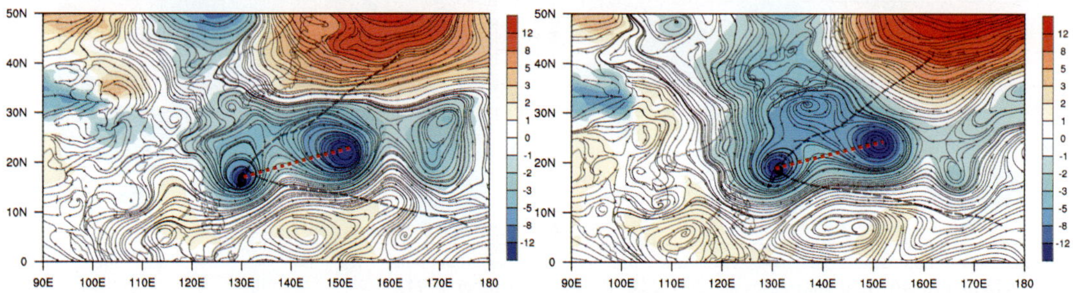

850 hPa flow/height anomalies at 1200 UTC 27 and 28 October 1994.

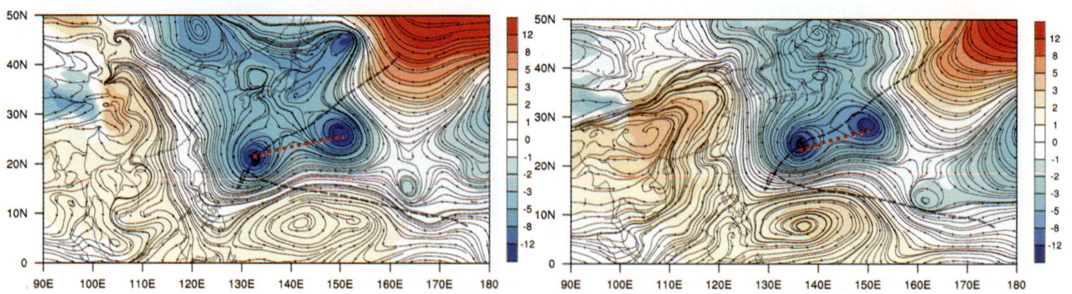

850 hPa flow/height anomalies at 1200 UTC 29 and 30 October 1994.

Figure 7.67. *(Continued)*

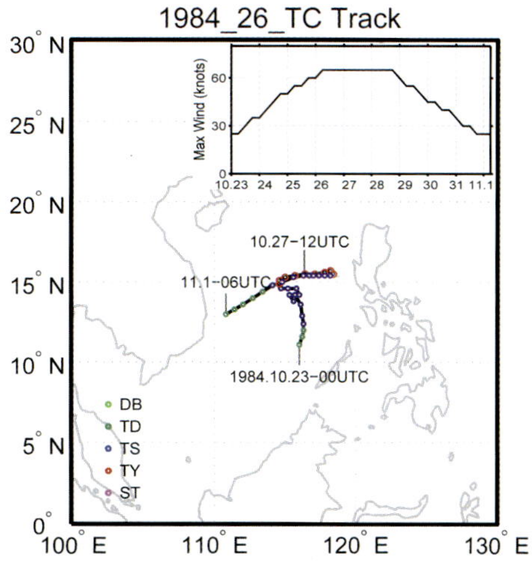

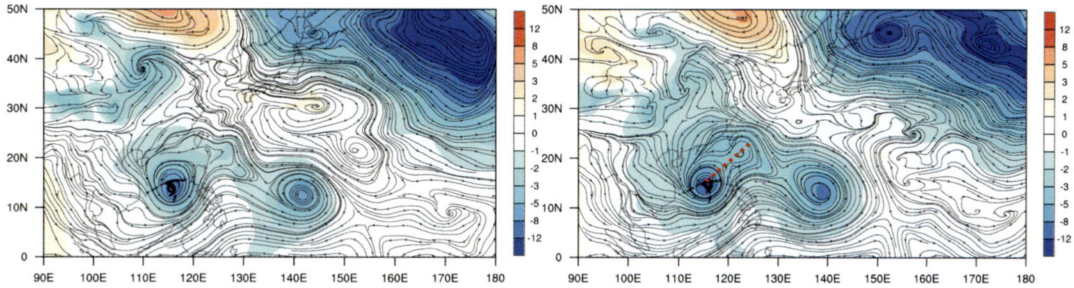

850 hPa flow/height anomalies at 0000 UTC and 1200 UTC 25 October 1984.

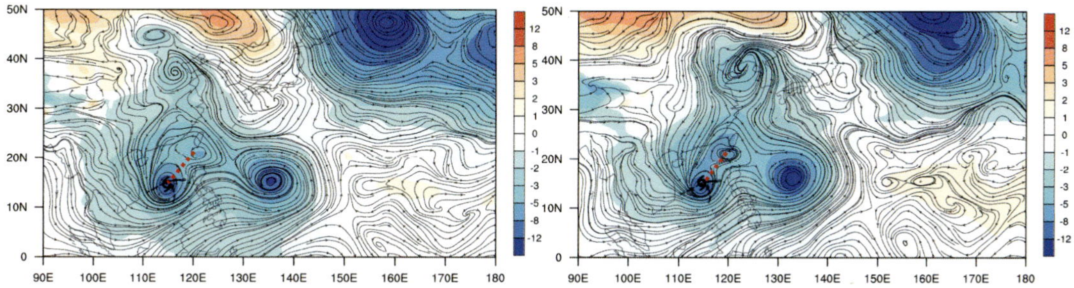

850 hPa flow/height anomalies at 0000 UTC and 1200 UTC 26 October 1984.

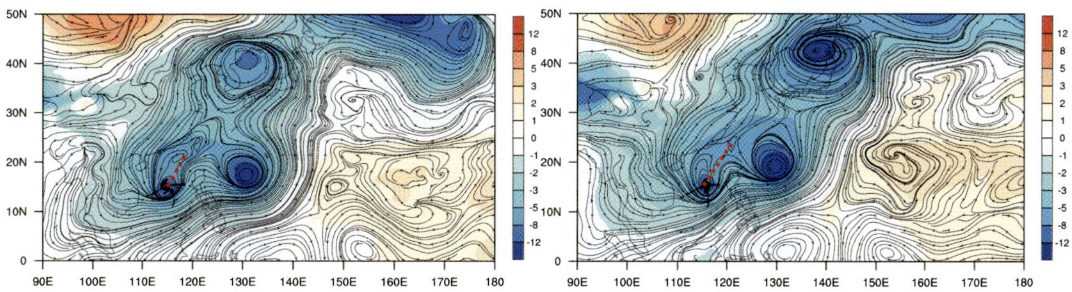

850 hPa flow/height anomalies at 0000 UTC and 1200 UTC 27 October 1984.

Figure 7.68. The typhoon (1984_No.26) generated in the central SCS on 23 October 1984. It had a "T"-type track and reached the peak intensity in the central SCS on 27 October 1984 when it was approaching the climatic monsoon trough. The "T"-type track was influenced by anomalous vortices at 850 hPa.